Reviewed by PSC

Agriculture, Ecosystems

and Environment

1985

29.2.83

Pest Resistance to Pesticides

Pest Resistance to Pesticides

Edited by
George P. Georghiou
University of California, Riverside
Riverside, California
and
Tetsuo Saito
Nagoya University
Nagoya, Japan

PLENUM PRESS · NEW YORK AND LONDON

Library of Congress Cataloging in Publication Data

Main entry under title:

Pest resistance to pesticides.

"Proceedings of a U.S.-Japan Cooperative Science Program Seminar on Pest Resistance to Pesticides: Challanges and Prospects, held December 3-7, 1979, in Palm Springs, California"—T.p. verso.

Includes bibliographical references and index.

1. Pesticide resistance—Congresses. I. Georghiou, George P. II. Saitō, Tetsuo, 1924- . III. U.S.-Japan Cooperative Science Program Seminar on Pest Resistance to Pesticides: Challenges and Prospects (1979: Palm Springs, Calif.)

SB957.P46 1983 632′.95 82-22369

ISBN 0-306-41246-2

Based on the Proceedings of a U.S.—Japan Cooperative Science Program Seminar on Pest Resistance to Pesticides: Challanges and Prospects, held December 3-7, 1979, in Palm Springs, California

A Division of Plenum Publishing Corporation
233 Spring Street, New York, N.Y. 10013

Printed in the United States of America

PREFACE

The development of resistance to pesticides is generally acknowledged as one of the most serious obstacles to effective pest control today. Since house flies first developed resistance to DDT in 1946, more than 428 species of arthropods, at least 91 species of plant pathogens, five species of noxious weeds and two species of nematodes were reported to have developed strains resistant to one or more pesticides. A seminar of U.S. and Japanese scientists was held in Palm Springs, California, during December 3-7, 1979, under the U.S.-Japan Cooperative Science Program, in order to evaluate the status of research on resistance and to discuss directions for future emphasis. A total of 32 papers were presented under three principal topics: Origins and Dynamics of Resistance (6), Mechanisms of Resistance (18), and Suppression and Management of Resistance (8). The seminar was unique in that it brought together for the first time researchers from the disciplines of entomology, plant pathology and weed science for a comprehensive discussion of this common problem.

Significant advances have been identified in (a) the development of methods for detection and monitoring of resistance in arthropods (electrophoresis, diagnostic dosage tests) and plant pathogens, (b) research on biochemical and physiological mechanisms of resistance (cytochrome P^{450}, sensitivity of target site, gene regulation), (c) the identification and quantification of biotic, genetic and operational factors influencing the evolution of resistance, and (d) the exploration of pest management approaches incorporating resistance-delaying measures. The participants also discussed recent progress in the synthesis of new pesticides that possess novel modes of action and are affected only minimally by cross resistance. The seminar delegates concluded that the increasing impact of pesticide resistance in the agricultural and public health fields requires substantial expansion of research with emphasis on the discovery of compounds with negatively correlated toxicity or multi-site mode of action and the formulation of pest control practices that minimize unidirectional selection pressure.

It was not the intention of the seminar to compile a formal list of research topics for future emphasis. However, many recommendations on promising areas for investigation were made by the participants, and there is no doubt that many of these will be pursued in the future. These areas include:

- The significance of enzyme induction in the development of resistance.

- Gene amplification as a mechanism contributing higher levels of resistance.

- Gene duplication as a source of resistance toward related compounds.

- Genetic studies of resistance in principal lepidopterous pests.

- Practical methods for detecting the presence of resistant mutants at low frequencies.

- Biotic fitness of resistant mutants as affecting the stability of resistance.

- The multiplicity and specificity of detoxifying oxidases, esterases and glutathione transferases.

- Purification of cytochrome P^{450} from resistant and susceptible strains and comparative studies on these.

- Basis for reduced sensitivity of acetylcholinesterase toward organophosphates and carbamates in resistant strains.

- Synthesis of chemicals with higher inhibitory activity on acetylcholinesterase of resistant than of susceptible strains.

- A better understanding of the mechanism of "reduced nerve sensitivity" in pyrethroid-resistant strains.

- Sequence of development of resistance in non-target organisms according to their status in the trophic chain.

- Discovery and utilization of pairs of chemicals which demonstrate negatively correlated cross resistance in pests.

- Search for chemicals with distinctly novel modes of action.

- Assessment of methods for inhibiting the evolution of resistance utilizing joint or rotational uses of chemicals.

- Search for synergists that may inhibit the evolution of resistance through suppression of detoxification mechanisms.

- Competitiveness of insecticide-resistant parasites and predators and stability of their resistance in the absence of insecticidal selection.

Each author was given the opportunity to revise his paper following the symposium. Research data are included for illustration, but the emphasis is on the comprehensive coverage of the principles and concepts of resistance, the critical appraisal of current knowledge, and prospects for future breakthroughs.

Thanks are due to the National Science Foundation of the U.S. and the Japan Society for the Promotion of Science for their sponsorship of this seminar. We are grateful to Miss Roni Mellon for patient and valuable editorial assistance. Of course, this volume could not have been prepared without the collaboration of the contributing authors to whom we wish to express our sincere appreciation.

George P. Georghiou
Tetsuo Saito

TABLE OF CONTENTS

I. ORIGINS AND DYNAMICS OF RESISTANCE

II. MECHANISMS OF RESISTANCE

PESTICIDE RESISTANCE IN TIME AND SPACE*

George P. Georghiou and Roni B. Mellon

Division of Toxicology and Physiology
Department of Entomology
University of California
Riverside, California 92521

INTRODUCTION

Within the evolutionarily insignificant period of just 65 years, beginning when the first case of resistance to a pesticide was reported (Melander, 1914), the phenomenon of resistance has proliferated exponentially so as to constitute today an indispensable consideration in nearly every pest control program.

Resistance is not limited to insects and insecticides (Fig. 1). It occurs in such relatively simple forms as bacteria and sporozoa, as well as in such advanced forms as mammals and plants. It affects a variety of toxicants, including antibiotics, antimalarials, coccidiostats, insecticides, rodenticides, etc. It is also evident, however, that while the capacity to develop resistance to chemicals is universal, it is those compounds that are employed as insecticides that have exceeded their targets and have selected for resistance in practically every type of organism, from bacteria to mammals. Since resistance cannot be induced by any other means except by lethal action, this undoubtedly implies a special responsibility for the pest control professional.

Questions regarding actual or potential resistance arise in the course of evaluating a new compound both in the laboratory and in the field; they are crucial during the marketing assessment of the

*The records of resistance contained in this review extend to December 1980.

	ANTIBIOTICS	ANTIMALARIALS	COCCIDIOSTATS	FUNGICIDES	INSECTICIDES	CHEMOSTERILANTS	NEMATOCIDES	RODENTICIDES	HERBICIDES
BACTERIA	●				●				
SPOROZOA		●	●						
FUNGI				●	●				
NEMATODES					●		●		
ACARINA				●	●				
INSECTA				●	●	●			
CRUSTACEA					●				
FISH					●				
FROGS					●				
RODENTS					●			●	
WEEDS					●				●

Figure 1. The occurrence of resistance to xenobiotics in various types of organisms.

new product, and occur frequently during monitoring of the degree of effectiveness that is obtained with continued use. Symptomatic of the impact of resistance in this respect is the considerable agonizing that we have witnessed recently on the part of industry over the question of whether or not pyrethroids should be developed commercially -- a question that two decades earlier would have been answered in the affirmative without hesitation. This is understandable in view of the considerably increased costs involved in the discovery and commercialization of a new insecticide, from less than \$10 million in 1972 to more than \$20 million in 1980 (Braunholtz, 1981). The phenomenon of resistance is a strong component of such costs. Many new compounds that are intrinsically highly active on "normal" strains of pests have been found to be affected by cross-resistance when tested against resistant strains. W. Mullison (in Lewert 1976) estimated that the number of chemicals screened to obtain one commercial success has increased from 1,800 in 1956 to 15,000 in 1975.

The cost of resistance is evident at the field level as well, in terms of more frequent applications, higher dosages, and changes to new, more expensive compounds. For the control of cotton pests in the Imperial Valley of California, which utilizes mainly pyrethroids and organophosphates, the cost is estimated at \$200-\$300/acre. In the public health sector, resistance in Anopheline mosquitoes has hindered the continuation of the initially highly successful malaria eradication program of the World Health Organization. Recently, concern regarding the outcome of the huge onchocerciasis control program in Africa has grown as a result of the detection of resistance to temephos in the vectors *Simulium sanctipauli* and *S. soubrense* (Guillet et al., 1980). In monetary terms, Pimentel et al. (1980) estimated the increased costs of pest control in the U.S. due to resistance at \$133.09 million annually, a figure that does not include the indirect costs of resistance on industrial research and development.

A seminar as comprehensive as this one requires that the subject be introduced in its historical perspective. Although in this initial

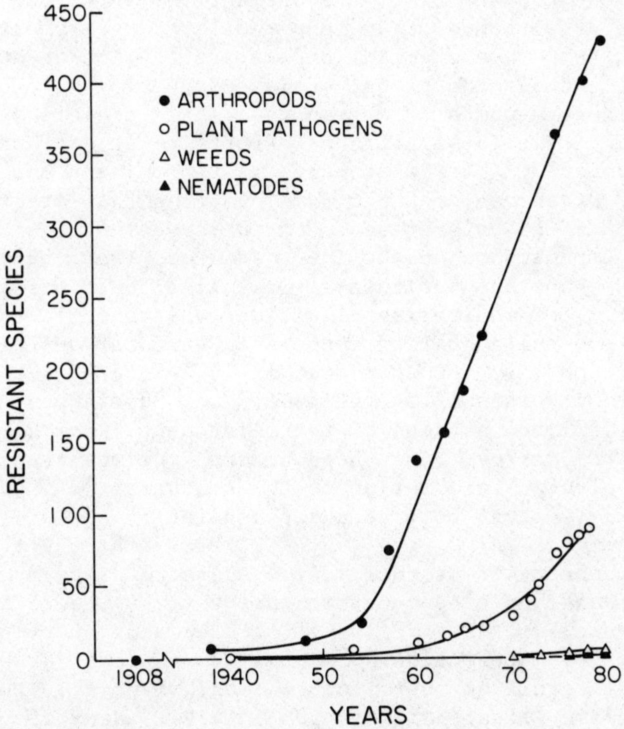

Figure 2. Chronological increases in numbers of species of arthropods, plant pathogens, weeds and plant parasitic nematodes with resistance to pesticides.

chapter we are emphasizing the current status of resistance, this overall view of the subject is established by examining the chronological, worldwide increase in the number of species known to be resistant (Fig. 2). Information is also provided on the extent of geographical distribution of resistance for a small number of important pests on which such information is available. Emphasis is placed on arthropods since information on the status of resistance in plant pathogens and weeds is covered in other chapters of this volume (see papers by Ogawa et al. and by Radosevich, respectively). For additional reading on the occurrence of resistance, the reader is referred to earlier general or specialized reviews on species of medical importance (Brown and Pal, 1971; WHO, 1976, 1980), on stored grain (Champ and Dyte, 1976) and agricultural pests (Brown, 1958, 1971), and on the ecology and dynamics of resistance (Georghiou, 1972a; Georghiou and Taylor, 1976).

SURVEILLANCE AND DOCUMENTATION OF CASES OF RESISTANCE

The early detection of resistance is of considerable importance to organized pest control campaigns as well as to individual growers since it allows for the timely adjustments of supplies, equipment, and training of personnel. By thus facilitating a smoother transition to an alternative chemical or the application of supplementary measures, it reduces the extent of losses that might otherwise be incurred. Thus, a number of organized, large-scale pest control programs have included significant efforts to detect and monitor the eventual development of resistance. Most prominent has been the World Health Organization's program involving the detection and documentation of cases of resistance in insects of medical importance, with emphasis on Anopheline mosquitoes (WHO, 1976, 1980). Likewise, the Food and Agriculture Organization is pioneering a global program for monitoring acaricide resistance in animal ticks by a network of collaborating centers in various countries (FAO, 1979). Additionally, FAO sponsored in 1972-73 a global field survey of resistance in pests of stored food, encompassing the collection and testing of 1,684 strains of the major pests from 85 countries (Champ and Dyte, 1976). This huge undertaking resulted in the collection of valuable information on the status of resistance to malathion, lindane, methyl bromide and phosphine.

At the country or state level, systematic programs for monitoring resistance have been carried out for many years in Egypt on cotton leafworm *Spodoptera littoralis* (El-Guindy et al., 1975), in Australia on the cattle tick *Boophilus microplus* (Roulston et al., 1977), in California on mosquitoes (Womeldorf et al., 1972), and in Denmark on house flies (Keiding, 1963, 1979). Many other extensive surveys of resistance on specific pests have been published as a result of individual studies involving a variety of pests, such as *Myzus persicae* in the U.K. (Sawicki et al., 1978), *Phorodon humuli* in Czechoslovakia (Hrdý and Kriz, 1976), *Diabrotica virgifera* in the U.S.A. (Ball and Weekman, 1963; Chio et al., 1978), rice leafhoppers in Japan (Ozaki and Kurosu, 1967; Hiramatsu et al., 1973; Ozaki and Kassai, 1971), *Plutella xylostella* in Malaysia (Sudderuddin and Kok, 1978), and many others.

Standardized Tests

Of considerable importance in resistance detection activities has been the development of standardized tests for various types of pests. Test procedures and kits for detecting resistance in several species of medically important pests have been provided by WHO (1970, 1976, 1980). Likewise, 23 standardized test methods involving some 36 species of agricultural importance have been formulated by FAO (Busvine, 1980), and a smaller number has been provided by the Entomological Society of America (Anonymous, 1968, 1970, 1972).

More recently, the surveillance for resistance has been facilitated by the introduction of the concept of "diagnostic doses" ("discriminating doses"), namely the use of a single dose that is lethal to normal, susceptible insects and sufficiently elevated (by a factor of 2- or 3-fold) to allow only individuals that are beyond the upper limits of confidence intervals of the LD_{99} to survive. Such diagnostic doses have been prescribed for adult and larval mosquitoes of *Anopheles*, *Culex* and *Aedes* groups (WHO, 1976, 1980) and thus have enabled the testing of larger numbers of individuals at these critical doses. This, in turn, has enhanced the validity of the conclusions. The designation of such diagnostic doses is based upon prior accurate determination of an LD_{99} of a field population and the confidence intervals of susceptibility at that level. As such, it requires the establishment of a complete dose-response curve by bioassay.

Further innovations in resistance surveillance have been introduced in the form of simplified biochemical diagnostic tests for specific resistance mechanisms that may be conducted at field laboratories. In most cases the tests involve the monitoring of organophosphorus insecticide resistance through detection of esterases that show high activity in degrading α- or β-naphthyl acetate. As with the designation of diagnostic doses by bioassay, the development of biochemical tests for resistance involves extensive and detailed investigations of esterases and requires the provision of evidence of co-identity or inseparability of high esterase activity and organophosphate resistance. Simple tests involving the detection of high esterase activity in single insects that are crushed on filter paper or in staining tiles for further processing have been introduced for detection of organophosphate resistance in a number of species, including *Nephotettix cincticeps* and *Laodelphax striatellus* (Ozaki, 1969), *Myzus persicae* (Sawicki et al., 1977), and *Culex quinquefasciatus* (Pasteur and Georghiou, 1981). Analogous tests have also been devised for detection of carbamate resistance in *Nephotettix cincticeps* that is due to reduced sensitivity of acetylcholinesterase toward carbamate insecticides (Miyata et al., 1980) and the detection of triazine resistance in weeds (Ali and Machado, 1981). It must be pointed out that these tests provide information on generic, rather than specific resistance, and also that the degree of resistance is not determined. Thus, these tests supplement, rather than replace the usual bioassay tests.

Computerization of Resistance Records

Since 1972 a systematic program of cataloging cases of resistance in insects and mites worldwide has been conducted by one of us (Georghiou) at the University of California, Riverside. This information is computerized and can be retrieved according to a number of parameters including species, chemical, host, country, locale, degree of resistance, year of detection, and source of information. The data presented in this paper have been obtained selectively from that program (Table 1). A more detailed presentation of the data will be given elsewhere (Georghiou, 1982). The data concern cases of resistance that have arisen as a result of the field application of insecticides; the numerous cases of resistance that were developed in the laboratory through simulated selection pressure are not included.

The criterion employed for designation of a population as resistant is the detection by means of sensitive tests of a significant decrease in the normal level of response of a population toward an insecticide. Such a change in sensitivity is understood to have resulted in diminished control or to be of such extent as to have created grounds for substitution of the chemical. This criterion is consistent with definitions of resistance proposed by the WHO Expert Committee on Insecticides (WHO, 1957) and by the FAO Working Party of Experts on Resistance (FAO, 1967).

Of much value to this documentation effort have been three surveys of resistance conducted by FAO in 1965, 1968 and 1974 by means of questionnaires distributed to researchers in various countries (FAO, 1967, 1969 and unpublished). Several earlier reviews were also consulted (Asakawa, 1975; Brown, 1958, 1971; Brown and Pal, 1971; Champ and Dyte, 1976; Croft, 1977; Dittrich, 1975; Kerr, 1977; Metcalf, 1980; WHO, 1970, 1976, 1980). In as far as it was possible, original reports and published papers were examined. Included also is information obtained through personal communications or from manuscripts in press provided by the authors. Although no longer of practical significance, certain cases of resistance involving obsolete chemicals, e.g. arsenicals, hydrogen cyanide, selenium, lime sulfur, tartar emetic ($KSbOC_4H_4O_6 \cdot \frac{1}{2}H_2O$), etc, have been retained as evidence of the ability of arthropods to develop resistance to these toxicants as well. It is also realized that resistance in those populations that have not remained under continued selection

Table 1. Number of Species of Arthropods with Reported Cases of Resistance to Pesticides through 1980

Order	Pesticide group[a/]							Importance[b/]		Total (%)
	DDT	Cyclod.	OP	Carb.	Pyr.	Fumig.	Other	Med./ Vet.	Agr.	
Acarina	17	15	42	6	1		30	15	38	53 (12.4)
Anoplura	4	4	2	1				6		6 (1.4)
Coleoptera	24	55	26	9	3	14	5		64	64 (14.9)
Dermaptera	1	1							1	1 (0.2)
Diptera	106	107	60	11	6		1	130	23	153 (35.8)
Ephemeroptera	2								2	2 (0.5)
Hemiptera/ Heteroptera	8	16	6					4	16	20 (4.7)
Hemiptera/ Homoptera	13	13	28	9	3	3	1		42	42 (9.8)
Hymenoptera	1	3							3	3 (0.7)
Lepidoptera	40	40	31	14	8		2		64	64 (14.9)
Mallophaga		2						2		2 (0.5)
Orthoptera	3	3	2	1	1			3		3 (0.7)
Siphonaptera	7	5	2					8		8 (1.9)
Thysanoptera	3	5	1				2		7	7 (1.6)
Total	229	269	200	51	22	17	41	168	260	428
(%)	(53.5)	(62.9)	(46.7)	(11.9)	(5.1)	(4.0)	(9.6)	(39.3)	(60.7)	(100)

a/ Cyclod.=cyclodiene, OP=organophosphate, Carb.=carbamate, Pyr.=pyrethroid, Fumig.=fumigant.
b/ Med./Vet. = medical/veterinary, Agr. = agricultural.

by the same or related chemicals may have regressed by now to non-detectable levels. However, in view of the known tendency of regressed resistance to be rapidly reselected, it was considered essential to retain all such cases regardless of their present levels.

STATUS OF RESISTANCE

The number of species of insects and acarines in which resistant strains appeared by the end of 1980 has reached a total of 428 (Table 1). Of these, 260 (60.7%) are of agricultural importance, and 168 (39.3%) are of medical or veterinary importance, or are considered a nuisance to man due to their biting habits. To enable comparisons with previous years, the numbers of species reported as resistant in various earlier reviews have been graphed in Figure 2. Also presented is information on the evolution of cases of resistance in plant pathogens (91 species), weeds (5), and plant parasitic nematodes (2).

The 12 cases of resistance in arthropods that were reported previous to 1946, when the first case of DDT resistance was detected at Arnas, Sweden (Wiesmann, 1947), have involved mainly inorganic chemicals such as arsenicals, hydrogen cyanide, lime sulfur, cryolite, selenium, and tartar emetic (see detailed reviews by Babers, 1949; Babers and Pratt, 1951). A similar record is also seen in the history of resistance in plant pathogens, i.e. a low incidence of cases of resistance to copper, sulfur, and mercury fungicides, but a considerable increase in reports of resistance upon introduction of benzimidazoles (see papers by Ogawa et al., Gilpatrick, and Eckert and Wild in this volume). It has been suggested that the relatively small number of species that developed resistance in the pre-DDT and pre-organic fungicide era can be attributed to the multisite mode of action of inorganic pesticides, i.e. their effect at several sensitive loci. In view of the initially rare occurrence of resistance genes in unselected populations, it was reasoned that it would be statistically very unlikely for such genes to exist together in the same individual. There is little doubt that the multiplicity of action sites for the inorganic compounds has been one of the causes of this limited resistance, but other factors must also be important, such as the ionic nature of the toxic principle of some of these compounds, which excludes the possibility of detoxication by metabolic enzymes.

More importantly, the quantities of pesticides employed in the post-World War II era have increased steadily with no signs of abating. Worldwide sales of insecticides, herbicides and fungicides grew from $1.1 billion in 1960 to $3.6 billion in 1970 and $9.7 billion in 1979 (Braunholtz, 1981)., indicating a tremendous increase in selection pressure being applied against pest species. It is of interest to note that the number of arthropod species in which resistance has been reported has nearly doubled during the last decade, rising from 224 species in 1970 to 428 in 1980.

Of these 428 resistant species, the majority are Diptera (153 species, i.e. 35.7%; Table 1), reflecting the strong chemical selection pressure that has been applied against mosquitoes throughout the world. Substantial numbers of resistant species are also evident in such agriculturally important Orders as the Lepidoptera and Coleoptera (64 species each, 14.9%), Acarina (53, 12.4%), Homoptera (42, 9.8%) and Heteroptera (20, 4.7%). The resistant species include many of the major pests since it is against these that chemical control is mainly directed.

An even more substantial increase in cases of resistance is evident when one considers the number of resistant species x pesticide groups that can be resisted (Table 2). On this basis, the number of cases of resistance has increased from 313 in 1970 to 829 in 1980, a

Table 2. Increases in Numbers of Species of Resistant Arthropoda During the Decade 1971-80

	1970[a]	1980	Fold increase
Species with reported resistance[b]	224	428	1.91
Cases[c] of resistance by pesticide group			
DDT	98	229	2.34
Cyclodiene	140	269	1.92
Organophosphate	54	200	3.70
Carbamate	3	51	17.00
Pyrethroid	3	22	7.33
Fumigant	3	17	5.67
Other	12	41	3.42
Total for all pesticide groups	313	829	2.65

[a] Calculated from data in Brown, 1971.
[b] Denotes number of species irrespective of number of chemical groups that are resisted.
[c] Denotes number of species resisting compounds within the indicated group.

2.65-fold increase. Additionally, if the number of species x individual insecticides that can be resisted is considered, the total number of cases reported through 1980 rises to 1,640. As might be expected, greater percentage increases in cases of resistance have occurred toward the relatively recently introduced classes of chemicals, i.e. carbamates (17-fold) and pyrethroids (7.33-fold), than toward organophosphates (3.7-fold), DDT (2.34-fold) and cyclodienes (1.92-fold). It must be emphasized that the statistics concerning increases in cases of resistance must be viewed only as indicators of broad trends, since exact numbers are undoubtedly influenced by the amount of research that is conducted in each area and time period as well as by the accessibility of the data that are generated. There is little doubt that the available data understate the extent of resistance since many cases probably either are not investigated or remain unreported.

Geographical Distribution of Resistance

The inclusion of a case of resistance in this report does not imply that the pesticide group or the specific insecticide involved is no longer effective against that species throughout its area of distribution. There are many examples of continued effectiveness of the same chemical in areas where selection pressure has been less severe or has not yet been applied. For example, organophosphate-resistant citrus red scale, *Aonidiella aurantii*, is found in a limited area of South Africa, although good control continues to be obtained elsewhere in that country (Georgala, 1975; Nel et al., 1979). Carbamate-resistant Colorado potato beetle, *Leptinotarsa decemlineata*, is found on Long Island, New York (Semel, 1980), in New Jersey (Forgash, 1981), and in Sherbrooke, Quebec, but not in London, Ontario (Harris and Svec, 1981) or in Maine (Forgash, 1981). Organophosphate and carbamate resistance is common in populations of *Anopheles albimanus* along the Pacific coast of Nicaragua, Honduras, El Salvador, Guatemala and the State of Chiapas, Mexico, but not on the Atlantic coast of the region (Georghiou, 1972b and unpublished). In contrast, some species have developed resistance to certain insecticides almost throughout the world, as in the case of the house fly toward DDT (WHO, 1980) and of *Sitophilus oryzae* toward lindane (Fig. 3) (Champ and Dyte, 1976). Records of development of resistance in a species toward a certain chemical should be considered as a forewarning of the possibility of similar developments elsewhere. They also indicate conditions under which the species in question can develop resistance and thus provide clues for the application of suitable resistance-delaying or resistance-avoiding control tactics.

The most frequently observed pattern of distribution of resistance is one in which isolated cases appear in areas where high pressure has been applied, as has been noted with DDT resistance in the Colorado potato beetle in various states (Brown, 1968). These islands gradually enlarge through natural dispersion of resistant

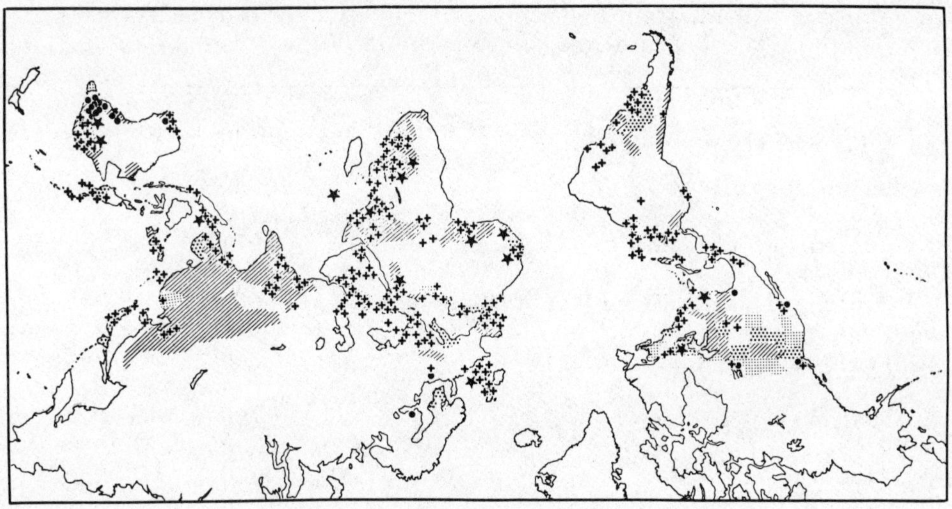

Figure 3. The occurrence of lindane resistance in *Tribolium castaneum*. • , susceptible; +, resistant; ★ , older cases of resistance. (After Champ and Dyte, 1976.)

individuals into surrounding areas, where they encounter further selection pressure. Even in the case of well-established and broadly occurring resistance, however, the frequency of resistance genes in each area is roughly commensurate with the degree of selection pressure that has been applied. Man-assisted accidental movement of resistant pests is a cause of increasing concern, especially in tropical countries. This is well illustrated by the case of organochlorine-resistant *Aedes aegypti*, which was re-established in countries of Central America and the Carribean following its eradication from the area. Thus, the possibility of introduction of resistant Anopholines and other important pests of man or crops cannot be overemphasized. The worldwide distribution of lindane-resistant strains of *Tribolium castaneum* believed to have occurred mainly through international trade serves as an illustration of the risks involved (Fig. 1).

The names of all species of insects and acarines reported to have developed resistance and the chemical group to which the compounds affected belong are presented alphabetically in taxonomic order in the Appendix. It is beyond the scope of this paper to review individually the resistance situation in each pest species or in each crop. It would be of interest, however, to examine briefly the status of resistance in species affecting cotton and rice. These constitute the two most important single crop markets for insecticides, as shown in 1979 estimates of end-user dollar value of leading crop pesticide sectors (Table 3).

Table 3. Estimated End-user Dollar Value of the Leading Crop Pesticide Sectors -- 1979

	- million dollars -
Maize herbicides	1,050
Cotton insecticides	975
Fruit and vegetable insecticides	900
Fruit and vegetable fungicides	860
Soybean herbicides	760
Rice insecticides	420
Maize insecticides	360
Cotton herbicides	350

Source: Braunholtz, 1981.

Resistance in Pests of Cotton

In no other crop is pest resistance to pesticides as widespread as it is in cotton. Serious control difficulties are being experienced in Central and South American countries, in the U.S., Mexico, Turkey, Egypt, the Sudan, and elsewhere. Partly for economic reasons but undoubtedly also because of resistance problems, cotton crops are treated frequently with a large variety of pesticides over the relatively long growing season of four to six months. In many cases, insecticides are applied weekly, or more frequently, so that as many as 30 applications are not unusual. In Central American countries where cotton is grown during the wet season, repeat applications are often necessitated by rain. Indicative of the high degree of selection pressure that is exerted on pests of cotton in the U.S. are statistics which show that 40% of the insecticide used in 1980 was applied on cotton (Eichers, 1981). Equally revealing are the records on the types and quantities of insecticides applied. In the area of Tapachula, Chiapas, Mexico, during the 1979 and 1980 growing seasons, 24 different insecticides were used to treat 28,000 hectares of cotton at an average rate of some 29 liters active ingredient per hectare (Table 4). Although methyl parathion and toxaphene were employed in the largest quantities, substantial amounts of parathion, monocrotophos, profenofos, methamidophos, mevinphos, sulprofos, mephospholan, DDT, and chlordimeform, and smaller amounts of carbamates and pyrethroids were also used.

Table 4. Insecticides Applied on Cotton in Tapachula, Mexico, 1979-81

Insecticide Class	Compound	1979 (Liters, a.i.)	1980 (Liters, a.i.)
Organo-phosphates	methyl parathion	369,626	340,800
	parathion	60,091	50,000
	monocrotophos	35,771	30,350
	profenofos	30,344	30,000
	methamidophos	14,441	21,880
	mevinphos	7,380	15,000
	sulprofos	7,589	14,400
	mephosfolan	1,773	10,000
	azinphosmethyl	2,595	4,000
	EPN	1,441	4,500
	dicrotophos	1,687	3,496
	dimethoate	684	
	omethoate		500
	Total	533,422	524,926
Cyclodienes	toxaphene	209,009	153,300
	endrin	4,896	3,797
	endosulfan	232	
	Total	214,137	157,097
Carbamates*	carbaryl	7,420	15,560
	bufencarb	688	
	Total	8,108	15,560
Pyrethroids	permethrin	2,314	5,200
	cypermethrin	660	1,300
	fenvalerate	529	690
	deltamethrin	60	50
	Total	3,563	7,240
DDT	DDT	44,388	60,000
Other	chlordimeform	24,450	25,500
GRAND TOTAL (liters)		828,068	789,823
Hectares treated		28,000	27,000
Liters a.i./HA		29.57	29.25

*Also methomyl, 7,740 Kg. 1979, 6,750 Kg. 1980.

Table 5. Insecticide Resistance in Pests of Cotton

Species	Pesticide Group*	Country
Acarina		
Tetranychus spp.	DDT OP	most countries
Coleoptera		
Anthonomus grandis	DDT;DL	USA; Mexico: Venezuela
Eutinobothrus brasiliensis	DL	Brazil
Graphognathus spp.	DL	USA
Heteroptera		
Lygus hesperus	DDT;DL;OP	USA
Dysdercus peruvianus	DL	Peru
Homoptera		
Bemisia tabaci	OP	Sudan
Trialeurodes abutilonea	OP	USA
Aphis craccivora	OP	USSR
A. gossypii	DL	Madagascar; USA; GDR; Peru
" "	OP	China; Japan; USSR; French Polynesia; Zambia
Empoasca biguttula	DDT;DL; Carb.	Taiwan
Lepidoptera		
Estigmene acrea	DDT;DL	USA
Pectinophora gossypiella	DDT	USA; Mexico
" "	OP;Carb.	Egypt
Bucculatrix thurberiella	DDT	USA; Peru
" "	DL	USA
" "	OP	USA; Mexico; Peru
" "	Carb.	USA
Alabama argillacea	DDT	USA
" "	DL	USA; Colombia; Venezuela
Anomis texana	DL	Peru
Cosmophila flava	DDT;DL	Ivory Coast; Mali; Niger; Upper Volta
Earias biplaga	DDT;DL	Madagascar
E. insulana	DDT	Madagascar
" "	DL	Israel; Spain; Madagascar

Species	Pesticide Group*	Country
Lepidoptera (continued)		
Heliothis armigera	DDT	Australia; USSR; Thailand
" "	DL	Australia; Portugal; Thailand
" "	OP	Canada; Portugal; Australia
" "	Carb.	Canada; USSR
H. virescens	DDT;DL	USA; Colombia; Mexico; Peru
" "	OP	USA; Mexico; Colombia
" "	Carb.;Pyr.	USA
Plusia brassicae	DDT; Carb	Barbados
Spodoptera exigua	DDT;DL;OP	USA
" "	Pyr.	El Salvador; Guatemala; Nicaragua
S. frugiperda	DDT	USA; Bolivia; Venezuela
" "	DL	Bolivia; Paraguay
" "	OP;Carb.;Pyr.	USA
S. littoralis	DDT	Egypt; Taiwan
" "	DL	Egypt; India; Turkey
" "	OP	Japan; Egypt; Cyprus; Turkey; Israel
" "	Carb.	Egypt; Japan
" "	Pyr.	Egypt
Thysanoptera		
Frankliniella occidentalis	DL	USA

*DL = cyclodienes, OP = organophosphates, Carb. = carbamates, Pyr. = pyrethroids.

The occurrence of resistance in pests of cotton is summarized in Table 5. At least 26 species are reported to be resistant, of which a high proportion (14 species) are Lepidoptera. Prominent among these are *Heliothis virescens*, *Spodoptera frugiperda* and *Spodoptera littoralis*, all displaying resistance to insecticides in the five principal classes (DDT, cyclodiene, organophosphate, carbamate and pyrethroid). Also notable is the high resistance to organophosphates in the cotton leaf perforator *Bucculatrix thurberiella* in the U.S. and the increasingly serious resistance to these insecticides in white flies (*Bemisia* and *Trialeurodes*) in a number of countries.

Resistance in Pests of Rice

The available records indicate that at least 12 species of insects attacking rice have developed strains resistant to insecticides

Table 6. Insecticide Resistance in Pests of Rice

Species	Pesticide Group*	Country
Diptera		
Agromyza oryzae	DDT;DL	Japan
Heteroptera		
Leptocorisa acuta	DL;OP	Thailand
L. varicornis	DL	Sri Lanka; Thailand
" "	OP	Thailand
Homoptera		
Delphacodes striatella	DL	Japan
Inazuma dorsalis	DDT	Taiwan
Laodelphax striatellus	DL Carb.	Japan
" "	OP	Japan; Korea
Nephotettix bipunctatus	DDT	Viet Nam
N. cincticeps	OP	Japan;Korea;China;Taiwan
" "	Carb.	Japan; Taiwan
Nilaparvata lugens	DDT	Viet Nam
" "	DL	Fiji; Taiwan; Japan
" "	OP	Philippines; Viet Nam
" "	OP;Carb.	Japan; Taiwan
Chilo suppressalis	DL	Japan; Taiwan
" "	OP	Japan; Taiwan; Korea
Lepidoptera		
Tryporyza incertula	DL	China
" "	OP	China; Taiwan

*See footnote, Table 5.

(Table 6). All such records concern countries of the Far East, but it is likely that resistance in rice pests also occurs in Central America but has not been investigated. Homoptera are by far the most important pests of rice and have been the target of intensive chemical control. The most serious resistance problems are found in the green rice leafhopper, *Nephotettix cincticeps*, involving organophosphates (Japan, Korea, China, Taiwan) and carbamates (Japan, Taiwan). In Japan, where resistance in rice pests has been studied in considerable detail, field populations of *N. cincticeps* show resistance to parathion, methyl parathion, EPN, fenitrothion, diazinon, dimethoate, malathion, carbaryl and propoxur (Ozaki and Kassai, 1971; Iwata and Hama, 1971) (Fig. 4). Somewhat lower resistance to organophosphates is found in the smaller brown planthopper, *Laodelphax striatellus*, in Japan and Korea. Another important pest of rice, the brown planthopper, *Nilaparvata lugens*, occurs in China and southeast Asia, but migrants reach Japan annually where they reproduce and cause economic losses. Organophosphate- and carbamate-resistant populations of this

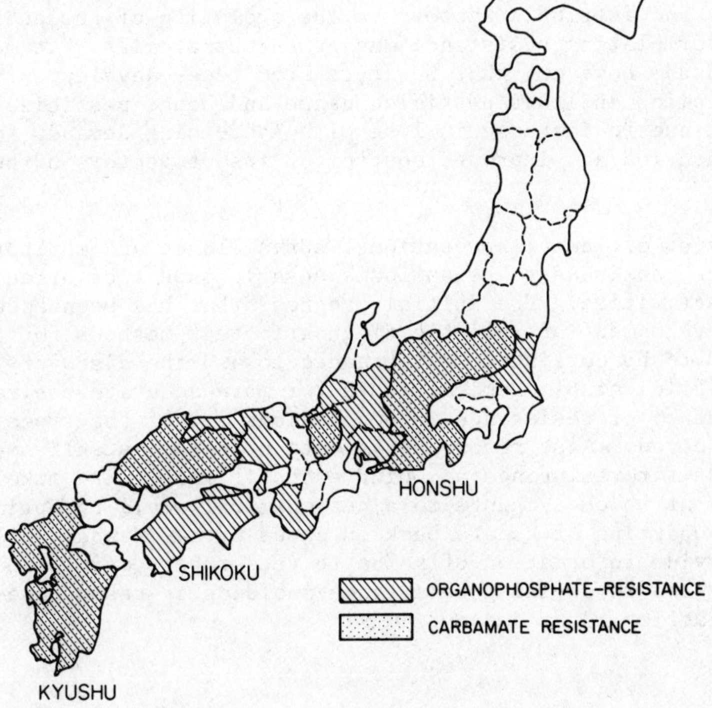

Figure 4. Geographical distribution of resistance to organophosphorus and carbamate insecticides in *Nephotettix cincticeps* in Japan. (After Asakawa, 1976.)

species were detected in Japan in 1977 (Ozaki, 1978) following the appearance of such resistance in Taiwan (Ku et al., 1977; Lin et al., 1979) and the Philippines (Heinricks and Valencia, 1978).

CONCLUSION

Despite greater costs brought about by the energy crisis and general inflation, the amount of pesticides being used is steadily increasing with concomitant increases in the number of reported cases of resistance. Such cases now involve at least 428 species of arthropods, 91 species of plant pathogens, 5 species of weeds and 2 species of plant parasitic nematodes.

There are encouraging signs that the formulation of integrated pest management programs for major crops could reduce the number of pesticide applications necessary for control and thus diminish the extent of selection pressure. Likewise, recent discoveries of new pesticides with unique modes of action, e.g. juvenile hormone mimics, chitin synthesis inhibitors, GABA inhibitors, bacterial toxins, etc., might contribute to a reduction of unidirectional selection pressure. Furthermore, increased use of computers in modelling for pesticide resistance has enhanced the capabilities of assessing the effects of extrinsic and intrinsic factors in the evolution of resistance and thus of formulating resistance management strategies. It must be acknowledged, however, that benefits from these developments will be slow in coming and that pesticide usage and hence pesticide resistance will continue to increase in line with increasing demands for higher food production and improved control of insect vectors of human diseases.

In view of these expectations, surveillance and monitoring of resistance continues to be an indispensable aspect of organized pest control activities. The initial progress that has been accomplished in the development of reliable diagnostic test methods for resistance will need to be continued and expanded toward the discovery of practical and efficient biochemical tests for more accurate estimation of the frequency of resistance in populations and of the types of mechanisms involved. Also requiring attention is the establishment of thresholds for resistance at which economic losses are likely to occur and at which a change to a new chemical would be optimal. Finally, updating of a data bank on cases of resistance worldwide would provide information of value to researchers on alternative chemicals and in devising resistance-avoiding or resistance-delaying control tactics.

REFERENCES

Ali, A., and Souza Machado, V., 1981, Rapid detection of "triazine resistant" weeds using chlorophyll fluorescence, *Weed Res.*, 21:191.

Anonymous, 1968, First conference on test methods for resistance in insects of agricultural importance. Method for the boll weevil and tentative method for spider mites, *Bull. Ent. Soc. Amer.*, 14:31.

Anonymous, 1970, Second conference on test methods for resistance in insects of agricultural importance. Standard method for detection of insecticide resistance in *Heliothis zea* (Boddie) and *H. virescens* (F.); tentative methods for detection in *Diabrotica* and *Hypera*, *Bull. Ent. Soc. Amer.*, 16:147.

Anonymous, 1972, Standard methods for detection of insecticide resistance in *Diabrotica* and *Hypera* beetles, *Bull. Ent. Soc. Amer.*, 18:179.

Asakawa, M., 1975, Insecticide resistance in agricultural insect pests of Japan, *Japan Pesti. Inf., No. 23*, pp. 5-8.

Asakawa, M., 1976, Resistance to carbamate insecticides in the green rice leafhopper, *Nephotettix cincticeps* Uhler, *Rev. Plant Protect. Res.*, 9:101.

Babers, F. H., 1949, Development of insect resistance to insecticides, U.S. Dept. Agri., Bureau of Entomol. and Plant Quarantine, mimeographed document E-776, 31 pp.

Babers, R. H., and Pratt, J. J. Jr., 1951, Development of insect resistance to insecticides, II. A critical review of the literature up to 1951. U. S. Dept. Agri., Agri. Res. Admin., Bureau Entomol. and Plant Quarantine, Report E-818.

Ball, H. J., and Weekman, G. T., 1963, Differential resistance of corn rootworms to insecticides in Nebraska and adjoining states, *J. Econ. Entomol.*, 56:553.

Braunholtz, J. T., 1981, Crop protection: The role of the chemical industry in an uncertain future, *Phil. Trans. Res. Soc. Lond.*, B 295:19.

Brown, A. W. A., 1958, The spread of insecticide resistance in pest species, *in:* "Advances in Pest Control Research," R. L. Metcalf, ed., Interscience Publishers, Inc., New York, pp. 351-414.

Brown, A. W. A., 1968, Insecticide resistance comes of age, *Bull. Entomol. Soc. Amer.*, 14:3.

Brown, A. W. A., 1971, Pest resistance to pesticides, *in:* "Pesticides in the Environment," Vol. 1, Part II, R. White-Stevens, ed., Marcel Dekker, New York, pp. 457-552.

Brown, A. W. A., and Pal, R., 1971, Insecticide resistance in arthropods, *WHO Monograph Series No. 38*, Geneva.

Busvine, J. R., 1980, Recommended methods for measurement of pest resistance to pesticides, *FAO Plant Production Protect. Paper No. 21*, FAO, Rome, 132 pp.

Champ, B. R., and Dyte, C. E., 1976, Report of the FAO global survey of pesticide susceptibility of stored grain pests, *FAO Plant Production Protect. Paper No. 5*, Food and Agriculture Organization of the United Nations, Rome.

Chio, H., Chang, C.-S., Metcalf, R. L., and Shaw, J., 1978, Susceptibility of four species of *Diabrotica* to insecticides, *J. Econ. Entomol.*, 71:389.

Croft, B. A., 1977, Resistance in arthropod predators and parasites, *in:* "Pesticide Management and Insecticide Resistance," D. L. Watson and A. W. A. Brown, eds., Academic Press, New York, pp. 337-393.

Dittrich, V., 1975, Acaricide resistance in mites, *Z. angew. Entomol.*, 78:28.

Eichers, T. R., 1981, Farm pesticide economic evaluation, 1981, USDA Economics and Statistics Service, Agri. Econ. Rep. No. 464.

El-Guindy, M. A., El-Sayed, G. N., and Madi, S. M., 1975, Distribution of insecticide resistant strains of the cotton leafworm *Spodoptera littoralis* in two governorates of Egypt, *Bull. Entomol. Soc. Egypt Econ. Ser.*, 9:191.

FAO, 1967, "Report of the First Session of the FAO Working Party of Experts on Resistance of Pests to Pesticides, 1965," PL/1965/18, FAO, Rome, 106 pp. (mimeo.).

FAO, 1969, "Report of the Fourth Session of the FAO Working Party of Experts on Resistance of Pests to Pesticides, 1968," PL/1968/M/10, FAO, Rome, 45 pp. (mimeo.).

FAO, 1979, Pest resistance to pesticides and crop loss assessment -- 2, *FAO Plant Production Protect. Paper 6/2*, FAO, Rome, 41 pp.

Forgash, A. J., 1981, Insecticide resistance of the Colorado potato beetle, *Leptinotarsa decemlineata* (Say), *in:* "Advances in Potato Pest Management," J. H. Lashomd and R. Casagrande, eds., Hutchinson Ross Publishing Co., Strausburg, PA.

Georgala, M. B., 1975, Possible resistance of the red scale *Aonidiella aurantii* Mask to corrective spray treatments, *Citrus Sub-Trop. Fruit J.*, 504:5.

Georghiou, G. P., 1972a, The evolution of resistance to pesticides, *Ann. Rev. Ecol. Systematics*, 3:133.

Georghiou, G. P., 1972b, Studies on resistance to carbamate and organophosphorus insecticides in *Anopheles albimanus*, *Am. J. Trop. Med. Hyg.*, 21:797.

Georghiou, G. P., 1982, "The Occurrence of Resistance to Pesticides in Arthropods -- An Index of Cases Reported Through 1980," FAO, Rome, in press.

Georghiou, G. P., and Taylor, C. E., 1976, Pesticide resistance as an evolutionary phenomenon, *Proc. XV Int. Cong. Entomol.*, Washington, D. C., August 19-27, 1976, Entomol. Soc. Amer., pp. 759-785.

Guillet, P., Escaffre, H., Ouedraogo, M., and Quillévéré, D., 1980, Note preliminaire sur une resistance au temephos dans le complexe *Simulium damnosum* (*S. sanctipauli* et *S. soubrense*) en Cote d'Ivoire (zone du programme de lutte contre l'onchocercose dans la region du bassin de la Volta), WHO/VBC/80.784, WHO, Geneva.

Harris, C. R., and Svec, H. J., 1981, Colorado potato beetle resistance to carbofuran and several other insecticides in Quebec, *J. Econ. Entomol.*, 74:421.

Heinricks, E. A., and Valencia, S. L., 1978, Resistance of the brown planthopper to carbofuran at IRRI, *IRRN* 3:4.

Hiramatsu, T., Tsuboi, A., and Kobayashi, M., 1973, Geographical distribution of the resistant smaller brown planthopper to organophosphorus insecticides in Okayama Prefecture, *Chugoku Agr. Res.*, 47:131.

Hrdý, I., and Kríz, J., 1976, Problems of hop pests control (in Czech), *Agrochemia*, 16:37-42.

Iwata, T., and Hama, H., 1971, Green rice leafhopper, *Nephotettix cincticeps*, resistant to carbamate insecticides, *Botyu-kagaku*, 36:174.

Keiding, J., 1963, Investigations of housefly resistance to insecticides in Danish farms, *Danish Pest Infestation Lab. Annu. Rep.*, 1961 & 1962, pp. 29-39.

Keiding, J., 1979, Insecticide resistance in houseflies, *Danish Pest Infestation Lab. Annu. Rep.*, 1978, pp. 43-59.

Kerr, R. W., 1977, Resistance to control chemicals in Australian arthropod pests, *J. Aust. ent. Soc.*, 16:327.

Ku, T. Y., Wang, S. C., and Hung, P. N., 1977, Resistance status in the brown planthopper, *Nilaparvata lugens* Stal, to commonly used insecticides in Taiwan, *Taiwan Agric. Quaterly*, 13:9.

Lewert, H. V., 1976, "A Closer Look at the Pesticide Question for Those Who Want the Facts," Dow Chemical Co., 41 pp.

Lin, Y.-H., Sun, C.-N., and Feng, H.-T., 1979, Resistance of *Nilaparvata lugens* to MIPC and MTMC in Taiwan, *J. Econ. Entomol.*, 72:901.

Melander, A. L., 1914, Can insects become resistant to sprays? *J. Econ. Entomol.*, 7:167.

Metcalf, R. L., 1980, Changing role of insecticides in crop protection, *Annu. Rev. Entomol.*, 25:219.

Miyata, T., Saito, T., Hama, H., Iwata, T., and Ozaki, K., 1980, A new and simple detection method for carbamate resistance in the green rice leafhopper, *Nephotettix cincticeps* Uhler, *Appl. Ent. Zool.*, 15:351.

Nel, J. J. C., de Lange, L., and van Ark, H., 1979, Resistance of citrus red scale, *Aonidiella aurantii* (Mask.) to insecticides, *J. Entomol. Soc. So. Africa*, 42:275.

Ozaki, K., 1969, The resistance to organophosphorus insecticides of the green rice leafhopper, *Nephotettix cincticeps* Uhler, and the smaller brown planthopper, *Laodelphax striatellus* Fallén, *Rev. Plant Protect. Res.*, 2:1.

Ozaki, K., 1978, Development of insecticide resistance in the brown planthopper, *Symp. 22nd Ann. Meet. Jap. Soc. appl. Ent. Zool.*, (abstract).

Ozaki, K., and Kassai, T., 1971, Patterns of insecticide resistance in field populations of the smaller brown planthopper, *Proc. Assoc. Pl. Prot. Shikoku*, 6:81.

Ozaki, K., and Kurosu, Y., 1967, Resistance pattern in four strains of insecticide-resistant green rice leafhopper collected in field, *Jap. J. appl. Ent. Zool.*, 11:145.

Pasteur, N., and Georghiou, G. P., 1981, Filter paper test for rapid determination of phenotypes with high esterase activity in organophosphate-resistant mosquitoes, *Mosq. News*, 41:181.

Pimentel, D., Androw, D., Dyson-Hudson, R., Gallahan, D., Jacobson, S., Irish, M., Kroop, S., Moss, A., Schreiner, I., Shepard, M., Thompson, T., and Vinzant, B., 1980, Environmental and social costs of pesticides: A preliminary assessment, *Oikos*, 34:126.

Roulston, W. J., Schuntner, C. A., Schnitzerling, H. J., Wilson, J. T., and Wharton, R. H., 1977, Characterization of three strains of organophosphorus resistant cattle ticks, *Boophilus microplus*, from Bajool, Tully and Ingram, *Aust. J. Agric. Res.*, 28:345.

Sawicki, R. M., Devonshire, A. L., and Rice, A. D., 1977, Detection of resistance to insecticides in *Myzus persicae* Sulz., *Meded. Rijksfac. Landbouwwet. Gent*, 42/2:1403.

Sawicki, R. M., Devonshire, A. L., Rice, A. D., Moores, G. D., Petzing, S. M., and Cameron, A., 1978, The detection and distribution of organophosphorus and carbamate insecticide-resistant *Myzus persicae* (Sulz.) in Britain in 1976, *Pestic. Sci.*, 9:189.

Semel, M., 1980, personal communication.

Sudderuddin, K. I., and Kok, P.-F., 1978, Insecticide resistance in *Plutella xylostella* collected from the Cameron Highlands of Malaysia, *FAO Plant Protect. Bull.*, 26:53.

WHO, 1957, Seventh Report of the Expert Committee on Insecticides, *WHO Tech. Rept. Ser.*, No. 125.

WHO, 1970, Insecticide resistance and vector control, Seventeenth Report of the WHO Expert Committee on Insecticides, *WHO Tech. Rept. Ser.*, No. 443, 279 pp.

WHO, 1976, Resistance of vectors and reservoirs of disease to pesticides, Twenty-second Report of the WHO Expert Committee on Insecticides, *WHO Tech. Rept. Ser.*, No. 585, 88 pp.

WHO, 1980, Resistance of vectors of disease to pesticides, Fifth Report of the WHO Expert Committee on Vector Biology and Control, *WHO Tech. Rept. Ser.*, No. 655, 82 pp.

Wiesmann, R. von, 1947, Untersuchungen uber das physiologische Verhalten von *Musca domestica* L. verschiedener Provenienzen, *Schweiz. Ent. Gesell. Mitt.*, 20:484.

Womeldorf, D. J., Gillies, P. A., and White, K. E., 1972, Insecticide susceptibility of mosquitoes in California, illustrated distribution of organophosphorus resistance in larvae *Aedes nigromaculis* and *Culex tarsalis*, *Proc. Pap. 40th Ann. Conf. Calif. Mosq. Control Assoc.*, 40:17.

APPENDIX

Cases of Resistance to Pesticides in Arthropoda

Vet. Med.	Agr.		DDT	Cyclod.	OP	Carb.	Fumigant	Other
		ACARINA						
		Acaridae						
	*	*Acarus siro*		*				
		Eriophyidae						
	*	*Aculus cornutus*			*			
	*	*A. fockeui*			*			
	*	*A. malivagrans*			*			
	*	*A. pelekassi*	*					
	*	*A. schlechtendali*			*			
	*	*Phyllocoptruta oleivora*				*		
		Ixodidae						
*		*Amblyomma americanum*		*				
*		*A. hebraeum*		*	*			sodium arsenite
*		*A. variegatum*		*				sodium arsenite
*		*Boophilus decoloratus*	*	*	*	*		sodium arsenite
*		*B. microplus*	*	*	*	*		sodium arsenite
*		*Dermacentor variabilis*	*	*				
*		*Haemaphysalis leachii*		*				

Vet. Med.	Agr.		DDT	Cyclod.	OP	Carb.	Fumigant	Other
		Ixodidae (cont.)						
*		*Hyalomma marginatum*		*				
*		*H. rufipes*		*				sodium arsenite
*		*H. truncatum*		*				sodium arsenite
*		*Ixodes rubicundus*						sodium arsenite
*		*Rhipicephalus appendiculatus*		*	*			sodium arsenite
*		*R. evertsi*		*	*			sodium arsenite
*		*R. sanguineus*		*	*			
		Macronyssidae						
*		*Ornithonyssus sylviarum*			*			
		Phytoseiidae						
	*	*Amblyseius andersoni*			*			
	*	*A. chilenensis*			*			
	*	*A. fallacis*	*		*	*		propargite
	*	*A. hybisci*			*			
	*	*Metaseiulus occidentalis*	*		*			sulfur
	*	*Phytoseiulus persimilis*			*			
	*	*Typhlodromus pyri*			*			
		Stigmaeidae						
	*	*Zetzellia mali*			*			
		Tenuipalpidae						
	*	*Brevipalpus chilensis*			*			

	Tetranychidae				
*	*Bryobia praetiosa*		*		
*	*Oligonychus pratensis*		*		propargite
*	*Panonychus citri*	*	*		ovex, tetradi- fon, quinomethi- onate, thioqui- nox, binapacryl, 'Nissol'
*	*P. ulmi*	*	*	*	chlorfenson, chlorfensulfide, fenson, Genite, ovex, propar- gite, tetradi- fon, tetrasul, binapacryl, chlordimeform, quinomethionate
*	*Tetranychus altheae*		*		
*	*T. arabicus*	*	*		
*	*T. atlanticus*		*		
*	*T. bimaculatus*		*		
*	*T. canadensis*		*		
*	*T. cinnabarinus*	*	*		quinomethionate
*	*T. crataegi*		*		
*	*T. cucurbitae*	*	*		
*	*T. desertorum*		*		
*	*T. hydrangaea*		*		
*	*T. kanzawai*	*	*		chlorfensulphide
*	*T. ludeni*		*		
*	*T. mcdanieli*	*	*		ovex, sulphe- none, tetradi- fon, binapacryl

Vet. Med.	Agr.		DDT	Cyclod.	OP	Carb.	Fumigant	Other
		Tetranychidae (cont.)						
	*	*Tetranychus pacificus*	*		*			
	*	*T. schoenei*	*		*			
	*	*T. tumidus*	*		*			
	*	*T. urticae*	*	*	*	*		chlorfensulphide, ovex, tetradifon, azobenzene, selenium, quinomethionate, binapacryl, chlordimeform, pyrethroids
	*	*T. viennensis*			*			
		ANOPLURA						
		Haematopinidae						
*		*Haematopinus eurysternus*		*	*			
		Linognathidae						
*		*Linognathus africanus*		*				
*		*L. stenopsis*	*					
*		*L. vituli*	*					
		Pediculidae						
*		*Pediculus humanus capitus*	*	*				
*		*P. h. humanus*	*	*	*	*		

	<u>COLEOPTERA</u>						
	<u>Bostrichidae</u>						
*	*Rhyzopertha dominica*		*	*			phosphine
	<u>Byturidae</u>						
*	*Byturus tomentosus*	*					
	<u>Carabidae</u>						
*	*Clivina impressifrons*		*				
	<u>Chrysomelidae</u>						
*	*Aulacophora femoralis*			*			
*	*Brontispa longissima*		*				
*	*Diabrotica balteata*	*	*				
*	*D. longicornis*		*	*			
*	*D. undecimpunctata*	*	*	*	*		
*	*D. virgifera*		*	*	*		
*	*D. vittata*	*	*	*			
*	*Epitrix atomaria*	*	*				
*	*E. cucumeris*	*	*				
*	*E. hirtipennis*	*	*				
*	*E. tuberis*	*	*				
*	*Galerucella birmanica*	*	*				Thanite
*	*Leptinotarsa decemlineata*	*	*	*	*	HCN	pyrethroids
*	*Oulema oryzae*	*	*				
*	*Phyllotreta striolata*	*	*				

Vet. Med.	Agr.		DDT	Cyclod.	OP	Carb.	Fumigant	Other
		Coccinellidae						
	*	*Coleomegilla maculata*	*	*	*			
	*	*Epilachna varivestis*						rotenone
		Cucujidae						
	*	*Cryptolestes ferrugineus*			*			
	*	*Oryzaephilus mercator*		*	*			
	*	*O. surinamensis*		*	*		EDB	
		Curculionidae						
	*	*Anthonomus eugenii*			*	*		
	*	*A. grandis*	*	*				
	*	*Bothynoderus punctiventris*		*			HCN	
	*	*Ceutorrhynchus assimilis*		*	*			
	*	*C. quadridens*		*	*			
	*	*Cosmopolites sordidus*		*				
	*	*Cylas formicarius*	*	*				
	*	*Eutinobothis brasiliensis*		*				
	*	*Graphognathus leucoloma*		*				
	*	*G. minor*		*				
	*	*G. peregrinus*		*				
	*	*Hypera postica*	*	*				
	*	*Lissorhoptrus oryzaephilus*		*				
	*	*Listronotus oregonensis*		*				
	*	*Otiorhynchus sulcatus*		*				
	*	*Sitophilus granarius*	*	*	*	*	EDB, MB, phosphine	pyrethroids

*	*S. oryzae*	*	*	*	*	phosphine, chloropicrin	
*	*S. zeamais*	*	*	*	*		
*	*Tychius picirostris*	*					
	Dermestidae						
*	*Dermestes maculatus*		*	*			
*	*Trogoderma granarium*			*			
	Elateridae						
*	*Agriotes lineatus*		*				
*	*A. obscurus*		*				
*	*Conoderus falli*	*	*	*			
*	*C. vespertinus*		*				
*	*Limonius californicus*		*				
*	*L. canus*		*				
*	*Melanotus tamsuyensis*		*	*			
	Mycetophagidae						
*	*Typhaea stercorea*			*			
	Nitidulidae						
*	*Meligethes aeneus*	*	*	*	*		Kelevan
	Scarabaeidae						
*	*Anomala orientalis*		*				*Bacillus popilliae*

Vet. Med.	Agr.		DDT	Cyclod.	OP	Carb.	Fumigant	Other
		Scarabaeidae (cont.)						
	*	*Ataenius spretulus*		*				
	*	*Cochliotis melalonthoides*		*				
	*	*Costelytra zealandica*	*	*				
	*	*Cyclocephala immaculata*		*				
	*	*Phyllophaga crinita*		*				
	*	*Popillia japonica*		*				*Bacillus popilliae*
	*	*Rhizotrogus majalis*		*				
		Tenebrionidae						
	*	*Gnathocerus cornutus*			*			
	*	*Tribolium castaneum*	*	*	*	*	MB, EDB, phosphine	bioresmethrin pyrethrins
	*	*T. confusum*		*	*		MB, phosphine	
		DERMAPTERA						
		Labiduridae						
	*	*Labidura riparia*	*	*				
		DIPTERA						
		Agromyzidae						
	*	*Agromyza oryzae*	*	*				
	*	*Liriomyza archiboldi*		*				

	*	*L. flaveola*	*		*	
	*	*L. minutiseta*			*	
	*	*L. munda*			*	*
	*	*Phytomyza atricornia*		*		
		<u>Anthomyiidae</u>				
	*	*Hylemya antiqua*	*	*	*	*
	*	*H. brassicae*	*	*	*	
	*	*H. floralis*		*		
	*	*H. florilega*		*		
	*	*H. platura*		*		*
		<u>Calliphoridae</u>				
*		*Chrysomya bezziana*		*		
*		*C. putoria*	*	*	*	
*		*Lucilia cuprina*		*	*	*
*		*L. sericata*		*		
*		*Protophormia terraenovae*	*	*	*	
		<u>Cecidomyiidae</u>				
	*	*Aphidoletes aphidomyza*			*	
	*	*Dasineura pyri*			*	
		<u>Ceratopogonidae</u>				
*		*Culicoides furens*		*		
*		*Leptoconops kerteszii*	*	*		
		<u>Chaoboridae</u>				
*		*Chaoborus astictopus*	*		*	

Vet. Med.	Agr.		DDT	Cyclod.	OP	Carb.	Fumigant	Other
		Chironomidae						
*		*Chironomus decorus*			*			
*		*C. zealandicus*		*				
*		*Glyptotendipes paripes*		*	*			
*		*Procladius freemani*			*			
*		*P. sublettei*			*			
		Chloropidae						
*		*Hippelates collusor*	*	*				
		Culicidae						
*		*Aedes aegypti*	*	*	*			pyrethroids
*		*Ae. albopictus*	*	*	*			
*		*Ae. atropalpus*	*					
*		*Ae. canadensis*			*			
*		*Ae. cantans*	*	*				
*		*Ae. cantator*	*	*				
*		*Ae. caspius*	*		*			
*		*Ae. detritus*	*	*	*			
*		*Ae. dorsalis*	*		*			
*		*Ae. fijiensis*	*					
*		*Ae. melanimon*	*	*	*			
*		*Ae. nigromaculis*	*	*	*			
*		*Ae. polynesiensis*	*					
*		*Ae. pseudoscutellaris*	*					
*		*Ae. sierrensis*	*					
*		*Ae. sollicitans*	*	*	*			

*	*Ae. taeniorhynchus*	*	*	*		
*	*Ae. togoi*	*		*		
*	*Ae. triseriatus*	*				
*	*Ae. vexans*	*		*		
*	*Ae. vittatus*	*				
*	*Anopheles aconitus*	*	*			
*	*An. albimanus*	*	*	*	*	
*	*An. albitarsis*	*	*			
*	*An. annularis*	*	*			
*	*An. aquasalis*		*			
*	*An. arabiensis*	*	*	*		pyrethroids
*	*An. atroparvus*	*	*	*	*	
*	*An. barbirostris*	*	*			
*	*An. coustani*		*			
*	*An. crucians*		*			
*	*An. culicifacies*	*	*	*		
*	*An. culicifacies adenensis*	*				
*	*An. donaldi*	*	*			
*	*An. d'thali*		*			
*	*An. farauti* (sp. No. 1)		*			
*	*An. farauti* (sp. No. 2)		*			
*	*An. filipinae*		*			
*	*An. flavirostris*		*			
*	*An. fluviatilis*	*	*			
*	*An. funestus*		*			
*	*An. gambiae*	*	*			
*	*An. hyrcanus*	*	*	*		
*	*An. labranchiae*	*	*			
*	*An. maculipennis*	*	*	*	*	
*	*An. messeae*	*	*	*		
*	*An. minimus*		*			
*	*An. multicolor*	*	*			
*	*An. neomaculipalpis*		*			

Vet. Med.	Agr.		DDT	Cyclod.	OP	Carb.	Fumigant	Other
		Culicidae (cont.)						
*		*Anopheles nigerrimus*	*	*				
*		*An. nili*		*				
*		*An. peditaeniatus*	*					
*		*An. pharoensis*	*	*				
*		*An. philippinensis*	*	*				
*		*An. pseudopunctipennis*	*	*				
*		*An. pulcherrimus*	*	*				
*		*An. quadrimaculatus*	*	*				
*		*An. rangeli*		*				
*		*An. rufipes*		*				
*		*An. sacharovi*	*	*	*	*		pyrethroids
*		*An. sergenti*		*				
*		*An. sinensis*	*	*	*			
*		*An. splendidus*		*				
*		*An. stephensi*	*	*	*			pyrethroids
*		*An. strodei*		*				
*		*An. subpictus*	*	*				
*		*An. sundaicus*	*	*				
*		*An. superpictus*	*					
*		*An. tesselatus*	*	*				
*		*An. triannulatus*	*	*				
*		*An. turkhudi*	*					
*		*An. vagus*	*	*				
*		*Armigeres subalbatus*	*	*	*			
*		*Culex annulus*	*	*	*			
*		*Cx. coronator*	*					
*		*Cx. erythrothorax*	*					
*		*Cx. fuscocephalus*	*	*	*			

*		*Cx. gelidus*	*	*			
*		*Cx. nebulosus*		*			
*		*Cx. nigripalpus*	*		*		
*		*Cx. peus*	*	*	*		
*		*Cx. pipiens pallens*	*	*	*		
*		*Cx. p. pipiens*	*	*	*	*	
*		*Cx. poicilipes*	*				
*		*Cx. quinquefasciatus*	*	*	*	*	
*		*Cx. restuans*	*	*			
*		*Cx. salinarius*	*	*			
*		*Cx. tarsalis*	*	*	*		
*		*Cx. tritaeniorhynchus*	*	*	*		
*		*Cx. vishnui*	*	*	*		
*		*Culiseta inornata*	*	*	*		
*		*Mansonia annulifera*	*	*			
*		*M. indiana*	*				
*		*M. uniformis*		*			
*		*Psorophora confinnis*	*	*			
*		*P. discolor*		*			
		<u>Drosophilidae</u>					
	*	*Drosophila melanogaster*	*				
	*	*D. virilis*	*				
		<u>Muscidae</u>					
*		*Fannia canicularis*	*	*	*		
*		*F. femoralis*	*	*	*		
*		*Haematobia irritans*	*	*	*		
*		*Musca domestica*	*	*	*	*	pyrethroids
*		*M. sorbens*	*	*			
*		*Ophyra leucostoma*	*	*			
*		*Stomoxys calcitrans*	*	*	*		

Vet. Med.	Agr.		DDT	Cyclod.	OP	Carb.	Fumigant	Other
		Otitidae						
	*	*Euxesta notata*	*	*				
	*	*Tetanops myopaeformis*		*				
		Phoridae						
	*	*Megaselia halterata*			*			pyrethrins
		Psilidae						
	*	*Psila rosae*		*	*			
		Psychodidae						
*		*Phlebotomus papatasi*	*					
*		*Psychoda alternata*	*	*				
		Simuliidae						
*		*Cnephia mutata*	*					
*		*Prosimulium mixtum*	*					
*		*Simulium aokii*	*					
*		*S. damnosum s. l.*	*					
*		*S. fuscum*	*					
*		*S. hargreavesi*	*					
*		*S. ornatum*	*		*			
*		*S. sanctipauli*			*			
*		*S. soubrense*			*			
*		*S. tuberosum*	*					
*		*S. venustum*	*					

		<u>Syrphidae</u>				
	*	*Eumerus striagatus*		*		
	*	*Merodon equestris*		*		
		<u>Tephritidae</u>				
	*	*Ceratitis capitata*	*	*	*	
	*	*Rhagoletis completa*				cryolite
		<u>EPHEMEROPTERA</u>				
		<u>Ephemeridae</u>				
	*	*Stenonema fuscum*	*			
		<u>Heptageniidae</u>				
	*	*Heptagenia hebe*	*			
		<u>HEMIPTERA/HETEROPTERA</u>				
		<u>Cimicidae</u>				
*		*Cimex hemipterus*	*	*		
*		*C. lectularius*	*	*	*	
		<u>Coreidae</u>				
	*	*Leptocoris acuta*		*	*	
	*	*L. varicornis*		*	*	
	*	*Leptoglossus membranaceus*	*	*		

Vet. Med.	Agr.		DDT	Cyclod.	OP	Carb.	Fumigant	Other
		Lygaeidae						
	*	*Blissus insularis*	*	*	*			
	*	*B. leucopterus hirtus*		*				
	*	*B. pulchellus*		*				
		Miridae						
	*	*Distantiella theobroma*		*				
	*	*Lygus elisus*	*					
	*	*L. hesperus*	*	*	*			
	*	*L. lineolaris*			*			
	*	*Psallus seriatus*		*				
	*	*Sahlbergella singularis*		*				
		Pentatomidae						
	*	*Nezara viridula*	*					
	*	*Rhynchocoris humeralis*	*					
	*	*Scotinophora lurida*		*				
		Pyrrhocoridae						
*		*Dysdercus peruvianus*		*				
		Reduviidae						
*		*Rhodnius prolixus*		*				
*		*Triatoma maculata*		*				

	HEMIPTERA/HOMOPTERA					
	Aleyrodidae					
*	*Dialeurodes citri*			*		
*	*Trialeurodes abutilonea*			*		
*	*T. vaporariorum*	*		*		pyrethroids
	Aphididae					
*	*Aphis citricola (=spiraecola)*			*		
*	*A. craccivora*			*		
*	*A. fabae*			*	*	
*	*A. gossypii*		*	*		
*	*A. pomi*			*		
*	*Aulacorthum solani*			*		
*	*Chaetosiphon fragaefolii*		*			
*	*Chromaphis juglandicola*		*	*		
*	*Dysaphis plantaginea*			*		
*	*Eriosoma lanigerum*		*			
*	*Hyalopterus pruni*			*		
*	*Myzus cerasi* (includes *M. umefoliae*)			*		
*	*M. persicae*	*	*	*	*	pyrethroids
*	*Phorodon humuli*		*	*	*	
*	*Rhopalosiphum padi*			*		
*	*Sapaphis pyri*			*		
*	*Schizaphis graminum*			*		
*	*Therioaphis maculata*			*		
	Cercopidae					
*	*Aeneolamia varia*	*	*			

Vet. Med.	Agr.		DDT	Cyclod.	OP	Carb.	Fumigant	Other
		Cicadellidae						
	*	*Delphacodes striatella*		*				
	*	*Empoasca biguttula*	*	*		*		
	*	*Erythroneura comes*	*					
	*	*E. elegantula*	*		*			
	*	*E. lawsoniana*	*					
	*	*E. variabilis*	*					
	*	*Inazuma dorsalis*	*					
	*	*Laodelphax striatellus*		*	*	*		
	*	*Nephotettix bipunctatus*	*					
	*	*N. cincticeps*			*	*		
	*	*Typhlocyba pomaria*	*		*			
		Coccidae						
	*	*Coccus pseudomagnoliarum*					HCN	
	*	*Saissetia oleae*					HCN	
		Delphacidae						
	*	*Nilaparvata lugens*	*	*	*	*		
		Diaspididae						
	*	*Aonidiella aurantii*			*	*	HCN	
	*	*Chrysomphalus aonidum*			*			
	*	*Quadraspidiotus perniciosus*						lime sulfur

	<u>Fulgoridae</u>					
*	*Pyrilla perpusilla*		*			
	<u>Pseudococcidae</u>					
*	*Planococcus ficus*			*		
	<u>Psyllidae</u>					
*	*Psylla pyricola*	*	*	*	*	pyrethroids
	<u>HYMENOPTERA</u>					
	<u>Apidae</u>					
*	*Apis mellifera*	*	*			
	<u>Formicidae</u>					
*	*Atta sexdens*		*			
*	*Iridomyrmex humilis*		*			
	<u>LEPIDOPTERA</u>					
	<u>Arctiidae</u>					
*	*Estigmene acrea*	*	*			
	<u>Gelechiidae</u>					
*	*Anarsia lineatella*	*				lead arsenate
*	*Pectinophora gossypiella*	*		*	*	
*	*Phthorimaea operculella*	*	*	*		

Vet. Med.	Agr.		DDT	Cyclod.	OP	Carb.	Fumigant	Other
		Gelechiidae (cont.)						
	*	*Sitotroga cerealella*		*				
	*	*Stegasta bosqueella*		*				
		Gracillariidae						
	*	*Lithocolletis ringoniella*			*			
		Hepialidae						
	*	*Oncopera intricata*	*					
	*	*O. rufobrunnea*	*					
		Lymantriidae						
	*	*Porthetria dispar*	*	*				
		Lyonetiidae						
	*	*Bucculatrix thurberiella*	*	*	*	*		
	*	*Lyonetia clerkella*			*			
	*	*Phyllocnistis citrella*			*			
		Noctuidae						
	*	*Agrotis ipsilon*	*	*	*			
	*	*Alabama argillacea*	*	*				
	*	*Anomis texana*		*				
	*	*Cosmophila flava*	*	*				

*	*Earias biplaga*	*	*			
*	*E. insulana*	*	*			
*	*Euxoa detersa*		*			
*	*E. messoria*		*			
*	*E. ochrogaster*	*	*			
*	*E. scandens*		*			
*	*Heliothis armigera*	*	*	*	*	
*	*H. assulta*	*				
*	*H. virescens*	*	*	*	*	pyrethroids
*	*H. zea*	*	*	*	*	
*	*Mamestra brassicae*			*		
*	*Phlogophora meticulosa*			*		
*	*Plusia brassicae*	*			*	
*	*P. eriosoma*	*	*	*		
*	*P. gamma*		*	*		
*	*Pseudoplusia includens*	*	*	*		
*	*Scrobipalpula obsoluta*					pyrethroids
*	*Spodoptera exigua*	*	*	*		pyrethroids
*	*S. frugiperda*	*	*	*	*	pyrethroids
*	*S. littoralis*	*	*	*	*	pyrethroids
*	*Trichoplusia ni*	*	*	*		
	Olethreutidae					
*	*Grapholitha molesta*	*				
*	*Laspeyresia pomonella*	*		*		lead arsenate
*	*Paralobesia botrana*	*				
	Pieridae					
*	*Pieris rapae*	*		*	*	

Vet. Med.	Agr.		DDT	Cyclod.	OP	Carb.	Fumigant	Other
		Pyralidae						
	*	*Chilo suppressalis*		*	*			
	*	*Diatraea saccharalis*		*	*	*		
	*	*Ephestia cautella*	*	*	*			pyrethrins
	*	*Margaronia hyalinata*	*		*	*		
	*	*Plodia interpunctella*	*	*	*			pyrethrins
	*	*Tryporyza incertula*		*	*			
		Pyraustidae						
	*	*Evergestis pallidata*	*	*				
	*	*Loxostege bifidalis*		*				
	*	*Zinckenia perspectalis*		*				
		Sphingidae						
	*	*Manduca quinquemaculata*	*					
	*	*M. sexta*		*				
		Tineidae						
	*	*Syringopais temperatella*		*				
	*	*Tinea pellionella*		*				
	*	*Tineola bisselliella*		*				
		Tortricidae						
	*	*Adoxophyes orana*			*	*		
	*	*Archips argyrospilus*			*			

	*	*Argyrotaenia velutinana*	*				
	*	*Choristoneura fumiferana*	*				
	*	*Epiphyas postvittana*	*		*	*	
	*	*Tortrix viridana*	*	*			
		<u>Yponomeutidae</u>					
	*	*Plutella xylostella*	*	*	*	*	pyrethroids
	*	*Prays oleae*	*				
		<u>MALLOPHAGA</u>					
		<u>Trichodectidae</u>					
*		*Bovicola caprae*		*			
*		*B. limbatus*		*			
		<u>ORTHOPTERA</u>					
		<u>Blattellidae</u>					
*		*Blattella germanica*	*	*	*	*	pyrethrins
		<u>Blattidae</u>					
*		*Blatta orientalis*	*	*			
*		*Periplaneta brunnea*		*			
		<u>SIPHONAPTERA</u>					
		<u>Certophyllidae</u>					
*		*Ceratophyllus fasciatus*	*				

Vet. Med.	Agr.		DDT	Cyclod.	OP	Carb.	Fumigant	Other
		Pulicidae						
*		*Ctenocephalides canis*	*	*				
*		*C. felis*	*	*				
*		*Pulex irritans*	*	*				
*		*Stivalius cognatus*			*			
*		*Xenopsylla astia*	*	*				
*		*X. brasiliensis*	*					
*		*X. cheopis*	*	*	*			
		THYSANOPTERA						
		Thripidae						
	*	*Chaetanaphothrips orchidii*		*				
	*	*Diarthrothrips coffeae*	*					
	*	*Frankliniella occidentalis*		*				
	*	*F. tritici*		*				
	*	*Scirtothrips citri*	*	*	*			tartar emetic
	*	*Taeniothrips simplex*						tartar emetic
	*	*Thrips tabaci*	*	*				

GENETIC ORIGINS OF INSECTICIDE RESISTANCE

Frederick W. Plapp, Jr. and T. C. Wang

Insecticide Toxicology Laboratory
Department of Entomology
Texas A&M University
College Station, Texas 77843

INTRODUCTION

Resistance to insecticides has been demonstrated in most important pest insect species, and this widespread occurrence is thought to provide proof of Darwinian evolutionary theory. That is, exposure to insecticides has acted as a powerful selecting force, which concentrates the various preexisting genetic factors that confer resistance.

The next step in the process, a careful evaluation of the nature of these genetic factors, remains to be taken. We shall try in this report to provide some insight into the problem and, hopefully, identify some of the approaches that can be taken to elucidate more fully the nature of resistance.

Given the widespread occurrence of insectide resistance, it is discouraging to have to report that we know very little about the basis for it in genetic terms. Our understanding of resistance biochemistry is in somewhat better shape, but it, too, suffers from a lack of comprehension of the genetic factors actually involved.

Entomologists have gained considerable knowledge of the biochemical mechanisms involved in metabolic resistance to insecticides. First, we know the detoxifying enzymes conferring metabolic resistance are remarkably nonspecific; that is, they have very broad ranges of activity. Secondly, we know these enzymes are inducible. In susceptible insects, detoxifying enzymes are present at relatively low levels, but the level can be increased, often by several fold, if the insects are exposed to appropriate inducers of enzyme activity. In resistant insects, amounts of detoxifying enzymes are often much greater. These, too, are frequently inducible but often to a lesser degree, at least on a percent basis, than in susceptible insects. Surprisingly, it is unclear if the higher levels of activity of resistant insects are due to better detoxifying enzymes (higher specific activity), to more enzyme (increased enzyme synthesis), or to a combination of both.

Thus, we do not know if resistance to insecticides is due to changes in structural genes, i.e., those genes actually specifying the nature of the enzymes, or to changes in regulatory genes or regulatory loci associated with the structural genes, i.e., changes that control the rates of synthesis of the enzymes. It will be the hypothesis of this paper that at least as far as metabolic resistance to insecticides is concerned, both types of change occur and resistance is due to the interaction of the two types of gene effects.

Fortunately, enough is known concerning the effects of the different types of genetic changes that, by means of appropriate genetic and biochemical assays, it is now possible to gain a good idea of the nature of the changes that may be involved in resistance.

If changes occur in a structural gene resulting in a detoxifying enzyme with higher specific activity, one would expect the enzyme to be as inducible as the enzyme in susceptible strains. Also, in a cross with susceptible flies, enzyme activity in the F_1 should be intermediate and show no change in inducibility. Note, however, that experience suggests that changes in structural genes are usually associated with a loss, not an increase, in enzyme activity. The best example of this in insecticide resistance is the mutant aliesterase of certain organophosphate-resistant house fly strains. Here, activity towards substrates such as methyl butyrate shows a significant decline (Van Asperen and Oppenoorth, 1959). The theory is that the mutant enzyme has lost activity towards normal substrates and has gained in its ability to metabolize certain toxic organophosphates (Oppenoorth, 1965).

Other possibilities for resistance involve changes that are regulatory in nature. These changes may occur with a separate regulatory gene or at a regulatory site associated with the structural gene. They may be transcriptional, posttranscriptional or translational in nature. Whatever their precise nature, they should lead to synthesis of increased amounts of detoxifying enzymes with no changes in specific activity levels. Unlike structural genes, the inheritance of regulatory gene effects is frequently either dominant or recessive, rather than intermediate in the F_1, and many differ in control and induced flies. However, for unknown reasons this model of inheritance that is derived from analogy with studies in microorganisms may not hold for many regulatory genes in eukaryotes (Paigen, 1979).

The recent discoveries of insertion elements in the DNA of many organisms (see reviews by Nebert, 1978 and Crick, 1979) suggest additional possibilities for regulatory control in eukaryotes. It is unclear just how these findings may relate to insecticide resistance.

Lastly, there is the possibility of gene amplification or redundancy. This possibility was first suggested in insecticide resistance by Walker and Terriere (1970). It has recently been substantiated by the studies of Blackmum et al. (1978) and Devonshire and Sawicki (1970) on esterase resistance to organophsophate insecticides in the aphid *Myzus persicae*, and by Alt et al. (1978) for drug metabolism in human cell cultures. Such a genetic change, at least if it involves mutant structural genes, should result in increased basal levels of enzyme activity in resistant strains with no loss in specific activity levels and with no loss in inducibility.

In this report we shall present evidence that several types of genetic change occur in metabolic resistance to insecticides in insects and that the total resistance present in many strains is due to the interaction of multiple genes. New information presented will include evidence for the occurrence of allelic genes of chromosome II that exhibit pleiotropic effects on resistance levels and detoxifying gene activity, evidence for position effects involving changes in gene order and distances on chromosome II that are associated with changes in levels of resistance, evidence for both dominant and recessive inheritance of chromosome II resistance depending on the position effects, and finally, preliminary evidence that at least one and probably additional structural genes controlling the detoxifying enzymes involved in resistance are located elsewhere than on chromosome II. The major finding described here will relate to the demonstration that position effects appear to play a major role in determining the total expression of insecticide resistance in the house fly.

REVIEW OF RELATED WORK

For comparative purposes it is pertinent to review what is known concerning resistance to xenobiotics in other organisms. The examples to be discussed are those cases involving multiple drug resistance or the ability to metabolize multiple xenobiotics that appear to be associated with a common genetic locus. In these examples, there are striking similarities to what occurs in insect populations resistant to a wide range of insecticides, although the mechanisms involved are certainly not totally identical.

Multiple Drug Resistance in Bacteria

The best known system of multiple xenobiotic resistance concerns the transmissible drug resistance occurring in Enterobacteriaceae (Mitsuhashi, 1971; Davies and Rownd, 1972). Current genetic theory holds that in bacterial strains exposed to drugs, resistance should occur to only one antibiotic at a time and multiple resistance should occur only by the accumulation of successive mutations. However, in a large number of cases resistance to many antibiotics appears simultaneously in bacterial populations. Genetic studies have revealed that the resistance is controlled by an extrachromosomal R (for resistance) factor that harbors a series of drug resistance genes and can be transferred from one bacterium to another by conjugation. The genetic material conferring resistance is capable of multiplying independently of its host and can be transferred not only to other bacteria of the same species, but also to other bacterial species. The occurrence of this process in bacteria poses a serious problem for continued use of antibiotics for disease control in much the same way that the occurrence of multiple insecticide resistance in insects is a problem for insect control by insecticides.

Multiple Drug Resistance in Yeast

A similar antibiotic resistance system is known from the yeast *Saccharomyces cerivisiase*. In a recent study (Cohen and Eaton, 1979) evidence was presented that a single nuclear gene controls resistance to a wide range of antibiotics. In addition, a series of genes located in the mitochondria interact with the nuclear gene and determine the response to specific antibiotics. Thus, resistance is controlled by the interaction of genes at two distinct loci in the organism. Implicit in these findings is the idea that the nuclear gene is regulatory in nature while the mitochondrial genes are structural genes associated with specific biochemical processes conferring resistance.

Genetics of Multiple Insecticide Resistance in *Drosophila melanogaster*

Early genetic studies on DDT resistance in *Drosophila* were done in Japan (Tsukamoto and Ogaki, 1953; Ogaki and Tsukamoto, 1953). In tests with strains designated Fukuoka and Hikone, evidence was obtained that the major gene for DDT resistance was dominant and located on chromosome II near the marker vestigial wing (*vg*), which is at locus 67.0.

Tsukamoto and Ogaki (1954) also tested for BHC resistance in *Drosophila* and proposed that the gene was the same as the DDT resistance gene. According to their study resistance to both DDT and BHC was controlled mainly by a single gene at locus 66.0 on the second chromosome.

The genetics of resistance to organophosphates and carbamates have also been studied (Kikkawa, 1964a,b). Resistance to both parathion and carbaryl was found to be controlled by a dominant gene located near 64.5 on the second chromosome.

Thus, these early studies indicated that the same or a series of very closely related genes conferred resistance to all types of insecticide in *D. melanogaster*. At that time our knowledge of insecticide resistance biochemistry was much less extensive than at present and no biochemical studies were done on the nature of the resistance mechanisms involved.

The cited studies were all performed with larvae as test insects. Experiments by other investigators who used adults as test insects yielded different results. These investigators (Crow, 1954; King, 1954; Oshima, 1954) all reported complex polygenic systems of genetic factors for DDT resistance.

The cited research failed to answer clearly the question of whether resistance in this species is caused by simple or complex systems. However, it is very interesting to note the many examples in which resistance to different types of insecticides is concentrated at a single locus, 64.5 on the second chromosome.

Coordinate Regulation of Drug-metabolizing Enzymes in the Mouse

Oxidative detoxifying enzymes in mammals, as in insects, are readily inducible by a wide range of xenobiotics including many insecticides. As recently reviewed by Nebert and Jensen (1979) studies with laboratory mouse populations have revealed that at least 20 different monooxygenase activities are associated with a common genetic locus that has been designated *Ah* (for aryl hydrocarbon hydroxylase). In addition, the induction of microsomal UDP glucuronosyl transferase, cytoxolic reduced NAD(P):menadione oxidoreductase, and cytosolic ornithine decarboxylase appear to be associated with the same genetic locus.

According to the authors, the regulatory *Ah* locus produces a cytosolic receptor, i.e., its protein product, which binds a variety of inducers of the system. The hypothesis is that a wide variety of inducers binds to a common cytosolic receptor, thereby invoking a pleiotropic response, which is the coordinate induction of a variety of detoxifying enzymes.

It is worth noting, as Nebert and Jensen have done, that not all oxidative detoxifying activities are under the regulation of the *Ah* locus. Detoxifying enzymes not related to this locus include NADPH-cytochrome c reductase, NADPH-cytochrome P-450 reductase, epoxide hydrase, and GSH transferases.

In summary, the chief significance of the Nebert studies in relation to insecticide resistance genetics is the occurrence in mice of a single genetic locus, regulatory in nature, that is a common factor in the metabolism of multiple xenobiotics and may interact with multiple structural genes. The implication is that changes in regulatory mechanisms rather than changes in structural genes may prove to be the major event taking place in insect populations responding to selection with insecticides.

INSECTICIDE RESISTANCE IN THE HOUSE FLY

Types of Resistance Mechanisms

There are three known types of insecticide resistance mechanisms that have been subjected to genetic and/or biochemical study. These include decreases in the ability of flies to absorb insecticides, increases in their ability to metabolize insecticides, and changes leading to a decrease in sensitivity of the target site or sites to the effects of insecticides.

With some insecticides, such as the hard-to-metabolize chlorinated hydrocarbons and cyclodienes, step two of this process has been bypassed. In these cases the insecticides are so difficult to metabolize that it has proven biologically more feasible to respond by means of changes in target sites.

Genetic studies have been made of all the described types of resistance. We know, for example, that in the house fly the gene *pen* (for decreased rate of penetration) is located on chromosome III and is inherited as a simple recessive factor (Sawicki and Farnham, 1968; Hoyer and Plapp, 1968). The gene confers a low level resistance to most insecticides, but is most important as a modifier of other resistance genes.

Metabolic resistance to insecticides is most important with the biodegradable insecticides, such as organophosphates and carbamates, and probably with synthetic pyrethroids. It also may be important in resistance to DDT. Detoxifying enzymes involved include aliesterases or carboxylesterases, mixed-function oxidases, GSH transferases, and DDT dehydrochlorinase. Extensive genetic and biochemical studies have resulted in findings that resistance associated with all these processes is controlled primarily by genes on chromosome II (Oppenoorth et al., 1977, 1979; Motoyama et al., 1977; Plapp, 1976) and appears to be inherited in an intermediate to incompletely dominant manner. From these studies we have assumed that the gene or genes measured on chromosome II are structural genes dealing with changes in each biochemical process involved. So far alternate explanations have not received adequate investigation.

Several house fly strains, notably SKA as studied by Sawicki and Farnham (1967) and R-Fc as studied by Oppenoorth (1967) and various others, have been found to have metabolic resistance factors controlled primarily by genes on chromosome V and to a lesser extent, by genes on chromosome II. In a recent study of strain R-Fc (Terriere and Schonbrod, 1976), however, evidence was obtained that resistance in the test population was located primarily on chromosome II and to a lesser extent on chromosome I. Reasons for the observed changes from earlier findings are not known.

Lastly, three target resistance mechanisms are known. These include the recessive gene for knockdown resistance to DDT and pyrethroids, which is located on chromosome III, a chromosome IV gene for resistance to cyclodiene insecticides, and a chromosome II gene for an altered acetylcholinesterase, the presumed target of organophosphate and carbamate insecticides. The location of the latter gene on chromosome II has only recently been confirmed (Oppenoorth, 1979). In cases of knockdown resistance to DDT and cyclodiene resistance, inheritance of resistance is recessive to susceptibility. This is in contrast to metabolic resistance in which resistance inheritance is more or less dominant.

Location of Resistance Genes on Chromosome II

Efforts to determine the precise location on chromosome II of various metabolic resistance genes have yielded inconsistent results. Part of the reason for this may be that the biochemical factors involved were not always studied in conjunction with measurements of the inheritance of resistance. As a result, the precise genetic and biochemical relationships have remained unclear.

For example, studies on the inheritance of resistance to diazinon have given contradictory results. Franco and Oppenoorth (1962) reported that cross overs between diazinon resistance and the mutant aristapedia (*ar*) were rare or absent. In contrast, Tsukamoto and Suzuki (1965) reported that diazinon resistance was located about 30 units from *ar*. Khan et al. (1973) used biochemical assay methods and reported that a gene for high levels of microsomal oxidase activity (*Ox-2*) was 40 units from stubby wing (*stw*) and 32 units from carnation eye (*car*). These results are in quite close agreement with those of Tsukamoto and Suzuki.

Variable results have also been reported for the location of a gene for DDT resistance, presumably due to DDT dehydrochlorinase. Lichtwardt (1964) reported the gene was located 7.35 units from *car* while Khan et al. (1973) reported it was very close to *car*. In another study, Tsukamoto and Suzuki (1964) studies the location of genes for DDT resistance in two strains in relation to the markers *ar* and *cm* (for carmine eye). In the two strains studied, the loci for DDT resistance were 5.8 and 3.1 units from the marker *ar*. Inversion or transposition of the chromosome was their explanation for the observed variability; however, there is no cytological evidence that this is the case.

Numerous other reports of the location of chromosome II metabolic resistance genes have been made, but will not be reviewed here. Overall, the data are similar to those already reported; resistance is apparently located either very close to the mutant *ar*, or else 20 to 30 units from it. As we shall show later, both estimates are correct.

Coordinate Regulation of Chromosome II Metabolic Resistance Genes

The chromosome II resistance genes of the house fly have a number of common properties. At least three of them, DDT dehydrochlorinase, mixed-function oxidase, and glutathione transferase occur at low levels in susceptible strains and at high levels in resistant strains. All are inducible enzymes; that is, exposure to insecticide substrates or to other known inducers can result in increases in enzyme activity.

Less appreciated is the fact that these enzymes are coordinately inducible. Chemicals that induce increased levels of one also induce increases in levels of the others, even in cases where the inducers are not substrates for the enzymes. For example, Terriere and Yu (1973) reported that a juvenile hormone analog induced DDT dehydrochlorinase as well as mixed-function oxidase. The inducer is a known substrate for the latter enzyme, but not for the former.

Results from our own laboratory (J. A. Ottea and F. W. Plapp, unpublished data) have shown that phenobarbital is as good an inducer of GSH transferase as it is of mixed-function oxidase. Again the inducer is a substarte for the latter, but not for the former.

The point is, induction of insecticide detoxifying enzymes appears to be a pleiotropic effect. Chemicals that induce one appear to induce several if not all. This implies that the inducers have a common site of action; that is, they react with the same enzyme or at least with the same gene product, which may then in turn act in some fashion on structural genes for specific detoxifying enzymes. The findings point out the similarities between chromosome II resistance in the house fly, the common gene site of resistance in *Drosophila* and the common metabolic locus, *Ah*, in mice. The data suggest that what we measure, at least in part, is the effect of genes regulating the quantity of detoxifying enzymes. This is in addition to the effects of structural genes controlling the quality of these enzymes.

EXPERIMENTAL

The remainder of this report will deal with recent investigations from our laboratory on the genetic origins of metabolic resistance to insecticides in the house fly. Specifically, we have investigated the genetics of chromosome II resistance and done some biochemical measurements related to these genetic studies.

Crosses Between Resistant Strains and Susceptible Strains with Single Chromosome II Visible Recessive Mutants

The first study we undertook was an investigation of the location of chromosome II resistance genes in relation to several chromosome II visible recessive mutants. This would, we hoped, give us a better idea as to the precise location of the various types of metabolic resistance genes associated with chromosome II.

The resistant (R) strains tested are listed in Table 1. All have been reported previously as having the major portion of their resistance controlled by chromosome II genes. Three susceptible (S) strains, each containing a chromosome II visible mutant, were used. The mutants were aristapedia (*ar*), stubby wing (*stw*) and carnation eye (*car*). All have been used in previous studies on the location of chromosome II resistance genes, either in our laboratory or by other investigators.

The test procedure consisted of crossing R strain females with S mutant strain males, followed by backcrossing F_1 females to S mutant males. Recombination between R genes and the mutants was then determined. The strain having only DDT resistance was not included in these experiments.

The results of these studies showed that in all cases resistance genes in the test strains were located in approximately the same area of chromosome II. In each case the gene order appeared to be *ar*-R-*stw*-*car* and the distances between genes were about five, five, and three crossover units, respectively. In other words, these preliminary tests indicated chromosome II resistance genes were at nearly the same location regardless of the differing metabolic resistance mechanisms present in the different strains.

Recombination of Resistance with Chromosome II Mutants

The next step was to recombine resistance from the R strains with the chromosome II mutants and compare the derived strains with their parents. Measurements of resistance levels in these R mutant strains gave unexpected results. As summarized in Figure 1, they showed that different levels of resistance appeared in each R mutant strain. For the R-Parathion series, only small amounts of resistance

Table 1. Resistant Strains and Resistance Mechanisms Present

Strain	Resistance Mechanism[a]				
	DDT dehydro-chlorinase	Altered ali-esterase	Mixed-function oxidase	GSH transferase	Altered AChE
R-DDT	Yes	No	No	No	No
R-Parathion	Yes	Yes	No	No	No
R-Diazinon	Yes	Yes	Yes	Yes	No
R-Propoxur	Yes	Yes	Yes	Yes	No
R-Tetrachlor-vinphos	Yes	Yes	Yes	Yes[b]	Yes

[a]Resistance mechanisms involve higher than normal levels of DDT dehydrochlorinase, mixed-function oxidase, and glutathione transferase, abnormally low levels of ali-esterase as measured with methyl n-butyrate as substrate, and an AChE of reduced sensitivity to inhibition by organophosphate insecticides.

[b]Glutathione transferases higher in this strain than in R-Diazinon and R-Propoxur.

were present in the *ar* strain, while the *stw* strain was about as resistant as the R parent strain. The *car* strain was even more resistant than the original R parent.

Similar results occurred with the R-Tetrachlorvinphos strains. Again, only a small amount of resistance was present in the *ar* strain, an intermediate amount was present in the *stw* strain, and a high level of resistance was present in the *car* strain. In this case resistance in the *stw* strain was less than in the R parent strain while the *car* strain resistance level was similar to that of the R parent.

From these tests it is evident that chromosome II resistance is inherited in a complex manner. However, the similarities in R levels in the mutant R strains derived from R-Parathion to those from R-Tetrachlorvinphos indicate that the genetic factor(s) involved affect both R strains in a similar way. This suggests that they are regulatory in nature or function, rather than structural genes for specific detoxification enzymes.

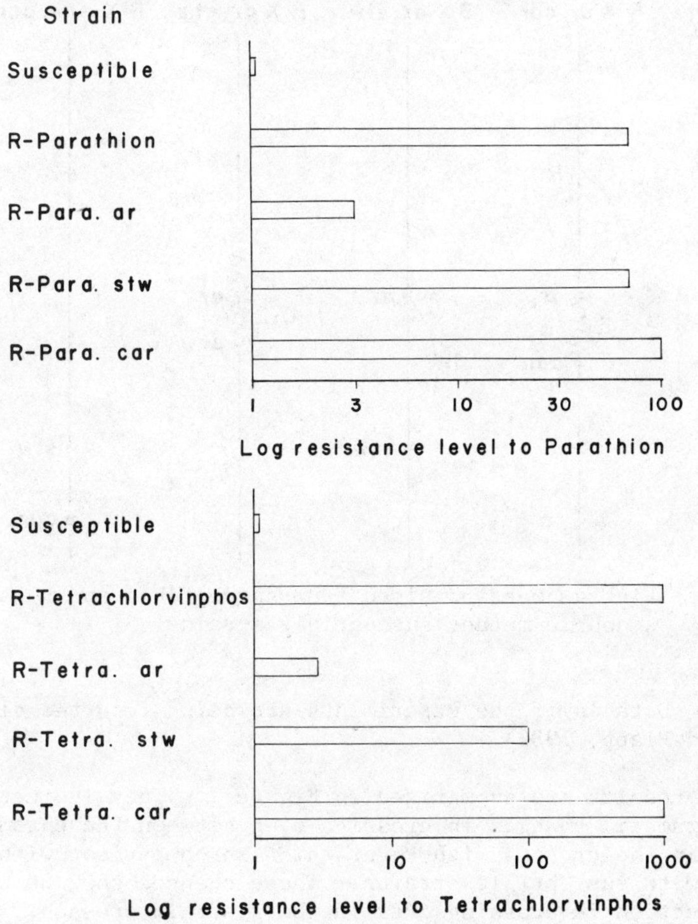

Figure 1. Resistance levels in parent resistant strains and substrains.

Crosses Between R-Parathion and Double Mutant Susceptible Strains

The previously described experiments were inadequate for precisely locating resistance genes in the various strains tested. Better for this purpose are 3-point test crosses. From such crosses it should be possible to locate precisely each gene in relation to the other genes tested.

For these experiments we prepared three double mutant susceptible strains, *ar car*, *ar stw*, and *stw car*. We then crossed these strains with both susceptible and R-Parathion strains and determined the recombination values and the gene order in the F_1 female back-

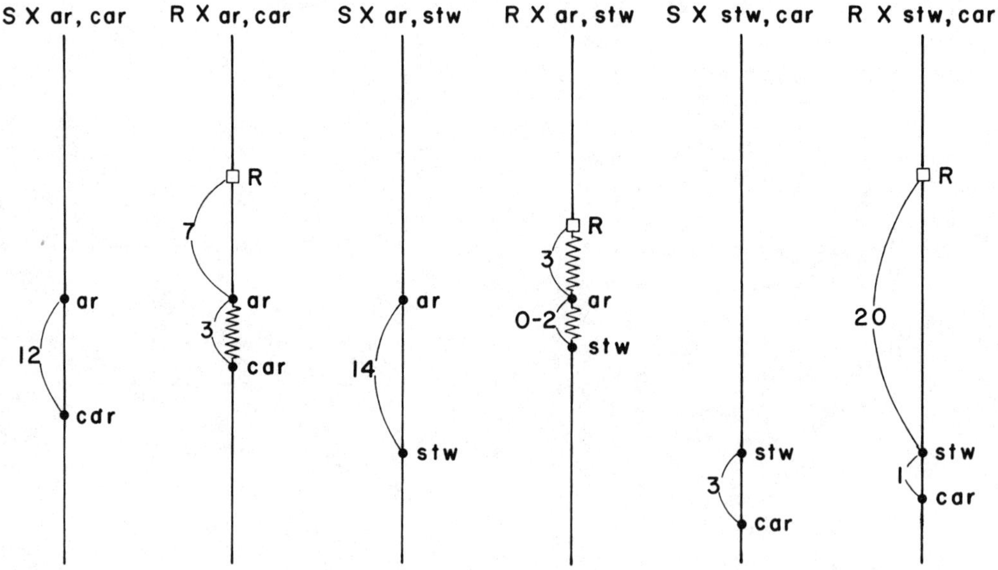

Figure 2. Linkage maps obtained from susceptible and R-Parathion x double mutant susceptible strains.

crosses. Details of the experiments are being reported elsewhere (Wang and Plapp, 1980).

The results are summarized in Figure 2. They indicate some rather dramatic changes in crossing over between the chromosome II mutants in the crosses with R-Parathion in comparison with similar crosses with susceptible strains. These changes show up primarily as a reduction in crossing over in the R strain crosses. For example, the 12% recombination between *ar* and *car* in S x *ar car* was reduced to 3% in R-Parathion x *ar car*. Similarly the 14% recombination between *ar* and *stw* in S x *ar stw* was reduced to from 0 to 2% in R x *ar stw*. Even with the closely linked mutants *stw* and *car* recombination was reduced from 3% to 1%.

The R gene did not appear to be at a constant position in relation to the visible mutants. In the cross with *ar car*, the distance from R to *ar* is 7 units. In the cross with *ar stw*, the R to *ar* distance is 3 units. Finally, in the cross with *stw car*, the R to *stw* distance is 20, a value much greater than in the other test crosses.

In several respects these results are quite confusing. What they show, however, is that chromosome II of the R-Parathion strain differs considerably from chromosome II of susceptible strains in the relative locations of several genes. The results suggest that changes in gene position on the chromosome are a factor in the expression of resistance. The precise nature of these changes has not so far been determined.

Crosses Between Resistant Strains and a Triple Mutant Susceptible Strain

The R-Parathion x double mutant tests indicated the occurrence of unexpected complexities in the inheritance of chromosome II resistance genes. In an attempt to resolve these difficulties we combined all three visible recessive chromosome II mutants into one strain, *ar stw car*. This strain was then crossed with the four organophosphate- and carbamate-resistant strains tested previously and also with the R-DDT strain. Recombination values and gene order were determined in F_1 female backcrosses. A preliminary report on these findings was made earlier (Wang and Plapp, 1978).

Data from these studies are summarized in Figure 3. In R-DDT, chromosome II is apparently similar in structure to that of the susceptible strain and the R gene is located 24 units from the marker *ar*. This distance is close to that reported earlier for R-Diazinon to *ar* (Tsukamoto and Suzuki, 1965).

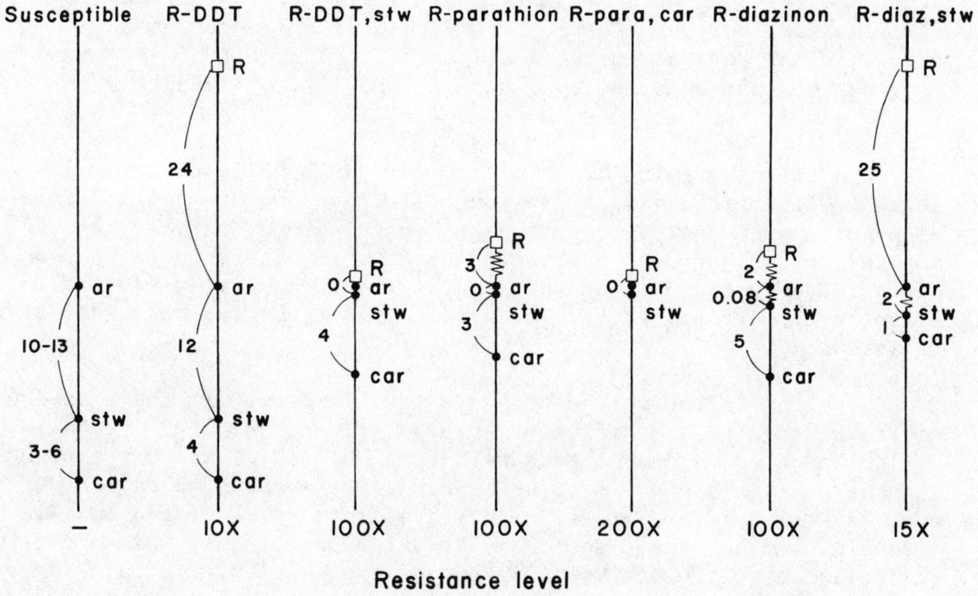

Figure 3. Linkage maps of house fly strains crossed with susceptible triple mutant strain.

DDT resistance from this strain was introduced into *stw* via recombination. Resistance in the DDT *stw* strain was higher than in R-DDT and so, the gene loci were remapped. This time the R gene showed no crossing over with the mutants *ar* and *stw* and recombination between the mutants themselves was absent. Thus, high level resistance to DDT in DDT *stw* is associated with a reduction in crossing over as compared to that of the parent DDT-R strain. In DDT *stw*, the gene locus for resistance is close to that reported by Lichtwardt (1964) and Tsukamoto and Suzuki (1964).

The R-Parathion x *ar stw car* test cross also showed a complete absence of crossing over between *ar* and *stw* as well as a small, but measurable amount of crossing over between R and *ar*. Introducing resistance into the mutant *car* (R-Parathion *car*) resulted in a strain with more resistance than in the parent R-Parathion strain and also, with a recombination pattern identical to that of DDT *stw*, i.e., no recombination between R and the mutants *ar* and *stw*. This complete suppression of crossing over between R and *ar* is identical to that reported earlier for diazinon resistance by Franco and Oppenoorth (1962).

The R-Diazinon x *ar stw car* cross gave results very similar to R-Parathion x *ar stw car*, i.e., greatly reduced crossing over between *ar* and *stw* and only a little crossing over between R and *ar*. When diazinon resistance was introduced into *stw* and then tested, the results were somewhat different. Here recombination between *ar* and *stw* remained low, but the distance from R to *ar* was 25 units, a value similar to that in the R-DDT strain. This R to *ar* distance is close to that reported by Tsukamoto and Suzuki (1965) who used a very similar test procedure.

From these results it is clear that the high levels of resistance to insecticides that are controlled by chromosome II genes are associated with marked reductions in crossing over between R genes and the mutants used as markers. Furthermore, the greater the level of resistance, the greater the reduction in crossing over; that is, position effects appear to play a role in determining the level of resistance.

A second factor worth noting is the similarity in the location of the R genes in the different strains. Thus, 24 to 25% recombination between R and *ar* occurs with both DDT and diazinon resistance in moderately resistant strains. In all the strains we have studied that have high levels of resistance associated with chromosome II, recombination between R and *ar* is very low or absent. This occurs regardless of the biochemical makeup of the particular strain tested.

A proposed evolution of chromosome II in terms of resistance development is shown schematically in Figure 4. The loci of the

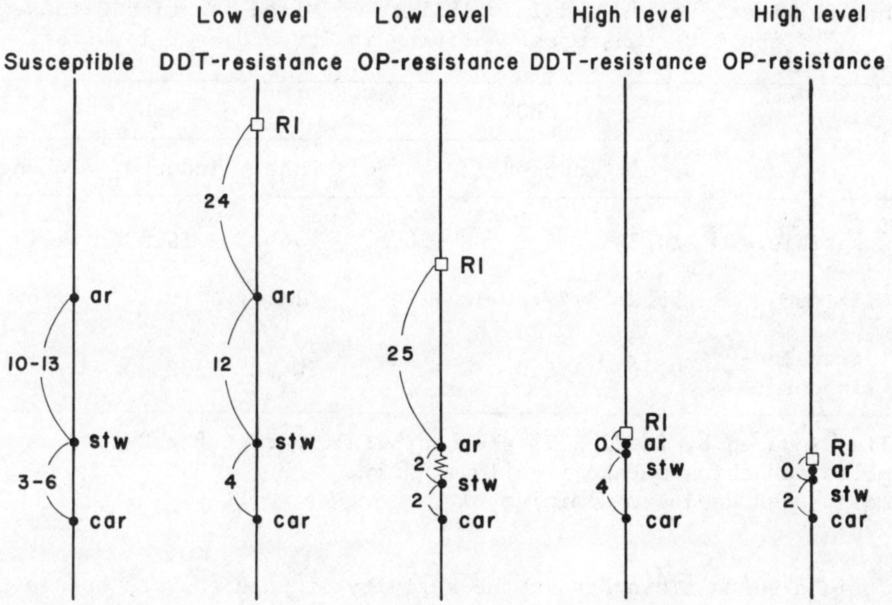

Figure 4. Evolution of chromosome II in relation to resistance.

three mutants in susceptible flies is similar to the map of chromosome II of Hiroyoshi (1977), and based on the results obtained, we propose that the resistance genes studied are in fact identical in all strains tested. This common gene appears to occur in one of two positions and the position determines the level of resistance present. The similarities between the gene positions in different strains are apparently independent of the metabolic resistance patterns present in the different strains.

Based on the above data we propose that the chromosome II gene we have studied is regulatory in nature. We propose to name it RI (for resistance to insecticides), a name identical to that given to the similar gene in *Drosophila* many years ago.

Inducibility of Detoxifying Enzymes in Susceptible and Resistant Strains

The genetic studies just described point out strong similarities between the inheritance of chromosome II resistance genes in all strains studied. To further investigate these similarities we measured the inducibility of mixed-function oxidases and GSH transferases in several of the tested strains. Results of these measurements are presented in Table 2, and the data show very similar levels of induction of both enzymes within each strain. They also show differences in inducibility between strains.

Table 2. Basal Levels and Inducibility of Mixed-function Oxidase and GSH-transferase Activity in Three House Fly Strains[a]

Strain	MFO[b]			GSH[c]		
	Basal	Induced	% change	Basal	Induced	% change
Susceptible *stw*	14.7	56.9	+289	4.8	16.5	+244
R-Diazinon	85.8	149.2	+ 74	11.3	19.1	+ 69
R-Tetrachlorvinphos	91.0	116.0	+ 27	204.3	202.4	- 1

[a]Flies induced by feeding 1% phenobarbital in diet for 2 to 3 days.
[b]pmoles parathion metabolized/15 min/fly.
[c]pmoles DCNB conjugated/min/gm fly.

In a susceptible strain the activity of both enzymes was increased by two to three-fold over that of control values by the phenobarbital treatment used. In R-Diazinon the percent change was less than two-fold. In R-Tetrachlorvinphos only a low level of mixed-function oxidase induction and no induction of GSH transferase was observed.

Clearly, induction is greater in the susceptible strain than in either resistant strain. Furthermore, there is less induction in R-Tetrachlorvinphos than in R-Diazinon, and based on numerous bioassay tests, R-Tetrachlorvinphos is a more highly resistant strain than R-Diazinon. From these findings it follows that basal enzyme activity is less inducible in R strains than in the susceptible strain.

The similarities in percent induction between each enzyme within each strain suggest a pleiotropic mechanism is active, i.e., the two enzymes may be coordinately regulated by the same process. While the precise regulatory mechanism is not understood, it may possibly be the previously described chromosome II regulatory resistance gene. However, the hypothesis has not yet been adequately tested.

Inheritance of Resistance and Inducibility in R-Diazinon and R-Diazinon *stw* Crosses

Previous results showed that the high level of resistance present in organophosphate-resistant strains was frequently reduced when resistance was introduced into the mutant *stw* by means of crossing over. In a study of this phenomenon we measured resistance

Table 3. Resistance Levels and MFO Activity in Several House Fly Strains and Crosses

Fly Strain or Cross	Resistance Level[a]	MFO Activity[b]	
		Control	Induced
Susceptible *stw*	1	15	57
F_1	27	34	63
R-Diazinon	100	88	140
Susceptible *stw*	1	15	57
F_1	2	13	50
R-Diazinon *stw*	15	53	121
R-Diazinon *stw*	15	53	121
F_1	16	50	75
R-Diazinon	100	88	140

[a] LC_{50} = 3.5 μg diazinon/450 ml glass jar. All values are ratio of response of indicated strain to response of susceptible *stw*.
[b] pmoles parathion metabolized/15 min/fly.

levels and both basal and induced mixed-function oxidase (MFO) levels in R-Diazinon and R-Diazinon *stw* strains and in crosses between these strains and a susceptible *stw* strain. The results are presented in Table 3.

In the cross between R-Diazinon and susceptible *stw*, both resistance and MFO activity were intermediate in the F_1, and induced MFO activity was similar to that in the susceptible parent.

In the R-Diazinon *stw* x susceptible *stw* cross, resistance and both control and induced MFO activity were similar to levels in susceptible *stw*; that is, they were inherited in a recessive manner.

In the cross between R-Diazinon *stw* and R-Diazinon, both resistance and oxidase activity in the F_1 were similar to levels in the less resistant R-Diazinon *stw* parent and inducibility was less than in either parent.

These results are the first demonstration we have made of a change in the inheritance of chromosome II resistance from intermediate or incompletely dominant to recessive. It is worth noting that both resistance and high oxidase activity disappeared in the F_1 of R-Diazinon *stw* x susceptible *stw*, while in the F_1 of R-Diazinon x R-Diazinon *stw* both resistance and oxidase activity levels were like those of the lower activity parent. This suggests that chromosome II resistance to diazinon is due to changes in regulatory genes rather than structural genes. This is true because if only structural genes were involved there should be some degree of intermediate level of both resistance and enzyme activity in the F_1. The results support the hypothesis that changes in regulation of detoxifying enzyme activity and resistance are important in chromosome II resistance.

Evidence That Structural Genes for Resistance Are Not Located on Chromosome II

The studies reported so far have dealt solely with the role of chromosome II in resistance in the house fly. They have suggested that the role of gene(s) on this chromosome is largely regulatory, controlling rates of enzyme synthesis.

However, regulatory factors cannot be the sole cause of resistance. If they were, then susceptible flies would become resistant to insecticides when appropriately induced. Numerous experiments in our laboratory, which were designed to test this hypothesis, have all failed to show significant increases in resistance in susceptible insects exposed to inducers. Clearly other factors must be involved.

The data suggest, then, that structural genes determining the specific nature of each detoxifying enzyme also play a role in resistance involving insecticide metabolism, and furthermore, that these genes may not be located on chromosome II. We describe here preliminary experiments that support this hypothesis.

For this work we have utilized the R-Tetrachlorvinphos strain because it seems to have all the known metabolic resistance factors. We crossed the R strain with the susceptible *ar stw car* strain described previously, which contains three chromosome II recessive mutants. We used an F_1 male backcross so that we could isolate flies homozygous for chromosome II from the susceptible parent and heterogeneous for all other chromosomes from both parents. The resulting R-Tetrachlorvinphos *ar stw car* strain was then selected for resistance to tetrachlorvinphos.

Data on resistance levels of R-Tetrachlorvinphos and R-Tetrachlorvinphos *ar stw car* to tetrachlorvinphos and several other insecticides are presented in Table 4. R-Tetrachlorvinphos *ar stw car* proved to be homozygous for a low level of resistance to tetrachlorvinphos and heterogeneous for low levels of resistance to

Table 4. Resistance Spectra of R-Tetrachlorvinphos and R-Tetrachlorvinphos *ar stw car* Strains

Insecticide	Resistance Level in Comparison with Susceptible *ar stw car*[a]	
	R-Tetrachlorvinphos	R-Tetrachlorvinphos *ar stw car*
Tetrachlorvinphos	>1000	5
Parathion	116	Heterogeneous, 1-5
Propoxur	> 500	Heterogeneous, 2-10
Trichlorfon	100	5

[a]Resistance level is based oncomparative 24 hr toxicity measurements in µg insecticide/ 20 ml glass vial required for 50% mortality.

parathion and propoxur. Resistance to trichlorfon also appeared to be homogeneous. We suggest that resistance to tetrachlorvinphos and possibly trichlorfon is controlled mainly by GSH transferases (and possibly by esterases) as described by Oppenoorth et al. (1977) while resistance to parathion and propoxur may involve mixed-function oxidases as well. The data indicate that a small, but measurable proportion of the resistance of the parent strain is controlled by genes not located on chromosome II.

We then measured GSH transferase in R-Tetrachlorvinphos *ar stw car* in comparison with the parent strains (Table 5). GSH transferase activity in R-Tetrachlorvinphos *ar stw car* was five times greater than in the susceptible parent strain and about 12 times lower than in the parent R strain. The results indicate that at least a portion of the high GSH transferase of R-Tetrachlorvinphos is controlled by genes not located on chromosome II. We hypothesize that the non-chromosomal II portion of high GSH transferase activity may relate to a structural gene for the enzyme, and the effects of this gene may possibly interact with the effects of a chromosome II regulatory gene to yield the extremely high enzyme level present in the R-Tetrachlorvinphos strain. Further studies are in progress to determine the chromosomal location of this gene in the house fly.

DISCUSSION

The data reviewed and presented in this paper offer significant extensions to our understanding of the genetic origins of insecticide

Table 5. GSH Transferase Activity of Several House Fly Strains

House Fly Strain	GSH Transferase Activity[a]
Susceptible *stw*	0.42
R-Tetrachlorvinphos	24.68
R-Tetrachlorvinphos *ar stw car*	2.10

[a]pmoles DCNB conjugated/min/gm fly.

resistance. The new work described here deals only with the house fly, but there is no reason to believe that it is not typical of resistance in other insect species as well.

The major new findings concern the nature of resistance factors on chromosome II. The data strongly suggest a high level of allelism, i.e., the genetic factor or factors involved in chromosome II resistance in one strain or to one insecticide are very similar to those conferring resistance in other strains to other insecticides.

The data also show the importance of position effects. Resistance is high when there is no recombination between resistance and the visible mutant aristapedia, and lower when R is located at a distance of approximately 25 crossing-over units from *ar*. Furthermore, the inheritance of both resistance and high levels of detoxifying enzyme activity are intermediate to semidominant when recombination between R and *ar* is reduced, and recessive when recombination is high. Thus, the position effect may be important because it increases the dominance of resistance inheritance.

The results fit well with the reports of other investigators who have dealt with similar problems. The finding of a common gene locus, *RI*, at 64.5 on the second chromosome in *D. melanogaster* that confers resistance to multiple insecticides agrees well with present evidence for a common gene locus, *RI*, in house flies. The results suggest that similar R mechanisms operate in both *Drosophila* and *Musca*.

The data also show similarities to the findings of Nebert and co-workers with the *Ah* (for aryl hydrocarbon hydroxylase) locus in mice. Both *RI* in house flies and *Ah* in mice are pleiotropic in effect and both are somewhat variable in inheritance (usually dominant, but sometimes recessive) depending on the particular cross involved.

Nebert and Jensen (1979) reported that an important product of the *Ah* locus is a cytosolic receptor capable of binding numerous inducers. Furthermore, they reported that the receptor:inducer complex in some manner activates structural genes that metabolize these inducers.

Our data agree well with this model, although to date there is no evidence on the nature of the product of the *RI* gene in insects. Furthermore, the genetic variation between strains of house flies should prove extremely useful in testing such a model of regulatory gene action.

A problem that has plagued researchers dealing with genetic variation in eukaryotes has been to determine if regulatory genes affecting enzyme activity exist, and if so, where they are located in relation to the structural genes whose action they modify. Specifically, are the regulator genes adjacent to the structural genes as in the *lac* operon or are they located elsewhere in the genome? Paigen (1979) summarized evidence for both types of regulation in eukaryotes, i.e., the occurrence of both near and distant regulators.

McDonald and Ayala (1978), working with alcohol dehydrogenase in *D. melanogaster*, and Korochkin et al. (1978), working with an esterase in *D. virilis*, presented evidence for the occurrence of regulatory genes on chromosomes other than those containing the structural genes whose action they modify in insects. McDonald and Ayala pointed out how variations and evolution in gene regulation can play an important role in adaptation. Korochkin et al. described their model as convenient for studying mechanisms of gene activity regulation in eukaryotes.

Our findings agree with both reports in that regulation of several metabolic resistance mechanisms in the house fly is apparently accomplished by genes at a common locus, a locus almost certainly different from the structural genes whose action it modifies. The precise locations of structural genes for the different detoxifying enzymes in the house fly remain to be determined, but the experiments described here on GSH transferase in R-Tetrachlorvinphos provide evidence that the structural gene for this enzyme may be located elsewhere than on chromosome II. Identifying structural gene loci and determining how they interact with what appear to be regulatory mechanisms on chromosome II is the next major problem to be investigated as we strive to increase our understanding of the genetic origins of insecticide resistance.

Acknowledgements: We thank P. J. McKinney, J. A. Ottea, C. P. Chang, L. Pavlas and W. H. Vance of this laboratory for technical assistance with experiments reported in this paper and G. Valenta and L. Mynar for typing the manuscript. We thank geneticists J. D. Smith and S. Johnston, Plant Sciences Department, Texas A&M University and R. H. Richardson, Zoology Department, University of Texas, Austin, for helpful consultations concerning various aspects of the genetic problems encountered during the course of this work. The majority of the research reported here was supported financially by the Texas Agricultural Experiment Station.

REFERENCES

Alt, F. W., Kellems, R. E., Bertino, J. R., and Schimke, R. T., 1978, Selective multiplication of dihydrofolate reductase genes in methotrexate-resistant variants of cultured murine cell, *J. Biol. Chem.*, 253:1357.

Blackmun, R. L., Takada, H., and Kawakami, K., 1978, Chromosomal rearrangement involved in insecticide resistance of *Myzus persicae*, *Nature*, 271:450.

Cohen, J. D., and Eaton, N. R., 1979, Genetic analysis of multiple drug cross resistance in *Saccharomyces cerevisiae:* A nuclear-mitochondrial gene interaction, *Genetics*, 91:19.

Crick, F., 1979, Split genes and RNA splicing, *Science*, 204:264.

Crow, J. F., 1954, Analysis of a DDT-resistant strain of *Drosophila*, *J. Econ. Entomol.*, 47:383.

Davis, J. E., and Rownd, R., 1972, Transmissible multiple drug resistance in Enterobacteriaceae, *Science*, 176:758.

Devonshire, A. L., and Sawicki, R. M., 1979, Insecticide resistant *Myzus persicae* as an example of evolution by gene duplication, *Nature*, 280:140.

Franco, M. G., and Oppenoorth, F. J., 1962, Genetic experiments on the gene for low aliesterase and organophosphate resistance in *Musca domestica* L., *Entomol. Exp. Appl.*, 5:119.

Hiroyoshi, T., 1977, Some new mutants and revised linkage maps of the housefly, *Musca domestica* L., *Japan J. Genetics*, 52:275.

Hoyer, R. F., and Plapp, F. W., 1970, Insecticide resistance in the house fly: Identification of a gene that confers resistance to organotin insecticides and acts as an intensifier of parathion resistance, *J. Econ. Entomol.*, 63:787.

Khan, M. A. Q., Morimoto, R. I., Bederka, J. T., and Runnels, J. M., 1973, Control of the microsomal mixed-function oxidase by Ox^2 and Ox^5 genes in houseflies, *Biochem. Genet.*, 10:243.

Kikkawa, H., 1964a, Genetical analysis on the resistance to parathion in *Drosophila melanogaster*. II. Induction of a resistance gene from its susceptible allele, *Botyu-Kagaku*, 29:37.

Kikkawa, H., 1964b, Genetical studies on the resistance to Sevin in *Drosophila melanogaster*, *Botyu-Kagaku*, 29:42.

King, J. C., 1954, The genetics of resistance to DDT in *Drosophila melanogaster*, *J. Econ. Entomol.*, 47:389.

Korochkin, L. I., Matveeva, N. M., Kuzin, B. A., Karasik, G. I., and Maximovsky, L. F., 1978, Genetics of esterases in

Drosophila. VI. Gene system regulating the phenotypic expression of the organ-specific esterase in *Drosophila virilis*, *Biochem. Genet.*, 16:709.

Lichtwardt, E. T., 1964, A mutant linked to the DDT-resistance of an Illinois strain of house flies, *Entomol. Exp. Appl.*, 7:296.

McDonald, J. R., and Ayala, F. J., 1978, Genetic and biochemical basis of enzyme activity variation in natural populations. I. Alcohol dehydrogenase in *Drosophila melanogaster*, *Genetics*, 89:371.

Mitsuhashi, S., 1972, "Transferable drug resistance factor R," University Park Press, Baltimore.

Motoyama, N., Dauterman, W. C., and Plapp, F. W., 1977, Genetic studies on glutathione-dependent reactions in resistant strains of the house fly, *Musca domestica* L., *Pestic. Biochem. Physiol.*, 7:443.

Nebert, D. W., 1978, Genetic aspects of enzyme induction by drugs and chemical carcinogens, *in:* "The Induction of Drug Metabolism," R. N. Eastabrook and E. Lindenlaub, eds., pp. 419-452, Schattauer Verlag, Stuttgart.

Nebert, D. W., and Jensen, N. M., 1979, The *Ah* locus: Genetic regulation of the metabolism of carcinogens, drugs, and other environmental chemicals by cytochrome P-450-mediated monooxygenases, *CRC Critical Reviews in Biochemistry*, 6:401.

Ogaki, M., and Tsukamoto, M., 1953, Genetical analysis of DDT-resistance in some Japanese strains of *Drosophila*, *Botyu-Kagaku*, 18:100.

Oppenoorth, F. J., 1965, Biochemical genetics of insecticide resistance, *Ann. Rev. Entomol.*, 10:185.

Oppenoorth, F. J., 1967, Two types of sesamex-suppressible resistance in the housefly, *Entomol. Exp. Appl.*, 10:75.

Oppenoorth, F. J., 1979, Localization of the acetylcholinesterase gene in the housefly, *Musca domestica*, *Entomol. Exp. Appl.*, 25:115.

Oppenoorth, F. J., Smissaert, H. J., Welling, W., van der Pas, L. J. T., and Hitman, K. T., 1977, Insensitive acetylcholinesterase, high glutathione-S-transferase, and hydrolytic activity as resistance factors in a tetrachlorvinphos-resistant strain of houseflies, *Pestic. Biochem. Physiol.*, 7:34.

Oppenoorth, F. J., van der Pas, L. J. T., and Houx, N. W. H., 1979, Glutathione-S-transferase, and hydrolytic activity as resistance factors in a tetrachlorvinphos-resistant strain of houseflies, *Pestic. Biochem. Physiol.*, 7:34.

Oshima, C., 1954, Genetical studies on DDT-resistance in populations of *Drosophila melanogaster*, *Botyu-Kagaku*, 19:93.

Paigen, K., 1979, Acid hydrolases as models of genetic control, *Ann. Rev. Genetics*, 13:417.

Plapp, F. W., 1976, Biochemical genetics of insecticide resistance, *Ann. Rev. Entomol.*, 21:179.

Sawicki, R. M., and Farnham, A. W., 1967, Genetics of resistance in the SKA strain of *Musca domestica*. II. Isolation of the dominant factors of resistance to diazinon, *Entomol. Exp. Appl.*, 10:363.

Sawicki, R. M., and Farnham, A. W., 1968, Genetics of resistance to insecticides of the SKA strain. III. Location and isolation of the factors of resistance to dieldrin, *Entomol. Exp. Appl.*, 11:133.

Terriere, L. C., and Schonbrod, R. D., 1976, Arguments against a fifth chromosomal factor in control of aldrin epoxidation and propoxur resistance in the Fc strain of the housefly, *Pestic. Biochem. Physiol.*, 6:551.

Terriere, L. C., and Yu, S. J., 1973, Insect juvenile hormones: Induction of detoxifying enzymes in the housefly and detoxification by housefly enzymes, *Pestic. Biochem. Physiol.*, 3:96.

Tsukamoto, M., and Ogaki, M., 1953, Inheritance of resistance to DDT in *Drosophila melanogaster*, *Botyu-Kagaku*, 18:39.

Tsukamoto, M., and Ogaki, M., 1954, Gene analysis of resistance to DDT and BHC in *Drosophila melanogaster*, *Botyu-Kagaku*, 19:25.

Tsukamoto, M., and Suzuki, R., 1964, Genetic analysis of DDT-resistance in two strains of the housefly, *Musca domestica* L., *Botyu-Kagaku*, 29:76.

Tsukamoto, M., and Suzuki, R., 1965, Genetic analysis of diazinon resistance in the house fly, *Botyu-Kagaku*, 31:1.

Van Asperen, K., and Oppenoorth, F. J., 1959, Organophosphate resistance and esterase activity in houseflies, *Entomol. Exp. Appl.*, 2:48.

Walker, C. R., and Terriere, L. C., 1970, Induction of microsomal oxidases by dieldrin in *Musca domestica*, *Entomol. Exp. Appl.*, 13:260.

Wang, T. C., and Plapp, F. W., 1978, Genetics of resistance to organophosphate insecticides and DDT in the house fly, presented at national meetings, Entomol. Soc. Amer., Houston, Texas, November, 1978.

Wang, T. C., and Plapp, F. W., 1980, Genetic studies on the location of a chromosome II gene conferring resistance to parathion in the house fly, *J. Econ. Entomol.*, 73: 200.

METHODS OF GENETIC ANALYSIS OF INSECTICIDE RESISTANCE

Masuhisa Tsukamoto

Department of Medical Zoology, School of Medicine
University of Occupational and Environmental Health
Iseigaoka, Yahata-nishiku, Kitakyushu 807, Japan

INTRODUCTION

Insecticide resistance in insects was first genetically analyzed nearly 40 years ago when Dickson (1941) reported that resistance to HCN fumigation in the California red scale, *Aonidiella aureantii*, was inherited as a sex-linked, incompletely dominant character. Shortly following this, Yust et al. (1943) obtained similar results. In a modern sense, however, one may say that the age of insecticide resistance actually started with the initiation of worldwide usage of synthetic chlorinated hydrocarbon insecticides.

Since Harrison (1951) first reported that knockdown resistance to DDT in the house fly gave a clear-cut pattern of Mendelian segregation, numerous investigators have attempted to carry out similar genetic experiments, such as crossing susceptible (S) and resistant (R) strains. Sometimes, results obtained from those crossings gave them rather contradictory conclusions, mainly due to the differences in the levels of resistance in the strains used. Some investigators reported results of non-Mendelian inheritance or a polygenic system in which no major R gene was detected, but other investigators obtained rather simple segregation ratios, such as 3:1 in F_2 progenies or 1:1 in backcross progenies. At that time, the levels of resistance to an insecticide in the R strains used were still generally low compared to those of the more recent R strains, and hence, ranges of the dosage-mortality curve for R and S strains were mostly overlapping. In such cases genetic analysis was difficult even when visible mutant markers were available for crossing experiments because no effective discriminating dose could be drawn between R and S phenotypes.

In *Drosophila melanogaster*, there are a variety of visible mutant markers for which gene loci have been determined (Bridges and Brehme, 1944; Lindsley and Grell, 1967). Therefore, it is no wonder that the first precise genetic analysis of insecticide resistance was carried out using a DDT-resistant strain and mutant marker strains of *Drosophila melanogaster* (Tsukamoto and Ogaki, 1953; Crow, 1954). In the house fly, however, the mode of action of insecticides, metabolism of insecticides, and mechanisms of insecticide resistance had been investigated extensively by various workers. It was, at that time, rather common knowledge that insecticides with completely different chemical structures could possess different modes of action, and hence, could require different mechanisms for resistance. Indeed, a DDT-resistant strain of an insect species is not necessarily resistant to dieldrin or BHC, and vice versa. Similar situations were also apparent in other groups of insecticides, such as organophosphorus (OP) or carbamate insecticides, and therefore, results obtained from precise genetic analyses in *Drosophila melanogaster* appeared to be rather unusual when compared with information obtained from the house fly, with which insect toxicologists were well acquainted. That is, in *Drosophila,* the same single R gene is probably responsible for resistance to various insecticides with different chemical structures, such as DDT, BHC, parathion and carbaryl, while resistance to one of the botanical insecticides, nicotine sulfate, is controlled by a completely different gene (see Tsukamoto, 1969). Thus, Hoskins and Gordon (1956) stated in their review that "this is contrary to all previous findings that these toxicants (DDT and BHC) are in different groups." Furthermore, phenylthiourea (PTU) and its halogenated analogs showed negatively correlated susceptibility to insecticide resistance in *Drosophila* (Ogita, 1958). This negative correlation between susceptibility to PTU and resistance to other insecticides would be useful if it could be applied to field populations of other insect species, but unfortunately, PTU and its analogs were not effective in killing other resistant pests of medical or agricultural importance. This evidence also indicates that the mechanism of resistance in *Drosophila* might be quite different from that found in other insect species, especially the house fly. Indeed, it was soon demonstrated in *Drosophila* that the metabolic pathway of DDT was completely different from that found in the house fly, i.e., there was oxygenation instead of dehydrochlorination (Tsukamoto, 1959). However, a question arises again as to why mechanisms of DDT resistance in *Drosophila* are effective in conferring resistance to other groups of insecticides. The mechanisms of insecticides other than DDT have not yet been investigated in this interesting species.

Nevertheless, genetic analysis is now considered to be a useful method of identifying or distinguishing insecticide resistance mechanisms, and the purpose of this paper is to describe the way in which genetic analyses are utilized in studies of insecticide resistance.

HOMOGENEITY OF STRAINS USED

Before proceeding to genetic analyses by crossing R and S strains, the homogeneity of both strains should first be examined; otherwise, analytical examinations of the resulting data could lead to errors in interpreting the mode of inheritance of resistance. The log dosage-probit mortality regression line (ld-pm line) is usually employed for expressing toxicological data in insecticide susceptibility tests and for the concurrent examination of the homogeneity of the populations used.

In a homogeneous strain, dosage-mortality data may offer a straight line with a steep slope. However, in a heterogeneous population, such as a field population and/or a field-collected laboratory colony, the ld-pm relationship will represent a more or less curved line, depending upon the components mixed and their progenies. In earlier studies of insecticide resistance, such a curve was usually considered to represent a straight regression line with a shallow slope. Therefore, selection pressure with an insecticide in this population gave rise to a steeper ld-pm line (Hoskins and Gordon, 1956; Brown, 1959). Tsukamoto (1963) discussed how to read ld-pm lines in detail.

From the statistical point of view, based on the probit transformation of percentage data, only the major part of the ld-pm line corresponding to about 10 to 90 percent mortality is effective in interpreting the straightness of the line or, in other words, the homogeneity of the population. Thus, leveling-off of the line or deviation of the mortality data at less than 10 percent or more than 90 percent was usually ignored by most toxicologists. From the genetic point of view, based on the segregation of progenies with various phenotypes, however, shallow portions of the slope or leveling-off of the curve at the lower or upper portion must be emphasized.

When a resistant strain tested does not show a straight ld-pm line, selection pressure for resistance should be continued for several generations to eliminate the possibility of the strain being heterogeneous. Then the crossing experiment may be started. If the continuous selection pressure still does not effectively purify the strain, it might be assumed that some special genetic mechanism is involved, such as a homozygous lethal gene linked with the R gene, or that the insecticides used were not pure, and so on. Continuous sib-matings combined with the insecticidal bioassay for subsequent generations are also effective in establishing homogeneous strains for either susceptibility or resistance.

CROSSING EXPERIMENTS

Virgin Females

The use of virgin females in crossing experiments is one of the essential things that must be strictly observed; otherwise the ld-pm line for the F_1 progeny from the cross between R and S strains may not be straight even if both parent strains are genetically pure.

Males and females should be emerged separately from individual pupae in small-scale crossing experiments. Sometimes it may be troublesome to do so, especially in large-scale crossing experiments, and in such cases, virgin females must be collected as a group. First of all, no emerged adult should be retained in a container of pupae. Then, only newly emerged females should be collected within a limited period. The maximum permissible period before the first emerged male reaches maturity depends on the species of insect and the temperature under which the pupae have been kept. In the case of the house fly, for example, the safety period is 8 hours at 25°C, and hence, adult flies emerging during an over-night period should not be used, unless the pupae are kept at a lower temperature during the night.

Reciprocal Crosses

Results of reciprocal crosses between R and S strains will provide at least the following three important pieces of information regarding the genetics of resistance: 1) dominance or recessivity, 2) cytoplasmic or chromosomal influence, and 3) sex chromosomal or autosomal inheritance.

Dominance. When the F_1 progenies obtained from reciprocal crosses possess almost the same high resistance level as the parental R strain, the resistance factor is said to be dominant over the susceptible one. On the other hand, when the F_1 progenies are just as susceptible to the insecticide as the parent S strain, susceptibility is dominant over resistance, or in other words, the resistance character is recessive to the susceptibility character. However, such complete dominance or complete recessivity is not often the case, and in most instances, resistance levels of F_1 progenies are found to be intermediate between those of S and R strains. These cases reveal either incomplete dominance or incomplete recessivity, respectively. Such usage of the term "incomplete" is rather ambiguous and Stone (1968) and Dittrich (1972) proposed a digital expression for the degree of dominance in insecticide resistance.

Genetic Symbols. According to genetics textbooks, the symbol for a dominant factor is expressed as a capital letter, such as *A*, and a small letter, *a*, for its corresponding recessive allele. Thus, *aa*, *Aa* and *AA* express the three possible combinations of genotypes at a single locus. This symbolic system is rather simple to understand even for beginners. For many applications, however, this is inconvenient and the symbol + has been adopted by geneticists for expressing a wild or normal standard allele, with a character dominant over the wild type symbolized by the capital letter and a recessive gene indicated by the small letter:

recessive	wild type	dominant
a	+ (or a^+)	
	+ (or A^+)	A

In this system, the *a* gene is not necessarily an allele of the *A* gene, and the wild type, expressed by +, is sometimes either dominant or recessive. The wild type allele for the gene *a* is a^+, or more simply +, unless confusion arises. For multiple alleles, numerals or letters are usually attached to the shoulder of the symbol. For example, in *Drosophila*, *bw*, bw^2, bw^D and bw^{au} are used to indicate *brown*, *brown-2*, *brown-dominant* and *brown-auburn*, respectively. For independent resistance genes, linkage group numbers should not be attached to the shoulder because they are not alleles of each other.

In genetic studies of insecticide resistance, the progenies of crossing experi nts are divided into only two phenotypic groups -- the resistant or the susceptible individuals classified as "alive" or "dead" after an insecticidal treatment. For convenience, the symbols R and S are used regardless of dominance and/or allelism by most insect toxicologists to express either the resistant or susceptible phenotypes or genotypes, respectively. Thus, the symbols RR, RS and SS are used to express the resistant homozygotes, the heterozygotes and the susceptible homozygotes even if the resistance character is recessive. Such a vague usage is sometimes very convenient for insect toxicologists, whereas strictly speaking, this is incorrect according to the conventional rule that is generally adopted by geneticists. In the present paper, the R and S system will be adopted in some cases because of its convenience and familiarity to insect toxicologists. The semicolon between mutant symbols should be used only when these genes belong to different linkage groups, for example *Deh;kdr* (2;3).

Cytoplasmic Influence. In usual susceptibility tests for an insect species, females show more or less higher resistance levels than males. In the reciprocal crosses, the F_1 progenies from both R♀ x S♂ and S♀ x R♂ have the same genotype except for sex chromosomes. Thus, differences observed in the F_1 progenies of these reciprocal crosses indicate the presence of either a cytoplasmic genetic factor or a maternal influence. Examples of cytoplasmic inheritance are known in CO_2 sensitivity and abnormal sex ratios in *Drosophila*, but no such obvious examples have ever been reported for insecticide resistance. When no significant differences are observed in the susceptibility levels in either sex from the reciprocal crosses (for example, F_1 females from the R♀ parent and those from the S♀ parent, or F_1 males from the R♀ and F_1 males from the S♀), resistance can be attributed to a chromosomal genetic factor.

Sex-linked vs. Autosomal Inheritance. When a resistance factor is linked to a sex chromosome, the sex ratio of the F_1 progeny will be largely affected by insecticide treatment. In most species of insects, except for Lepidoptera and Trichoptera, females are homozygous (XX) in the sex chromosome and males are heterozygous (XY). Therefore, the comparison of XX and XY individuals of the F_1 progeny tells us whether or not a sex chromosomal resistance factor had been involved in the parental R strain. When there are no significant differences between sexes in F_1 progenies of the reciprocal crosses, the R factor may be located on the autosome(s). If all the male progenies are resistant and all the female progenies are susceptible to a particular insecticide, some sex chromosomal abnormality or an autosomal male-determining factor may be involved. The first example of this was the sex-limited DDT resistance in house flies demonstrated by Kerr (1960).

BACKCROSSES

When only a single dominant or recessive gene is involved in insecticide resistance, the familiar 3:1 segregation ratio of R and S phenotypes would be expected in the F_2 progeny obtained from the cross between R and S strains. However, if dominance is incomplete, interpretation of the ld-pm curve for a mixture of SS, RS and RR genotypes in the F_2 progenies is sometimes very difficult, and hence, this has often misled earlier insect toxicologists to the assumption that resistance is due to a complicated polygenic mode of inheritance.

If resistance is assumed to be controlled by a single gene, interpretation of the 1:1 segregation ratio is easier in the backcross of the F_1 progeny to one of the parent strains. Of course, the F_1 progeny should be backcrossed to the parental R strain when the resistance character is recessive and to the parental S strain when the resistance character is dominant. The ld-pm curves for such backcross progenies will show a plateau corresponding to about 50 percent mortality. If the position of the plateau deviates far from 50 percent mortality, viability of the RS progenies may be higher or lower than that of the parental homozygotes. Sometimes another plateau or disorder of the ld-pm curve may be detected by a careful examination at an upper portion of the curve, although such a small change of curve is easily overlooked. If this phenomenon occurs consistently, it may indicate the involvement of some other minor R gene.

Therefore, examination of the backcross data is quite important and gives useful information on the mode of inheritance of the resistance character even if no visible mutant marker is available in crossing experiments.

FORMAL GENETICS

Although analytical examination of the ld-pm curves for progenies of crossing experiments is an important tool, we cannot draw any more information without utilizing visible mutant markers. Twenty-five years ago, practically no mutant markers were available for genetic analyses for insecticide resistance in insect pests of medical or agricultural importance, but now linkage groups have been estabilshed in several insect pests and a number of mutations have been located on the chromosome.

Mutant and Linkage Group in the House Fly

Since Bodenstein (1939) first described numerous variations in the house fly, more than 100 morphological mutants and aberrant forms have been recorded, and have been listed by Milani (1967, 1975), Nickel and Wagoner (1970) and Hiroyoshi (1977). Because of their lower viability, poor penetrance or varied expression, many mutant strains have been discarded before or after their locations on the chromosome were determined. At present, therefore, only a limited number of "good" markers are available for analysis of insecticide resistance.

The house fly has six pairs of chromosomes, as shown by Perje (1948), and six linkage groups have been established by Hiroyoshi (1960, 1961). During the early stages of study on house fly genetics, each investigator proposed a different numbering system for the various linkage groups because they found mutants independently; this caused some confusion not only among the house fly geneticists but also among toxicologists. In 1967, Wagoner was able to relate the genetic linkage groups to the cytological chromosomes as they were first numbered by Perje (1948). At present, all the house fly investigators have adopted the Wagoner's linkage numbers to avoid unnecessary confusion and to respect Perje's priority. Readers who refer to papers written before 1970, therefore, should be careful to avoid this confusion. The historical examples for transition of the numbering to the same linkage groups of the house fly are shown in Table 1.

For linkage group analysis of insecticide resistance factors, several combinations of "good" mutant markers are necessary. Furthermore, in order to carry out a three-point method for determining a locus for an R gene on a particular chromosome map, the shortage of good mutants (or a combination of good mutants) is

Table 1. Numbering System for Linkage Groups of the House Fly Reported by Various Investigators

Author (Year)	Chromosome or Linkage Group					
Cytological investigation						
Perje (1948)	1	2	3	4	5	XY
Genetic investigation						
Hiroyoshi (1960, 1961)	6	5	2	4	3	1
Milani (1961)	3	5	2	6	4	1
Tsukamoto et al. (1961)	4	5	2	6	3	1
Milani (1965*, 1967)	3	5	2	4	6	1
Genetic and cytological investigations						
Wagoner (1967, 1969)	1	2	3	4	5	XY
Representative marker	*ac*	*ar*	*bwb*	*ct*	*ocra*	sex

*Proposed in personal communication.

keenly felt. Much more effort should be made to discover a number of "good" visible markers for studying the relation between the mechanism of an R gene and its percise location on the chromosome.

Mosquito Genetics

Mosquitoes are one of the largest groups of insect pests of medical importance in relation to the global control of insect-born diseases. Most of them belong to three major genera: *Anopheles*, *Aedes* and *Culex*. Numerous visible mutants of mosquitoes have been reported and are still being reported by various investigators (for earlier mutants, see the book edited by Wright and Pal, 1967). Three linkage groups are known in mosquitoes where the first linkage group is generally allotted for the sex chromosomes. At that time, limited numbers of mutants and linkage groups were available in only a few mosquito species, such as *Culex pipiens* and *Aedes aegypti* (Laven, 1967; Graig and Hickey, 1967), but since then a number of mutants of various species have been found. Especially, information on mutants of *Anopheles stephensi, Anopheles culicifacies* and *Culex tritaeniorhynchus* has been rapidly accumulated by a group of mosquito geneticists working in Paksitan, and Kitzmiller (1976) has reviewed recent advances in mosquito genetics. Insecticide resistance genes are also being investigated, using these visible mutant markers.

GENETICS ANALYSIS OF RESISTANCE

Linkage Group Analysis for Resistance Factors

When the results of reciprocal crosses have indicated the presence of the autosomal genetic factor(s) for insecticide resistance, the next step of the genetic analysis is to determine which chromosome is responsible. For this purpose, however, the use of several visible mutant markers is an absolute requirement. This is a bottleneck for the study of most insect pests of medical or agricultural importance.

Synthesis and utilization of a multiple marker strain produced an advantage in the genetic analysis of insecticide resistance. Therefore, it should be well understood that availability of various combinations of "good" mutant markers that correspond to each linkage group is tremendously important. At present, only a few "good" multiple marker strains, such as *ac;ar;bwb;ocra* (1;2;3;5), are available for linkage group analysis in the house fly. In mosquitoes the number of visible mutants discovered is still increasing and the linkage relationship is being firmly established among several species, as mentioned above.

Results of a linkage group analysis statistically based on a factorial analysis can indicate the presence of R gene(s) both qualitatively and quantitatively; that is, we can learn with which chromosome the R gene is associated and whether that genetic factor is major or minor. Details on practical crossing experiments for the linkage group analysis in which multiple marker strains are utilized in the house fly have been described by Tsukamoto (1964), Sawicki and Farnham (1967) and Georghiou (1969). Fundamentally, similar technical procedures are applicable to mosquitoes or other insects if mutant markers are available.

Location of Resistance Genes on Chromosome Maps

After determining the linkage relationship between the R gene and the chromosome, the next step in genetic analysis is locating the locus on the chromosome to which the gene links. Again, the design for crossing experiments is based on the backcross but it is different from that for linkage group analysis: here females of the F_1 progeny are backcrossed to males of one of the parent marker strains, because in the house fly crossing over occurs almost exclusively in females. When the R gene in question is dominant, males of the S marker strains should be used. When the R gene is recessive, males of the R marker strain should be crossed to the F_1 females.

For determining the locus for the R gene on the chromosome in question, at least two "good" mutant markers are necessary for the three-point method, whereas only one marker is necessary for each linkage group in the case of linkage group analysis. These two markers are usually recessive genes and it is desirable that they be an appropriate distance from each other on the chromosome, such as a 10 to 30 map distance. Practical crossing schemes and formulas for calculating recombination values are given by Tsukamoto (1965). If the distance between the R gene and a marker gene is too short, the experiment may be a time-consuming and difficult task. However, as will be explained later, determination of a precise locus for each R gene is extremely important, especially for cases in which allelism of different R genes is suspected by similarity of the gene action, for example, *kdr* (Milani, 1956), *kdr-o* (Milani and Franco, 1959), *r-DDT* (Tsukamoto and Suzuki, 1964), a gene for low nerve sensitivity (Tsukamoto et al., 1965), a pyrethroid-R gene (Tsukamoto, 1969), etc., in the third chromosome of the house fly.

POPULATION GENETICS

Except for extremely inbred "isogenic" strains as in *Drosophila melanogaster*, we must recognize that any laboratory strain or actual field population of insect pests is not a genetically pure population but is a mixture of various genotypes. By successive selection with an insecticide even a standard laboratory S strain, such as NAIDM and CSMA, can develop resistance, rapidly or slowly, to a selective agent. The insecticide resistance problem, therefore, should be considered as one of the subjects of population genetics. By introducing formulas of population genetics based on the Hardy-Weinberg law, with some modifications, increases in resistance levels within a population under selection pressure and maintenance or decreases in resistance levels after removing selection pressure can be calculated by given information about the initial frequency of R and S genes in a given population, population size, frequency of inbreeding, strength of selection pressure, number of generations, fitness of R and S genes, etc.

Rapid advances in computer technology have made it possible to predict the fate of an R gene in natural or laboratory populations by inputting the various parameters mentioned above. Recent contributions to the study of computer simulations in insecticide resistance problems by Georghiou and Taylor (1977a,b) and Taylor and Georghiou (1979) are quite remarkable. In the near future such computer simulations will become more readily available as technological developments in electronics for microcomputers occur and as it becomes possible to simply select one model from various ready-made computer programs.

RESISTANCE GENES AND MECHANISMS IN THE HOUSE FLY

Information on biochemical and physiological mechanisms of resistance to insecticides has been rapidly accumulating, especially during this decade, as summarized in the excellent reviews by Oppenoorth and Welling (1976) and Plapp (1976). As the most extensive studies on mechanisms of insecticide resistance have been done on the house fly, only this insect will be considered here as an actual example. Linkage groups for major R factors and mechanisms are summarized in Table 2.

Linkage Group 1

Concerning Linkage Group 1, it seems that no important resistance gene is involved, although a minor factor for BHC resistance was indicated (Tsukamoto, in Georghiou, 1965). This, however, does not mean that the first chromosome is genetically inert regarding insecticide resistance, because this chromosome is cytologically the largest one among the house fly autosomes (Wagoner, 1967) and many visible and biochemical mutants are known to be located along a large portion of the chromosome map.

Recently one exception was reported by Rupeš and Pinterová (1975) in the case of fenitrothion resistance in a DDT-resistant strain in which the R factor of the first chromosome behaved as dominant for fenitrothion resistance and incompletely dominant for DDT resistance. However, levels of resistance (LD_{50}) to these insecticides in this strain are quite a bit lower than those reported in the usual R strains. The mechanisms of these R genes are completely unknown.

Linkage Group 2

The second chromosome is very interesting in insecticide resistance in the house fly. Several major R genes and important mechanisms assumed for R genes are associated with this linkage group. These include genes for resistance to DDT, BHC, diazinon, malathion, dimethoate, pyrethrins, Isolan, carbaryl, propoxur, Matacil, MPMC, naphthalene, juvenile hormone analogs, piperonyl butoxide, etc., and the mechanisms reported are DDT-dehydrochlorinase, altered aliesterase, glutathione S-transferase, phosphatase, insensitive acetylcholinesterase, microsomal mixed-function oxidases, P-450, etc.

Table 2. Linkage Groups for Major Resistance Factors and Mechanisms in the House Fly

I	II	III	IV	V
	DDT	DDT		DDT
	R: Lichtwardt (1964)	*kdr* (knockdown-r): Milani (1956)		*md* (microsomal detoxication): Oppenoorth & Houx (1968)
	R-DDT: Tsukamoto & Suzuki (1964)	*kdr-o*: Milani & Franco (1959)		Plapp & Casida (1969)
	Deh: Hoyer & Plapp (1966)	r-DDT: Tsukamoto & Suzuki (1964)		
			Dieldrin	
			Dld: Tsukamoto & Suzuki (1963)	
	BHC	BHC		BJC
	R-BHC: Tsukamoto (1969)	r-BHC: Tsukamoto (1969)	R-Dieldrin: Oppenoorth & Nasrat (1966)	R-BHC: Tsukamoto (1969)
			DR_4: Sawicki & Farnham (1968)	
	Pyrethroids	Dieldrin		
	R-Pyr: Tsukamoto (1969)	dld_3: Georghiou (1969)	Dld_4: Georghiou (1969)	
				Pyrethroids
	R-Allethrin: Plapp & Casida (1969)			R-Pyr: Tsukamoto (1969)
		Pyrethroids		
	py-ex: Farnham (1973)	r-pyr: Tsukamoto (1969)		*py-ses*: Farnham (1973)

OP	OP			OP
R-Fenitrothion: Rupeš & Pinterová (1975)	*a*: Oppenoorth (1959) *Dz*: Tsukamoto & Suzuki (1966) Plapp & Casida (1969) *g*: Oppenoorth & Rupeš (1972) *M*: Sawicki (1974) *D*: Sawicki (1974)			R-Diazinon: Tsukamoto & Suzuki (1966) Plapp & Casida (1969)
	Carbamates	Carbamates		Carbamates
	R: Hoyer et al. (1965) Tsukamoto et al. (1968) Georghiou (1969) Plapp & Casida (1969)	R: Tsukamoto et al. (1968) Georghiou (1969)		R: Tsukamoto et al. (1968) Georghiou (1969) Plapp & Casida (1969)

(continued)

Table 2. (continued)

I	II	III	IV	V
	Miscellaneous			
	R-Naphthalene: Schafer & Terriere (1970)			
	R-JHA: Vinson & Plapp (1974)			
	Yu & Terriere (1978)			
	Microsomal Oxidation	Decreased Penetration		Microsomal Oxidation
	Ox-2: Khan et al. (1973)	*Pen*: Sawicki & Farnham (1967)		*Ses*: Sawicki & Farnham (1967)
	P-450: Tate et al. (1974)	*tin*: Hoyer & Plapp (1968)		*Ox-5*: Khan et al. (1973)
		R_2: Sawicki & Farnham (1968)		R_5: Sawicki (1974)
	Insensitive AChE			
	Devonshire & Sawicki (1975)			
	Motoyama et al. (1977)	Nerve Insensitivity		
	Plapp & Tripathi (1978)	Tsukamoto et al. (1965)		

DDT-dehydrochlorination. Prior to precise genetic analysis with mutant markers in several house fly strains, it had been assumed that a major mechanism of DDT resistance was dehydrochlorination of DDT to DDE (Lovell and Kearns, 1959). That the enzyme DDT-dehydrochlorinase is involved in the mechanism of the second chromosomal major R gene has been demonstrated by using some inhibitors of DDT-dehydrochlorinase, such as DMC and WARF Antiresistant (Tsukamoto and Suzuki, 1964; Ogita and Kasai, 1965c; Grigolo and Oppenoorth, 1966; Sawicki and Farnham, 1967). Gene loci for R-DDT in a Japanese strain and in an American strain were slightly different both in map position and in recombination value (Tsukamoto and Suzuki, 1964).

Metabolism of Organophosphorus Insecticides. Oppenoorth (1959) and Oppenoorth and van Asperen (1960) adopted a symbol *a* for both aliesterase activity and the organophosphorus (OP) resistance gene. Subsequently, Franco and Oppenoorth (1962), using the visible mutant markers *ar* and *cm*, searched for the location of *a*, but crossing over between the markers and the *a* gene was either absent or rare. Oppenoorth and his co-workers suggested a hydrolyzing mechanism for the action of this resistance gene, but no biochemical evidence has been given yet for any actual metabolite yielded by this enzyme system. This has prompted studies on OP resistance, but at the same time, it has caused some confusion concerning the nature of second chromosomal metabolic resistance to OP compounds: phosphorylation or desalkylation.

According to Lewis (1969), at least two diazinon-detoxifying mechanisms are involved in the second chromosomal resistance. One is a microsomal cleavage that requires O_2 and NADPH or GSH, and another is a desethylation that requires only GSH. Similar results on the second chromosomal R gene were obtained by Oppenoorth et al. (1972) and this GSH-dependent R gene was named *g*. Motoyama et al. (1977) also showed the second chromosomal control of a GSH-dependent enzyme system.

Using a Japanese diazinon-selected Hokota strain, Tsukamoto and Suzuki (1966) determined the locus for a diazinon-R gene, *Dz*, to be near one end of the second chromosome, far from the *ar* and *cm* markers. At present the exact gene action of *Dz* is not known. According to results of in vitro experiments by Shono (1974a,b) in this strain, diazinon is metabolized through diazoxon, which is further metabolized to a water-soluble metabolite, diethylphosphoric acid, by a phosphatase. Neither a microsomal mixed-function oxidase system nor a GSH S-transferase system seems to play an important role in diazoxon metabolism in the Hokota strain, but diazinon is metabolized by the microsomal mixed-function oxidase to diazoxon and to diethylphosphorothioic acid (Shono, 1974c). Thus, more precise genetic analyses are necessary to correlate these metabolic enzyme systems and the diazinon R gene, *Dz*.

Sawicki (1974) reported that at least five R genes are involved in a dimethoate-selected strain originating from a Danish field population. Three of them, *D*, *M* and *Pb*, are on the second chromosome. Gene *D* confers stronger resistance to dimetoxon than to the original phosphorothionate, dimethoate, and probably controls mixed-function oxidase(s) because the resistance mechanism is suppressed by sesamex. Gene *M* is a major R gene to malathion, malaoxon and tetrachlorvinphos, but not to dimethoate. This gene is not a modified aliesterase gene *a* because the esterase activity to α-naphthyl acetate is normal. Malathion resistance is not affected by pretreatment with TBTP, a carboxylesterase inhibitor. Gene *Pb* controls resistance to methylenedioxyphenyl-type synergists, such as piperonyl butoxide (PB), and to a PB + pyrethrum combination, but not to pyrethrum alone. Loci on the chromosome map are still unknown but gene *M* is located near a marker gene *ar*, and gene *D* is close to gene *Pb* and is located about 20 map units from *ar*.

Microsomal Oxidation. The second chromosomal mixed-function oxidase(s) that requires NADPH as a cofactor for microsomal fractions can oxidize a variety of insecticides, not only carbamates (Tsukamoto et al., 1968; Plapp and Casida, 1969) but also OP compounds (Lewis, 1969; Plapp and Casida, 1969; Tsukamoto, 1969; Sawicki, 1973), naphthalene (Schafer and Terriere, 1970), and cyclodienes (Khan, 1970; Khan et al., 1970, 1973). High levels of cytochrome P-450, which has an important role in microsomal oxidation, are also controlled by a second chromosomal gene (Tate et al., 1974). It would be very interesting to know whether all these microsomal mixed-function oxidases are controlled by a single resistance gene or by several distinct genes.

Modified Acetylcholinesterase. Several examples of insensitive or altered acetylcholinesterase (AChE) have been reported, mainly in pests of agricultural importance. In the case of the house fly, Devonshire and Sawicki (1975) reported that the genetic factor for the decreased susceptibility of AChE in OP-resistant strains is associated with the second chromosome and the locus has a map distance of about 20 units from the mutant marker, *ar*. Plapp (1976) also confirmed the second chromosomal genetic factor for altered AChE, but the map distance reported by Plapp and Tripathi (1978), being 5 percent for *stw* (stubby wing) and AChE genes, is quite different from that by Oppenoorth (1979), being 23.1 percent. O'Brien et al. (1978) showed differences in substrate preferences between normal and insensitive AChE mutant strains of the house fly.

Resistance to Juvenile Hormone Analogs. Resistance to juvenile hormone analogs (JHA) is usually observed in house fly strains with an increased oxidative detoxifying enzyme system (Plapp and Vinson, 1973). Results obtained from crossing experiments with a multichromosomal mutant marker strain and an increased oxidase R strain have indicated that the JHA resistance is controlled by a genetic factor on the second chromosome. Thus, an increase in mixed-function oxidase seems to be responsible for JHA resistance (Cerf and Georghiou, 1974; Plapp, 1976). However, it should be noticed that Yu and Terriere (1978) have recently shown that microsomal fractions contain at least three detoxifying enzyme systems: esterase, oxidase, and epoxide hydrase which, like oxidase, required NADPH. Therefore, the relationship between the second chromosomal genetic factor and the microsomal oxidation should be reexamined on the basis of such advanced information.

Resistance to Pyrethroid Insecticides. Farnham (1973), utilizing mutant marker strains and a strain highly resistant to pyrethroids, demonstrated that four R genes are involved in the R strain used. Among them, the *py-ex* gene is linked to the second chromosome, two others are linked to the third chromosome, and the last one is linked to the fifth chromosome. In the house fly, pyrethroid chemicals are also oxidized by an abdominal microsomal system, as shown by Yamamoto and Casida (1966). The resistance mechanism controlled by this second chromosomal gene is unknown, but it might be that no oxidase system is involved because the gene action is not affected by synergists.

Linkage Group 3

This linkage group of the house fly is also very important in relation to incompletely recessive major genes for insecticide resistance.

Delayed penetration of various insecticides is controlled by the *Pen* (= *tin*) gene, and this gene acts as a strong enhancer in the presence of other major R genes (Hoyer and Plapp, 1968; Sawicki and Lord, 1970). The famous *kdr* (knockdown-resistance) gene, which was named by Milani (1956), is located near one end of this linkage map. Another similar gene, *kdr-o*, was found in an American R strain (Milani and Franco, 1959). In a Japanese DDT-R strain, an incompletely recessive gene, *r-DDT*, has been linked to this chromosome (Tsukamoto and Suzuki, 1964), and the mechanism of this gene seems to be the control of low nerve sensitivity to DDT (Tsukamoto et al., 1965). Recessive major resistance genes against γ-BHC and pyrethrum are also linked to this third chromosome (Tsukamoto, 1969). Thus, to confirm the relationship of the loci for these phenotypically similar genes is important, considering the new knowledge on resistance mechanisms.

Linkage Group 4

This is a peculiar linkage group because relatively few visible mutant markers are available, and the only known R genes on this chromosome involve dieldrin resistance: *Dld* (Tsukamoto and Suzuki,

1963); *R-dieldrin* (Oppenoorth and Nasrat, 1966); DR_4 (Sawicki and Farnham, 1968); and Dld_4 (Georghiou, 1969). Although various symbols have been used by different investigators and although there is no direct evidence based on crossing experiments, all these R genes might be alleles at a single locus near the end of the chromosome. The resistance mechanism of this gene still remains unsolved.

Linkage Group 5

Major dominant genes for resistance to γ-BHC and to pyrethrum are involved in this linkage group (Tsukamoto, 1969). Microsomal oxidation mechanisms other than that of the second chromosome are also associated with this group. This oxidation is controlled by the *Ox-5* gene for which the map position is still unknown, apart from the fact that the recombination value is about 40 units from a marker gene, *ocra* (Khan et al., 1973). The methylenedioxyphenyl synergist-suppressive mechanism controlled by the *Ses* gene (Sawicki and Farnham, 1967) may be the same as the fifth chromosomal mixed-function oxidase mechanism controlled by the *Ox-5* gene. Oppenoorth and Houx (1968) also reported a similar R gene for microsomal detoxication of DDT, and the gene action is suppressed by methylenedioxyphenyl synergists but not by DDT-dehydrochlorinase synergists. Furthermore, this linkage group is associated with dominant genes for resistance to OP and carbamate insecticides (Tsukamoto and Suzuki, 1966; Tsukamoto et al., 1968; Plapp and Casida, 1969). Thus, some of those resistance genes mentioned above may be synonymous with *Ses*.

SEX CHROMOSOMES AND SEX DETERMINATION

Fundamentally the mode of sex determination in the house fly is XX females and XY males. Both sex chromosomes are thought to be genetically inert except for the male-determining factor, M, located on the Y chromosome, and viability factor(s). Thus, no visible sex-linked mutants have been found, and the M factor behaves as a dominant gene in the house fly. However, phenomena of abnormal sex ratios and sex-limited inheritance of DDT resistance have been frequently reported by several investigators from various areas of the world (Milani, 1954, 1961; Kerr, 1960, 1961, 1970; Sullivan, 1961; etc.). Sullivan (1961) and Hiroyoshi (1964) have explained these phenomena by the translocation of the Y chromosome, or at least a part of the Y chromosome together with the M factor, to one of the autosomes. The autosome to which the M factor can attach is not limited except for the fourth chromosome, whereas the second or third chromosomal cases are most frequently reported. According to Hiroyoshi and Fukumori (1977, 1978), the attachment of the M factor to the autosome seems to act as a mutator and/or enhancer of crossing over in males. This might explain in part why such abnormal sex determination and biased sex ratios occur frequently in insecticide-resistant populations in the field. Cytological examinations show abnormalities of the sex chromosome; for example, in Japan, most highly resistant field populations of the house fly possess the XX karyotype in both males and females.

Furthermore, such highly resistant strains or field populations sometimes involve a feminizing factor, F, whose gene action is superior to that of M factors. Thus, curiously enough, the presence of XY females and XX males is possible. The F factor might also enhance crossing over (Hiroyoshi, personal communication). Earlier attempts to associate the F factor with any chromosome were not successful, but recently McDonald et al. (1978) reported that the dominant female-determining factor is linked to the fourth chromosome, near *Ba* (Bald abdomen).

In mosquitoes, it is well known that the M factor is located on the first chromosome (i.e., the sex chromosome). Sometimes this male-determining factor is located on one of the autosomes, and Baker and Sakai (1976) have shown such a case in *Culex tritaeniorhynchus*.

ENZYME POLYMORPHISM AS A BIOCHEMICAL MARKER

Most of the resistance mechanisms demonstrated or inferred in the house fly and mosquitoes involve enzyme systems such as DDT-dehydrochlorinase, microsomal mixed-function oxidases, carboxylesterase, GSH-transferase, and so on. The hypothesis proposed by Oppenoorth (1959) and co-workers that aliesterase is a major cause of OP resistance stimulated various insect toxicologists to further research the enzyme systems involved in resistance. The importance of the nature of enzyme proteins at a molecular level has also been recognized.

It is now well established that many insects possess various types of enzyme polymorphism. Electrophoresis has proven to be a useful and effective method for distinguishing an enzyme protein from a mixture of proteins or for comparing similar enzyme molecules or proteins obtained from R and S strains of insects. For example, Stanton et al. (1978) reported the presence of multiple forms of cytochrome P-450 in different strains of the house fly by utilizing SDS-gel electrophoresis.

Velthuis and van Asperen (1963) and van Asperen (1964) first studied agar gel electrophoresis of house fly esterases in relation to biochemical genetics of insecticide resistance. Since then, various workers have investigated isoenzyme patterns and some of these enzymes have been associated with special linkage groups or have been located on the chromosome map. In the house fly, the acid phosphatase locus and esterase loci have been located on the second chromosome (Ogita and Kasai, 1965a,b), lactate dehydrogenase (LDH) isoenzymes on the third chromosome (Agatsuma et al, 1977), amylase loci on the fifth chromosome (Ogita and Hiroyoshi, 1964), and so on. Similar experiments have also been carried out in some mosquito species: acid phosphatase and alcohol dehydrogenase loci are on the second chromosome and esterase on the third chromosome in *Anopheles stephensi* (Iqbal et al., 1973); alkaline phosphatase gene on the second chromosome in *Culex tritaeniorhynchus* (Sakai et al., 1973); esterases on the third chromosome in *Culex pipiens* (s.l.) (Narang et al., 1977; Gargan and Barr, 1977); hexokinase activity on linkage group 3 in *Aedes aegypti* (Tabachnick and Powell, 1978), and so on. Steiner and Joslyn (1979) have recently reviewed a number of enzymes and have cited much of the literature dealing with electrophoretic methods and genetic studies on enzyme polymorphism in mosquitoes.

CONCLUDING REMARKS

Genetic analysis has been shown to be a useful method for resolving various problems in insecticide resistance, not only in determining a mode of inheritance of resistance to an insecticide, but also in distinguishing differences in the mechanisms involved in resistance to either a single insecticide or to different insecticides.

For these purposes, morphologically visible mutant markers have been used in crossing experiments. Due to the development of many biochemical techniques, especially various kinds of electrophoresis and high speed liquid chromatography, some biochemical mutants, such as genetic polymorphism in isoenzymes, can now be utilized as markers as well. Purification of enzymes or metabolites is now easier to do than it was just one decade ago because of recent advances in analytic methods, especially in micro-scale studies using sophisticated analytical instruments.

Our knowledge of resistance mechanisms has also increased. For example, it was previously considered quite a strange phenomenon that in *Drosophila melanogaster* only a single gene locus at 66± on the second chromosome was responsible for resistance to various types of synthetic insecticides whose chemical structures were completely distinct from each other. At present, however, it is rather commonly accepted that this R gene may control high microsomal oxidase activity.

Before, it was considered adequate to use a highly resistant strain that had been selected with a certain insecticide for several generations in order to biochemically investigate resistance mechanisms. Now we have learned that several resistance mechanisms might be involved in such strains, and therefore, to investigate resistance mechanisms for each R gene, it is necessary to prepare special R strains in which only a single R gene is involved in the background of the S strain. Such a coisogenic strain is already used in biochemical investigations of *Drosophila* to reduce unnecessary variable factors that might disturb analyses. Even in the house fly or in mosquitoes, other chromosomes should be replaced by those of the standard S strain. Much more precise genetic analysis will be required in the future to distinguish several similar mechanisms and to know whether these mechanisms are controlled by alleles or by distinct loci. Therefore, it should be emphasized agains that the currently available number of visible mutants, especially "good" markers, is still very limited and efforts to seek and to establish such "good" mutant markers should be continued; otherwise, modern biochemical instruments will not be able to display their real abilities.

Developments in population genetics and their application, especially by computer simulation, will soon bring us more useful and practical information, not only for geneticists but also for insect toxicologists, in forecasting the fate of an R gene in a given field population under environments with or without insecticidal pressure.

REFERENCES

Agatsuma, T., Shiroishi, T., and Takeuchi, T., 1977, Genetic studies on LDH isozymes in the house fly, *Musca domestica*, *Japan. J. Genet.*, 52:149.

Baker, R. H., and Sakai, R. K., 1976, Male-determining factor on chromosome 3 in the mosquito, *Culex tritaeniorhynchus*, *J. Hered.*, 67:289.

Baker, R. H., and Rabbani, H. G., 1970, Complete linkage in females of *Culex tritaeniorhynchus* mosquitoes, *J. Hered.*, 61:59.

Baker, R. H., Sakai, R. K., and Mian, A., 1971, Linkage group-chromosome correlation in *Culex tritaeniorhynchus*, *Science*, 171:585.

Bodenstein, G., 1939, Die Auslöslung von Modifikationen und Mutationen bei *Musca domestica* L., *Roux' Arch. Entwickl.*, 140:614.

Bridges, C. B., and Brehme, K. S., 1944, "The Mutants of *Drosophila melanogaster*," Carnegie Inst. Washington Publ. 552, Washington, D.C.

Brown, A. W. A., 1959, Inheritance of insecticide resistance and tolerance, *Misc. Publ. Entomol. Soc. Amer.*, 1:20.

Cerf, D. C., and Georghiou, G. P., 1974, Cross resistance to juvenile hormone analogs in insecticide-resistant strains of *Musca domestica* L., *Pestic. Sci.*, 5:759.

Craig, G. B., and Hickey, W. A., 1967, Genetics of *Aedes aegypti*, *in:* "Genetics of Insect Vectors of Disease," Wright, J. W., and Pal, R., ed., pp. 67-131, Elsevier, Amsterdam.

Crow, J. F., 1954, Analysis of a DDT-resistant strain of *Drosophila*, *J. Econ. Entomol.*, 47:393.

Devonshire, A. L., and Sawicki, R. M., 1975, The importance of the decreased susceptibility of acetylcholinesterase in the resistance to organophosphorus insecticides, *in:* "Environmental Quality and Safety," Special Issue, Proc. 3rd Internat. Congr. Pestic. Chemists, Helsinki, p. 441.

Dickson, R. C., 1941, Inheritance of resistance to HCN fumigation in the California red scale, *Hilgardia,* 13:515.

Dittrich, V., 1972, Phenotypic expression of gene OP for resistance in two-spotted spider mites tested with various organophosphates, *J. Econ. Entomol.,* 65:1248.

Farnham, A. W., 1973, Genetics of resistance of pyrethroid-selected houseflies, *Musca domestica* L., *Pestic. Sci.,* 4:513.

Franco, M. G., and Oppenoorth, F. J., 1962, Genetical experiments on the gene for low aliesterase activity and organophosphate resistance in *Musca domestica* L., *Entomol. Exp. Appl.,* 5:119.

Gargan, T. S., II, and Barr, A. R., 1977, Inheritance of an esterase locus in *Culex pipiens, Ann. Entomol. Soc. Amer.,* 70:402

Georghiou, G. P., 1965, Genetic studies on insecticide resistance, *Adv. Pest Control Res.,* 6:171.

Georghiou, G. P., 1969, Genetics of resistance to insecticides in houseflies and mosquitoes, *Exp. Parasitol.,* 26:224.

Georghiou, G. P., and Pasteur, N., 1978, Electrophoretic esterase patterns in insecticide-resistant and susceptible mosquitoes, *J. Econ. Entomol.,* 71:201.

Georghiou, G. P., and Taylor, C. E., 1977a, Genetic and biological influences in the evolution of insecticide resistance, *J. Econ. Entomol.,* 70:319

Georghiou, G. P., and Taylor, C. E., 1977b, Operational influences in the evolution of insecticide resistance, *J. Econ. Entomol.,* 70:653.

Grigolo, A., and Oppenoorth, F. J., 1966, The importance of DDT-dehydrochlorinase for the effect of the resistance gene *kdr* in the housefly *Musca domestica* L., *Genetica,* 37:159.

Harrison, C. M., 1951, Inheritance of resistance to DDT in the housefly, *Musca domestica* L., *Nature,* 167:855.

Hiroyoshi, T., 1960, Some new mutants and linkage groups of the house fly, *J. Econ. Entomol.,* 53:985.

Hiroyoshi, T., 1961, The linkage map of the house fly, *Musca domestica* L., *Genetics,* 46:1373.

Horoyoshi, T., 1964, Sex-limited inheritance and abnormal sex ratio in strains of the housefly, *Genetics,* 50:373.

Hiroyoshi, T., 1977, Some new mutants and revised linkage maps of the housefly, *Musca domestica* L., *Japan. J. Genet.,* 52:275.

Hiroyoshi, T., and Fukumori, Y., 1977, On the III^{M}-type houseflies frequently appeared in Japan, *Japan. J. Genet.,* 53:443 (Abstr. in Japanese).

Hiroyoshi, T., and Fukumori, Y., 1978, On the sex-determination in wild populations of the housefly, *Japan. J. Genet.,* 54:420 (Abstr. in Japanese).

Hoskins, W. M., and Gordon, H. T., 1956, Arthropod resistance to chemicals, *Ann. Rev. Entomol.*, 1:89.

Hoyer, R. F., and Plapp, F. W., Jr., 1966, A gross genetic analysis of two DDT-resistant house fly strains, *J. Econ. Entomol.*, 59:495.

Hoyer, R. F., and Plapp, F. W., Jr., 1968, Insecticide resistance in the house fly: Identification of a gene that confers resistance to organotin insecticides and acts as an intensifier of parathion resistance, *J. Econ. Entomol.*, 61:1269.

Hoyer, R. F., Plapp, F. W., and Orchard, R. D., 1965, Linkage relationships of several insecticide resistance factors in the housefly (*Musca domestica* L.), *Entomol. Exp. Appl.*, 8:65.

Iqbal, M. P., Tahir, M. K., Sakai, R. K., and Baker, R. H., 1973, Linkage groups and recombination in the malaria mosquito, *J. Hered.*, 64:133.

Kerr, R. W., 1960, Sex-limited DDT-resistance in houseflies, *Nature*, 185:868.

Kerr, R. W., 1961, Inheritance of DDT resistance involving the Y chromosome in the housefly (*Musca domestica* L.), *Aust. J. Biol. Sci.*, 14:605.

Kerr, R. W., 1970, Inheritance of DDT resistance in a laboratory colony of the housefly, *Musca domestica*, *Aust. J. Biol. Sci.*, 23:377.

Khan, M. A. Q., 1970, Genetic and biochemical characteristics of cyclodiene epoxidase in the housefly, *Biochem. Pharmacol.*, 19:903.

Khan, M. A. Q., Chang, J. L., Sutherland, D. J., Rosen, J. D., and Kamal, A., 1970, House fly microsomal oxidation of some foreign compounds, *J. Econ. Entomol.*, 63:1807.

Khan, M. A. Q., Morimoto, R. I., Bederka, J. P., Jr., and Runnels, J. M., 1973, Control of the microsomal mixed-function oxidase by Ox^2 and Ox^5 genes in houseflies, *Biochem. Genet.*, 10:243.

Kitzmiller, J. B., 1976, Genetics, cytogenetics, and evolution of mosquitoes, *Adv. Genet.*, 18:315.

Laven, H., 1967, Formal genetics of *Culex pipiens*, *in:* "Genetics of Insect Vectors of Disease," Wright, J. W., and Pal, R., ed., pp. 17-65, Elsevier, Amsterdam.

Lewis, J. B., 1969, Detoxication of diazinon by subcellular fractions of diazinon-resistant and susceptible houseflies, *Nature*, 224:917.

Lichtwardt, E. T., 1964, A mutant linked to the DDT-resistance of an Illinois strain of house flies, *Entomol. Exp. Appl.*, 7:296.

Lindsley, D. L., and Grell, E. H., 1967, "Genetic Variations of *Drosophila melanogaster*," Carnegie Inst. Washington Publ. 627, Washington, D. C.

Lovell, J. B., and Kearns, C. W., 1959, Inheritance of DDT-dehydrochlorinase in the house fly, *J. Econ. Entomol.*, 52:931.

McDonald, I. C., Evenson, P., Nickel, C. A., and Johnson, O. A., 1978, House fly genetics: Isolation of a female determining factor on chromosome 4, *Ann. Entomol. Soc. Amer.*, 71:692.

Milani, R., 1954, The genetics of the house fly. Preliminary note, *Atti IXth Congr. Internat. Genet., Bellagio, 1953, Caryologia, Suppl.*, p. 791.

Milani, R., 1956, Mendelian inheritance of knock-down resistance to DDT and correlation between knock-down and mortality in *Musca domestica* L., *Selected Sci. Papers Istit. Super. Sanità*, I, Part 1, p. 176.

Milani, R., 1961, Results of genetic research on *Musca domestica* L., *Atti Assoc. Genet. Ital.*, 6:427.

Milani, R., 1967, The genetics of *Musca domestica* and of other muscoid flies, *in:* "Genetics of Insect Vectors of Disease," Wright, J. W., and Pal, R., ed., pp. 315-369, Elsevier, Amsterdam.

Milani, R., 1975, The housefly, *Musca domestica, in:* "Handbook of Genetics. Vol. 3," King, R. C., ed., pp. 377-399, Plenum Press, New York.

Milani, R., and Franco, M. G., 1959, Comportamento ereditario della resistenza al DDT in incroci tra il ceppo Orlando-R e ceppo *kdr* e kdr^+ di *Musca domestica* L., *Symp. Genet. Biol. Ital.*, 6:269.

Motoyama, N., and Dauterman, W. C., 1978, Molecular weight, subunits, and multiple forms of glutathione S-transferase from the house fly, *Insect Biochem.*, 8:337.

Motoyama, N., and Plapp, F. W., Jr., 1977, Genetic studies on glutathione-dependent reactions in resistant strains of the house fly, *Musca domestica* L., *Pestic. Biochem. Physiol.*, 7:443.

Narang, S., Bhalla, S. C., and Narang, N., 1977, Isozymes of *Culex p. fatigans*: I. An esterase locus in linkage group III and its variability in natural populations, *J. Hered.*, 68:95.

Nickel, C. A., and Wagoner, D. E., 1970, Some new mutants of house flies and their linkage groups and map positions, *J. Econ. Entomol.*, 63:1385.

O'Brien, R. D., Tripathi, R. K., and Howell, L. L., 1978, Substrate preference of wild and mutant house fly acetylcholinesterase and a comparison with the bovine erythrocyte enzyme, *Biochim. Biophys. Acta*, 526:129.

Ogita, Z., 1958, The genetical relation between resistance to insecticides in general and that to phenylthiourea (PTU) and phenylurea (PU) in *Drosophila melanogaster*, *Botyu-Kagaku*, 23:188.

Ogita, Z., and Hiroyoshi, T., 1965, Further genetico-biochemical study on amylase-isoenzymes in the house fly, *Japan. J. Genet.*, 40:411.

Ogita, Z., and Kasai, T., 1965a, Genetic control of multiple molecular forms of the acid phosphomonoesterases in the housefly, *Musca domestica*, *Japan. J. Genet.*, 40:185.

Ogita, Z., and Kasai, T., 1965b, Genetic control of multiple esterases in *Musca domestica*, *Japan. J. Genet.*, 40:1.

Ogita, Z., and Kasai, T., 1965c, A genetic analysis of synergistic action of sulfonamide derivatives with DDT against houseflies (*Musca domestica*), *Botyu-Kagaku*, 30:119.

Oppenoorth, F. J., 1959, Genetics of resistance to organophosphorus compounds and low ali-esterase activity in the housefly, *Entomol. Exp. Appl.*, 2:304.

Oppenoorth, F. J., 1979, Localization of the acetylcholinesterase gene in the housefly, *Musca domestica, Entomol. Exp. Appl.*, 25:115.

Oppenoorth, F. J., and Houx, N. W. H., 1968, DDT-resistance in the housefly caused by microsomal degradation, *Entomol. Exp. Appl.*, 11:81.

Oppenoorth, F. J., and Nasrat, G. E., 1966, Genetics of dieldrin and γ-BHC resistance in the housefly, *Entomol. Exp. Appl.*, 9:223.

Oppenoorth, F. J., and van Asperen, K., 1960, Allelic genes in the housefly producing modified enzymes that cause organophosphate resistance, *Science*, 132:298.

Oppenoorth, F. J., and Welling, W., 1976, Biochemistry and physiology of resistance, *in:* "Insecticide Biochemistry and Physiology," Wilkinson, C. F., ed., pp. 507-551, Plenum Press, New York.

Oppenoorth, F. J., Rupeš, V., El Bashir, S., Houx, N. W. H., and Voerman, S., 1972, Glutathione-dependent degradation of parathion and its significance for resistance in the housefly, *Pestic. Biochem. Physiol.*, 2:262.

Oppenoorth, F. J., Smissaert, H. R., Welling, W., van der Pas, L. T. J., and Hitman, K. T., 1977, Insensitive acetylcholinesterase, high glutathione-S-transferase, and hydrolytic activity as resistance factors in a tetrachlorvinphos-resistant strain of house fly, *Pestic. Biochem. Physiol.*, 7:34.

Perje, A. M., 1948, Studies on the spermatogenesis in *Musca domestica, Hereditas*, 34:209.

Plapp, F. W., Jr., 1970, Inheritance of dominant factors for resistance to carbamate insecticides in the house fly, *J. Econ. Entomol.*, 63:138.

Plapp, F. W., Jr., 1976, Biochemical genetics of insecticide resistance, *Ann. Rev. Entomol.*, 21:179.

Plapp, F. W., Jr., and Casida, J. E., 1969, Genetic control of house fly NADPH-dependent oxidases: Relation to insecticide chemical metabolism and resistance, *J. Econ. Entomol.*, 62:1174.

Plapp, F. W., Jr., and Tripathi, R. K., 1978, Biochemical genetics of altered acetylcholinesterase resistance to insecticides in the house fly, *Biochem. Genet.*, 16:1.

Plapp, F. W., Jr., and Vinson, S. B., 1973, Juvenile hormone analogs: Toxicity and cross-resistance in the housefly, *Pestic. Biochem. Physiol.*, 3:131.

Plapp, F. W., Jr., Tate, L. G., and Hodgson, E., 1976, Biochemical genetics of oxidative resistance to diazinon in the house fly, *Pestic. Biochem. Physiol.*, 6:175.

Rupes, V., and Pinterová, J., 1975, Genetic analysis of resistance to DDT, methoxychlor and fenitrothion in two strains of housefly (*Musca domestica*), *Entomol. Exp. Appl.*, 18:480.

Sakai, R. K., Iqbal, M. P., and Baker, R. H., 1973, Genetics of alkaline phosphatase in a mosquito *Culex tritaeniorhynchus*, *Ann. Entomol. Soc. Amer.*, 66:913.

Sawicki, R. M., 1973a, Resistance to insecticides in the SKA strain of houseflies, Report Rothamsted Exp. Station for 1972, pp. 168-181.

Sawicki, R. M., 1973b, Resynthesis of multiple resistance to organophosphorus insecticides from strains with factors of resistance isolated from the SKA strain of house flies, *Pestic. Sci.*, 4:171.

Sawicki, R. M., 1973c, Recent advances in the study of the genetics of resistance in the housefly, *Musca domestica*, *Pestic. Sci.*, 4:501.

Sawicki, R. M., 1974, Genetics of resistance of a dimethoate-selected strain of houseflies (*Musca domestica* L.) to several insecticides and methylenedioxyphenyl synergists, *J. Agr. Food Chem.*, 22:344.

Sawicki, R. M., and Farnham, A. W., 1967, The use of visible mutant markers in the study of resistance of house flies to insecticides, *Proc. 4th British Insectic. Fungic. Confer.*, 1967:355.

Sawicki, R. M., and Farnham, A. W., 1968, Genetics of resistance to insecticides of the SKA strain of *Musca domestica*, III. Location and isolation of the factors of resistance to dieldrin, *Entomol. Exp. Appl.*, 11:133.

Sawicki, R. M., and Lord, K. A., 1970, Some properties of a mechanism delaying penetration of insecticides into houseflies, *Pestic. Sci.*, 1:213.

Schafer, J. A., and Terriere, L. C., 1970, Enzymatic and physical factor in house fly resistance to naphthalene, *J. Econ. Entomol.*, 63:787.

Shono, T., 1974a, Studies on the mechanism of resistance to diazinon-resistant Hokota strain of houseflies. II. In vitro degradation of diazoxon, *Botyu-Kagaku*, 39:54.

Shono, T., 1974b, Studies on the mechanism of resistance in diazinon-resistant Hokota strain of houseflies. III. Diazinon degradation by glutathione-S-transferase, *Botyu-Kagaku*, 39:75.

Shono, T., 1974c, Studies on the mechanism of resistance in diazinon-resistant Hokota strain of houseflies. IV. Diazinon metabolism by mixed-function oxidase, *Botyu-Kagaku*, 39:80.

Stanton, R. H., Plapp, F. W., Jr., White, R. A., and Agosin, M., 1978, Induction of multiple cytochrome P-450 species in housefly microsomes: SDS-gel electrophoresis studies, *Comp. Biochem. Physiol., B. Comp. Biochem.*, 61:297.

Steiner, W. W. M., and Joslyn, D. J., 1979, Electrophoretic techniques for the genetic study of mosquitoes, *Mosq. News*, 39:35.

Stone, B. F., 1968, A formula for determining degree of dominance in cases of monofactorial inheritance of resistance to chemicals, *Bull. WHO*, 38:325.

Sullivan, R. L., 1961, Linkage and sex limitation of several loci in the housefly, *J. Hered.*, 52:282.

Tabachnick, W. J., and Powell, J. R., 1978, Genetic structure of the East African domestic populations of *Aedes aegypti*, *Nature*, 272:535.

Tate, L. G., Plapp, F. W., Jr., and Hodgson, E., 1974, Genetics of cytochrome P-450 in two insecticide-resistant strains of the housefly, *Musca domestica* L., *Biochem. Genet.*, 11:49.

Taylor, C. E., and Georghiou, G. P., 1979, Suppression of insecticide resistance by alteration of gene dominance and migration, *J. Econ. Entomol.*, 72:105.

Tripathi, R. K., Telford, J. N., and O'Brien, R. D., 1978, Molecular and structural characteristics of house fly brain acetylcholinesterase, *Biochem. Biophys. Acta*, 525:103.

Tsukamoto, M., 1959, Metabolic fate of DDT in *Drosophila melanogaster*. I. Identification of a non-DDE metabolite, *Botyu-Kagaku*, 24:141.

Tsukamoto, M., 1963, The log dosage-probit mortality curve in genetic research of insect resistance to insecticides, *Botyu-Kagaku*, 28:91.

Tsukamoto, M., 1964, Method for the linkage-group determination of insecticide-resistance factors in the housefly, *Botyu-Kagaku*, 29:51.

Tsukamoto, M., 1965, The estimation of recombination values in backcross data when penetrance is incomplete, with a special reference to its application to genetic analysis of insecticide-resistance, *Japan. J. Genet.*, 3:159.

Tsukamoto, M., 1969, Biochemical genetics of insecticide resistance in the housefly, *Residue Rev.*, 25:289.

Tsukamoto, M., and Ogaki, M., 1953, Inheritance of resistance to DDT in *Drosophila melanogaster*, *Botyu-Kagaku*, 18:39.

Tsukamoto, M., and Suzuki, R., 1963, Communication to World Health Organization, *WHO Inf. Cir. Insectic. Resist.*, 46:20.

Tsukamoto, M., and Suzuki, R., 1964, Genetic analyses of DDT-resistance in two strains of the housefly, *Musca domestica* L., *Botyu-Kagaku*, 29:76.

Tsukamoto, M., and Suzuki, R., 1966, Genetic analyses of diazinon-resistance in the house fly, *Botyu-Kagaku*, 31:1.

Tsukamoto, M., Baba, Y., and Hiraga, S., 1961, Mutation and linkage groups in Japanese strains of the housefly, *Japan. J. Genet.*, 36:168.

Tsukamoto, M., Narahashi, T., and Yamasaki, T., 1965, Genetic control of low nerve sensitivity to DDT in insecticide-resistant houseflies, *Botyu-Kagaku*, 30:128.

Tsukamoto, M., Shrivastava, S. P., and Casida, J. E., 1968, Biochemical genetics of house fly resistance to carbamate insecticide chemicals, *J. Econ. Entomol.*, 61:51.

van Asperen, K., 1964, Biochemistry and genetics of esterases in houseflies (*Musca domestica*) with special reference to the development of resistance to organophosphorus compounds. *Entomol. Exp. Appl.*, 7:205.

Velthuis, H. H., and van Asperen, K., 1963, Occurrence and inheritance of esterases in *Musca domestica, Entomol. Exp. Appl.*, 6:79.

Wagoner, D. E., 1967, Linkage group-karyotype correlation in the house fly determined by cytological analysis of X-ray induced translocations, *Genetics*, 57:729.

Wagoner, D. E., 1969, Linkage group-karyotype correlation in the house fly, *Musca domestica* L., confirmed by cytological analysis of X-ray induced Y-autosomal translocations, *Genetics*, 62:115.

Wright, J. W., and Pal, R., ed., 1967, "Genetics of Insect Vectors of Disease," Elsevier, Amsterdam.

Yamamoto, I., and Casida, J. E., 1966, *O*-Dimethyl pyrethrin II analogs from oxidation of pyrethrin I, allethrin, dimethrin, and phthalthrin by house fly enzyme systems, *J. Econ. Entomol.*, 59:1542.

Yu, S. J., and Terriere, L. C., 1978, Metabolism of juvenile hormone I by microsomal oxidase, esterase, and epoxide hydrase of *Musca domestica* and some comparisons with *Phormia regina* and *Sarcophaga bullata, Pestic. Biochem. Physiol.*, 9:237.

Yust, H. R., Nelson, H. D., and Busbey, R. L., 1943, Comparative susceptibility of two strains of California red scale to HCN, with special reference to the inheritance of resistance, *J. Econ. Entomol.*, 36:744.

DETECTION AND MONITORING METHODS FOR RESISTANCE IN ARTHROPODS BASED ON BIOCHEMICAL CHARACTERISTICS

Tadashi Miyata

Laboratory of Applied Entomology and Nematology
Faculty of Agriculture, Nagoya University
Chikusa, Nagoya 464, Japan

INTRODUCTION

Resistance of insect pests (including mites) to insecticides is one of the most serious problems in pest control. According to Georghiou (1981), the number of insect pests showing resistance to insecticides increased to 414 in 1979. It is difficult to imagine the cessation of the development of resistance, because resistance is acknowledged as one of the evolutionary products of pesticide application. However, it is possible to reduce the rate of development of resistance or to control resistant (R) insect pests if insecticides are utilized ideally.

The most accurate way to monitor the resistance is through bioassays, and many of these methods, such as topical application, dipping, spraying, etc., can be found in the literature. These methods are useful in determining resistance levels in the laboratory. Large numbers of insect pests or long periods of time are required to obtain results, which even then can be ambiguous, especially when resistance is light and the population is heterogeneous. In such instances, a simple biochemical monitoring method for resistance is particularly valuable.

Van Asperen and Oppenoorth (1959) and Oppenoorth and Van Asperen (1960) reported that organophosphate-R house flies showed lower phenylbutyrate hydrolyzing activity due to the alternation of a specific enzyme. In other house fly strains, however, there was no correlation between insecticide resistance and esterase activity (Motoyama and Dauterman, 1974). Motoyama and Dauterman (1974) summarized the relationship between nonspecific esterase activity and organophosphorus insecticide resistance. In this chapter, detection and monitoring methods for resistance based on biochemical characteristics are discussed.

NATURE OF NONSPECIFIC ESTERASES

Rice Leafhopper and Planthopper

Resistance of the green rice leafhopper (*Nephotettix cincticeps* Uhler) to malathion was first observed in 1961 in Kochi Prefecture, Japan (Kojima et al., 1963). Malathion-R green rice leafhoppers had a high nonspecific esterase for methyl *n*-butyrate (Hayashi and Hayakawa, 1962; Kojima et al., 1963). Ozaki and Koike (1965) reported that a nonspecific esterase for β-naphthyl acetate of the malathion-R green rice leafhopper was significantly more active than that of the susceptible (S) strain. They did not find similar relationships between esterase activity and resistance levels to methyl parathion; however, later investigations proved that all organophosphate-R green rice leafhoppers have high esterase activity (Ozaki and Kassai, 1970a).

Kasai and Ogita (1965) and Ozaki et al. (1966) separated β-naphthyl acetate hydrolyzing esterases of the green rice leafhopper by thin layer agar gel electrophoresis. Ozaki et al. (1966) also studied the relationship between malathion resistance and nonspecific esterase activity of the S, R and F_1 progeny of R and S strains. Both the resistance levels and the nonspecific esterase activity of the F_1 progeny of R and S were intermediate in both the males and females. The cummulative dose-mortality curves and the nonspecific esterase activity were closely correlated among R, S, F_1 and backcross offspring (Figure 1). From these results Ozaki et al. (1966) suggested that inheritance of malathion resistance and esterase activity was controlled by the same codominant factor(s) located on an autosomal chromosome.

The relationship between high esterase activity and organophosphate resistance of the green rice leafhopper was reconfirmed in many populations from various parts of Japan (Tsuboi et al., 1973; Miyata and Saito, 1976; Miyata et al., 1977; Miyata and Saito, 1978b).

Resistance of the green rice leafhopper to carbamates also developed in districts where organophosphate resistance had developed in this insect. It is now difficult to locate green rice leafhopper populations resistant to carbamates but susceptible to organophosphates. In the laboratory, the Rmc strain was obtained from S and R strains by repeated backcrossings under selection with a carbamate insecticide, Bassa® (Hama, 1975; Hama and Iwata, 1978). The Rmc strain showed almost the same resistance levels to carba-

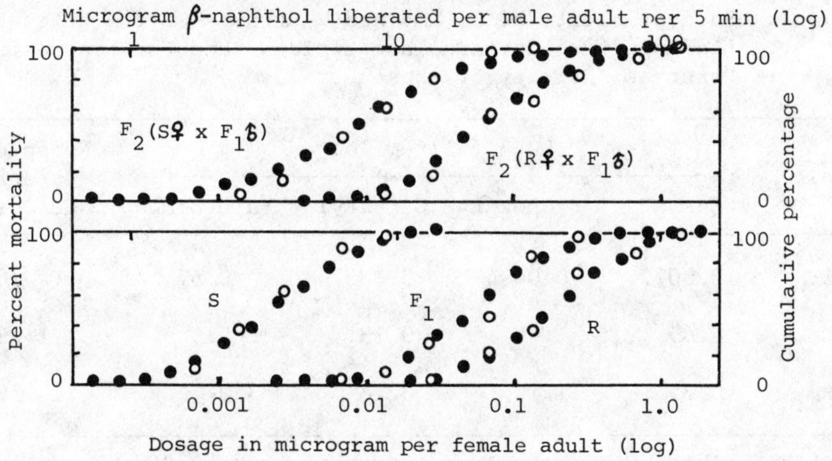

Figure 1. Relationship between the cumulative diagram of esterase activity (•) and the log-dosage curves (o) of R and S strains and their offspring for malathion (from Ozaki et al., 1966).

mates as the Nakagawara (R) strain, but showed a high susceptibility to many organophosphates. The nonspecific esterase of the Rmc strain was almost as active against methyl *n*-butyrate as that of the S strain (Table 1). It was concluded that there was no correlation between carbamate resistance and nonspecific esterase activity in the green rice leafhopper and that nonspecific esterase activity, therefore, cannot be used to monitor carbamate resistance (Miyata et al., 1978a).

Ozaki and Kassai (1970b) reported that there is a close correlation between a nonspecific esterase for β-naphthyl acetate of the E_7 band separated by thin layer agar gel electrophoresis and malathion resistance in the smaller brown planthopper (*Laodelphax striatellus* Fallèn). The nonspecific esterase activity of the E_7 band was thought to be controlled by codominant factors on the autosomal chromosome(s) (Figure 2, Table 2). The nonspecific esterase of the E_7 band of the smaller brown planthopper strains selected by malathion and fenitrothion was significantly more active than that of an S strain (Okuma and Ozaki, 1969). A similar phenomenon was observed in the malathion-R or fenitrothion-R brown planthopper (*Nilaparvata lugens* Stål) (Hasui and Ozaki, 1977; Miyata et al., 1978b).

Green Peach Aphid

The green peach aphid (*Myzus persicae* Sulz.) is resistant to organophosphorus and carbamate insecticides worldwide (Sudderuddin,

Table 1. Susceptibility of S, Rmc and R Strains of the Green Rice Leafhopper to Certain Insecticides, and the Nonspecific Esterase Activity of These Strains

Strain[a]	LD_{50} (μg/female)				Nonspecific esterase activity[b]
	Bassa	Propoxur	Carbaryl	Malathion	
S	0.0072	0.0094	0.0027	0.57	0.70
Rmc	0.65	1.40	0.13	5.5	1.13
R	0.75	1.15	0.18	330	15.59

[a]S, susceptible (Miyagi); R, resistant to both organophosphates and carbamates.
[b]μmol methyl-*n*-butyrate hydrolyzed/10 males/30 minutes at 37°C.
(From Hama, 1977; Hama and Iwata, 1978).

1973; Sawicki et al., 1978). Needham and Sawicki (1971) reported that there is a good correlation between nonspecific esterase for α-naphthyl acetate and resistance levels to organophosphorus insecticides (Table 3). This correlation was confirmed in many aphid populations throughout the United Kingdom (Beranek, 1974; Needham and Devonshire, 1975; Sawicki et al., 1978).

Beranek (1974) separated nonspecific esterases of the green peach aphid by starch gel electrophoresis and found that all R clones were more active in esterase 2 than several S clones. Devonshire (1975) separated the green peach aphid nonspecific esterases by polyacrylamide gel electrophoresis. From the seven esterase bands detected, the esterase 4 (equivalent to the esterase 2 described by Beranek, 1974) showed higher activity in R clones than in S clones.

When sexuales of an organophosphate-R, high esterase clone (R) of the green peach aphid were crossed with those from an S, i.e., low esterase clone (J), both esterase activity and resistance to dimethoate of F_1 progeny were intermediate between those of the parents. The F_1 clones fell into two distinct groups in the esterase activity, indicating that the R parent might be heterozygous for two alleles coding for decreased esterase activity (Table 4) (Blackman et al., 1977). Blackman and Devonshire (1978) later showed that a mutation at a single regulatory locus caused lower intermediate esterase activity, and that a second locus may be involved in the expression of higher esterase activity.

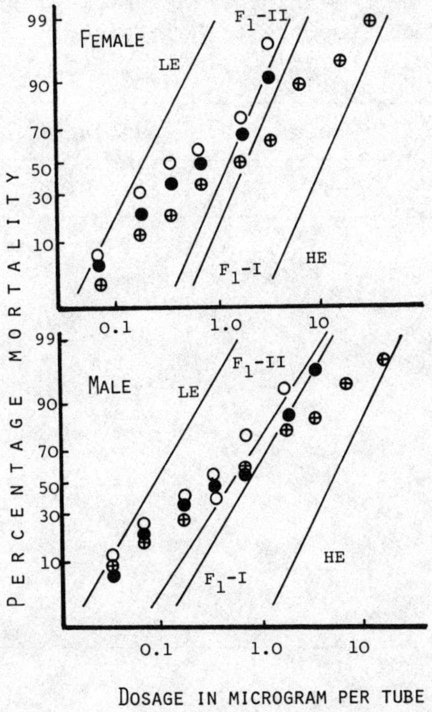

Figure 2. Log dosage-probit mortality lines of F_2 and backcross offspring for malathion, compared with those of low esterase (LE), high esterase (HE) and F_1 (⊕ = F_2; ● = backcross I; o = backcross II) (from Ozaki and Kassai, 1970b).

Other Insect Pests

For mosquitoes, Yasutomi (1970, 1971) has shown a good correlation between nonspecific esterase activity for β-naphthyl acetate and resistance levels to various organophosphates (see Yasutomi, this volume). Similar phenomena have been reported with the citrus red mite, *Panonychus citri* McGregor (Takehisa and Tanaka, 1967), predacious mite, *Neoseiulus fallacis* Garman (Motoyama et al., 1971) and the Indian meal moth, *Plodia interpunctella* Hübner (Zettler, 1974). Further investigations will be necessary to confirm the correlation between esterase activity and resistance in these insect pests, except for mosquitoes.

Table 2. Inheritance of β-Naphthyl Acetate Hydrolyzing Esterase Activity and Response to Malathion in the Smaller Brown Planthopper

Strain	Frequency of E_7 band activity			LD_{50} for malathion (μg/tube)
	Low	Middle	High	
HE	0	0	140	11.17
LE	150	0	0	0.187
F_1, I	0	140	0	1.879
F_1, II	0	100	0	1.279
F_2	76	145	62	
Backcross, I	140	154	0	
Backcross, II	120	121	0	

(From Ozaki and Kassai, 1970b.)

ROLE OF HIGH NONSPECIFIC ESTERASE ACTIVITY IN RESISTANCE

In organophosphate-R green rice leafhoppers, nonspecific esterases were active in the E_2, E_3 and/or E_4 bands when separated by thin layer agar gel electrophoresis (Ozaki et al., 1966; Ozaki and Kassai, 1970a). ^{14}C-Methyl malathion was degraded by E_1, E_2 and E_3 bands with the highest activity on the E_2 band (Miyata and Saito, 1976, 1978b). In the smaller brown planthopper, in vitro degradation of ^{14}C-methyl malathion was seven and five times as high in malathion-R (Rm) and fenitrothion-R (Rf) strains as in the S strain, respectively. Malathion degradation was detected at the E_7 band (Miyata et al., 1976). In the brown planthopper, ^{14}C-methyl malathion degraded in vitro was almost eight times higher in Rm and Rf strains than in S strains. Degradation was detected around the E_1, E_2, E_3 and E_4 bands with the highest activity on the E_2 band (Figure 3). These malathion degrading activities correlated well with esterase activity and resistance levels (Miyata et al., 1978b).

These results indicate that the high esterase activity of the green rice leafhopper, smaller brown planthopper and brown planthopper may play an important role in the degradation of malathion. However, it is not easy to determine whether or not these esterase bands represent more than one enzyme for it could always be possible

Table 3. Hydrolysis of α-Naphthyl Acetate by Several Strains of *Myzus persicae*[a]

Strain	Origin	Amount of α-naphthyl acetate hydrolyzed in 30 minutes	Response of strains to OPs
Rothamsted (S)	Rothamsted	140 nmol/mg aphid	Susceptible
GR (R)	UK	860	Very resistant
GRUS (R)	Unselected substrain of GR	840	Very resistant
		230	Slightly resistant
Ferrara	Italy	1,490	Resistant
East Malling	UK	470	Moderately resistant
		420	Resistant

[a]Figures were corrected according to Oppenoorth and Voerman (1975).
(From Needham and Sawicki, 1971. Reprinted by permission from *Nature*, Vol. 230, p. 125. Copyright (c) 1971 Macmillan Journals Limited.)

that a number of enzymes may not be separated in the electrophoretic experiment.

Beranek and Oppenoorth (1977) reported that an enzyme hydrolyzing methyl paraoxon in vitro in an organophosphate-R strain of the green peach aphid is present in the same electrophoretic fraction as a nonspecific esterase (esterase 2). This esterase was characteristically more active in organophosphate-R strains (Table 5). Devonshire (1977) reported that the nonspecific esterase associated with resistance could not be separated from paraoxon-hydrolyzing enzymes by electrophoresis or ion-exchange chromatography. He has also shown that one enzyme hydrolyzes both α-naphthyl acetate and paraoxon.

NATURE OF ACETYLCHOLINESTERASE

The mechanism of carbamate resistance in the green rice leafhopper is reported to be reduced sensitivity of acetylcholinesterase

Table 4. Relationship Between Total Carboxylesterase Activity, Esterase-4 Activity and Resistance to Dimethoate in Parents and F_1 Progeny of Cross Between Resistant (French R) and Susceptible (J) *Myzus persicae*

Clone		Total carboxyl-esterase activity (μmol/h/mg aphid) (± s.e.)	Esterase band 4 (peak area/mg aphid) (± s.e.)	Resistance factor
Parent (French R)		0.96 ± 0.08	69.8 ± 1.0	14.3
Parent (J)		0.26 ± 0.02	-	1
F_1 (French RxJ)	1	0.42 ± 0.03	24.0 ± 0.5	3.0
	4	0.41 ± 0.02	20.4 ± 0.9	2.0
	5	0.44 ± 0.02	28.9 ± 0.7	2.0
	6	0.44 ± 0.03	18.7 ± 1.2	2.5
	11	0.41 ± 0.03	23.4 ± 0.2	-
Mean		0.42	23.1	2.4
F_1 (French RxJ)	3	0.75 ± 0.04	45.7 ± 2.6	10.5
	7	0.72 ± 0.07	53.5 ± 1.5	8.0
	9	0.68 ± 0.04	51.4 ± 1.5	7.0
	10	0.60 ± 0.03	59.6 ± 1.0	7.0
	12	0.73 ± 0.02	58.9 ± 0.8	13.5
	13	0.75 ± 0.07	55.8 ± 1.0	10.0
Mean		0.71	54.6	9.3
F_1 (JxFrench R)		0.66 ± 0.05	55.8 ± 1.0	12.5

(From Blackman et al., 1977.)

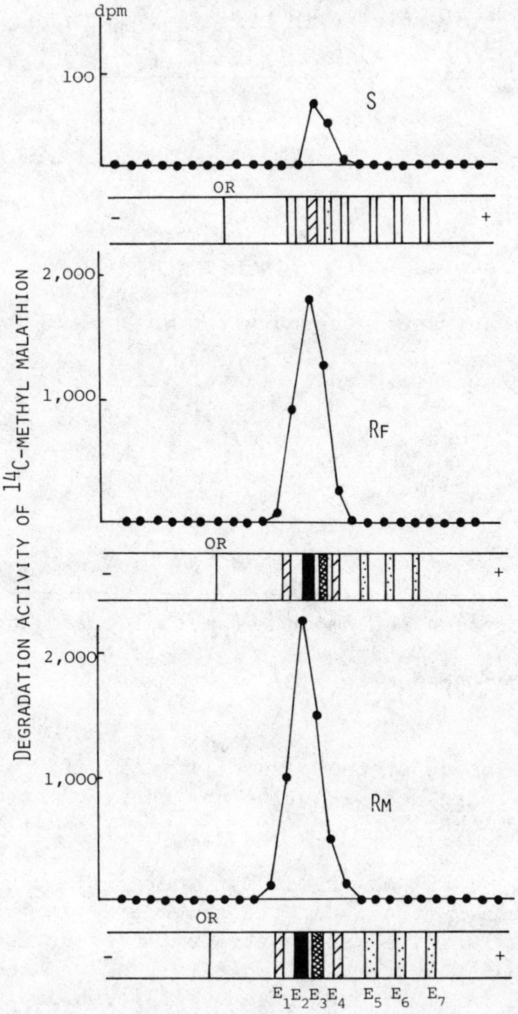

Figure 3. Degradation of ^{14}C-methyl malathion by the brown planthopper enzymes separated by thin layer agar gel electrophoresis. The corresponding zymogram for β-naphthyl acetate indicates the relative position of esterase bands against malathion degrading enzyme (from Miyata et al., 1978b).

(AChE) to insecticides (Hama and Iwata, 1971, 1973; Iwata and Hama, 1972; Yamamoto et al., 1977). The same resistance mechanism has been shown in other insect pests (reviewed by Plapp, 1976). To determine AChE activity, the Ellman method (Ellman et al., 1961) is widely used due to its rapidity and high sensitivity. Miyata and

Table 5. Hydrolytic Activity in Different Parts of Electropherogram of E and M1 Aphid Homogenates[a]

Slice number	Esterase present	E	M1
1 (origin)	None	0.0	0.0
2	Esterase 1 (cholinesterase)	0.4	0.0
3	Esterase 2 (carboxylesterase)	1.8	0.0
4	Esterase 3 and 4 (carboxylesterase)	0.0	0.0
5	Esterase 5 (unknown)	0.0	0.0
6 (front)	None	0.0	0.0
	Whole aphid homogenate	4.4	0.0

[a]The figures represent the amount of methyl-paraoxon hydrolyzed in ng/mg aphid/h.
(From Beranek and Oppenoorth, 1977.)

Saito (1978a) pointed out the possibility of differences in activity determination between acetylcholine and acetylthiocholine when used as substrates of AChE.

Hama et al. (1979) observed that there is a big difference in AChE between S and R strains of the green rice leafhopper. AChE from the S strain was inhibited as substrate concentration was increased; however, modified AChE from the Rmc strain showed inhibitory action of acetylcholine (ACh) but not of acetylthiocholine (ATCh) when the substrate concentration was increased (Figure 4). This indicates that a high concentration of ATCh does not inhibit AChE activity in carbamate-R strains but does in S strains. Individual green rice leafhoppers from the Miyagi (S), Nakagawara (R) and Rmc (R) strains were crushed on the filter paper, which was sprayed with ATCh (2×10^{-2}, 2×10^{-3} and 2×10^{-4}M). After being kept at 37°C for 30 minutes, the filter paper was sprayed with eserine solution to stop the reaction. High concentrations of ATCh resulted in a large difference in AChE activity between S and R strains (Table 6) (Miyata et al., 1979).

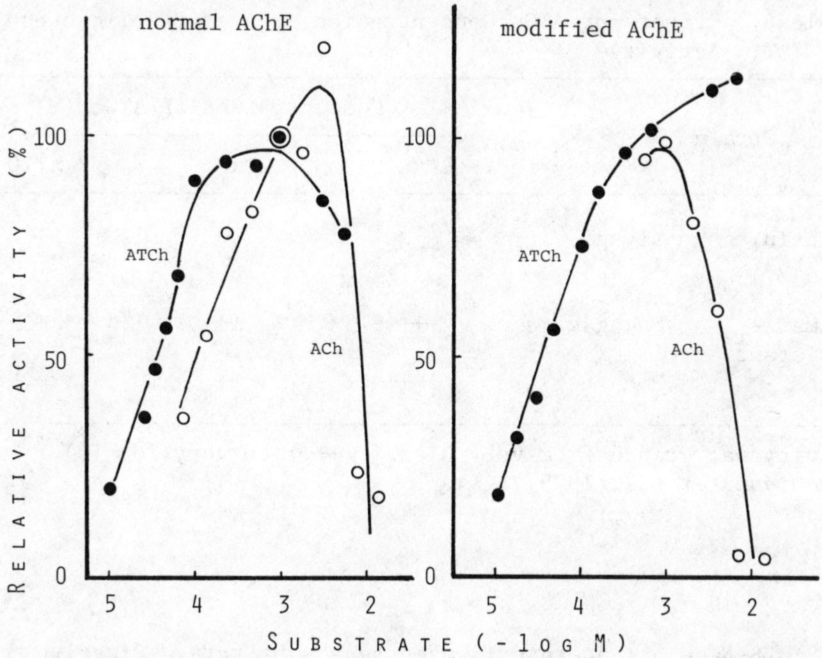

Figure 4. Effect of substrate concentration on normal and modified AChE activity (from Hama et al., 1979).

SIMPLE TECHNIQUE FOR ASSAY OF RESISTANT POPULATIONS

Organophosphate-R Populations

As described previously, in several insect species a degree of resistance to organophosphates was closely correlated with nonspecific esterase activity, suggesting that the organophosphate resistance of field populations could be determined by examining their esterase activity either colorimetrically (Ozaki and Koike, 1965: Ozaki and Kassai, 1970b; Needham and Sawicki, 1971), manometrically (Hayashi and Hayakawa, 1962) or electrophoretically (Kasai and Ogita, 1965; Ozaki and Kassai, 1970b; Beranek, 1974). Ozaki (1969) and Sawicki et al. (1978) have reported simple assay methods to monitor organophosphate resistance.

Filter Paper Method (Ozaki, 1969).

1) Toyo No. 51 filter paper or equivalent, specifically available for chromatographic analysis, should be used in order to assay esterase activity of the green rice leafhopper and the smaller brown planthopper. The filter paper should be cut to 2 x 2 cm squares.

Table 6. Effects of ATCh Concentration on Acetylcholinesterase Activity[a]

Strain		Concentration of ATCh (M)		
		2×10^{-2}	2×10^{-3}	2×10^{-4}
Susceptible	Miyagi	±	++	+
Carbamate-resistant	Nakagawara	+++++	++	+
	Rmc	+++++	++	+

[a]Activity was graded from very high (+++++) to very low (±). (From Miyata et al., 1979.)

2) β-Naphthyl acetate is used as a substrate. Dissolve 5 mg β-naphthyl acetate in 2 ml acetone, and then 50 ml of pH 7.0 phosphate buffer. Fill container to 100 ml with distilled water. This substrate solution should be prepared immediately before the experiment.

3) Collect adult green rice leafhoppers and smaller brown planthoppers in fields and keep them in a refrigerator at 0-5°C. Adult males should be used for assay because of their uniform body size compared to that of adult females. Place the insect on filter paper. Crush it with a glass rod so that the enzymes of the insect are absorbed into the filter paper. Keep the paper in a refrigerator at 4-5°C after complete removal of wings, legs and integument from the filter paper.

4) Immerse the filter paper into substrate solution at 37°C for 15 sec and then put it on a glass plate. Put about 0.3 ml of 0.4% naphthanil diazo blue B on the filter paper, and after 1 min, add a few drops of 25% trichloroacetic acid.

5) Assay at least 100 adult males for each population. Whether an insect is S or R can be determined by the degree of activity of esterase shown on the filter paper. Individuals whose reaction color is purplish and particularly remarkable at a portion in and around the enzyme-absorbing part of the filter paper should be rated to be organophosphate-R. Individuals whose reaction color is not apparent are rated S.

Agar Gel Plate Method (Ozaki, 1969).

1) To prepare agar gel plates, dissolve 0.75 g agar and 1.5 g polyvinylpyrolidon (K-90) in 100 ml phosphate buffer at pH 6.8 and ionic strength 0.025μ. Pour 20 ml of hot solution on a 10 x 17 cm glass plate on which a thin layer of gel has been settled.

2) Dissolve 500 mg β-naphthyl acetate in 100 ml acetone (0.5%). This substrate solution should be kept at $0^{o}C$.

3) Collect adult males from the field and keep them in a refrigerator until testing.

4) Crush a single male with a glass rod on a glass plate in distilled water, 0.025 ml for the green rice leafhopper, or 0.01 ml for the smaller brown planthopper. Immerse a strip of 1 x 5 mm filter paper in the resultant brei.

5) Place 50 strips of the filter paper on an agar plate at regular intervals. After a 30 min conditioning in a refrigerator at 4-$5^{o}C$, remove the strips, and spray the plate with a 0.5% acetone solution for substrate. Incubate the plate at $37^{o}C$ for 20 min. Add a few drops of a 0.4% aqueous solution of naphthanil diazo blue B. Wash the plate in running water and dry at $50^{o}C$.

6) Assay at least 100 adult males for each population. Esterase activity of the insect is classified into three groups based on a standard of esterase activity set in advance, i.e., low activity for S individuals, middle for hybrid, and high for R. In the case of the green rice leafhopper, the standard of esterase activity is determined using S populations and organophosphate-R field populations because most R insects have two or more high esterase activity bands in the electrophoretic zymogram, as described above.

Tile Test (Sawicki et al., 1978).

1) Phosphate buffer, 0.02M, pH 7.0: Dissolve disodium hydrogen phosphate dodecahydrate (900 mg) and potassium dihydrogen phosphate (340 mg) in distilled water (250 ml).

2) α-Naphthyl acetate in acetone (6 g/liter).

3) α-Naphthol in acetone (300 mg/liter).

4) Aqueous sodium dodecyl sulphate (50 g/liter).

5) Substrate: Mix 0.5 ml of α-naphthyl acetate in acetone with 50 ml of phosphate buffer (pH 7.0). Coupling reagent: Dissolve 150 mg of fast blue B salt (Searle, High Wycombe) in 15 ml

water and add 35 ml of aqueous sodium dodecyl sulphate. Immediately prior to assay, freshly prepare these two reagents from the above stock solutions (1 through 4).

6) Pipette the substrate (0.2 ml) into each of the 12 wells of a staining tile and add one apterous adult aphid to each of the first eleven wells. Then crush the aphids in the substrate at 10 sec intervals with a glass rod and mix well. After the last aphid is crushed, mix the coupling reagent (0.05 ml) with homogenate in the first well, then at 10 sec intervals in the remaining eleven wells. In this way, incubate each aphid homogenate with the substrate for 2 min. Finally, add 0.05 ml of α-naphthol using a "Microcap" to the twelfth well to be used as a standard. (It is sufficiently accurate when dispensing the substrate and coupling reagent to consider that one drop from a Pasteur pipet is 0.05 ml.)

7) A red color develops, which turns to blue after standing for 10 min, and the intensity of this color gives an estimate of the resistance of the aphid. A color less intense than the standard will form with those homogenates from aphids that are S or only slightly R to dimethoate and demeton-S-methyl, and those with strong color are from highly resistant R_2 aphids.

A quantitative total esterase assay and semi-quantitative technique by electrophoresis would be useful.

Carbamate-R Populations of the Green Rice Leafhopper

1) ATCh iodide, 4 x 10^{-2}M. Keep this substrate solution in a refrigerator.

2) Dithiobisnitrobenzoic acid (DTNB) reagent solution (2 x 10^{-2}M): Prepare with pH 7.0 phosphate buffer.

3) Crush a single adult male on the filter paper (Toyo No. 2) with a glass rod.

4) Mix an equal volume of ATCh and DTNB solutions. Immediately spray a mixed solution of substrate and reagent solutions onto the filter paper and keep it at 37°C for 30 min.

5) Spray a 10^{-3}M eserine sulfate solution to stop the reaction. Samples with strong yellow should be rated as carbamate-R. Samples whose reaction color is not apparent are rated as S (Miyata et al., 1979).

CONCLUDING REMARKS

In many insect species, nonspecific esterase activity is significantly high in organophosphate-R populations. Genetic studies have revealed that there is a close correlation between nonspecific esterase activity and organophosphate resistance in several insect species. This enhanced nonspecific esterase activity contributes to degradation of organophosphorus insecticides; however, there is no correlation between nonspecific esterase and carbamate resistance. Only organophosphorus insecticide resistance may be monitored by assaying nonspecific esterase activity in some insect species. Also, a simple biochemical method to monitor carbamate resistance in the green rice leafhopper has been described.

These methods are quite simple for assaying individual insects within the population. Extension personnel can easily determine insecticide resistance in the laboratory or in the field, and these methods will contribute to the monitoring of the development of insecticide resistance.

Acknowledgement: This work was supported in part by Grant-in-Aid for Scientific Research Nos. 456037 and 544008 from the Ministry of Education, Science and Culture, Japan.

REFERENCES

Beranek, A. P., 1974, Stable and non-stable resistance to dimethoate in the peach-potato aphid (*Myzue persicae*), *Ent. Exp. & Appl.*, 17:381.

Beranek, A. P., and Oppenoorth, F. J., 1977, Evidence that the elevated carboxylesterase (Esterase 2) in organophosphorus-resistant *Myzus persicae* (Sulz.) is identical with organophosphate-hydrolyzing enzyme, *Pestic. Biochem. Physiol.*, 7:16.

Blackman, R. L., and Devonshire, A. L., 1978, Further studies on the genetics of the carboxylesterase regulatory system involved in resistance to organophosphorus insecticide in *Myzus persicae* (Sulzer), *Pestic. Sci.*, 9:517.

Blackman, R. L., Devonshire, A. L., and Sawicki, R. M., 1977, Co-inheritance of increased carboxylesterase activity and resistance to organophosphorus insecticides in *Myzus persicae* (Sulzer), *Pestic. Sci.*, 8:163.

Devonshire, A. L., 1975, Studies on the carboxylesterase of *Myzus persicae* resistant and susceptible to organophosphorus insecticides, *Proc. 8th Brit. Insecti. and Fungi. Conf.*, p. 67.

Devonshire, A. L., 1977, The properties of carboxylesterase from the peach-potato aphid, *Myzus persicae* (Sulz.), and its role in conferring insecticide resistance, *Biochem. J.*, 167:675.

Ellman, G. L., Courtney, K. D., Andres, V. Jr., and Featherstone, R. M., 1961, A new and rapid colorimetric determination of acetylcholinesterase activity, *Biochem. J.*, 7:88.

Georghiou, 1980, Resistance in time and space, this volume.

Hama, H., 1975, Toxicity and anticholinesterase activity of propaphos, 0,0-di-(*n*)-propyl-0-4-methylthiophenyl phosphate, against the resistant green rice leafhopper, *Nephotettix cincticeps* Uhler, *Botyu-Kagaku*, 40:14.

Hama, H., and Iwata, T., 1971, Insensitive cholinesterae in the Nakagawara strain of the green rice leafhopper, *Nephotettix cincticeps* Uhler (Hemiptera: Cicacelidae), as a cause of resistance to carbamate insecticides, *Appl. Ent. Zool.*, 6:183.

Hama, H., and Iwata, T., 1973, Resistance to carbamate insecticides and its mechanism in the green rice leafhopper, *Nephotettix cincticeps* Uhler, *Jap. J. Appl. Ent. Zool.*, 17:154.

Hama, H., and Iwata, T., 1978, Studies on the inheritance of carbamate resistance in the green rice leafhopper, *Nephotettix cincticeps* Uhler (Hemiptera: Cicadelidae). Relationship between insensitivity of acetylcholinesterase and cross-resistance to carbamate and organophosphate insecticides, *Appl. Ent. Zool.*, 13:190.

Hama, H., Miyata, T., and Saito, T., 1979, unpublished data.

Hasui, H., and Ozaki, K., 1977, Studies on the resistance to organophosphorus insecticides and esterase activity of the brown planthoppers, *Ann. Meet. Jap. Soc. Appl. Ent. Zool.*, (Abst.)

Hayashi, M., and Hayakawa, H., 1962, Malathion tolerance in *Nephotettix cincticeps* Uhler, *Jap. J. Appl. Ent. Zool.*, 6:250.

Iwata, T., and Hama, H., 1972, Insensitivity of cholinesterase in *Nephotettix cincticeps* resistant to carbamate and organophosphorus insecticides, *J. Econ. Entomol.*, 65:643.

Kasai, T., and Ogita, Z., 1965, Studies on malathion resistance and esterase activity in green rice leafhopper, *SABCO J.*, 1:130.

Kojima, K., Isizuka, T., and Kitakata, S., 1963, Mechanism of resistance to malathion in the green rice leafhopper, *Nephotettix cincticeps*, *Botyu-Kagaku*, 28:17.

Miyata, T., and Saito, T., 1976, Mechanism of malathion resistance in the green rice leafhopper, *Nephotettix cincticeps* Uhler (Hemiptera: Deltocephalidae), *J. Pestic. Sci.*, 1:23.

Miyata, T., and Saito, T., 1978a, Adaptation and resistance to insecticides in the planthopper and leafhopper, *J. Pestic. Sci.*, 3:179.

Miyata, T., and Saito, T., 1978b, The role of enhanced nonspecific esterase activity in the insecticide resistance of the green rice leafhopper, *4th Int. Cong. Pestic. Chem. (IUPAC)*, (Abst.)

Miyata, T., Haonda, H., Saito, T., Ozaki, K., and Sasaki, Y., 1976, In vitro degradation of ^{14}C-methyl malathion by organophosphate susceptible and resistant smaller brown planthopper, *Laodelphax striatellus* Fallèn, *Botyu-Kagaku*, 41:10

Miyata, T., Saito, T., and Honda, H., 1977, Studies on insecticide resistance III. Relationship between patterns of aliesterase zymogram and resistance to insecticides in the green rice leafhopper, *Ann. Meet. Jap. Soc. Appl. Ent. Zool.*, (Abstr.)

Miyata, T., Hama, H., and Saito, T., 1978a, Studies on insecticide resistance IV. Relationship between carbamate resistance and aliesterase activity in the green rice leafhopper, *Ann. Meet. Jap. Soc. Appl. Ent. Zool.*, (Abst.)

Miyata, T., Saito, T., Kassai, T., and Ozaki, K., 1978b, Studies on insecticide resistance V. Aliesterase activity and metabolism of ^{14}C-methyl malathion of organophosphorus insecticide resistant brown planthoppers, *Ann. Meet. Pestic. Sci. Soc. Japan*, (Abst.)

Miyata, T., Saito, T., Hama, H., Iwata, T., and Ozaki, K., 1979, unpublished data.

Motoyama, N., and Dauterman, W. C., 1974, The role of non-oxidative metabolism in organophosphorus resistance, *J. Agr. Food Chem.*, 22:350.

Motoyama, N., Rock, G. C., and Dauterman, W. C., 1971, Studies on the mechanism of azinphosmethyl resistance in the predacious mite, *Neoseiulus (T.) fallacis* (Family: Phytosiidae), *Pestic. Biochem. Physiol.*, 1:205.

Needham, P. H., and Devonshire, A. L., 1975, Resistance to some organophosphorus insecticides in field populations of *Myzus persicae* from sugar beet in 1974, *Pestic. Sci.*, 6:547.

Needham, P. H., and Sawicki, R. M., 1971, Diagnosis of resistance to organophosphorus insecticides in *Myzus persicae* (Sulz.), *Nature*, 230:125.

Okuma, M., and Ozaki, K., 1969, Development of resistance to sumithion and malathion in the smaller brown planthopper, *Laodelphax striatellus* Fallen, *Proc. Assoc. Pl. Prot. Shikoku*, 4:45.

Oppenoorth, F. J., and Van Asperen, K., 1960, Allelic genes in the housefly producing modified enzymes that cause organophosphate resistance, *Science*, 132:298.

Oppenoorth, F. J., and Voerman, S., 1975, Hydrolysis of paraoxon and malaoxon in three strains of *Myzus persicae* with different degrees of parathion resistance, *Pestic. Biochem. Physiol.*, 5:431.

Ozaki, K., 1969, The resistance to organophosphorus insecticides of the green rice leafhopper, *Nephotettix cincticeps* Uhler and the smaller brown planthopper, *Laodelphax striatellus* Fallen, *Rev. Plant Protec. Res.*, 2:1.

Ozaki, K., and Kassai, K., 1970a, Relationship between patterns of esterase zymogram and resistance to insecticides in the green rice leafhopper, *Nephotettix cincticeps* Uhler, *Rep. Kagawa Agr. Exp. Sta.*, 20:62.

Ozaki, K., and Kassai, K., 1970b, Biochemical genetics of malathion resistance in the smaller brown planthopper, *Laodelphax striatellus*, *Ent. Exp. & Appl.*, 13:162.

Ozaki, K., and Koike, H., 1965, Naphthyl acetate esterase in the green rice leafhopper, *Nephotettix cincticeps* Uhler, with special reference to the resistant colony to the organophosphorus insecticides, *Jap. J. Appl. Ent. Zool.*, 9:53.

Ozaki, K., Kurosu, Y., and Koike, H., 1966, The relation between malathion resistance and esterase activity in the green rice leafhopper, *Nephotettix cincticeps* Uhler, *SABCO J.*, 2:98.

Plapp, F. W., 1976, Biochemical genetics of insecticide resistance, *Ann. Rev. Ent.*, 21:179.

Sawicki, R. M., Devonshire, A. L., Rice, A. D., Moores, G. D., Petzing, S. M., and Cameron, A., 1978, The detection and distribution of organophosphorus and carbamate insecticide-resistant *Myzus persicae* (Sulz.) in Britain in 1976, *Pestic. Sci.*, 9:189.

Sudderuddin, K. I., 1973, Studies of insecticide resistance in *Myzus persicae* (Sulz.) (Hem., Aphididae), *Bull. Ent. Res.*, 62:533.

Takehisa, T., and Tanaka, M., 1967, Detection of insecticide resistance of citrus red mite by thin layer agar gel electrophoresis, *Proc. Assoc. Pl. Prot. Kyushu*, 13:126.

Tsuboi, A., Hiramatsu, T., and Kobayashi, M., 1973, Geographical distribution of resistant green rice leafhopper, *Nephotettix cincticeps* Uhler, to organophosphorus-insecticide in Okayama prefecture, *Chyugoku Nogyo Kenkyu*, 47:97.

Van Asperen, K., and Oppenoorth, F. J., 1959, Organophosphate resistance and esterase activity in houseflies, *Ent. Exp. & Appl.*, 2:48.

Yamamoto, I., Kyomura, N., and Takahashi, Y., 1977, Aryl *N*-propylcarbamates, a potent inhibitor of acetylcholinesterase from the resistant green rice leafhopper, *Nephotettix cincticeps*, *J. Pestic. Sci.*, 2:463.

Yasutomi, K., 1970, Studies on organophosphate-resistance and esterase activity in the mosquitoes of the *Culex pipiens* group, *Jap. J. Sanit. Zool.*, 21:41.

Yasutomi, K., 1971, Studies on diazinon-resistance and esterase activity in *Culex tritaeniorhynchus*, *Jap. J. Sanit. Zool.*, 22:8.

Yasutomi, K., 1980, this volume.

Zettler, J. L., 1974, Esterases in a malathion-susceptible and a malathion-resistant strain of *Plodia interpunctella* (Lepidoptera: Phycitidae), *J. Georgia Entomol.*, 9:207.

METHODS FOR DETECTING AND MONITORING THE RESISTANCE OF PLANT PATHOGENS TO CHEMICALS

J. M. Ogawa, B. T. Manji, C. R. Heaton,
J. Petrie* and R. M. Sonoda†

Department of Plant Pathology
University of California
Davis, California 95616

INTRODUCTION

The detection and monitoring of strains of plant pathogens resistant to chemicals became of concern only recently, with the introduction of fungicides and bactericides that are highly specific, partially systemic, and very effective. Agriculturists have been fortunate in having had no resistance problems to contend with during the long period when sulfurs, copper, mercuries, dithiocarbamates, captan and similar nonsystemic fungicides and soil fumigants were in extensive use. Temporary resistance or adaptation of plant pathogens to copper was detected, but these resistant pathogens lost their resistance once they were allowed to grow on media free of the chemical (Parry and Wood, 1959). Thus, temporary adaptation of plant pathogens to some fungicides was not related to or correlated with any loss in their efficacy in controlling diseases. Decades and even centuries of repeated applications with these chemicals produced no evidence that they would fail to protect crops against loss to diseases because of the development of resistance. Concern about fungicide resistance and methods for determining resistance came with the discovery by Harding (1962) that control of *Penicillium* on citrus fruit was being lost through the development of resistance to orthophenylphenate and biphenyl. Ogawa et al. (1963) later detected dicloran resistance in *Gilbertella*. Review articles by Ashida (1965) on the adaptation of fungi to metal toxicants and by Georgopoulos

*Decco-Tiltbelt Division, Pennwalt Corp., Monrovia, CA 91016.
†IFAS, Agricultural Research Center, University of Florida, Ft. Pierce, FL 33450.

and Zaracovitis (1967) on the tolerance of fungi to organic fungicides soon followed.

In recent years, resistance of pathogens to newly introduced chemicals resulted in crop losses, and resistance is now of much concern in recommendations of chemical use. In estimating the expected life of newly introduced fungicides and bactericides, the questions that must be explored include: (1) can we determine from laboratory studies the odds that the pathogen will develop resistance to the chemical from use under field conditions; (2) can feasible and economically sound systems be developed to detect resistant populations to fungicides before economic losses occur, and (3) will early detection and monitoring of low-level resistant populations of the pathogens be an important tool in fungicide management programs that will assure a maximum useful life of a fungicide? The discussion of programs for detection and monitoring necessarily rests on information learned from all past and present experiences. Thus, new research programs on resistance in plant pathogens should refer to the direction of resistance research taken by entomologists, weed scientists and medical researchers through discussions of mutual problems.

Terminology must be considered first. According to Sakurai (1977), "resistance" (in Japanese, "Teikosei") is the preferred term to indicate inherited, decreased fungicide sensitivity ("kanjusei") in a population of plant pathogens. At the 1979 meeting of the FAO of the United Nations Committee on Pest Resistance to Pesticides and Crop Loss Assessments, it was recommended that "resistance" apply only to heritable resistance in fungi and bacteria, as it does in insects and nematodes. Other decreases in sensitivity to fungicides would be reported by such terms as training, adaptation, insensitivity, tolerance (taisei), etc. Many researchers object to the use of the term "resistance" because it can refer to any physiological or anatomical reaction of crops to diseases and insects. Sakurai (1977) noted that "kanjusei" corresponds to "sensitivity" or "susceptibility" to any outside stimulus and is not limited to a response to chemical treatments. In the United States, Japan and Europe, the terms "resistance" and "tolerance" are generally used interchangeably. In a computer search of the literature we found that the most common terms for decreased sensitivity of organisms to toxic compounds were "resistance," "tolerance" and "insensitivity." The terms "training" and "adaptation" have been reserved for instances in which the influence of the chemical is temporary. The term "resistance" will be used throughout the following text.

The term "fungicides" is used herein to include all chemicals used against any plant pathogen, including fungi, bacteria, mycoplasmas and viruses. In Japan, the term "drug" is often used, as in "drug-resistant strain."

"Detection method" refers to any procedure that finds or verifies chemical resistance. The procedures may range from detecting reduced disease control in practice to the development of laboratory or greenhouse techniques to differentiate fungicide-resistant strains from sensitive strains. It is important that the method be scientifically sound, the pathogen correctly identified, the resistance related to the chemical in question, and the mechanism of resistance heritable (not one of individual adaptation). The method must take into account that often low levels of resistance are temporary (Schroth et al., 1979) and that higher levels of resistance to the same fungicide are usually stable.

Monitoring procedures can be used to determine the degree and distribution of a plant pathogen's resistance to a chemical in a given population. The technique may sample the plant pathogen population within a greenhouse, packing shed, field, orchard or region, and must critically assess whether or not resistant populations are present.

REPORTS OF FUNGICIDE RESISTANCE IN PLANT PATHOGENS

The objective of detecting and monitoring fungicide-resistant plant pathogens is to prevent crop losses and to prolong use of the chemicals. There are two separate monitoring strategies, one of which focuses on detecting the presence, distribution and level of resistance in order to minimize short-term losses. This type of program is predicated on the assumption that, given such information, growers can adjust their programs (e.g., by planting alternative crops or using alternative fungicides) to reduce losses foreshadowed by the presence of fungicide-resistant pathogens. The second involves tracking the development and distribution of resistant (R) species over time, thus yielding data essential to the design of effective, long-term countermeasures. Prescriptions for countering the effects of existing fungicide-R pathogens and preventing further increases in resistance levels and the numbers of R pathogens will remain haphazard until the major environmental and biological factors influencing the development and evolution of fungicide-R pathogens are more clearly identified and the nature of their interaction better defined.

The experience has been somewhat different in chemical control of insects. Although with insects, resistance has developed to almost all insecticides introduced, this is not true for fungicides, which can be divided into three categories: (1) chemicals to which no resistance has been detected; (2) chemicals to which resistance has been detected only in the laboratory, and (3) chemicals to which resistance has been detected under field use.

Chemicals To Which No Resistance Has Been Detected

Table 1 lists over 100 fungicides (with year introduced) to which no resistance has been detected. Many of these chemicals are important to plant pathologists because they provide tools for the control of plant diseases that are not often controlled by other means, and it is believed that most of these chemicals have modes of action that do not lead to selection for R strains. Included are such important chemicals as sulfurs (1800), coppers (1887), maneb (1950), fenaminosulf (1960) and captafol (1961). Because of the extensive use of these chemicals it is unlikely that resistance to them would have gone undetected. Sulfurs and dithiocarbamates are used most widely throughout the world. Table 1 also includes chemicals used for soil treatment, such as methyl bromide, chloropicrin and carbon disulfide. Although many of the fungicides listed have limited usage because they are less effective than newer compounds, plant pathologists are fortunate to have available these compounds to which resistance has been undetected after such long periods of use.

The fungicides introduced in the 1970s, listed in Table 1, are effective in controlling a broad spectrum of diseases. Of note are metaxanin, for the downy mildews and soil-borne *Pythium* and *Phytophthora*, and CGA 64251, biloxazol and fenarimol, general fungicides for powdery mildew and nonpythiaceous-induced diseases. After limited use, resistance to these chemicals has not been reported, and they are potentially useful alternatives for fungicides with resistance problems.

Chemicals To Which Resistance Has Been Detected Only in the Laboratory

Table 2 lists fewer than 27 chemicals that are of special interest, since resistance to them has been demonstrated in the laboratory tests but not in the field. Many of these fungicides, such as Bordeaux mixture, captan, carboxin, chlorneb, chlorothalonil, cycloheximide, dichlorofluanid, ethazol, ferbam, quintozene and thiram have been used extensively in practice, and some of them control strains resistant to other chemicals, such as oxytetracycline for streptomycin-R *Erwinia amylovora* on pear, and chlorothalonil for several pathogens resistant to benomyl. Knowledge of why this important group of fungicides has not encountered resistance under field use should help in developing new fungicides. At this point we assume that plant pathogens will not readily develop high levels of resistance to captan, carboxin, chlorothalonil, quintozene, and thiram, but may become resistant to the antibiotics listed. For new introductions with limited use in the field, such as iprodione, imazalil, procymidione and vinclozolin, continued monitoring may determine whether or not resistance will develop in practice.

Table 1. Fungicides and Bactericides to Which No Resistance Has Been Detected in Plant Pathogens, the Year Introduced, and Examples of Plant Pathogens Controlled, in Order of Introduction

Common or trivial name (trade name)	Year introduced	Examples of disease pathogen, or use
Sulfur	1800 BC	Powdery mildews, *Monilinia*
Hypochlorite (chlorine)	1798	General disinfestant
Lime sulfur (Ca polysulfide)	1852	General fungicide, powdery mildews
Carbon disulfide (carbon bisulfide)	1854	*Armillaria*, soil treatment
Copper carbonate (basic cupric carbonates)	1887	Seed treatment
Formaldehyde (Formalin)	1888	Soil treatment General disinfestant
Sorbic acid	1895	Control molds on processed food
Copper oxychloride (basic cupric chlorides)	1900	General
Petroleum oils (oils)	1922	Used as adjuvant, Sigatoka disease
Cheshunt compound (Cu sulfate + NH_4 carbonate)	1922	Seedling diseases
Mercurous chloride (Calomel)	1929	Turf fungicide
Ziram (Zerlate)	1930	General, *Coryneum*, *Taphrina*
Salicylanilide (Shirlan)	1931	Mildew proofing fabrics

(continued)

Table 1. (Continued)

Common or trivial name (trade name)	Year introduced	Examples of disease pathogen, or use
Cuprous oxide (Perenox)	1932	General
Methyl bromide (Bromomethane)	1932	Soil-borne diseases
Methoxyethylmercury chloride (Agallol)	1935	Soil-borne diseases
Chinosol (8-hydroxyquinoline sulphate)	1936	Seedling diseases, *Botrytis*
Gliotoxin (Bliotoxin)	1936	Antibiotic
Pentachlorophenol (Dowicide 7)	1936	Wood-rotting fungi
Chloranil (Spergon)	1937	Seed treatment, foliar
Methoxyethylmercury acetate (Panogen)	1938	Seed dressing
Griseofulvin (Fulsin)	1939	Antibiotic, *Botrytis*, *Cladosporium*, powdery mildews
Nabam (Dithane D-14)	1943	General
Zineb (Dithane Z-78)	1943	General
2,4-Dinitrophenyl-thiocyanate (nirit)	1946	*Peronospora*
Oxine-copper (Oxine)	1946	Treatment of fruit-handling equipment
Chloropicrin (Trichloronitromethane)	1948	Soil fungi, *Verticillium*

Common or trivial name (trade name)	Year introduced	Examples of disease pathogen, or use
Folpet (Phaltan)	1949	General
Sodium pentachlorophenate (Niagara dormant)	--	*Monilinia*, wood preservative
Dehydroacetic acid (DHA)	1950	Postharvest diseases
Maneb (Dithane M-22, Manzate)	1950	General fungicide
Sodium arsenite	--	*Phomopsis* on grapes
Sodium lauryl sulfate	--	*Uncinula necator*
Dazomet (Mylone)	1952	Soil fungi
bis (Dimethylthiocarbamoyl thio) methyl arsine (Tuzet, Urbacid)	1953	Coffee diseases and apple scab
Glyodin (Crag fungicide 341)	1954	General and as surfactant
Benquinox (Ceredon Special)	1955	Seed-borne pathogens
Sodium-N-methyldithiocarbamate (Vapam)	1955	Soil-borne fungi and bacteria
Methyl-arsinsulfide (Rhizoctol)	1958	*Rhizoctonia*, *Fusarium*, *Helminthosporium*
Binapacryl (Morocide 40)	1960	Powdery mildew
Copper ammonium carbonate (Copper-Count N)	1960	General use
Fenaminosulf (Dexon)	1960	Seedling diseases, phycomycetes

(continued)

Table 1. (Continued)

Common or trivial name (trade name)	Year introduced	Examples of disease pathogen, or use
Triamphos (Wepsyn 1555)	1960	Powdery mildew
Captafol (Difolatan)	1961	General
Hexachlorophene (Isobac)	1962	*Rhizoctonia*
1-Chlor-2-nitropropane (Lanstan)	1962	Soil fungi (*Pythium*, *Rhizoctonia*, *Fusarium*, *Verticillium*)
Dithianon (Delan)	1962	General, downy mildew
Propineb (Antracol)	1963	Downy mildew
Copper hydroxide complex (Kocide)	1964	General use
Copper naphthenates	1964	Mildew on wood and fabrics
N-Cyanathyl-chloracetamide (Udonkor)	1964	Mildews
Dinobuton (Acrex)	1965	Powdery mildew
Dodemorph (BAF 3101F)	1965	Powdery mildew
Piperalin (pipron)	1965	Powdery mildew
Dinocton (Dinocton-4)	1966	Powdery mildew
Drazoxolon (Ganocide, Mil-Col)	1966	Powdery mildew, *Fusarium*, *Pythium*, *Ganoderma*

Common or trivial name (trade name)	Year introduced	Examples of disease pathogen, or use
Fuberidazole (Voronit)	1966	Snow mold and stinking smut of cereals
Parinol (Parnon)	1966	Powdery mildew
2,4-Xylenol plus metacresol (Bacticin)	1967	*Agrobacterium*
Chloraniformethan (Imugan, Milfaron)	1968	Powdery mildew
Guazatine (Panoctine)	1968	*Fusarium*, *Septoria*, *Helminthosporium*
Methyl-metiram (Basfungin)	1968	*Peronospora*, *Botrytis*
Dithiolphosphospate-butyl-S-diethyl-S-benzylester (Conen)	1969	*Pyricularia*
Dimethachlor (Ohric)	1969	*Helminthosporium* *Botrytis*, *Corticium*
Pyrazophos (Afugan)	1969	Powdery mildew
Tridemorph (Calixin)	1969	Powdery mildew of cereals
Chlorquinox (Lucel)	1970	Mildews
Hymexazol (Tachigaren)	1970	*Fusarium*, *Pythium*, *Aphanomyces*
2-Brom-2-nitro-1,3-propandiol (Bronopol)	1971	*Xanthomonas*
Pyracarbolid (Sicarol)	1971	Rust fungi, *Rhizoctonia*

(continued)

Table 1 (Continued)

Common or trivial name (trade name)	Year introduced	Examples of disease pathogen, or use
Tetrachlorphthalide (Rabcide, TCP)	1971	*Pyricularia*
Validamycin A (Validacin)	1972	Rice sheath blight, *Rhizoctonia*
Quinacetol sulphate (Fongoren)	1973	*Rhizoctonia, Penicillium*
Benodanil (Calirus)	1974	Rust fungi, *Rhizoctonia*
Fluoroimid (Spaticide, Sparticide)	1974	*Septoria,* apple scab
Bupirimate (Nimrod)	1975	Powdery mildew
n-Butyl-1,2,4-triazol (Indar)	1975	Rusts
Prothiocarb (Previcur)	1975	*Phycomycetes*
3-Carboxanilido-2-4,5 trimethyl furan (Granovax, UNI H 719)	1976	*Rhizoctonia*
Fenfuram (Panoram)	1976	*Rhizoctonia*
Ferric methanearsonate (ferric ammonium salt) (Neo-asozin)	--	*Rhizoctonia, Botrytis*
Fluotrimazol (Persulon, Bay 6660)	1976	Powdery mildew
Lecithin (Lecithinon)	1976	Powdery mildew
Tricyclazole (Beam 75)	1976	*Pyricularia*

Common or trivial name (trade name)	Year introduced	Examples of disease pathogen, or use
Aluminum ethylphosphate (Aliette)	1977	Downy mildew, *Phytophthora*, *Alternaria*, *Phomopsis*
2-Cyan-N-ethylaminocarbonyl-2-(methyoxyimino-)acetamide (Curzate)	1977	Downy mildew
Furalazyl (Fongarid)	1977	Downy mildew, *Pythium*, *Phytophthora*
Metalaxyl (Ridomil, CGA 48988)	1977	*Phytophthora*, *Plasmopara*, *Peronospora*, *Pythium*, *Pseudoperonospora*
Nickel-dimethyl-dithiocarbamate (Sankel)	--	*Xanthomonas oryzae*
Nuarimol (Trimidal, EL 228)	1977	Powdery mildew
CGA 64251	1976	General, Ascomycetes, basidiomycetes, and deuteromycetes
Biloxazol (Baycor, BAY KWG 0599)	--	Powdery mildew, *Monilinia*, *Cercospora*

Table 2. Fungicides and Bactericides for Which Resistance in Plant Pathogens Has Been Detected in Laboratory Tests but Not in the Field

Common or trivial name[a] (trade name)	Year reported or introduced	Examples of pathogens with resistance (Reference)[b]
Antimycin A	1957	*Venturia* (Leben et al., 1955)
S-Benzlethylphenylphos-phonothiolate (Inezin)	1967	*Pyricularia* (Uesugi, 1969)
Bordeaux mixture (Copper sulphate plus lime)	1885	*Physalospora* (Taylor, 1953)
Boron (Na tetraborate decahydrate)	1938	*Pyricularia* (Yamasaki et al., 1964)
Captan	1949	*Botrytis cinerea* (Parry and Wood, 1959)
		Puccinia graminis (Polyakov and Mende, 1965)
Carboxin	1966	*Ustilago* (Ben-Yepeth et al., 1974)
		Aspergillus (Van Tuyl, 1975)
Chloramphenicol	1948	*Aspergillus* (Gunatilleke et al., 1975)
Chloroneb (Demosan)	1967	*Ustilago* (Tillman and Sisler, 1973)
Chlorothalonil (Bravo 6F)	1963	*Botrytis* (Sakurai, 1977)
Copper sulphate (Pentahydrate of cupric hydrate)	1807	*Monilinia* (Mader and Schneider, 1948)
Cycloheximide (Actidione)	1948	*Cochliobolus* (MacKenzie et al., 1971)

Common or trivial name[a] (trade name)	Year reported or introduced	Examples of pathogens with resistance (Reference)[b]
Cycloheximide (Actidione)	1948	*Schizophyllum* (Pikalek et al., 1978)
Dichlozoline (Sclex)	1967	*Aspergillus* (Dekker, 1972)
		Botrytis (Leroux, 1977)
Ethazol (Truban, Terrazole)	1969	*Pythium* (Halos and Huisman, 1976)
Fenarimol	1977	*Aspergillus* (De Waard et al., 1977)
Fentin chloride (Tinmate)	1960	*Cercospora* (Giannopolitis, 1978)
Ferbam (Fermate)	1940	*Botrytis* (Parry and Wood, 1959)
Imazalil (Fungaflor)	1974	*Aspergillus* (Van Tuyl, 1977)
Mancozeb (Dithane M-45)	1961	*Sclerotium rolfsii* (Anilkmar, 1976)
Oxytetracycline (Terramycin)	1950	*Erwinia* (English and Van Helsema, 1954)
Phenylmercury acetate (Tag HL 331)	1914	*Pyricularia* (Yoshii et al., 1958)
Pimaricin (Myprozine)	1958	*Aspergillus* (Van Tuyl, 1977)
		Cladosporium (Dekker et al., 1979)
		Fusarium (Dekker et al., 1979)
Polyram (Metiram)	1958	*Monilinia* (Good and MacNeill, 1972)

(continued)

Table 2. (Continued)

Common or trivial name[a] (trade name)	Year reported or introduced	Examples of pathogens with resistance (Reference)[b]
TCMTB (Busan-72)	1970	*Fusarium* (Vanachter, 1977)
		Rhizoctonia (Vanachter, 1977)
Triarimol (Trimidal, EL 273)	1969	*Sclerotinia* (Wolfe, 1971)
		Sphaerotheca (Gilpatrick and Provvidenti, 1973)
Triforine (Cela 524, Saprol)	1969	*Cladosporium* (Fuchs et al., 1977)
Vinclozolin (Ronilan)	1976	*Botrytis* (LeRout, 1977)
		Monilinia (Sztejnberg and Jones, 1978)

[a]Chemical name used when common or trivial name is not available.
[b]References not listed in literature cited. List obtained from Review of Plant Pathology and review published by Ogawa et al., 1977.

Chemicals To Which Resistance Has Been Detected Under Field Use

Table 3 lists 34 chemicals of great concern because of reported failures in disease control. By far the most numerous examples are provided by the benzimidazole fungicides (benomyl, thiophanate methyl, carbendazim and TBZ). Of these, the greatest number of resistant fungus genera have been reported for benomyl, including *Botrytis, Cercospora, Erysiphe, Monilinia, Sphaerotheca* and *Penicillium* and *Venturia*. Other important chemicals include diphenyl and orthophenylphenate, both used effectively for control of postharvest decay in citrus (Harding, 1962). The pyrimidines (dimethirimol and ethirimol) when first introduced provided excellent control of mildews but soon encountered resistance problems, and the antibiotics kasugamycin and blasticidin-S were developed for control of rice blast caused by *Pyricularia oryzae,* but again, their potential was not realized due to the development of resistance. Most of these chemicals are still used in some areas where resistance has not yet occurred or losses have not become of economic importance in the presence of resistance biotypes. For example, benomyl is still used against *Monilinia* on stone fruits and *Botrytis* on grapes; pyrimidines are used against barley mildew; dodine against *Venturia;* and sodium orthophenylphenol, biphenyl, 2-aminobutane and thiabendazole against postharvest citrus decay.

Resistance in plant pathogens to more than one fungicide of different modes of action is difficult to detect. To date, multiple resistance has been found in laboratory tests of field isolates resistant to one fungicide, but not to fungicide combinations applied in the field. An example of multiple resistance is in *Botrytis cinerea* in which some (or all) strains resistant to benomyl were resistant to dicloran (Chastagner and Ogawa, 1979). This can be of practical importance since benomyl is used in the tomato field and dicloran can be used as a postharvest treatment (Chastagner and Ogawa, 1979). Surprisingly, strains resistant to carbendazim (benzimidazole) also developed resistance to chlorothalonil (Sakurai et al., 1975). Chlorothalonil recently replaced benomyl for control of *Botrytis* on tomatoes as well as *Cercospora* on peanuts and *Septoria* on celery, but no failures in control of these diseases have been related to cross resistance to chlorothalonil. Chastagner (1979) reported cross tolerance of two new fungicides being tested in the United States: glycophene and vinocolozolin. *Pyricularia* strains resistant to kasugamycin were also found to be resistant to both blasticidin-S (Sakurai et al., 1976) and IBP (Yaoita et al., 1979); in another report a kasugamycin-R strain of *Pyricularia* was resistant not only to blasticidin-S but also to polyoxin D (Sakurai et al., 1975).

From these reports, one should look into the possibility that susceptibility to the development of resistance to a second

Table 3. Fungicides and Bactericides For Which Resistance to Plant Pathogens Has Been Detected under Commercial Usage

Common or trivial name[a] (trade name)	Year introduced	Examples of pathogens showing resistance (Ref.)[b]
2-Aminobutane (Tutane, Frucote)	1962	*Penicillium digitatum* (Harding, 1976)
		Penicillium italicum (Harding, 1976)
Anilazine (Dyrene)	1955	*Sclerotinia homoeocarpa* (Nicholson et al., 1971)
Benomyl (Benlate)	1967	*Ascochyta chrysanthemi* (Steekelenburg, 1973)
		Aspergillus nidulans (Hastie, 1970)
		Botrytis cinerea (Bollen et al., 1971)
		Ceratocystis ulmi (Nishijima et al., 1971)
		Cercospora apii (Berger, 1973)
		Cercospora arachidicola (Littrell, 1974)
		Cercospora beticola (Georgopoulos et al., 1973)
		Cercosporella herpotrichoides (Rashid et al., 1975)
		Cercosporidium personatum (Clark et al., 1974)
		Cladosporium carpophilum (Chandler et al., 1978)
		Cladosporium cucumerinum (Dekker, 1972)

Common or trivial name[a] (trade name)	Year introduced	Examples of pathogens showing resistance (Ref.)[b]
Benomyl (cont.)		*Cladosporium effusum* (Littrell, 1976)
		Colletotrichum coffeanum (Okioga, 1976)
		Colletotrichum musae (Griffee, 1973)
		Colletotrichum lindemuthianum (Meyer, 1975)
		Didymella ligulicola (Krijthe and Hoorst, eds., 1973)
		Erysiphe cichoracearum (Iida, 1975)
		Erysiphe graminis (Vargas, 1973)
		Erythronium sp. (Duineveld, 1975)
		Fusarium oxysporum f. sp. *dianthi* (Tramier et al., 1974)
		Fusarium oxysporum f. sp. *gladioli* (Magie et al., 1974)
		Fusarium oxysporum f. sp. *lycopersici* (Thanassoulopoulos et al., 1971)
		Fusarium oxysporum f. sp. *melonis* (Bartels-Schooley et al., 1971)
		Fusarium roseum (Hoitink et al., 1970)

(continued)

Table 3. (Continued)

Common or trivial name[a] (trade name)	Year introduced	Examples of pathogens showing resistance (Ref.)[b]
Benomyl (cont.)		*Fusarium solani* (Richardson, 1973)
		Fusicaldium effusum (Littrell, 1976)
		Fulvia fulva (Jordan et al., 1975)
		Glicladium sp. (Lyon, 1978)
		Monilinia fructicola (Whan, 1976)
		Monilinia fructigena (Abelentsev, 1973)
		Monilinia laxa (Abelentsev, 1973)
		Mycospharella fragaries (Remior et al., 1974)
		Neurospora crassa (Sisler, 1972)
		Oidiopsis taurica (Netzer et al., 1970)
		Oidium begoniae (Strider, 1978)
		Penicillium brevicompactum (Bollen, 1971)
		Penicillium corymbiferum (Bollen, 1971)
		Penicillium digitatum (Muirhead, 1974)
		Penicillium expansum (Ogawa and Manji, 1975)

Common or trivial name[a] (trade name)	Year introduced	Examples of pathogens showing resistance (Ref.)[b]
Benomyl (cont.)		*Penicillium italicum* (Muirhead, 1974)
		Penicillium verrucosum (Penrose et al., 1978)
		Rhizoctonia solani (Kataria et al., 1974)
		Sclerotinia homoeocarpa (Goldenberg et al., 1973)
		Sclerotinia sclerotiorum (Netzer et al., 1970)
		Sclerotium rolfsii (Anilkumar, 1976)
		Septoria leucanthemi (Paulus et al., 1974)
		Septoria nodorum (Fehrmann, 1977)
		Septoria passiflora (Peterson, 1977)
		Sphaerotheca fuligenea (Schroeder et al., 1969)
		Sphaerotheca pannosa (Yoshii et al., 1975)
		Uncinula necator (Nagler et al., 1977)
		Ustilago hordei (Ben-Yepeth et al., 1974)
		Venturia inaequalis (Wicks, 1974)
		Venturia nashicola (Ishii et al., 1977)

(continued)

Table 3. (Continued)

Common or trivial name[a] (trade name)	Year introduced	Examples of pathogens showing resistance (Ref.)[b]
Benomyl (cont.)		*Venturia pirina* (Shabi, 1976)
		Verticillium dahliae (Talboys et al., 1976)
		Verticillium fungicola (Bollen et al., 1975)
		Verticillium tricorpus (Locke et al., 1977)
		Verticillium malthousei (Wuest, 1974)
S-Benzyl O,O-diisopropyl phosphorothioate (Kitazin, IBP)	1968	*Pyricularia* (Katagiri, in press)
Blasticidin-S (BLA-S)	1959	*Pyricularia oryzae* (Sakurai, 1974)
Cadmium succinate (Cadminate)	1948	*Aspergillus nidulans* (Davidse, 1976)
		Cochliobolus carbonum (MacKenzie et al., 1971)
		Pyricularia oryzae (Yamasaki et al., 1964)
		Sclerotinia homoeocarpa (Cole et al., 1968)
Carbendazim (Bavistan, MBC)	1973	*Botrytis cinerea* (Geeson, 1978)
		Cercosporella herpotrichoides (Fehrmann, 1977)
		Colletotrichum coffeanum (Okioga, 1976)

Common or trivial name[a] (trade name)	Year introduced	Examples of pathogens showing resistance (Ref.)[b]
Carbendazim (cont.)		*Septoria nodorum* (Fehrmann et al., 1977)
		Sporobolomyces roseus (Nachmias et al., 1976)
Dichlone (Phygon)	1946	*Puccinia graminis* (Polyakov and Mende, 1964)
Dicloran	1959	*Botrytis cinerea* (Webster et al., 1968)
		Gilbertella persicaria (Ogawa et al., 1963)
		Rhizopus stolonifer (Webster et al., 1968)
		Sclerotium cepivorum (Locke, 1969)
Dimethirimol (Milcurb)	1968	*Sphaerotheca fuligenea* (Bent et al., 1971)
Dinocap (Karathane)	1946	*Sphaerotheca fuligenea* (Iida, 1975)
Diphenyl (Biphenyl)	1944	*Diplodia* sp. (Littnauer and Gutter, 1953)
		Penicillium digitatum (Harding, 1959)
		Penicillium italicum (Harding, 1959)
Dodine (Cyprex, Melprex)	1956	*Venturia inaequalis* (Szkolnik and Gilpatrick, 1969)
Edifenphos (Hinosan EDDP)	1968	*Pyricularia oryzae* (Katagiri and Uesugi, in press)

(continued)

Table 3. (Continued)

Common or trivial name[a] (trade name)	Year introduced	Examples of pathogens showing resistance (Ref.)[b]
Edifenphos (cont.)		*Septoria nodorum* (Fehrmann, 1977)
Ethirimol (Milstem)	1969	*Erysiphe graminis* (Bent, 1971)
Fentin acetate (Brestan)	1954	*Cercospora beticola* (Giannopolitis, 1978)
Fentin hydroxide (Du-Ter)	1963	*Cercospora beticola* (Giannopolitis, 1978)
Hexachlorobenzene (Perchlorobenzene, HCB)	1949	*Tilletia foetida* (Kuiper, 1965)
Isoprothiolane (Fuji-One, IPT)	1974	*Pyricularia oryzae* (Uesugi, 1980)
Iprodione (Glycophene, Rouval)	1976	*Botrytis cinerea* (Kotani, 1979)
		Monilinia fructicola (Sztenberg and Jones, 1978)
Kasugamycin (Kasumin)	1965	*Pyricularia oryzae* (Miura et al., 1973)
Methoxyethylmercury silicate (Cerasan)	1935	*Pyrenophora avenae* (Noble et al., 1966)
Oxycarboxin (Plantvax, F461)	1966	*Puccinia horiana* (Abiko et al., 1977)
2-Phenylphenol (O-Phenylphenol, Dowcide A)	1936	*Penicillium digitatum* (Harding, 1962)
Polyoxin B	1965	*Alternaria kikuchiana* (Shimada et al., 1972)
		Alternaria mali (Iida, 1975)

Common or trivial name[a] (trade name)	Year introduced	Examples of pathogens showing resistance (Ref.)[b]
Polyoxin B (cont.)		*Botrytis cinerea* (Iida, 1975)
Polyram		*Monilinia fructicola* (Good et al., 1972)
Pyrazophos (Curamil, Afugan)	1969	*Sphaerotheca fuligenea*
Quinomethionate (Morestan)	1962	*Botrytis alii* (Priest et al., 1961)
		Sphaerotheca fuligenea (Iida, 1975)
Quintozene (Terraclor, PCNB)	1930	*Botrytis alii* (Priest and Wood, 1961)
		Hypomyces solani f. sp. *cucurbitae* (Georgopoulos, 1963)
		Rhizoctonia solani (Shatla and Sinclair, 1962)
		Sclerotium rolfsii (Georgopoulos, 1964)
Streptomycin (Agrimycin 100)	1952	*Erwinia amylovora* (Schroth et al., 1969)
		Pseudomonas apii (Thayer, 1963)
		Pseudomonas mori (Iida, 1975)
		Pseudomonas syringae (Dye, 1958)
		Pseudomonas tabaci (Cole, 1960)

(continued)

Table 3. (Continued)

Common or trivial name[a] (trade name)	Year introduced	Examples of pathogens showing resistance (Ref.)[b]
Streptomycin (cont.)		*Xanthomonas diffenbachiae* (Knauss, 1972)
		Xanthomonas phaseoli (Zaumeyer et al., 1953)
		Xanthomonas vesicatoria (Stall and Thaner, 1962)
Tecnazene (Folosan, TCNB)	1946	*Botrytis cinerea* (Esuruoso et al., 1971)
		Fusarium coeruleum (Makee, 1951)
Thiabendazole (mertect, TBZ)	1962	*Colletotrichum musae* (Griffee, 1973)
		Fusarium oxysporum f. sp. *gladioli* (Magie et al., 1974)
		Penicillium digitatum (Harding, 1972)
		Penicillium italicum (Harding, 1972)
Thiophanate methyl (Topsin M, Cercobin)	1969	*Botrytis cinerea* (Bollen et al., 1971)
		Colletotrichum coffeanum (Okiaga, 1975)
		Colletotrichum musae (Griffee, 1973)
		Erysiphe cichoracearum (Iida, 1975)
		Erysiphe graminis (Vargas, 1973)

Common or trivial name[a] (trade name)	Year introduced	Examples of pathogens showing resistance (Ref.)[b]
Thiophanate methyl (cont.)		*Fusarium oxysporum* f. sp. *dianthi* (Tramier et al., 1974)
		Penicillium digitatum (Kuramoto, 1976)
		Penicillium italicum (Kuramoto, 1976)
		Septoria leucanthemi (Paulus et al., 1974)
		Sphaerotheca fuliginea (Iida, 1975)
		Sphaerotheca humuli (Iida, 1975)
		Venturia inaequalis (Tate and Samuels, 1976)
		Venturia nashicola (Ishii et al., 1977)
Thiram (Arasan, Tersan)	1931	*Cochliobolus carbonum* (MacKenzie et al., 1971)
		Fusarium solani (Abdalla, 1975)
		Rhizoctonia solani (Abdalla, 1975)
		Sclerotinia homoeocarpa (Cole, 1968)

Includes plant pathogens for which resistance has been detected in the field and laboratory.
[a]Chemical name used when common or trivial name is not available.
[b]References not listed in literature cited. References obtained from Review of Plant Pathology and review published by Ogawa et al., 1977.

fungicide may become greater after resistance has developed to the first fungicide. These experiences suggest that procedures for detection of resistance should not be limited only to the fungicide in use but to fungicides formerly used and those that may be used in the future.

In Table 4 the numbers of pathogens detected with resistance are listed with the number of sites (crop/pathogen combinations) registered for some of the important fungicides used in the United States. Over 543 sites were registered for the EDBC's, captan, DMDC, sulfur and copper fungicides, and resistance has been detected in only six sites. The benzimidazoles have 45 registered sites, but 61 plant pathogens have been found resistant primarily to benomyl. Of the other 20 fungicides listed with 366 sites registered, serious crop losses due to resistance have been reported only with the tin compounds and streptomycin. Chlorine is probably the most widely used fungicide in postharvest treatments, but no resistance has yet been detected.

Thus, fungicide resistance of plant diseases in the United States involves principally the benzimidazoles, diphenyl, sodium orthophenylphenol, streptomycin, dodine and cadmium (the last is no longer recommended). In Japan, where antibiotics are widely used, resistance problems have occurred primarily with blasticidin-S, kasugamycin and, more recently, the organic phosphates, such as IBP.

DESIGNING EFFECTIVE PROGRAMS FOR DETECTION

Numerous factors must be considered in designing a detection program. Many of these have been discussed elsewhere (Sakurai, 1977; Littrell, 1976; Shepherd et al., 1975; Ogawa et al., 1978), but a further elaboration is warranted.

Sample Location, Time and Size

Location. Disease specimens should be sampled where the probability of detecting fungicide-R plant pathogens is high, e.g., where large quantities of fungicide have been applied for extended periods. Such locations may include areas where the disease causes the largest crop losses. Sampling for fungicide resistance in plant pathogens should be more fruitful for those pathogens in which one expects the greatest number of secondary infections and regeneration of secondary spores. The first reports on fungicide resistance listed in Table 3 were for pathogens of this type, such as *Venturia*, *Monilinia*, *Botrytis*, *Penicillium*, *Cercospora* and *Sphaerotheca*.

Time. To demonstrate resistance, the optimum time for collection is when the disease develops after fungicide application. The highest population of R strains would be expected after several

Table 4. Number of Fungicide-resistant Pathogens Detected on Fungicides Used Most Commonly in the United States, Based on the Number of Sites Registered on all Crops

Chemical	Number of sites registered[a]	Number of pathogens with resistance detected
EDBD's (maneb, zineb, etc.)	122	2
Captan	112	2
DMDC (ziram, ferbam)	110	1
Sulfur	105	0
Copper	88	1
Benzimidazole	45	81
Chlorine	b	0
Thiram	55	4
Quintozene	44	4
Folpet	36	0
Captafol	27	0
Dinocap	30	1
Dichlone	31	1
Dicloran	24	4
Chlorothalonil	24	1
Ethazol	19	1
Fenaminosulf	18	0
Anilazine	15	1
Cycloheximide	15	2
Streptomycin	13	8
Dodine	9	1

(continued)

Table 4. (Continued)

Chemical	Number of sites registered	Number of pathogens with resistance detected
Chloroneb	8	2
Carboxin	5	2
Tin	6	3
Oxytetracycline	3	1
Cadmium	1	4

[a]APS Committee Report, 1979. Contemporary control of plant diseases with chemicals. Sites refer to crop/pathogen combinations.
[b]General use as postharvest treatment on crops.

fungicide applications, and if sexual spores are sampled, the life cycle of the pathogen must be considered. With *Monilinia fructicola* on stone fruits, the apothecia develop during bloom and sampling must be conducted at this time. With *Venturia inequalis*, mature ascospores are discharged from the green-tip stage until early summer, and with the powdery mildews the cleistothecia develop at a phenological stage of the host that depends on mildew species. In the laboratory, sexual fruiting bodies have been developed for *Monilinia* spp. and *Botrytis* spp., although the necessary conditions are difficult to reproduce, but for certain fungal pathogens, such as *Sclerotinia sclerotiorum*, such development is not difficult at all.

Size. In designing a program to monitor fungicide resistance in a given field or area, the effort and expense of each level of sampling precision should be weighed against the economic losses incurred from failure to detect R populations at each density. Theoretically, sampling numbers can be increased until the cost of detection equals the revenue that would be saved by detection, but unfortunately, until more is known about the magnitude of crop losses due to the presence of different frequencies and levels of fungicide-R pathogens, an optimal program cannot be assured.

The tedium and expense of monitoring can be reduced by employing a sequential sampling scheme, and a suggested strategy is as follows:

1. Select an appropriate form for the population distribution and reasonable values for any population parameters not subject to sampling. Data for such use might come from preliminary sampling in other areas that are believed to have similar populations of R strains.

2. Establish the levels of resistance associated with the null (H_0) and the alternative hypothesis (H_1), the former being the level below which detection need not occur (i.e., the maximum tolerable level) and the latter being the level above which detection is desired.

3. Select the probability levels associated with type I (i.e., accepting H_0 when H_1 is true) errors.

4. On the basis of decisions made in the first three steps, calculate the acceptance and rejection regions as a function of sample size. These can then be plotted on a graph or tabulated for easy reference.

As might be expected, the exact calculation and procedure associated with a given sequential sampling regime will depend on the hypothesized distribution of the target populations (e.g., random, negative binomial, Poisson). Presented here for illustration is a numerical example based on a negative binomial distribution,* which is frequently used to model populations in which units tend to clump or aggregate, and hence might be appropriate in sampling *Monilinia*-decayed fruit from trees in a given area.

The exact form of a negative binomial distribution is defined by two values, the mean and a dispersion parameter. The latter is a measure of aggregation or clumping. Only one of these parameters can be sampled in a single experiment. Thus, a researcher wishing to determine the average number of R pathogens in a particular orchard would specify an *a priori* value for the dispersion parameter.

Given the assumption of a negative binomial distribution the formula for the acceptance and rejection regions as a function of sample size has been proven to be:

$$y_o = \Theta_n + b \qquad (1)$$

$$y_1 = \Theta_n + b_1 \qquad (2)$$

---

*This example is adapted from Southwood, J. R. E., Ecological Methods, London: Meuthen and Company.

where

$$\Theta = \gamma \frac{\log (q_1/q_0)}{\log (p_1 q_0/p_0 q_1)} \tag{3}$$

$$b_0 = \frac{\log [\beta/(1-\alpha)]}{\log (p_1 q_0/p_0 q_1)} \tag{4}$$

$$b_1 = \frac{\log [(1-\beta)/\alpha]}{\log (p_1 q_0/p_0 q_1)} \tag{5}$$

and y = the number of R strains isolated; n = the number of trees samples; p_0 = the population level if H_0 is true; p_1 = the population level if H_1 is true; $p_0 = 1 + p_0$; $q_1 = 1 + p_1$, and γ = the dispersion parameter.

Setting $\alpha = \beta = .05$, selecting a value of two for γ, and specifying $p_0 = 1$ and $p_1 = 10$, equations (1) and (2), respectively, become

$$y_0 = 5.7\ n - 4.90 \tag{6}$$

$$y_1 = 5.7\ n + 73.2 \tag{7}$$

Equations (6) and (7) define the lower and upper limits, respectively, of the uncertainty region. They can be used to prepare a table summarizing the appropriate decision for various sample sizes. Several such calculations are presented in Table 5.

Thus, in this example, if two trees are sampled and no R strains are isolated one would reject H_1, and conversely, if 100 R isolates are obtained with a sample size of two trees, one would reject H_0. However, if the number of R isolates identified after sampling two trees lies between 6 and 85, neither hypotheses can be rejected -- and the researcher should proceed with additional sampling.

Thus far we have focused on the advantages of a sequential sampling strategy. Researchers should be aware, however, that there are also several drawbacks to this method. First, it is possible that even after numerous samples one can reject neither H_0 nor H_1 with the predetermined level of confidence on the basis of the cumulated data.* Secondly, the validity of the results obtained depends

*To avoid such a difficulty, Southwood (op. cit.) recommends setting an "arbitrary upper limit" on the number of samples to be taken, with expense and effort likely to be prime factors in deciding this limit.

Table 5. Criteria for Distinguishing Between Alternative Hypotheses When Using a Sequential Sampling Strategy

Number of trees in sample (n)	Number of resistant isolates from sampled trees (cumulative total)	
	Reject H_1[a] (i.e., p = 10)	Reject H_0[b] (i.e., p = 1)
1	< .8	> 78.9
2	< 6.5	> 84.6
3	< 12.2	> 90.3
4	< 17.9	> 96
5	< 23.6	> 101.7

[a]Figures in this column are obtained by substituting the appropriate value for the sample size into equation (6).

[b]Figures in this column are obtained by substituting the appropriate value for the sample size into equation (7).

upon the accuracy of the *a priori* assumptions regarding the form of the population distribution. Finally, the effectiveness of the method depends upon the limits established for H_0 and H_1, yet little information is available to help the researcher select appropriate values for these parameters.

Collection, Transport and Storage of Samples

Collection. The method of collection must ensure that individual specimens are not contaminated. One way is to use new plastic bags as gloves in taking samples of infected shoots, blossoms, fruit, etc. Each specimen is then stored in the same plastic bag used for collection. In addition, hands should be rinsed in dilute chlorine solutions before other sites are sampled. Samples collected in summer should be refrigerated to prevent loss in viability of spores. A high percentage of viable spores is ensured by collecting only samples of newly infected tissue.

Transport. The transport of R isolates may not present serious problems within areas where they are already established, but movement of R strains into areas where there is no resistance would be unwise. Although pathogens are extremely unlikely to escape from the laboratory to commercial fields, even that remote risk should be avoided since inadvertent release could be disastrous to local agricultural crops.

Storage. Collected samples should be stored for only a limited period and under conditions that ensure the viability of the pathogen.

Isolation and Pure Culture of Sample Pathogens

Many tests for resistance in plant pathogens have not been conducted with pure cultures. That failure may not have serious consequences when the sole objective is the detection of R pathogens or if isolations are from either surface-sterilized tissue or a heavily sporulating structure of the pathogen, such as a sporodochium, or if spores were ejected from a fruiting structure. However, sporulating structures or infected tissues are not always free from contaminating fungi, bacteria or yeasts, and these organisms could invalidate test results by detoxifying the active ingredient of the fungicide, by inhibiting growth, or by inducing abnormal growth of the plant pathogen. Thus, purified cultures should be used whenever possible. Transfer of single spores onto hosts to establish a culture is not feasible with obligate parasites, so with the host-specific obligate parasites, we recommend that at least one or more transfers be made onto uncontaminated host tissue, such as xenobiotically produced seedlings or surface-sterilized, excised plant parts.

Identification of the Pathogen

Classifying pathogens to genera is usually rather simple, but correct speciation and race determination of certain pathogens can be difficult. The procedures used to classify and identify pathogens can be learned by obtaining and studying known isolates. The importance of this step cannot be overemphasized since it appears that misidentification of even clearly distinguishable fungi such as *Rhizopus* spp. is possible. One method of separating *R. stolonifer* from *R. circinans*, *R. arrhizus* and *R. oryzae*, all of which appear similar in gross characteristics, is by the use of the fungicide dicloran, to which *R. stolonifer* is sensitive (Ogawa et al., 1963). In working with dicloran-R *R. stolonifer* (Webster et al., 1968), care must be taken to avoid contamination by other species to prevent erroneous conclusions. The identification of *Penicillium* spp. is more difficult; although *P. digitatum* and *P. italicum*, agents of citrus fruit decay, are distinguishable in pure culture, they are difficult to identify when mixed with other *Penicillium* spp., even on special medium. The key to success in identifying bacteria has been selective media, such as prepared for *Erwinia amylovora* (Schroth et al., 1979) and *Agrobacterium tumefaciens* (Schroth et al., 1965). For the powdery mildews, classification on the basis of the conidia is not easy, and cleistothecia are not always found; in such cases, confirmation of identity by a specialist is suggested.

Preparing Test Media

In preparing test media, use formulations that give normal vegetative growth unless resting bodies or fruiting structures are desired. The media reported most often in the literature are potato-dextrose agar (PDA) or amended PDA, made from fresh potatoes or prepared from commercial mixes. A recent report (Ogawa et al., 1978), however, indicates that certain commercial preparations of PDA will not support fungal vegetative growth and will curtail the production of sclerotia, conidia and pycnidia, and media should therefore be pretested with several isolates of the pathogen under study. Fungicide compatibility with test media should also be investigated. With obligate parasites it is important that the host tissue be susceptible at the time of the test, and the development of the pathogen in this case must be examined before significant changes in susceptibility occur. Excised host parts, grown under light and in contact with sterile water, have been used for culturing downy and powdery mildews.

Solid, semi-synthetic media. Most tests employ synthetic media because they are easily prepared and can support the growth of a wide range of plant pathogens. The proprietary formulation of the fungicide is used routinely, but more precise tests require the use of the more pure technical compounds. Solvents are often needed to incorporate the fungicide into the medium, and when they are used, a solvent control is required to determine if the solvent itself affects the plant pathogen or if it converts the fungicide to breakdown products that could be either more or less active than the original fungicide.

Germination and germ-tube elongation of fungal spores are readily evaluated on solid agar with various fungicide concentrations. The spores are dispersed in sterile distilled water (wetting agents may be necessary), sprayed or streaked onto the solid medium, and incubated. Fungicides such as benomyl and dicloran do not inhibit spore germination, so germ-tube growth is compared. Another method of testing for resistance is to put the spores on a cellophane strip that is placed on a fungicide-agar medium. Fungicides are difficult to evaluate exclusively in liquid suspension because certain pathogens require stimulatory compounds for uniform spore germination. Furthermore, uniform distribution of spores is difficult to obtain in a microbeaker or water droplet since spores tend to clump or since some may be submerged while others float. Clumped spores do not germinate uniformly.

Test concentrations should be chosen on the basis of the activity of the fungicides on specific pathogens. This information can be obtained from technical data available from the manufacturer, or one can test a logrithmic series of concentrations (e.g., 0, 1, 10, 100 and 1,000 μg/ml) followed by a more narrow range of concentrations. The two-fold serial dilution with 100 μg/ml base for testing antibiotics against bacteria (Sakurai, 1977) also can be used for fungi. The final range of concentration should encompass values that both allow and inhibit the growth of the pathogen. The lowest concentration at which the pathogen does not grow has been called the minimal inhibitory concentration (MIC) (Sakurai, 1977), and for bacteria, it can be difficult to determine since bacteria mutate readily and are likely to develop individuals capable of growth and multiplication at concentrations that prevent the growth and multiplication of the originally introduced cells (Sakurai, 1977).

Liquid medium. Liquid media can be used to test the growth rate of bacteria, fungi and yeasts if increases in population numbers or the dry weight are desired. A specific amount of propagule is added and the experiment is terminated after a definite period. Thus, daily increments of growth are not determined. This technique requires additional equipment for aeration and for distribution of the suspended fungicide, but it allows for a better adjustment of pH and allows evaluation of mycelial growth without the presence of spores.

Spore germination in liquid media provides a rapid evaluation for fungicide activity, but is difficult to conduct with pathogens such as the powdery mildews.

Zoospore motility time and germination percentage in a liquid medium was used to test fungicides for activity against certain *Phytophthora* (Sonoda and Ogawa, 1972) and *Pythium* spp. The test offers much promise for the detection of fungicide resistance since it requires only a few minutes for critical examination of zoospore motility, and less than 24 hr are needed to test germination.

Incorporation and Application of Fungicides

Incorporation. The stability of fungicides during autoclaving of media, exposure to light, or exposure to changes in acidity must be known. Inactivation of captan is prevented if it is added after the medium is cooled but before solidification of the agar. Media with fenaminosulf must be kept in the dark, since the active compound is photosensitive. Sodium orthophenylphenate must be buffered to prevent its conversion to orthophenylphenol, which is insoluble in water. On the other hand, the activity of benomyl incorporated in the medium is enhanced during autoclaving.

In preparing low concentrations of highly active compounds, large volumes of media (up to a liter) should be prepared to ensure accurate concentrations of the fungicides in the media, unless tests are conducted in serial dilutions.

Treatment of soil and seed. When applying systemic fungicides to protect seedlings, the chemicals may be either added to the soil or planting mixes, or applied to the seed. The chemical concentration in growing plants should be monitored by chemical analyses, bioassay or the addition of radioactive tracers.

Application of protectant fungicide on host. Fungicides should be applied uniformly on plants or plant parts. The smallest possible host unit, such as potted plants, or detached fruit, leaf or blossoms on a shoot, should be used. To test benomyl-R *M. fructicola* on peach fruit, harvest the mature fruit, surface sterlize it with sodium hypochlorite, air dry, and then spray benomyl on the styler end of the fruit with an atomizer using constant air pressure. Szkolnik (1978) sprayed blossoms on peach trees in pots on a rotating table. In field tests, the natural spread of pathogens such as *Monilinia* (Wilson and Ogawa 1979) and *Cercospora* (Berger, 1973) must be considered in designing an experiment.

Inoculating Test Media

Inoculum. The uniformity of inoculation is dependent on the age of the culture or spores, their wettability, and the amount of inoculum applied. The inoculum is usually grown on a medium free of inhibitory chemicals, but with tests in which resistance is being checked, we advise using inoculum grown on the same concentration of chemical to prevent the initial growth of the pathogen from the inoculum medium onto the diluted medium. Such precautions are necessary when a large mycelial plug is used (4 to 5 mm in diameter).

The age of the culture has a bearing on the inhibition of growth for some fungi. Taking plugs from margins of colonies will initiate uniform colony growth of *Monilinia laxa* from one- to two-day-old cultures, but will not always do so with cultures that are one to two weeks old. When spores are used, special harvest methods are needed to prevent transfer of the substrate. Spores of *M. fructicola* may lose viability if collected dry with a device such as the cyclone spore collector for too long a period. To remove mycelial fragments, *M. fructicola* spores are usually rinsed in distilled water by filtering through a sterile cheesecloth and centrifuged once or twice. This procedure could remove nutrients required for germination if spores are to be sown on water agar.

Since sexual spores such as the ascospores of *M. fructicola* and *V. inaequalis* are difficult to suspend in water, it is best to have them eject directly onto the test medium.

Obligate parasites. Spores of obligate parasites such as the powdery mildew fungi can be dusted into an inoculation settling chamber with a host plant on a rotating wheel to ensure uniform deposit. With the downy mildews, however, sporangia must be exposed to free moisture for germination. Sporangia can be suspended in water and sprayed onto plants, or the plants can be sprayed with water and sporangia released onto the plants from diseased leaves. Sporangia of hop downy mildew were found to have very short survival periods under low relative humidity conditions (Sonoda and Ogawa, 1972); however, a drop in relative humidity was required for sporangial release.

Replication

Replications are required for reliable experiments. In vitro fungicide testing is considered less variable than in vivo testing since environmental conditions are presumably more uniform and more stable in vitro and host influences are removed. For in vitro tests, three replications should be adequate, but if variability is high, replications can be increased in later experiments. For in vitro tests, costs associated with increasing the number of replications can be reduced by using petri plates of 50 mm instead of the standard 100 mm diameter or (if the chemical has no vapor action) by using a divided plate with one side serving as control with no chemical. In tests with plant materials (e.g., seedlings, growing plants or fruit) more than three replications are recommended to reduce the effects of variability in inoculation and of any natural contamination.

Incubation

The length of the incubation period for resistance tests should be closely monitored since extended incubation could invalidate test results. There is also a possibility of partial inactivation of the fungicide or an increase in fungicide concentration as moisture in the medium evaporates.

Since the host plant of an obligate parasite changes continuously in susceptibility, it is important that readings are made during a time span when treated parts are susceptible. As plants grow, they continue to produce new tissue, which may not be covered by protective fungicide. Plant growth may also dilute the concentration of systemic fungicides.

We feel that the incubation temperature should be at or near the optimum for growth of the pathogen and normal growth habits of the host. For postharvest studies on most fruits, the best incubation temperature for ripening is near 20°C. Optimum storage temperatures range from near 0°C (for stone fruits) to slightly below freezing (for pome fruits) and above 13°C (for tomatoes and sweet potatoes). The incubation of postharvest-treated crops should be at the optimum for growth of the pathogen unless that temperature damages the crop, and in testing systemic fungicides, the incubation time should be sufficient to allow the chemical to move to the site of action.

Recording Results and Analyses

The response of sensitive and resistant strains must be compared under identical test conditions. Too often, reports of resistance are based on claims of reduced sensitivity in a few field isolations when the previous level of resistance in the field is unknown. Data on vegetative growth should be taken at regular intervals and for a range of fungicide concentrations. To test the sensitive wild type for minimal inhibitory concentrations the number of isolates must be an adequate sample of the population, and the method for determining the presence of R isolates is discussed in the previous section on designing effective programs for detection. Statistical analysis is necessary to differentiate small differences.

Test for Pathogenicity

Testing the pathogenicity of an isolate is an important part of the procedure for detecting fungicide resistance. Before resistance screening certain pathogens, such as *Geotrichum* or *Fusarium*, should be tested on the host to ascertain pathogenicity. Such precautions have been so far unnecessary with *Monilinia* and *Venturia*, as isolates from infected plant tissue have all retained their pathogenicity.

Preserving the Isolates

Pathogens should be preserved in a manner that minimizes the chance of variation. Several techniques are available, such as lyophilization, storage in liquid nitrogen, and other long-term storage methods that temporarily inhibit or reduce growth. For some organisms, especially obligate parasites, the most reliable method is to grow them on host parts and then transfer them at intervals. None of the methods of storage absolutely insure against the occurrence of variations, but some are more reliable than others. The selected means of storage will depend upon the equipment available, the pathogen and length of the study.

MONITORING STRATEGIES USED IN PREVIOUS RESEARCH

Monitoring for Streptomycin-resistant *Erwinia amylovora*

The first chemicals used to control bacterial diseases were metals, such as coppers and mercury. Bacterial resistance to these materials was not a problem, but after antibiotics were introduced, fungicide-resistant bacteria were reported.

In vitro bacterial resistance to streptomycin and oxytetracycline was reported in California by English and Van Helsema in 1954. The detection technique used involved incorporation of antibiotics in the culture medium followed by introduction of suspensions of bacterial cells. When streptomycin and oxytetracycline were combined, fewer R strains developed; however, this forewarning was ignored. On the basis of performance data from field tests for disease control, the cost of application, and the ease of registration, the recommendation was to use only streptomycin. In vivo streptomycin resistance was detected in 1971 (Schroth et al., 1972) in a few orchards where 17% streptomycin sulfate had been applied at up to 21 kg/ha in 760 liters of water every other day without success.

An orchard-monitoring program was established with the aid of a selective medium for *Erwinia amylovora* (Miller and Schroth, 1972). The procedure involved collecting healthy or diseased blossoms, washing them in 10 ml of sterile distilled water, and spreading 0.1 ml of the resulting suspension on the special medium with a bent glass rod. Plates were incubated for 96 hr at 28°C before colonies were detected. These colonies were tested on streptomycin at replicated concentrations of 0 to 500 μg/ml, and the levels of resistance were usually high (500 μg/ml). Strains with low-level resistance, 1 μg/ml, died within a few days. In another test, disks were impregnated with streptomycin at 10, 25 and 100 μg/ml, and no inhibition of the R strain was detected although small zones (2 to 4 mm) developed when 500 μg per disk was applied and placed on Mueller-Hinton medium with streptomycin-R *Erwinia*. The pathogenicity of the isolates can be confirmed by inoculation of succulent Bartlett pear shoot tips in the field or greenhouse, or inoculation of green fruit or flowers in the laboratory (Schroth et al., 1979).

The unanswered question is whether there is a ratio of sensitive to resistant strains below which streptomycin sprays can be used and above which they should be abandoned. In orchards where antibiotics have failed to control disease, growers have reverted to copper fungicides. In 1978 oxytetracycline was approved for use as an alternate for control of fire blight of pears, and field monitoring for *Erwinia* resistant to oxytetracycline is in progress.

Greenhouse and Field Monitoring for Fungicide-resistant Powdery Mildews

The first report of powdery mildew resistance to fungicide was by Schroeder and Provvidenti (1969), who detected benomyl resistance in *Sphaerotheca fuliginea* on curcubits. To confirm their findings, they inoculated leaf disks floated on water and benomyl suspension. Similar studies enabled Bent et al. (1971) to detect *S. fuliginea* resistance to dimethirimol. After the use of alternate fungicides, such as benomyl, dinocap, quinomethionate and pyrazophos, resistance was not detected in greenhouse cucumbers; therefore dimethirimol is again used as a soil application, followed by alternate protective sprays. Multiple resistance has not been reported from this spray program.

Pyrimidine compounds, introduced in 1970, have been used widely to control powdery mildew on barley, caused by *Erysiphe graminis* DC f. sp. *hordei*. Ethirimol resistance was detected in 1971 but no yield losses were experienced. By 1973, reliable methods for detecting and monitoring ethirimol resistance were developed by Imperial Chemical Industries, Ltd., and by 1974, this chemical had been used to treat over 0.6 million ha of spring barley in the United Kingdom. Shepherd et al. (1975) established a unique detecting and monitoring program. Their first problem was to culture an obligate parasite without cross contamination. Proctor barley seeds, untreated and treated (100 to 8000 μg a.i. ethirimol per g seed), were grown in a controlled environment for 10 to 11 days; then, two leaf pieces 2.5 cm long were cut from each prophyll of the seedlings. Five replicate leaf pieces for each treatment were mounted with one end embedded in tap water agar in large plastic petri dishes. Spores (24 hr old) of test isolates were blown into a settling tower and allowed to settle for 1 min before the lid was replaced. After six days, infection on each leaf piece was recorded according to a 0 to 4 grading system (0 = highest infection). The experimental design, including the grading system, is most important in obtaining critical analyses of mildew sensitivity and resistance to ethirimol.

To correlate field and laboratory tests, similar grading systems were used to record disease levels in the field before sampling. Five to six mildewed leaves (5 cm pieces) from each of 25 sites were collected twice from sample fields. A large number of fields were sampled in each year before conclusions were made (400 fields in 1974 and 90 in 1975). Sampling continued for six years. In the latter years, samples were mainly from the mildew-susceptible barley cultivars, Proctor, Julia and Golden Promise. The preceding information serves as the basis of a program whereby a single ethirimol seed treatment is used in Scotland and England, and a single spray of ethirimol in Europe. This was found possible only after Holloman (1977) found that a resistant population of *E. graminis* did not build up uniformly but oscillated from planting to harvest. Therefore, a single application of ethirimol on the seed or at the beginning stage of mildew development does not affect its use in the following year.

Wolfe and Minchen (1976) reported another method of monitoring powdery mildew resistance. When the prophylls were fully extended, barley seedlings previously treated with ethirimol or tridemorph were left in the field for 24 hr, and disease was evaluated after seven to eight days of incubation in the greenhouse. Fungicide resistance can be monitored for the life of the fungicide on the seedling.

Orchard Monitoring for Benomyl-resistant Brown Rot Fungi

Brown rot of stone fruit is a difficult disease to control with protectant fungicides because repeated applications are needed, although resistance has not been a problem with repeated applications of coppers, sulfurs, dithiocarbamates and captan. Because benzimidazole fungicides (benomyl and thiophanate M) provided outstanding disease control with fewer applications (Ogawa et al., 1969; Ramsdell and Ogawa, 1973), benomyl has been used in most of the stone fruit acreage in California. After reports of resistance in other pathogens, a monitoring program was started to detect benomyl resistance in *Monilinia fructicola* and *Monilinia laxa*. Ten blighted shoots and/or mummies were collected from various stone fruit (almond, apricot, cherries, peaches, plums, nectarines), fungus pathogens were isolated, grown on PDA in pure culture, and mycelial plugs were placed on PDA medium containing benomyl at 1.0 μg/ml (Tate et al., 1974). No resistance to benomyl was detected until 1977, when resistance to benomyl was detected in *M. fructicola* but not in *M. laxa*. Because *M. laxa* is the predominate species on almond and apricot, growers may continue to use benomyl effectively in a program of one spray only during bloom.

Many difficulties are encountered in detecting resistance. In a recent report on the detection of benomyl resistance in *Monilinia* (Ogawa et al., 1978), it was suggested that the fungus be cultured first on media free of benomyl. The fungus should then be tested at various concentrations of proprietary Benlate 50W. Resistance levels of 4 μg/ml can quickly lead to higher levels of resistance in populations if benomyl use is continued. If benomyl resistance levels are above 10 μg/ml (Whan, 1976; Szkolnik and Gilpatrick, 1977) the chemical is ineffective, with serious crop losses possible. Early detection and monitoring of low-level resistance (1 μg/ml) is critical in planning a strategy to minimize buildup. Moreover, benomyl can still be used for control but must be managed according to monitoring data. At present, benomyl is still effective in most California orchards with resistance detected in only three areas.

Field Monitoring for *Pyricularia* Resistance to Various Fungicides

Rice blast caused by *Pyricularia oryzae* Cavara has been controlled in Japan with mercuries, antibiotics and, most recently, phosphates (Brooks and Buckely, 1977). A method was developed for detecting and monitoring resistance (Sakurai, 1977) that included testing large numbers of isolates (over 100). A rice plant juice medium incorporated with a two-fold serial dilution of the antibiotics starting at about 0.05 μg/ml for blasticidin-S and 0.4 μg/ml for kasugamycin is used. Mycelial fragments of the test isolates are placed on the medium and incubated for 48 hr at 28°C. The lowest concentration of fungicide at which the bacterial or fungal isolate is inhibited is the minimal inhibitory concentration (MIC) for that isolate. Each MIC value is plotted with precent isolation frequency on one axis and chemical concentration on the other, which differentiates isolates into groups called sensitive, intermediate and resistant. With kasugamycin, large differences in MIC values were found between physiologic races and strains. Additional monitoring methods are under study in Japan to serve as a basis of fungicide recommendation to delay the development of resistant *Pyricularia* (Sakurai, 1977).

Postharvest Monitoring for *Penicillium* spp. Resistance to Fungicides Used in Controlling Citrus Fruit Decay

Fungicides play a major role in controlling postharvest decay during storage, transport and marketing. Since postharvest fungicides on citrus are not applied in the field, resistance is not a problem until the fruit is brought into the packing shed. The two major pathogens on citrus are *Penicillium digitatum* (green mold) and *P. italicum* (blue mold), and nonpathogenic *Penicillium* strains are common contaminants.

There are isolates of *Penicillium* species resistant to one or more of the following fungicides currently used for postharvest use on citrus: sodium orthophenylphenate, biphenyl, TBZ, benomyl and 2-aminobutane. A monitoring program was developed to detect resistance to these products so that their use in situations where R isolates were present could be avoided. The procedures developed by Decco Division of Pennwalt are as follows:

Detection medium. The detection medium is PDA fortified with neopeptone for better growth of *Penicillium* and easier detection of nonpathogenic *Penicillium*. The formulations are as follows:

Procedure for 20 μg/ml thiabendazole (TBZ):

1) Add 0.203 g of 98.5% TBZ to 100 ml volumetric flask.
2) Add ethanol (absolute) to make 100 ml.

3) The stock solution contains 0.2% TBZ.
4) Store in refrigerator at around 5°C.
5) Add 1 ml of stock solution to 100 ml of Harding's medium (Harding, 1962) that has been melted and cooled to 55°C. If *Rhizopus* contamination exists, also all 0.1 ml of 3,000 μg/ml dicloran stock (dicloran crystals dissolved in acetone). Mix and pour from 10-15 ml into petri dishes.

Procedure for 10 μg/ml benomyl:

1) Add 0.1 ml of 1% benomyl to 100 ml of Harding's agar. To make the 1% benomyl, add the following to 50 ml methanol: 1.05 g or 95% benomyl, 10 to 15 ml of concentrated HCl (12 N) and methanol to 100 ml volume.
2) The final concentration of benomyl and methanol is 10 μg/ml.
3) Dicloran is added to control any *Rhizopus* contamination.

Procedure for 20 μg/ml sodium orthophenylphenate SOPP:

1) Prepare 0.44 g of Dowicide A (71% SOPP anhydrous) dissolved in 100 ml of isopropanol.
2) Add 1.3 ml of above solution to 200 ml of cooled agar to make 20 μg/ml SOPP anhydrous.
3) Dicloran to control *Rhizopus* is not necessary.

Procedure for 500 μg/ml of 2-aminobutane:

1) To 100 ml of Harding's medium (sterile, melted, and cooled to 55° C) add aseptically the following: 0.2 ml 25% 2-aminobutane phosphate and 0.1 ml dicloran solution (optimal for control of *Rhizopus*).
2) Mix well and pour into petri plates.

Procedure for control plates with no chemicals:

1) Add 0.1 ml of stock 300 μg/ml dicloran for *Rhizopus* control.

Exposure and incubation. Petri dishes with the appropriate fungicide are exposed by opening the Petri dish lid for 10 to 20 seconds up to as long as 1 minute (or longer for storage areas) for deposit of fewer than 20 to 30 spores per plate, differing with the location of spore collection and intensity of the spore load anticipated in the air. Monitoring should include the packout dump, box-cleaning area, washing and color-sorting area, packaging area and storage room. The plates are incubated for three to four days before reading.

Testing for diphenyl resistance. For diphenyl resistance, commercially used diphenyl pads are placed in lids of the control plates after visible growth is detected and circled. *Penicillium digitatum* and *P. italicum* mycelial growth inhibition is recorded.

Identification of pathogen. Identification is critical because *P. citrinum*, *P. corylophilum* and *P. frequentans* are common contaminants. Change in substrate color during the three to four days of incubation provides an aid for identifying some *Penicillium* species. *P. italicum*, blue mold, characteristically produces a dark substrate, whereas green mold would express a light substrate.

Since stone fruits are shipped almost immediately to markets and not usually stored for extended periods this citrus monitoring program is not applicable to stone fruits.

DISCUSSION

Detecting and monitoring programs for fungicide resistance can allow for the development of strategies that might delay or prevent the onset and spread of resistance such as that experienced with: (1) streptomycin against fireblight disease of pears; (2) benomyl against brown rot of sweet cherries and peaches; (3) sodium orthophenylphenate and biphenyl against blue and green mold of citrus fruit, and (4) kasugamycin against rice blast. Detection of resistance before the occurrence of crop losses and a switch to alternate fungicides as a disease management technique could maintain the usefulness of the chemical for which resistance has developed, as with pyrimidines in powdery mildew.

In establishing monitoring programs, program costs must be compared with the revenues to be gained. Considering fungicide development costs, in most cases the gains will outweigh the cost, as was found with dimethirimol use on barley.

Although the development of fungicide-R strains in vitro has not always been expressed in the field, this does not imply that laboratory tests are meaningless: They play a major role in research on mechanisms and provide us with a valuable guide. To add a cautionary note, however, we are not confident that proper detection methods have been used and tested for the fungicides listed as not having resistance problems (Tables 1 and 2), since in certain instances baseline sensitivities were not established before the fungicide was introduced in the field. In other cases, experimental techniques were inadequate (e.g., the manner in which the chemical was incorporated and the concentration levels were inappropriate).

Future research should increase efforts to assess the stability of detected field-R pathogens and to determine the probability that such R strains will develop multiple resistance. Also needing attention are the competitive nature of the various sensitive and R strains and their relationship to pathogenicity. More accurate, more precise, and quicker methods for detecting low levels and low frequencies of fungicide resistance must be developed. Such techniques will be invaluable in fungicide-disease management programs to help prevent or delay increases in populations of R strains. Further work must be done to determine the occurrence of, the conditions necessary for, and the frequency of reversion of pathogens resistant to a fungicide back to sensitive ones as was found with *Erwinia amylovora* (Schroth et al., 1979). It is our hope that information obtained from these studies will serve as the basis for techniques to increase the useful life span of fungicides that performed so effectively before resistance to them became a problem.

REFERENCES

Ashida, J., 1965, Adaptation of fungi to metal toxicants, *Annu. Rev. Phytopathol.*, 3:153.

Bent, K. J., Cole, A. M., Turner, J. A. W., and Woolner, M., 1971, Resistance of cucumber mildew to dimethirimol, *Proc. 6th Brit. Insecti. and Fungi. Conf.*, p. 274.

Berger, R. D., 1973, Disease progress of *Cercospora apii* resistant to benomyl, *Plant Dis. Rep.*, 57:837.

Brooks, D. H., and Buckley, N. G., 1977, Results in practice 1. Cereals and grasses, *in:* "Systemic Fungicides," R. W. Marsh, ed., p. 233, Butler and Tanner, Ltd.

Chastagner, G. A., 1980, Tolerance of *Botrytis tulipae* to glycophene and vinclozolin, *Phytopathology* (in press).

Chastagner, G. A., and Ogawa, J. M., 1979, DCNA-benomyl multiple tolerance in strains of *Botrytis cinerea*, *Phytopathology*, 69:699.

Chastagner, G. A., and Ogawa, J. M., 1979, A fungicide-wax treatment to suppress *Botrytis cinerea* and protect fresh-market tomatoes, *Phytopathology*, 56:59.

Dekker, J., and Gielink, A. J., 1979, Acquired resistance to pimaricin in *Cladosporium cucumerinum* and *Fusarium oxysporum* f. sp. *narcissi* associated with decreased virulence, *Neth. J. Pl. Pathol.*, 85:67.

English, A. R., and Van Helsema, G., 1954, A note on the emergence of resistant *Xanthomonas* and *Erwinia* strains by the use of streptomycin plus Terramycin combinations, *Plant Dis. Rep.*, 38:429

FAO Plant Production and Protection Paper on Pest Resistance to Pesticides and Crop Loss Assessment, 1979, AGP:1979/M/2. 41 p.

Georgopoulos, S. G., and Zaracovitis, C., 1967, Tolerance of fungi to organic fungicides, *Annu. Rev. Phytopathol.*, 5:109.

Harding, P. R., 1962, Differential sensitivity to sodium orthophenylphenate by biphenyl-sensitive and biphenyl-resistant strains of *Penicillium digitatum*, *Plant Dis. Rep.*, 46:100.

Hills, F. J., and Leach, L. D., 1962, Photochemical decomposition and biological activity of p-dimethylaminobenzenediazo sodium sulfonate (Dexon), *Phytophathology*, 52:51.

Holloman, D. W., 1977, Laboratory evaluation of ethirimol activity, *in:* "Crop Protection Agents--Their Biological Evaluation," N. R. McFarlane, ed., pp. 505-515, Academic Press, London.

Littrell, R. H., 1976, Techniques in monitoring for resistance in plant pathogens, *Proc. Am. Phytopathol. Soc.*, 3:90.

Miller, T. D., and Schroth, M. N., 1972, Monitoring the epiphytic populations of *Erwinia amylovora* on pear with a selective medium, *Phytopathology*, 62:1175.

Ogawa, J. M., Mathre, J. H., Weber, D. J., and Lyda, S. D., 1963a, Effects of 2,6-Dichloro-4-nitroaniline on *Rhizopus* species and its comparison with other fungicides on control of *Rhizopus* rot of peaches, *Phytopathology*, 53:950.

Ogawa, J. M., Ramsey, R. H., and Moore, C. J., 1963b, Behavior of variants of *Gilbertella persicaria* arising in medium containing 2,6-dichloro-4-nitroaniline, *Phytopathology*, 53:97.

Ogawa, J. M., Manji, B. T., and Bose, E., 1968, Efficacy of fungicide 1991 in reducing fruit rot of stone fruit, *Plant Dis. Rep.*, 52:722.

Ogawa, J. M., Gilpatrick, J. D., and Chiarappa, L., 1977, Review of plant pathogens resistant to fungicides and bactericides, *FAO Plant Protection Bull.*, 25(3):97.

Ogawa, J. M., Gilpatrick, J. D., Uyemoto, J. K., and Abawi, G. S., 1978a, Variations in fungal growth on various preparations of potato-dextrose agar media, *Plant Dis. Rep.*, 62:437.

Ogawa, J. M., Manji, B. T., Bose, E. A., Szkolnik, M., and Frate, C. A., 1978b, Methods for detection of benomyl-tolerant *Monilinia fructicola*, *Phytopathol. News* 12(9):248 (Abstr.).

Parry, K. E., and Wood, R. K. S., 1959, The adaptation of fungi to fungicides: Adaptation of thiram, ziram, ferbam, nabam and zineb, *Ann. Appl. Biol.*, 47:10.

Ramsdell, D. C., and Ogawa, J. M., 1973, Systemic activity of methyl-2-benzimidazolecarbamate (MBC) in almond blossoms following prebloom sprays of benomyl MBC, *Phytopathology*, 63:959.

Sakurai, H., 1977, Methods of determining the drug-resistant strains in phytopathogenic bacteria and fungi and its epidemiology in the field, *Rev. J. Pestic. Sci.*, 2:177.

Sakurai, H., and Naito, H., 1976, A cross resistance of *Pyricularia oryzae* Cavara to kasugamycin and blasticidin-S, *J. Antibiotics*, 29:1341.

Sakurai, H., Naito, H., and Yoshida, K., 1975, Studies on cross resistance to antifungal antibiotics in kasugamycin-resistant strains of *Pyricularia oryzae* Cavara, *Bull. Agr. Chem. Inspect. Stn. (Tokyo)*, 15:82.

Schroeder, W. T., and Provvidenti, R., 1969, Resistance to benomyl in powdery mildew of cucurbits, *Plant Dis. Rep.*, 53:271.

Schroth, M. N., Thompson, J. P., and Hildebrand, D. C., 1965, Isolation of *Agrobacterium tumefaciens* - *A. radiobacter* group from soil, *Phytopathology*, 55:645.

Schroth, M. N., Beutel, J. A., Moller, W. J., and Beil, W. O., 1972, Fireblight of pears in California, current status and research progress, *Calif. Plant Pathol.*, 7(1):1.

Schroth, M. N., Thomson, S. V., and Moller, W. J., 1979, Streptomycin resistance in *Erwinia amylovora*, *Phytopathology*, 69:565.

Shepherd, M. C., Bent, K. J., Woolner, M., and Cole, A. M., 1975, *Proc. 8th Brit. Insecti. and Fungi. Conf.*, 59.

Sonoda, R. M., and Ogawa, J. M., 1972, Ecological factors limiting epidemics of hop downy mildew in arid climates, *Hilgardia*, 41:457.

Sonoda, R. M., Ogawa, J. M., Lyons, T., and Hansen, J. A., 1970, Correlation between immobilization of zoospores by fungicides and the control of *Phytophthora* root and crown rot of transplanted tomatoes, *Phytopathology*, 60:783.

Szkolnik, M., 1978, Techniques involved in greenhouse evaluation of deciduous tree fruit fungicides, *Annu. Rev. Phytopathol.*, 16:103.

Szkolnik, M., and Gilpatrick, J. D., 1977, Tolerance of *Monilinia fructicola* to benomyl in western New York State orchards, *Plant Dis. Rep.*, 61:654.

Tate, K. G., Ogawa, J. M., Manji, B. T., and Bose, E., 1974, Survey for benomyl-tolerant isolates of *Monilinia fructicola* and *Monilinia laxa* in stone fruit orchards of California, *Plant Dis. Rep.*, 58:663.

Webster, R. K., Ogawa, J. M., and Moore, C. J., 1968, The occurrence and behavior of variants of *Rhizopus stolonifer* tolerant to 2,6-dichloro-4-nitroaniline, *Phytopathology*, 58:997.

Whan, J. H., 1976, Tolerance of *Sclerotinia fructicola* to benomyl, *Plant Dis. Rep.*, 60:200.

Wilson, E. E., and Ogawa, J. M., 1979, Fungal, bacterial and certain nonparasitic diseases of fruit and nut crops in California, Div. Agric. Sciences, Univ. Calif., 190 pp.

Wolfe, M. S., and Minchin, P. M., 1976, Quantitative assessment of variations in field populations of *Erysiphe graminis* f. sp. *hordei* using mobile nurseries, *Trans. Br. Mycol. Soc.*, 66:332.

Yaoita, T., Goh, N., Aoyagi, K., Iwano, M., and Sakurai, H., 1979, Studies on drug-resistant strain of rice blast fungus, *Pyricularia oryzae* Cavara 1. The occurrence of the multi-resistant strains of *Pyricularia oryzae* and its epidemiology in the field, *J. Niigata Agr. Exp. Sta.*, No. 28 (March).

EVOLUTION OF RESISTANCE TO INSECTICIDES:

THE ROLE OF MATHEMATICAL MODELS AND COMPUTER SIMULATIONS

Charles E. Taylor

Department of Biology
University of California
Riverside, California 92521

INTRODUCTION

Computers have become an important tool for scientific investigation in nearly all areas of science. In particular, the studies of ecology, resource management and pest control have been much affected by mathematical modeling and computer simulation. The widely heralded IBP and integrated pest management projects are only some of the most visible aspects of a transformation that is certain to continue. As yet, the study of resistance to insecticides has been less affected by mathematical modeling than these other areas, but several laboratories are now actively engaged in computer analysis of how such resistance has evolved and how it might be controlled. The number is certain to grow. Below I will review the principal studies of the evolution of insecticide resistance that have been published up to now and then speculate briefly where I think future work in this area might be most usefully directed.

Kenneth Watt (1968) has classified the models used for ecology and resource management into four types, depending on how complex they are. As they would apply to insecticide resistance Watt's categories are:

1. Simple models capable of straightforward mathematical analysis.
2. Unrealistic models using empirically derived relations.
3. Complex models incorporating intrinsic factors.
4. Complex models incorporating extrinsic factors.

It will be helpful to adopt this scheme for the discussion below, although some models do not fit exactly within any one category.

It should be stressed at the outset that no type of model is necessarily worse or better than the others. Instead, there is a trade-off: The simpler models allow one to manipulate and understand better what is going on, while the more complex models allow for more interactions that may be important for mimicking reality. This was stated nicely by Comins (1979): "It is as good to provide an exact answer to an approximate problem as to provide an approximate answer to the exact problem." Each approach has its special use.

SIMPLE MODELS CAPABLE OF STRAIGHTFORWARD ANALYSIS

The simplest, most general models attempt to describe evolution and population growth solely on the basis of a relation between allele frequency and population size. Much of the mathematical theory of population genetics is of this sort.

It may be assumed, for example, that there are two alleles, R (resistant) and S (susceptible), with the genotypes (RR, RS, SS) and with the fitnesses (W_{RR}, W_{RS}, W_{SS}) at frequencies (p_t^2, $2p_tq_t$, q_t^2) in generation t. The average fitness of the population would then be $\overline{W} = p_t^2W_{RR} + 2p_tq_tW_{RS} + q_t^2W_{SS}$. Let the population size be N_t, the intrinsic rate of increase be r, and the carrying capacity be K. Then a typical model of Watt's first type would be

$$p_{t+1} = (p_t^2W_{RR} + p_tq_tW_{RS})\overline{W}^{-1}$$

$$N_{t+1} = N_t \exp\{r(K - N_t\overline{W})/K\}$$

Starting with arbitrary parameters the course of evolution in such simplified circumstances is well understood (see, for example, Crow and Kimura, 1970; Spiess, 1977; or Wright, 1969).

With such equations MacDonald (1959) demonstrated that evolutionary rates depended on whether the alleles conferring resistance were dominant or recessive. On the basis of this analysis he proposed that the joint use of DDT and dieldrin on anopheline mosquitoes should slow the evolution of resistance to DDT but not to dieldrin. MacDonald's paper is especially remarkable because it antedated other mathematical analyses of this problem by more than 15 years.

More recently, models of this category, which are simple but tractable, have been used by Comins (1977a, 1979) and Taylor and Georghiou (1979) to better understand how immigration of susceptible individuals might dilute selection and affect the rate at which resistance to pesticides can evolve. The approaches taken and the parameters studied were different, but both analyses indicated that immigration of susceptibles retarded evolution by the target population. More important, they pointed to the existence of a critical set of genotype fitnesses and migration rates that permit such evolution to be stopped or even reversed. The next step toward developing those ideas into a control strategy has been to try them out in more complex but realistic models, described below.

UNREALISTIC MODELS USING EMPIRICALLY DERIVED RELATIONS

When modeling for commercially important applications, such as fisheries, it is common to use regression equations for predicting population growth. For example, yield in one year might be predicted from a linear combination of yields during the four or five years previous. The biological relation among these variables in the regression equation might be unclear and the mathematical relations might be quite unrealistic, but for commercial purposes this is unimportant if the yield can be predicted fairly accurately. While providing little insight into the biology of resistance, such analyses may nonetheless be useful in deciding whether or not to use an insecticide in any particular case.

At this time there is still little impact of pest modeling on actual control practices (Norton, 1979), so the incentive (primarily economic) for models of Watt's second type is lacking. But the situation is changing, and fairly accurate regression models exist for some species where resistance is a problem (e.g., *Culex pipiens*, see Hacker et al., 1973). It will probably be only a matter of time before adaptation is included into their formulation.

COMPLEX MODELS CONTROLLED BY INTRINSIC FACTORS

Models typically become more complex as they become more realistic. Insect life histories and economic thresholds for insecticide application will, for example, influence the rate at which resistance evolves; yet these two factors alone can make a complete mathematical analysis next to impossible. Some of the most useful and widely studied models are too complicated to be manipulated without the aid of a computer, and they are often simplified in an attempt to describe population growth and evolution only with factors that are intrinsic to the population. Extrinsic factors like the weather or competing species are ignored.

The models that have been developed so far and that fall into Watt's category of intermediate complexity are directed principally toward one or another of two objectives: (1) to better understand the biological nature of resistance; or (2) to develop the most economical method of control (Norton, 1979). Depending on which objective is favored the models employ somewhat different mathematical techniques and can best be considered separately.

Table 1. Known or Suggested Factors Influencing the Selection of Resistance to Insecticides in Field Populations[a]

A. Genetic

1. Frequency of R alleles
2. Number of R alleles
3. Dominance of R alleles
4. Penetrance; expressivity; interactions of R alleles
5. Past selection by other chemicals
6. Extent of integration of R genome with fitness factors

B. Biological

a. Biotic

1. Generation turn-over
2. Offspring per generation
3. Monogamy/polygamy; parthenogenesis

b. Behavioral

1. Isolation; mobility; migration
2. Monophagy/polyphagy
3. Fortuitous survival; refugia

C. Operational

a. The chemical

1. Chemical nature of pesticide
2. Relationship to earlier used chemicals
3. Persistance of residues; formulation

b. The application

1. Application threshold
2. Selection threshold
3. Life stage(s) selected
4. Mode of application
5. Space-limited selection
6. Alternating selection

[a]Adapted from Georghiou and Taylor, 1976.

Biological Nature of the Problem

Almost every facet of a pest's biology influences the rate at which it can adapt to pesticides. These have been systemized by Georghiou as in Table 1 (from Georghiou and Taylor, 1976). We have conducted a series of computer simulations in order to learn how each of those factors might retard or enhance adaptation to insecticides, and the results of these studies are reported in Georghiou and

Taylor (1977a, 1977b) and Taylor and Georghiou (1980). It was found that each factor studied did, indeed, influence the rate at which resistance evolved. For example, it was found that populations recovered from insecticide more slowly when there were refugia available, when resistance was recessive, and when the initial frequency of the R allele was low. There was typically a trade-off: When the population size was more suppressed in the early generations, then it evolved resistance more quickly, and when less suppressed it evolved more slowly.

Two factors stood out in these simulations as being especially important for controlling the rate of evolution. One was the influx of susceptible migrants. In line with the simpler models discussed above, we found that migrants could not only slow the build-up of resistance, but with the appropriate combination of dosage and application schedules evolution could even be reversed (Taylor and Georghiou, 1980). Fabio Quaglia has followed up these computer simulations with population-cage experiments intended to mimic conditions in which resistance was predicted to evolve more or less slowly. His results, as yet unpublished, are exciting, for they show a remarkable concordance between the population size and R allele frequencies predicted before the experiment began and those which were actually observed. He is finding, as the models predicted, that an inward flux of migrants may have a deciding effect on the rate at which resistance evolves.

The second important factor to stand out from our simulations was the integration of resistance with other aspects of fitness. Does an increase in one fitness component, resistance, also entail a loss in another component, reproductive potential? The possibility had been suggested earlier by Brown (1971) and Georghiou (1972) but its theoretical importance had not been appreciated. In all our simulations such a trade off could be vital for control. Following up these simulations a student in our laboratory, James Ferrari, began a careful study of how resistance was integrated into the genome. His results with *C. pipiens* indicated that the resistant mosquitos were at a surprisingly large reproductive disadvantage when the insecticide was absent. If these results are found to be general, then we will have a much better chance of containing the resistance problem.

It would be expected that the integration of R alleles with fitness would depend on interactions among several loci -- what Dobzhansky (1972) has termed *coadaptation*. Ferrari (1980) showed further that there is a close correspondence between what we know of selection for resistance in pest populations and what has been observed in selection experiments in laboratory populations of *Drosophila*. The subsequent analyses of these *Drosophila* populations has invariably indicated that coadaptation among several loci is a major part of evolution. The same may be true for resistance. In our laboratory we are now following up this possibility with simulations of multi-locus selection, the object being to better understand this coadaptation. The only published analysis of multi-locus selection for insecticide resistance of which I am aware is that by Plapp et al. (1979). They studied a model wherein several loci contributed to quantitative variation for resistance. Their model was fairly simple (it did not include coadaptation), and the principal results were not surprising -- resistance evolved faster when the selection differential was greater and when the heritability was higher. Nonetheless, their work represents a step in the right direction, and indicates where further work can be expected. Multi-locus selection is currently receiving much attention from theoretical population geneticists: The opportunity for cooperation among population geneticists and people concerned with pest control seems especially promising here.

Economic Optimization

The biological analyses just described are for the most part just more complicated versions of the simpler models of Watt's first category. The mathematics used is principally differential and difference equations. Separate from this are the models for pest control that have been developed by the agricultural economists and which employ an analysis called non-linear programming. A typical problem of this sort is to construct a vector $\underset{\sim}{x} = (x_1, x_2, \ldots, x_n)$ of insecticide doses, where x_t equals the dose applied at time t. There is a cost, $C(\underset{\sim}{x})$, and a benefit, $B(\underset{\sim}{x})$, associated with each vector. There is a series of constraints that apply to the population which must be written in the form $f_i(\underset{\sim}{x}) < \xi_i$. Then the problem is to find that vector which gives

$$\max \{B(\underset{\sim}{x}) - C(\underset{\sim}{x})\}$$

subject to

$$f_i(\underset{\sim}{x}) \leqslant \xi_i \qquad\qquad i = 1, \ldots, k.$$

Such problems and the technique for solving them have found extensive use in nearly all areas of industry and economics. While shedding less light into the biological principles of resistance, they are likely to be helpful in finding the most economical solutions to problems of pest control.

The first models of this sort were developed by Taylor and Headley (1975) and by Hueth and Regev (1974). The principal value of these early studies was to show that resistance could be incorporated into economic analyses and that resistance could substantially modify which control strategy was most desirable. More recently Comins (1977b) showed that resistance increased the cost of control, and that this increased cost could be incorporated into other models merely by changing the cost of insecticides.

A potentially important result of Comins' study is that when resistance might evolve then the schedule of application for each grower individually may not be the best schedule for all growers collectively. (This conclusion would also follow from the studies by Georghiou and Taylor (1977a) on refugia and immigration.) The net benefit is often greater when refugia exist or when some areas are left without insecticide because evolution is then slower, but of course, each individual would prefer that it were his neighbor's field that was left without insecticide and not his own. Comins suggested that this may point toward the desirability for further governmental regulation of pesticide use.

As yet, models such as these have had little direct application, but further refinements, similar to those reported by Guttierez et al. (1979), are likely, I think, to someday be an integral part of agricultural decision-making.

COMPLEX MODELS INCORPORATING EXTRINSIC FACTORS

It is unlikely that any computer model will ever exactly predict the response of pest numbers to insecticides. The closest approximations will come from computer models that consider as many factors as possible, including extrinsic factors -- such as weather, competitors and predators. Typically these simulations are expensive to run, they are so involved it is difficult to see what is really happening, and because they are only so good as the weakest part of their database, they may often be unreliable. It seems likely, however, that simulations have the best chance to approximate actual population responses closely only when all factors controlling resistance are incorporated, and too, such models represent the best objective test of our understanding of what really is happening. Their deficiencies point toward those areas in which additional information is needed most.

Greever and Georghiou's (1979) simulation of *C. tarsalis* populations exemplifies what may go into such a model. They have concentrated on the influence that weather has on the population growth and the evolution of resistance. In particular, they were interested in better understanding why the largest increases in resistance had occurred in just those years of exceptionally abundant rainfall. They divided the year into half-day periods and then used a data file of daily minimum and maximum temperatures. Survival of different age classes, stochastic rates of development, diapause and autogeny rates, and insecticide decay schedules were among the many variables of the model. They were able to reproduce the yearly fluctuations of a population at Bakersfield, California, moderately well, and suggested that migration among local populations is an underestimated factor for control. The most important aspect of their work, however, is probably their indication of those areas in which knowledge is most sparse and in which further effort at data collection would be most meaningfully directed.

An ambitious attempt to incorporate resistance into a full-blown simulation has been made by Guttierez's group at Davis and Berkeley (Guttierez et al., 1976, 1979). They have been simulating control of the Egyptian alfalfa weevil and have incorporated into their model several aspects of the problem that other simulations have not yet studied. These include the physiology and life history of the alfalfa, harvest time, the weather (including solar radiation, temperature and rainfall), the economic effects of secondary pests, etc. The emphasis on this effort has been directed toward finding the most economically beneficial plan for control.

It is somewhat ironic that this most intricate model of control should be performed with a pest that has not yet evolved resistance and for which adaptation to insecticide is not a problem. This does temper its usefulness and certainly prevents its being tested against reality. These simulations are probably most useful for integrating what is thought to be true about resistance into an economic analysis. As such, the alfalfa weevil model has supported Comins' conclusion cited above, that the collective good and the individual good are not served by the same strategy -- and that the theoretically optimal control strategy can be achieved only through a central body like the government (Guttierez et al., 1979).

Finally, an interesting experiment should be mentioned. Acknowledging the overwhelming complexity of pest control in an evolving population is tantamount to implying control is more an art than a science. In order to develop a feeling for the problem and how best to deal with it, a group at Michigan State University developed an interactive program in which farmers or extension workers, unskilled at programming, could sit at a terminal and try out different strategies of control on simulated populations. The objective was to let them sharpen their skills at control on imaginary populations in much the same way that airplane pilots improve their flying skills in simulated cockpits. I am aware of no demonstration that this practice actually does sharpen instincts and lead to more efficient control, but certainly that seems likely. Such experience may someday become a part of all agricultural training.

SPECULATION ON FUTURE MODELING AND COMPUTER SIMULATIONS

There are several reasons that computer simulations have become so widespread in ecology-related fields of study. Important among these are the difficulty of experimentation, the malevolent effects the experiments might have, and the overwhelming complexity of the subject itself. Experimentally selecting for resistance, especially in the field, may take many years and is costly of man-hours and land. Replication is difficult because every location and population is peculiar. Then too, the end product of a successful experiment, another insecticide-resistant population, is hardly desirable. Simulating such selection on a computer is easier, faster, cheaper and less dangerous. But computer or no computer, the models are only as good as the experiments testing their results. It is illusory to believe that an experiment on a computer is even remotely as meaningful as the real thing.

I believe that a problem is developing. Models are developed and simulations run when no effort is made to assess their accuracy experimentally. Dobzhansky (personal communication) has referred to this proliferation of models as "mental masturbation." A little bit may be all right, but it should not be overdone.

I believe that the basic sciences of population genetics and population ecology have already suffered from a pathological reliance on mathematical models: Empiricism is being abandoned in favor of scholasticism (Stearns, 1976). It would be a pity if that were to happen with pest control, yet, the only attempts to rigorously reproduce experimentally the results of simulations of which I am aware are the studies of MacDonald (1959) and of Quaglia (as yet unpublished). Computer models will, I believe, be no more useful than idle speculation if more testing is not performed. In the past five years, the development of models has far outstripped their verification. Of course some modeling is called for, as I have indicated above, but the principle problem now is different: It is now most necessary to test such models against actual populations in the laboratory and in the field. I hope that is where more future effort in this area will be directed. /al

This work has been supported by NIH grant AI15543-01.

REFERENCES

Brown, A.W.A., 1971, Pest resistance to pesticides, *in:* "Pesticides in the Environment," R. White-Stevens, ed., Vol. I, Part II, p. 457-552, Marcel Dekker, N.Y.

Comins, H., 1977a, The development of insecticide resistance in the presence of migration, *J. Theor. Biol.*, 64:177.

Comins, H., 1977b, The management of pesticide resistance, *J. Theor.*

Biol., 65:399.
Comins, H., 1979, The control of adaptable pests, G. A. Norton and C.S. Holling, eds., Pest management, Proceedings of an International Conference. Pergamon Press, Oxford, p. 217-226.
Crow, J. F., and Kimura, M., 1970, "An Introduction to Population Genetics Theory," Harper and Row, N.Y.
Dobzhansky, M., 1972, "Genetics of the Evolutionary Process," Columbia University Press, N.Y.
Ferrari, J., 1980, Effects of Insecticidal Selection and Treatment on the Reproductive Potential of a Resistant, Susceptible Strain of *Culex pipiens quinquefasciatus* (Say), M.S. thesis, University of California, Riverside.
Georghiou, G. P., 1972, The evolution of insecticide resistance, *Ann. Rev. Ecol., Syst.*, 3:133.
Georghiou, G. P., and Taylor, C. E., 1976, Pesticide resistance as an evolutionary phenomenon, p. 759-785, Proc. 15th Int. Congr. Entomol., Washington, D.C.
Georghiou, G. P., and Taylor, C. E., 1977a, Genetic and biological influences in the evolution of insecticide resistance, *J. Econ. Entomol.*, 70:319.
Georghiou, G. P., and Taylor, C. E., 1977b, Operational influences in the evolution of insecticide resistance, *J. Econ. Entomol.*, 70:653.
Greever, J., and Georghiou, G. P., 1979, Computer simulations of control strategies for *Culex tarsalis* (Diptera: Culicidae), *J. Med. Ent.*, 16:180.
Guttierez, A. P., Regev, U., and Summers, C. G., 1976, Computer model aids in weevil control, *Calif. Agr.*, April, 1976:8.
Guttierez, A. P., Regev, U., and Shalet, H., 1979, An economic optimization model of pesticide resistance: Alfalfa and Egyptian alfalfa weevil -- an example, *Environmental Ent.*, 8:101.
Hacker, C. S., Scott, D. W., and Thompson, J. R., 1973, A forecasting model for mosquito population densities, *J. Med. Ent.*, 10:544.
Hueth, D., and Regev, U., 1974, Optimal agricultural pest management with increasing pest resistance, *Am. J. Agr. Econ.*, 543.
MacDonald, G., 1959, The dynamics of resistance to insecticides by anophelines, *Rivista di Parassitologia*, 20:305.
Norton, G. A., 1979, Background to agricultural pest management modeling, *in:* "Pest Management, Proceedings of an International Conference," G. A. Norton and C. S. Holling, eds., p. 161-176, Paragon Press, Oxford.
Plapp, F. W., Browning, C. R., and Sharpe, P. J. H., 1979, Analysis of rate of development of insecticide resistance based on simulation of a genetic model, *Environmental Ent.*, 8:494.
Spiess, E. B., 1977, "Genes in Populations," Wiley and Sons, N.Y.
Stearns, S., 1976, Life-history tactics: A review of the ideas, *Quart. Rev. Biol.*, 51:3.

Taylor, C. E., and Georghiou, G. P., 1979, Suppression of insecticide resistance by alteration of gene dominance and migration, *J. Econ. Ent.*, 72:105.

Taylor, C. E., and Georghiou, G. P., 1980, The influence of pesticide persistence in the evolution of resistance, *Environmental Ent.*, (in press).

Taylor, C. R., and Headley, J. C., 1975, Insecticide resistance and the evolution of control strategies for an insect population, *Can. Entomologist*, 107.

Watt, K. E. F., 1968, "Ecology and Resource Management," McGraw-Hill, N.Y.

Wright, S., 1969, "Genetics and the Evolution of Populations," Vol. II, University of Chicago Press, Chicago.

ROLE OF MIXED-FUNCTION OXIDASES IN INSECTICIDE RESISTANCE

C. F. Wilkinson

Department of Entomology
Cornell University
Ithaca, New York 14853

INTRODUCTION

The realization that insects were able to metabolize modern synthetic organic insecticides, and that insect resistance to these insecticides was often associated with an enhanced metabolic detoxication capability, came initially as something of a shock during the late 1940s and early 1950s. It was difficult to comprehend how any organism could possibly attain the complex enzymatic machinery necessary to metabolize such a chemical, which until a few short years earlier had never even seen the light of day and the development of which clearly could not have been foreseen by the insect at which it was directed. Could it really be that the chemical itself was in some way dictating the synthesis of new enzymes, a heretical evolutionary thought, or was it possible that the insects were preadaptively equipped to metabolize the insecticide? The latter was obviously a more comfortable concept from a genetic, evolutionary standpoint, although it posed some difficult questions regarding the natural substrates and catalytic functions of the insecticide-metabolizing enzymes.

What we failed to appreciate at that time was that "there is no new thing under the sun" (Ecclesiastes, 1:9) and that modern synthetic insecticides can be viewed simply as more recently contrived additions to the host of naturally occurring toxicants against which insects and other organisms have been in constant battle throughout the course of evolutionary history. It is now recognized that as a result of these battles, insects and other organisms have developed a variety of mechanisms that have allowed them to survive exposure to a vast number of potentially hazardous chemicals of both biotic and abiotic origin. Probably the most important of these mechanisms is a remarkably effective biochemical defense system that is as proficient against man-made chemicals as it is against the naturally occurring chemicals for which it was selected (Wilkinson, 1980). Insects, therefore, were forewarned and forearmed to meet the challenges posed to them by the advent of modern synthetic organic insecticides. Viewed in this light, the development of insect resistance to insecticides through enhanced enzymatic detoxication should be considered as an accelerated version of the natural selection process that has been operating since life first began.

METABOLISM

General Aspects

Of the many types of chemicals to which organisms are exposed, it is the lipophilic compounds that represent the greatest potential threat, since these have the ability to penetrate the outer protective barriers of the organism and distribute themselves in the tissues. Furthermore, such fat-soluble materials tend to accumulate in the tissues, since their physico-chemical characteristics preclude their ready removal from the body in the aqueous or polar media in which excretion usually occurs. To counter this problem organisms have evolved a process whereby lipophilic foreign compounds are converted to more polar, hydrophilic materials that can be removed through the normal excretory mechanisms. The process typically occurs in two major steps, often referred to as primary and secondary metabolism (Figure 1). Primary metabolism usually involves an oxidative, reductive, or hydrolytic biotransformation in which a polar, reactive group is added to or uncovered in the molecule. In some cases the products of primary metabolism are excreted directly, but

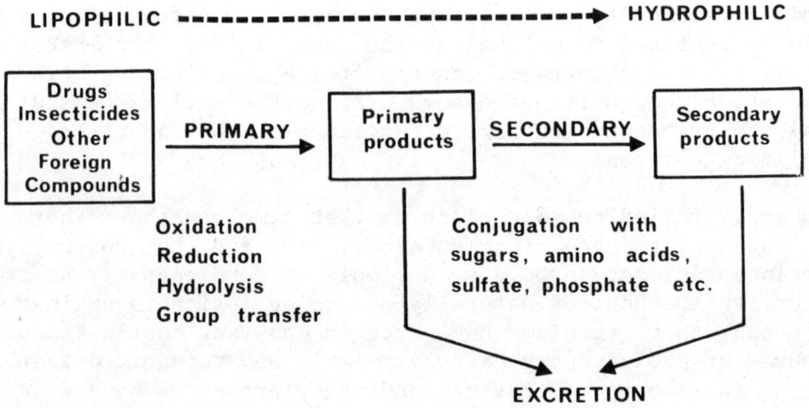

Figure 1. Metabolism of lipophilic foreign compounds.

more often they undergo further secondary metabolism resulting in the formation of water-soluble conjugates with endogenous materials such as glucose, glucuronic acid, sulfate, phosphate or amino acids.

The overall process by which fat-soluble compounds are converted to hydrophilic, readily-excretable metabolites is typically associated with a decrease in biological activity or toxicity, and is often termed detoxication. However, it should be emphasized that since organisms cannot prejudge the toxicity of a given material, detoxication has to be considered secondary to the function of facilitating removal of the chemical from the tissues.

The Importance of Oxidative Metabolism

Of the several types of reactions affecting the primary metabolism of insecticides and other foreign compounds, oxidation by the so-called mixed-function oxidases is of considerable importance and often plays a dominant role in determining the biological activity or toxicity of a given material. Consequently, it is not surprising that animals possessing high titers of the mixed-function oxidases exhibit a high degree of tolerance to many chemicals, and that in insects it constitutes a potentially important mechanism through which resistance can occur.

Early studies on the mixed-function oxidases were conducted almost exclusively in mammalian species where they occur mainly in the liver (Mannering, 1971; Nakatsugawa and Morelli, 1976; Testa and Jenner, 1976) and to a lesser extent in extrahepatic tissues such as lung, kidney, small intestine and skin (Bend and Hook, 1977). More recent comparative studies have established the presence of the enzymes in a large number of other organisms representing diverse phylogenetic groups (Khan and Bederka, 1974; Brattsten, 1979a; Wilkinson, 1980). In the animal kingdom, oxidases have been described and partially characterized in most of the major chordate classes, including mammals, birds (Pan et al., 1975), fish (Bend and James, 1978), reptiles and amphibia (Machinist et al., 1968). Invertebrates, too, have been the object of much recent interest, and mixed-function oxidases with properties basically similar to those found in vertebrate liver are known to occur in the tissues of several classes of arthropods, including insects (Wilkinson, 1979; Wilkinson and Brattsten, 1972; Hodgson and Plapp, 1970; Agosin and Perry, 1974), millipedes, centipedes and terrestrial and aquatic crustacea (Neuhauser and Hartenstein, 1976); they have also been reported in earthworms (phylum Annelida) (Nelson et al., 1976; Neuhauser and Hartenstein, 1976) and several species of the phylum Mollusca (Neuhauser and Hartenstein, 1976), including both terrestiral (snails and slugs) and aquatic species. Distribution of the mixed-function oxidases is not limited to members of the animal kingdom. They occur in the tissues of higher plants (Lamoureux and Frear, 1979), yeasts and fungi, and mechanistic studies on the oxidases of aerobic bacteria such as *Pseudomonas putida* have contributed greatly to our understanding of the mammalian hepatic enzymes (Gunsalus, 1972). Thus, mixed-function oxidases with remarkably similar properties are distributed ubiquitously throughout the plant and animal kingdoms as well as in a variety of primitive prokaryotic organisms. This immediately indicates the early evolutionary development of the enzymes (Wilkinson, 1980; Wickramasinghe and Villee, 1975) and suggests that they play a common functional role of considerable importance to a large number of living organisms.

The objective of this presentation is to discuss in rather general terms some of the major characteristics of the mixed-function oxidases, and to consider their importance in insecticide metabolism and insect resistance to insecticides. Details on more specific aspects of mixed-function oxidation are covered in other presentations at this symposium (Hodgson, 1980; Terriere, 1980).

Mixed-function Oxidases

In all vertebrate and invertebrate species so far examined, the mixed-function oxidases involved in foreign compound metabolism are associated with the microsomal fraction of tissue homogenates, which is derived from the endoplasmic reticulum of the intact cell. They exhibit an unusual degree of nonspecificity and a predilection for fat-soluble compounds, which they metabolize through reactions involving numerous functional groups. Among these reactions are aromatic, alicyclic and aliphatic hydroxylation, dealkylation of ethers and substituted amines, oxidation of thioethers to sulfoxides and sulfones, epoxidation of aromatic and olefinic double bonds, and desulfuration (Testa and Jenner, 1976; Nakatsugawa and Morelli, 1976).

Role in insecticide metabolism. The extensive involvement of mixed-function oxidation in insecticide metabolism is well known and has been the subject of several excellent reviews (Brooks, 1972; Nakatsugawa and Morelli, 1976). The following summary will serve to illustrate some of the major types of reactions that have been established for different groups of insecticides.

One of the first demonstrations of mixed-function oxidase activity in insects was the hydroxylation of DDT (Figure 2) to a dicofol-like derivative (Agosin et al., 1961), and it is of interest that the subsequent development of the more biodegradable analogs of DDT (Metcalf, 1976; Metcalf et al., 1971) has been based largely on the intentional introduction into the molecule of groups that are particularly susceptible to oxidative metabolism (e.g., alkoxy or alkyl) (Figure 2). Microsomal epoxidation of the unchlorinated double bond of several cyclodiene insecticides (e.g., chlordene, aldrin, isodrin, heptachlor) (Figure 2) is a well established reaction and leads to

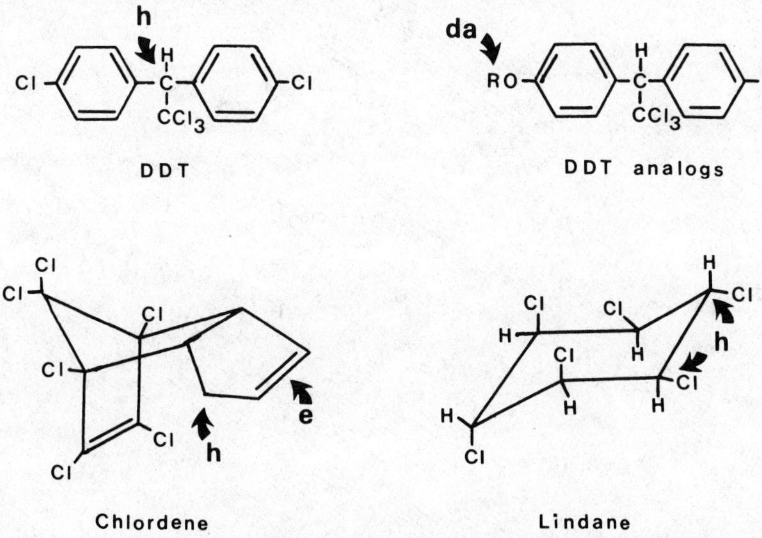

Figure 2. Mixed-function oxidation of organochlorine insecticides. da = dealkylation, e = epoxidation, h = hydroxylation.

the formation of *trans*-diols following subsequent cleavage by epoxide hydratases (Brooks, 1972, 1974; Nakatsugawa and Morelli, 1976). Hydroxylation of several cyclodienes may also occur (Figure 2) depending on the structure of the molecule, and recent studies have established that the metabolism of lindane (Figure 2) and other hexachlorocyclohexane isomers to trichlorophenols is also initiated by direct microsomal hydroxylation of the cyclohexane ring (Tanaka et al., 1979).

If our assumption is correct that the microsomal enzymes have evolved as a protective mechanism against naturally occurring toxicants, natural product insecticides should be prime condidates for oxidative metabolism. This is indeed the case. Nicotine (Papadopoulos and Kintzios, 1963), rotenone (Fukami et al., 1969) and the natural pyrethroids (Yamamoto et al., 1969; Elliott et al., 1972) are all metabolized by a variety of oxidative reactions (Figure 3), and in most insect species, the anti-juvenile hormones, precocenes I and II, are so metabolically labile (Burt et al., 1978) that they are unable to exert their potential biological activity. While the structures of some of the recently developed synthetic pyrethroids (e.g., permethrin, Table 1) preclude some of the reactions observed in the original natural product (e.g., the oxidation of the isobutenyl group of pyrethrin I) they provide additional sites for oxidative attack. Indeed, in the case of the synthetic juvenoids such as methoprene (Figure 3) and the geranyl phenyl ether, R-20458

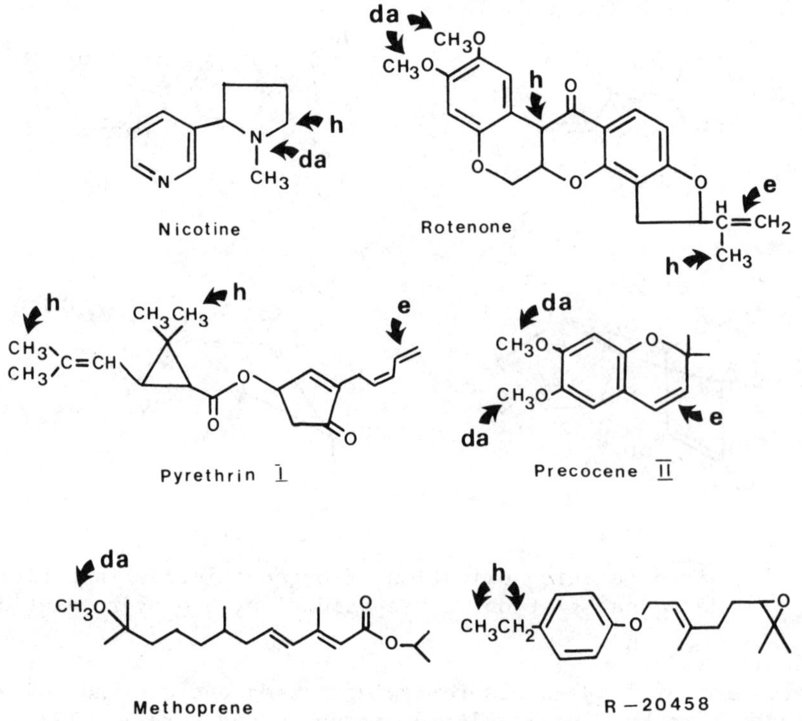

Figure 3. Mixed-function oxidation of some natural product insecticides and synthetic analogs. da = dealkylation, ds = sulfuration, h = hydroxylation.

(Figure 3), metabolism appears to occur primarily by oxidation (Hammock et al., 1977; Bigley and Vinson, 1979), whereas most reports indicate that the natural juvenile hormones on which these structures are based are degraded mainly by hydrolytic mechanisms (Slade and Zibitt, 1972).

The structural diversity of the carbamate insecticides provides a large number of sites at which metabolic attach can occur (Figure 4), and these compounds appear to be metabolized almost exclusively by mixed-function oxidation (Wilkinson, 1971; Nakatsugawa and Morelli, 1976). This may occur through dealkylation of the methycarbamoyl or dimethylcarbamoyl moieties, mbut most of the major metabolic pathways are initiated by oxidation at other sites in the molecule. Examples of all types of microsomal oxidation have been demonstrated with the

Table 1. Preferred Sties of Oxidative Attack on Permethrin by Different Species[a]

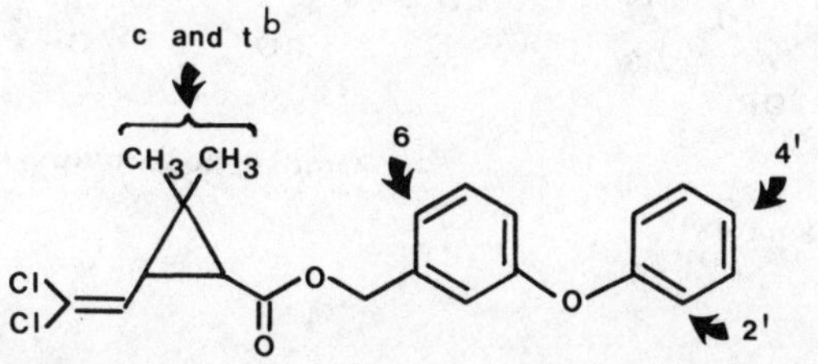

Species	Permethrin isomer	
	trans	*cis*
American cockroach	4'>t>>c	4'>t>>c
House fly	4'>t>6	4'>t>6>c
Cabbage looper	4'>>t	4'>>t
Rat	4'>>c>t	4'>c≥t>2'
Cow	t>c>4'	t>4'>c

[a]Data from Shono et al. (1978).
[b]c and t refer to the methyl groups *cis* and *trans*, respectively, to the carboxyl.

wide variety of aromatic substituents found in the phenyl N-methylcarbamates, and these typically result in the rapid detoxication of the insecticide.

Many organophosphorus insecticides are also extensively metabolized by the mixed-function oxidases (Figure 4), although in this case the results are complex and difficult to predict because the reactions may lead to either an increase or decrease in toxicity depending on the insecticide in question (Wilkinson, 1971; Nakatsugawa and Morelli, 1976). Thus, while oxidation of various structural components of the leaving group of the insecticide may promote rapid detoxication, the oxidative desulfuration of phosphorothionates to the corresponding phosphate anticholinesterases is an essential step in the activation of insecticidal activity. Evaluation of the role of oxidation in the metabolism of organophosphorus compounds is further complicated by the fact that in many cases (e.g., ester cleavage,

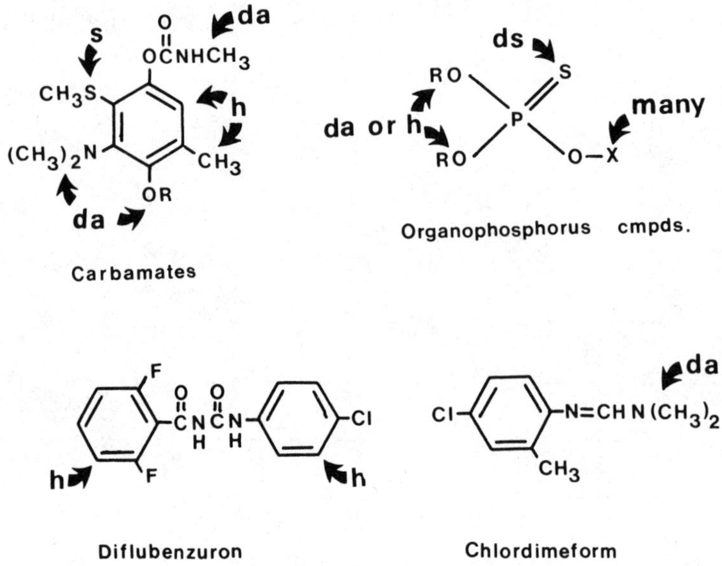

Figure 4. Mixed-function oxidation of insecticides: da = dealkylation, ds = sulfuration, h = hydroxylation, s = sulfoxidation.

dealkylation), the products generated by oxidase action are the same as those resulting from the action of other types of enzymes, such as esterases and glutathione transferases.

To conclude this brief survey, the hydroxylation of diflubenzuron (Pimprikar and Georghiou, 1979), the demethylation of chlordimeform (Benezet et al., 1978) (Figure 4) and the several metabolic reactions involving methylenedioxyphenyl synergists (Casida, 1970) provide additional examples illustrating the importance of mixed-function oxidation in the metabolism of the many types of chemicals employed in insect control.

Mechanism of Mixed-function Oxidation. In vitro studies with preparations from both insects and mammals indicate that the enzymes responsible for these reactions require NADPH and O_2 and are associated with the microsomal fraction of tissue homogenates. Since in all cases investigated in detail, one atom of molecular oxygen is incorporated into the substrate and the other reduced to water, the enzymes are by definition classified as mixed-function oxidases and can be represented by the general equation:

$$RH + O_2 + NADPH + H^+ \rightarrow ROH + H_2O + NADP^+$$

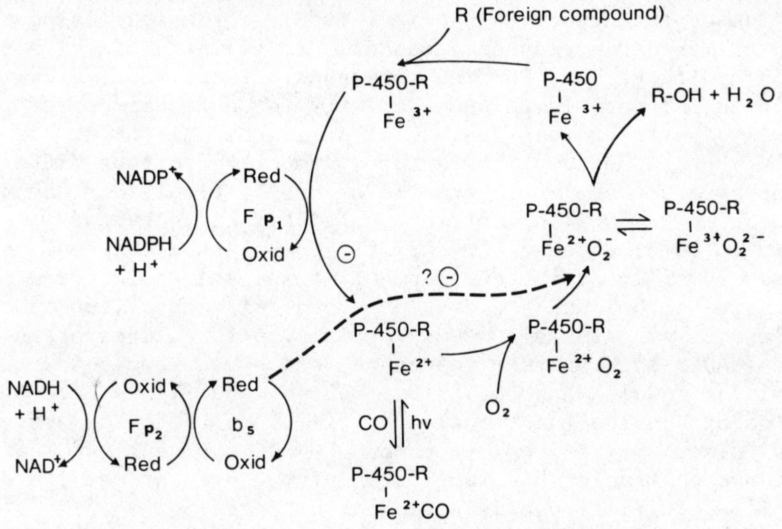

Figure 5. Microsomal electron transport pathway. From Wilkinson (1980).

The key to the system is a common electron transport pathway (Figure 5) that transfers reducing equivalents from NADPH through a flavoprotein (NADPH-cytochrome *c* reductase, Fp_1), which is the terminal oxidase of the chain. Cytochrome P-450 is an unusual *b*-type cytochrome, which derives its name from the fact that its reduced form combines with CO to produce a complex absorbing at 450 nm. The overall mechanism through which oxidation occurs is quite well understood (Estabrook et al., 1979). The foreign compound substrate first forms a complex with the oxidized form of cytochrome P-450 and the complex is reduced by one electron passing down the chain from NADPH. The reduced cytochrome P-450/substrate complex then reacts with and activates molecular oxygen and the resulting oxygenated complex breaks down to yield the product and water. Reduction, therefore, occurs in two separate one-electron steps. Although the first of these certainly arises from NADPH, it is still uncertain whether this is true of the second or whether this originates from NADH and passes down another microsomal electron transport pathway involving cytochrome b_5 and the flavoprotein NADH-cytochrome b_5 reductase (Fp_2) (Figure 5).

Multiple Forms of Cytochrome P-450. The apparent ability of the mixed-function oxidase system to metabolize almost any foreign compound substrate with which it is provided has been at least partially explained in recent years by discovery of the existence of multiple forms of cytochrome P-450 (Lu et al., 1976; Coon et al., 1977).

Evidence to support this came initailly from studies on the induction of mixed-function oxidase activity following in vivo treatment of animals with various foreign compounds. Induction typically results in an increase in oxidase activity through enhanced *de novo* synthesis of cytochrome P-450 and other microsomal enzyme components (Cooney, 1967). However, it has been demonstrated that different inducing agents cause the synthesis of qualitatively different forms of cytochrome P-450 with different catalytic and structural properties (Lu et al., 1976; Coon et al., 1977). Comparisons between these have been made possible by the development of suitable solubilization and purification techniques and by the successful reconstitution of catalytically active oxidase systems requiring only a lipoprotein, flavoprotein (NADPH-cytochrome *c* reductase) and cytochrome P-450 (Lu and Levin, 1974; Kawalek and Lu, 1975). These experiments have clearly established that the functional specificity of the oxidase system resides primarily with the cytochrome P-450 component and that there exist immunochemically distinct forms of the cytochrome with different catalytic activities towards various substrates.

The complex, yet fascinating picture that seems to be emerging is of a common flavoprotein transferring electrons from NADPH to a family of different cytochrome P-450s (Figure 6), each catalyzing a limited, perhaps overlapping spectrum of oxidative reactions with a variety of substrates, and each exhibiting varying degrees of sensitivity to different inducing agents. It is probable that the composition of the cytochrome "P-450 family" will be found to vary, not only between different animal species, but also, perhaps, between the tissues of a single animal. This could explain some of the subtle variations observed between species with respect to the preferred sites of oxidase attack on insecticides and other foreign compounds containing more than one oxidizable group. A recent example of this comes from metabolic studies showing that microsomal

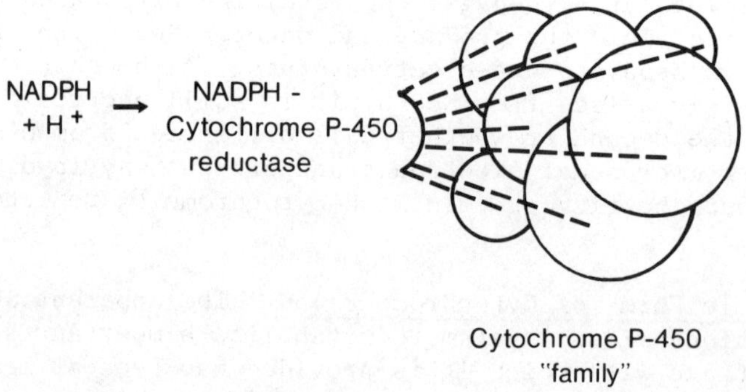

Figure 6. Multiple forms of cytochrome P-450. From Wilkinson (1980).

hydroxylation of *cis* and *trans*-permethrin (Shono et al., 1978) can occur at four different sites in the molecule (2', 4', 6 and the geminal-dimethyl group) and that there are species-dependent differences in the preferred site of attack in house flies, American cockroaches, cabbage loopers, rats and cows (Table 1). In all species except cows, 4'-hydroxylation of the phenoxybenzyl group was the preferred site of attack. However, house flies were the only species able to hydroxylate the 6 position of the alcohol moiety, and 2'-hydroxylation occurred only in rats. Hydroxylation at the geminal-dimethyl group occurred primarily at the methyl group *trans* to the carboxyl (except in rats) and was the preferred site of attack in cows. There is currently no information available that will allow the prediciton of preferred sites of oxidation of any molecule by any species.

The probable existence of multiple forms of cytochrome P-450 in insects (Hodgson, 1976; Stanton et al., 1978) could have an important bearing on our understanding of hitherto unexplainable facets of selective toxicity and insect resistance to insecticides and is discussed further elsewhere in this volume (Hodgson, 1980).

ROLE OF MIXED-FUNCTION OXIDASES IN INSECT RESISTANCE

In view of the ubiquitous distribution of the mixed-function oxidases, their apparently long-established natural protective function in living organisms, and their remarkable catalytic versatility towards a large number of naturally occurring and synthetic insecticides, it is not surprising that they are now known to play an important role in many cases of insect resistance (Perry and Agosin, 1974; Oppenoorth and Welling, 1976).

In vivo Evidence

Evidence to support this, however, came initially in a rather indirect way through the combined results of several separate studies on the mode of action of methylenedioxyphenyl synergists such as piperonyl butoxide (PB) and sesamex. Following the early discoveries that these compounds were able to inhibit aldrin epoxidation and phosphorothionate activation in vivo in insects (Sun and Johnson, 1960) and that they inhibited carbamate oxidation in vitro in mammalian liver preparations (Hodgson and Casida, 1960), it was established that carbaryl resistance in house flies could be abolished when the flies were treated with a carbaryl-sesamex combination (Eldefrawi et al., 1960). The strong suggestion that resistance to carbaryl was due to enhanced oxidative metabolism, and that this was being blocked by the synergist, was strengthened by subsequent studies demonstrating that the reversal of house fly resistance to *m*-isopropylphenyl *N*-methylcarbamate by PB was in fact due to metabolic inhibition (Georghiou and Metcalf, 1961). Since that time, the ability of methylenedioxyphenyl and other groups of compounds (propynyl ethers, 1,2,3-benzothiadiazoles, imidazoles, etc.) to inhibit microsomal oxidation had been clearly established as the primary mechanism through which they exert their synergistic activity (Casida, 1970; Hodgson, 1976; Wilkinson, 1976; Hodgson and Philpot, 1974). As a result, synergists of this type have become widely used as in vivo indicators of the involvement of mixed-function oxidase activity in insecticide metabolism, and in conjunction with the synergists known to inhibit other types of enzymes (e.g., hydrolases, glutathione transferases, etc.), they provide valuable information on the relative importance of different metabolic resistance mechanisms (Oppenoorth, 1971; Wilkinson, 1968, 1971).

A recent example of the type of information obtained from such investigations is provided by the results of studies on the high levels of resistance to diflubenzuron in two strains of house flies, R-OMS-12 and R-diflubenzuron, selected for several years with *O*-ethyl-*O*-(2,4-dichloro-phenyl) phosphoramidothioate and diflubenzuron, respectivley (Pimprikar and Georghiou, 1979) (Table 2). The high levels of synergism observed with PB (77.6) and sesamex (36.3) in the R-diflubenzuron flies clearly indicate that mixed-function oxidases play a significant role in the resistance mechanism of this strain: This was confirmed by metabolic studies. Conversely, the relatively low synergistic ratios with *S,S,S*-tributyl phosphorotrithioate (DEF), an esterase inhibitor, and diethylmaleate (DEM), an inhibitor of glutathione transferase, suggest that metabolism by these pathways is of little importance to either of the R strains, although activities in each are slightly higher than in the NAIDM susceptible flies. The large differences observed in the synertistic efficacy of PB and sesamex in the R-OMS-12 and NAIDM strains is unusual and emphasizes the danger in assuming that all methylenedioxyphenyl synergists have a similar inhibitory effect on all types of mixed-function oxidation. If we return for a moment to the concept of multiple forms of cytochrome P-450 it is possible to understand how the composition of these could vary between different strains and how they could exhibit different sensitivities to inhibitors. The other danger inherent in these in vivo synergist-indicator studies is, of course, the assumption that the synergist is a specific inhibitor of a given enzyme system. Thus, from the data in Table 2, is it reasonable to assume that the small degree of synergism observed with DEM reflects a real contribution of glutathione transferase to the resistance mechanism, or is DEM a weak oxidase inhibitor? Or is it possible that some of the effects of PB or sesamex might result from inhibition of esterase activity? Although in some cases (e.g., with the house cricket, *Acheta domesticus*) excellent correlations have been observed between in vitro mixed-function oxidase activity and in vivo carbaryl synergism with PB (Benke and Wilkinson, 1971), and although the latter has been used as a measure of mixed-function oxidase activity in many different insect species

Table 2. Effect of Synergists on Toxicity of Diflubenzuron to Various Strains of House Flies[a]

Treatment	Toxicity (μg/larva) for different strains[b]					
	S-NAIDM		R-OMS-12		R-Diflubenzuron	
	ED_{50}	SR[d]	ED_{50}	SR	ED_{25}	SR
Diflubenzuron						
Alone	0.059	--	22.67	--	17.08	--
+ Piperonyl butoxide	0.023	2.6	15.32	1.5	0.22	77.6
+ Sesamex	0.004	14.8	0.77	29.4	0.47	36.3
+ DEF	0.038	1.6	8.06	2.8	5.31	3.2
+ DEM	0.061	1.0	9.2	2.5	8.69	2.0

[a]Data from Pimprikar and Georghiou (1979).
[b]NAIDM was susceptible, R-OMS-12 was selected since 1960 with *O*-ethyl-*O*-(2,4-dichlorophenyl) phosphoramidothioate and R-diflubenzuron was selected with diflubenzuron.
[c]Diflubenzuron:synergist ratio, 1:5.
[d]ED_{50} diflubenzuron alone/ED_{50} diflubenzuron plus synertist.

(Brattsten and Metcalf, 1970), it is clear that data relating to resistance mechanisms should be interpreted with caution and should be confirmed by in vivo and in vitro metabolic studies.

In spite of some of the problems involved, the reversal of insecticide resistance by methylenedioxyphenyl synergists constitutes a useful in vivo indicator of the extent of mixed-function oxidase involvement in insect resistance. Using this approach mixed-function oxidases have been shown to be of importance in the resistance of various strains of insects to DDT (Oppenoorth, 1965; Sawicki, 1973), the pyrethrins (Farnham, 1973), the carbamates (Georghiou et al., 1961; Metcalf and Fukuto, 1965; Shrivastava et al., 1969; Wilkinson, 1971), several organophosphorus compounds (Wildinson, 1971) and some of the newer groups of compounds, including the synthetic juvenoid methoprene (Hammock et al., 1977) and the chitin synthetase inhibitor diflubenzuron (Pimprikar and Georghiou, 1979).

The effect is particularly well illustrated in the various strains of house flies that have developed high levels of resistance

Table 3. Piperonyl Butoxide Synergism of Two Carbamates Against Susceptible and Resistant Strains of the House Fly[a]

Strain	Topical LD_{50} of *N*-methylcarbamate (μg/female fly)			
	m-isopropylphenyl		*o*-isopropoxyphenyl	
	alone	with PB[b]	alone	with PB
Susceptible				
NAIDM	1.8	0.18	0.51	0.14
LAB	2.0	0.33	0.47	0.19
SCR	-	-	0.20	0.12
SCRS	-	-	0.30	0.08
Resistant (carbamate selection)				
MIP-1[c]	100	1.0	100	0.56
MIP-2[c]	100	1.2	100	0.64
Baygon[d]	-	-	15	0.6
Hokota[d]	-	-	100	1.0
Resistant (selection with other insecticides)				
Super Pollard[3]	20	0.86	-	-
Stauffer Chlor[f]	8.3	0.63	-	-
Ronnel[f]	100	2.6	100	0.96

[a]Data from Georghiou et al. (1961), Metcalf and Fukuto (1965) and Shrivastava et al. (1969).
[b]Carbamate:piperonyl butoxide (PB) ratio, 1:5.
[c]Selected with *m*-isopropylphenyl *N*-methylcarbamate.
[d]Selected with *o*-isopropoxyphenyl *N*-methylcarbamate.
[e]Selected with DDT and lindane.
[f]Selected with chlorthion.

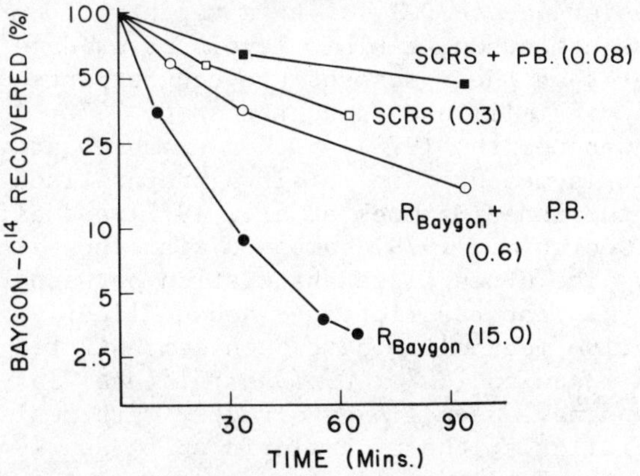

Figure 7. Effect of piperonyl butoxide (PB) on the metabolism of propoxur in susceptible (SCRS) and resistant [R-Baygon (propoxur)] house flies. Figures in parentheses are LD_{50} values for propoxur (μg/fly). Data from Shrivastava et al. (1969).

to the carbamates as a result of either selection pressure with the carbamates themselves or through cross resistance resulting from selection with other groups of insecticides (Table 3). In all cases PB almost completely abolishes the resistance, strongly suggesting that carbamate resistance results almost exclusively from metabolism due to enhanced mixed-function oxidase activity. This has been amply confirmed by in vivo experiments demonstrating the ability of PB to stabilize various carbamates to metabolic attack. Pretreatment of both susceptible (SCRS) and resistant (R-propoxur) house flies with PB impaired their ability to metabolize propoxur (Figure 7), and this correlated with the toxicity of synergized and unsynergized propoxur to these two strains (Shrivastava et al., 1969). Similar results have been obtained with other synergists known to inhibit the mixed-function oxidases (Sacher et al., 1968).

In Vitro Evidence

More direct evidence of the importance of the mixed-function oxidases in many cases of insect resistance comes from the results of in vitro assays of the enzymes in insect tissues. These usually involve direct measurement of the oxidative metabolism of insecticides or of a variety of model substrates, or determination of the levels of important microsomal enzyme components such as cytochrome P-450. Despite the many technical problems encountered in in vitro microsomal enzyme studies in insects (Wilkinson and Brattsten, 1972; Wilkinson, 1979), it has now been amply demonstrated that enzyme preparations from numerous strains of insects (mostly house flies) selected for resistance to DDT, or to a variety of carbamates and organophosphorus compounds, exhibit levels of oxidase activity considerably higher than their susceptible counterparts (for many references see reviews by Wilkinson and Brattsten, 1972; Oppenoorth and Welling, 1976; Oppenoorth, 1976). Other examples are found in the high oxidase levels measured in in vitro preparations from house flies resistant to methoprene (Hammock et al., 1977) and diflubenzuron (Pimprikar and Georghiou, 1979) compared with those in preparations from S strains. The often close association between resistance and oxidase activity is not restricted to house flies. High in vitro levels of oxidative metabolism have been demonstrated in preparations from propoxur-R mosquito (*Culex pipiens fatigans*) larvae (Shrivastava et al., 1970) and have been linked directly with resistance to this insecticide. Similarly, the high aldrin epoxidase levels observed in the gut tissues of field-collected strains of the tobacco budworm (*Heliothis virescens*) (Williamson and Schecter, 1970) have been associated with resistance to DDT and methyl parathion. Kuhr (1971) also showed that the levels of cytochrome P-450 and carbaryl oxidation activity in homogenates of the guts and fat bodies of strains of cabbage loopers (*Trichoplusia ni*) resistant to DDT, or to parathion and other organophosphorus compounds, were considerably higher than those measured in an insecticide-susceptible strain.

In most of these strains, the developemnt of resistance seemed to be associated with a rather general increase in all types of microsomal oxidation. Thus, enzyme preparations from strains of house flies selected with carbamates showed increased rates of hydroxylation, *O*- and *N*-dealkylation, epoxidation and desulfuration (Tsukamoto and Casida, 1967; Casida, 1969), and similar in vitro studies with the F_c strain of flies, selected with the organophosphorus compound diazinon, showed not only high DDT oxidase activity (Oppenoorth and Houx, 1968) but also a high capacity to hydroxylate naphthalene and epoxidize aldrin (Schonbrod et al., 1968). Microsomal fractions from another strain of flies selected with dimethoate exhibit higher dimethylase activity towards methoprene and *p*-nitroanisole, as well as higher levels of aldrin epoxidase and dihydroisodrin hydroxylase, than those from the NAIDM susceptible strain (Hammock et al., 1977).

Consequently, in addition to enhancing the metabolism of the insecticide with which selection is achieved, resistance due to increased mixed-function oxidase activity also facilitates the metabolism of a large number of other unrelated (even unsynthesized!) insecticides by one or more of the many pathways involving microsomal oxidation. This provides a ready explanation of the phenomenon of cross resistance, which is, of course, one of the most threatening aspects of insect resistance involving this mechanism.

On the other hand, in spite of the difficulties inherent in extrapolating in vitro enzymatic data to the results of in vivo toxicity studies, and the fact that in most cases resistance is due to a combination of several factors (e.g., metabolism, penetration), there are indications that resistance due to enhanced oxidative metabolism is not always associated with a general increase in all types of microsomal oxidation. Thus, although enhanced oxidase activity towards carbamates, pyrethrins, organophosphorus compounds and aldrin was observed in the R-Propoxur strain of house flies (Plapp and Casida, 1969), this did not extend to DDT, and the F_c strain showed high oxidase activity towards DDT, aldrin and some organophosphorus compounds, but not to other insecticides. The absence of any real correlation between in vitro oxidase (naphthalene hydroxylase and aldrin epoxidase) levels and insect resistance patterns, or susceptibility to naphthalene vapor, was also demonstrated by Schonbrod et al. (1968) in a survey of 14 strains of house flies, despite earlier indications that such a correlation might exist (Schonbrod et al., 1965). In conclusion, therefore, it appears that insecticide selection can lead either to a simultaneous increase in several types of mixed-function oxidations causing a broad spectrum of cross resistance, or it can lead to more selective increases in qualitatively different forms of the enzyme resulting in more restricted patterns of cross resistance. Such a situation is consistent with the proposed existence of multiple forms of cytochrome P-450 in insects and, moreover, suggests that these are under separate genetic control (Oppenoorth and Welling, 1976).

Natural Tolerance in Insects

The considerable variations observed in mixed-function oxidase levels in the tissues of natural populations of insect (i.e., those not selected with insecticides) are of interest, since clearly natural tolerance represents resistance attained through the process of natural selection. It also may provide some clues to what we can expect to find following further selection with insecticides. As discussed earlier, one of the first observations to be made is that all insects studied possess some level of mixed-function oxidase activity, and consequently are preadapted for further selection. However, the actual content and activity of the mixed-function oxidases vary considerably between different species and appear to be related to food preferences and the life-style of the insect (Brattsten, 1979a,b; Wilkinson, 1980).

Krieger et al. (1971) found that aldrin epoxidase levels in the gut tissues of 35 species of lepidopterous larvae vary substantially in a way reflecting the number of plant families that could be utilized as host plants. Thus, polyphagous species have higher levels of oxidase activity than those with more restricted feeding habits (Table 4) and are presumably better equipped to metabolize the wide variety of plant toxicants they are likely to encounter. As a result of the overall nonspecificity of the enzymes, such polyphagous

Table 4. Relationship Between Midgut Epoxidase Activity and Feeding Range in Lepidopterous Larvae[a]

Feeding range (No. of plant families)	Species tested	Aldrin epoxidase (pmole/min/mg protein)
Monophagous (1)	8	20.4 ± 9.1
Oligophagous (2-10)	15	90.0 ± 33.6
Polyphagous (11 or more)	12	297.4 ± 65.9

[a]Data from Kreiger et al. (1971).

species are also likely to show a higher natural tolerance to synthetic organic insecticides. Gordon (1961) specifically suggested this possibility several years ago, and it is of interest to note that as early as 1939, before the advent of synthetic organic insecticides, and prior to any knowledge of the mixed-function oxidases, Swingle (1939) foresightedly predicted a relationship between insecticide tolerance and feeding behavior in lepidopterous larvae.

Perhaps more important than the fact that polyphagous larvae possess higher levels of one particular type of oxidase activity (in this case epoxidation) than monophagous or oligophagous species, is the possibility that they possess a much greater degree of metabolic versatility. Polyphagous species clearly show what we can term natural cross resistance through selection with many different types of secondary plant substances and if, indeed, there exist multiple forms of cytochrome P-450 with qualitatively different catalytic activities, it is more likely that they will be found in polyphagous species than in those with more restricted host plant ranges. Consequently, we might predict that polyphagous insect species have a greater preadaptive potential to develop resistance to any of a large number of different insecticides. Conversely, we could speculate that insect species that have more specialized feeding habits and have consequently undergone natural selection with a more restricted group of natural toxicants, will possess less metabolic versatility, and that their ability to develop resistance to a particular insecticide will vary with the qualitative make-up of the cytochrome P-450 they possess. However, it must be noted as pointed out by Georghiou and Taylor (1976), that polyphagous behavior may tend to delay resistance if a population is only partially selected by allowing the insects to feed on non-treated hosts.

The possibility that insecticide resistance may indeed involve the selection of specific forms of cytochrome P-450 will be discussed further by Dr. Hodgson and could well provide an explanation for some of the subtle and unexplained cases of cross resistance that have been observed.

ADAPTATION OF INSECT OXIDASES FOR THEIR PROTECTIVE ROLE

In view of the importance of the mixed-function oxidases in insecticide metabolism and resistance, it may be informative to review some of the characteristics of the system that make it so beautifully adapted for its protective role. At the same time we can consider the implications of some of these characteristics with respect to insect control with insecticide chemicals.

Substrate Nonspecificity

As we have discussed already, the system is able to accept and metabolize any of a large number of lipophilic foreign compounds, i.e., it shows a remarkable degree of nonspecificity. Consequently, it is likely to continue to present a major obstacle to the effectiveness of all current and future insecticides and will continue to be involved in the development of insect resistance to these chemicals.

Strategic Localization

Consideration of the distribution of oxidase activity in the organs of different animals reveals that the enzymes are located primarily in tissues associated with major portals of entry into the body (Wilkinson and Brattsten, 1972; Nakatsugawa and Morelli, 1976). In mammals, for example, they are located in the lung, skin and intestinal tract and it should be noted that the liver, the major site of oxidation in mammals, is derived embryologically from the intestine; a similar pattern of strategic location occurs in fish, where the enzymes are found mainly in liver, kidney and gill (Bend and James, 1978). In insects, oxidase activity is usually found in either the gut or fat body (occasionally the Malphighian tubules), presumably tissues representing the first lines of defense for compounds entering the body in the diet or by cuticular penetration (Figure 8). Thus, the enzymes appear to be located in tissues where they can function to maximum advantage.

Inducibility

The ability of the oxidase system to be induced by a wide variety of foreign compounds allows it to respond rapidly to periods of unusually severe environmental (chemical) stress, and in some instances this may prove sufficient to ensure the immediate survival of the animal. The phenomenon of induction is now well established in insects (Table 5) (Brattsten and Wilkinson, 1973; Brattsten et

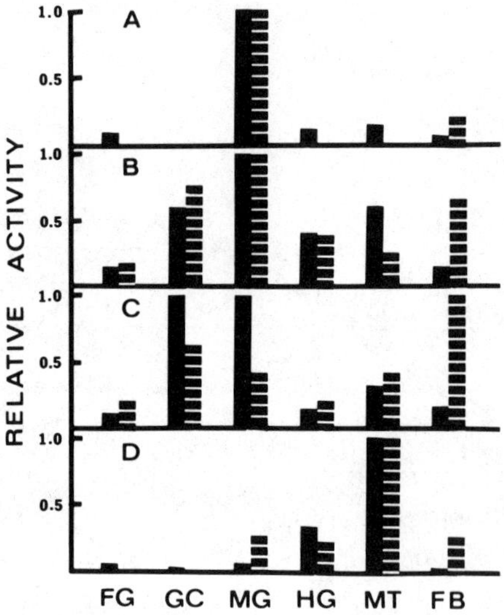

Figure 8. Distribution of microsomal epoxidase activity in insect tissues: A = southern armyworm (*Spodoptera eridania*); B = Madagascar cockroach (*Gromphadorhina portentosa*); C = American cockroach (*Periplaneta americana*); D = house cricket (*Acheta domesticus*). Solid bars represent specific activity (per mg protein); striped bars represent activity per insect. From Wilkinson (1979).

al., 1977; Yu and Terriere, 1971, 1972) although its role, if any, in insect resistance has not yet been established. Although the induced state *per se* is not inheritable, it is possible that "inducibility" (i.e., the capacity to be effectively and rapidly induced) could be closely linked with genes for high oxidase activity and that selection for both may occur concurrently. Even if induction occurs to the same extent in both low oxidase and high oxidase strains, the net increase of the latter would result in a considerably greater total oxidative capacity (Terriere et al., 1971) and would provide the insect with a more effective measure of protection.

One factor that should be stressed is the amazing rapidity with which induction can occur in insects. Studies with the southern armyworm (*Spodoptera eridania*) have shown a dramatic four-fold increase in gut oxidase activity within the first five hours of dietary exposure to pentamethylbenzene (0.2%) (Brattsten and Wilkinson, 1973) and a significant increase in activity has been observed only 30 minutes after ingestion of low levels of the natural products α-pinene and sinigrin (Brattsten et al., 1977). The fact that enzyme activity returns to control levels equally rapidly following the cessation of inducer exposure further illustrates the remarkable

Table 5. Effect of Induction on In Vitro Midgut Microsomal Oxidase Activity and In Vivo Tolerance of Southern Armyworm Larvae to Orally Administered Carbaryl[a]

Inducer treatment	Percent control activity		LD_{50} carbaryl (μg/g)
	Epoxidase	Cytochrome P-450	
Control	100	100	30
Hexamethylbenzene (2,000 ppm in diet)	225	219	67
Pentamethylbenzene (2,000 ppm in diet)	314	299	350

[a]Data from Brattsten and Wilkinson (1973).

flexibility of the system, a characteristic that serves to minimize the energy load imposed by enhanced protein synthesis.

With respect to insect control it might be mentioned that many of the solvents commonly used in insecticide formulations are not "inert" as claimed, but are in fact very effective oxidase-inducing agents (Brattsten and Wilkinson, 1977). In view of the rapidity with which induction can occur, one wonders whether such solvent effects may be the cause of some of the variations in insecticidal efficacy observed with different formulations. Are we using some formulations that help the insects to survive insecticide treatment?

The extent of the specificity of induction remains to be established, although there is mounting evidence to indicate that, as in mammals, induction of the insect mixed-function oxidase system with different compounds does not result in the simultaneous increase in all types of cytochrome P-450, but rather results in an increase in specific types of the cytochrome (Stanton et al., 1978). This may provide at least a partial explanation for the obvious qualitative differences observed in in vitro epoxidase and *N*-demethylase activities in gut tissues of southern armyworm larvae feeding on the foliage of different plant species (Table 6) (Brattsten, 1979). These data certainly suggest that insects feeding on different host plants can be expected to vary in thier tolerance to insecticides.

Further details on mixed-function oxidase induction and its possible relationship to insect resistance will be discussed elsewhere in this volume (Terriere, 1980).

Table 6. Mixed-function Oxidase Activity in Gut Tissues of Southern Armyworm Larvae Feeding on Different Host Plants[a]

Foodplant	Specific activity (nmole/min/mg protein)[b]	
	N-Demethylase	Epoxidase
Lima bean	0.29 ± 0.03	0.26 ± 0.02
Carrot	2.73 ± 0.55	1.08 ± 0.14
Parsley	1.24 ± 0.16	0.10 ± 0.02
Tomato	0.46 ± 0.06	0.20 ± 0.02

[a]Data from Brattsten (1979b).
[b]Activity measured in crude midgut homogenates of sixth-instar larvae feeding for 24 hours on plants indicated.

Synchronization with Development

The strict conservatism of animals with regard to energy expenditure on the mixed-function oxidases is further shown by the fact that the enzymes are synthesized and maintained only during those periods when they are required. In insects it is interesting to note that mixed-function oxidase activity is found only in those developmental stages that are actively involved in feeding (Wilkinson and Brattsten, 1972). This is particularly clear in lepidopterous larvae such as the southern armyworm, where within a few hours of finishing feeding in preparation for pupation, oxidase activity in the gut virtually disappears (Krieger and Wilkinson, 1969). It is further shown in the absence of activity during larval and nymphal molts of various species, and by the general absence of activity in pupae and in some adult insects that do not feed. In all cases, the presence of the enzymes in the tissue is synchronized with periods of maximum foreign compound exposure.

Most of our current pest control effects are directed at the insect stages responsible for maximum crop damage (e.g., lepidopterous larvae) but unfortunately these are the stages likely to be most tolerant to insecticides; they also constitute the stages at which the insect is best prepared to develop resistance through selection for high oxidase activity. Perhaps more effort should be made to develop pest control practices that are directed at the non-feeding, developmental stages in which the insect is likely to be most vulnerable to insecticidal action.

COUNTERMEASURES FOR RESISTANCE

Having established the importance of the mixed-function oxidases in insect resistance, and having considered some of the characteristics of the enzyme system that make it such an effective protective mechanism for insects and other organisms, it is now necessary to address the question of what, if any, measures exist to counter this class of insect resistance.

The design of new types of insecticides not susceptible to oxidase attack would clearly be one approach. However, in view of the tremendous versatility of the enzymes, this is not a very realistic goal and, indeed, there are only very few compounds, such as various perfluoro derivatives, known to resist oxidative metabolism. Furthermore, if we were to be successful in developing such compounds, they would undoubtedly pose severe environmental problems, since in addition to being resistant to the insect mixed-function oxidases, they would probably also prove resistant to oxidative degradation by man and other species. One quite attractive possibility is to design new groups of compounds that are activated but not degraded by mixed-function oxidases. In theory such so-called "negatively correlated" insecticides (Casida, 1969) would prove most effective against strains of insects resistant through enhanced mixed-function oxidase activity, although again they might be hazardous to man and other non-target species.

Since we now know of many types of compounds that inhibit the mixed-function oxidases in vitro, and that, as a consequence, are effective in vivo synergists for several insecticides towards both S and R strains of insects, the use of synergized insecticide formulations has frequently been advocated as a possible means of counteracting insect resistance (Wilkinson, 1968, 1971, 1976; Oppenoorth, 1971, 1976; Oppenoorth and Welling, 1976). It is also well known that the development of carbamate resistance is often greatly reduced, or almost completely eliminated, if insects are selected with carbamate-synergist combinations (Moorefield, 1960; Georghiou et al., 1961; Georghiou, 1962). Thus, Moorefield (1960) found that house flies selected with a combination of carbaryl and PB exhibited only a five-fold increase in resistance to the combination over 50 generations of selection (Table 7), and Georghiou (1962) showed a similar effect using flies selected with a combination of *m*-isopropylphenyl *N*-methylcarbamate and PB. It appears, however, that the situation with the carbamates may prove to be the exception rather than the rule and might result from the fact that metabolism occurs almost exclusively via mixed-function oxidation. In the case of insecticides that can be metabolized by several alternative routes, selection with a combination of the insecticide and a synergist blocking one metabolic pathway simply selects for resistance by another; the result is often a rapid build-up of resistance to the synergized

Table 7. Development of House Fly Resistance to Carbaryl and Carbaryl Plus Piperonyl Butoxide[a]

Generations of selection	μg/fly carbaryl for 70% mortality	
	Carbaryl alone	Carbaryl plus piperonyl butoxide[b]
0	2.7	0.12
2	5.0	0.13
4	15.0	0.16
6	40.0	0.19
8	70.0[c]	0.22
10	70.0	0.24
20	70.0	0.38
30	--	0.48
40	--	0.50
50	--	0.50

[a]Data from Moorefield (1960).
[b]Carbaryl:piperonyl butoxide ratio, 1:5.
[c]Limit of solubility.

mixture. This has been demonstrated in the development of house fly resistance to DDT/WARF-anti-resistant mixtures (Brown and Rogers, 1950) and more recently in studies of the resistance of *Culex pipiens fatigans* to organophosphorus insecticides (Ranasinghe and Georghiou, 1980). In the latter case, selection of mosquito larvae with either temephos alone or in combination with the synergists DEF or PB gave strains with different resistance characteristics and indicated that, although the synergist prevented enhancement of those detoxication mechanisms it normally inhibits, it allowed selection by the alternative detoxication pathway. Consequently, in a population of insects capable of metabolizing an insecticide by several different enzymatic pathways, the use of synergized combinations will serve only to offset resistance temporarily and ultimately may lead to a much more serious resistance problem.

CONCLUSION

In conclusion, the future problem of insect resistance involving mixed-function oxidation must be viewed with a considerable degree of pessimism. There are no simple answers with respect to how to avoid its development or how to overcome it once it has developed. We must remember that we are attempting to combat a biochemical, protective mechanism that has been continuously and severely tested throughout the course of evolutionary history. The fact that insects continue to thrive in a host of most unlikely (chemically unfriendly) ecological niches is itself witness to the effective protection it has provided. It is almost certain that insect resistance of this type will remain a major problem in insect control as long as we continue to use insecticide chemicals in the way they have been used until now.

REFERENCES

Agosin, M., and Perry, A. S., 1974, Microsomal mixed-function oxidases, *in:* "The Physiology of Insecta," Vol. V., M. Rockstein, ed., pp. 537-596, Academic Press, New York.

Agosin, M., Michaeli, D., Miskus, R., Nagasawa, S., and Hoskins, W. M., 1961, A new DDT-metabolizing enzyme in the German cockroach, *J. Econ. Entomol.*, 54:340.

Bend, J. R., and Hook, G. E. R., 1977, "Handbook of Physiology," Section 9, pp. 419-440, American Physiological Society, Washington, D.C.

Bend, J. R., and James, M. O., 1978, Xenobiotic metabolism in marine and freshwater species, *in:* "Tiochemical and Biophysical Perspectives in Marine Biology," D. C. Malins and J. R. Sargent, eds., Vol. 4., pp. 125-188, Academic Press, New York.

Benezet, H. J., Chang, K. M., and Knowles, C. O., 1978, Formamidine pesticides — metabolic aspects, *in:* "Pesticide and Venom Neurotoxicity," D. L. Shankland, R. M. Hollingworth and T. Smyth, Jr., eds., pp. 189-206, Plenum Press, New York.

Benke, G. M., and Wilkinson, C. F., 1971, In vitro mecrosomal epoxidase activity and susceptibility to carbaryl and carbaryl-piperonyl butoxide combinations in house crickets of different age and sex, *J. Econ. Entomol.*, 64:1032.

Bigley, W. S., and Vinson, S. B., 1979, Degradation of (^{14}C)methoprene in the imported fire ant, *Solenopsis invicta, Pestic. Biochem. Physiol.*, 10:1.

Brattsten, L. B., 1979a, Biochemical defense mechanisms in herbivores against plant allelochemicals, *in:* "Herbivores: Their Interactions with Secondary Plant Metabolites," G.A. Rosenthal and D. H. Janzen, eds., pp. 199-270, Academic Press, New York.

Brattsten, L. B., 1979b, Ecological significance of mixed-function oxidation, *Drug Metab. Revs.*, 10:35.

Brattsten, L. B., and Metcalf, R. L., 1970, The synergistic ratio of carbaryl and piperonyl butoxide as an indicator of the

distribution of multifunction oxidases in the Insects, *J. Econ. Entomol.*, 63:101.
Brattsten, L. B., and Wilkinson, C. F., 1973, Induction of microsomal enzymes in the southern armyworm (*Prodenia eridania*), *Pestic. Biochem. Physiol.*, 3:393.
Brattsten, L. B., and Wilkinson, C. F., 1977, Insecticide solvents: Interference with insecticidal action, *Science*, 196:1211.
Brattsten, L. B., Wilkinson, C. F., and Eisner, T., 1977, Herbivore plant interactions: Mixed-function oxidases and secondary plant substances, *Science*, 196:1349.
Brooks, G. T., 1972, Pathways of enzymatic degradation of pesticides, *in:* "Environmental Quality and Safety," F. Coulston and F. Korte, eds., pp. 106-164, Academic Press, New York.
Brooks, G. T., 1974, "Chlorinated Insecticides," Vol. II, CRC Press, Cleveland.
Brown, H. D., and Rogers, E. F., 1950, The insecticidal activity of 1,1-dianisyl neopentane, *J. Am. Chem. Soc.*, 72:1864.
Burt, M. E., Kuhr, R. J., and Bowers, W. S., 1978, Metabolism of precocene II in the cabbage looper and European corn borer, *Pestic. Biochem. Physiol.*, 9:300.
Casida, J. E., 1969, Insect microsomes and insecticide chemical oxidations, *in:* "Microsomes and Drug Oxidations," J. R. Gillette et al., eds., pp. 517-531, Academic Press, New York.
Casida, J. E., 1970, Mixed function oxidase involvement in the biochemistry of insecticide synergists, *J. Agr. Food Chem.*, 18:753.
Conney, A. H., 1967, Pharmacolgical implications of microsomal enzyme induction, *Pharmacol. Rev.*, 19:317.
Coon, M. J., Vermillion, J. L., Vatsis, K. P., French, J. S., Dean, W. L., and Haugen, D. A., 1977, Biochemical studies on drug metabolism: Isolation of multiple forms of liver microsomal cytochrome P-450, *in:* "Drug Metabolism Concepts," D. M. Jerina, ed., pp. 46-71, A.C.S. Symposium Series No. 44, Washington, D.C.
Eldefrawi, M. E., Miskus, R., and Sutcher, V., 1960, Methylenedioxyphenyl derivatives as synergists for carbamate insecticides on susceptible, DDT- and parathion-resistant house flies, *J. Econ. Entomol.*, 53:231.
Elliott, M., Janes, N. F., Kimmel, E. C., and Casida, J. E., 1972, Metabolic fate of pyrethrin I, pyrethrin II and allethrin administered orally to rats, *J. Agr. Food Chem.*, 20:300.
Estabrook, R. W., Wrringloer, J., and Peterson, J. A., 1979, The use of animal subcellular fractions to study type I metabolism of xenobiotics, *in:* "Xenobiotic Metabolism: In Vitro Methods," G. D. Paulson, D. S. Frear and E. P. Marks, eds., pp. 149-179, A.C.S. Symposium Series No. 97, Washington, D.C.
Farnham, A. W., 1973, Genetics of resistance of pyrethroid-selected houseflies, *Musca domestica* L., *Pestic. Sci.*, 4:513.
Fukami, J., Shishido, T., Fukunaga, K., and Casida, J. E., 1969, Oxidative metabolism of rotenone in mammals, fish and insects and its relation to selective toxicity, *J. Agr. Food Chem.*, 17:1217.

Georghiou, G. P., 1962, Carbamate insecticides: The cross-resistance spectra of four carbamate-resistant strains of the house fly after protracted selection pressure, *J. Econ. Entomol.*, 55:494.

Georghiou, G. P., and Metcalf, R. L., 1961, The absorption and metabolism of 3-isopropylphenyl *N*-methylcarbamate by susceptible and carbamate-selected strains of houseflies, *J. Econ. Entomol.*, 54:231.

Georghiou, G. P., Metcalf, R. L., and March, R. B., 1961, The development and characterization of resistance to carbamate insecticides in the housefly, *Musca domestica, J. Econ. Entomol.*, 54:132.

Georghiou, G. P., and C. E. Taylor, 1976, Pesticide resistance as an evolutionary phenomenon, Proc. XV Int. Congr. Entomol., pp. 759-785.

Gordon, H. T., 1961, Nutritional factors in insect resistance to insecticides, *Ann. Rev. Entomol.*, 6:27.

Gunsalus, I. C., 1972, Early reactions in the degradation of camphor: P-450 hydroxylase, *in:* "Degradation of Synthetic Organic Molecules in the Biosphere," pp. 137-145, National Academy of Sciences, Washington, D.C.

Hammock, B. D., Mumby, S. M., and Lee, P. W., 1977, Mechanisms of resistance to the juvenoid methoprene in the housefly, *Musca domestica* L., *Pestic. Biochem. Physiol.*, 7:261.

Hodgson, E., 1976, Cytochrome P-450 interactions, *in:* "Insecticide Biochemistry and Physiology," C. F. Wilkinson, ed., pp. 115-148, Plenum Press, New York.

Hodgson, E., 1980, This volume.

Hodgson, E., and Casida, J. E., 1960, Biological oxidation of *N,N*-dialkyl carbamates, *Biochem. Biophys. Acta*, 43:184.

Hodgson, E., and Philpot, R. M., 1974, Interaction of methylenedioxyphenyl (1,3-benzodioxole) compounds with enzymes and their effect, in vivo, on animals, *Drug Metab. Revs.*, 3:323.

Hodgson, E., and Plapp. F. W., Jr., 1970, Biochemical Characteristics of insect microsomes, *J. Agr. Food Chem.*, 18:1048.

Kawalek, J. C., and Lu, A. Y. H., 1975, Reconstituted liver microsomal enzyme system that hydroxylates drugs, other foreign compounds and endogenous substrates, *Mol. Pharmacol.*, 11:201.

Khan, M. A. Q., and Bederka, J. P., Jr., (eds.), 1974, "Survival in Toxic Environments," Academic Press, New York, 553 pp.

Krieger, R. I., and Wilkinson, C. F., 1969, Microsomal mixed-function oxidases in insects. I. Localization and properties of an enzyme system effecting aldrin epoxidation in larvae of the southern armyworm (*Prodenia eridania*), *Biochem. Pharmacol.*, 18:1403.

Krieger, R. I., Feeny, P. P., and Wilkinson, C. F., 1971, Detoxication in the guts of caterpillars: An evolutionary answer to plant defenses? *Science*, 172:579.

Kuhr, R. J., 1971, Comparative metabolism of carbaryl by resistant and susceptible strains of the cabbage looper, *J. Econ. Entomol.*, 64:1373.

Lamoureux, G. L., and Frear, D. S., 1979, Pesticide metabolism in higher plants: In vitro enzyme studies, *in:* "Xenobiotic Metabolism: In Vitro Methods," G. D. Paulson, D. S. Frear and E. P. Marks, eds., pp. 77-128, A.C.S. Sy-posium Series No. 97, Washington, D.C.

Lu, A. Y. H., and Levin, W., 1974, The resolution and reconstitution of the liver microsomal hydroxylation system, *Biochim., Biophys. Acta,* 344:205.

Lu, A. Y. H., Ryan, D., Kawalek, J., Thomas, P., West, S. B., Huang, M. T., and Levin, W., 1976, Multiplicity of liver microsomal cytochrome P-450: Separation, purification and characterization, *Biochem. Soc. (London) Trans.,* 4:169.

Machinist, J. M., Dehner, E. W., and Ziegler, D. M., 1968, Microsomal oxidases. III. Comparison of species and organ distribution of dialkylarylamine *N*-oxide dealkylase and dialkylamine *N*-oxidase, *Arch. Biochem. Biophys.,* 125:854.

Mannering, G. J., 1971, Microsomal enzyme systems which catalyze drug metabolism, *in:* "Fundamentals of Drug Metabolism and Drug Disposition," B. N. LaDu, H. G. Mandel and E. L. Way, eds., pp. 206-252, Williams and Wilkins, Baltimore.

Metcalf, R. L., 1976, Organochlorine insecticides, survey and prospects, *in:* "Insecticides for the Future: Needs and Prospects," R. L. Metcalf and J. J. McKelvey, Jr., eds., pp. 223-285, John Wiley and Sons, New York.

Metcalf, R. L., and Fukuto, T. R., 1965, Carbamate insecticides: Effect of chemical structure on intoxication and detoxication of phenyl *N*-methyl-carbamates in insects, *J. Agr. Food Chem.,* 13:220.

Metcalf, R. L., Kapoor, I. P., and Hirwe, A. S., 1971, Biodegradable analogues of DDT, *Bull. WHO,* 44:363.

Moorefield, H. H., 1960, Resistance of carbamate insecticides, *Misc. Publ. Entomol. Soc. Am.,* 2:151.

Nakatsugawa, T., and Morelli, M. A., 1976, Microsomal oxidation and insecticide metabolism, *in:* "Insecticide Biochemistry and Physiology," C. F. Wilkinson, ed., pp. 61-114, Plenum Press, New York.

Nelson, P. A., Stewart, R. R., Morelli, M. A., and Nakatsugawa, T., 1976, Aldrin spoxidation in the earthworm, *Lumbricus terrestris* L., *Pestic. Biochem. Physiol.,* 6:243.

Neuhauser, E., and Hartenstein, R., 1976, On the presence of *O*-demethylase activity in invertebrates, *Comp. Biochem. Physiol.,* 53C:37.

Oppenoorth, F. J., 1965, DDT-resistance in the housefly dependent on different mechanisms and the action of synergists, *Mededeel, Landbouwhogeschool Opzoekingsstat. Gent.,* 30:1390.

Oppenoorth, F. J., 1971, Resistance in insects: The role of metabolism and the possible use of synergists, *Bull. WHO,* 44:195.

Oppenoorth, F. J., 1976, Development of resista-ce to insecticides, *in:* "Insecticides for the Future: Needs and Prospects," R. L. Metcalf and J. J. McKelvey, Jr., eds., pp. 41-59, John Wiley

and Sons, New York.
Oppenoorth, F. J., and Houx, N. W. H., 1968, DDT resistance in the house fly caused by microsomal degradation, *Entomol. Exp. Appl.*, 11:81.
Oppenoorth, F. J., and Welling, W., 1976, Biochemistry and Physiology of resistance, *in:* "Insecticide Biochemistry and Physiology," C. F. Wilkinson, ed., pp. 507-551, Plenum Press, New York.
Pan, H. P., Hook, G. E. R., and Fouts, J. R., 1975, The liver parenchyma and foreign compound metabolism in red-winged blackbird compared with rat, *Xenobiotica*, 5:17.
Papadopoulos, N. M., and Kintzios, J. A., 1963, Formation of metabolites from nicotine by a rabbit liver preparation, *J. Pharmacol. Exp. Ther.*, 140:269.
Perry, A. S., and Agosin, M., 1974, The physiology of insecticide resistance by insects, *in:* "The Physiology of Insecta," Vol. VI, M. Rockstein, ed., pp. 3-124, Academic Press, New York.
Pimprikar, G. D., and Georghiou, G. P., 1979, Mechanisms of resistance to diflubenzuron in the house fly, *Musca domestica* (L.), *Pestic. Biochem. Physiol.*, 12:10.
Plapp, F. W., Jr., and Casida, J. E., 1969, Genetic control of house fly NADPH-dependent oxidases: Relation to insecticide chemical metabolism and resistance, *J. Econ. Entomol.*, 62:1174.
Ranasingh, L. E., and Georghiou, G. P., 1980, Comparative modifications of insecticide-resistance spectrum of *Culex pipiens fatigans* Wied. by selection with temephos and temephos/synergist combinations, *Pestic. Sci.*, submitted.
Sacher, R. M., Metcalf, R. L., and Fukuto, R. R., 1968, Propynyl naphthyl ethers as selective cambamate synergists, *J. Agr. Food Chem.*, 16:779.
Sawicki, R. M., 1973, Recent advances in the study of the genetics of resistance in the housefly, *Musca domestica*, *Pestic. Sci.*, 4:501.
Schonbrod, R. D., Philleo, W. W., and Terriere, L. C., 1965, Hydroxylation as a factor in resistance in houseflies and blow flies, *J. Econ. Entomol.*, 58:74.
Schonbrod, R. D., Khan, M. A. Q., Terriere, L. C., and Plapp, F. W., Jr., 1968, Microsomal oxidases in the housefly: A survey of fourteen strains, *Life Sci.*, 7:681.
Shono, T., Unai, T., and Casida, J. E., 1978, Metabolism of permethrin isomers in American cockroach adults, housefly adults and cabbage looper larvae, *Pestic. Biochem. Physiol.*, 9:96.
Shrivastava, S. P., Tsukamoto, M., and Casida, J. E., 1969, Oxidative metabolism of C^{14}-labeled Baygon by living houseflies and by housefly enzyme preparations, *J. Econ. Entomol.*, 62:483.
Shrivastava, S. P., Georghiou, G. P., Metcalf, R. L., and Fukuto, T. R., 1970, Carbamate resistance in mosquitoes: The metabolism of propoxur by susceptible and resistant larvae of *Culex pipiens fatigans*, *Bull. WHO*, 42:931.
Slade, M., and Zibitt, C. H., 1972, Metabolism of Cecropia juvenile

hormone in insects and in mammals, *in:* "Insect Juvenile Hormones: Chemistry and Action," J. J. Menn and M. Beroza, eds., pp. 155-176, Academic Press, New York.

Stanton, R. H., Plapp. F. W., Jr., White, R. A., and Agosin, M., 1978, Induction of multiple cytochrime P-450 species in house fly microsomes — SDS gel electrophoresis studies, *Comp. Biochem. Physiol.,* 61B:297.

Sun, Y. P., and Johnson, E. R., 1960, Synergistic and antagonistic actions of insecticide-synergist combinations and their mode of action, *J. Agr. Food Chem.,* 8:261.

Swingle, M. C., 1939, The effect of previous diet on the toxic action of lead arsenate to a leaf-feeding insect, *J. Econ. Entomol.,* 32:884.

Tanaka, K., Kurihara, N., and Nakajima, M., 1979, Oxidative metabolism of lindane and its isomers with microsomes from rat liver and housefly abdomen, *Pestic. Biochem. Physiol.,* 10:96.

Terriere, L. C., 1980, This volume.

Terriere, L. C., Yu, S. J., and Hoyer, R. F., 1971, Induction of microsomal oxidase in F_1 hybrids of a high and a low oxidase housefly strain, *Science,* 171:581.

Testa, B., and Jenner, P., 1976, "Drub Metabolism: Chemical and Biochemical Aspects," Dekker, New York, 500 pp.

Tsukamoto, M., and Casida, J. E., 1967, Metabolism of methylcarbamate insecticides by the $NADPH_2$-requiring enzyme system from houseflies, *Nature,* 213:49.

Wickramasinghe, R. H., and Villee, C. A., 1975, Early role during chemical evolution for cytochrome P-450 in oxygen detoxification, *Nature,* 256:509.

Wilkinson, C. F., 1968, The role of insecticide synergists in resistance problems, *Wrld. Rev. Pest Control,* 7:155.

Wilkinson, C. F., 1971, Effects of synergists on the metabolism and toxicity of anticholinesterases, *Bull. WHO,* 44:171.

Wilkinson, C. F., 1976, Insecticide synergism, *in:* "Insecticides for the Future: Needs and Prospects," R. L. Metcalf and J. J. McKelvey, Jr., eds., pp. 195-218, John Wiley and Sons, New York.

Wilkinson, C. F., 1979, The use of insect subcellular components for studying the metabolism of Xenobiotics, *in:* "Xenobiotic Metabolism: In Vitro Methods," G. D. Paulson, D. S. Frear and E. P. Marks, eds., pp. 249-284, A.C.S. Symposium Series No. 97, American Chemical Society, Washington, D.C.

Wilkinson, C. F., 1980, The metabolism of xenobiotics: A study in biochemical evolution, *in:* "The Scientific Basis of Toxicity Assessment," H. R. Witschi, ed., pp. 251-268, Elsevier, North Holland.

Wilkinson, C. F., and Brattsten, L. B., 1972, Microsomal drug metabolizing enzymes in insects, *Drug Metab. Revs.,* 1:153.

Williamson, R. L., and Schecter, M. S., 1970, Microsomal epoxidation of aldrin in lepidopterous larvae, *Biochem. Pharmacol.,* 19:1719.

Yamamoto, I., Kimmel, E. C., and Casida, J. E., 1969, Oxidative metabolism of pyrethroids in houseflies, *J. Agr. Food Chem.*, 17:1227.

Yu, S. J., and Terriere, L. C., 1971, Induction of microsomal oxidases in the housefly and the action of inhibitors and stress factors, *Pestic. Biochem. Physiol.*, 1:173.

Yu, S. J., and Terriere, L. C., 1972, Enzyme induction in the housefly: The specificity of the cyclodiene insecticides, *Pestic. Biochem. Physiol.*, 2:184.

CHARACTERIZATION OF CYTOCHROME P-450

IN STUDIES OF INSECTICIDE RESISTANCE

Ernest Hodgson and Arun P. Kulkarni

Department of Entomology
North Carolina State University
Raleigh, North Carolina 27650

INTRODUCTION

The importance of cytochrome P-450 (P-450*)-dependent monooxygenase systems in the metabolism of xenobiotics, including insecticides, is well established and has been frequently reviewed (e.g., Estabrook et al., 1973; Hodgson and Tate, 1976; Hodgson and Dauterman, 1980; Kulkarni and Hodgson, 1980a; Nakatsugawa and Morelli, 1976; Ullrich, 1977). Although reviews are often restricted to investigations carried out on mammals, such systems have been described in many insects, and their importance is widely recognized (for references see reviews by Agosin and Perry, 1974; Agosin, 1976; Hodgson, 1976; Kulkarni and Hodgson, 1976d; Wilkinson and Brattsten, 1972). In one or two species of insect, such as the house fly, *Musca domestica*, and the southern armyworm, *Spodoptera eridania*, P-450 and its related enzymes have been described in some detail.

Since these foreign compounds enter the insect body by virtue of their lipophilicity, and since the lack of specificity of the microsomal P-450-dependent monooxygenase system is such that many lipophilic substrates are attacked, it is not surprising that this system is of primary importance for xenobiotic metabolism in insects. Since increased metabolism of the toxicant has long been known to be involved in resistance, it follows that this system would also be involved. While this should not be construed as decreasing the

*The following abbreviations are used: P-450 for cytochrome P-450; R for resistant; S for susceptible; EtNC for ethyl isocyanide; P-420 for cytochrome P-420.

Thus, for studies of resistance due to oxidative metabolism (essentially the comparative biochemistry of P-450 in closely related strains), we must not only exercise careful technique, but we must also question the underlying assumptions on which our techniques are based.

PREPARATIVE TECHNIQUES

Much has already been written on techniques for the preparation of insect microsomes (Hanson and Hodgson, 1971; Kulkarni and Hodgson, 1975), and this will not be pursued further except to stress that there are many variables that may affect the ultimate oxidative activity and the stability of P-450. These include not only the chemical nature and pH of the buffer, but also its ionic strength, as well as the temperature at which all operations are carried out and the nature of the homogenization technique.

By way of summary, the following points are worthy of emphasis:

1. Variables such as the above should be examined with each newly investigated species and each new strain, both susceptible and resistant.

2. These variables should also be examined in each organ and subcellular fraction in which P-450 is to be investigated.

3. The optimum conditions for homogenization may not be the same as those for resuspension and subsequent determination of enzyme activity.

4. The effect and optimum concentration of any co-factors or protectants should likewise be determined for each species, strain, organ and subcellular organelle.

Less frequently stressed is the need to observe particular care in the selection of living material. In this regard, we may stress the following:

1. Field-caught material is seldom useful. Induction, particularly in polyphagous lepidopterous larvae, is both rapid and related to feeding (Brattsten et al., 1977), and thus the levels of monooxygenase activity and P-450 are subject to unknown variables. Furthermore, the effects of disease and physiological stress are largely unknown, and it is almost impossible to determine the age of such material, relative to the previous molt, with any accuracy.

2. Since the monooxygenase activity and P-450 levels are usually close to zero at the molt, rise to a maximum during the intermolt period, and fall to a low level at the next molt

importance of cuticular characteristics affecting penetration, altered target sites (Oppenoorth and Welling, 1976), or other xenobiotic-metabolizing enzymes such as the glutathione S-transferases (Motoyama and Dauterman, 1980), it is clear that microsomal monoxygenase systems are, in many cases, the single most important factor in insecticide resistance. Studies of the mechanism of this system (Gillette et al., 1972; Estabrook et al., 1973; Estabrook and Werringloer, 1979) make it clear that although reduced pyridine nucleotides and flavoprotein reductases are necessary for the supply of electrons, it is P-450, the site of interaction with xenobiotic substrates, that is the key component.

This brief review is a consideration of the proper characterization of P-450 and, to a lesser extent, the monooxygenase system dependent on it, in the hope that the results obtained from such characterization studies may increase our knowledge of the mechanisms involved in resistance to insecticides.

P-450 possesses a number of features that complicate investigations and make analytical errors difficult to avoid. These include the following:

1. The membrane-bound location of P-450 complicates enzyme preparation as well as subsequent spectral and kinetic measurements.

2. Foreign compounds not only serve as substrates but can also serve as inducers and/or inhibitors, frequently performing more than one of these roles, either simultaneously or in sequence.

3. Variations are known to exist in insects, as in mammals, between species, strains, organs, and subcellular organelles (Hodgson, 1974; Smith et al., 1979; Wilkinson and Brattsten, 1972).

4. Primarily from studies of mammalian liver, we know that multiple forms of the P-450, which differ in both their physical properties and in their substrate specificity, exist within the same subcellular organelle of the same cell type (Guengerich, 1979; Johnson, 1979). It is now becoming abundantly clear that similar multiple forms exist in insects and, moreover, that they vary between resistant (R) and susceptible (S) strains. A suggestion, made somewhat casually at an AAAS symposium in early 1979 (Hodgson, 1979), that P-450 might have a polyphyletic origin is already receiving some supporting evidence from studies of multiplicity. On the basis of partial sequences, Levin and co-workers (1979) have shown little or no homology among several purified P-450s from mammalian liver microsomes (Levin et al., 1979).

5. Last, but by no means least, insect P-450 appears to be more labile than that of mammals.

(Wilkinson and Brattsten, 1972), it is essential that the age of the insects, relative to the previous molt, be accurately known. In our recent studies on *Heliothis virescens* larvae (Gould and Hodgson, 1980) we found it necessary to use insects that had molted within four hours of each other.

3. Whole insect homogenates are rarely suitable for examination of oxidase activity due to the probability of mixing endogenous inhibitors from one organ and/or organelle with enzymes from another. Endogenous inhibitors of at least three different types have been described from insects (Schonbrod and Terriere, 1971; Wilson and Hodgson, 1972; Krieger and Wilkinson, 1969; Orrenius et al., 1971; Gilbert and Wilkinson, 1975) and, at the very least, their presence should be suspected. There are several very simple tests that should be applied to any previously uninvestigated enzyme preparation: First, does dialysis increase activity; second, does the 100,000 x g supernatant inhibit monooxygenase activity in microsomes from other species, and third, when a common enzyme activity (e.g., O-demethylation of *p*-nitroanisole) is measured in a mixed preparation (i.e., the microsomes from the test species in question plus those from rat or mouse liver), is the activity less than additive. Although this has not yet been intensively investigated, the presence of endogenous ligands would have similar complicating effects on spectral studies of insect P-450.

It should be noted that this problem is not restricted to monooxygenase enzymes since we have recently demonstrated that quinones, of wide occurrence in insects, are inhibitors of glutathione S-transferases (Kulkarni et al., 1978; Motoyama et al., 1978).

PURIFICATION VERSUS MICROSOMAL PREPARATIONS: THE PROBLEM POSED BY MULTIPLICITY

Assets and Liabilities

In studies of resistance, as in any physiological or biochemical study, the ultimate goal of experiments conducted in vitro is to provide information by which we can more realistically interpret events in vivo. While the need to solubilize and purify P-450 from insects was recognized early (Hodgson et al., 1974; Capdevila et al., 1974), the results to date have been of limited utility. Although they do indicate multiplicity in a variety of house fly strains (Agosin, 1976; Capdevila et al., 1975; Schonbrod and Terriere, 1975), the low yields and low purities achieved lessen their utility as tools for the interpretation of events in vivo.

For example, the most comprehensive purification studies on an insect reported to date (Agosin, 1976; Capdevila et al., 1975) were carried out on *Musca domestica* and yielded two P-450 fractions, one with a specific content of 0.83 nmoles/mg protein, the other 0.123 nmoles/mg protein. A simple calculation based on the assumption that the molecular weight is c. 50,000 daltons indicates that these are 4.15% and 0.615% pure, respectively. This compares with a specific content of 0.265 nmoles/mg protein (1.325% pure) for the original microsomes. Later studies by the same authors (Capdevila and Agosin, 1977), while indicating considerably more purification, are not reported in enough detail to permit proper comparison. The only other purification study of note (Schonbrod and Terriere, 1975), while again showing multiple forms of P-450 in a strain of *Musca domestica*, did not supply the data necessary to calculate specific contents. Although these studies are important, until purification procedures can be improved dramatically both in yield and purity, comparisons between R and S strains of insects must be made on microsomal preparations and thus the problem of multiplicity must be addressed if such comparisons are to be valid. We have examined multiplicity of P-450 in microsomal membranes by three techniques and a fourth has been used by others.

Off-balance Spectra

The principle of this method is the comparison by difference spectroscopy of microsomes from different sources, such as R and S strains of an insect species or microsomes from a specific organ of induced and uninduced mammals, after suitable dilutions to equalize the amount of cytochrome b_5. In comparisons of microsomes from S and R strains of the house fly, we modified the procedure to take into account all dithionite-reduceable components of the microsomes, not just cytochrome b_5. In a typical experiment (Kulkarni and Hodgson, 1980b), dithionite-reduced minus oxidized spectra of the microsomal samples to be compared were first recorded and the ΔAs thus determined were used to dilute the appropriate oxidized samples in such a way that, if reduced, they would have been of equal absorbance. The apparent absolute absorption spectrum of the oxidized P-450 was then recorded. Both samples were subsequently reduced to check the accuracy of the dilutions.

The microsomes from all possible pairs of the following strains were compared: Rutgers (Diazinon-R); Fc; Baygon; CSMA; SBO (stubby-winged, brown-bodied, ocra-eyed, susceptible). When microsomes from different strains are compared in this way, one of several results is possible. A flat baseline results if both strains have the same amount of qualitatively similar P-450s. However, if one strain or the other has an excess of a spectrally dissimilar P-450, or class of P-450s, a peak or trough appears at the wavelength at which the cytochromes differ from each other.

It was apparent that at least two P-450s, or classes of P-450, were present in all of these strains, one causing a peak at about

394 nm and the other at about 412 nm. In general, the 394 nm type of P-450 predominated in the R strains and the 412 nm type in the S.

Tryptic Digestion

Since different forms of P-450 presumably differ in their protein moieties and in their topology, the rate at which they are degraded by proteolytic enzymes offers a potential for their characterization. In experiments to test this possibility, microsomal suspensions from R and S house flies were digested with trypsin at 37°C for varying time periods (Kulkarni and Hodgson, 1980b). Digestion was stopped by the addition of soybean trypsin inhibitor and the spectral characteristics determined. Thus, it was possible to obtain a time course for each spectral characteristic as digestion proceeded.

In a number of parameters, changes were observed that varied from one strain to another. For example, type I binding in microsomes from Fc flies was abolished much more rapidly than was the P-450, while in microsomes from Rutgers flies, these two characteristics were abolished at about the same rate. Thus, it appeared that the P-450 of Fc flies that showed marked type I binding was either more labile or more accessible than that of the Rutgers strain.

In general, the results from measurements of CO, ethyl isocyanide (EtNC), n-octylamine, and benzimidazole binding supported the conclusion that the ligand binding spectra characteristic of undigested microsomes from R flies were gradually lost as digestion proceeded, and the surviving P-450 species exhibited features that increasingly resembled those of S flies. The data obtained suggest that not only are there multiple forms of P-450 in these strains, but that they are not randomly arranged in the microsomal vesicles.

Density Gradient Fractionation

The two-step discontinuous gradient centrifugation procedure developed by Dallner (1963) for the separation of rough and smooth microsomes from mammalian liver has been modified in our laboratory to permit microsomes of several different densities to be separated and their spectral characteristics determined. This method has proven valuable in studies of the multiplicity of P-450 in rodent liver (Mailman et al., 1975) and has enabled us to demonstrate that the multiple forms of P-450 characteristic of this organ are not randomly distributed but that different cytochromes are characteristic of microsomes of different density.

Similar experiments have been carried out on microsomes from R (Fc) and S (CSMA) house flies (Kulkarni and Hodgson, 1980b). In these experiments, microsomal suspensions in 0.25 M sucrose containing 15 mM cesium chloride were gently layered onto a lower layer of 1.0 to 1.5 M sucrose containing 15 mM cesium chloride. After centrifugation at 104,000 x g for 140 min, a heavy fraction was found at the bottom of the centrifuge tube and a light fraction at the interphase between the two layers. These fractions, whose relative densities varied as the density of the lower layer was varied, can be collected and resuspended for spectral examination.

Variations in the λ_{max} of the CO binding spectra, in the ratio of type I to type II binding and in the nature of the n-octylamine, benzimidazole and pyridine spectra, were seen between subfractions of different density in both strains. This indicates that in the house fly, as in mammalian liver, not only is there multiplicity of P-450, but its distribution within the endoplasmic reticulum is not random, since microsomes of different densities differ in their spectral characteristics. In addition, variations were seen between fractions of the same density from different strains, indicating that the multiplicity seen in one strain did not resemble that of the other. The typical spectral characteristics described previously for the microsomes of S flies (Philpot and Hodgson, 1971; Tate et al., 1973, 1974) were generally retained in the light subfractions. Thus, with few exceptions, the light fractions showed absorption maxima in the pyridine, benzimidazole and n-octylamine difference spectra at longer wavelengths than did the heavy fractions. Similarly, the 390 nm trough of the benzimidazole spectrum was not seen using light fractions. In contrast, the results obtained with microsomes of R flies suggest that, in general, the characteristics typical of these strains were retained in the heavy fractions.

SDS-Polyacrylamide Gel Electrophoresis

This technique, while potentially of considerable utility, is frequently used in an uncritical manner. The principle of the method is the electrophoretic separation, according to size, of completely denatured polypeptides invested with sodium dodecyl sulfate. If the polypeptide is not completely denatured, the relationship between electrophoretic mobility and molecular weight is lost. If the protein is completely denatured, the heme of P-450, which is not covalently bound, is lost and heme staining of gels, which depends on residual heme, cannot be used. An extensive investigation of this methodology, using mouse liver microsomes (Bell and Hodgson, 1977a,b), showed that rigorous pretreatment completely abolished heme staining.

The problems may be summarized as follows:

1. Lack of adequate pretreatment.
2. Uncritical use of heme staining techniques.

3. Failure to recognize that enzymes other than cytochrome, but of similar molecular weight, may also be induced.
4. Lack of adequate standards. Partially purified cytochrome P-450 preparations are inadequate unless essentially pure.
5. Quantitation, by densitometry, of extensively overlapping bands.

Much of the work carried out on mammalian microsomes is flawed by one or more of these problems, and similar problems were encountered in the only comprehensive study of insect microsomes that has been reported (Stanton et al., 1978).

SPECTRAL CHARACTERIZATION

Here we are confronted with the most useful technique to date for the characterization of P-450s, a technique that, although requiring a sophisticated spectrophotometer, is basically simple both in conception and execution. However, for reasons that are entirely unclear, the use of optical difference spectroscopy has added at least as much confusion as light to studies of insecticide resistance. Results in this area are often conspicuous examples of a failure to take advantage of advances made in mammalian studies.

Principles

Spectroscopy of particulate preparations, such as microsomes, is complicated by the fact that turbid samples scatter light and that such light scattering is a function of wavelength. Thus, changes in the absolute spectrum caused by the addition of a ligand are seen as variations in a sloping baseline, the principal component of which is not due to the cytochrome but rather the physical properties of the particles. Optical difference spectroscopy avoids the problem of light scattering and nonspecific absorption by recording only the changes in light absorption caused by the ligand interaction and not the absolute spectra themselves. This is accomplished by placing the microsomal suspension in both the sample and reference cuvettes of a high resolution split-beam spectrophotometer, thus balancing these effects and permitting the recording of a flat baseline. On placing the ligand in question in the sample cuvette, a difference spectrum can be recorded that shows only the perturbations in the absolute spectrum caused by the ligand and not the absolute spectrum itself. Problems associated with the determination of difference spectra of insect P-450 have already been discussed (Hodgson, 1974; Hodgson et al., 1974; Kulkarni and Hodgson, 1975).

Difference Spectra of Oxidized Cytochrome P-450

Type I spectra, with a peak at about 385 nm and a trough at about 420 nm, are caused by ligand binding to a lipophilic site adjacent to the heme and are believed to be a manifestation of substrate binding to the oxidized form of the cytochrome. Type I spectra can be readily demonstrated by standard techniques using microsomes from a number of R strains of the house fly. These spectra are typically identical to those of mammalian liver microsomes. The apparent lack of type I binding in microsomes from CSMA house flies was first demonstrated with benzphetamine (Philpot and Hodgson, 1971; Tate et al., 1973) and subsequently, with a large number of pesticidal (Kulkarni et al., 1975) and non-pesticidal (Kulkarni et al., 1974) chemicals, all of which are type I ligands with mammalian liver microsomes. This apparent lack of type I binding was also seen in three species of *Hippelates* and in *Manduca sexta* (Kulkarni et al., 1976). Subsequent studies, however, revealed a significant type I spectrum with the insecticide synergist sulfoxide, and very small type I spectra with other methylenedioxyphenyl compounds (Kulkarni and Hodgson, 1976a). Moreover, benzphetamine gave type I spectrum with some microsomal subfractions from S house flies (Kulkarni and Hodgson, 1980b). These results suggest that, although a type I binding site(s) exists in the P-450 species of S house flies, it is qualitatively different, probably being more remote from the heme than that in the P-450 of R strains such as Fc and Diazinon-R.

Capdevila et al. (1973a,b, 1974) failed to observe normal type I spectral interactions with oxidized P-450 from insecticide-resistant house flies of the Fc strain. Subsequently, Agosin (1976) did report type I spectra from microsomes of the Rutgers (Diazinon-R) strain. These authors also reported that type I difference spectra could be detected in such microsomes pretreated with EtNC or reduced with either dithionite or NADPH. Difference spectra obtained under these conditions exhibited a peak at 410 nm and a trough at 428 nm, wavelengths that are not typical of normal type I difference spectra from either mammalian (Schenkman et al., 1967) or insect (Kulkarni and Hodgson, 1975, 1976a, 1980b; Kulkarni et al., 1974, 1975, 1976) P-450. The spectral magnitudes of type I difference spectra in the presence of NADPH are different from those with oxidized P-450 and may represent binding of both ligand and metabolite(s). Mammalian P-450 has been found to be only partially reduced by NADPH under aerobic conditions (Omura and Sato, 1964; Hlavica, 1972). In view of these facts, type I spectra using reduced microsomes are of questionable value. Those determined in the presence of EtNC may represent partial displacement of the oxidized spectrum of EtNC but, in any case, do not appear to be type I spectra.

Detailed consideration of the type I spectrum is essential to any study of insecticide resistance due to high oxidase activity. Not only is it related to substrate specificity, but in all the genetic studies of resistance in the house fly carried out in our laboratory, it is the only characteristic that always accompanies high oxidase activity (Tate et al., 1973, 1974; Plapp, 1976). By crossing over, strains can be developed in which the P-450 levels are identical to those of S strains and that differ only in the characteristics of their type I binding site.

Type II spectra appear to be caused by binding of the ligand to the heme iron of the oxidized P-450. Characteristically, they have a trough at about 395 nm and a peak at 430 nm and are due primarily to nitrogen compounds, such as pyridine and n-octylamine, with spatially accessible sp^2 and sp^3 non-bonded electrons. Such spectra have been seen in all insect species examined (Kulkarni and Hodgson, 1976c; Kulkarni et al., 1976).

The type II spectrum formed by n-octylamine is of special interest since it occurs in two forms, one with a double trough at 410 nm and 394 nm and the other with a single trough at 390 nm. The former occurs in S strains of the house fly and the latter is formed by R strains with high oxidase activity (Philpot and Hodgson, 1971; Tate et al., 1973).

It is unwise to place too much reliance on the n-octylamine spectrum as a means of characterizing microsomal P-450. The two spectral forms appear to reflect the relative amounts of low spin and high spin forms of the cytochrome. Since each P-450 species can exist in either spin state, the transition being a function of the binding of ligands, particularly type I ligands, the two spectra do not necessarily reflect two different cytochromes but may simply reflect the magnitude of binding of endogenous substrates. Prudence dictates that n-octylamine spectra be used as a tool for the characterization of cytocrhome P-450 from R and S insects only in conjunction with other spectral characteristics, particularly type I spectra.

The characterization of P-450 by difference spectroscopy is further compounded by the facts that certain ligands produce either a concentration-dependent shift in spectral type or give rise to mixed spectra. In addition, it is also known that a ligand may exhibit different spectral responses with microsomes from different species (Kulkarni et al., 1975).

Difference Spectra of Reduced Cytochrome P-450

The CO-reduced difference spectrum is the basis for the quantitation of P-450 and its denatured form, P-420. The most reliable estimates of these hemoproteins are obtained when CO treatment of microsomes is followed by dithionite reduction. Reversal of this sequence or the use of NADPH as a reductant results in erroneously lower values, especially for P-420 (Kulkarni and Hodgson, 1975, 1980b) (see Table 1). Although oxidized P-420 does not exhibit normal type I and II difference spectra, it does interact with EtNC, and its presence seriously affects the evaluation of the EtNC type III difference spectrum and its pH equilibrium point (Kulkarni and Hodgson, 1975, 1976b). When present in relatively large amounts, P-420 may even cause a shift in the absorption maximum of the P-450 CO-difference spectra and be responsible for a 10% to 50% underestimation of P-450 (Kulkarni and Hodgson, 1976d). The use of NAD(P)H as a reductant in the measurement of insect microsome cytochromes should be avoided for yet another reason: As shown in Figure 1, the reduction of both cytochrome b_5 and P-450 by reduced pyridine nucleotides is a two-step process, the second step appearing after a variable time delay, frequently of several minutes, a time longer than that used in most cytochrome determinations. While the reasons for this are not yet obvious, the second step is correlated with physical changes in the microsomal vesicles (e.g., there is a simultaneous change in light scattering), presumably permitting complete reduction of cytochrome thus rendered accessible.

Therefore, while NADPH is unsuitable as a reductant for the measurement of the amount of P-450 (ΔO.D.), it should be emphasized that it is the most appropriate reductant for the determination of absorption maximum (λ_{max}) of P-450-CO complex. Since NADPH does not reduce P-420 the slight shift in λ_{max} at 450 nm, caused by a significant peak at 420 nm, does not occur.

Type III difference spectra caused by insecticide synergists of the methylenedioxyphenyl group and EtNC have been demonstrated using either NADPH- or dithionite-reduced house fly microsomes, respectively. The former case has been related to the mechanism of synergism (Hodgson and Philpot, 1974; Kulkarni and Hodgson, 1976a) while the magnitude of the 455 nm peak in the latter case has been used to characterize strain or species differences in P-450 (Philpot and Hodgson, 1971; Tate et al., 1973; Kulkarni et al., 1976). Accurate estimation of the magnitude of the 430 nm peak of type III difference spectra of EtNC is significantly affected by several factors, such as concentration of EtNC, time and P-420 content (Kulkarni and Hodgson, 1975, 1976b). Furthermore, dithionite reduction prior to CO treatment results in underestimation of P-420, and increased denaturation of P-450 to P-420 at low pH conditions results from the addition of excess dithionite to poorly buffered microsomal suspensions. In view of these factors, data reported (Capdevila et al., 1973a, 1973b, 1974; Agosin, 1976) on EtNC pH equilibrium points to demonstrate qualitative differences between control and induced P-450 from house flies should be regarded as

Table 1. Reduction of Microsomal Pigments by Various Reducing Agents

	Reduced minus Oxidized				Reduced CO Difference Spectrum			
	Fc		CSMA		Fc		CSMA	
Reductant	ΔA^a	%	ΔA^a	%	ΔA^a	%	ΔA^a	%
Dithionite	0.094	100	0.061	100	0.077	100	0.060	100
NADPH	0.086	91	0.057	93	0.063	82	0.055	92
NADH	0.036	38	0.026	43	0.050	65	0.035	58
Reduced GSH	0.014	15	0.008	13	0.000	0	0.000	0
Dithiothreitol[b]	0.080	85	0.055	90	0.051	66	0.034	57
L-Cysteine[b]	0.083	88	0.015	25	0.000	0	0.00	0
L-Ascorbate[b]	0.071	76	0.040	66	0.000	0	0.00	0
Succinate	0.000	0	0.00	0	0.000	0	0.00	0

The values represent typical results of a single experiment (N = 3 or more).

[a]A = peak minus trough.
[b]Less reliable data owing to very slow reduction of microsomal pigments. See text for further details.

tentative until confirmed by other methods. Because of these problems related to the 430-433 nm peak, we have reported spectral data on this ligand only with regard to the 455 nm peak (Philpot and Hodgson, 1971; Tate et al., 1973; Kulkarni and Hodgson, 1976b; Kulkarni et al, 1976) and have avoided the calculation of pH equilibrium points.

The argument that the 455 nm peak should not be used in the characterization of P-450 in the presence of P-420 (Capdevila et al., 1974) is specious since the value is used relative to the magnitude of the CO spectrum at 450 nm, a value that can be measured separately from P-420. Thus, the ratio EtNC(455nm)/CO(450nm) is a characteristic of that P-450 which has not been broken down to P-420.

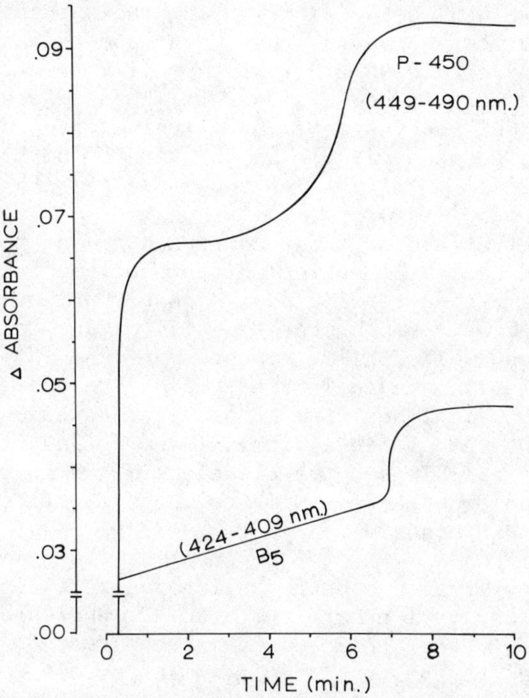

Figure 1. Reduction of microsomal cytochromes from the house fly. Cytochrome b_5, from abdominal microsomes obtained from house flies of the CSMA strain, was reduced with NADH. Cytochrome P-450, from abdominal microsomes obtained from house flies of the Rutgers strain, was reduced with NADPH.

New and Less-commonly Used Ligands

It should be emphasized that ligands other than those commonly used in studies of mammalian P-450 are available and have considerable potential for the characterization of P-450 from R and S insects.

For example, the type II ligand, benzimidazole, shows an interesting concentration-dependent shift with microsomes from Fc house flies (Kulkarni et al., 1974). At low ligand concentrations, the type II spectrum shows a single trough at 410 nm. As the ligand concentration is increased, an additional trough appears at 392 nm until at saturation there is a single, broad trough with a minimum at 392 nm. This is not seen with microsomes from the CSMA strain, which shows only a single, symmetrical trough at 392 nm at all ligand concentrations.

Recently, we have been investigating the structure-function relationships between compounds that cause spectral shifts at or around 450 nm and/or a double Soret, type III, spectrum. These studies included certain new classes of ligands, such as the group IVB dihalides (Dahl and Hodgson, 1977), trivalent phosphorus compounds (Dahl and Hodgson, 1978) and substituted dioxolanes (Dahl and Hodgson, 1979).

It appears that all of those compounds giving rise to a peak at or around 450 nm in the optical difference spectrum of reduced P-450 are strong π-acceptor ligands. Those that have an additional peak at a shorter wavelength, the type III or double Soret spectrum, have, in addition, a lipophilic group. For example, stannous fluoride, the widely used anticaries toothpaste additive, and other tin and germanium dihalides form complexes with hemoproteins such as hepatic P-450, hemoglobin and peroxidase. These complexes are characterized by visible spectra closely similar in shape, molar absorptivity, and absorbance maxima to those obtained with analogous complexes of carbon monoxide.

Trivalent oxygenated phosphorus ligands were also tested, including alkyl and aryl phosphites, $(RO)_3P$, phosphonites, $(RO)_2PR$, and phosphinites, $ROPR_2$. All such compounds tested, with the exception of triphenyl phosphite, interact with ferrous P-450 and its denatured form, P-420, to produce complexes having two peaks in the Soret region of their optical difference spectra. Careful evaluation of these spectra indicate that both of these peaks originate from each of the cytochromes. The evidence clearly shows that P-450 is not denatured by these ligands nor is P-420 converted to P-450. The high affinity of these ligands for heme iron is indicated by small K_s values.

The spectra resulting from the interaction of a series of substituted dioxolanes with microsomal P-450 or P-420, as well as purified P-450, were also measured. With the exception of dioxolane, 4-methyldioxolane and 4-ethyldioxolane, these compounds (e.g., 4-n-butyldioxolane and 4-n-pentyldioxolane) interacted with ferric P-450 to give complexes exhibiting type I optical difference spectra, and after incubation with NADPH, spectra with peaks at about 430 nm. These complexes, as well as those formed from dioxolanes in the presence of cumene hydroperoxide, inhibit the binding of CO to the cytochrome. Although the purpose of these studies was to elucidate the nature of the type III interactions of P-450 and although only mouse liver microsomes were used, we are currently investigating the effects of these compounds on microsomes from S and R house flies. The point to be made at the present time is that there are numerous ligands of potential worth in resistance studies, and we should not restrict our methods to those commonly used in studies of mammals.

GENETIC STUDIES

Since these studies have been reviewed a number of times (Hodgson et al., 1974; Hodgson, 1976; Plapp, 1976), a comprehensive review will not be attempted here. However, it can be said that examination of a number of R strains of the house fly has led to the conclusion that those showing a high oxidase activity generally have an increased titer of P-450, and moreover, this cytochrome differs qualitatively from that of S strains (Tate et al., 1973). The "resistant P-450" as found in diazinon-R, Fc and dimethoate-R strains differs from the susceptible P-450 as follows:

1. The λ_{max} of the CO-reduced cytochrome P-450 spectrum is shifted several nm toward the blue.
2. Type I binding is present.
3. Type II binding is increased relative to the CO spectrum.
4. The EtNC spectrum is reduced relative to the CO spectrum.

Other R strains may have some, but not all, of these characteristics. It is also apparent (Chang and Hodgson, 1975) that the catalytic properties of these cytochromes toward xenobiotic substrates are different.

The genes controlling these characteristics are found primarily on chromosome II, except in the Fc strain in which the gene that apparently controls type I binding is found on chromosome V. In all of the crosses, backcrosses and crossing over experiments using R strains and S strains with genetic markers, type I binding is the only spectral characteristic that is always associated with high oxidase activity.

These studies, insofar as they refer to the Fc strain, have recently been criticized (Terriere and Schonbrod, 1976) on several grounds. Some of these criticisms refer to measurements of substrate level oxidations carried out earlier in another laboratory (Plapp and Casida, 1969) and will not be commented on here. The more recent work involving P-450 is criticized on two principal grounds: first, that multiplicity is not adequately taken into account; second, that if aldrin epoxidation is linked to chromosome II, then chromosome V cannot be involved in resistance. While these criticisms are valid in part, the discrepancies can be explained by a relatively simple hypothesis.

With regard to the first problem, the bulk of the work was carried out before the extent of multiplicity of house fly P-450 was known. It now appears clear that, although final proof is lacking, the genes that appeared to control spectral characteristics actually control the synthesis of different forms of the cytochrome and that the spectral characteristics by which the predominant cytochrome differs from the others is the one that appeared to be under the control of that particular gene. All of the previous results can be explained by this simple hypothesis.

The second criticism, that aldrin epoxide activity is associated with chromosome II and not chromosome V, is presumably related to the multiplicity of P-450 and the substrate specificity of the different forms. Since the DDT resistance that is associated with chromosome V is oxidative, it must be assumed that the association between type I binding and chromosome V is, in fact, an association with a particular form of cytochrome that has high type I binding and which oxidizes DDT at a much faster rate than aldrin. Aldrin, on the other hand, appears to be most efficiently oxidized by a P-450, the gene for which is on chromosome II. This is in agreement with the fact that chromosome II is associated with a high P-450 level in the Fc strain.

Whatever the final resolution of this controversy, it must take into account the objective and reproducible finding that type I binding is associated with chromosome V and with high oxidase activity.

INDUCTION

While the relationship between high, but uninduced, levels of P-450 and/or different types of P-450 appears to be established, the relationship between induction and resistance is not yet clear. Although it will not be discussed in detail, it is clear that induction of cytochrome P-450 does occur in insects such as the house fly, *Musca domestica* (Capdevila et al., 1973a,b), and the southern armyworm, *Spodoptera eridania* (Brattsten et al., 1976), for example. While it is also clear that induction frequently involves changes in the qualitative characteristics of the P-450 complex, this is not always the case (Tate et al., 1973), and due to the problems associated with preparation, purification and spectral characterization referred to above, comparisons between different laboratories are difficult to make with any degree or certitude.

Although induction may be important in tolerance, if it is to be demonstrated that it is important in resistance then the extent of induction must be shown to be under genetic control, and furthermore, it must be shown that high induction is selected for under conditions of insecticide pressure.

SUMMARY AND CONCLUSIONS

Several factors have now been clearly established concerning resistance to insecticides and P-450. High oxidase activity is the factor most frequently associated with resistance, and this high oxidase activity is due to the microsomal P-450-dependent monooxygenase system. It is also clear, although the details are still obscure, that several forms of P-450 exist in both S and R insects and that the qualitative characteristics of the complex of P-450s vary between one strain and another. Furthermore, these cytochromes can be induced.

From the methodological point of view much confusion still remains in the areas of selection of experimental material, preparation of microsomes, presence of endogenous inhibitors and ligands, polyacrylamide gel electrophoresis in the presence of sodium dodecyl sulfate, and in the most important area of spectral characterization.

Several recommendations can be made:

1. Established methods should be standardized among laboratories and particular attention should be paid to developments in the area of instrumental analysis.

2. Care must be taken not to uncritically adapt methods developed as a result of investigations on other organisms, although after proper analysis, such methods may be adopted.

3. New methods should be explored, e.g., the use of new ligands in spectral characterization.

4. Purification studies should be extended until several forms of the cytochrome from S and R strains of the same species can be obtained essentially pure. This would open numerous avenues for investigation of resistance mechanisms. Findings to date appear to indicate that insect P-450 is more labile than that of mammalian liver microsomes, and it may be that we need to develop special protectants, solubilization agents, affinity columns, etc. It should also be recalled, however, that less than a decade ago the microsomal P-450 of rabbit and rat liver was believed to be labile and difficult to purify and today several essentially pure forms have been isolated from each other.

5. Studies in biochemical genetics should be extended, particularly in the areas of induction and gene expression. The development of micro-methods suitable for the examination of P-450 from single insects would be of inestimable value.

6. From a practical point of view, the study of the inhibition of P-450-dependent oxidations should be rewarding.

REFERENCES

Agosin, M., 1976, Insect cytochrome P-450, *Mol. Cell. Biochem.*, 12:33.

Agosin, M., and Perry, A. S., 1974, Microsomal mixed-function oxidases, *in:* "The Physiology of Insecta," M. Rockstein, ed., Vol. 5, pp. 537-596, Academic Press, New York.

Bell, D. Y., and Hodgson, E., 1977a, Methods for polyacrylamide gel electrophoresis of mammalian liver cytochrome P-450 in the presence of sodium dodecyl sulfate: A critique, *Gen. Pharmacol.*, 8:113.

Bell, D. Y., and Hodgson, E., 1977b, SDS-Polyacrylamide gel electrophoresis of hepatic cytochrome P-450 from normal, 3-methylcholanthrene and phenobarbital treated mice, *Gen. Pharmacol.*, 8:121.

Brattsten, L. B., Wilkinson, C. F., and Root, M. M., 1976, Microsomal hydroxylation of aniline in the southern armyworm, *Spodoptera eridania, Insect Biochem.*, 6:615.

Brattsten, L. B., Wilkinson, C. F., and Eisner, T., 1977, Herbivore-plant interactions: Mixed-function oxidases and secondary plant substances, *Science*, 196:1349.

Capdevila, J., and Agosin, M., 1977, Multiple forms of house fly cytochrome P-450, *in:* "Microsomes and Drug Oxidations," V. Ullrich, ed., pp. 144-151, Pargamon Press, Oxford.

Capdevila, J., Morello, A., Perry, A. S., and Agosin, M., 1973a, Effect of phenobarbital and naphthalene on some components of the electron transport system and hydroxylating activity of housefly microsomes, *Biochem.*, 12:1445.

Capdevila, J., Perry, A. S., Morello, A., and Agosin, M., 1973b, Some spectral properties of cytochrome P-450 from microsomes isolated from control, phenobarbital and naphthalene treated houseflies, *Biochem. Biophys. Acta*, 314:93.

Capdevila, J., Perry, A. S., and Agosin, M., 1974, Spectral and catalytic properties of cytochrome P-450 from a diazinon-resistant housefly strain, *Chem.-Biol. Interact.*, 9:105.

Capdevila, J., Ahmad, N., and Agosin, M., 1975, Soluble cytochrome P-450 from housefly microsomes. Partial purification and characterization of two hemoprotein forms, *J. Biol. Chem.*, 250:1048.

Chang, L. L., and Hodgson, E., 1975, Biochemistry of detoxication in insects. Microsomal mixed function oxidase activity in the housefly, *Musca domestica, Insect Biochem.*, 5:93.

Dahl, A. R., and Hodgson, E., 1977, Complexes of stannous fluoride and other group IVB dihalides with mammalian hemoproteins, *Science*, 197:1376.

Dahl, A. R., and Hodgson, E., 1978, Complexes of trivalent oxygenated phosphorus compounds with cytochrome P-450 and cytochrome P-420: The origin of double Soret spectra, *Chem.-Biol. Interact.*, 21:137.

Dahl, A. R., and Hodgson, E., 1979, The interaction of aliphatic analogs of methylenedioxyphenyl compounds with cytochromes

P-450 and P-420, *Chem.-Biol. Interact.*, 27:163.
Dallner, G., 1963, Studies on the structural and enzymic organization of the membraneous elements of liver microsomes, *Acta Pathol. Microbiol. Scand.*, Suppl. 166.
Estabrook, R. W., and Werringloer, J., 1979, The microsomal enzyme system responsible for the oxidative metabolism of many drugs, *in:* "The Induction of Drug Metabolism," R. W. Estabrook and E. Lidenlaub, eds., pp. 187-199, Schattauer Verlag, Stuttgart and New York.
Estabrook, R. W., Gillette, J. R., and Leibman, K. N., editors, 1973, "Microsomes and Drug Oxidations," Williams and Wilkins, Baltimore, Maryland.
Gilbert, M. D., and Wilkinson, C. F., 1975, An inhibitor of microsomal oxidation from gut tissues of the honey bee, *Apis mellifera*, *Comp. Biochem. Physiol.*, 50:613.
Gillette, J. R., Davis, D. C., and Sasame, H. A. 1972, Cytochrome P-450 and its role in drug metabolism, *Ann. Rev. Pharmacol.*, 12:57.
Gould, F., and Hodgson, E., 1980, Mixed-function oxidase and glutathione transferase activity in last instar *Heliothis virescens* larvae, *Pestic. Biochem. Physiol.*, in press.
Guengerich, F. P., 1979, Isolation and purification of cytochrome P-450 and the existence of multiple forms, *Pharmacol. Therap.*, 6:99.
Hansen, L. G., and Hodgson, E., 1971, Biochemical characterization of insect microsomes: N- and O-demethylation, *Biochem. Pharmacol.*, 20:1569.
Hlavica, P., 1972, Interaction of oxygen and aromatic amines with hepatic microsomal mixed-function oxidases, *Biochem. Biophys. Acta*, 273:318.
Hodgson, E., 1974, Comparative studies of cytochrome P-450 and its interaction with pesticides, *in:* "Survival in Toxic Environments," M. A. Q. Khan and J. P. Bederka, eds., pp. 213-260, Academic Press, New York.
Hodgson, E., 1976, Comparative toxicology: Cytochrome P-450 and mixed-function oxidase activity in target and nontarget organisms, *Essays in Toxicology*, 7:73.
Hodgson, E., 1979, Comparative aspects of the distribution of cytochrome P-450-dependent monooxygenase systems: An overview, *Drug Metabol. Rev.*, 10:15.
Hodgson, E., and Dauterman, W. C., 1980, Metabolism of toxicants: Phase one reactions, *in:* "Introduction to Biochemical Toxicology," E. Hodgson and F. E. Guthrie, eds., Elsevier/North Holland, New York.
Hodgson, E., and Philpot, R. M., 1974, Interaction of methylenedioxyphenyl (1,3-benzodioxole) compounds with enzymes and their effects on mammals, *Drug Metabol. Rev.*, 3:231.
Hodgson, E., and Tate, L. G., 1976, Cytochrome P-450 interactions, *in:* "Insecticide Physiology and Biochemistry," C. F. Wilkinson,

ed., pp. 115-148, Plenum Press, New York.

Hodgson, E., Tate, L. G., Kulkarni, A. P., and Plapp, F. W., 1974, Microsomal cytochrome P-450: Characterization and possible role in insecticide resistance in *Musca domestica*, *J. Ag. Food Chem.*, 22:360.

Johnson, E. F., 1979, Multiple forms of cytochrome P-450: Criteria and significance, *Rev. Biochem. Toxicol.*, 1:1.

Krieger, R. I., and Wilkinson, C. F., 1969, Microsomal mixed-function oxidases in insects, 1. Localization and properties of an enzyme system effecting aldrin epoxidation in larvae of the southern armyworm, *Biochem. Pharmacol.*, 18:1403.

Kulkarni, A. P., and Hodgson, E., 1975, Microsomal cytochrome P-450 from the housefly, *Musca domestica*: Assay and spectral characterization, *Insect Biochem.*, 5:679.

Kulkarni, A. P., and Hodgson, E., 1976a, Spectral interactions of insecticide synergists with microsomal cytochrome P-450 from insecticide-resistant and susceptible houseflies, *Pestic. Biochem. Physiol.*, 6:183.

Kulkarni, A. P., and Hodgson, E., 1976b, Effect of storage on the biochemistry of cytochrome P-450 from the housefly, *Musca domestica*, *Insect Biochem.*, 6:89.

Kulkarni, A. P., and Hodgson, E., 1976c, Spectral characterization of microsomal cytochrome P-450 from the midgut of the tobacco hormworm, *Manduca sexta*, *Insect Biochem.*, 6:385.

Kulkarni, A. P., and Hodgson, E., 1976d, Microsomal electron transport in the housefly, *Musca domestica*: A model for the study of detoxication systems in insects, *Israel J. Entomol.*, 11:93.

Kulkarni, A. P., and Hodgson, E., 1980a, Metabolism of insecticides by microsomal mixed-function oxidase systems, *Pharmacol. Therap.*, in press.

Kulkarni, A. P., and Hodgson, E., 1980b, Multiplicity of cytochrome P-450 in microsomal membranes from the housefly, *Musca domestica*, submitted to *Biochem. Biophys. Acta*.

Kulkarni, A. P., Mailman, R. B., Baker, R. C., and Hodgson, E., 1974, Cytochrome P-450 difference spectra: Type II interactions in insecticide-resistant and susceptible houseflies, *Drug Metabol. Disp.*, 2:309.

Kulkarni, A. P., Mailman, R. B., and Hodgson, E., 1975, Cytochrome P-450 optical difference spectra of insecticides: A comparative study, *J. Ag. Food Chem.*, 23:177.

Kulkarni, A. P., Smith, E., and Hodgson, E., 1976, Occurrence and characterization of microsomal cytochrome P-450 in several vertebrate and insect species, *Comp. Biochem. Physiol.*, 54B:509.

Kulkarni, A. P., Motoyama, N., Dauterman, W. C., and Hodgson, E., 1978, Inhibition of glutathione S-transferase by catecholamines and related compounds in the housefly, *Bull. Environ. Contam. Toxicol.*, 20:277.

Levin, W., Botelho, L. M., Thomas, P. E., and Ryan, D. E., 1979, Purification and characterization of multiple forms of rat

liver cytochrome P-450, *in:* "Fourth International Symposium Microsomes and Drug Oxidations," pp. 1-106, Ann Arbor, Michigan.
Mailman, R. B., Tate, L. G., Muse, K. E., Coons, L. B., and Hodgson, E., 1975, The occurrence of multiple forms of cytochrome P-450 in hepatic microsomes from untreated rats and mice, *Chem.-Biol. Interact.*, 10:215.
Motoyama, N., and Dauterman, W. C., 1980, Glutathione S-transferases: Their role in the metabolism of organophosphorus insecticides, *Rev. Biochem. Toxicol.*, 2:249.
Motoyama, N., Kulkarni, A. P., Hodgson, E., and Dauterman, W. C., 1978, Endogenous inhibitors of glutathione S-transferases in houseflies, *Pestic. Biochem. Physiol.*, 9:255.
Nakatsugawa, T., and Morelli, M. A., 1976, Microsomal oxidation and insecticide metabolism, *in:* "Insecticide Biochemistry and Physiology," C. F. Wilkinson, ed., pp. 61-114, Plenum Press, New York.
Omura, T., and Sato, R., 1964, The carbon monoxide binding pigment of liver microsomes. I. Evidence for its hemoprotein nature, *J. Biol. Chem.*, 239:2370.
Oppenoorth, F. J., and Welling, W., 1976, Biochemistry and physiology of resistance, *in:* "Insecticide Biochemistry and Physiology," C. F. Wilkinson, ed., pp. 507-551, Plenum Press, New York.
Orrenius, S., Berggen, M., Moldeus, P., and Krieger, R. I., 1971, Mechanism of inhibition of microsomal mixed-function oxidases by the gut contents inhibitor of the southern armyworm *(Prodenia eridania)*, *Biochem. J.*, 124:427.
Philpot, R. M., and Hodgson, E., 1971, Differences in cytochrome P-450s from resistant and susceptible houseflies, *Chem.-Biol. Interact.*, 4:399.
Plapp, F. W., 1976, Biochemical genetics of insecticide resistance, *Ann. Rev. Entomol.*, 21:179.
Plapp, F. W., and Casida, J. E., 1969, Genetic control of housefly NADPH-dependent oxidases: Relation to insecticide chemical metabolism and resistance, *J. Econ. Entomol.*, 62:1174.
Schenkman, J. B., Remmer, H., and Estabrook, R. W., 1967, Spectral studies of drug interaction with hepatic microsomal cytochrome, *Mol. Pharmacol.*, 3:113.
Schonbrod, R. D., and Terriere, L. C., 1971, Inhibition of housefly microsomal epoxidase by the eye pigment, xanthommatin, *Pestic. Biochem. Physiol.*, 1:409.
Schonbrod, R. D., and Terriere, L. C., 1975, The solubilization and separation of two forms of microsomal cytochrome P-450 from the housefly, *Musca domestica* L., *Biochem. Biophys. Res. Commun.* 64:829.
Smith, S. L., Bollenbacher, W. C., Cooper, D. Y., Schleyer, H., Wielgus, J. J., and Gilbert, L. I., 1979, Ecdysone 20-monoxygenase: Characterization of an insect cytochrome P-450-dependent steroid hydroxylase, *Molec. Cell. Endocrinol.*, 15:111.
Stanton, R. H., Plapp, F. W., White, R. A., and Agosin, M., 1978,

Induction of multiple cytochrome P-450 species in housefly microsomes SDS gel electrophoresis studies, *Comp. Biochem. Physiol.*, 61B:297.

Tate, L. G., Plapp, F. W., and Hodgson, E., 1973, Cytochrome P-450 difference spectra of microsomes from several insecticide-resistant and susceptible strains of housefly, *Musca domestica* L., *Chem.-Biol. Interact.*, 6:237.

Tate, L. G., Plapp, F. W., and Hodgson, E., 1974, Genetics of cytochrome P-450 in two insecticide-resistant strains of the housefly, *Musca domestica* L., *Biochem. Genetics*, 11:49.

Terriere, L. C., and Schonbrod, R. D., 1976, Agruments against a fifth chromosomal factor in control of aldrin epoxidation and propoxur resistance in the Fc strain of the housefly, *Pestic. Biochem. Physiol.*, 6:551.

Ullrich, V., 1977, "Microsomes and Drug Oxidations," Pergamon Press, Oxford.

Wilkinson, C. F., and Brattsten, L. B., 1972, Microsomal drug metabolizing enzymes in insects, *Drug Metabol. Rev.*, 1:153.

Wilson, T. G., and Hodgson, E., 1972, Mechanism of microsomal mixed-function oxidase inhibitor from the housefly, *Musca domestica* L., *Pestic. Biochem. Physiol.*, 2:64.

ROLE OF HYDROLASES AND GLUTATHIONE S-TRANSFERASES IN INSECTICIDE RESISTANCE

Walter C. Dauterman

Toxicology Program, Department of Entomology
North Carolina State University
Raleigh, North Carolina 27650

INTRODUCTION

During the last few decades, many species of insects have acquired resistance to insecticides. Resistance is inherited and has proved to be one of the major obstacles in the successful contol of insects. Factors recognized to contribute to insecticide resistance include changes in the rate of metabolism, decreased cuticular penetration of the insecticide, and modifications in the target site (Oppenoorth, 1971: Oppenoorth and Welling, 1976).

The present review is concerned with detoxication involving phosphorotriester hydrolases and glutathione S-transferases and the role of these enzymes in insecticide resistance. Many insect species have a complexity of mechanisms responsible for resistance that may vary from species to species and between strains as well as with the insecticide stressor. In most cases, it has been impossible to separate the effect of the various mechanisms and their components and relate them individually to the total end product, resistance.

HYDROLASES

General Reaction

The hydrolysis of organophosphorus insecticides is mediated by a number of enzymes that are responsible for the cleavage of the phosphorus ester or anhydride bond (Dauterman, 1971, 1976). A variety of names have been used to identify and describe the enzymes that catalyze these ractions, such as DFP-ase, paraoxonase, A-esterase, phosphorylphosphatase, aryl esterase, phosphatase, etc., but for the present discussion, the general term "phosphorotriester hydrolase" will be used. Triester hydrolysis of organophosphate compounds results in the formation of phosphorus-containing metabolites, which are ionized at neutral pH and are extremely poor cholinesterase inhibitors. The overall effect of hydrolysis is to detoxify the parent compound (O'Brien, 1960; Heath, 1961; Eto, 1974).

Phosphorotriester hydrolases may attack the intact organophosphate insecticide molecule at two sites:

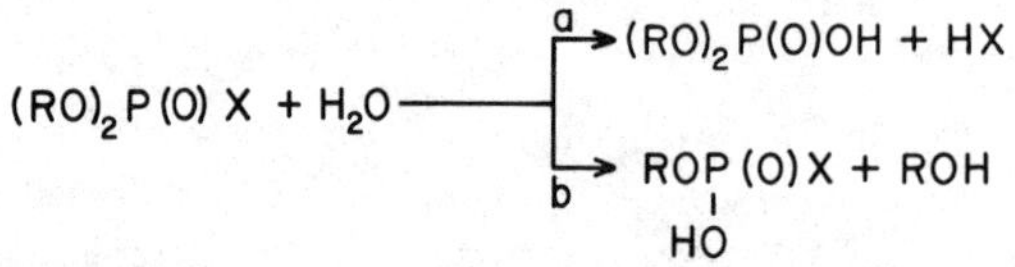

One reaction (a) leads to the formation of dialkyl phosphoric acid and HX, and the other (b) results in the formation of a desalkyl derivitive and an alcohol. Both reactions may be considered to be detoxication.

Phosphorotriester Hydrolases

A relationship between hydrolase (phosphatase) activity and organophosphorus insecticide resistance was initially suggested as a result of the low carboxylesterase (aliesterase) levels found in several strains of resistant house flies (van Asperen and Oppenoorth, 1959). These findings led to the so-called "mutant aliesterase theory," which suggested that the increased phosphatase activity in organophosphate-resistant insects was due to a mutant form of the aliesterase normally found in the susceptible strain and was responsible for the degradation of the insecticide and the development of resistance (Oppenoorth and van Asperen, 1960). However, much of the evidence for this theory relied on indircet measurements of the degradation of the oxygen analogs based on a decrease in anticholinesterase activity (Oppenoorth and van Asperen, 1961). Even with direct measurements of the metabolites, it is not possible to infer the nature of the detoxication mechanisms since the same metabolic products may be generated by a number of different enzyme systems (Dauterman, 1976). In order to distinguish one enzyme system from another, it is necessary to examine the subcellular localization of the enzyme, its cofactor requirements, and its response to specific inhibitors. A complete analysis and identification of all the metabolic components formed from a particular substrate is also required (Motoyama and Dauterman, 1974, 1980).

In 1964, Matsumara and Hogendijk studied phosphatase activities in resistant (R) and susceptible (S) strains of house flies and partially purified enzymes for R house flies that degraded parathion, a phosphorothioate, to diethyl phosphorothioic acid. However, subsequent attempts by both Nakatsugawa et al. (1969) and Welling et al. (1971) failed to demonstrate the presence of phosphatases that were capable of degrading parathion. At present, one must assume that insect phosphorotriester hydrolases have a substrate specificity similar to that of the mammalian phosphorotriester hydrolases and only phosphate triesters are hydrolyzed (Dauterman, 1976).

Welling et al. (1971) investigated the in vitro metabolism of paraoxon in preparations from R and S strains of house flies. Microsomes from the R strain (E_1) degraded paraoxon and formed diethyl phosphoric acid and two minor metabolites. Although this strain had enhanced mixed-function oxidase activity that contributed to the total resistance, inhibitor studies demonstrated that the reaction leading to diethyl phosphoric acid was not catalyzed by the mixed-function oxidases but was in fact due to a hydrolase (phosphatase). The production of diethyl phosphoric acid was higher in whole fly homogenates from the R strain than from the S strain, strongly suggesting that enhanced phosphatase activity was responsible for the observed resistance.

Lewis and Sawicki (1971) reported that microsomes from R and S house flies yielded diethyl phosphoric acid form both diazoxon and paraoxon in the absence of NADPH and oxygen; they observed some interstrain difference in hydrolase activity, which suggested that the enzyme was involved in the resistance mechanism.

Nolan and O'Brien (1970) reported that ^{3}H-ethoxy paraoxon was metabolized in vivo in R and S house flies to labeled ethanol and derivatives and not to acetaldehyde or ethyl S-glutathione. This strongly suggests the involvement of a hydrolytic reaction (reaction b above). While the formation of diethyl phosphoric acid was higher in R than in S strains, the formation of ^{3}H-ethanol was highest in the S strain. Based on this evidence, it appears that hydrolysis of the alkoxy groups of phosphorotriester insecticides does not constitute an important resistance mechanism.

High phosphorotriester hydrolase activity has also been demonstrated in a number of R strains of the green peach aphid, *Myzus persicae* (Sulz.) (Oppenoorth and Voerman, 1975). Homogenates from R strains were from 2.6 to 7.8-fold more active in the hydrolysis of paraoxon, methyl paraoxon and malaoxon than that from the S strain (Table 1). The radioactive hydrolysis products were identified as the dialkyl phosphoric acids.

Table 1. Phosphorotriester Hydrolysis of Insecticides by Aphid Homogenates[a]

Substrate[b]	Strain	Degradation Rate pmol/mg aphid/hr
	A	2.3
Paraoxon	R	4.7
	E	8.6
	A	4.0
Methyl paraoxon	R	7.5
	E	10.5
	A	2.4
Malaoxon	R	11.9
	E	18.8

[a]Data from Oppenoorth and Voerman (1975).
[b]Homogenates incubated at pH 7.0 at 27°C.

In summary, the degradation of phosphate insecticides by hydrolases occurs in only a limited number of strains of R house flies and in only a few species of insects. Consequently, hydrolases appear to play a smaller role in resistance than might be expected and typically are associated with other more important resistance mechanisms.

GLUTATHIONE S-TRANSFERASES

General Reactions

The initial step in the biosynthesis of mercapturic acids involves the conjugation of electrophilic xenobiotics with endogenous nucleophilic glutathione. This reaction is catalyzed by an elaborate complex of enzymes, the glutathione S-transferases (EC 2.5.1.18), which exist in different forms and whose exact number and specificities remain to be established (Boyland and Chasseaud, 1969; Arias and Jakoby, 1976). The generalized scheme of mercapturic acid

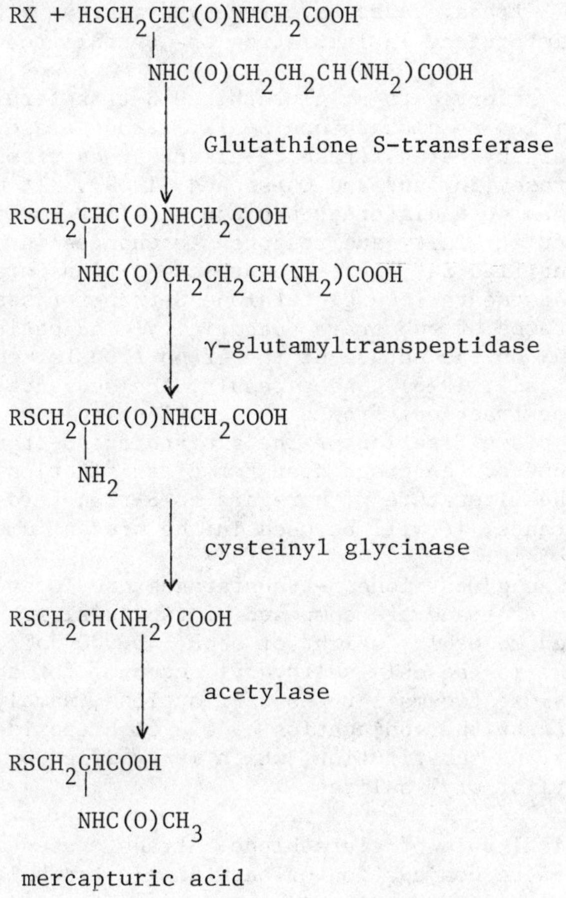

Figure 1. Mercapturic acid biosynthesis.

biosynthesis is presented in Figure 1. The overall reaction involves conjugation of the foreign compound with reduced glutathione, followed by transfer of the glutamate group, loss of glycine, and finally, acetylation. As can be seen, a number of enzymes catalyze the various steps in the biosynthesis of mercapturic acid, and in contrast to other major conjugation processes, such as glucosidation, sulfate formation, etc., glutathione conjugation does not require a high energy intermediate involving ATP. Mercapturic acid biosynthsis has been demonstrated to occur in vivo in insects (Gessner and Smith, 1960; Cohen and Smith, 1964; Dykstra and Dauterman, 1978). In some insect species, only the cysteine conjugate has been isolated, while with others, the acetyl cysteine conjugate has been found. Of the enzymes involved in mercapturic acid biosynthesis, the glutathione S-transferases have received maximum attention because of their importance in initiating the overall reaction.

The enzymes referred to as glutathione S-transferases were initially classified as glutathione S-alkyl transferase, S-aryl transferase, S-aralkyl transferase, S-alkene transferase and S-epoxide transferase (Boyland and Chasseaud, 1969). It was suggested that these enzymes were different because of differences in substrate specificity, heat stability and response to changes in pH. Recent findings with purified rat liver glutathione S-transferases have demonstrated that the various glutathione S-transferases have broad overlapping patterns of substrate specificity, indicating that classification based on the nature of the group transferred is inadequate (Jakoby et al., 1976). As a result, the sequence of elution of various enzyme fractions from a CM cellulose column has been utilized in the classification of the glutathione S-transferase. However, since use of the group transfer classification is well entrenched in the literature with regard to S-transferases and insecticide resistance, it will be used in the present discussion.

The mammalian glutathione S-transferases are located in the 100,000 g supernatant and are composed of two equal subunits, having a combined molecular weight of about 46,000 daltons. Reduced glutathione (GSH) is the only sulfhydryl acceptor for this system. The enzymes catalyze two main types of reactions, namely addition, which is typified by the conjugation of GSH with epoxides or unsaturated compounds, and substitution, which is typified by conjugation of GSH with alkyl or aryl halides.

The biological role of glutathione S-transferases appears to be twofold since the conjugation of harmful electrophiles with the endogenous nucleophile GSH not only provides protection for biological macromolecules such as nucleic acids and proteins from the potential toxic consequences of a covalent reaction, but also provides an effective means of excretion of the electrophile through the formation of an anionic, water-soluble product (Chasseaud, 1973).

The involvement of glutathione S-transferases in the metabolism of organophosphorus insecticides was first reported by Shishido and Fukami (1963) and Fukami and Shishido (1963, 1966), who found that methyl parathion was dealkylated by a soluble enzyme that required GSH and exhibited a preference for methyl groups. A number of reviews are available concerning the role of glutathione S-transferases in the metabolism of insecticides (Hollingworth, 1970; Dauterman, 1971; Motoyama and Dauterman, 1980; Fukami, 1980), so this will not be further discussed.

Distribution and Amount of Glutathione

Glutathione (γ-glutamyl L-cysteinyl glycine) is widely distributed in living cells and is an important component in a number of enzyme reactions, either as a substrate or as a coenzyme. Since GSH is a substrate for the glutathione S-transferase system, the tripeptide may become rate-limiting when an organism is exposed to a large amount of xenobiotic. Therefore, low or decreased levels of GSH during certain developmental stages of an insect or in certain insect species would be expected to result in a decrease in the level of protection afforded by a functional glutathione S-transferase system.

The amount of GSH present in various stages of house fly development has been determined by Saleh et al. (1978). It remained fairly constant during the larval and pupal stages, but increased with age in the adult insect (Figure 2). It should be noted that a difference in the amount of GSH was found in individual male and female house flies, probably due to the larger size of the female, but on a protein or wet weight basis, a slightly larger amount of GSH was found in the male adult than in the female.

The distribution of GSH in various body regions of the house fly was also investigated (Table 2). Approximately 60 to 68 percent of the GSH was localized in the thorax whereas 60 to 65 percent of the glutathione S-transferase activity occurred in the abcomen. It would appear that the GSH localized in the throax may have a function other than that of a substrate for the glutathione S-transferase system.

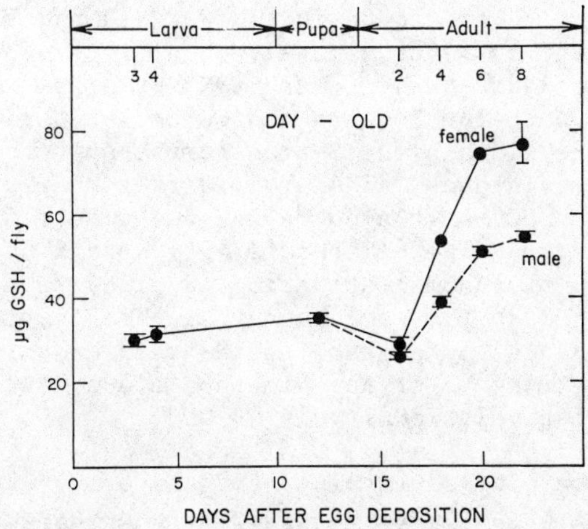

Figure 2. Micrograms of glutathione per individual during various stages of development of the CSMA strain of house flies (adapted from Saleh et al., 1978).

Table 2. Localization of Glutathione in Six-day-old House Flies (Cornell-R Strain)[a]

Body Region	Reduced glutathione μg/body region (%)[b]	
	Male	Female
Head	2.7 (3.3%)	4.4 (3.9%)
Thorax	56.9 (68.4%)	68.4 (60.2%)
Abdomen	23.4 (28.3%)	40.0 (35.9%)

[a]Data from Saleh et al. (1978).
[b]Average of two separate experiments.

Larvae and adults of DDT-R strains of house fly were found to have slightly lower levels of GSH than those of a susceptible strain (Cotty and Henry, 1958; Lipke and Chalkely, 1962). DDTase, which is responsible for DDT resistance in the strains employed, is a GSH-dependent enzyme; but since GSH is not depleted during dehydrochlorination (Lipke and Kearns, 1960), increased DDTase activity in the resistant strains may not require additional levels of GSH.

Saleh et al. (1978) determined the amount of GSH in five R strains of house flies. GSH levels in insects increased with age, and the amount of GSH was determined at four different ages (Table 3). High GSH levels were found in the Rutgers and the Cornell-R strains, and lower, but approximately equal levels occurred in the Orlando DDT, dimethoate $49r_2$, Fc and CSMA strains. This is not surprising since both the Rutgers and Cornell-R strains utilize GSH as part of the glutathione S-transferase system, which accounts for part of the resistance. The small difference in GSH content in the Orlando DDT and CSMA strains may be due to the function of GSH in the DDTase system. The dimethoate $49r_2$ strain appears to owe its resistance to microsomal oxidation (Suplicy et al., 1972; Sawicki, 1974) in much the same manner as the Fc strain (Oppenoorth et al., 1972). Since the glutathione S-transferase systems are not operable in either the Fc or the dimethoate 49r2 strains, this may explain their relatively low levels of GSH.

The high levels of glutathione that have been reported in several R strains of house flies appears to be associated with high levels of glutathione S-transferase activity (Motoyama and Dauterman, 1975). Therefore, it would appear that GSH is present in sufficient quantities and is not a rate-limiting component in the resistance to insecticides in these house fly strains. One would

Table 3. Reduced Glutathione in Various Strains of the Adult House Fly[a,b]

Strain	2-day-old Adult		4-day-old Adult		6-day-old Adult		8-day-old Adult	
	Male	Female	Male	Female	Male	Female	Male	Female
Orlando DDT	18.4±1.0	18.4±2.5	40.5±0.6	31.9±0	53.7--	31.3±1.7	55.8--	38.7±1.6
Dimethoate 49r_2	23.1±2.9	19.6±1.3	37.6±1.9	23.7--	31.5--	32.1±2.7	38.2±2.2	32.5±0.9
Rutgers	33.9±0.6	30.0±2.0	40.2±0.6	45.2±5.1	77.7±1.8	48.4±2.0	71.7±5.1	59.2±1.9
Cornell-R	50.8±1.7	40.5±8.1	48.0±1.2	52.5±1.3	59.1±0.9	47.6±0.9	43.3±3.6	45.5±0
Fc	19.7±0.3	26.2±1.4	49.1±1.3	40.6±1.3	41.3±2.2	35.3±0.8	40.1±1.1	33.6±1.1
CSMA	13.8±0.7	14.2±0.6	21.7±0.5	20.5±0	31.7±0.6	27.9±0	37.6±0.3	27.2±1.7

[a] µg GSH/mg protein ± S.E.
[b] Data from Saleh et al. (1978).

also expect that the amount of GSH found in an insect strain is genetically determined in a similar manner to that of the glutathione S-transferases.

Glutathione S-Transferases and Resistance

Metabolism mediated by glutathione S-transferases has recently been implicated as a biochemical mechanism that contributes to organophosphorus insecticide resistance. Lewis (1969) first reported that several diazinon-resistant house fly strains that were known to have a gene "a" for low aliesterase activity also had high glutathione S-transferase activity toward diazinon and diazoxon, the major metabolites being identified as desethyl diazinon and desethyl diazoxon, respectively. Lewis and Sawicki (1971) confirmed that certain R house fly strains with low aliesterase and high phosphatase activity associated with chromosome II also had high glutathione S-transferase activity. The soluble enzymes formed desethyl diazinon as well as diethyl phosphorothioic acid from diazinon in the presence of GSH. Yang et al. (1971) found that diazinon was degraded faster by the soluble fraction from the Rutgers multiresistant house fly strain than from the CSMA house fly strain in the presence of GSH. The metabolites identified were diethyl phosphorothioic acid and, indirectly, desethyl diazinon. Subsequently, Dauterman (1971) postulated that diethyl phosphorothioic acid was formed from diazinon as a result of the transfer of the pyrimidinyl group to GSH by the glutathione S-transferase(s). The glutathione conjugate, S-(6-ethoxy-2-ethyl-4-pyrimidinyl) glutathione, was isolated and characterized by Shishido et al. (1972). This finding demonstrated that organophosphate insecticides could undergo two types of transfer, one reaction (a) involving alkyl transfer and the other (b) involving aryl transfer.

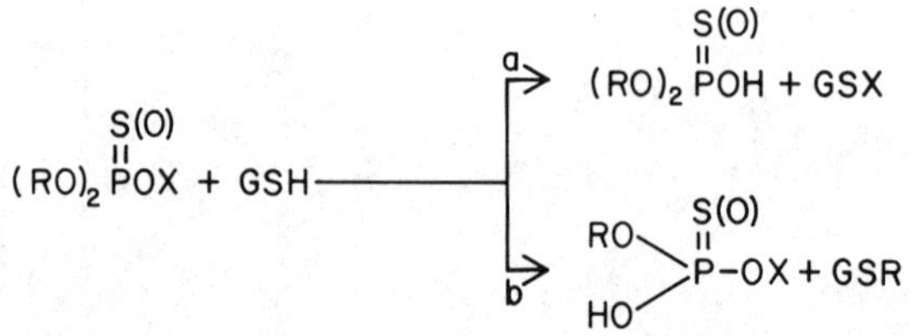

Oppenoorth et al. (1972) compared the glutathione-dependent metabolism of parathion by the soluble fraction from several R and S house fly strains. Both *O*-dealkylation and *O*-dearylation were demonstrated to occur, since diethyl phosphorothioic acid, desethyl parathion and ethyl glutathione were detected. Whereas Lewis and Sawicki (1971) found higher levels of desethyl diazinon than diethyl phosphorothioic acid, Oppenoorth et al. (1972) reported higher levels of diethyl phosphorothioic acid than desethyl parathion. The contribution of glutathione S-transferase to parathion resistance was concluded to be of minor importance because the level of transferase activity was not proportional to the level of resistance.

Motoyama and Dauterman (1972, 1974) observed a significant interstrain difference in the metabolism of azinphosmethyl by the soluble fraction from the Rutgers R strain and the S house fly strain in the presence of GSH (Figure 3). High glutathione S-transferase activity was found for azinphosmethyl, diazinon, methyl iodide and 3,4-dichloronitrobenzene; the latter two compounds are known substrates for the glutathione S-transferase system. With azinphosmethyl as the substrate, the only metabolite detected was desmethyl azinphosmethyl. No dimethyl phosphorothioic acid was found, which would indicate that the benzazimide moiety was not a suitable substrate for dearylation by the glutathione S-transferases.

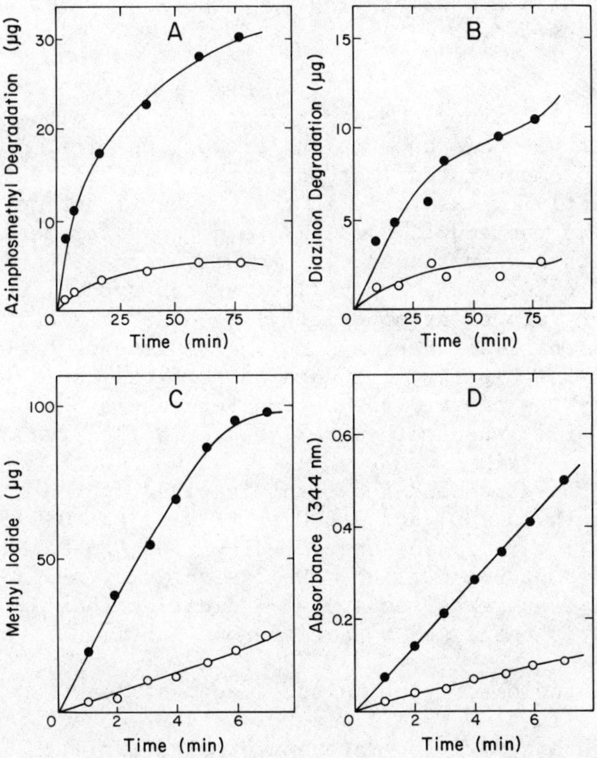

Figure 3. Glutathione S-transferase activity in the soluble fraction from resistant (●) and susceptible (o) house flies with the following substrates: A, azinphosmethyl; B, diazinon; C, methyl iodide, and D, 3,4-dichloronitrobenzene (adapted from Motoyama and Dauterman, 1974).

Table 4. Azinphosmethyl Degradation by Homogenates of Resistant (R) and Susceptible (S) Predacious Mite, *Neoseiulus fallacis*[a]

Cofactor	Degradation[b]	
	S	R
None	0	6.0
GSH	10.8	61.2
NADPH	9.8	17.2

[a]Data from Motoyama et al. (1971).
[b]μg of azinphosmethyl equivalents/100 mg of mites/2 hr.

In a parallel study the importance of glutathione S-transferases in azinphosmethyl resistance in the predacious mite, *Neoseiulus fallacis*, was demonstrated (Motoyama et al., 1971) (Table 4). High degradative activity was found in the soluble fraction. The activity was enhanced by the addition of GSH, and the major metabolite isolated was desmethyl azinphosmethyl. No difference in cholinesterase inhibition or cuticular permeability was observed. Substitution of the dimethoxy group of azinphosmethyl by other dialkoxy groups resulted in a significant decrease in the resistance factor (Motoyama et al., 1977b) (Table 5). The data support the conclusion that desmethylation by the glutathione S-transferases is the only mechanism responsible for azinphosmethyl resistance in the predacious mite.

Glutathione-dependent reactions involved in the metabolism of diazinon, parathion, DDT and γ-BHC paralleled the GSH-conjugation activity with methyl iodide and 3,4-dichloronitrobenzene in the various strains of house flies. DDTase activity did not parallel alkyl or aryl transferase activity, indicating that dehydrochlorination of DDT is catalyzed by a different enzyme.

Genetic studies of glutathione-dependent reactions were conducted with a diazinon R house fly strain and a susceptible house fly strain with mutant markers (Motoyama et al., 1977a). High levels of glutathione-dependent enzyme activity towards 3,4-dinitrobenzene, methyl iodide and γ-benzene hexachloride were controlled by gene(s) on chromosome II, which contrasted with previous studies (Lewis and Sawicki, 1971) that suggested that glutathione transferase activity was associated with gene "g" on the same chromosome. It was also

Table 5. Toxicity of Azinphosmethyl Analogs to a Resistant (R) and a Susceptible (S) Strain of Predacious Mite[a]

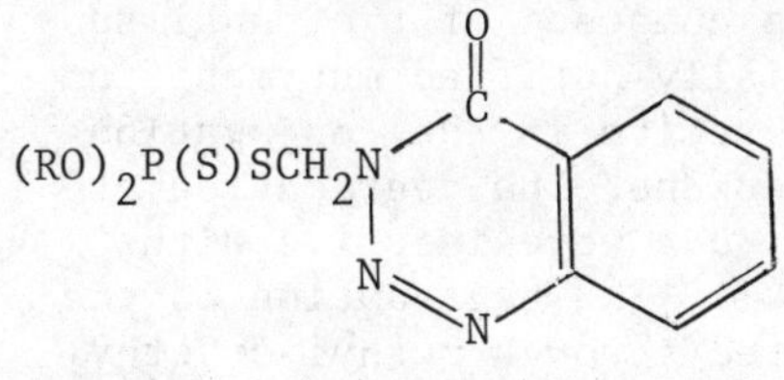

R	LC_{50}^{b} ($\mu g/cm^2$) S	LC_{50}^{b} ($\mu g/cm^2$) R	Resistance Factor R/S
CH_3	0.127	100,688	79,287
C_2H_5	0.750	7.96	11
n-C_3H_7	3.959	42.89	11
i-C_3H_7	>4,210	>4,210	--

[a]Data from Motoyama et al. (1977a).
[b]Determined 12 hours after exposure to treated filter paper.

shown that in the house fly, transferase activities towards 3,4-dichloronitrobenzene and methyl iodide were genetically inseparable and, therefore, were catalyzed by the same enzyme.

A tetrachlorvinphos-resistant house fly strain (Cornell-R strain) shown to have an acetylcholinesterase with reduced sensitivity to tetrachlorvinphos (Tripathi and O'Brien, 1973) also possessed high glutathione S-transferase and hydrolase activities. The high resistance exhibited by the Cornell-R strain appeared to be due to the combined action of these three different resistance mechanisms (Oppenoorth et al., 1977). The existence of high glutathione S-transferase activity in the Cornell-R strain was verified in another study (Motoyama et al., 1978). In subsequent studies with this strain, it was shown that tetrachlorvinphos was exclusively demethylated by the transferase and that although the hydrolases did not contribute significantly to tetrachlorvinphos resistance, they were important in the resistance to other OP compounds (Oppenoorth et al., 1979).

Properties of Insect Glutathione S-Transferase

Glutathione S-transferase was partially purified from the Rutgers-R and the susceptible CSMA strains of house flies. The purification procedures utilized DEAE-cellulose chromatography, ammonium sulfate fractionation, Sephadex G-200 and hydroxylapatite chromatography (Motoyama and Dauterman, 1977). The molecular weight was estimated to be 50,000 daltons by gel filtration, and SDS electrophoresis revealed that the enzyme was composed of two equal subunits of 23,000 daltons each. The partially purified enzymes from both the R and S house fly strains were active in the conjugation of methyl iodide and 3,4-dichloronitrobenzene, the degradation of organophosphorus insecticides and γ-BHC, but were inactive with respect to the dehydrochlorination of DDT. The degradation of the organophosphorus insecticides was effected through methyl or ethyl conjugation and/or "leaving group" conjugation (Table 6). It appears that the house fly enzyme catalyzes both alkyl and "leaving group" conjugation at relative rates that vary according to the structure of the insecticide, the concentration of both the substrate and glutathione, and possibly the reaction conditions. Dimethoxy compounds formed only the alkyl conjugates, while the diethoxy insecticides formed both types of conjugates.

Further studies with the glutathione S-transferase from the Cornell-R strain resulted in a 106-fold purification, and polyacrylamide gel electrophoresis of the purified enzyme resolved multiple activity bands (Motoyama and Dauterman, 1978). The molecular weights of the native enzyme and its primary dissociated form were estimated to be 54,000 daltons and 37,000 daltons, respectively. SDS gel

Table 6. O-Alkyl and "Leaving Group" Conjugation of Organophosphorus Insecticides by a Partially Purified Glutathione S-Transferase from the Rutgers Strain of House Flies[a]

Substrate	% Conjugation	
	Alkyl	"Leaving Group"
Dimethoxy		
malathion	100	0
methyl parathion	100	0
Diethoxy		
ethyl parathion	28	72
diazinon	63	37
diazoxon	0	100

[a]Data from Motoyama and Dauterman (1977b).

electrophoresis of the native enzyme and of individual electrophoretic bands resulted in the separation of a predominant subunit with an approximate molecular weight of 22,700 daltons.

NONSPECIFIC ESTERASES

The term "nonspecific esterases" will be used in the following discussion to refer to enzymes that have been detected with certain aliphatic or aromatic esters as substrates, but whose enzymological identity and function with regard to resistance have not been established. Numerous studies on insecticide resistance have suggested that it may be associated with both low and high levels of nonspecific esterase activity (Motoyama and Dauterman, 1974). In the course of purifying glutathione S-transferases from the house fly, it was noted that nonspecific esterase activity towards α-naphthyl acetate typically paralleled glutathione S-transferase activity. In other studies, extra electrophoretic esterase bands stained with α-naphthyl acetate were found in the R strain in which resistance was shown to be due to glutathione S-transferase activity (Motoyama et al., 1971). (See also papers by T. Miyata and K. Yasutomi in this volume.)

Recently, Keen and Jakoby (1978) reported that incubation of various purified liver glutathione S-transferases with *p*-nitrophenyl acetate in the presence of glutathione resulted in the formation of *p*-nitrophenol and the thiol ester; they also catalyzed the formation of *p*-nitrophenol from *p*-nitrophenyl trimethylacetate.

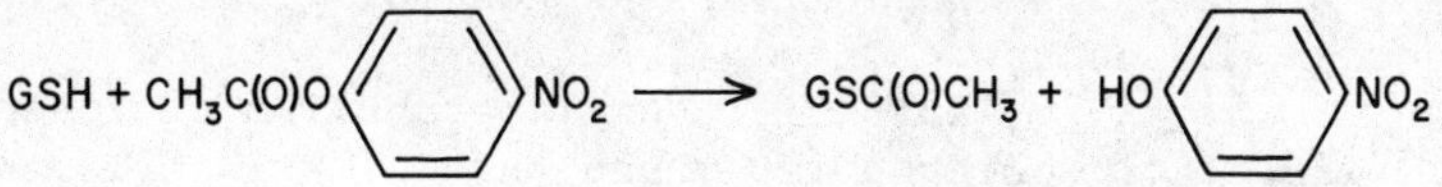

The formation of an alcohol or phenol from an ester is normally considered to be the result of hydrolysis, and it would be natural to assume that the reaction was catalyzed by nonspecific esterases. Sufficient endogenous glutathione is present in house fly homogenates (Saleh et al., 1978) and other insect preparations to mediate this type of reaction, and it is possible that the high nonspecific esterase activity often found to be associated with OP-resistance is in reality due to high glutathione S-transferase activity. Further studies are needed to verify this suggestion.

SUMMARY

Both the phosphorotriester hydrolases and glutathione S-transferases are biological mechanisms responsible for organophosphate insecticide resistance in certain species and certain strains of insects. They both detoxify the insecticide and often function in concert with other resistance mechanisms, such as the mixed-function oxidases, altered cholinesterase and penetration. The importance of triester hydrolases and glutathione S-transferases in resistance is dependent upon the structure of the insecticide, the concentration of the insecticide and possibly the stage of development of the insect.

Acknowledgement: Paper No. 6225 of the Journal Series of the North Carolina Agricultural Research Service, Raleigh, North Carolina. Work supported in part by PHS Research Grant ES-00044 from the National Institute of Environmental Health Science.

REFERENCES

Arias, I. M., and Jakoby, W. B., 1976, "Glutathione: Metabolism and Function," Raven Press, New York.

Boyland, E., and Chasseaud, L. F., 1969, The role of glutathione and glutathione S-transferase in mercapturic acid biosynthesis, *Advan. Enzymol.,* 32:173.

Chasseaud, L. F., 1973, The nature and distribution of enzymes catalysing the conjugation of glutathione with foreign compounds, *Drug Metab. Rev.,* 2:185.

Cohen, A. J., and Smith, J. N., 1964, The metabolism of some halogenated compounds by conjugation with glutathione in locusts and other insects, *Biochem. J.,* 90:457.

Cotty, V. H., and Henry, S. M., 1958, Glutathione metabolism in DDT resistant and susceptible house flies, *Contribs. Boyce Thompson Inst.,* 19:393.

Dauterman, W. C., 1971, Biological and nonbiological modifications of organophosphorus compounds, *Bull. WHO,* 44:133.

Dauterman, W. C., 1976, Extramicrosomal metabolism of insecticides, *in:* "Insecticide Biochemistry and Physiology," C. F. Wilkinson, ed., pp. 149-176, Plenum Press, New York.

Dyskstra, W. G., and Dauterman, W. C., 1978, Excretion, distribution and metabolism of S-(2,4-dinitrophenyl) glutathione in the American cockroach, *Insect Biochem.*, 8:263.

Eto, M., 1974, "Organophosphorus Pesticides: Organic and Biological Chemistry," CRC Press, Cleveland, Ohio.

Fukami, J., 1980, The metabolism of several insecticides by glutathione S-transferases, *in:* "Pharmacology and Theurapeutics," F. Matsumura, ed., in press, Pergamon Press, New York.

Fukami, J., and Shishido, T., 1963, Studies on the selective toxicities of organic phosphorus insecticides (III). The characters of enzyme system in cleavage of methyl parathion to desmethyl parathion in the supernatant of several species of homogenate, *Botyu-Kagaku,* 28:77.

Fukami, J., and Shishido, T., 1966, Nature of a soluble glutathione-dependent enzyme system active in cleavage of methyl parathion to desmethyl parathion, *J. Econ. Entomol.,* 59:1338.

Gessner, T., and Smith, J. N., 1960, The metabolism of chlorobenzene in locust. Excretion of unchanged chlorobenzene and cysteine conjugates, *Biochem. J.*, 75:165.

Heath, D. F., 1961, "Organophosphorus Poisons," Pergamon Press, New York.

Hollingworth, R. M., 1970, The dealkylation of organophosphorus triesters by liver enzymes, *in:* "Biochemical Toxicology of Insecticides," R. D. O'Brien and I. Yamamoto, eds., pp. 75-92, Academic Press, New York.

Jakoby, W. B., Ketley, J. N., and Habig, W. H., 1976, Rat glutathione S-transferases: Binding and physical properties, *in:* "Glutathione: Metabolism and Function," I. M. Arias and W. B. Jakoby, eds., pp. 213-223, Raven Press, New York.

Keen, J. H., and Jakoby, W. B., 1978, Glutathione transferases: Catalysis of nucleophilic reactions of glutathione, *J. Biol. Chem.*, 253:5654.

Lewis, J. B., 1969, Detoxication of diazinon by subcellular fractions of diazinon resistant and susceptible house flies, *Nature* (London), 224:917.

Lewis, J. B., and Sawicki, R. M., 1971, Characterization of the resistance mechanism to diazinon, parathion, and diazoxon in the organophosphorus SKA strain of house flies (*Musca domestica* L.), *Pestic. Biochem. Physiol.*, 1:275.

Lipke, H., and Chalkey, J., 1962, Glutathione, oxidized and reduced in some dipterans treated with 1,1,1-trichloro-2,2-di-(p-chlorophenyl) ethane, *Biochem. J.*, 85:104.

Lipke, H., and Kearns, C. W., 1960, DDT dehydrochlorinase, *in:* "Advances in Pest Control Research," Vol. III, R. L. Metcalf, ed., pp. 253-287, Interscience, New York.

Matsumura, F., and Hogendijk, C. J., 1964, The enzymatic degradation of parathion in organophosphate-susceptible and resistant house flies, *J. Agr. Food Chem.*, 12:447.

Motoyama, N., and Dauterman, W. C., 1972, In vitro metabolism of azinphosmethyl in susceptible and resistant house flies, *Pestic. Biochem. Physiol.*, 2:113.

Motoyama, N., and Dauterman, W. C., 1974, The role of nonoxidative metabolism in organophosphate resistance, *J. Agr. Food Chem.*, 22:350.

Motoyama, N., and Dauterman, W. C., 1975, Interstrain comparison of glutathione dependent reactions in susceptible and resistant houseflies, *Pestic. Biochem. Physiol.*, 5:489.

Motoyama, N., and Dauterman, W. C., 1977, Purification and properties of house fly glutathione S-transferase, *Insect Biochem.*, 7:361.

Motoyama, N., and Dauterman, W. C., 1978, Molecular weight, subunits and multiple forms of glutathione S-transferase from the house fly, *Insect Biochem.*, 8:337.

Motoyama, N., and Dauterman, W. C., 1980, Glutathione S-transferases: Their role in the metabolism of organophosphorus insecticides, *Rev. Biochem. Toxicol.*, 2:49.

Motoyama, N., Rock, G. C., and Dauterman, W. C., 1971, Studies on the mechanism of azinphosmethyl resistance in the predaceous mites, *Neoseiulus* (T.) *fallacis* (Family: Phytoseiidae), *Pestic. Biochem. Physiol.*, 1:205.

Motoyama, N., Dauterman, W. C., and Plapp, F. W., Jr., 1977a, Genetic studies of glutathione-dependent reactions in resistant strains of the house fly, *Musca domestica* L., *Pestic. Biochem. Physiol.*, 7:433.

Motoyama, N., Dauterman, W. C., and Rock, G. C., 1977b, Toxicity of O-alkyl analogues of azinphosmethyl and other insecticides to resistant and susceptible predacious mites, *Amblyseius fallacis*, *J. Econ. Entomol.*, 70:475.

Motoyama, N., Kulkarni, A. P., Hodgson, E., and Dauterman, W. C., 1978, Endogenous inhibitors of glutathione S-transferase in house flies, *Pestic. Biochem. Physiol.*, 9:225.

Nakatsugawa, T., Tolman, N. M., and Dahm, P. A., 1969, Metabolism of S^{35} parathion in the house fly, *J. Econ. Entomol.*, 62:408.

Nolan, J., and O'Brien, R. D., 1970, Biochemistry of resistance to paraoxon in strains of housefly, *J. Agr. Food Chem.*, 18:802.

O'Brien, R. D., 1960, "Toxic Phosphorus Esters: Chemistry, Metabolism and Biological Effects," Academic Press, New York.

Oppenoorth, F. J., 1971, Resistance in insects: The role of metabolism and the possible use of synergists, *Bull. WHO*, 44:311.

Oppenoorth, F. J., and van Asperen, K., 1960, Allelic genes in the house fly producing modified enzymes that cause organophosphate resistance, *Science*, 132:298.

Oppenoorth, F. J., and van Asperen, K., 1961, The detoxication enzymes causing organophosphate resistance in the house fly; properties, inhibition and the action of inhibitors as synergists, *Entomol. Exp. Appl.*, 4:311.

Oppenoorth, F. J., and Voerman, S., 1975, Hydrolysis of paraoxon and malaoxon in three strains of *Myzus persicae* with different degrees of parathion resistance, *Pestic. Biochem. Physiol.*, 5:431.

Oppenoorth, F. J., and Welling, W., 1976, Biochemistry and physiology of resistance, *in:* "Insecticide Biochemistry and Physiology," C. F. Wilkinson, ed., pp. 507-551, Plenum Press, New York.

Oppenoorth, F. J., Rupes, V., El Bashir, S., Houx, N. W. H., and Voerman, S., 1972, Glutathione-dependent degradation of parathion and its significance for resistance in the housefly, *Pestic. Biochem. Physiol.*, 2:262.

Oppenoorth, F. J., Smissaert, H. R., Welling, W., van der Pas, L. J. T., and Hitman, K. J., 1977, Insensitive acetylcholinesterase, high glutathione S-transferase and hydrolytic activity as resistance factors in a tetrachlorvinphos-resistant strain of house fly, *Pestic. Biochem. Physiol.*, 7:34.

Oppenoorth, F. J., van der Pas, L. J. T., and Houx, N. W. H., 1979, Glutathione S-transferase and hydrolytic activity in a tetrachlorvinphos-resistant strain of house fly and their influence on resistance, *Pestic. Biochem. Physiol.*, 11:176.

Saleh, M. A., Motoyama, N., and Dauterman, W. C., 1978, Reduced glutathione in the house fly: Concentration during development and variation in strains, *Insect Biochem.*, 8:311.

Sawicki, R. M., 1974, Genetics of resistance to a dimethoate-selected strain of house flies (*Musca domestica* L.) to several insecticides and methylene dioxyphenyl synergists, *J. Agr. Food Chem.*, 22:344.

Shishido, T., and Fukami, J., 1963, Studies on the selective toxicities of organic phosphorus insecticides (II). The degradation of the ethyl parathion, methyl parathion, methyl paraoxon and sumithion in mammal, insect and plant, *Botyu-Kagaku*, 28:69.

Shishido, T., Usui, K., Sato, M., and Fukami, J., 1972, Enzymatic conjugation of diazinon with glutathione in rat and American cockroach, *Pestic. Biochem. Physiol.*, 2:51.

Suplicy, N., Guthrie, F. E., and Dauterman, W. C., 1972, Studies on the toxicity of a series of dimethoate analogs to a resistant and susceptible strain of house flies, *J. Econ. Entomol.*, 65:1585.

Tripathi, R. K., and O'Brien, R. D., 1973, Insensitivity of acetylcholinesterase as a factor in resistance of house flies to the organophosphate Rabon, *Pestic. Biochem. Physiol.*, 1:61.

van Asperen, K., and Oppenoorth, F. J., 1959, Organophosphate resistance and esterase activity in house flies, *Entomol. Exp. Appl.*, 2:48.

Welling, W., Blaakmeer, P., Vink, G. J., and Voerman, S., 1971, In vitro hydrolysis of paraoxon by parathion resistant house flies, *Pestic. Biochem. Physiol.*, 1:61.

Yang, R. S. H., Hodgson, E., and Duaterman, W. C., 1971, Metabolism in vitro of diazinon and diazoxon in susceptible and resistant house flies, *J. Agr. Food Chem.*, 19:14.

ROLE OF DETOXICATION ESTERASES IN INSECTICIDE RESISTANCE

Kazuo Yasutomi

Pest Control Laboratory
Department of Medical Entomology
National Institute of Health
Kamiosaki, Shinagawa-ku, Tokyo 141, Japan

INTRODUCTION

Organophosphate (OP) resistance among Japanese mosquitoes was first observed in 1967 in the city of Amagasaki in the larvae of *Culex pipiens pallens*, and since then resistant populations of the *Culex pipiens* complex have spread widely to many parts of the country. Recent tests indicated that several populations of *Culex tritaeniorhynchus*, the vector of Japanese B-encephalitis, are also resistant to organophosphorus insecticides.

The mechanism of OP resistance is mainly detoxication by the enzymes hydrolyzing the insecticides. Malathion-resistant strains of insects are characterized by increased carboxyesterase, which attacks the unique weak point of this OP molecule. Matsumura and Brown (1961, 1963) studied malathion resistance in *Culex tarsalis* and found a considerable increase in carboxyesterase activity and a slight increase in phosphatase activity in the resistant (R) strains compared with the susceptible (S) ones.

By using the thin layer electrophoresis method, Ogita and Kasai (1965) discovered that soluble esterases of house flies separated into 24 bands that hydrolyzed β-naphthylacetate, and that no esterase band was related to diazinon resistance.

Ozaki (1964), Kasai and Ogita (1965), Ozaki et al. (1966) and Ozaki (1969) studied esterase activities in the green rice leafhopper, *Nephotettix cincticeps* Uhler, or the smaller brown planthopper, *Laodelphax striatellus* Fallen. They reported that esterase bands that hydrolyze β-naphthylacetate were higher in the malathion-R strains than in the S strains.

Since 1950, the author has been engaged in the research of insecticide resistance development in medical insects in Japan, the evaluation of the resistance level, cross resistance, and resistance mechanisms. The present paper deals with the results of experiments using thin-layer electrophoresis on the relationship between OP resistance and esterase activity in mosquitoes.

MATERIALS AND METHODS

Insects

The following colonies of test insects were obtained from the sources indicated:

(A) *Culex pipiens pallens*

(1) Amagasaki: open sewer in Amagasaki (1969).
(2) Tsunashima: open sewer in Tsunashima district, Yokohama (1976).
(3) Susceptible: standard strain maintained in our laboratory since 1965.

(B) *Culex pipiens fatigans*

(1) Okinawa: open sewer in Koza, Okinawa (1977).
(2) Susceptible: open sewer in Chichijima, Ogasawara (1968).

(C) *Culex tritaeniorhynchus*

(1) Kyoto: paddy field in the suburbs of Kyoto (1966).
(2) Yokohama: paddy field in the suburbs of Yokohama (1970).
(3) Susceptible: standard strain, maintained in our laboratory since 1960.

(D) *Aedes aegypti*

(1) Bangkok: a field colony sent by Dr. Boonluan Phanthumachinda, collected from Bangkok, Thailand, in 1977.
(2) Susceptible: standard strain, collected from Bangkok in 1967 by Dr. Manabu Sasa.

The larvae were fed dry brewer's yeast and were reared at a temperature of 25±1°C with 16 hr of illumination per day. Fourth instar larvae and five-day-old adults were used in the test. Adult mosquitoes were fed a 3% sugar solution.

Insecticides

Ethanol solutions of insecticides were prepared as the sample for testing the resistance levels of larvae, and acetone solutions were prepared for adult mosquitoes.

Test Methods for Resistance Levels

The percentage mortalities were recorded according to the WHO method; the early fourth instar larvae were exposed to one of several concentrations of diazinon in 250 ml dishes for 24 hr at a temperature of 25°C. In order to measure the resistance levels of adults, 0.25 μl of acetone solution was applied topically to the prothorax by means of a microsyringe. The treated mosquitoes were then transferred into the clean tubes for a 24 hr mortality count. From these data the regression of the probit mortality on log of dosage was computed and the LC_{50} and LD_{50} were obtained.

Crossing Experiments

Crossing experiments were carried out in breeding cages (30 x 30 x 30 cm) at a temperature of 25 ± 1°C, and the adults were fed a 3% sugar solution. The female mosquitoes were also fed on mice. Reciprocal crosses were made between the OP-resistant and S strains, and the susceptibility tests and thin-layer eletrophoresis experiments were carried out on early fourth instar larvae of F_1, F_2 and back-crossed offspring.

Electrophoresis

The esterase activity of mosquitoes was measured by using the thin-layer electrophoresis method described by Ogita (1963, 1964). The medium containing 0.7 g of agar and 2 g of polyvinyl pyrrolidone in 100 ml of potassium phosphate buffer (pH 6.8, ionic strength 0.025μ) was used for the separation of esterase, and 20 ml of the agar solution were placed on a glass plate (16.5 x 15 cm) to form a thin layer 0.8 mm thick.

A single insect was crushed with a glass rod on a glass plate in a drop of deionized water and a strip of filter paper 1 mm x 5 mm was immersed into the resultant brei and placed on an agar plate. After the samples infiltrated the agar gel layer in a freezing box, strips were removed and the agar gel plate was connected by filter paper to a buffer solution of ionic strength 0.05μ in the electrode vessels. A constant current of 1.7 mA per cm width of the agar gel plate was applied for 80 min at 5°C in a cold box. After electrophoresis, the agar gel was sprayed with 1% β-naphthylacetate acetone solution and incubated for 30 min at 37-38°C. The esterase activities were then revealed by spraying with 2% aqueous solution of naphthanil diazoblue B. These plates were washed with running water and dried at 50°C.

Specific esterases were revealed by immersing the agar gel plates after electrophoresis into the solutions consisting of 0.2-1% substrate (acetylcholine chloride 0.2%, methyl-*n*-butyrate and phenylacetate 1%), 21 mg of $NaHCO_3$ and 2 ml of 1% bromthymol blue ethanol solution in 100 ml of distilled water for 10-15 min and incubating for 5-10 min at 37°C.

RESULTS

Culex pipiens pallens

Amagasaki colony. The esterases of *Culex pipiens pallens* larvae hydrolyzing β-naphthylacetate were separated into four electrophoretic bands, which were designated as E_1 to E_4 from the anodal to the cathodal side. The esterase zymograms of test colonies are shown in Figure 1, and the relationship between esterase activity and the level of diazinon resistance is shown in Tables 1 and 2. It is evident that the R Amagasaki colony has high esterase activity of

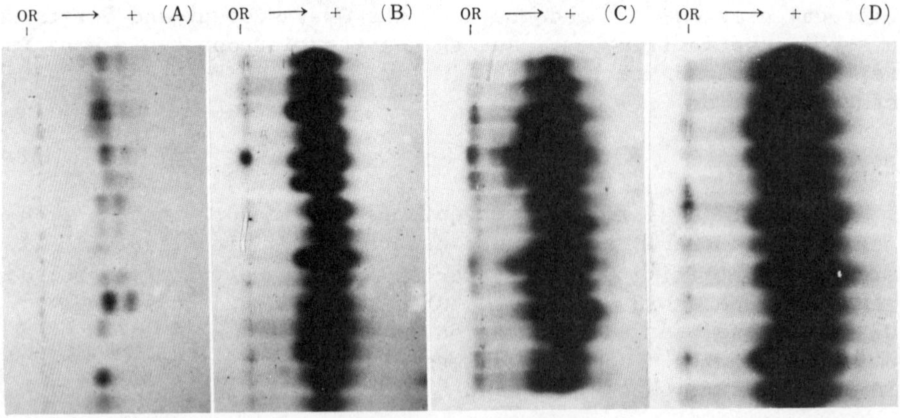

Figure 1. Esterase zymograms for four colonies of *Culex pipiens pallens*. The esterase activity was detected by β-naphthylacetate and naphthanil diazoblue B. OR shows the line of origin in electrophoresis (Yasutomi, 1970); A, susceptible strain; B, Amagasaki colony; C, Amagasaki colony selected further with diazinon pressure for two generations; D, Amagasaki colony selected further with diazinon pressure for five generations.

Table 1. Relationship Between Activity of the E_2 Esterase and Resistance to Diazinon in *Culex pipiens pallens* Larvae

Colony	No. of insects used	Activity of the E_2 band				% frequency of high activity	Levels of resistance LC_{50} (ppm)
		High	Inter-mediate	Low	No.		
Susceptible	50	0	2	23	25	0	0.037
Amagasaki	50	46	3	1	0	92.0	0.46
Selected Amagasaki (with diazinon pressure for two generations)	30	29	1	0	0	96.7	0.85
Selected Amagasaki (with diazinon pressure for five generations)	50	50	0	0	0	100.0	1.27

Table 2. Relationship Between Activity of the E_3 Esterase and Resistance to Diazinon in *Culex pipiens pallens* Larvae

Colony	No. of insects used	Activity of the E_3 band				% frequency of high activity	Levels of resistance LC_{50} (ppm)
		High	Inter- mediate	Low	No.		
Susceptible	50	0	15	29	6	0	0.037
Amagasaki	50	16	17	11	6	32.0	0.46
Selected Amagasaki (with diazinon pressure for two generations)	30	14	14	2	0	46.7	0.85
Selected Amagasaki (with diazinon pressure for five generations)	50	50	0	0	0	100.0	1.27

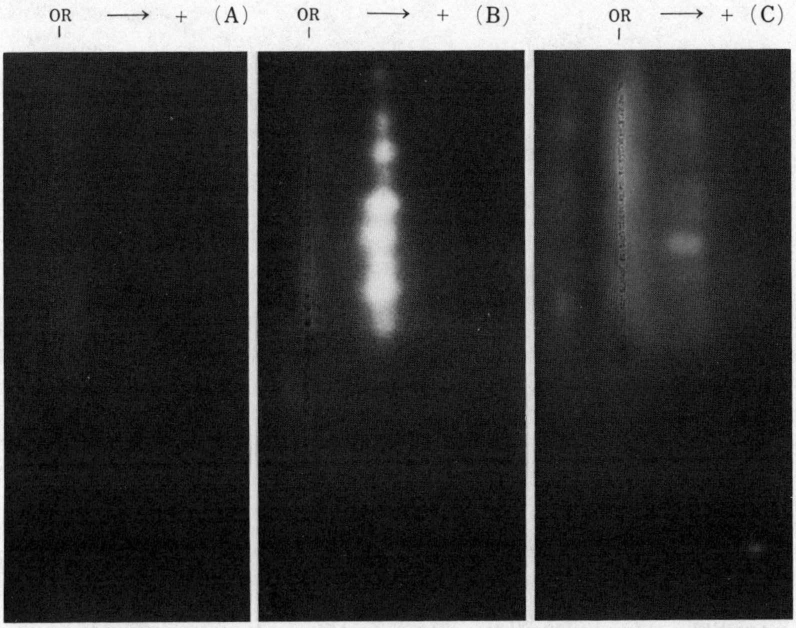

Figure 2. Specific esterase zymograms detected by the pH-indicator method using the following substrates: A, acetylcholine chloride; B, phenylacetate; C, methyl-*n*-butyrate. Upper part (10 insects) of each zymogram indicates esterase of the Amagasaki colony and lower part (10 insects) indicates that of the susceptible strain of *Culex pipiens pallens* (Yasutomi, 1971).

the E_2 and E_3 bands in comparison with the S strain. This colony, on further selection with diazinon pressure for five generations in our laboratory, showed the highest LC_{50} value and markedly high esterase activities of both the E_2 and E_3 bands.

In order to characterize these enzymes, the substrate specificity and inhibitor susceptibility of the E_2 or E_3 bands were studied. Esterase zymograms revealed by the pH-indicator method are shown using acetylcholine chloride (Figure 2A), phenylacetate (Figure 2B) and methyl-*n*-butyrate as substrates (Figure 2C). The esterase that hydrolyzes β-naphthylacetate also has the ability to hydrolyze phenylacetate and methyl-*n*-butyrate (Figure 2B,C). However, there were no significant differences in cholinesterase activity between the R and S strains, as shown in Figure 2A.

These results are quite different from those of the house fly, in which methyl-*n*-butyrate-splitting or phenylacetate-splitting

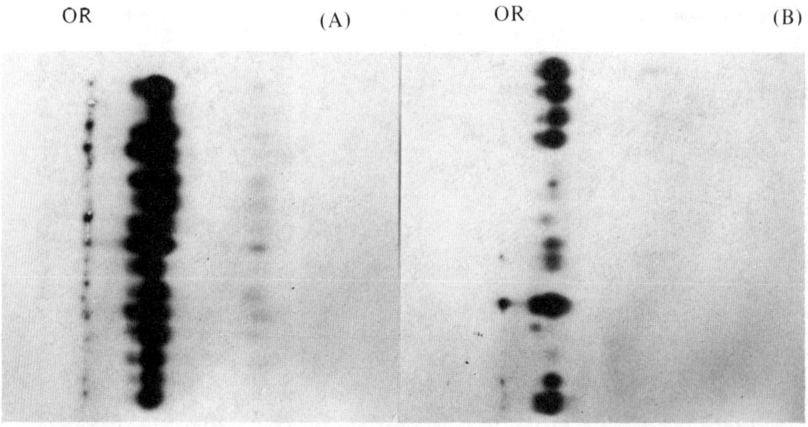

Figure 3. Esterase zymograms for larvae of *Culex pipiens pallens*. A, Tsunashima A; B, Tsunashima B.

esterases were somewhat lower in organophosphate-R strains than in S strains, and these esterases were not revealed by β-naphthylacetate (van Asperen and Oppenoorth, 1959; Ogita and Kasai, 1965a).

Tsunashima colonies. In Tsunashima A district, fenitrothion became ineffective after five years of heavy use, and increased resistance to fenitrothion has been discovered, the LC_{50} value being 0.153 ppm (22.4 times that of the S strain). This colony has high esterase activity of the E_2 and E_3 bands, as shown in Figure 3A. In contrast, the Tsunashima B population, with a moderate increase in OP resistance, contains individuals of high and low activity of the E_2 band, as shown in Figure 3B.

In the green rice leafhopper, *Nephotettix cincticeps* Uhler, ali-esterase activity was higher in malathion-R strains than in S strains, and no significant difference in cholinesterase activity was found (Hayashi and Hayakawa, 1962; Kojima et al., 1963). Kasai and Ogita (1965) studied the esterases of green rice leafhoppers by thin-layer electrophoresis and reported that the malathion-R strains showed a markedly higher E_9 esterase band, which hydrolyzes β-naphthylacetate, and that the E_9 esterase had the ability to hydrolyze methyl-*n*-butyrate, tributyrin and phenylacetate, but not acetylcholine. Osaki et al. (1966) reported that the malathion-R colonies of green rice leafhoppers showed high activity of the E_2 (reported as E_9 by Kasai and Ogita), E_3 and/or E_4 bands.

The results of the present study suggest that the mechanism of organophosphate resistance in *Culex pipiens pallens* may resemble that of green rice leafhoppers.

Table 3. LC_{50} Values in ppm of *Culex p. fatigans*

Insecticides	Ogasawara (S)	Okinawa (R)
Malathion	0.013	2.05 (158)[a]
Diazinon	0.024	0.152 (6.3)
Fenitrothion	0.0037	0.032 (8.6)
Fenthion	0.0028	0.057 (20.4)
Temephos	0.0019	0.045 (23.7)

[a]R/S

Crossing experiment. The relationship between esterase activities and log dosage-mortality curves of fenitrothion was tested in the R Tsunashima A strain, S strain, F_1, F_2 and backcross offspring. The results of the crossing experiment suggest that inheritance of fenitrothion resistance and esterase activity is controlled by the same factor of intermediate dominance located on the autosomal chromosome.

Culex pipiens fatigans

Development of resistance in the Okinawa strain. Table 3 shows LC_{50} values of the fourth instar larvae of the S Ogasawara strain and the OP-R Okinawa strain of *C. p. fatigans* for various OP compounds. The resistance ratio of the Okinawa strain to the Ogasawara strain was about 160 for malathion.

Thin-layer agar gel electrophoresis revealed four bands of esterases hydrolyzing β-naphthylacetate. Of these, the esterase activities of the E_2 and E_4 bands were higher in the R Okinawa strain than in the S strain, and the frequency of individuals with high esterase activity was 88.3% in the Okinawa strain. Selection pressure using malathion had been applied to the larvae of the Okinawa strain in the laboratory. The resistance levels attained were up to 692 with high esterase activities in all individuals by the fifth generation, as shown in Figure 4. An increase in the esterase activity was closely correlated with malathion resistance.

Resistance mechanism. The relationship between the carboxylester degrading enzyme and the esterase bands of the malathion-selected strain was studied by means of the agar gel electrophoresis method. These high esterase bands disappeared completely with treatment by

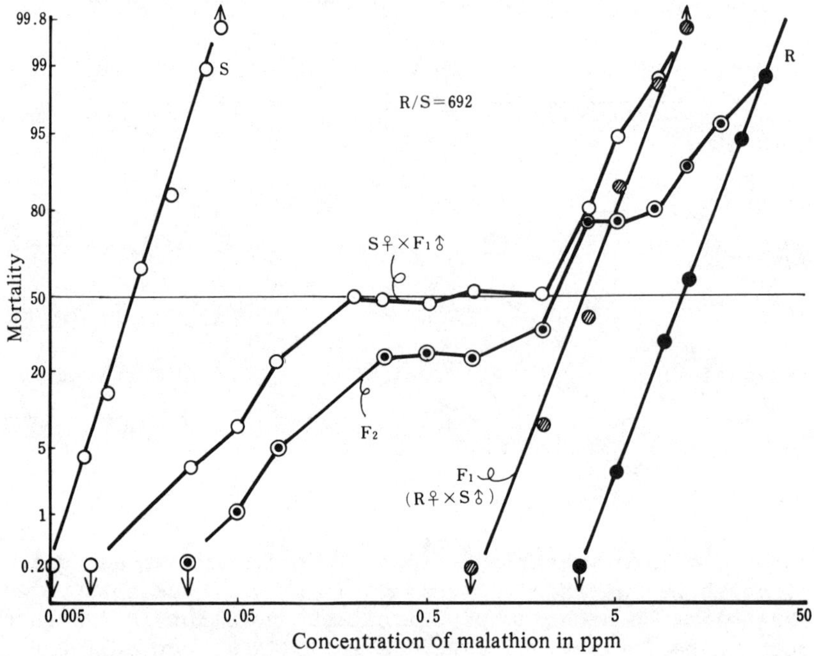

Figure 4. Concentration-mortality lines of the fourth instar larvae of *Culex p. fatigans* for malation.

K_2 (2-phenoxy-4*H*-1,3,2-benzodioxaphosphorin-2-oxide) and TPP (triphenyl phosphate) at 1 ppm for 4 hr. These synergists of malation are effective against the R strain: dipping tests using the WHO method on the fourth instar larvae, K_2 and TPP showed remarkable synergistic activities towards malathion (Table 4). The synergism of K_2 and TPP may be at least partly due to the inhibition of esterase. The results suggest that the malathion resistance of this strain is closely

Table 4. Effect of Synergists on the Fourth Instar Larvae of *Culex p. fatigans*

Strain	LC_{50} (ppm)		
	Malathion alone	+TPP (1:1)	+K_2 (1:1)
Selected Okinawa	9.0	0.126	0.132
Ogasawara (Susceptible)	0.013	0.012	0.014

correlated with the increase of carboxylesterase activity hydrolyzing malathion.

Crossing experiment. Females of the selected Okinawa strain, which was highly resistant to malathion, were crossed with males of the susceptible Ogasawara strain, and the F_1 offspring were then reciprocally backcrossed with the S strain. The F_1 hybrid was resistant but showed a somewhat lower degree of resistance than the R strain. The F_2 hybrid obtained from mass mating of the F_1 was estimated to be composed of a mixture of SS, SR and RR in the ratio of 1:2:1, as shown by the presence of a plateau at the 25% mortality level and an inflection point at the 75% level of the log dosage-probit mortality line

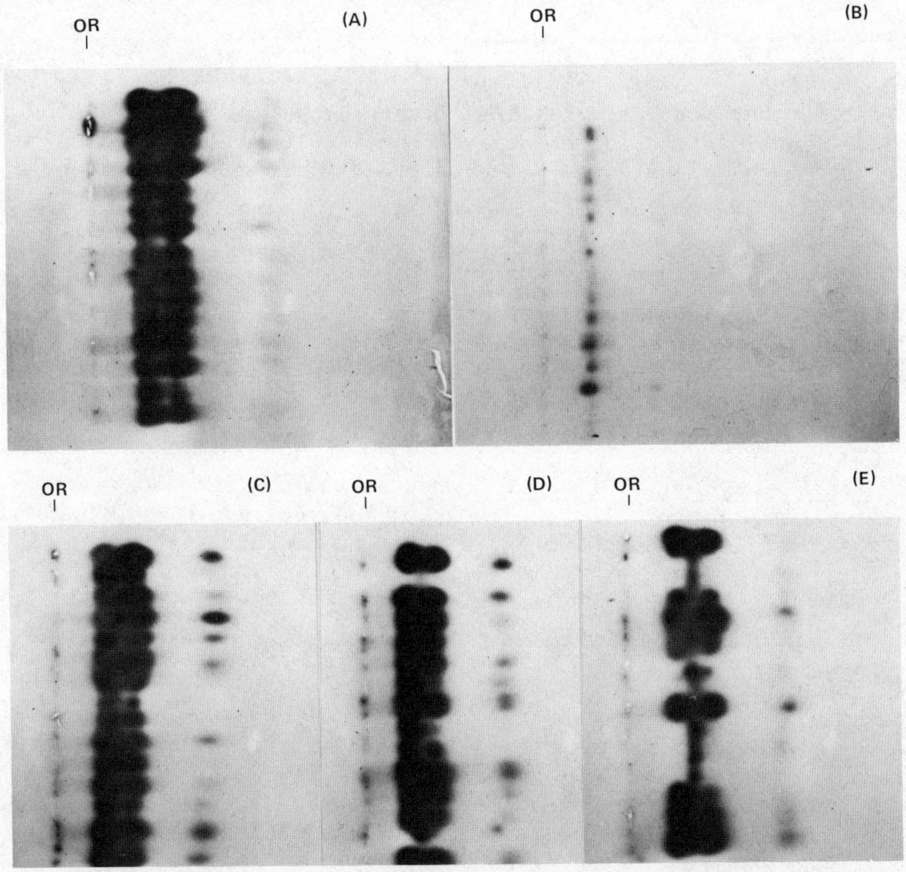

Figure 5. Esterase zymograms for larvae of *Culex p. fatigans*. A: Resistant strain. B: Susceptible strain. C: F_1 (R♀ x S♂). D: F_2. E: Backcross (S♀ x F_1♂).

Table 5. Resistance Levels of *Culex tritaeniorhynchus summorosus*

Colony	LC_{50} fourth instar larvae (ppm)	LD_{50} of adults (microgram/)
Susceptible	0.0055	0.0049
Yokohama	0.026	-
Kyoto	0.021	0.028
Diazinon-selected Kyoto	0.034	0.052

(Figure 4). There was a plateau or an inflection at the 50% mortality level in the log dosage-probit line for each of the backcrosses, indicating that monofactorial inheritance by an autosomal gene of partial dominance exists in resistance to malathion. This conclusion is supported by esterase zymograms obtained by agar gel electrophoresis (Figure 5).

Culex tritaeniorhynchus summorosus

Resistance levels of larvae. As shown in Table 5, the LC_{50} values of the field colonies (0.021 ppm for the Kyoto colony, 0.026 ppm for the Yokohama colony and 0.034 ppm for the diazinon-selected Kyoto colony) for diazinon were about 5 to 6 times as high as those of the S laboratory strain (0.0055 ppm).

Resistance levels of adults. Resistance to diazinon was discovered in the Kyoto colony, the LD_{50} values being 0.028 to 0.052 μg per female. These adults were 6 to 10 times more resistant to diazinon than normal insects (Table 5).

Diazinon resistance and esterase activity. When the esterases of *Culex tritaeniorhynchus summorosus* were subjected to thin layer electrophoresis, the esterases of larvae and adults hydrolyzing β-naphthylacetate were separated into several electrophoretic bands. Although there were remarkable individual variations in their esterase patterns, the E_1, E_3 and E_4 bands were commonly observed for many individuals of the R colonies. Therefore, the E_1, E_3 and E_4 bands, which were designated as E_1 to E_4 from the anodal to the cathodal side, were analyzed in the present study. Each esterase band was divided into four types, according to its degree of staining, i.e., high, intermediate, low and no activity. The rate of occurrence of each band is shown in Tables 6 and 7. It is evident that the R

Table 6. Percent Frequency of Individuals of Various Esterase Activities in *Culex tritaeniorhynchus summorosus* Larvae

Colony	E_1 band			E_3 band			E_4 band		
	H	I	L or N	H	I	L or N	H	I	L or N
Susceptible	0	0	100.0	0	18.3	81.7	0	0	100.0
Yokohama	15.0	5.0	88.0	45.0	15.0	40.0	30.0	15.0	55.0
Kyoto	11.7	8.3	80.0	28.3	31.7	40.0	11.7	15.0	73.3
Diazinon-selected Kyoto	23.3	33.3	43.3	53.3	30.0	16.7	33.3	30.0	36.7

H, high activity; I, intermediate activity; L, low activity; N, no detectable esterase activity.

Table 7. Percent Frequency of Individuals of Various Esterase Activities in *Culex tritaeniorhynchus summorosus* Adults

Colony	E_1 band			E_3 band			E_4 band		
	H	I	L or N	H	I	L or N	H	I	L or N
Susceptible	0	10.0	90.0	0	31.7	68.3	0	0	100.0
Kyoto	6.7	13.3	80.0	26.7	20.0	53.3	13.3	33.3	53.3
Diazinon-selected Kyoto	100.0	0	0	61.7	25.0	13.3	31.7	33.3	35.0

H, high activity; I, intermediate activity; L, low activity; N, no detectable esterase activity.

colonies have high esterase activity of these bands in comparison with the S strain. The Kyoto strain further selected with diazinon pressure in our laboratory for five generations showed the highest resistance level and high esterase activities of the E_1, E_3 and E_4 bands. In contrast to the R colony, no individual from the S strain showed high activity of its esterase bands. These results seem to suggest that the mechanism of OP resistance in *C. tritaeniorhynchus summorosus*, as in *C. p. pallens*, is similar to that of the green rice leafhopper, *Nephotettix cincticeps* Uhler.

Aedes aegypti

Development of resistance to malathion and esterase activity have been studied with the fourth instar larvae of *Aedes aegypti*.

Malathion pressure has been applied to a field strain collected in Bangkok, Thailand. This strain had 8.3 times the malathion resistance of the S strain, and the resistance level was increased to 22.7 times that of the S strain with higher esterase activities in most individuals by the fifteenth generation, as shown in Figure 6.

CONCLUSIONS AND DISCUSSION

The esterase zymograms of house flies revealed by β-naphthylacetate indicated different patterns among the testing strains, and no esterase band was correlated with OP resistance (Ogita, 1963; Ogita and Kasai, 1965). On the contrary, the results of recent studies showed that higher esterase zymograms were parallel to malathion resistance in the green rice leafhoppers and the smaller brown planthoppers (Ozaki et al., 1966; Ozaki, 1969).

Among the culicine mosquitoes, four species and subspecies of *Culex* have developed resistance to organophosphorus insecticides, and the failure to control them has been reported from many districts in Japan. In order to pursue the relationship between OP resistance and esterase activity in mosquitoes, the agar gel thin layer electrophoresis method was carried out by using *Culex pipiens pallens*, *C. p. fatigans*, *C. tritaeniorhynchus* and *Aedes aegypti* as testing insects.

In the present study on four species of culicine mosquitoes, esterase bands that hydrolyze β-naphthylacetate showed higher activities in the resistant colonies compared with the susceptible strains. The β-naphthylacetate-hydrolyzing esterase bands had an ability to hydrolyze organic acid esters, such as methyl-*n*-butyrate and phenylacetate.

These results suggest that the mechanism of OP resistance in culicine mosquitoes may be different from that in house flies, and it may be closely similar to that of green rice leafhoppers or smaller brown planthoppers. Genetic studies in the *Culex pipiens* complex have shown that OP resistance and high esterase activity were inherited as a single factor with intermediate dominance.

REFERENCES

Hayashi, M., and Hayakawa, M., 1962, Malathion tolerance in *Nephotettix cincticeps* Uhler, *Jap. J. Appl. Ent. Zool.*, 6:250.

Kasai, T., and Ogita, Z., 1965, Studies on malathion-resistance and esterase activity in green rice leafhoppers, *SABCO J.*, 1:130.

Kojima, K., Ishizuka, T., and Kitakata, S., 1963, Mechanism of resistance to malathion in the green rice leafhopper, *Nephotettix cincticeps*, *Botyu-Kagaku*, 28:17.

Matsumura, F., and Brown, A. W. A., 1961, Biochemistry of malathion resistance in *Culex tarsalis*, *J. Econ. Ent.*, 54:1176.

Matsumura, F., and Brown, A. W. A., 1963, Studies on carboxylesterase in malathion-resistant *Culex tarsalis*, *J. Econ. Ent.*, 56:381.

Ogita, Z., 1963, Separation and revelation of isozymes by means of the thin layer electrophoresis, *Nucleus and Cytoplasm*, 5:7.

Ogita, Z., and Kasai, T., 1964, Separation and revelation of specific esterase in thin layer electrophoresis by means of a pH-indicator method, *SABCO J.*, 1:37.

Ogita, Z., 1964, Improved agar gel media for thin layer electrophoresis, *Med. J. Osaka Univ.*, 15:141.

Ozaki, K., 1969, The resistance to organophosphorus insecticides of the green rice leafhopper, *Nephotettix cincticeps* Uhler, and the smaller brown planthopper, *Laodelphax striatellus* Fallen, *Rev. Plant Prot. Res.*, 2:1.

Ozaki, K., Kurosu, Y., and Koike, H., 1966, The relation between malathion resistance and esterase activity in the green rice leafhopper, *Nephotettix cincticeps* Uhler, *SABCO J.*, 2:28.

van Asperen, K., and Oppenoorth, F. J., 1959, Organophosphate resistance and esterase activity in houseflies, *Ent. Exp. Appl.*, 2:48.

Yasutomi, K., 1970, Studies on organophosphate-resistance and esterase activity in the mosquitoes of the *Culex pipiens* group, *Jap. J. Sanit. Zool.*, 21:41.

Yasutomi, K., 1971, Studies on diazinon-resistance and esterase activity in *Culex tritaeniorhynchus*, *Jap. J. Sanit. Zool.*, 22:8.

ENZYME INDUCTION, GENE AMPLIFICATION AND INSECT RESISTANCE TO INSECTICIDES

Leon C. Terriere

Department of Entomology
Oregon State University
Corvallis, Oregon 97331

INTRODUCTION

Perhaps the most fully understood mechanism of insecticide resistance in insects is that due to increased metabolism of the toxicant. We will refer to this as biochemical resistance. It is obvious that any metabolism that inactivites an insecticide will be beneficial to the insect and that such traits will be transmitted genetically. What is not so obvious, however, is how the organism achieves the observed increase in enzyme activity. We will consider some possibilities in this chapter.

The first possibility to be mentioned is the production of a mutant enzyme capable of attacking different substrates than had been attacked before. This was observed with the altered aliesterase discussed elsewhere in this book and will not be mentioned further here.

Other possibilities include gene amplification, i.e., the presence of multiple copies of the structural genes that direct the synthesis of enzymes, and induction, the phenomenon wherein exogenous or endogenous chemicals stimulate the production of additional enzyme. Both possibilities are supported by experimental evidence. This evidence and its bearing on the phenomenon of biochemical resistance will be discussed in the following paragraphs.

GENE AMPLIFICATION

Gene amplification refers to the process by which variously postulated alterations of gene linkage result in the multiplication of the normal complement of genes that code for a particular enzyme or enzyme system. This "amplification" could provide as little as one duplication of a gene to as much as several hundred copies (Alt et al., 1978; Schimke et al., 1978). Whether the amplification is a result of a genetic accident, such as uneven crossing over or translocations, or of some evolutionary adaptation process, such as DNA insertion, is not known. It is clear, however, that the magnification of gene capacity could result in a larger supply of the enzymes that metabolize insecticides.

The possibility that gene amplification might be involved in biochemical resistance to insecticides was suggested by Walker and Terriere (1970), who developed their hypothesis from the results of enzyme induction studies with the house fly. They found that dieldrin induced a microsomal oxidase system, heptachlor epoxidase, to differing extents in three dieldrin-resistant strains. When the induction was expressed in terms of the net increase in enzyme activity, it was observed that the strain with the highest basal epoxidase activity, Isolan-B, was induced approximately three times as much as the low oxidase strain, dieldrin-curly wing (Dld:cyw), while the strain intermediate in microsomal oxidase activity, Orlando-DDT, had a net increase twice that of the low oxidase strain. Assuming a genetic model similar to that of the lac operon of *E. coli*, these authors reasoned that the mixed-function oxidase of the Isolan-B strain must exist at triple the level of the operon of the Dld:cyw strain. It was assumed that the inducer completely de-repressed the structural genes, resulting in the maximum enzyme producing capabilities (Figure 1).

In a second experiment Terriere et al. (1971) induced the heptachlor epoxidase system of the Dld:cyw and Isolan-B strains and their F_1 hybrids. It was found that the net increase in epoxidase activity of the F_1 population was almost exactly midway between that of the parents, indicating that the tendency to be induced to a greater extent was inheritable (Figure 2) and thus supporting the gene amplification hypothesis.

The diagrams in Figure 1 illustrate the genetic model for gene amplification and how induction could be an indicator of multiple copies of genes. In the normal case, e.g., a susceptible house fly, it is supposed that a basal level of enzyme is produced, either through partial inhibition by the repressor or because some constitutive structural genes direct enzyme synthesis at all times. This is depicted in Part A of the diagram. On induction (Part B) repressor activity is prevented by an inducer and the maximum amount of enzyme is produced. Gene amplification could result in several different arrangements of the multiple genes, two of which are shown. In Part C, it is supposed that each set of structural genes is accompanied by its regulator genes, while in Part D the structural genes are amplified, but not the regulator genes.

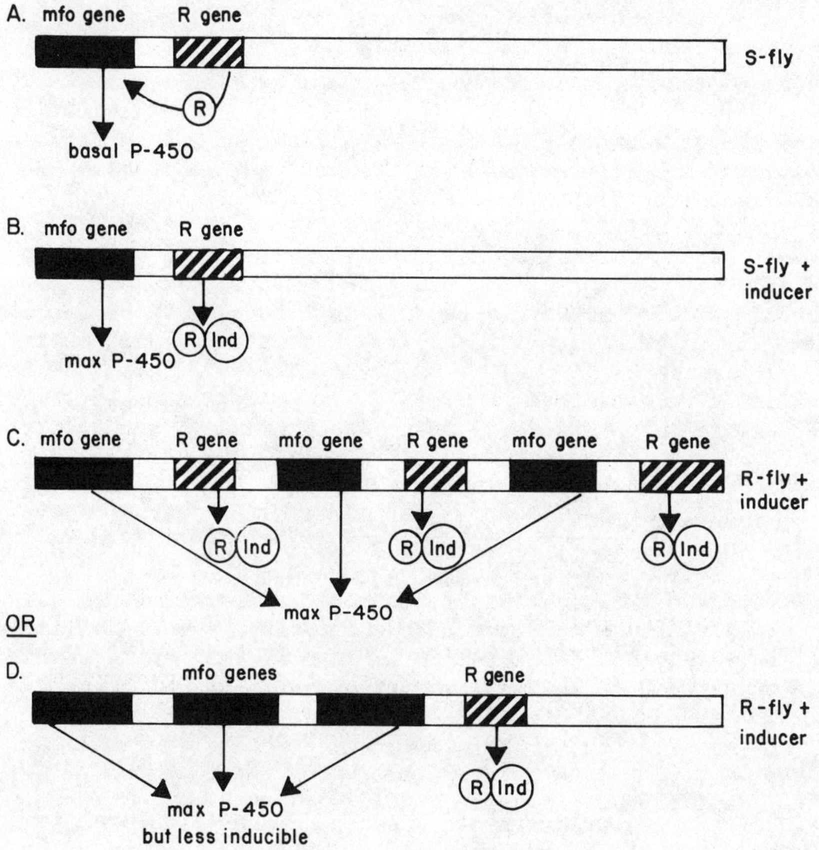

Figure 1. Model for gene amplification and the effect of an inducer on cytochrome P-450 production in resistant and susceptible house flies.

In both C and D the capacity for enzyme production is the same, but C would appear to be more susceptible to an inducer. This is because of the larger amount of repressor expected in C and hence the greater response to induction. In the few experiments that have been done (Table 3), however, resistant strains of house flies appear to be readily induced. Hence, the evidence at hand seems to favor arrangement C.

These ideas are supported by experiments reported by Ishaaya and Chefurka (1971), who compared the production of RNA by the microsomal fraction of DDT-resistant and susceptible house flies after treatment with DDT. With the resistant strain they found an 88% increase in RNA, compared to untreated controls. In the susceptible strain, there was only a 15% increase in RNA. This result would be

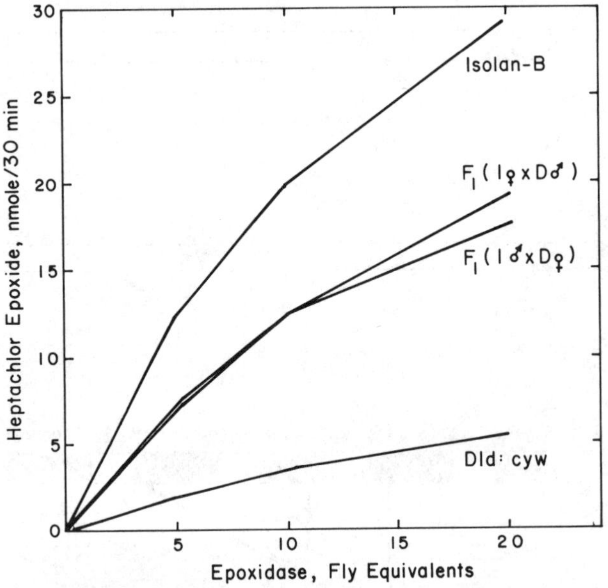

Figure 2. Dieldrin induction of microsomal heptachlor epoxidase in a high oxidase house fly strain (Isolan-B), a low oxidase strain (Dld:cyw), and their F_1 hybrids (from Terriere et al., 1971; copyright 1971 by the American Association for the Advancement of Science).

expected if the resistant strain possessed more structural genes for the DDT-metabolizing enzymes.

Another study of resistance, involving mouse liver sarcoma cells resistant to the anti-cancer drug methotrexate, resulted in more evidence of gene redundancy (Alt et al., 1978; Schimke et al., 1978). These cells are resistant to the drug by virtue of increased production of the enzyme dihydrofolate reductase (DHFR), the target of the methotrexate. The most resistant cell line produced approximately 200 times more DHFR than the susceptible lines, and this has been related to a 200-fold increase in m-RNA production. The additional enzyme appears to be identical with the native enzyme and only the DHFR is increased in the resistant cells. To account for these facts the authors postulated a 200-fold increase in the DHFR gene.

The most recent evidence that gene amplification is involved in insecticide resistance is that of Devonshire and Sawicki (1979), who postulated its occurrence in resistant aphids (*M. persicae*). Resistance in this case is due to an increased titer of an enzyme that hydrolyzes esters such as paraoxon. The activity in the case of the resistant aphids has been traced to an electrophoretically

Table 1. Concentrations of the Insecticide-hydrolyzing Esterase, E4, in Seven Aphid Variants

Variant	Clone	pmole E-4 per mg aphid[a] ± s.d. (no. of independent determinations)
V1(S)	USIL	0.37 ± 0.20(9)
V2	24ON	0.85 ± 0.18(2)
V4	MSIG	1.78 ± 0.75(7)
V8	French R	4.80 ± 0.60(3)
V16[b]	TIV	6.70 ± 0.70 (2)
V32[b,c]	PirR	11.80 ± 1.20(5)
V64[b,c]	G6	24.70 ± 0.20 (2)

[a]Calculated from the rate of paraoxon hydrolysis and the catalytic centre activity of E-4 for this substrate.
[b]Karyotype with A1,3 translocation.
[c]E4 reverts to low activity in the absence of insecticide selection.
(Data from Devonshire and Sawicki, 1979.)

distinct isozyme known as E-4. The results of assays of E-4 activity in seven variants of aphids of differing degrees of resistance are shown in Table 1 from Devonshire and Sawicki (1979).

It will be seen that the enzyme titers of the seven variants are related in a geometric progression from 1 to 64. The authors have interpreted this to indicate a similar progression in the number of copies of the gene coding for E-4.

The resistance in two of the aphid variants in the Devonshire and Sawicki study, those with enzyme titers 32 and 64 times that of the susceptible variant, was unstable, disappearing when the insecticidal pressure was removed. The resistance of the other variants was retained in the absence of insecticide. The authors attribute this difference in resistance stability to differing mechanisms of gene amplification and suggest that it is a survival mechanism in the case of the highly resistant strains. It is reasoned that excess levels of the esterase may be disadvantageous in environments free of insecticide and that a means of rapidly losing this production capacity is necessary to overcome this disadvantage. A similar lack of stability in the resistance mechanism was reported in the methotrexate-resistant mouse sarcoma cells (Alt et al., 1978; Schimke et al., 1978).

This is an attractive theory of biochemical resistance, for it explains how the additional enzyme could be produced. Also, it is believed that redundancy of genes is common in higher animals (Britten and Davidson, 1969; Britten and Kohne, 1968). However, there are some questions arising from the facts known at present. Why, for example, in the case of the aphid, does the enzyme titer, and presumably the gene copy number, increase by doubling, i.e., in a geometric progression? It would seem that unequal crossing over could result in odd numbers of genes as well and in arithmetical progressions during the amplification. Furthermore, since it is believed that such genes are controlled by regulator genes, it becomes necessary to postulate some degree of amplification of these as well. This may be possible if both structural and regulator genes occur in tandem arrays and can thus move around as a unit, but it would seem very unlikely if structural and regulator genes are separated on the chromosome. As for the mechanism of the amplification, we will leave the subject to better qualified authors.

ENZYME INDUCTION

Chemicals that cause increased activity in the enzymes of treated organisms are known as enzyme inducers. Various types of structures, such as chlorinated and polycyclic hydrocarbons, steroids, barbiturates, terpenoids, and several insecticides, have shown such activity in both insects and higher animals. Some examples of insecticides and related compounds as inducers in vertebrate animals are presented in Table 2.

These data have been selected, where possible, to show the lowest dose causing a significant increase in enzyme activity, rather than the maximum increase observed. It can be seen that some compounds, such as mirex, dieldrin and especially TCDD, are rather potent inducers while others, e.g., DDT, DDE, and carbaryl, must be given in large doses to achieve any induction.

Most studies of induction have involved the microsomal oxidase system with measurements of increases in cytochrome P-450 and of the several biochemical reactions it catalyzes (O-dealkylation, N-dealkylation, hydroxylation, epoxidation, and sulfoxidation). However, other detoxifying enzymes, such as epoxide hydrase (Oesch et al., 1973), glutathione S-transferase (Kaplowitz et al., 1975; Baars et al., 1978) and UDP glucuronyl transferase (Aitio, 1973; Bock et al., 1973), are also known to be inducible. The enzyme 5-ALA synthetase, shown in Table 2, is involved in the biosynthesis of the heme moiety of cytochrome P-450.

Table 2. Examples of Enzyme Induction in Pesticide-treated Vertebrates

Compound	Species	Enzyme Induced[a]	Treatment[b]	Increase x Control	Acute Oral LD_{50}[c]	Reference
carbaryl	mouse	mfo	5-13 g/kg, diet, 14 da	1.4	–	Cress & Strother (1974)
	cockerel	P-450	6-300 mg/kg, diet, 6 da	1.8	–	Puyear & Paulson (1972)
		mfo	3-200 mg/kg, diet 3 da	1.4	–	Puyear & Paulson (1972)
	rat	P-450	7-800 mg/kg/da, IP	0.7	–	Madhukar & Matsumura (1979)
		mfo	7-800 mg/kg/da, IP	4.2		Madhukar & Matsumura (1979)
chlordane	rat	mfo	1-25 mg/kg, 8 da	2.0	430	Hart et al. (1963)
		P-450	7-65 mg/kg/da, IP	1.6	–	Madhukar & Matsumura (1979)
		mfo	7-65 mg/kg/da, IP	6.6	–	Madhukar & Matsumura (1979)
chlordimeform	rat	P-450	7-40 mg/kg/da, IP	0.9	–	Madhukar & Matsumura (1979)
		mfo	7-40 mg/kg/da, IP	1.7	–	Madhukar & Matsumura (1979)

(continued)

Table 2. (Continued)

Compound	Species	Enzyme Induced[a]	Treatment[b]	Increase x Control	Acute Oral LD_{50}[c]	Reference
DDT	rat	P-450	7-25 mg/kg/da, IP	2.5	118	Bickers et al. (1974)
		mfo	7-26 mg/kg/da, IP	3.5	118	Bickers et al., (1974)
		P-450	7-100 mg/kg/da, IP	2.2	-	Madhukar & Matsumura (1979)
		mfo	7-100 mg/kg/da, IP	6.6	-	Madhukar & Matsumura (1979)
	chicken	P-450	100 ppm diet, 8 wk	1.4	-	Sell & Davison (1973)
		mfo	100 ppm diet, 8 wk	1.4	-	Sell & Davison (1973)
	duck	P-450	100 ppm diet, 8 wk	3.0	-	Sell & Davison (1973)
		mfo	100 ppm diet, 8 wk	1.8	-	Sell & Davison (1973)
	J. quail	P-450	200 ppm diet, 22 da	1.6	-	Sell & Davison (1973)
DDE	rat	P-450	150 ppm diet, 21 da	2.0	1240	Bunyan & Page (1973)
		mfo	150 ppm diet, 21 da	3.0	-	Bunyan & Page (1973)
	J. quail	P-450	150 ppm diet, 21 da	2.0	-	Bunyan & Page (1973)
		mfo	150 ppm diet, 21 da	1.1	-	Bunyan & Page (1973)
dieldrin	chicken	P-450	10 ppm diet, 21 da	1.5	-	Sell & Davison (1973)
		mfo	10 ppm diet, 8 wk	1.7	-	Sell & Davison (1973)

Compound	Species	Enzyme Induced[a]	Treatment[b]	Increase x Control	Acute Oral LD_{50}[c]	Reference
dieldrin (cont.)	duck	P-450	10 ppm diet, 8 wk	1.9	–	Sell & Davison (1973)
		mfo	20 ppm diet, 8 wk	1.4	–	Sell & Davison (1973)
	rat	P-450	7–5 mg/kg/da, IP	1.8	46	Madhukar & Matsumura (1979)
		mfo	7–5 mg/kg/da, IP	10.2	–	Madhukar & Matsumura (1979)
diazinon	rat	P-450	7–50 mg/kg/da, IP	0.8	285	Madhukar & Matsumura (1979)
		mfo	7–50 mg/kg/da, IP	1.5	–	Madhukar & Matsumura (1979)
heptachlor	rat	mfo	1–80 mg/kg, oral 24 hr	2.4	–	Krampl et al. (1973)
hexachloro-benzene	rat	P-450	10–333 ppm diet, 10 da	1.8	–	Turner & Green (1974)
		mfo	10–333 ppm diet, 10 da	2.4	–	Turner & Green (1974)
Kepone	mouse	mfo	10 ppm diet, 14 da	1.5	–	Fabacher & Hodgson (1976)
	rat	P-450	7–15 mg/kg/da, IP	2.1	125	Madhukar & Matsumura (1979)

(Continued)

Table 2. (Continued)

Compound	Species	Enzyme Induced[a]	Treatment[b]	Increase x Control	Acute Oral LD_{50}[c]	Reference
kepone (cont.)	rat	mfo	7-15 mg/kg/da, IP	3.9	-	Madhukar & Matsumura (1979)
lindane	rat	P-450	7-25 mg/kg/da, IP	0.6	91	Madhukar & Matsumura (1979)
		mfo	7-25 mg/kg/da, IP	6.3	-	Madhukar & Matsumura (1979)
malathion	rat	P-450	7-250 mg/kg/da, IP	0.9	1000	Madhukar & Matsumura (1979)
		mfo	7-250 mg/kg/da, IP	2.3	-	Madhukar & Matsumura (1979)
mirex	mouse	P-450	1-10 mg/kg, IP	2.0	-	Baker et al. (1972)
	rat	P-450	7-40 mg/kg/da, IP	1.7	600	Madhukar & Matsumura (1979)
		mfo	7-40 mg/kg/da, IP	3.3	-	Madhukar & Matsumura (1979)
		P-450	1-25 mg/kg, IP	1.5	-	Baker et al. (1972)
TCDD	chick embryo	mfo	1-1.55 x 10^{-10} mole/egg, 1 da	10.0	-	Poland et al. (1976)
		5-ALA-synth	1-4.66 x 10^{-12} mole/egg, 1 da	2.0	-	Poland & Glover (1973)

Compound	Species	Enzyme Induced[a]	Treatment[b]	Increase x Control	Acute Oral LD_{50}[c]	Reference
TCDD (cont.)	chick embryo	5-ALA-synth	1-1.55 x 10^{-9} mole/egg, 1 da	35.0	-	Poland & Glover (1973)
	rat	mfo	1-0.031 μmole/kg/da, IP	13.00	-	Poland & Glover (1973)
		UDPG-T	1-25 μg/kg, oral, 6 da	6.0	-	Lucier et al. (1975)
		P-450	1-25 mg/kg, IP 6 da.	1.3	-	Madhukar & Matsumura (1979)
		mfo	1-25 mg/kg, IP 6 da	3.5	-	Madhukar & Matsumura (1979)

[a]Mfo, mixed-function oxidase; P-450, cytochrome P-450; 5-ALA-synth, 5-aminolevulinic acid synthetase; UDPG-T, uridinediphospho-glucuronyl transferase.
[b]Refers to number and duration of treatments. Lowest effective dose cited when several doses were tested.
[c]For female rats (Gaines, 1969).

Induction occurs slowly in vivo, but it can be detected in minutes in cell culture systems (Nebert et al., 1972), indicating that the response is rapid once the demands of transport and accumulation are met. The effect, i.e., the increased activity of enzymes, remains as long as the inducer is present and disappears at a rate dependent upon the rate of metabolism and/or excretion of the inducer.

From the nature of the experiments summarized in Table 2 as well as the toxicity data for rats, it appears that these inducers are effective at sublethal doses. As will be seen later, this is not often the case when these compounds are inducers of insect enzymes.

The first reports of induction in insects were those of Agosin and colleagues who reported that DDT induced NAD kinase, resulting in increased levels of NADP in *Triatoma infestans* (Agosin and Dinamarca, 1963; Ilevicky et al., 1964) and suggesting that this induction might be related to DDT resistance. In additional work on the phenomenon they found that DDT-treated insects did metabolize DDT more rapidly (Agosin et al., 1969). It was also reported that DDT was an inducer of oxidative mechanisms in certain resistant strains of the house fly (Gil et al., 1968), and a connection between induction and resistance, especially cross-resistance, was again suggested. The early work on induction in insects has been reviewed by Terriere and Yu (1974).

Most of the more recent examples of enzyme induction in insects are summarized in Table 3. At least 12 species have been shown to be susceptible to induction, with a variety of chemical structures exerting the inductive action. It is clear from the examples shown that most of the enzyme systems involved in the metabolism of insecticides are inducible, including cytochrome P-450 and several P-450-dependent systems, phosphotransferase, carboxyesterase, DDT-dehydrochlorinase, epoxide hydrase and others (see Table 3 for references).

On the other hand, though induction is possible, it appears that rather high doses of inducer are required, higher, in the case of the insecticides tested, than their toxic dose for the insect species being induced. This is one reason to doubt that induction by an insecticide could be involved in the development of resistance in a population.

The Mechanism of Induction

Our first understanding of enzyme induction was provided by the work of Jacob and Monod (Jacob, 1966) whose experiments revealed how the sugar lactose induced the enzymes necessary for its metabolism in *E. coli*. In the absence of lactose, a regulator gene in the

Table 3. Examples of Enzyme Induction in Insects

Species	Compound	Enzyme[a]	Treatment[b]	Increase x Control	Reference
M. domestica	phenobarb	epoxidase	0.1% in diet, 3 da	6	Yu & Terriere (1973)
			1% in diet, 3 da	25	Yu & Terriere (1973)
			1% in diet, 3 da	4.1	Yu & Terriere (1973)
*[c]		O-demethylase	1% in diet, 3 da	4.7	Yu & Terriere (1973)
		DDT-ase	1% in diet, 3 da	1.7	Yu & Terriere (1973)
*			1% in diet, 3 da	1.9	Yu & Terriere (1973)
*	DDT	DDT-ase	10 μg/fly, 10 hr	1.5	Capdevila et al. (1973)
*	dieldrin	epoxidase	50 μg/100 flies, contact	3.1	Terriere et al. (1971)
*	phenobarb	N-demethylase	0.5% in diet, 2 da	1.8	Capdevila et al. (1973)
*	naphthalene	N-demethylase	5% in diet, 2 da	2.1	Capdevila et al. (1973)
	hydroprene	carboxy-esterase	5 μg/insect, 36 hr	2.2	Maa & Terriere (unpublished)
	phenobarb	P-450	0.25% in diet, 3 da	1.4	Perry et al. (1971)
		epoxidase	0.25% in diet, 3 da	4.4	Perry et al. (1971)
†[d]	piperonyl butoxide	epoxidase	0.5% in diet, 3 da	11	Yu & Terriere (1974)

(Continued)

Table 3. (Continued)

Species	Compound	Enzyme[a]	Treatment[b]	Increase x Control	Reference
M. domestica (cont.)	phenobarb	epoxide hydrase	1% in diet, 3 da	1.8	Yu & Terriere (1978)
†	difluo-benzuron	epoxidase	0.3 ppm, diet, 2 da	1.9	Yu & Terriere (1977)
†	TH-6038	epoxidase	16 ppm, diet, 2 da	11	Yu & Terriere (1975)
*	dieldrin	epoxidase	0.5 μg/fly, 12 hr	4	Walker & Terriere (1970)
S. bullata	phenobarb	epoxidase	0.1% in diet, 3 da	1.4	Terriere & Yu (1976)
			1% in diet, 3 da	5.8	Terriere & Yu (1976)
†	difluo-benzuron	epoxidase	1 ppm, diet, 2 da	1.5	Yu & Terriere (1977)
†	phenobarb	epoxidase	0.01% in diet, 3 da	2.5	Terriere & Yu (1976)
†			0.5% in diet, 3 da	56	Terriere & Yu (1976)
	hydroprene	carboxy-esterase	10 μg/fly, 2 da	3.5	Maa & Terriere (unpublished)
P. regina	phenobarb	epoxidase	0.1% in diet, 3 da	2.4	Terriere & Yu (1976)
			1% in diet, 3 da	13.6	Terriere & Yu (1976)
†			0.01% in diet, 3 da	18.5	Terriere & Yu (1976)
†			0.1% in diet, 3 da	60	Terriere & Yu (1976)

Species	Compound	Enzyme[a]	Treatment[b]	Increase x Control	Reference
P. regina (cont.)	phenobarb	P-450	0.5% in diet, 3 da	7	Rose et al. (unpub.)
		epoxidase	0.5% in diet, 3 da	10	Rose et al. (unpub.)
	hydroprene	carboxy-esterase	5 μg/fly, 2 da	2.3	Maa & Terriere (unpublished)
S. eridana	pentamethyl benzene	epoxidase	0.2% in diet, 3 da	3.1	Brattsten & Wilkinson (1973)
	pentamethyl benzene	N-demethylase	0.2% in diet, 3 da	3.1	Brattsten & Wilkinson (1973)
	pentamethyl benzene	RNA-polymerase III	2000 ppm in diet, 1 da	1.4	Elshourbagy & Wilkinson (1978)
	insecticide solv.	N-demethylase	0.05-0.25% in diet, 1 da	up to 4.5	Brattsten & Wilkinson (1977)
	pentamethyl benzene	5-ALA synth	2000 ppm in diet, 3 da	3	Brattsten & Wilkinson (1975)
	phenobarb	5-ALA synth	2500 ppm in diet, 3 da	3.7	Brattsten & Wilkinson (1975)
	DDTPC[e]	5-ALA synth	2000 ppm in diet, 3 da	7	Brattsten & Wilkinson (1975)
	AIA[f]	5-ALA synth	500 ppm in diet, 2 da	2	Brattsten & Wilkinson (1975)
	pentamethyl benzene	N-demethylase	2000 ppm in diet, 2 da	4.5	Brattsten & Wilkinson (1975)

(Continued)

Table 3. (Continued)

Species	Compound	Enzyme[a]	Treatment[b]	Increase x Control	Reference
S. eridana (cont.)	phenobarb	N-demethylase	2500 ppm in diet, 2 da	3	Brattsten & Wilkinson (1975)
	DDTPC	N-demethylase	2000 ppm in diet, 2 da	3.8	Brattsten & Wilkinson (1975)
	AIA	N-demethylase	1000 ppm in diet, 2 da	2.1	Brattsten & Wilkinson (1975)
	sec. plant subs.	N-demethylase	0.05%-0.20% in diet, 1 da	up to 3.9	Brattsten et al. (1977)
P. saucia	sec. plant subs.	epoxidase	0.1-0.2% in diet, 2.5 da	up to 25	Yu et al. (1979)
P. americana	chlor-cyclizine	O-demethylase	60 μg/g, inject 1 da	1.5	Turnquist & Brindley (1975)
	DDT	hydroxylase	0.01 μg/jar, contact	2	Khan & Matsumura (1972)
G. portentosa	phenobarb	hydroxylase	1.2 mg/insect, inject, 4 da	1.3	Gil et al. (1974)
		P-450	1.2 mg/insect, inject, 4 da	2	Gil et al. (1974)
		phospho-transferase	1.2 mg/insect, inject, 4 da	1.3	Gil et al. (1974)

Species	Compound	Enzyme[a]	Treatment[b]	Increase x Control	Reference
G. portentosa (cont.)	phenobarb	epoxidase	1.2 mg/insect, inject, 4 da	1.6	Gil et al. (1974)
B. germanica	dieldrin	hydroxylase	0.07 μg/jar contact	1.6	Khan & Matsumura (1972)
Lucilia illustris	phenobarb	P-450	1% in diet, 3 da	2	Rose et al. (unpub.)
		epoxidase	1% in diet, 3 da	7	Rose et al. (unpub.)
Eucalliphora lilaea	phenobarb	P-450	0.5% in diet, 3 da	2.5	Rose et al. (unpub.)
		epoxidase	0.5% in diet, 3 da	5	Rose et al. (unpub.)
Agrotis ypsilon	piperonyl butoxide	hydroxylase	1.4 mg/larvae, 15 hr	2.5	Thongsinthusak & Krieger (1974)
M. sexta[d]	mono-crotophos	carboxy-esterase	1 ppm diet, 1 da	2	Wongkobrat & Dahlman (1976)

[a]P-450, cytochrome P-450; 5-ALA synth, 5-aminolevulinic acid synthetase.
[b]Refers to number and duration of treatments.
[c](*) Resistant strain.
[d](†) Larvae.
[e]Diethyl-1,4-dihydro-2,4,6-trimethylpyridine-3,5-carboxylate.
[f]2-allyl-2-isopropylacetamide.

E coli operon produces a repressor protein that occupies the operator region of the structural genes that code for the necessary enzymes. In the presence of lactose, the repressor protein is derepressed, i.e., it is removed from the operator region, thus permitting the structural genes to be transcribed to m-RNA. This system, though from a single-celled organism, can be used as a model for enzyme induction in higher animals because it contains the essential features of gene regulation.

Induction of enzymes is probably a normal, physiologically important mechanism of gene regulation carried on by endogenous substances such as hormones. An example of such a system, provided by Gelehrter (1978), involves glucocorticosteroids and insulin and their induction of tyrosine aminotransferase, required in tyrosine metabolism. The glucocorticosteroids act at the transcription level and the insulin at the translational level to regulate supply of this enzyme.

Numerous studies of induction by such compounds as phenobarbital (PB) and 3-methylcholanthrene (3-MC) have shown that the main effect of such inducers is to stimulate protein synthesis through the increased transcription of structural genes. The action is more or less specific, depending upon the inducer, and the new or additional enzyme is identical to that which was present prior to induction (Omura, 1978).

Several other sites of action have also been indicated by experimental results or suggested on theoretical grounds. Insulin, for example, is known to stimulate tyrosine aminotransferase activity, acting within six minutes in cell culture systems, by stabilizing the enzyme (Gelehrter, 1978). A similar effect has been postulated for inducers of aromatic hydrocarbon hydroxylase, which has a half life of 4 to 8 hours (Gelboin and Whitlock, 1978). Other possibilities suggested by these authors include the modification of m-RNA, the stabilization of m-RNA, and the stimulation of one or more processes through cyclic-AMP. Still other possibilities remain, including increased activity of 5-aminolevulinic acid synthetase as well as various combinations of any of the above. It is clear that a full understanding of the mechanism is not yet at hand.

Another problem requiring explanation is how so many compounds, representing so many different structural types, can act as inducers. One answer that has some experimental support is that special receptor proteins bind the various inducers, transporting them to the nucleus for interaction with the gene regulation system (Gelboin and Whitlock, 1978; Gelehrter, 1978; Nebert, 1978).

RNA Production in Induced Insects

Prior to the report of Ishaaya and Chefurka (1971) indicating increased production of RNA in DDT-induced resistant and susceptible house flies (mentioned earlier under Gene Amplification), evidence of a similar nature was provided by Balazs and Agosin (1968), Litvak and Agosin (1968), Litvak et al. (1968). This increase in RNA production is preceded by an increase in activity of DNA-dependent RNA polymerase (Agosin, 1971). Tsang and Agosin (1976) extended their observations to obtain evidence that the phenobarbital induction of resistant (Fc) and susceptible (NAIDM) house flies resulted in several differences: greater RNA polymerase and chromatin template activity in the resistant strain, and different nucleotide ratios in the RNA produced in resistant and susceptible strains. These workers also believe that phenobarbital is bound to a low molecular weight protein in the cytosol and that this may be its mechanism of transport into the nucleus.

The *Ah* Gene in Inducer-responsive Mice

In extensive research on the genetic nature of 3-MC induction of the enzymes that hydroxylate polycyclic hydrocarbons in mice, Nebert and his colleagues have made considerable progress. The mice strains used in their research were of two general types with respect to such induction. Some strains produced only basal levels of cytochrome P-450 whether or not they were induced, i.e., they were non-responsive, while other strains produced additional P-450 when induced, i.e., they were responsive. This responsiveness to 3-MC induction has been traced to gene loci identified as *Ah* (for aromatic hydrocarbon hydroxylase). When the responsive mice were treated with 3-MC, there was a considerable increase in hydroxylase activity, and this was accompanied by increases in the activity of UDPG-transferase and at least 20 different microsomal oxidases (Nebert, 1978).

The *Ah* gene(s) is dominant and may control the production of several cytochrome P-450s, two of which, P_1-450 and P-448, have been partially characterized (Nebert, 1978). Indeed, Nebert has postulated dozens of such hemoproteins, which arise in response to exogenous inducers through cytosolic protein receptors that bind to the inducers and carry them into the nucleus. There is also evidence of temporal control of the genes at the *Ah* locus so that the response to inducers by different enzymes occurs at different times.

In the case of the 3-MC-induced cytochrome P-450, which hydroxylates benzo (a)pyrene, the newly formed hemoprotein is the high spin type of vastly greater catalytic activity (Nebert and Gielen, 1972). Cell culture studies indicate that the response to 3-MC occurs within

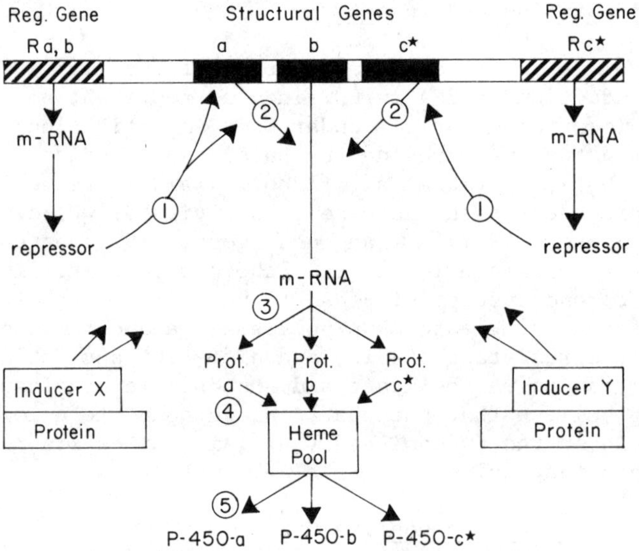

Figure 3. Model for gene-directed cytochrome P-450 and possible sites of induction.

30 minutes, with the production of m-RNA specific for the hydroxylase system (Nebert et al., 1972).

The ideas reviewed in the previous paragraphs are expressed in Figure 3, which depicts the gene-directed synthesis of cytochrome P-450. In this model it is supposed that there are multiple genes coding for the proteins of cytochrome P-450 and that these proteins couple with heme from the heme pool to produce the multiple forms that are probably present in most organisms. Two regulator genes are shown to account for the specificity of some inducers.

The numbers in the figure refer to the points at which the inducers might act, either alone, or more likely, as inducer-protein complexes. The methods involved at these points are: 1) prevention of repression by the repressor proteins; 2) stimulation of transcription; 3) stimulation of the translation of m-RNA; 4) and 5) stabilization of proteins before and after their link-up with heme.

It is assumed that heme is always available from a cellular pool, but this too could be a point of genetic control and of induction, i.e., the stimulation of 5-aminolevulinic acid synthetase (see Tables 2 and 3). The protein c* represents a hypothetical apoprotein that couples with heme and converts it to the high spin form, which is catalytically more active. This accounts for the production of high spin P-450 in animals induced with 3-MC (Nebert and Gielen,

Table 4. Cytochrome P-450 Level and Aldrin Epoxidase Activity of a Resistant and a Susceptible House Fly Strain

Property	S (NAIDM)	R (Rutgers)	R/S
P-450, nmole/g abdomen	4.7	8.1	1.7
aldrin epox, pmole/min/mg abdomen	32.2	326.4	10.1
LD_{50} propoxur, mg/g - males	.04	7.75	194
LD_{50} propoxur, mg/g - females	.05	>16.2	>324

[a]Average of 2 experiments.

1972) and in a high-oxidase house fly strain (Yu and Terriere, 1979).

Induction of Cytochrome P-450 in House Flies

There is evidence in Table 1 that the cytochrome P-450 of induced insects differs in catalytic activity from that of the non-induced control insect, i.e., the microsomal oxidase activity increases to a greater extent than the P-450 level. Similar results were obtained in our comparisons of cytochrome P-450 levels and enzyme activity in insecticide-resistant and susceptible house flies (Table 4) and in earlier reports (Perry et al., 1971; Matthews and Casida, 1970). These observations suggest the possibility that the hemoprotein produced as a result of induction in susceptible flies is the same as that already present in strains with resistance conferred by high oxidase activity. Further evidence of this is provided by results obtained in the author's laboratory.

Resistant and susceptible strains of the house fly were fed a diet containing phenobarbital for three days, after which microsomes were prepared from abdomen homogenates. Cytochrome P-450 was measured by the method of Omura and Sato (1964), and the type II binding spectra were determined by the method of Jefcoate et al. (1970). The latter method indicates the presence of high-spin heme iron in the cytochrome P-450 mixture and can be used to obtain an estimate of the relative amounts of the high and low spin forms present. This is done by calculating the ratio $\Delta A410\text{-}500/\Delta A932\text{-}500$. Ratios of 0.7, for example, represent P-450 mixtures that are approximately 1 part high spin to 3 parts low spin P-450 while ratios of 1.0 represent mixtures of approximately 1 part high spin to 5 parts low spin P-450. This ratio is of importance because of the greater enzymatic activity of the high spin form. Although the Jefcoate method has not been verified for insect cytochrome P-450, there seems little reason to doubt its general applicability.

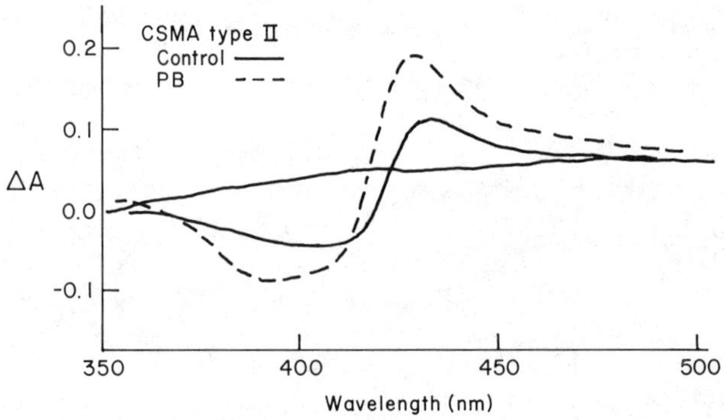

Figure 4. Type II difference spectra of the n-octylamine-microsomal cytochrome P-450 complex from phenobarbital-induced and control house flies of the CSMA strain.

Typical carbon monoxide and n-octylamine difference spectra of microsomes from control and phenobarbital-induced house flies are shown in Figures 4 through 7. Additional data for these and other house fly strains are presented in Table 5.

Two aspects of the susceptible fly P-450 spectra are of interest: the shift of the absorption maximum of the CO-P-450 complex after PB treatment (Figure 4) and the appearance of the absorption of the trough at 392 nm in the n-octylamine binding spectrum after PB treatment (Figure 5). As seen in the corresponding spectra of the R fly microsomes (Figures 6 and 7), these characteristics are already present in the control spectra and are not altered appreciably by PB treatment.

It is now believed that multiple forms of cytochrome P-450 are present in the microsomal membranes of most species and that spectra such as those shown here represent a composite of such forms. Thus, the lower λ_{max} values for the CO-P-450 complex of the R fly microsomes indicate that this P-450 mixture contains different forms or different proportions of the hemoprotein than the mixture from the S flies. A shift in the absorption maximum of this complex in the PB-treated S flies to a value closer to that of the untreated R flies may indicate that the P-450 mixture has a composition more like that of the R flies.

Similarly, in type II binding spectrum of P-450 from the PB-induced S insects, resulting in a spectrum resembling that of the P-450 of the R insects, the appearance of the 392 nm trough may indicate that induction results in the increased production of the hemoprotein already present in R insects. This type of spectrum is known to be formed with P-450 mixtures containing higher proportions

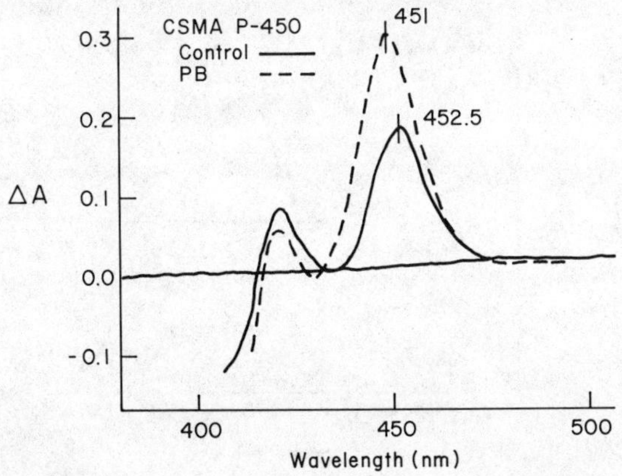

Figure 5. Difference spectra of the carbon monoxide-microsomal cytochrome P-450 complex from phenobarbital-induced and control house flies of the CSMA strain.

of high spin iron (Jefcoate et al., 1970). Therefore, it can be speculated that the newly synthesized hemoprotein found in the microsomes of PB-induced S flies is predominantly of the high spin form.

As summarized in Table 5, the spectral shifts depicted in Figures 4 through 7 have been observed in three reference strains. Using the Jefcoate method to calculate the ΔA410/Δ392 ratios, it is seen the PB treatment produces a marked change in this value, from about 1.5 to 1.0 or less, whereas the corresponding change in the ratio for the R flies is negligible.

The possible significance of these observations is seen in our attempts to solubilize the cytochrome P-450 of a susceptible and a highly resistant strain (Yu and Terriere, 1979). Methods developed for the solubilization and purification of hepatic cytochrome P-450 were applied to microsomes from the two strains. Six chromatographically distinct fractions contained active hemoprotein. The two containing the largest amounts of P-450 are compared in Table 6.

Of greatest interest is a comparison of the specific activities of the fractions prepared from the high oxidase Rutger strain and those of similar fractions from the NAIDM reference strain. It is clear that the P-450 in the B_1 fraction of the R strain is at least 200-fold more active than that from the S strain and its C_1 fraction is approximately 4 times more active. Only a small part of the P-450 of either strain is represented, but this indicates the potential for greatly increased activity of the enzyme in a resistant strain. Furthermore, there may be a relationship between these results and those obtained with phenobarbital-induced S insects (Table 5). This

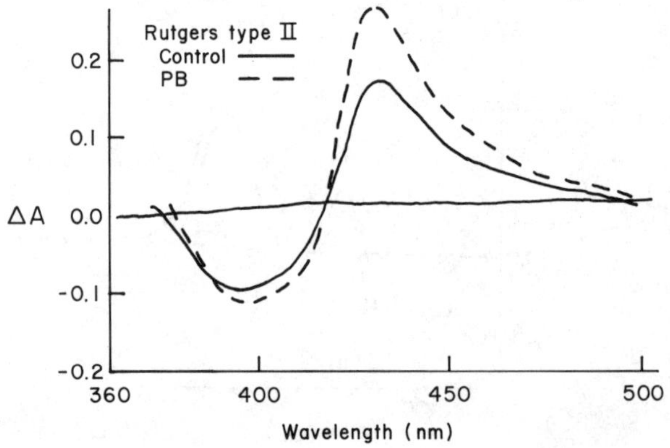

Figure 6. Type II difference spectra of the n-octylamine-microsomal cytochrome P-450 complex from phenobarbital-induced and control house flies of the Rutgers strain.

suggests that the same regulatory genes may be involved in both induction and biochemical resistance.

INDUCTION AND RESISTANCE

From the information reviewed in the previous paragraphs, it is clear that induction and biochemical resistance have several characteristics in common. There is no doubt that several of the important detoxifying enzymes of insects are inducible (Table 3) or that some insecticides are inducers of these enzymes (Tables 2 and 3). Also, there is little to dispute the fact that both phenomena are genetic in nature, and still other evidence indicates that both may originate at the same genetic site or with the same genetic mechanism.

There is one major difficulty with this theory: as pointed out earlier, the dose required for induction is high compared to that required for toxicity. Thus, it is difficult to see how selection of a resistant population could occur as a result of exposure to a toxic inducer. There remains the possibility that a highly active impurity or metabolite (such as TCDD) might provide the necessary stimulation of enzyme production.

As a practical point, it is not likely that experimental insecticides that are potent detoxication inducers would survive initial screening programs. In case the initial enzyme attack resulted in activation, as in the conversion of parathion to paraoxon, action as an inducer would have the opposite effect, increasing toxicity. It is very difficult to see how induction could be involved in resistance to this type of compound.

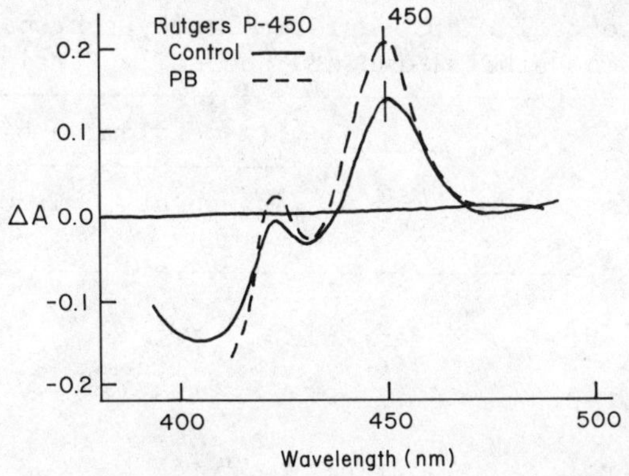

Figure 7. Difference spectra of the carbon monoxide-microsomal cytochrome P-450 complex from phenobarbital-induced and control house flies of the Rutgers strain.

If induction of detoxifying enzymes is of significance in resistance, it should result in some degree of protection against insecticides. There have been several attempts to demonstrate this, but the results are mixed. Meksongsee et al. (1967) found that for the most part, both DDT and phenobarbital induction increased the susceptibility of the house fly to several carbamate insecticides.

House flies induced by naphthalene were more tolerant of propoxur (twofold increase in LD_{50}), but phenobarbital treatment resulted in increased sensitivity to the insecticide (Capdevila et al., 1973). Walker and Terriere (1970) were able to demonstrate a slight decrease in susceptibility to carbaryl in house flies induced with dieldrin, while Yu and Terriere (1973) demonstrated a similar degree of protection against propoxur in house flies induced with phenobarbital.

In the case of the leaf cutter bee treated with inducing drugs (aminopyrine, phenobarbital and chlorcyclizine), there was evidence of protection against carbaryl by chlorcyclizine (a twofold increase in the LD_{50}), but the aminopyrine made the bees more sensitive to the insecticide, and phenobarbital had no effect (Lee and Brindley, 1974).

Results were more clearcut in experiments with Lepidoptera exposed to secondary plant substances. Brattsten and Wilkinson (1977) showed that the toxicity of nicotine to the southern armyworm was considerably less after the insects were fed α-pinene. The southern armyworm is also protected against carbaryl after previous exposure (one day) to various insecticide solvents known to be

Table 5. Spectral Characteristics of Cytochrome P-450 from Phenobarbital-induced House Fly Strains

Strain	PB Treatment	CO-P-450 λ_{max} nm	n-Octylamine binding spectrum		
			Trough at 410 nm	Trough at 392 nm	ΔA410/ΔA392
NAIDM	control	453.5	+	-	1.54
	treated	452.2	+	+	1.06
CSMA	control	452.5	+	-	1.16
	treated	451.0	+	+	0.93
SRS	control	452.0	+	-	1.29
	treated	451.0	+	+	0.95
Rutgers	control	450.0	-	+	0.72
	treated	450.0	-	+	0.75
Fc	control	452.0	±	+	0.77
	treated	451.0	±	+	0.73

inducers (Brattsten and Wilkinson, 1977). Yu et al. (1979) showed that cutworm larvae fed peppermint leaves (rich in secondary plant substances) were much less susceptible to carbaryl than larvae fed bean leaves.

To speculate further about induction as a mechanism of resistance, the possibility of a mutant regulator gene that produces a more inducer-susceptible gene product should be mentioned. Such mutations could result in greater sensitivity to either inducers or their early-appearing metabolites and could result in protection against toxic inducers. This mechanism is somewhat like that postulated for 3-MC induction in genetically responsive and non-responsive mice (Neberg, 1978).

There is another means by which induction might be viewed as part of the resistance phenomenon -- as a supplementary source of enzyme activity. Once a population has achieved some degree of resistance as a result of selection and is thus able to tolerate larger doses of the toxicant, those compounds that induce defensive

Table 6. Specific Activities of Cytochrome P-450 Fractions from Rutgers and NAIDM Flies and their *n*-Octylamine Binding Characteristics

House Fly Strain	P-450 Fraction	Aldrin Epoxidase (pmole/nmole P-450/min)[a]	*n*-Octylamine Difference Spectrum ($\Delta A410/\Delta A392$)[b]
Rutgers	B_1	297.9 ± 39.8	0.79 ± 0.05
	C_1	160.0 ± 38.4	0.64 ± 0.13
NAIDM	B_1	1.15[b]	1.98 ± 0.09
	C_1	41.6 ± 16.7	1.25 ± 0.06

[a]Mean ± S.E. of 4 to 5 assays.
[b]Based on a one-hr incubation with 1.5 nmoles of cytochrome P-450. No dieldrin could be detected after the standard incubations.
Data from Yu and Terriere (1979).

enzymes might be able to exert their effect. Furthermore, this action could prevail with resistance mechanisms that do not involve detoxication, such as insensitive nerve tissue, barriers to penetration, or altered target enzymes. In this way, the insect would be afforded the additional mechanism of increased detoxication as a result of the induction. For this to occur, it would seem necessary for the originally selected defenses to provide tolerances several times higher than those of the susceptible individuals.

These ideas may have some validity in explaining a related phenomenon: cross resistance. In this case we can imagine an inducer that stimulates the production of an enzyme capable of detoxifying a second toxicant, thereby broadening the insect's basis of resistance.

Reports that secondary plant substances such as mono-terpenes and substituted indoles are potent inducers of microsomal oxidases in insects (Brattsten and Wilkinson, 1977; Yu et al., 1979) and in rats (Pantuck, 1976) raise the possibility that induction might be mistaken for resistance in certain field situations. For example, the N-demethylase activity of midgut microsomes of southern armyworm larvae was increased 2- to 4-fold after a few hours of feeding on a diet containing α-pinene or other mono-terpenes. Furthermore, there was an increase in the tolerance of the larvae to the insecticide nicotine.

In the example reported by Yu et al. (1979), the variegated cutworm was more tolerant of the insecticide carbaryl after feeding on peppermint leaves than after feeding on bean leaves. This difference was associated with a several-fold increase in microsomal oxidase activity in the peppermint-fed larvae and various constituents of peppermint, including α-pinene, were shown to be inducers when added to the semi-defined artificial diet on which the cutworms were reared.

Effects such as these could conceivably result in a higher treatment rate being required for one crop than for another, thus indicating an early stage of resistance. On the other hand, such induction might increase susceptibilities to pesticides that are activated by the microsomal oxidase system.

REFERENCES

Agosin, M., 1971, Ribonucleic acid synthesis in nuclei isolated from *Musca domestica*, *Insect Biochem.*, 1:363.

Agosin, M., and Dinamarca, M. L., 1963, The effect of DDT on the level of di- and triphosphopyridine nucleotides in *Triatoma infestans*, *Exp. Parasitol.*, 13:199.

Agosin, M., Scaramelli, N., Gil, L., and Letelier, M. E., 1969, Some properties of the microsomal system metabolizing DDT in *Triatoma infestans*, *Comp. Biochem. Physiol.*, 29:785.

Aitio, A., 1973, Induction of UDP glucuronyltransferase in the liver and extrahepatic organs of the rat, *Life Sci.*, 13:1705.

Alt, F. W., Kellems, R. E., Bertino, J. R., and Schimke, R. T., 1978, Selective multiplication of dihydrofolate reductase genes in methotrexate-resistant variants of cultured murine cells, *Jour. Biol. Chem.*, 253:1357.

Baars, J. J., Jansen, M., and Breimer, D. D., 1978, The influence of phenobarbital, 3-methylcholanthrene and 2,3,7,8-tetrachlorodibenzo-p-dioxin on glutathione S-transferase activity of rat liver cytosol, *Biochem. Pharmacol.*, 27:2487.

Baker, R. C., Coons, L. G., Mailman, R. B., and Hodgson, E., 1972, Induction of hepatic mixed-function oxidases by the insecticide, mirex, *Environ. Res.*, 5:418.

Balazs, I., and Agosin, M., 1968, The effect of 1,1,1-trichloro-2,2-bis (p-chlorophenyl) ethane on ribonucleic acid metabolism in *Musca domestica* L., *Biochem. Biophys. Acta*, 157:1.

Bickers, R. D., Kappas, A., and Alvares, A.P., 1974, Differences in inducibility of cutaneous and hepatic drug metabolizing enzymes and cytochrome P-450 by polychlorinated biphenyls and 1,1,1-trichloro-2,2-bis(p-chlorophenyl) ethane (DDT), *J. Pharmacol. Exp. Therap.*, 188:300.

Bock, K. W., Frohling, W., Remmer, H., and Rexter, B., 1973, Effects of phenobarbital and 3-methylcholanthrene on substrate specificity of rat liver microsomal UDP-glucuronyltransferase,

Biochem. Biophys. Acta, 327:46.

Brattsten, L. B., and Wilkinson, C. F., 1973, Induction of microsomal enzymes in the southern armyworm (*Prodenia eridania*), *Pestic. Biochem. Physiol.,* 3:393.

Brattsten, L. B., and Wilkinson, C. F., 1975, Properties of 5-aminolaevulinate synthetase and its relationship to microsomal mixed-function oxidation in the southern armyworm (*Spodoptera eridania*), *Biochem. J.,* 150:97.

Brattsten, L. B., and Wilkinson, C. F., 1977, Insecticide solvents: Interference with insecticidal action, *Science,* 196:1211.

Brattsten, L. B., Wilkinson, C. F., and Eisner, T., 1977, Herbivore-plant interactions: Mixed-function oxidases and secondary plant substances, *Science,* 196:1349.

Britten, R. J., and Advidson, E. H., 1969, Gene regulation for higher cells: A theory, *Science,* 165:349.

Britten, R. J., and Kohne, D. E., 1968, Repeated sequences in DNA, *Science,* 161:529.

Bunyan, P. J., and Page, J. M. J., 1973, Pesticide-induced changes in hepatic microsomal enzyme systems: Some effects of 1,1-di(p-chlorophenyl)-2-chloroethylene (DDMU) in the rat and Japanese quail, *Chem. Biol. Interactions,* 6:249.

Capdevila, J., Morello, A., Perry, A. S., and Agosin, M., 1973, Effect of phenobarbital and naphthalene on some of the components of the electron transport system and the hydroxylating activity of house fly microsomes, *Biochemistry,* 12:1445.

Cress, C. R., and Strother, A., 1974, Effects on drug metabolism of carbaryl and 1-naphthol in the mouse, *Life Sci.,* 14:861.

Devonshire, A. L., and Sawicki, R. M., 1979, Insecticide-resistant *Myzus persicae* as an example of evolution by gene duplication, *Nature*, 280:140.

Elshourbagy, N. A., and Wilkinson, C. F., 1978, The role of DNA-dependent RNA polymerases in microsomal enzyme induction in southern armyworm (*Spodoptera eridania*) larvae, *Insect Biochem.*, 8:425.

Fabacher, D. L., and Hodgson, E., 1976, Induction of hepatic mixed-function oxidase enzymes in adult and neonatal mice by Kepone and mirex, *Tox. and Appl. Pharm.*, 38:71.

Gaines, T. B., 1969, Acute toxicity of pesticides, *Tox. and Appl. Pharm.,* 14:515.

Gelboin, H. V., and Whitlock, J. P., Jr., 1978, On the mechanism of mixed-function oxidase induction, *in:* "The Induction of Drug Metabolism," R. W. Estabrook and E. Lindenlaub, eds., pp. 67-79, F. K. Schattauer Verlag, Stuttgart - New York.

Gelehrter, T. D., 1978, Enzyme induction in mammals -- An overview, *in:* "The Induction of Drug Metabolism," R. W. Estabrook and E. Lindenlaub, eds., pp. 7-24, F. K. Schattauer Verlag, Stuttgart - New York.

Gil, D. L., Rose, H. A., Yang, R. S. H., Young, R. G., and Wilkinson, C. F., 1974, Enzyme induction by phenobarbital in the Madagascar cockroach, *Gromphadorhina portentosa, Comp. Biochem. Physiol.,* 47B:657.

Gil, L., Fine, B. B., Dinamarca, M. L., Balazs, I., Busvine, J. R., and Agosin, M., 1968, Biochemical studies on insecticide resistant *Musca domestica, Ent. Exp. and Appl.,* 11:15.

Hart, L. G., Shultice, R. W., and Fouts, J. R., 1963, Stimulatory effects of chlordane on hepatic microsomal drug metabolism in the rat, *Toxicol. and Appl. Pharmacol.,* 5:371.

Ilevicky, J., Dinamarca, M. L., and Agosin, M., 1964, Activity of NAD-kinase of nymph *Triatoma infestans* upon treatment with DDT and other compounds, *Comp. Biochem. Physiol.,* 11:291.

Ishaaya, I., and Chefurka, W., 1971, Induction of RNA and protein biosynthesis in the house fly microsomes after DDT treatment, *in:* "Insecticide Resistance, Synergism, Enzyme Induction," A. S. Tahori, ed., pp. 267-279, Gordon and Breach, New York.

Jacob, F., 1966, Genetics of the bacterial cell, *Science,* 152:1470.

Jefcoate, C. R. E., Calabrese, R. L., and Gaylor, J. L., 1970, Ligand interaction with hemoprotein P-450 III. The use of *n*-octylamine and ethyl isocyanide difference spectroscopy in the quantitative determination of high- and low-spin P-450, *Mol. Pharmacol.,* 6:391.

Kaplowitz, N., Kuhlenkamp, J., and Clifton, G., 1975, Drug induction of hepatic glutathione S-transferases in male and female rats, *Biochem. J.,* 146:351.

Khan, M. A. Q., and Matsumura, F., 1972, Induction of mixed-function oxidase and protein synthesis of DDT and dieldrin in German and American cockroaches, *Pestic. Biochem. Physiol.,* 2:236.

Krampl, V., Vargova, M., and Vladar, M., 1973, Induction of hepatic microsomal enzymes after administration of a combination of heptachlor and phenobarbital, *Bull. Environ. Contam. Tox.,* 9:156.

Lee, R. M., and Brindley, W. A., 1974, Synergist ratios, EPN detoxication, lipid, and drug-induced changes in carbaryl toxicity in *Megachile pacifica, Environ. Entomol.,* 3:899.

Litvak, S., and Agosin, M., 1968, Protein synthesis in polysomes from house flies and the effect of 2,2-bis (p-chlorophenyl)-1,1,1-trichloroethane, *Biochemistry,* 7:1560.

Litvak, S., Tarrago-Litvak, L., Poblete, P., and Agosin, M., 1968, Evidence for the DDT-induced synthesis of messenger ribonucleic acid in *Triatoma infestans, Comp. Biochem. Physiol.,* 26:45.

Lucier, G. W., McDaniel, O. S., and Hook, G. E. R., 1975, Nature of the enhancement of hepatic uridine diphosphate glucuronyltransferase activity by 2,3,7,8-tetrachlorodibenzo-p-diozin in rats, *Biochem. Pharmacol.,* 24:325.

Maa, W. C. J., and Terriere, L. C., (unpublished data).

Madhukar, B. V., and Matsumura, F., 1979, Comparison of induction patterns of rat hepatic microsomal mixed-function oxidases by pesticides and related chemicals, *Pestic. Biochem. Physiol.,* 11:301.

Matthews, H. B., and Casida, J. E., 1970, Properties of house fly microsomal cytochromes in relation to sex, strain, substrate specificity, and apparent inhibition and induction by synergist and insecticide chemicals, *Life Sci.,* (pt. 1) 9:989.

Meksongsee, B., Yang, R. S., and Guthrie, F. E., 1967, Effect of inhibitors and inducers of microsomal enzymes on the toxicity of carbamate insecticides to mice and insects, *J. Econ. Entomol.,* 60:1469.

Nebert, D. W., 1978, Genetic aspects of enzyme induction by drugs and chemical carcinogens, *in:* "The Induction of Drug Metabolism," R. W. Estabrook and E. Lindenlaub, eds., pp. 419-452, F. K. Schattauer Verlag, Stuttgart - New York.

Nebert, D. W., and Gielen, J. E., 1972, Genetic regulation of aryl hydrocarbon hydroxylase induction in the mouse, *Fed. Proc.,* 31:1315.

Nebert, D. W., Gielen, J. E., and Goujon, F. M., 1972, Genetic expression of aryl hydrocarbon hydroxylase induction III. Changes in the binding of n-octylamine to cytochrome P-450, *Mol. Pharmacol.,* 8:651.

Oesch, F., Morris, N., and Daly, J. W., 1973, Genetic expression of the induction of epoxide hydrase and aryl hydrocarbon hydroxylase activities in the mouse by phenobarbital or 3-methylcholanthrene, *Mol. Pharmacol.,* 9:692.

Omura, T., 1978, Biosynthesis and drug-induced increase of microsomal enzymes, *in:* "The Induction of Drug Metabolism," R. W. Estabrook and E. Lindenlaub, eds., pp. 161-175, F. K. Schattauer Verlag, Stuttgart - New York.

Omura, T., and Sato, R., 1964, The carbon monoxide-binding pigment of liver microsomes I. Evidence for its hemoprotein nature, *J. Biol. Chem.,* 239:2370.

Pantuck, E. J., Hsiao, K. C., Loub, W. D., Wattenberg, L. W., Kuntzman, R., and Conney, A. H., 1976, Stimulatory effect of vegetables on intestinal drug metabolism in the rat, *J. Pharm. Exper. Ther.*, 198:278.

Perry, A. S., Dale, W. E., and Buckner, A. J., 1971, Induction and repression of microsomal mixed-function oxidases and cytochrome P-450 in resistant and susceptible house flies, *Pestic. Biochem. Physiol.,* 1:131.

Poland, A., and Glover, E., 1973, 2,3,7,8-tetrachlorodibenzo-p-dioxin: A potent inducer of ∂-aminolevulinic acid synthetase, *Science,* 179:476.

Poland, A., and Glover, E., 1974, Comparison of 2,3,7,8-tetrachlorobenzo-p-dioxin, a potent inducer of aryl hydrocarbon hydroxylase, with 3-methylcholanthrene, *Mol. Pharmacol.,* 10:349.

Poland, A., Glover, E., Kende, A. S., DeCamp, M., and Giandomenico, C. M., 1976, 3,4,3',4'-tetrachloro azoxybenzene and azobenzene: Potent inducers of aryl hydrocarbon hydroxylase, *Science,* 194:627.

Puyear, R. L., and Paulson, G. D., 1972, Effect of carbaryl

(1-naphthyl N-methylcarbamate) on pentobarbital-induced sleeping time and some liver microsomal enzymes in white leghorn cockerels, *Toxicol. Appl. Pharmacol.*, 22:621.
Rose, H., Yu, S. J., and Terriere, L. C., (unpublished data).
Schimke, R. T., Kaufman, R. J., Alt, F. W., Kellems, R. F., 1978, Gene amplification and drug resistance in cultured murine cells, *Science*, 202:1051.
Sell, J. L., and Davison, K. L., 1973, Changes in the activities of hepatic microsomal enzymes caused by DDT and dieldrin, *Fed. Proc.*, 32:2003.
Terriere, L. C., and Yu, S. J., 1974, The induction of detoxifying enzymes in insects, *J. Agr. Food Chem.*, 22:366.
Terriere, L. C., and Yu, S. J., 1976, Microsomal oxidases in the flesh fly (*Sarcophaga bullata*, Parker) and the black blow fly (*Phormia regina*, Meigen), *Pestic. Biochem. Physiol.*, 6:223.
Terriere, L. C., Yu, S. J., and Hoyer, R. F., 1971, Induction of microsomal oxidase in F_1 hybrids of a high and a low oxidase house fly strain, *Science*, 171:581.
Thongsinthusak, T., and Krieger, R. I., 1974, Inhibitory and inductive effects of piperonyl butoxide on dihydroisodrin hydroxylation in vivo and in vitro in black cutworm (*Agrotis ypsilon*) larvae, *Life Sci.*, 14:2131.
Tsang, V. C. W., and Agosin, M., 1976, Phenobarbital stimulation of house fly chromatin template activity, *Insect Biochem.*, 6:425.
Turner, J. C., and Green, R. S., 1974, Effect of hexachlorobenzene on microsomal enzyme systems, *Biochem. Pharmacol.*, 23:2387.
Turnquist, R. L., and Brindley, W. A., 1975, Microsomal oxidase activities in relation to age and chlorcyclizine induction in American cockroach, *Periplaneta americana*, fat body, midgut, and hindgut, *Pestic. Biochem. Physiol.*, 5:211.
Walker, C. R., and Terriere, L. C., 1970, Induction of microsomal oxidases by dieldrin in *Musca domestica*, *Ent. Exp. Appl.*, 13:260.
Wongkobrat, A., and Dahlman, D. L., 1976, Larval *Manduca sexta* hemolymph carboxylesterase activity during chronic exposure to insecticide-containing diets, *J. Econ. Entomol.*, 69(2):237.
Yu, S. J., and Terriere, L. C., 1973, Phenobarbital induction of detoxifying enzymes in resistant and susceptible house flies, *Pestic. Biochem. Physiol.*, 3:141.
Yu, S. J., and Terriere, L. C., 1974, A possible role for microsomal oxidases in metamorphosis and reproduction in the house fly, *J. Insect Physiol.*, 20:1901.
Yu, S. J., and Terriere, L. C., 1975, Activities of hormone metabolizing enzymes in house flies treated with some substituted urea growth regulators, *Life Sci.*, 17:619.
Yu, S. J., and Terriere, L. C., 1977, Ecdysone metabolism by soluble enzymes from three species of Diptera and its inhibition by the insect growth regulator TH-6040, *Pestic. Biochem. Physiol.*, 7:48.

Yu, S. J., and Terriere, L. C., 1978, Juvenile hormone epoxide hydrase in house flies, flesh flies and blow flies, *Insect Biochem.*, 8:349.

Yu, S. J., and Terriere, L. C., 1979, Cytochrome P-450 in insects. I. Differences in the forms present in insecticide resistant and susceptible house flies, *Pestic. Biochem. Physiol.*, 12:239.

Yu, S. J., Berry, R. E., and Terriere, L. C., 1979, Host plant stimulation of detoxifying enzymes in a phytophagous insect, *Pestic. Biochem. Physiol.*, 12:280.

RESISTANCE TO INSECTICIDES DUE TO REDUCED SENSITIVITY OF ACETYLCHOLINESTERASE

Hiroshi Hama

National Institute of Agricultural Sciences

Yatabe, Tsukuba, Ibaraki, 305 Japan

INTRODUCTION

Acetylcholinesterase has been extensively investigated as the target of organophosphate and carbamate insecticides. Two types of cholinesterases are well known to be present in mammals; acetylcholinesterase, or true-cholinesterase, and butyrylcholinesterase, or pseudo-cholinesterase. The two cholinesterases are different in their function, distribution and properties, although the physiological function of pseudo-cholinesterase is still unknown (Augustinsson, 1948, 1971).

The properties of cholinesterase in insects are similar to those of true-cholinesterase rather than pseudo-cholinesterase in mammals, so it is therefore called either acetylcholinesterase or cholinesterase. Butyrylcholinesterase has not been proved to exist in insects (O'Brien, 1976). In the green rice leafhopper, *Nephotettix cincticeps*, a butyrylcholinesterase-like enzyme hydrolyzing propyonylthiocholine and butyrylthiocholine at a high rate was separated from acetylcholinesterase (Hama, 1978). However, this enzyme could not be separated from a nonspecific esterase (aliesterase, AliE) by Sepharose 6B gel filtration, DEAE-Sephadex or hydroxylapatite column chromatography (Hama, unpublished).

It has been reported that acetylthiocholine (ATCh), used as a substrate of acetylcholinesterase (AChE), could be hydrolyzed in the human brain by esterases other than AChE (Bernsohn et al., 1962). In the cattle tick, *Boophilus microplus*, ratios of AChE activity toward acetylcholine (ACh), its intrinsic substrate, and ATCh varied extremely among three strains (Roulston et al., 1968). AChE activity

Table 1. AChE Activity in Aqueous Head Extracts of Adult Females of Green Rice Leafhopper

Strain[a]	Amount substrates hydrolyzed[b]		ATCh/ACh
	ACh	ATCh	
S	1.42	4.06	2.86
M	2.28	5.37	2.36
Rmc	2.10	5.06	2.41
N	2.75	6.29	2.29

[a]S, susceptible strain with no altered AChE, collected in Miyagi; M, OP-resistant strain with no altered AChE, collected in Doi, Ehime, and selected with malathion in the laboratory; Rmc, carbamate-resistant strain with altered AChE, bred from S and N strains by repeated back-crossings to S strain under selection with a carbamate Bassa, *o-sec*-butylphenyl methylcarbamate; N, OP- and carbamate-resistant strain with altered AChE, collected in Nakagawara, Ehime. Rmc and N strains have been selected with a carbamate, Bassa, for several generations.
[b]n moles/head/min at 30°C.
Based on data from Hama (1977).

in the green rice leafhopper toward ATCh was higher than toward ACh, but ratios of each activity, i.e., ATCh/ACh ratios, were almost the same among the different strains (Table 1) (Hama, 1977).

It is well known that in most AChE sources activity is distributed in both sediment and supernatant, although more than half of the activity exists in the sediment. Table 2 shows the distribution of AChE activity to ATCh and AliE activity to methyl-*n*-butyrate between 105,000 *g* sediment and supernatant preparations from various strains of the green rice leafhopper, respectively (Hama, 1977). More than 70% of AChE activity was distributed in the sediment, whereas more than 85% of AliE activity was distributed in the supernatant in all strains. The result in the leafhopper is in contrast with the other insects. For example, in the blow fly, *Chrysomya putoria*, AliE activity was found mostly in a microsomal fraction along with AChE activity (Townsend and Busvine, 1969). Similar findings were reported in both fractions of the 100,000 *g* supernatant and pellet in the green peach aphid, *Myzus persicae*, and the house fly (Needham and Sawicki, 1971). The distribution of AChE and AliE activity in these insects may not necessarily show the true distri-

Table 2. AChE and AliE Activity in 105,000 *g* Sediment and Supernatant Prepared from Aqueous Whole Body Extracts of Adult Females of the Green Rice Leafhopper

Enzyme source[a]	Strain[b]	AliE activity methyl-*n*-butyrate hydrolyzed[c]	AChE activity ACh hydrolyzed[d]
Sediment	S	0.36	1.65
	M	0.18	1.83
	Rmc	0.18	1.80
	N	0.85	2.55
Supernatant	S	1.98 (85)[e]	0.58 (26)[e]
	M	8.55 (98)	0.58 (24)
	Rmc	2.37 (93)	0.73 (29)
	N	30.61 (97)	0.86 (25)

[a]Whole body homogenates of adult females were centrifuged at 700 *g* for 10 min, and the supernatant was further centrifuged at 105,000 *g* for 60 min. Both sediment and supernatant were used for assay.
[b]See Table 1.
[c]μ moles/equivalent to 10 females/30 min at 30° C.
[d]μ moles/equivalent to 10 females/40 min at 30° C.
[e]Figures in parentheses indicate precent activity in the supernatant of the total activity of both sediment and supernatant.
Based on data from Hama (1977).

bution of AChE and AliE because it is possible that each enzyme hydrolyzes the other substrates.

There were no appreciable differences in inhibition by propoxur and malaoxon between the sediment AChE and the supernatant AChE of the leafhopper (Table 3) (Hama, 1977). The elution patterns of AChE in the solubilized sediment AChE and the supernatant AChE using DEAE-cellulose column chromatography were similar (Hama, 1976). Therefore, it is likely that the sediment AChE and the supernatant AChE of this insect are not intrinsically different (Hama, 1976).

Table 3. Sensitivity of AChE to Propoxur and Malaoxon in 105,000 *g* Sediment and Supernatant Prepared from Aqueous Whole Body Extracts of Green Rice Leafhopper Adults

Enzyme source[a]	Strain[b]	I_{50} (M)[c]	
		Propoxur	Malaoxon
Sediment	S	2.3×10^{-6}	5.0×10^{-8}
	M	2.3×10^{-6}	4.0×10^{-8}
	Rmc	$> 10^{-4}$	4.0×10^{-7}
	N	2.2×10^{-5}	1.7×10^{-7}
Supernatant	S	1.0×10^{-6}	5.0×10^{-8}
	M	1.0×10^{-6}	5.0×10^{-8}
	Rmc	$> 10^{-4}$	5.0×10^{-7}
	N	5.0×10^{-5}	3.8×10^{-7}

[a]See Table 2.
[b]See Table 1.
[c]Pre-incubation of enzyme sources with propoxur for 15 min and with malaoxon for 10 min at 30°C. ATCh was used as the substrate.
Based on data from Hama (1977).

CASES OF RESISTANCE TO INSECTICIDES DUE TO REDUCED SENSITIVITY OF ACETYLCHOLINESTERASE

Reduced sensitivity of AChE to organophosphates (OP) was first found in resistant (R) spider mites, *Tetranychus urticae*, by Smissaert (1964); these spider mites had about one-third as active an AChE as susceptible (S) mites, and their AChE was much less sensitive to inhibition by paraoxon and diazoxon. Since then, similar phenomena have been observed in other R mites and in ticks and insects, as listed in Table 4. It has been accepted that reduced sensitivity of AChE is one of the main factors of resistance to OP in mites and ticks in addition to the increased activity of degradation of insecticides.

Sensitivity of AChE to inhibition by OP and carbamates has also been determined and compared among strains of insects as one

Table 4. Cases of Resistance to Insecticides Related to Reduced Sensitivity of AChE

Species	Resistance	Places	References
Tetranychus urticae	OP (parathion, diazinon)	West Germany	Smissaert, 1964
	(parathion)	New Zealand	Ballantyne and Harrison, 1967
Tetranychus pacificus	parathion	USA (California)[a]	Zon and Helle, 1966
Tetranychus telarius	malathion	Israel	Zahavi et al., 1970
Boophillus microplus	OP and carbamates (coumaphos, diazinon, carbaryl)	Australia	Lee and Batham, 1966; Roulston et al., 1968
Nephotettix cincticeps	carbamates and OP (propoxur, carbaryl, malathion)	Japan	Hama and Iwata, 1971, 1978; Iwata and Hama, 1972
Musca domestica	OP and carbamates (tetrachlorvinphos, dichlorvos, propoxur)	USA (New York)[a]	Tripathi and O'Brien, 1973b; Tripathi, 1976
	(dimethoate)	Denmark	Devonshire, 1975
Anopheles albimanus	OP and carbamates (parathion, propoxur)	El Salvador[a]	Ayad and Georghiou, 1975

[a]Populations selected with an insecticide further in laboratory.

of the suspected resistance factors, but no differences were found among the strains in their AChE sensitivity.

Mengle and Casida (1960) found that when whole body homogenates of the house fly were pre-incubated with paraoxon and malaoxon, the rate of AChE inhibition was more rapid in an S strain than in the other two OP-R strains. However, when head homogenates were used as the enzyme source, AChE sensitivity was almost equal among all the strains. It was suggested that an unknown factor protecting AChE from inhibition existed in the thorax and/or abdomen of the

R house fly. Similarly, the rate of AChE inhibition by diazoxon was 1.5 times smaller in the OP-R strain of the sheep blow fly, *Lucilia cuprina,* than the other two strains when thoracic AChE was used, but no difference was found in the rates of AChE inhibition among the strains when the head homogenates were used (Schuntner and Roulston, 1968).

Kojima et al. (1963) compared AChE sensitivity to inhibition by malaoxon between S and R strains of the green rice leafhopper, but found no sensitivity difference between the 10-times malathion-R strain and the S strain. Iwata and Hama (1977) selected malathion-R and methyl parathion-R strains of the leafhopper in the laboratory with malathion and methyl parathion, respectively. The insects exhibited extremely high resistance to many other OP insecticides, but did not show clear cross resistance to carbamate insecticides. Malaoxon and propoxur equally inhibited the AChE's of both S and OP-R strains (Hama, 1977; Hama et al., 1977) (See Table 3 for an example); however, AChE of the multiple-R leafhoppers, highly resistant to both carbamates and OP, was much less sensitive to inhibition by substituted phenyl methylcarbamates and carbaryl (Hama and Iwata, 1971, 1973). Most of the carbamate-R leafhoppers in the field are typically multiple-R because resistance to OP developed prior to carbamate resistance (Iwata and Hama, 1971, 1972). Acetylcholinesterases of head extracts of multiple-R strains were less sensitive to propoxur and carbaryl inhibition (Hama and Iwata, 1971), and those of whole body extracts of strains Rmc and N were inhibited by propoxur and malaoxon at a lesser degree (Table 3) (Hama, 1977). Consequently, it may be concluded that reduced sensitivity of AChE in the miltiple-R leafhoppers is attributed to a structural alteration of AChE.

Since then, much reduced AChE sensitivity to OP and carbamates inhibition has also been found in R strains of the house fly and the mosquito *Anopheles albimanus,* as listed in Table 4.

When there are small differences in sensitivity of AChE to inhibitors among strains of insects with crude enzyme preparations, it is necessary to clarify whether true alteration of the enzyme or some other factors are responsible for these differences.

EXISTENCE OF "ALTERED ACETYLCHOLINESTERASE"

It is considered that the structurally altered active site of AChE results in reduced sensitivity of AChE to inhibition. The kinetics studies of AChE with inhibitors have suggested that there are at least two different types of AChE in the spider mite (Smissaert, 1964), the cattle tick (Lee and Batham, 1966; Nolan et al., 1972; Schnitzerling et al., 1974; Nolan and Schnitzerling, 1975), the green rice leafhopper (Hama, 1977; Takahashi et al., 1978) and house fly (Tripathi and O'Brien, 1973b; Devonshire, 1975).

However, mutant-type enzymes in R strains have not been separated from wild-type enzymes, i.e., normal AChE, by the conventional methods, such as gel electrophoresis, gel filtration or various types of column chromatography (Nolan et al., 1972; Tripathi and O'Brien, 1975). Failure of separation may be attributed to the fact that alteration in altered AChE must be minor (Nolan et al., 1972).

In the green rice leafhopper, the mutant-type enzyme was separated from the normal AChE by DEAE-cellulose column chromatography (Hama, 1976). Since then, further studies have been made (Hama, unpublished) and those results are described here. Materials and methods are essentially the same as those of the previous paper (Hama, 1976), except for the column size in gel filtration and for the use of DEAE-Sephadex (Pharmacia Fine Chemical Inc.) instead of DEAE-cellulose.

Sepharose 6B Gel Filtration

Figures 1 and 2 show elution patterns of AChE activity in 105,000 *g* supernatant and solubilized sediment of leafhopper aqueous whole body homogenates using Sepharose 6B gel filtration. The insects studied were susceptible (S), carbamate-R (Rmc) and multiple-R (N) strains (data on Rmc strain omitted). Three active peaks were detected in the supernatant, i.e., a small peak (AChE 3) in the void

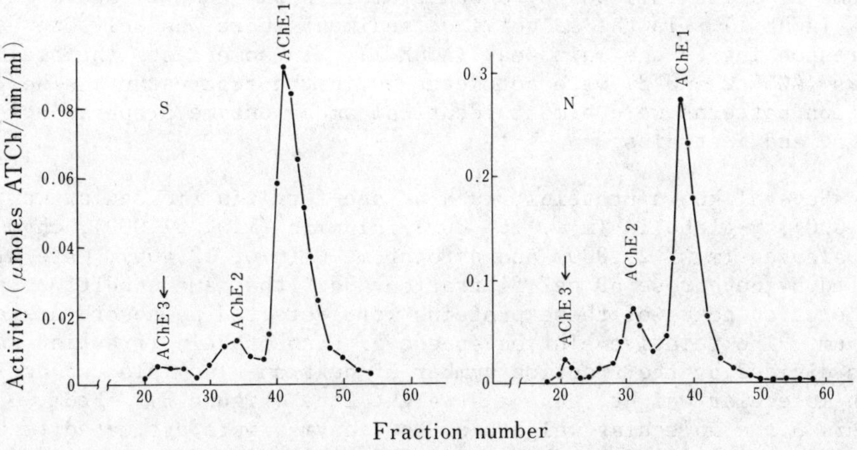

Figure 1. Elution patterns of AChE activity in 105,000 *g* supernatant prepared from aqueous whole body extracts of susceptible (S) and multiple-resistant (N) strains using Sepharose 6B gel filtration. Activity of enzyme sources introduced onto the column was not the same among the strains. Fraction volume: 8 ml/fraction, arrow: void volume.

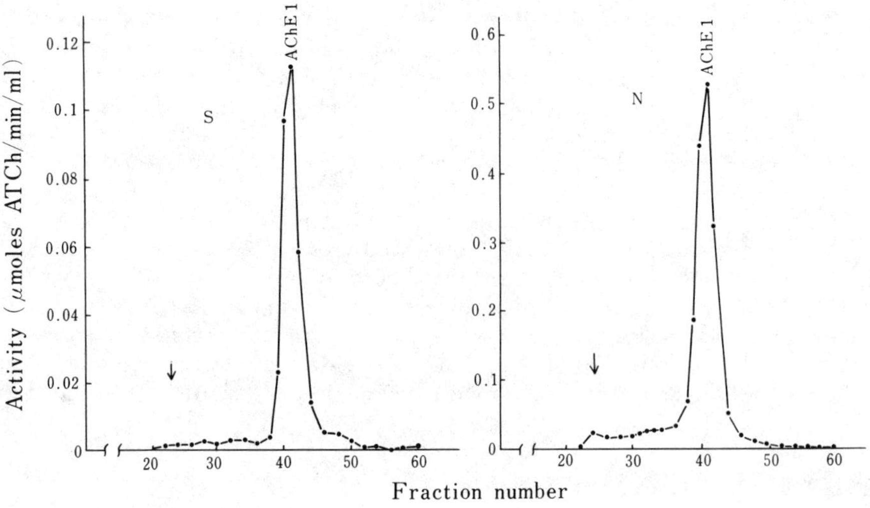

Figure 2. Elution patterns of AChE activity in 105,000 *g* solubilized sediment prepared from aqueous whole body extracts of susceptible (S) and multiple-resistant (N) strains using Sepharose 6B gel filtration. Activity of enzyme sources introduced onto the column was not the same among the strains. Fraction volume: 8 ml/fraction, arrow: void volume.

volume fraction followed by another small peak (AChE 2) and a main peak (AChE 1). In the solubilized sediment there was only one peak, corresponding to the main peak (AChE 1). In some cases the small peaks (AChE 2 and 3) were not detected in the supernatant. The elution patterns were almost identical among enzyme preparations of S, Rmc and N strains.

Several known proteins, such as apo-ferritin (molecular weight 480,000), γ-globulin (m.w. 160,000), albumin (m.w. 67,000), chymotrypsinogen (m.w. 25,000) and cytochrome C (m.w. 12,400), were also eluted by Sepharose 6B gel filtration under the same conditions and the elution peaks of these proteins were detected by absorbance at 280 nm. The relative elution speeds of each protein (fraction numbers divided by the fraction number of dextran blue) plotted against each molecular weight, gave a straight line (Figure 3). From Figure 3 the molecular weights of the enzymes were estimated to be 120,000 to 150,000 for AChE 1, 350,000 to 450,000 for AChE 2 and ≧ 2,000,000 for AChE 3.

It has been well known that AChE's of the electric eel, the torpedo, the mayfly, the honey bee and the house fly are oligomer

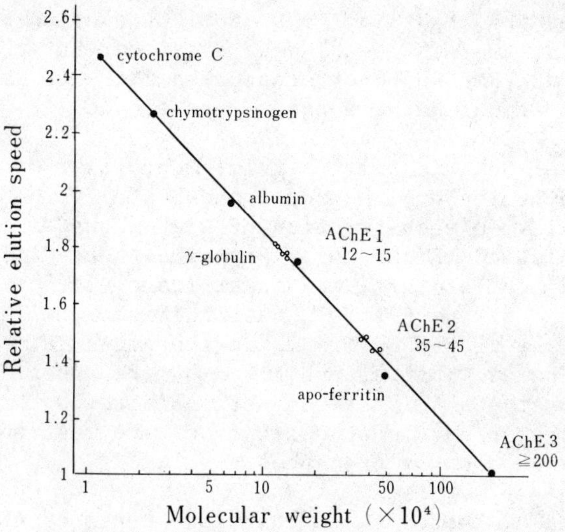

Figure 3. Relationship between molecular weight of proteins and their relative elution speed (fraction numbers divided by the fraction number of dextran blue) on Sepharose 6B gel filtration.

enzymes, and that aggregation and deaggregation of oligomers are strongly affected by various factors such as salt concentration, pH and detergents (Grafius and Millar, 1965, 1967; Massoulié and Rieger, 1969; Krysan and Kruckeberg, 1970; Dudai et al., 1972; Steele and Smallman, 1976b).

The results with the leafhopper, namely the distribution of molecular weights of AChE 1, 2 and 3, the relative instability of AChE 2 and 3 as compared with the main AChE 1, and the disappearance of AChE 2 and 3 in the sediment treated with Triton X-100, suggest that AChE 2 and 3 in the leafhopper are aggregating forms of AChE 1.

The molecular weight of the main AChE 1 of the leafhopper is almost the same as that of AChE of the house fly (Krysan and Chadwick, 1966; Krysan and Kruckeberg, 1970; Tripathi and O'Brien, 1977). It was reported that an AChE showing a molecular weight of 80,000 was the "fundamental unit" of the AChE of the house fly (Steele and Smallman, 1976a).

DEAE-Sephadex Column Chromatography

The AChE 1, a major fraction of the supernatant on Sepharose 6B gel, was rechromatographed using DEAE-Sephadex. Figure 4 shows the elution patterns. Although AChE was eluted as one peak in both the S and Rmc strains, in the S strain it was eluted around 0.44 M KCl, which was later than that in the 0.38 M KCl elution of the Rmc strain. In the N strain, two AChE peaks were observed that coincided with the AChE peaks of the Rmc and S strains, respectively, as judged by KCl concentrations at which the enzymes were eluted.

After pre-incubation with 2 x 10^{-5}M propoxur or 2 x 10^{-4}M propaphos (*p*-methylthiophenyl dipropyl phosphate) for 10 min at 30°C, the AChE activity in each fraction of DEAE-Sephadex column chromatography was plotted (Figure 4). Propaphos served as an inhibitor of the altered AChE of this leafhopper (Hama, 1975; Hama and Iwata, 1978). The AChE of the S strain was inhibited completely by propoxur, but not by propaphos, while in the Rmc strain it was not inhibited by propoxur, but was inhibited by propaphos. In the N strain, the first AChE eluate was not inhibited by propoxur but was inhibited by propaphos, and the second eluate was inhibited completely by propoxur but not by propaphos.

Consequently, there are at least two types of AChE's in the green rice leafhopper: the normal AChE and the altered AChE. It was proved that the AChE in the S strain was the normal AChE, most of the AChE in the Rmc strain was the altered AChE, and the N strain had both the normal AChE and the altered AChE with an activity ratio of about 1:1.

Since the elution patterns of AChE activity in the solubilized sediment and the supernatant were not different in each strain using DEAE-Sephadex column chromatography, AChE in the sediment is considered intrinsically identical to that in the supernatant.

PROPERTIES OF "NORMAL ACETYLCHOLINESTERASE" AND "ALTERED ACETYLCHOLINESTERASE"

The normal AChE and the altered AChE in the green rice leafhopper were purified from whole body homogenates of the S and Rmc strains by ultracentrifugation and solubilization with Triton X-100 followed by Sepharose 6B gel filtration and DEAE-cellulose column chromatography as described in a previous paper (Hama, 1976). A partially purified AChE preparation derived from the solubilized sediment was used for most of the experiments. A 70-fold purification of AChE was obtained with a 15% recovery.

Sensitivity of Acetylcholinesterase to Inhibitors

It has been shown that the AChE of the multiple-R leafhopper is much less sensitive to inhibition by many methylcarbamates, including eserine (Table 5), and dimethyl organophosphates, such as malaoxon and fenitroxon (Table 6). Several chemicals of diethyl and *n*-dipropyl organophosphates (Tables 6 and 7) and *n*-propylcarbamates (Yamamoto et al., 1977) more strongly inhibit the altered

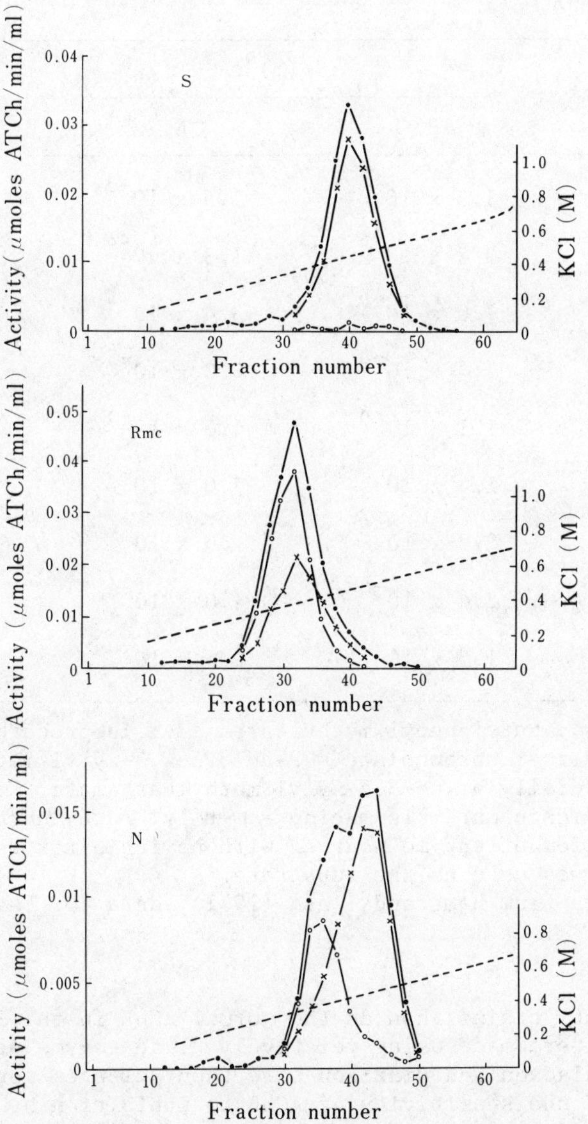

Figure 4. Elution patterns of a major fraction of the supernatant, AChE 1, by DEAE-Sephadex column chromatography following Sepharose 6B gel filtration. ●: AChE activity without an inhibitor, o: AChE activity after pre-incubation with 2 x 10^{-5}M propoxur for 10 min at 30°C, x: AChE activity after pre-incubation with 2 x 10^{-4}M propaphos for 10 min at 30°C. Fraction volume: 5 ml/fraction.

Table 5. Sensitivity of AChE to Methylcarbamates in Aqueous Whole Body Extracts of Adult Females of the Green Rice Leafhopper

Inhibitor[a]	I_{50} (M)[b]		Ratio of I_{50} N/S
	S	N	
propoxur	1.3×10^{-5}	1.5×10^{-3}	115
Bassa	2.5×10^{-6}	1.3×10^{-4}	52
isoprocarb	7.0×10^{-6}	3.4×10^{-4}	49
carbaryl	1.4×10^{-6}	6.0×10^{-5}	43
carbanolate	1.1×10^{-6}	4.0×10^{-5}	36
Hydrol	1.4×10^{-5}	7.0×10^{-5}	5
Meobal	2.7×10^{-5}	1.9×10^{-4}	7
Tsumacide	2.4×10^{-5}	4.0×10^{-4}	17
ererine	1.5×10^{-7}	2.5×10^{-6}	17

[a]Bassa = *o*-*sec*-butylphenyl methylcarbamate; isoprocarb = *o*-cumenyl methylcarbamate; carbanolate = 2-chloro-4,5-xylyl methylcarbamate; Hydrol = 4-diallylamino-3,5-xylyl methylcarbamate; Meobal = 3,5-xylyl methylcarbamate; Tsumacide = *m*-tolyl methylcarbamate.
[b]Pre-incubation of enzyme sources with an inhibitor for 30 min at 37°C. ACh was used as the substrate.
Based on data from Hama and Iwata (1971), data for Bassa and eserine were added.

AChE in the R strains than do the normal AChE in the S strain. The comparisons were made using relatively crude enzyme preparations. Propoxur, malaoxon and diazoxon were then selected for further studies, and the sensitivity of AChE to inhibition by these chemicals was determined using partially purified normal and altered AChE's (Hama et al., unpublished).

As shown in Figure 5, the altered AChE was much less sensitive to inhibition by propoxur and malaoxon but more highly sensitive to inhibition by diazoxon than was the normal AChE. These results were not very different from those listed in Tables 5 and 6.

Table 6. Sensitivity of AChE to Organophosphates in 105,000 *g* Sediment Prepared from Aqueous Whole Body Extracts of Green Rice Leafhopper Adults

Inhibitor[a]	I_{50} (M)[b]		Ratio of I_{50}
	S	Rmc	Rmc/S
malaoxon	2.5×10^{-8}	3.0×10^{-7}	12
fenitroxon	2.5×10^{-7}	2.0×10^{-6}	8
chlorfenvinphos	3.2×10^{-7}	1.1×10^{-6}	3.4
diazoxon	1.5×10^{-7}	3.0×10^{-8}	0.2
pyridafenoxon	1.2×10^{-7}	3.6×10^{-8}	0.3
propaphos	$> 2 \times 10^{-4}$	7.5×10^{-5}	< 0.4
acephate	$>> 2 \times 10^{-4}$	$>> 2 \times 10^{-4}$	-

[a]Pyridafenoxon = 2,3-dihydro-3-oxo-2-phenyl-6-pyridazinyl diethyl phosphate; propaphos = *p*-methylthiophenyl dipropyl phosphate; acephate = *O,S*-dimethyl *N*-acetylphosphoramidothiolate.
[b]Pre-incubation of enzyme sources with an inhibitor for 15 min at 30°C. ATCh was used as the substrate.
Data from Hama and Iwata (1978).

The bimolecular reaction constant, ki, between AChE and an inhibitor was calculated by the Aldridge method (1950). The ki values for malaoxon were $3 \times 10^{6}\ M^{-1}min^{-1}$ in the normal AChE and they were $2 \times 10^{5} M^{-1}min^{-1}$ in the altered AChE, respectively, whose ki was decreased by a factor of 15. The ki values for diazoxon, however, were $8.7 \times 10^{5} M^{-1}min^{-1}$ in the normal AChE and $1.7 \times 10^{6} M^{-1}min^{-1}$ in the altered AChE, respectively, whose ki was increased by a factor of 2.

Inhibition of AChE (E) by carbamate and OP chemicals (AX) progresses with time according to the following equation (O'Brien, 1967; Aldridge and Reiner, 1972):

Table 7. Sensitivity of AChE to Analogs of Propaphos in 105,000 *g* Sediment Prepared from Aqueous Whole Body Extracts of Green Rice Leafhopper Adults

Inhibitor	$(R_1O)_2P(O){\cdot}O{\cdot}C_6H_4{-}R_2$ Substituents		I_{50}(M)[a]		Ratio of I_{50} (Rmc/S)
	R_1	R_2	S	Rmc	
methyl propaphos	CH_3	$S{\cdot}CH_3$	7.5×10^{-7}	1.4×10^{-6}	1.9
ethyl propaphos	C_2H_5	$S{\cdot}CH_3$	3.0×10^{-6}	1.3×10^{-6}	0.43
propaphos	*n*-C_3H_7	$S{\cdot}CH_3$	$> 2 \times 10^{-4}$	7.5×10^{-5}	< 0.38
propaphos sulfoxide	*n*-C_3H_7	$SO{\cdot}CH_3$	1.4×10^{-5}	1.7×10^{-6}	0.12
propaphos sulfone	*n*-C_3H_7	$SO_2{\cdot}CH_3$	1.0×10^{-6}	2.0×10^{-7}	0.2

[a]Pre-incubation of enzyme sources with an inhibitor for 15 min at 30°C.
ATCh was used as the substrate. Data from Hama (unpublished).

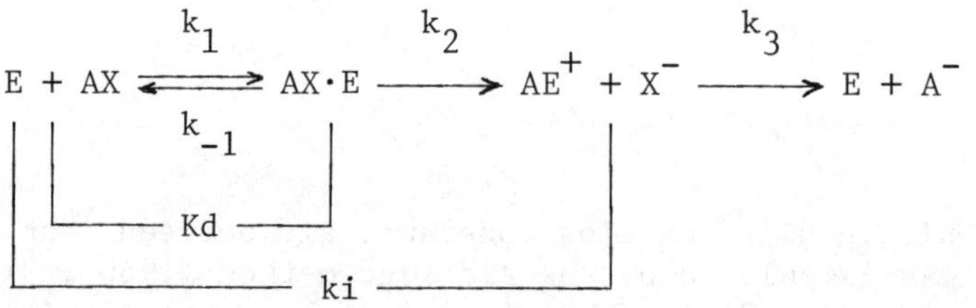

It has been shown that in the R strains of the house fly and the green rice leafhopper the much reduced sensitivity of the altered AChE to OP and carbamates inhibition is mainly caused by a reduced affinity for inhibitors, Kd, rather than by the rates of phosphorylation or carbamylation, k_2, followed by dephosphorylation or decarbamylation, k_3 (Tripathi and O'Brien, 1973b, 1975; Tripathi, 1976; Yamamoto et al., 1977).

It has been established that quaternary ammonium chemicals inhibit AChE activity due to binding at the "anionic site" of the

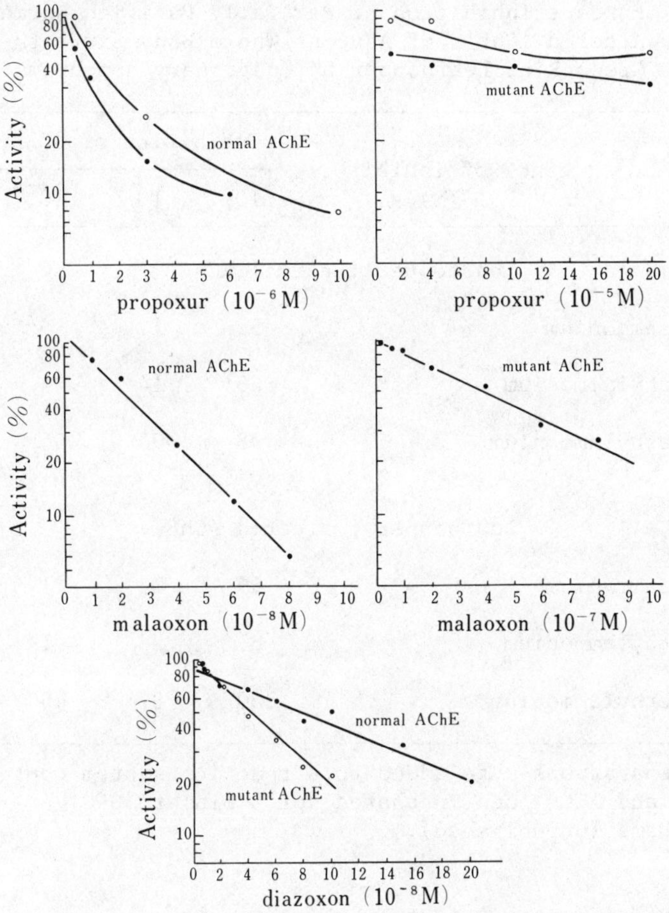

Figure 5. Sensitivity of partially purified normal and altered AChE's of aqueous whole body extracts of susceptible and carbamate-resistant (Rmc) strains to propoxur, malaoxon and diazoxon. Pre-incubation of enzyme preparations with an inhibitor for 10 min at 30°C. Hollow circles in data for propoxur indicate values determined by radioassay using ^{14}C-ACh. The other values were determined by Ellman's method using ATCh. (Mutant AChE = altered AChE.)

enzyme (Wilson, 1952), and Table 8 shows the sensitivity of purified normal AChE and altered AChE to these chemicals. Although phenyltrimethylammonium was the strongest inhibitor of the leafhopper AChE among the chemicals tested, there was no significant difference in their respective sensitivities between the normal and altered AChE's.

Table 8. Percent Inhibition of Partially Purified Normal and Altered AChE's of Aqueous Whole Body Extracts of the Green Rice Leafhopper by Quaternary Ammonium Chemicals

Inhibitor	Conc. of inhibitor (M)	% Inhibition of AChE activity[a] 10^{-2}	10^{-3}	10^{-4}	10^{-5}
Leafhopper : normal AChE					
tetramethylammonium		52	12	3	4
ethyltrimethylammonium		68	14	2	0
phenyltrimethylammonium		98	90	55	17
Leafhopper : altered AChE					
tetramethylammonium		69	17	1	0
ethyltrimethylammonium		81	28	1	0
phenyltrimethylammonium		100	87	46	11

[a]Enzyme preparations were added to a reaction system containing an inhibitor and ATCh, and incubated for 5 min at 30°C.
Data from Hama (unpublished).

Substrate Specificity of Acetylcholinesterase

Figure 6 shows the relative activity of partially purified normal AChE and altered AChE of the leafhopper toward acetylthiocholine (ATCh), propyonylthiocholine (PrTCh) and butyrylthiocholine (BuTCh), with bovine erythrocyte AChE (purchased from Sigma Chemical Co.) used as a standard enzyme (Hama, unpublished). The activity patterns of the normal AChE and the bovine erythrocyte AChE were similar, but that of the altered AChE was obviously different. The relative activity of the altered AChE toward PrTCh was lower than that of the normal AChE, but it was rather high toward BuTCh.

Also, the activity of both the normal AChE and the bovine erythrocyte AChE was suppressed or inhibited at more than 10^{-3}M ATCh or PrTCh, but in the altered AChE no apparent inhibition of activity toward ATCh, PrTCh and BuTCh occurred within the concentrations of substrates tested. In another test, however, the activity of the

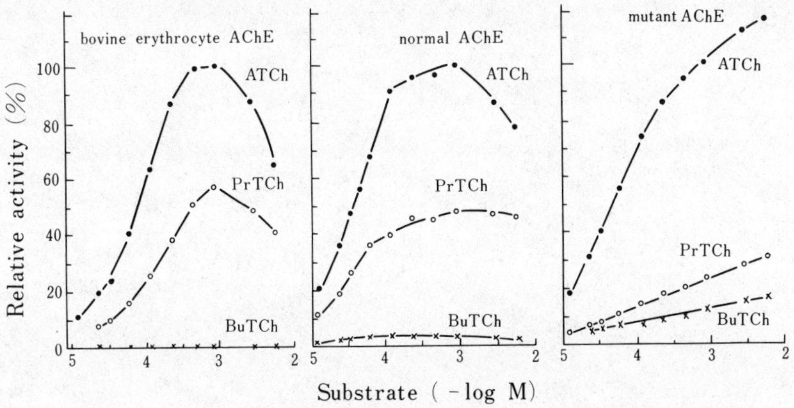

Figure 6. Relative activity of partially purified normal and altered AChE's of aqueous whole body extracts of susceptible and carbamate-resistant (Rmc) strains toward ATCh (o), PrTCh (o) and BuTCh (x) with bovine erythrocyte AChE. (Mutant AChE = altered AChE.)

altered AChE toward ATCh was suppressed at more than 10^{-2}M ATCh (data not shown).

Figure 7 shows the relative activity of the normal AChE and the altered AChE of the leafhopper toward ^{14}C-ACh, as determined by radiometric methods (Hama et al., unpublished). The optimum ACh concentration for the altered AChE was higher than that for the normal AChE.

Similar findings were reported in the dimethoate-R house fly (Devonshire, 1975) and in the Ridgelands R strain of the cattle tick, *Boophilus microplus*, whose altered AChE was not inhibited even at 1.73 x 10^{-2}M ATCh (Nolan and Schnitzerling, 1975).

The inhibition of AChE activity by an excess amount of substrates is well known as one of the critical properties of a true-cholinesterase (Augustinsson, 1948, 1971). It has been proposed that such an inhibition of AChE activity may be caused by the combining of the free substrates at the acylated AChE, which inhibits deacylation of the acylated AChE (Krupka and Laidler, 1961), and that this inhibition may be related to sites other than the active site, at which intrinsic substrate ACh binds (Aldridge and Reiner, 1969).

The affinity of each AChE of the leafhopper for ATCh and PrTCh was evaluated by the Linewever-Burk plots (Table 9). The Km values of both ATCh and PrTCh for the bovine erythrocyte AChE were in the order of 10^{-4}M, which were higher than those for the leafhopper

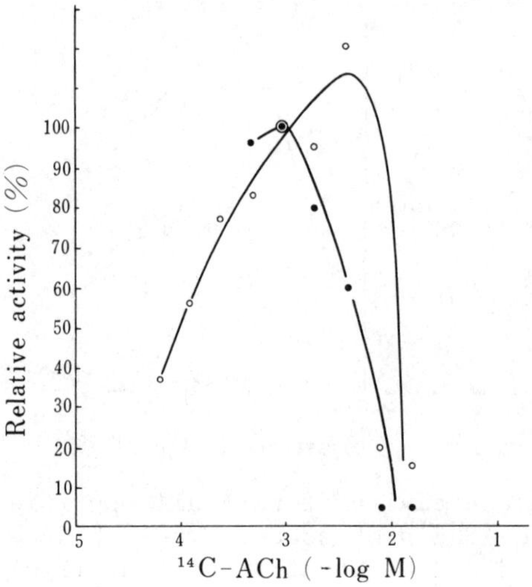

Figure 7. Relative activity of partially purified normal and altered AChE's of aqueous whole body extracts of susceptible and carbamate-resistant (Rmc) strains. ^{14}C-ACh was used as the substrate. (● = normal AChE, o = altered AChE.)

Table 9. Km Values of ATCh and PrTCh for Bovine Erythrocyte AChE and Partially Purified Normal and Altered AChE's of Aqueous Whole Body Extracts of the Green Rice Leafhopper

Enzyme source	Km(M)	
	ATCh	PrTCh
bovine erythrocyte AChE	1.72×10^{-4}	2.0×10^{-4}
leafhopper: normal AChE	4.76×10^{-5}	4.17×10^{-5}
leafhopper: altered AChE	7.14×10^{-5}	7.14×10^{-5}

Data from Hama (unpublished).

AChE, which were in the order of 10^{-5} M. The values of both ATCh and PrTCh for the altered AChE were less than twice as high as those for the normal AChE.

In the mite, *Tetranychus urticae*, the Km values of ATCh for the altered AChE and the normal AChE were not extremely different, the values being less than two-fold higher in the altered than in the normal, although the AChE activity of the R mites was one-third that of the S mites (Smissaert, 1964, 1970). On the other hand, the Km values of PrTCh and BuTCh for R AChE were much greater than were those for the susceptible AChE (Smissaert, 1970).

O'Brien et al. (1978) found that in the house fly the Km value of butyrylcholine for the altered AChE was 25 times greater than that for the normal AChE, although the values with the other substrates between the two enzymes were not significantly different.

Considerations on Modification of "Altered Acetylcholinesterase"

The fact that the Km values of ATCh and PrTCh for the altered AChE were less than twice those for the normal AChE (Table 9) strongly supports a hypothesis proposed by Tripathi and O'Brien (1973b, 1975) and O'Brien et al. (1978). They have stated that a binding site of AChE for OP and carbamates was different from the binding site for intrinsic substrate ACh. In relation to this hypothesis, it has been proposed that the binding site of AChE for OP and carbamates is "hydrophobic" rather than anionic (Casida, 1973; Aldridge, 1975; O'Brien, 1976).

It has been shown that for the house fly, the anticholinesterase activity of substituted phenyl methylcarbamates increases progressively with the size and branching of their alkyl and alkoxyl substituents until the maximum activity is attained by isopropyl and *sec*-butyryl (Metcalf et al., 1962). Such a tendency for an increased anticholinesterase activity with increased methylation was fully explained by van der Wall's force between the substituents and the anionic site of AChE (Wilson, 1952; Metcalf et al., 1962).

Similarly, the bulky alkyl- or alkoxyl-substituted carbamates, such as propoxur, isoprocarb and Bassa, exhibited stronger anticholinesterase activity for the S leafhopper than did methyl-substituted carbamates, such as Tsumacide and Meobal (Table 5). However, AChE in multiple-R leafhoppers, e.g., the N strain, was much less sensitive to inhibition by propoxur, isoprocarb and Bassa than was the S strain (Table 5). From these results it had been once postulated that some changes occurred at the anionic site of AChE in the multiple-R leafhopper (Hama and Iwata, 1971).

However, quaternary ammonium chemicals, which are also inhibitors of AChE due to binding at the anionic site, inhibited the altered AChE of the leafhopper to the same extent as they inhibited the normal AChE (Table 8). Therefore, the binding site of AChE for alkyl and alkoxyl substituents of phenyl methylcarbamates must be a "hydrophobic site" rather than an anionic site, and some changes must have occurred at the hydrophobic site of the altered AChE in the leafhopper.

Aldridge (1975) proposed that inhibition of AChE activity by an excess substrate may be due to binding of the substrate at both sites, anionic and hydrophobic, in order to prevent its hydrolysis. This may relate to the shift of the optimum concentration of ACh or ATCh to higher concentration in the altered AChE of the leafhopper (Figures 6 and 7).

It was mentioned above that the altered AChE of the leafhopper had reduced sensitivity to inhibition by methylcarbamates and dimethyl organophosphates, but increased sensitivity to inhibition by diethyl and *n*-dipropyl organophosphates and *n*-propylcarbamates when compared with the normal AChE. Additionally, relative activity of the altered AChE toward PrTCh and BuTCh was different from that of the normal AChE (Figure 6). Therefore, possibilities cannot be neglected that a structural change at a catalytic or esteratic site and a distance change between a catalytic site and binding sites are involved in the reduced sensitivity of the altered AChE.

RELATIONSHIP BETWEEN REDUCED SENSITIVITY OF ACETYLCHOLINESTERASE AND RESISTANCE LEVEL

Resistance factors in arthropods have been considered to be related to penetration, movement, metabolism, excretion and storage of insecticides, and inhibition of the enzyme at the target site, but it is very difficult to evaluate which factors contribute to the actual resistance level.

Since inhibition of an AChE by carbamates and OP progresses with time, even the altered AChE could be inhibited by these inhibitors, and R insects having the altered AChE could be killed after all, unless the amount of an inhibitor in the target site is reduced (Smissaert, 1964). In fact, the amount of an inhibitor was shown to be reduced by a detoxication capacity present even in most of the S insects so that it is likely that R strains having the altered AChE can survive as a result of a mechanism that allows sufficient time for the degradation of inhibitors (Smissaert, 1964; Oppenoorth and Welling, 1976).

The relationship between the reduced sensitivity of AChE and resistance to carbamate and OP insecticides in the green rice

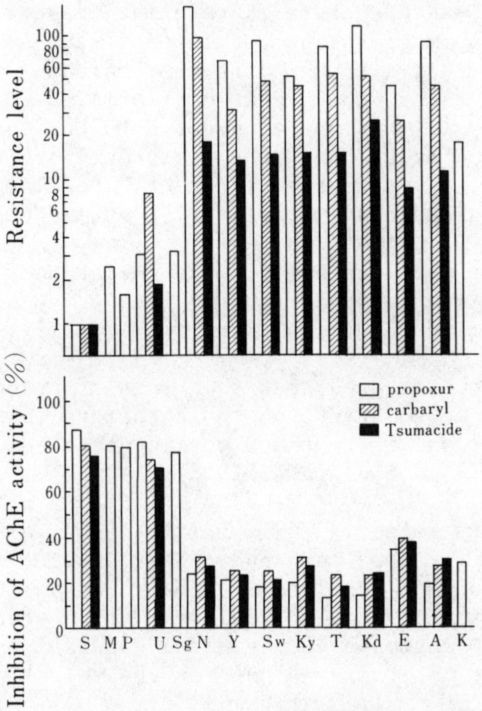

Figure 8. Relationship between a resistance level to methylcarbamates and inhibition of AChE activity by methylcarbamates among various populations of the green rice leafhopper. Inhibition of AChE activity was determined following preincubation with 10^{-4}M propoxur, 10^{-5}M carbaryl and 10^{-4}M Tsumacide for 30 min at 37°C. S, M and N, see Table 1; P, an OP-resistant strain selected with methyl parathion in the laboratory; U and Sa, OP-resistant insects showing a low level of resistance to carbamates; the other insects are multiple-resistant insects collected in Kyushu, except for Y and K, which were collected in Hiroshima and Kochi, respectively. Based on data from Hama and Iwata (1973).

leafhopper is as follows:

1) The rate of development of carbamate resistance was much more rapid than that of OP resistance. When field personnel reported a poor leafhopper control by carbamate insecticides, a few generation times usually elapsed in field populations before the actual level of resistance was determined. It had often been found that the populations had already developed extremely high levels of resistance to carbamate insecticides, as shown in Figure 8.

2) The resistance spectra to carbamate insecticides were very similar in pattern among the tested insects representing multiple resistance (Figure 8), although different carbamates had been used in each location. Also, the sensitivity of AChE to carbamates in the multiple-R populations was considered to be reduced uniformly (Figure 8). On the other hand, the resistance spectra to OP insecticides varied extremely with the different R populations (Ozaki and Kurosu, 1967; Yoshioka and Iwata, 1967; Iwata and Hama, 1971).

3) For eight methylcarbamates the resistance level of the multiple-R strain was in proportion to the degree of reduction in AChE sensitivity (Figure 9) (Hama and Iwata, 1971).

4) It was demonstrated that the trait of reduced AChE sensitivity could not be isolated from the carbamate resistance trait by successive back crossings to the S strain under selection pressure with a carbamate (Hama and Iwata, 1978).

5) Laboratory selection of a multiple-R strain K with propaphos, an inhibitor of the leafhopper altered AChE, was continued for several generations and resulted in a decrease in the level of resistance to propoxur instead. The level was one-tenth that of the original strain K, although the selected population had developed a four-fold increase in its resistance to the selecting agent (Iwata and Hama, 1976).

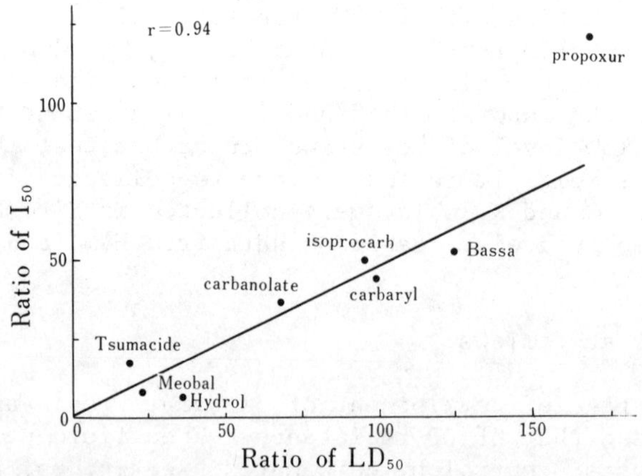

Figure 9. Relationship between a resistance level (ratio of LD_{50}) and reduction in sensitivity of AChE (ratio of I_{50}) to methylcarbamates. Data from Hama and Iwata (1971); data for Bassa were added.

The decrease in the level of resistance to propoxur after selection with propaphos was explained as an increase in the sensitivity of AChE to inhibition by propoxur (Hama and Iwata, 1979).

These characteristics indicate that carbamate resistance in the leafhopper is mainly due to reduced sensitivity of AChE, and that cross resistance to carbamate insecticides in the multiple-R populations can also be explained by the same mechanism. The rapid development of resistance to carbamates and the uniformity of the resistance spectra can well be attributed to one incompletely dominant factor (Hama and Iwata, 1978).

In order to clarify the relationship between reduced sensitivity of AChE and cross resistance to carbamates and OP in the multiple-R leafhopper, a sub-strain (Rmc) was used. This strain was created by introducing a factor for the reduced sensitivity of AChE into the background of the S strain by repeated back crossings under selection with a carbamate, Bassa (*o-sec*-butylphenyl methylcarbamate) (Hama, 1975; Hama and Iwata, 1978). By comparing the susceptibility of the strain Rmc to insecticides and the sensitivity of its AChE to inhibitors with those of the two parental strains, it was determined that resistance to methylcarbamate insecticides was mainly caused by a reduced sensitivity of AChE, as listed in Table 10. It was also determined that the altered AChE contributed to a part of the resistance to dimethyl OP insecticides such as malathion, phenthoate, fenitrothion and Fujithion (Table 10) (Hama and Iwata, 1978).

On the other hand, it was confirmed that in the multiple-R strain N in vivo, propoxur, Bassa and carbaryl were degraded twice as much as in the S strain (Kazano et al., 1978; Hama et al., 1979). It was observed that susceptibility to propoxur in the strain selected with *n*-dipropyl organophosphate propaphos, as mentioned above, remained about ten times lower than that of the S strain, although AChE activity of the selected leafhoppers and that of the S strain were inhibited equally by propoxur (Hama and Iwata, 1979).

These results suggest that in addition to alteration of the target enzyme, increased degradation activity of carbamate insecticides may be a minor carbamate resistance factor (Hama et al., 1979).

In the house fly, it was observed that in the Cornell R strain having the altered AChE resistance levels to OP and carbamates were approximately proportional to the degree of AChE sensitivity reduction to inhibition (Tripathi, 1976). On the other hand, Oppenoorth et al. (1977) found that although the gene for reduced sensitivity of AChE in the CH strain (Cornell R strain) was introduced into the genome of an S strain, resistance levels to tetrachlorvinphos and other OP chemicals became considerably lower in this sub-strain CH_2 than were those in the original R strain, and that other resistance

Table 10. Relationship between Reduced Sensitivity of AChE and Cross Resistance to Insecticides in the Multiple-Resistant N Strain of the Green Rice Leafhopper

Resistance mainly due to reduced sensitivity of AChE:

propoxur, Bassa, isoprocarb, carbaryl, terbam,[a] Tsumacide,

Macbal,[a] Meobal, Hopcide.[a]

Resistance partly due to reduced sensitivity of AChE:

carbanolate, Hydrol,

malathion, phenthoate, methyl parathion, fenitrothion,

Fujithion,[a] chlorfenvinphos.

Resistance not related to reduced sensitivity of AChE:

methomyl

diazinon, pyridafenthion, propaphos, acephate.

[a]terbam = *m*-*tert*-butylphenyl methylcarbamate; Macbal = 3,5-xylyl methylcarbamate; Hopcide = *o*-chlorophenyl methylcarbamate; Fujithion = dimethyl *S*-(*p*-chlorophenyl)phosphorothiolate.

factors, including increased degradation due to glutathione S-transferase and phosphatase, were involved in the resistance of the original strain. Therefore, it was concluded that the high resistance levels to tetrachlorvinphos and other OP chemicals in the original strain were due to the joint action of at least these three mechanisms (Oppenoorth et al., 1977).

Such a joint action of resistance factors has also been reported in the house fly in cases in which reduced penetration of insecticides through the cuticle and increased degradation of insecticides are causes of resistance (Plapp and Hoyer, 1968; Sawicki, 1970, 1973; Plapp, 1970; Georghiou, 1971).

GENETICS OF RESISTANCE AND REDUCED SENSITIVITY OF ACETYLCHOLINESTERASE

The analysis of dosage-mortality curves of the progenies derived from crossing the S strain and the multiple-R strain N suggests that in the green rice leafhopper, resistance to carbamates such as propoxur, Tsumacide and carbaryl is mainly controlled by an incompletely dominant autosomal factor (Hama and Iwata, 1978).

On the other hand, it was confirmed that the N strain used in the genetic study had both the normal AChE and the altered AChE with an activity ratio of 1:1, as previously described. According to genetic study, the R strain N appears homozygous for carbamate resistance as judged from the dosage-mortality curve, but not for reduced sensitivity of AChE. We then tried to determine the sensitivity to propoxur of the AChE's of individual insects in order to clarify the mode of inheritance of the reduced sensitivity.

Sensitivity of "individual AChE's" in heads of the leafhopper to propoxur is determined precisely by the technique of Oppenoorth et al. (1977) with only slight modifications (Hama, unpublished). An aqueous extract of a single head was incubated in a cuvette with ATCh (a final concentration of 5×10^{-4}M), followed by adding propoxur (a final concentration of 6.7×10^{-4}M) a few minutes later. The absorbance increase was correlated directly with time prior to the addition of propoxur, after which it correlated almost directly, with some variations among the samples tested. The initial slope (v_o) and that obtained after the addition of propoxur (v_i) were estimated and the ratio v_i/v_o was calculated. Some of the results are shown in Figure 10 (Hama and Yamazaki, unpublished). All individuals of the S strain had the normal AChE, which was inhibited completely by propoxur (ratio of 0). Sensitivity of individual AChE's in the N strain centered on the ratio of 0.5 with a small variation, whereas the ratio in the Rmc strain ranged from 0.5 to 1.0 (Figure 10). No individual of either the R strains N or Rmc had AChE as highly sensitive as the S strain.

From these results, it is unlikely that individuals with the normal AChE or the altered AChE co-exist in the N strain, and that the allele or multiple allele responsible for AChE in the leafhopper exists at a single locus on some autosome.

No individual having AChE with a ratio of more than 0.7 was detected in the populations collected in Nakagawara in 1970 (N) and 1975, where the N strain originated, but in the population collected there in 1979 individuals having AChE with a ratio greater than 0.7 appeared along with an increasing level of resistance to propoxur (Hama and Yamazaki, unpublished).

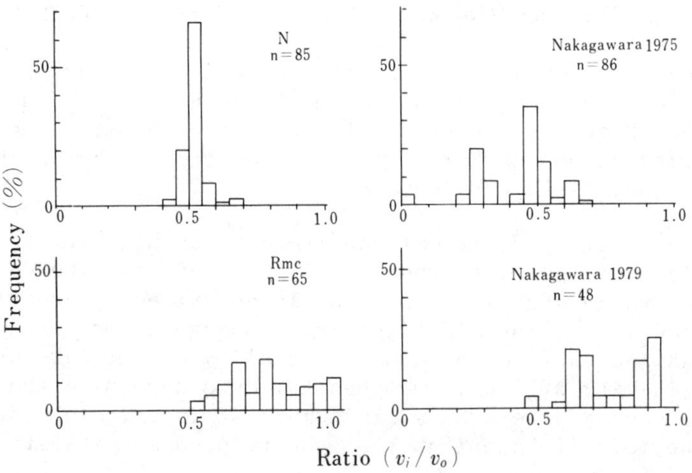

Figure 10. Frequency of sensitivity of AChE in individual head extracts of the green rice leafhopper. Ratio (v_i/v_o): v_o, absorbance change/min prior to addition of inhibitor; v_i, absorbance change/min after addition of propoxur (final concentration 6.7 x 10^{-4}M).

In the house fly, it has been found that three distinct phenotypes can be recognized (homozygote for normal AChE, and heterozygote and homozygote for resistant AChE), and that the reduced sensitivity of AChE is controlled by an incompletely dominant gene on the second chromosome (Oppenoorth et al., 1977; Plapp and Tripathi, 1978; Oppenoorth, 1979).

CONCLUDING REMARKS

Tripathi et al. (1973) indicated that there were several isozymes of AChE in the house fly that were separated by polyacrylamide gel electrophoresis, and that their reactivities to substrates and inhibitors were all different. They also reported that isozymes of AChE in the thorax of the house fly were different from those in the head (Tripathi and O'Brien, 1973a; Tripathi et al., 1973). This finding seems to support a view that the thoracic AChE is the target enzyme for OP poisoning of house flies rather than the head AChE (Mengle and Casida, 1960; Bigley, 1966; Farnham et al., 1966; Booth and Metcalf, 1970; Brady, 1970; Booth and Lee, 1971).

As previously described, AChE is a typical oligomer enzyme, although how isozymes detected by the gel electrophoresis are related to oligomers is not necessarily clear. Recent work (Steele and Smallman, 1976a, b, c; Steele and Maneckjee, 1979) suggests that all isozymes of head and thoracic AChE's in the house fly detected by gel electrophoresis constitute a size isomer family, and that the thoracic AChE is not biochemically different from the head AChE.

If isozymes detected in the house fly are of genetic origin, they may be very important in relation to the development of resistance to insecticides in the case of reduced AChE sensitivity. However, the origin of these isozymes in the house fly seems to be epigenetic rather than genetic (Tripathi and O'Brien, 1975).

In the green rice leafhopper, two or three bands of AChE activity were detected with polyacrylamide gel electrophoresis (Hama, 1976; Takahashi et al., 1978), but these bands were thought to be oligomers, as previously described. The altered AChE detected in the multiple-R leafhoppers seems to be a charge isomer of the wild-type enzyme since the altered AChE was not separated from the normal AChE by the gel filtration, but the two types of enzymes were distinguished by DEAE-cellulose column chromatography (Hama, 1976). The appearance of the altered AChE in the leafhopper is apparently genetic although the genetic background of reduced AChE sensitivity is not fully known.

It has been postulated that the reduced AChE sensitivity in the house fly appeared as a result of such excessive selection pressure that R populations exceeded the limit of their capacity for metabolic defense (Plapp, 1976; Georghiou and Taylor, 1977). In the green rice leafhopper, it is unlikely that resistance factors more important than the reduced sensitivity of AChE are involved in the carbamate resistance. Therefore, it may be concluded that the extensive use of carbamate insecticides since 1964 for the control of the OP-R leafhopper has led to the selection of insects with the altered AChE, which is much less sensitive to inhibition by methylcarbamates.

Acknowledgement: The author is grateful to Dr. C. Tomizawa, National Institute of Agricultural Sciences, for his valuable comments on the manuscript, and to Prof. T. Saito and Dr. T. Miyata, Nagoya University, for kindly allowing him the use of unpublished data.

REFERENCES

Aldridge, W. N., 1950, Some properties of specific cholinesterase with particular reference to the mechanism of inhibition by diethyl *p*-nitrophenyl thiophosphate (E 605) and analogues, *Biochem. J.*, 46:451.

Aldridge, W. N., 1975, Survey of major points of interest about reactions of cholinesterases, *in:* "Cholinesterases and Cholinergic Receptors," E. Reiner, ed., pp. 225-233, Croatica Chemica Acta, Zagreb.

Aldridge, W. N., and Reiner, E., 1969, Acetylcholinesterase. Two types of inhibition by an organophosphorus compound: One the formation of phosphorylated enzyme and the other analogous to inhibition by substrate, *Biochem. J.*, 115:147.

Aldridge, W. N., and Reiner, E., 1972, "Enzyme Inhibitors as Substrates," North-Holland Publishing Co., Amsterdam.

Augustinsson, K. B., 1948, Cholinesterases. A study in comparative enzymology, *Acta Physiol. Scand.*, 15:Suppl. 52:1.

Augustinsson, K. B., 1971, Comparative aspects of the purification and properties of cholinesterases, *Bull. Wld. Hlth. Org.*, 44:81.

Ayad, H., and Georghiou, G. P., 1975, Resistance to organophosphates and carbamates in *Anopheles albimanus* based on reduced sensitivity of acetylcholinestase, *J. Econ. Entomol.*, 68:295.

Ballantyne, G. H., and Harrison, R. A., 1967, Genetic and biochemical comparisons of organophosphate resistance between strains of spider mites (*Tetranychus* species: Acari), *Ent. Exp. Appl.*, 10:231.

Bernsohn, J., Barron, K. D., and Hess, A. R., 1962, Multiple nature of acetylcholinesterase in nerve tissue, *Nature*, 195:285.

Bigley, W. S., 1966, Inhibition of cholinesterase and ali-esterase in parathion and paraoxon poisoning in the house fly, *J. Econ. Entomol.*, 59:60.

Booth, G. M., and Lee, A. H., 1971, Distribution of cholinesterases in insects, *Bull. Wld. Hlth. Org.*, 44:91.

Booth, G. M., and Metcalf, R. L., 1970, Histochemical evidence for localized inhibition of cholinesterase in the house fly, *Ann. Ent. Soc. Amer.*, 63:197.

Brady, U. E., 1970, Localization of cholinesterase activity in housefly thoraces: Inhibition of cholinesterase with organophosphate compounds, *Ent. Exp. Appl.*, 13:423.

Casida, J. E., 1973, Insecticide biochemistry, *Ann. Rev. Biochem.*, 42:259.

Devonshire, A. L., 1975, Studies of the acetylcholinesterase from houseflies (*Musca domestica* L.) resistant and susceptible to organophosphorus insecticides, *Biochem. J.*, 149:463.

Dudai, Y., Silman, I., Kalderon, N., and Blumberg, S., 1972, Purification by affinity chromatography of acetylcholinesterases from electric organ tissue of the electric eel subsequent to tryptic treatment, *Biochim. Biophys. Acta*, 268:138.

Farnham, A. W., Gregory, G. E., and Sawicki, R. M., 1966, Bioassay and histochemical studies of the poisoning and recovery of house flies (*Musca domestica* L.) treated with diazinon and diazoxon, *Bull. Ent. Res.*, 57:107.

Georghiou, G. P., 1971, Isolation, characterization and re-synthesis of insecticide resistance factors in the housefly, *Musca domestica, in:* "Insecticide Resistance, Synergism, Enzyme Induction, Vol. II," A. S. Tahori, ed., pp. 77-94, Gordon and Breach Science Publishers, New York.

Georghiou, G. P., and Taylor, C. E., 1977, Pesticide resistance as an evolutionary phenomenon, Proc. XV Intern. Cong. Entomol.,

pp. 759-785.
Grafius, M. A., and Millar, D. B., 1965, Reversible aggregation of acetylcholinesterase, *Biochim. Biophys. Acta*, 110:540.
Grafius, M. A., and Millar, D. B., 1967, Reversible aggregation of acetylcholinesterase. II. Interdependence of pH and ionic strength, *Biochem.*, 6:1034.
Hama, H., 1975, Toxicity and anticholinesterase activity of propaphos, *o,o*-di-(*n*)-propyl-*o*-4-methylthiophenyl phosphate, against the resistant green rice leafhopper, *Nephotettix cincticeps* Uhler, *Botyu-Kagaku*, 40:14.
Hama, H., 1976, Modified and normal cholinesterases in the respective strains of carbamate-resistant and susceptible green rice leafhoppers, *Nephotettix cincticeps* Uhler (Hemiptera: Cicadellidae), *Appl. Ent. Zool.*, 11:239.
Hama, H., 1977, Cholinesterase activity and its sensitivity to inhibitors in resistant and susceptible strains of the green rice leafhopper, *Nephotettix cincticeps* Uhler, *Botyu-Kagaku*, 42:82.
Hama, H., 1978, Preliminary report on the existence of butyrylcholinesterase-like enzyme in the green rice leafhopper, *Nephotettix cincticeps* Uhler (Hemiptera: Cicadellidae), *Appl. Ent. Zool.*, 13:324.
Hama, H., and Iwata, T., 1971, Insensitive cholinesterase in the Nakagawara strain of the green rice leafhopper, *Nephotettix cincticeps* Uhler (Hemiptera: Cicadellidae), as a cause of resistance to carbamate insecticides, *Appl. Ent. Zool.*, 6:183.
Hama, H., and Iwata, T., 1973, Resistance to carbamate insecticides and its mechanism in the green rice leafhopper, *Nephotettix cincticeps* Uhler, *Jap. J. Appl. Ent. Zool.*, 17:154.
Hama, H., and Iwata, T., 1978, Studies on the inheritance of carbamate resistance in the green rice leafhopper, *Nephotettix cincticeps* Uhler (Hemiptera: Cicadellidae). Relationships between insensitivity of acetylcholinesterase and cross resistance to carbamate and organophosphate insecticides, *Appl. Ent. Zool.*, 13:190.
Hama, H., and Iwata, T., 1979, Selection of the resistant green rice leafhopper with propaphos and propoxur, Changes in resistance level and component of acetylcholinesterase, Abstract of Ann. Meeting of Pesticide Sci. Soc. Japan, Kyoto.
Hama, H., Iwata, T., Tomizawa, C., and Murai, T., 1977, Mechanism of resistance to malathion in the green rice leafhopper, *Nephotettix cincticeps* Uhler, *Botyu-Kagaku*, 42:188.
Hama, H., Iwata, T., and Tomizawa, C., 1979, Absorption and degradation of propoxur in susceptible and resistant green rice leafhoppers, *Nephotettix cincticeps* Uhler, *Appl. Ent. Zool.*, 14:333.
Iwata, T., and Hama, H., 1971, Green rice leafhopper, *Nephotettix cincticeps* Uhler, resistant to carbamate insecticides, *Botyu-Kagaku*, 36:174.

Iwata, T., and Hama, H., 1972, Insensitivity of cholinesterase in *Nephotettix cincticeps* resistant to carbamate and organophosphorus insecticides, *J. Econ. Entomol.*, 65:643.

Iwata, T., and Hama, H., 1976, Selection of the resistant green rice leafhopper with propaphos: Developing resistance to propaphos and simultaneous restoring susceptibility to carbamate insecticides, Abstract of Ann. Meeting of Japanese Soc. Appl. Ent. Zool., Kyoto.

Iwata, T., and Hama, H., 1977, Comparison of susceptibility to various chemicals between malathion-selected and methyl parathion-selected strains of the green rice leafhopper, *Nephotettix cincticeps* Uhler, *Botyu-Kagaku*, 42:181.

Kazano, H., Asakawa, M., Miyata, T., and Saito, T., 1978, Comparison of penetration rates of carbamate insecticides in carbamate-susceptible and -resistant green rice leafhoppers, Abstract of Ann. Meeting of Japanese Soc. Appl. Ent. Zool., Sendai.

Kojima, K., Ishizuka, T., Kitakata, S., 1963, Mechanism of resistance to malathion in the green rice leafhopper, *Nephotettix cincticeps*, *Botyu-Kagaku*, 28:17.

Krupka, R. M., and Laidler, K. J., 1961, Molecular mechanisms for hydrolytic enzyme action. II. Inhibition of acetylcholinesterase by excess substrate, *J. Amer. Chem. Soc.*, 83:1448.

Krysan, J. L., and Chadwick, L. E., 1966, The molecular weight of cholinesterase from the house fly, *Musca domestica* L., *J. Insect Physiol.*, 12:781.

Krysan, J. L., and Kruckeberg, W. C., 1970, The sedimentation properties of cholinesterase from a may fly (*Hexagenia bilineata* (Say); Ephemeroptera) and the honey bee (*Apis mellifera* L.), *Int. J. Biochem.*, 1:241.

Lee, R. M., and Batham, P., 1966, The activity and organophosphate inhibition of cholinesterases from susceptible and resistant ticks (Acari), *Ent. Exp. Appl.*, 9:13.

Massoulié, J., and Rieger, F., 1969, L'acétylcholinestérase des organes électriques de poissons (torpille et gymnote); complexes membranaires, *European J. Biochem.*, 11:441.

Mengle, D. C., and Casida, J. E., 1960, Biochemical factors in the acquired resistance of houseflies to organophosphate insecticides, *J. Agr. Food Chem.*, 8:431.

Metcalf, R. L., Fukuto, T. R., and Winton, M. Y., 1962, Insecticidal carbamates: Position isomerism in relation to activity of substituted phenyl *N*-methylcarbamates, *J. Econ. Entomol.*, 55:889.

Needham, P. H., and Sawicki, R. M., 1971, diagnosis of resistance to organophosphorus insecticides in *Myzus persicae* (Sulz.), *Nature*, 230:125.

Nolan, J., and Schnitzerling, H. J., 1975, Characterization of acetylcholinesterases of acaricide-resistant and susceptible strains of the cattle tick *Boophilus microplus* (Can) 1. Extraction of the critical component and comparison with enzyme from other sources, *Pestic. Biochem. Physiol.*, 5:178.

Nolan, J., Schnitzerling, H. J., and Schuntner, C. A., 1972, Multiple forms of acetylcholinesterase from resistant and susceptible strains of the cattle tick, *Boophilus microplus* (Can.), *Pestic. Biochem. Physiol.*, 2:85.

O'Brien, R. D., 1967, "Insecticides, Action and Metabolism," Academic Press, New York.

O'Brien, R. D., 1976, Acetylcholinesterase and its inhibition, *in:* "Insecticide Biochemistry and Physiology," C. F. Wilkinson, ed., pp. 271-296, Plenum Press, New York.

O'Brien, R. D., Tripathi, R. K., and Howell, L. L., 1978, Substrate preferences of wild and a mutant house fly acetylcholinesterase and a comparison with the bovine erythrocyte enzyme, *Biochim. Biophys. Acta,* 526:129.

Oppenoorth, F. J., 1979, Localisation of the acetylcholinesterase gene in the housefly, *Musca domestica, Ent. Exp. Appl.*, 25:115.

Oppenoorth, F. J., and Welling, W., 1976, Biochemistry and physiology of resistance, *in:* "Insecticide Biochemistry and Physiology," C. F. Wilkinson, ed., pp. 507-551, Plenum Press, New York.

Oppenoorth, F. J., Smissaert, H. R., Welling, W., Pas, L. J. T. van der, and Hitman, K. T., 1977, Insensitive acetylcholinesterase, high glutathione-*S*-transferase, and hydrolytic activity as resistance factors in a tetrachlorvinphos-resistant strain of house fly, *Pestic. Biochem. Physiol.*, 7:34.

Ozaki, K., and Kurosu, Y., 1967, Resistance pattern in four strains of insecticide-resistant green rice leafhopper, *Nephotettix cincticeps* Uhler, collected in fields, *Japan. J. Appl. Ent. Zool.*, 11:145.

Plapp, F. W., Jr., 1970, On the molecular biology of insecticide resistance, *in:* "Biochemical Toxicology of Insecticides," R. D. O'Brien and I. Yamamoto, eds., pp. 179-192, Academic Press, New York, London.

Plapp, F. W., Jr., 1976, Biochemical genetics of insecticide resistance, *Ann. Rev. Ent.*, 21:179.

Plapp, F. W., Jr., and Hoyer, R. F., 1968, Insecticide resistance in the house fly: Decreased rate of absorption as the mechanism of action of a gene that acts as an intensifier of resistance, *J. Econ. Entomol.*, 61:1298.

Plapp, F. W., Jr., and Tripathi, R. K., 1978, Biochemical genetics of altered acetylcholinesterase. Resistance to insecticides in the house fly, *Biochem. Genetics,* 16:1.

Roulston, W. J., Schnitzerling, H. J., and Schuntner, C. A., 1968, Acetylcholinesterase insensitivity in the Biarra strain of the cattle tick *Boophilus microplus,* as a cause of resistance to organophosphorus and carbamate acaricides, *Aust. J. Biol. Sci.*, 21:759.

Sawicki, R. M., 1970, Interaction between the factor delaying penetration of insecticides and the desethylation mechanism of resistance in organophosphorus-resistant houseflies, *Pestic. Sci.*, 1:84.

Sawicki, R. M., 1973, Resynthesis of multiple resistance to organophosphorus insecticides from strains with factors of resistance isolated from the SKA strain of house flies, *Pestic. Sci.*, 4:171.

Schnitzerling, H. J., Schuntner, C. A., Roulston, W. J., and Wilson, J. T., 1974, Characterization of the organophosphorus-resistant Mt. Alford, Gracemere and Silkwood strains of the cattle tick, *Boophilus microplus*, *Aust. J. Biol. Sci.*, 27:397.

Schuntner, C. A., and Roulston, W. J., 1968, A resistance mechanism in organophosphorus-resistant strains of sheep blowfly (*Lucilia cuprina*), *Aust. J. Biol. Sci.*, 21:173.

Smissaert, H. R., 1964, Cholinesterase inhibition in spider mites susceptible and resistant to organophosphate, *Science*, 143:129.

Smissaert, H. R., Voerman, S., Oostenbrugge, L., and Renooy, N., 1970, Acetylcholinesterases of organophosphate-susceptible and -resistant spider mites, *J. Agr. Food Chem.*, 18:66.

Steele, R. W., and Smallman, B. N., 1976a, Acetylcholinesterase from the housefly head, Molecular properties of soluble forms, *Biochim. Biophys. Acta*, 445:131.

Steele, R. W., and Smallman, B. N., 1976b, Acetylcholinesterase of the house-fly head, Affinity purification and subunit composition, *Biochim. Biophys. Acta*, 445:147.

Steele, R. W., and Smallman, B. N., 1976c, Organophosphate toxicity: Kinetic differences between acetylcholinesterase of the housefly thorax and head? *Life Sci.*, 19:1937.

Steele, R. W., and Maneckjee, A., 1979, Toxicological significance of acetylcholinesterase of the housefly thorax, *Pestic. Biochem. Physiol.*, 10:322.

Takahashi, Y., Kyomura, N., and Yamomoto, I., 1978, Mechanism of joint action of *N*-methyl and *N*-propylcarbamates for inhibition of acetylcholinesterase from resistant green rice leafhopper, *Nephotettix cincticeps*, *J. Pestic. Sci.*, 3:55.

Townsend, M. G., and Busvine, J. R., 1969, The mechanism of malathion-resistance in the blowfly *Chrysomya putoria*, *Ent. Exp. Appl.*, 12:243.

Tripathi, R. K., 1976, Relation of acetylcholinesterase sensitivity to cross-resistance of a resistant house fly strain to organophosphates and carbamates, *Pestic. Biochem. Physiol.*, 6:30.

Tripathi, R. K., and O'Brien, R. D., 1973a, Effect of organophosphates in vivo upon acetylcholinesterase isozymes from housefly head and thorax, *Pestic. Biochem. Physiol.*, 2:418.

Tripathi, R. K., and O'Brien, R. D., 1973b, Insensitivity of acetylcholinesterase as a factor in resistance of houseflies to the organophosphate Rabon, *Pestic. Biochem. Physiol.*, 3:495.

Tripathi, R. K., and O'Brien, R. D., 1975, The significance of multiple molecular forms of acetylcholinesterase in the sensitivity of houseflies to organophosphorus poisoning, *in:* "Isozymes II. Physiological Function," C. L. Makkert, ed., pp. 395-408, Academic Press, New York.

Tripathi, R. K., and O'Brien, R. D., 1977, Purification of acetylcholinesterase from house fly brain by affinity chromatography, *Biochim. Biophys. Acta*, 480:382.

Tripathi, R. K., Chiu, Y. C., and O'Brien, R. D., 1973, Reactivity in vitro toward substrate and inhibitors of acetylcholinesterase isozymes from electric eel electroplax and housefly brain, *Pestic. Biochem. Physiol.*, 3:55.

Wilson, I. B., 1952, Acetylcholinesterase XII. Further studies of binding forces, *J. Biol. Chem.*, 197:215.

Yamamoto, I., Kyomura, N., and Takahashi, Y., 1977, Aryl *N*-propylcarbamates, a potent inhibitor of acetylcholinesterase from the resistant green rice leafhopper, *Nephotettix cincticeps*, *J. Pestic. Sci.*, 2:463.

Yoshioka, K., and Iwata, T., 1967, Susceptibility to insecticides of the green rice leafhopper, *Nephotettix cincticeps* Uhler, collected from various localities, *Jap. J. Appl. Ent. Zool.*, 11:193.

Zahavi, M., Tahori, A. S., and Stolero, F., 1970, Sensitivity of acetylcholinesterase in spider mites to organo-phosphorus compounds, *Biochem. Pharmacol.*, 19:219.

Zon, A. Q. van and Helle, W., 1966, A search for linkage between genes for albinism and parathion resistance in *Tetranychus pacificus* McGregor, *Genetica*, 37:181.

RESISTANCE TO INSECTICIDES DUE TO REDUCED SENSITIVITY OF THE NERVOUS SYSTEM

Toshio Narahashi

Department of Pharmacology
Northwestern University Medical School
Chicago, Illinois 60611

INTRODUCTION

In order to understand the mechanism underlying the physiological resistance of an insect to an insecticide, we must know how the insecticide exerts its toxic action. Among a number of processes involved, three factors have been regarded as playing the most important roles in insect resistance to insecticides: penetration of insecticides through the cuticle, detoxication, and sensitivity of the target site. Penetration and detoxication mechanisms are discussed in other papers of this volume.

In some cases, resistant (R) insects surviving an application of an insecticide have been found to contain a larger amount of undetoxified insecticide than the susceptible (S) insects killed by the insecticide. In other words, the R insects tolerate, without showing any signs of symptoms of poisoning, a large internal amount of insecticide that would be more than enough to kill S insects (Babers and Pratt, 1953; Brooks, 1960; Perry and Hoskins, 1951; Sternburg et al., 1950; Tahori and Hoskins, 1953; see also Narahashi, 1964, 1971). These observations suggest that in certain R insects the target site is less sensitive to the insecticide than in S insects.

INSECT RESISTANCE AND LOW NERVE SENSITIVITY TO INSECTICIDES

The nervous systems of R strains have acquired resistance to the direct action of insecticides. This has been shown to be the case for several strains of house flies resistant to DDT, BHC, dieldrin and organophosphates (Browne and Kerr, 1967; Narahashi, 1964;

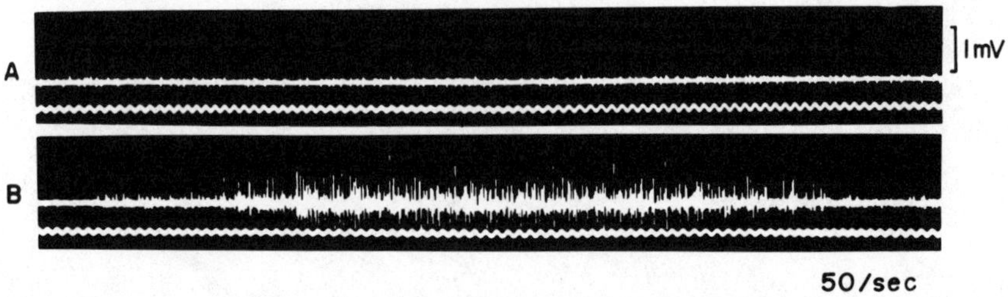

Figure 1. Repetitive discharges induced by application of DDT to the thoracic ganglion of a house fly. Recording was made from the femur by means of an extracellular wire electrode. A, control; B, 7 min after application of 1 x 10^{-6}M DDT (reproduced by permission from Yamasaki and Narahashi, 1962).

Pratt and Babers, 1953; Smyth and Roys, 1955; Tsukamoto et al., 1965; Weiant, 1955; Yamasaki and Narahashi, 1958, 1962).

Methods of Nerve Sensitivity Measurement

In order to compare the sensitivities of the nervous systems of S and R strains of insects to insecticides, a method must be developed to meet certain specific requirements. First, the method should permit measurement of the direct action of an insecticide on the nervous system. Second, the method should be applicable to various species of insects in which R strains have been developed. Third, the method should be simple enough to permit the performance of a large number of experiments for estimation of effective dose 50 (ED_{50}).

We developed a method for this purpose using house flies. A house fly is pinned on a board (wood, cork or Sylgard) with the ventral side up. The ventrum of the thorax is removed to expose the thoracic ganglion, and physiological saline solution with or without an insecticide is applied directly onto the ganglion. Muscle action potentials are recorded by inserting a fine silver or platinum wire electrode in the femur, and another wire is placed in the abdomen as the reference electrode. If the insecticide stimulates motor neurons to induce bursts of impulses, these impulses in turn elicit muscle action potentials and contractions (Figure 1). Thus, the frequency of the muscle action potentials can be taken as a measure of the general stimulating action of an insecticide.

The recorded muscle action potentials are amplified, displayed on an oscilloscope, and photographed. At the same time, the recorded action potentials are fed into an electronic counter to display the frequency per unit time in a digital form. Using two channels of

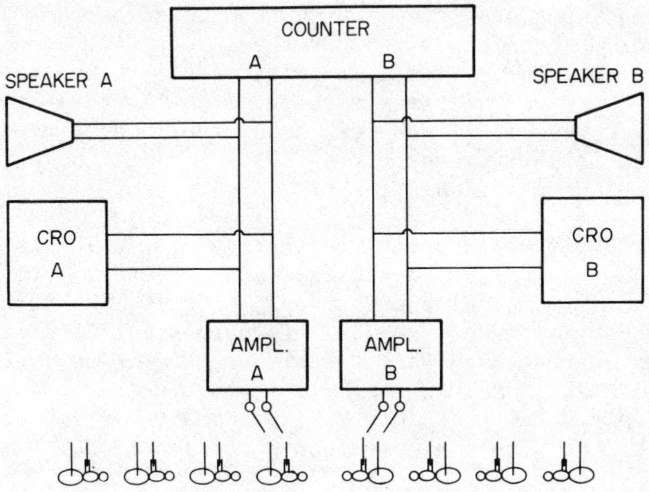

Figure 2. Schematic diagram of the recording system for the study of nerve sensitivity to insecticides. The action potentials recorded from the femur muscle while an insecticide solution was being applied to the exposed thoracic ganglion were amplified and were then fed into an oscilloscope (CRO), a speaker and a pulse counter. Two-channel system enables eight house flies to be studied in one experiment.

the recording system and switches, as many as eight house flies can be handled in one experiment (Figure 2). The digital data can be converted into analog form and registered on a chart recorder. Although this method was originally developed for house flies, it can be applied to other species of insects as well. The recorded action potentials may also be either fed into a computer for on-line analysis or stored on tape.

Studies on Nerve Sensitivity to Insecticides

Our early study performed with Takatsuki and Hikone strains of house flies showed some differences in nerve sensitivities to DDT, BHC and dieldrin (Yamasaki and Narahashi, 1958). Both strains were highly resistant (or tolerant) to DDT with LD_{50} values exceeding 5000 μg/g. When applied directly to the exposed thoracic ganglion, DDT caused bursts of muscle action potentials in both strains, and the ganglion of the Takatsuki strain was slightly more sensitive to the direct action of DDT than that of the Hikone strain, measured by the threshold concentration necessary to induce discharges. The Hikone strain was five times more resistant to the killing action of BHC than the Takatsuki strain, and the nerves of the Hikone strain were 1.8 times less sensitive to BHC than those of the Takatsuki strain, based on the measurement of the latent time for initiation of impulse discharges. Similar results were obtained for dieldrin: the Hikone strain was 40 times more resistant in terms of mortality, and its nerves were 1.5 times less sensitive. These results clearly indicate that low nerve sensitivity plays a direct role in insect resistance to insecticides.

Similar experiments were later carried out using three strains of house flies: NAIDM (susceptible), CSMA (moderately resistant) and DKM (resistant) (Yamasaki and Narahashi, 1962). The concentrations of DDT necessary to induce impulse discharges in 50% of the population were estimated. The insect resistance ratio as measured by the LD_{50} was NAIDM:CSMA:DKM = 1:17:217, and the nerve insensitivity ratio as measured by the ED_{50} was NAIDM:CSMA:DKM = 1:6.3:7.7. Thus, the nerves of the R strains are less sensitive to DDT than those of the S strain.

The study of nerve sensitivity was then extended to diazinon (Narahashi, 1964). Since the activated form, diazoxon, is more potent than diazinon in inhibiting cholinesterase in vitro, the nerve sensitivities to both diazinon and diazoxon were measured. The ratio of insect resistance to diazinon was estimated to be NAIDM:Hokota = 1:21, while the ratio of nerve insensitivity was NAIDM:Hokota = 1:0.6. The ratio of nerve insensitivity to diazoxon was NAIDM:Hokota = 1:0.5. Thus, there is a slighlty negative correlation between insect resistance and nerve sensitivity, and the mechanism of resistance to diazinon must lie in other factors.

Possible Mechanisms of Low Nerve Sensitivity

Two possible mechanisms are conceivable for the low sensitivity of nerves to insecticides. One is the high activity of detoxication mechanisms within the nervous system, and this may possibly be the case with a DDT-R strain of house fly (Miyake et al., 1957). However, since many R strains of house flies exhibit low nerve sensitivity when an insecticide is applied directly to the nerve in a large volume of bathing medium, it is unlikely that the detoxication of the insecticide within the nerve plays any major role in resistance. It is clearly shown that the detoxication of DDT is governed by a dominant gene located on the second chromosome (Tsukamoto and Suzuki, 1964), whereas nerve sensitivity is governed by a recessive gene located on the third chromosome (Tsukamoto et al., 1965). Thus, these observations do not support the notion that the low sensitivity of nerves to insecticides is due to the detoxication of the insecticides in the nerves.

The other mechanism is the intrinsic low nerve sensitivity to insecticides, which is our main concern. First, evidence will be presented in support of the notion that the target sites of DDT and pyrethroids are located in the nervous system, or to be more specific, in nerve membrane ionic channels. Step-by-step explorations of the mechanisms of action of these insecticides will be described, starting from the symptoms of poisoning. Second, possible approaches to the mechanism of this low nerve sensitivity will be given in light of this study of nerve membrane ionic channels.

THE SITE OF ACTION OF DDT AND PYRETHROIDS

Symptoms of Poisoning in Insects

The initial symptoms of poisoning by DDT and some pyrethroids in insects are characterized by hyperactivity, loss of coordination, ataxia and convulsions. This stage is followed by a stage of paralysis, which in turn leads to death. The time course varies greatly depending on the kind of insecticide, species of insect, dose and temperature. In testing the potency of insecticides, an index called "knockdown" is often used. This symptom reflects a certain stage of initial poisoning, but has little meaning in terms of the mechanism of action of the insecticide. The initial symptoms of hyperactivity and convulsions are shared by other types of insecticides, including cyclodienes, BHC, organophosphates and carbamates, all of which are potent neuropoisons.

Nervous Disorder

When an insect exhibits hyperactivity and convulsions as a result of intoxication by DDT or pyrethroids, hyperexcitability can be observed in various parts of the nervous system (Narahashi, 1971, 1976; Wouters and van den Bercken, 1978). The nerve hyperexcitability is manifested by repetitive discharges in sensory, motor and/or central neurons (Figure 1). Such repetitive discharges, either induced by a stimulus or spontaneous, lead to severe muscle convulsions. The overall intensity of repetitive discharges and the relative activity of various parts of the nervous system vary greatly depending on the kind of insecticide, dose, species of insects and temperature. It should be emphasized that the common denominator is repetitive discharges, which form the basis of the action of DDT and certain pyrethroids.

Careful recording of impulse discharges from various portions of the nerve and muscle system poisoned by DDT or some pyrethroids reveals that repetitive discharges are induced in the sensory neurons, motor neurons, interneurons, nerve fibers, synapses and neuromuscular junctions. It has been claimed that the nerve preparations containing synapses are more sensitive to DDT and certain pyrethroids than those containing only nerve fibers (Burt and Goodchild, 1971; Camougis, 1973; Narahashi, 1976). Thus, the next question to be asked is where the site of action of DDT and pyrethroids is located in the nervous system.

The Nerve Membrane as the Site of Action

Crayfish have been found to be highly sensitive to DDT and pyrethroids. For example, the LD_{50} value for (1*R*)-*trans*-tetramethrin has been estimated to be 0.02 μg/gm for crayfish (Narahashi and Lund, 1980) compared to 2 to 16 μg/gm for house flies (Adams and Miller, 1979; Briggs et al., 1974; Clements and May, 1977) and 37 μg/gm for locusts (Clements and May, 1977). The isolated abdominal nerve cord of crayfish responds to 1×10^{-8}M allethrin, giving rise to a high frequency of impulse discharges (Takeno et al., 1977). The neuromuscular junction of the crayfish is also sensitive to DDT-type and pyrethroid-type insecticides, producing repetitive muscle activity in response to a single nerve stimulus (Takeno et al., 1977; Farley et al., 1979). The crayfish system of nerves and muscles has been extensively used for the study of basic neurophysiology, and a wealth of information has been accumulated. The crayfish has giant nerve fibers, which are convenient for the detailed electrophysiological analysis of nerve membrane excitability. There is no such material in insects. The giant nerve fiber of the cockroach has been extensively used for intracellular microelectrode recording of resting potential and action potential (Yamasaki and Narahashi, 1957; Narahashi and Yamasaki, 1960a,b,c; Narahashi, 1962a,b; Boistel and Coraboeuf, 1954; Coraboeuf and Boistel, 1955), but the voltage clamp measurement of membrane ionic conductance using this preparation (Pichon, 1969a,b) is far less satisfactory than that performed with the giant axons of crayfish and squid, in terms of accuracy. Internal perfusion methods that provide great flexibility and accuracy in voltage clamp experiments are also applicable to these giant axons, and therefore the crayfish nerve and muscle systems have been used extensively for elucidation of the mechanism of action of DDT and pyrethroids.

In our recent study to determine the site of action of DDT-type insecticides, 2,2 bis(p-ethoxyphenyl)-3,3-dimethyloxetane (EDO or GH149) was used as a model compound (Farley et al., 1979). EDO is a biodegradable DDT analog (Holan, 1971) with an oxetane group substituted for a trichloro group of DDT (Figure 3), and greatly augments the muscle contraction induced by stimulation of the excitatory nerve (Figure 4). The excitatory postsynaptic potential (EPSP) recorded by means of an intracellular microelectrode is effectively potentiated as a result of the summation of repetitive EPSPs. In order to determine the site of origin of the EDO-induced repetitive EPSPs, an extracellular or focal recording by a microelectrode was attempted. When the tip of a recording microelectrode is brought very close to the end-plate region, local circuit currents due to the action potential at the nerve terminal and the EPSP of the

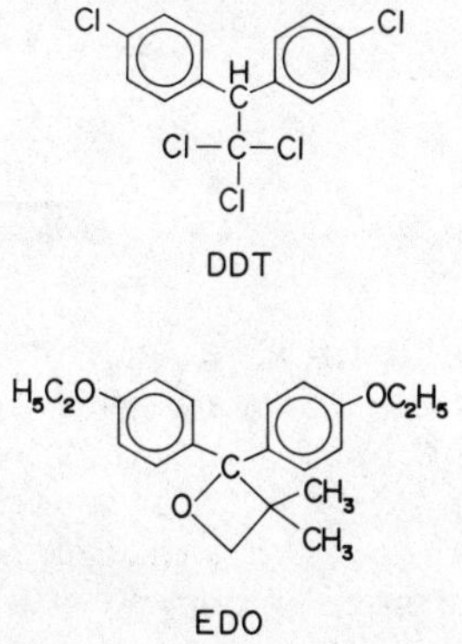

Figure 3. Structures of p,p'-DDT and EDO.

muscle can be recorded. Figure 5 illustrates an example of such a focal recording in the preparation poisoned by EDO. A single nerve stimulus induces repetitive EPSPs, and each EPSP is preceded by an action potential generated in the nerve. Thus, it can be concluded that the repetitive responses originate in the nerve rather than in the muscle. Similar observations were made in the frog neuromuscular junctions poisoned with pyrethroids (van der Bercken, 1977; Evans, 1976; Wouters et al., 1977).

In accordance with the above observation, the crayfish giant axon poisoned by pyrethroid- or DDT-type insecticides generates repetitive after-discharges in response to a single stimulus (Figure 6). Thus, it has become clear that the nerve membrane is the site of action of these insecticides, with the repetitive discharge there induced being responsible for synaptic and neuromuscular after-discharges.

Depolarizing After-potential Induces Repetitive After-discharges

The next question to be asked is how repetitive after-discharges are produced in the nerve fiber poisoned with DDT and pyrethroids.

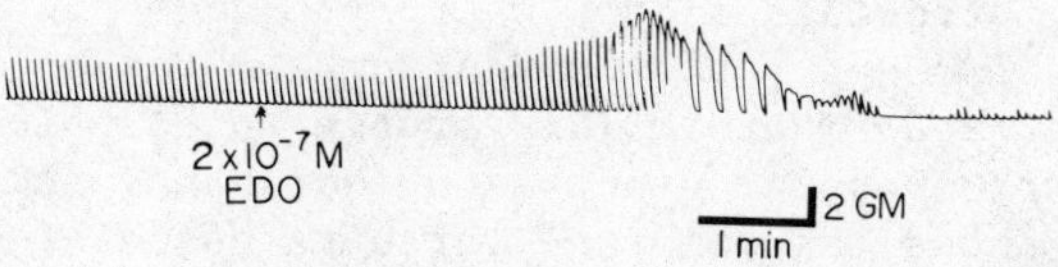

Figure 4. Effect of 2 x 10^{-7}M EDO on the crayfish muscle contraction induced by stimulation of the excitatory nerve. (Reproduced by permission from Farley et al., 1979.)

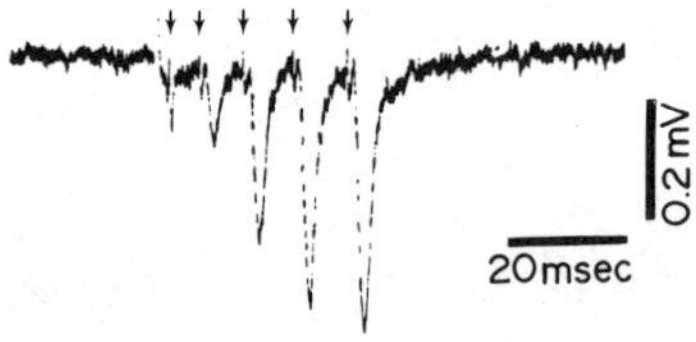

Figure 5. Focal (extracellular) recording of multiple end-plate potentials in the crayfish muscle induced by a single stimulus of the excitatory nerve in the presence of 2 x 10^{-6}M EDO. The initial upward deflection is the stimulus artifact; the small deflections indicated by the arrows prior to each end-plate potential are nerve terminal action potentials. (Reproduced by permission from Farley et al., 1979.)

Working with the cockroach nerve, Yamasaki and Ishii (1952) reported that the postsynaptic giant fiber response induced by a presynaptic cercal nerve stimulus was greatly facilitated in the presence of DDT, and a prolonged after-discharge was noted. During the after-discharge, the extracellularly recorded action potentials from individual giant fibers were clearly discernible, and the falling phase of the individual action potential was greatly prolonged after intoxication by DDT. Intracellular recording of the action potential from the cockroach giant axons gave more concrete support to the notion (Yamasaki and Narahashi, 1957). Figure 7 gives an example of such an experiment. More detailed analyses of the depolarizing after-potential in the cockroach giant axon led to the hypothesis that the inhibition of sodium inactivation and potassium activation, both of which are responsible for the falling phase of the action potential, cause the depolarizing after-potential to increase (Narahashi and Yamasaki, 1960b, 1960c). The elevated depolarizing after-potential serves as a stimulus and elicits action potentials when it exceeds the threshold depolarization for excitation.

The aforementioned hypotehsis was later demonstrated to be the case by voltage clamp experiments with lobster giant axons (Narahashi and Haas, 1967, 1968), frog nodes of Ranvier (Hille, 1968) and cockroach giant axons (Pichon, 1969a, 1969b). The voltage clamp is the most straightforward method whereby changes in membrane permeabilities to sodium and potassium during excitation can be measured as conductance, and Figure 8 illustrates an example of such an experiment with the lobster giant axon (Narahashi and Haas, 1968). The upper set shows the membrane currents associated with a step depolarization to -20 mV before (normal) and during application of saxitoxin. Since saxitoxin selectively blocks the sodium current without any effect on the potassium current (Narahashi et al., 1967b), the difference between the currents before and during

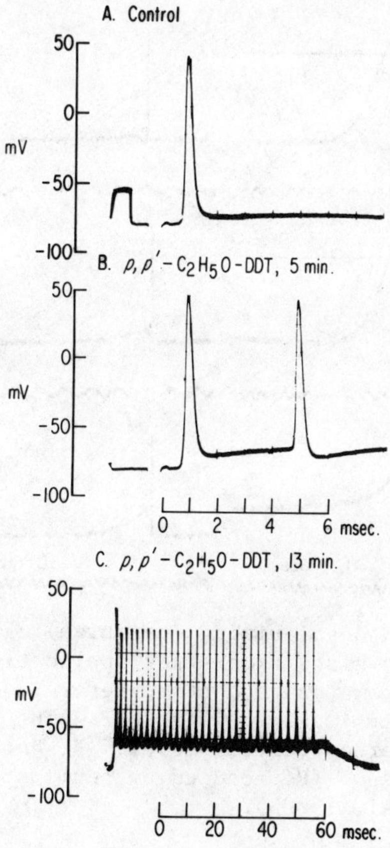

Figure 6. Repetitive after-discharges induced by a single stimulus in a crayfish giant axon exposed to 1 x 10^{-4}M p,p'-C_2H_5O-DDT. (Reproduced by permission from Wu et al., 1975.)

saxitoxin application represents the sodium current (I_{Na}), and the current remaining in the presence of saxitoxin represents the potassium current. It can be seen that the sodium current rises and falls quickly as a result of sodium activation and sodium inactivation during a step depolarization of the membrane, and that the potassium current rises more slowly and is maintained at a steady state during depolarization. The lower set of Figure 8 shows similar membrane current records in the DDT-poisoned axon. Tetrodotoxin, which also inhibits the sodium current selectively (Narahashi et al., 1964), was used here in place of saxitoxin. It is clearly seen that the sodium current rises normally but decays much more slowly in the presence of DDT and that the potassium current is suppressed. Thus, the two ionic conductance mechanisms responsible for the falling phase of the action potential, i.e., the sodium inactivation and the

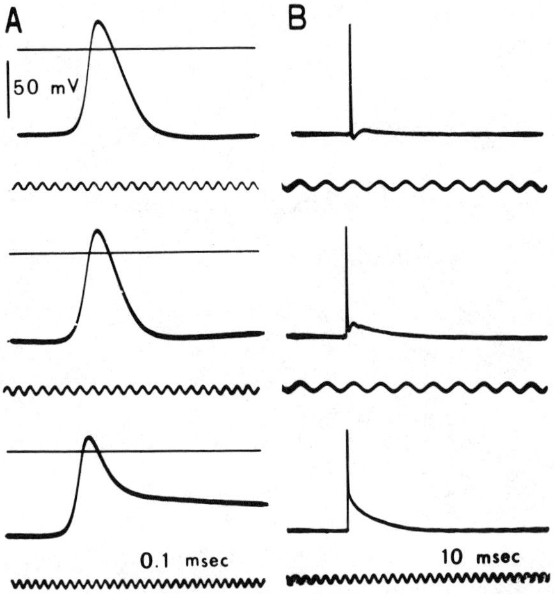

Figure 7. Increase in depolarizing (negative) after-potential of a cockroach giant axon after application of 1 x 10^{-4}M DDT. A, from top to bottom, before, 38 min after, and 90 min after treatment with DDT. The horizontal lines indicate zero potential level. B, as in A, but with slower sweep. (Reproduced by permission from Narahashi and Yamasaki, 1960b).

potassium activation, are inhibited by DDT, thereby causing a prolonged and elevated depolarizing after-potential, which in turn induces repetitive after-discharges. The prolongation of the sodium current appears to contribute more significantly to the increase in depolarizing after-potential than the suppression of the potassium current, since the former is more evident than the latter at the low concentrations of DDT that are high enough to increase the depolarizing after-potential. Similar results were obtained with EDO (Wu et al., 1980) and allethrin (Narahashi and Anderson, 1967; Wang et al., 1972).

The Sodium Channel as the Possible Target Site

The sites of increases in sodium and potassium conductance during excitation are called sodium and potassium channels, which are holes with gating mechanisms that open and close in a manner dependent upon the membrane potential. There is evidence to support the notion that the sodium channel and the potassium channel are separate entities. The densities of these channels have been

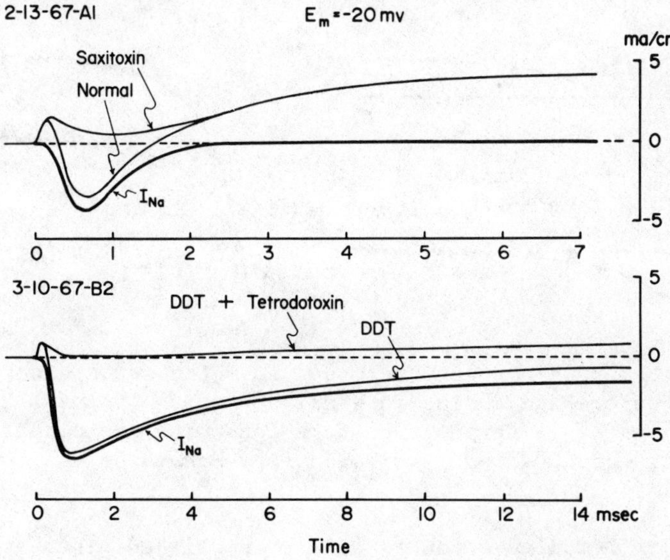

Figure 8. Effect of 3 x 10^{-7}M DDT on peak sodium current and steady-state potassium current in the lobster giant axon. Sodium current (I_{Na}) is obtained by subtracting the current in the presence of saxitoxin (upper set) or tetrodotoxin (lower set) from the total membrane current in a normal axon (upper set) or in an axon treated with DDT (lower set). The steady-state current remaining in saxitoxin or tetrodotoxin represents potassium current. DDT greatly prolongs the falling phase of the sodium current and suppresses the potassium current. (Reproduced by permission from Narahashi and Haas, 1968.)

estimated by various methods, including measurement of the binding of specific blockers such as tetrodotoxin by bioassay (Moore et al., 1967) and tritiated compounds (Ritchie, 1979; Ritchie et al., 1976), gating current measurement (Neumcke et al., 1978) and membrane fluctuation analysis (Conti et al., 1976; Neumcke et al., 1978). The sodium channel density is relatively low, ranging from about 30/μm^2 of membrane in garfish olfactory nerve fibers to about 400/μm^2 in squid axons. The frog nodes of Ranvier have a higher density of approximately 10,000/μm^2.

Our recent voltage clamp analyses for the action of synthetic pyrethroids on the channels of the crayfish and squid giant axons indicate that the pyrethroids modify a certain fraction of the sodium channel population to give rise to extremely slow kinetics in opening and closing (Lund and Narahashi, 1979; Narahashi and

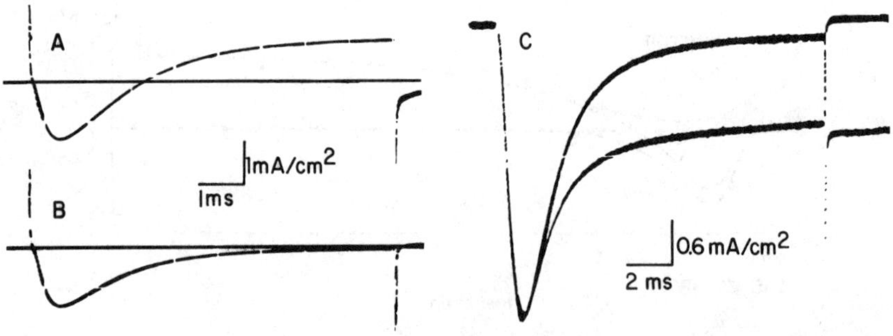

Figure 9. Membrane currents recorded from the squid giant axon before and during internal perfusion of 1 x 10^{-6}M (1*R*)-*trans*-allethrin. See text for further explanation. (Reproduced by permission from Narahashi and Lund, 1980.)

Lund, 1980). Detailed accounts will be published elsewhere, so only a few observations will be described here. Figure 9 illustrates the effect of 1 x 10^{-6}M (1*R*)-*trans*-allethrin on the sodium current recorded from the squid giant axon. Record A shows the normal membrane current associated with a step depolarization from -80 mV to 0 mV. The transient inward sodium current is followed by a steady-state outward potassium current. Record B is the membrane current from the same axon after external and internal potassium ions have been replaced by tetramethylammonium and cesium, respectively. These substitutes do not penetrate the potassium channels, so the observed ionic current represents the sodium current. Record C shows such sodium currents in response to a step depolarization from -100 mV to -20 mV before and after internal perfusion with 1 x 10^{-6}M (1*R*)-*trans*-allethrin. Before application of allethrin, the sodium current rises and falls in much the same way as that shown in Record B (note the difference in the ordinate scale). After introduction of allethrin, however, the sodium current decays much more slowly, exhibiting a steady-state current, and the rising phase of the sodium current is virtually unchanged. It should also be noted that the tail current associated with a step repolarization from -20 mV to -100 mV is greatly elevated in the presence of allethrin inside the axon. External perfusion of allethrin gave essentially the same results. Other pyrethroids, including (1*R*)-*trans*-tetramethrin, had the same effect as allethrin. Although not shown in the figure, the potassium current was not affected by 1 x 10^{-6}M allethrin. Thus, it is clear that the sodium inactivation is most sensitive to action of the pyrethroids, being remarkably inhibited at low concentrations.

Experiments with prolonged depolarizing pulses under voltage clamp conditions have revealed important features of pyrethroid-channel interactions. Figure 10 illustrates the sodium current from the crayfish giant axon perfused internally with 2 x 10^{-5}M

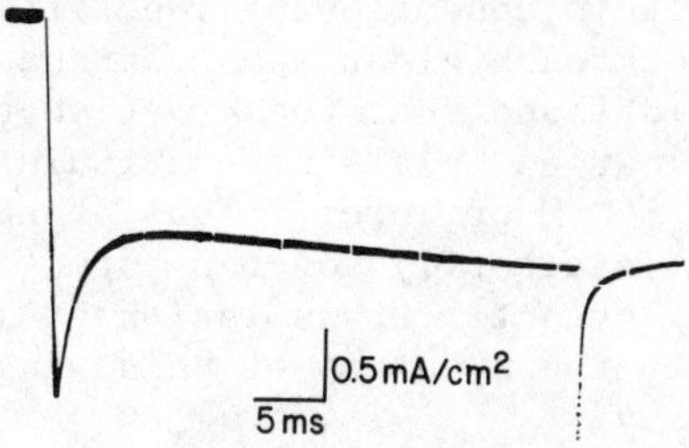

Figure 10. Membrane sodium current associated with a step depolarization in the crayfish giant axon perfused internally with 2 x 10^{-5}M (1*R*)-*trans*-tetramethrin. See text for further explanation. (Reproduced by permission from Narahashi and Lund, 1980.)

(1*R*)-*trans*-tetramethrin. Compared with the experiments shown in Figure 9, a longer depolarizing pulse is applied. The sodium current, after having risen normally to attain a peak (control record not shown), decays to a certain level and rises again, but very slowly. A more prolonged depolarizing pulse lasting as long as 10 seconds reveals that this secondary, slowly rising sodium current attains a maximum and then decays very slowly (not shown). Thus, the secondary sodium current also activates and inactivates in much the same way as the initial normal sodium current but with a much longer time course. This and several other observations in kinetic analysis of the secondary, slow sodium current are explained in terms of the existence of a population of the pyrethroid-modified sodium channels. In other words, the pyrethroids modify a population of the sodium channels to give rise to a slow sodium conductance change. This notion is also supported by the observation that the kinetics of the slow current are independent of the pyrethroid concentration, while its amplitude varies with the concentration. Thus, in the presence of the pyrethroids the sodium currents derived from the unmodified sodium channels and those from the modified sodium channels are recorded together, as shown in Figure 10.

The dose-response relationship for the modification of the sodium channels by allethrin and tetramethrin deserves further discussion. The experiments shown in Figures 9 and 10 were performed with relatively high concentrations of the pyrethroids, i.e., 1 x 10^{-6}M and 2 x 10^{-5}M, respectively. However, in order to elevate the depolarizing after-potential to the threshold for repetitive discharges, only a very small fraction of the sodium channels needs to be modified. Preliminary calculations indicate that modification of less than 1% of the sodium channels would be sufficient to increase the depolarizing after-potential for repetitive firing (Narahashi and Lund, 1980). Further increase in depolarizing after-potential would result in cessation of repetitive discharges in the allethrin-poisoned cockroach giant axon (Narahashi and Yamasaki, 1960b). It was indeed found that the crayfish giant axon generates repetitive after-discharges by a single stimulus at a pyrethroid concentration of 1 x 10^{-8}M or lower. Thus, these results with nerve membranes account satisfactorily for the potent action of the pyrethroids in facilitating synaptic transmission at low concentrations, which in turn produces the symptoms of poisoning in animals as manifested by hyperactivity.

THE LOW NERVE SENSITIVITY IN RESISTANT INSECTS

As was described earlier in this communication, the low nerve sensitivity to an insecticide is at least in part responsible for the resistance of the insect to the insecticide. Having established the notion that the sodium channels are the target site of the DDT- and pyrethroid-type insecticides, the next question to be asked is how the nerve of the R insect becomes less sensitive to the insecticide. Aside from the possible detoxication of the insecticide within the nerve tissue as discussed before, at least three mechanisms are conceivable for the intrinsic low nerve sensitivity.

First, the nerve of the R insect may be devoid of a binding site for the insecticide. Although no concrete evidence has been obtained, it is reasonable to assume that the insecticide binds to a site in the nerve membrane and thus modifies the sodium channels. The binding site may be either part of the sodium channels, which are thought to be composed of proteins, or part of an area outside of the sodium channels, which somehow influences the gating mechanism. Studies along this line have been in progress for the binding site of tetrodotoxin and saxitoxin, which are believed to bind to a site on or near the external mouth of the sodium channel (Kao and Nishiyama, 1965; Narahashi et al., 1967a; Hille, 1975). Local anesthetics such as lidocaine and procaine appear to bind to a site inside the sodium channel via its internal mouth or hydrophobic lipid phase (Hille, 1977; Narahashi et al., 1970).

Second, the sodium channel or the binding site of the R insect may not undergo change even when the insecticide binds to the site. This would result in no modification of the sodium channel function as a whole in the presence of the insecticide. In this case, however, one should be able to demonstrate the normal binding of the insecticide to the nerve membrane of the R strain of insect.

Third, the sodium channels of the R insect are modified by the insecticide to give rise to a slow sodium conductance change, but no repetitive discharges may be initiated. In this case, one should be able to see a slow sodium current under voltage clamp conditions and an elevation of the depolarizing after-potential without repetitive discharges. Since the initiation of repetitive after-discharges from the elevated depolarizing after-potential depends on a delicate balance among various electrical properties of the membrane, slight modifications of some of these properties could easily result in the cessation of repetitive after-discharge in the face of the elevated depolarizing after-potential.

Only a few studies have so far been undertaken to prove or disprove the validity of any of these three possibilities as the cause of the low nerve sensitivity to the insecticide. The nerve from the dieldrin-R strain of the German cockroach binds a smaller amount of dieldrin than that from the normal S strain (Matsumura and Hayashi, 1966, 1969; see also paper by Matsumura in this volume). Recently, Hall et al. (1979) have reported that in the tetrodotoxin-supersensitive strain of *Drosophila*, binding of the tritiated saxitoxin to the brain is the same as that found in the normal strain of *Drosophila*. This observation excludes the factor of binding as the cause of supersensitivity of *Drosophila* to saxitoxin. However, since the physiological sensitivity of the nerve to saxitoxin has not been compared between the normal and tetrodotoxin-supersensitive strains of *Drosophila*, it remains to be seen whether such supersensitivity is due to a nerve factor or to other factors, such as detoxication of tetrodotoxin.

SUMMARY AND CONCLUSION

A target site of DDT- and pyrethroid-type insecticides has been identified as the sodium channels in the nerve membrane. These insecticides modify a population of the sodium channels to give rise to an extremely slow opening and closing of the channels, leading to an increase in the depolarizing after-potential, which in turn induces repetitive after-discharges. Such repetitive activity in the nerve fibers causes disturbances in synaptic transmission and produces hyperactivity in the insect.

The mechanism underlying the reduced nerve sensitivity to an insecticide in an insecticide-R strain largely remains to be seen. The possible mechanisms include reduced binding of the insecticide to the nerve membrane, a smaller degree of modification of the sodium channels by the insecticide, and modification of the membrane electrical properties to reduce repetitive response in the face of the increased depolarizing after-potential. Each of these factors must be analyzed to understand the molecular nature of the lowered nerve sensitivity to the insecticide.

Acknowledgment: I wish to thank Caroline Myss and Zarin Karanjia for secretarial assistance. This study was supported by a grant from the National Institutes of Health, NS 14143.

REFERENCES

Adams, M. E., and Miller, T. A., 1979, Site of action of pyrethroids: Repetitive "backfiring" in flight motor units of house fly, *Pestic. Biochem. Physiol.*, 11:218.

Babers, F. H., and Pratt, J. J., Jr., 1953, Resistance of insects to insecticides: The metabolism of injected DDT, *J. Econ. Entomol.*, 46:977.

Boistel, J., and Coraboeuf, E., 1954, Potentiel de membrane et potentiels d'action de nerf d'insecte recueillis à l'aide de microélectrodes intracellulaires, *C. R. Acad. Sci.*, Paris, 238:2116.

Briggs, G. G., Elliott, M., Farnham, A. W., and Janes, N. F., 1974, Structural aspects of the knockdown of pyrethroids, *Pestic. Sci.*, 5:643.

Brooks, G. T., 1960, Mechanisms of resistance of the adult housefly (*Musca domestica*) to "cyclodiene" insecticides, *Nature*, 186:96.

Browne, L. B., and Kerr, R. W., 1967, The response of the labellar taste receptors of DDT-resistant and non-resistant houseflies (*Musca domestica*), *Entomol. Exp. Appl.*, 10:337.

Burt, P. E., and Goodchild, R. E., 1971, The site of action of pyrethrin I in the nervous system of the cockroach *Periplaneta americana* L., *Entomol. Exp. Appl.*, 14:179.

Camougis, G., 1973, Mode of action of pyrethrum on arthropod nerves, *in:* "Pyrethrum: The Natural Insecticide," J. E. Casida, ed., pp. 211-222, Academic Press, New York.

Clements, A. N., and May, T. E., 1977, The actions of pyrethroids upon the peripheral nervous system and associated organs in the locust, *Pestic. Sci.*, 8:661.

Conti, F., Hille, B., Neumcke, B., Nonner, W., and Stämpfli, R., 1976, Conductance of the sodium channel in myelinated nerve fibres with modified sodium inactivation, *J. Physiol.*, 262:729.

Coraboeuf, E., and Boistel, J., 1955, Quelques aspects de la microphysiologie nerveuse chez les insectes. Colloq. Internat. Cent. Nat. Rech. Sci. No. 67, Microphysiologie Comparée des Éléments Excitables, pp. 57-72.

Evans, M. H., 1976, End-plate potentials in frog muscle exposed to a synthetic pyrethroid, *Pestic. Biochem. Physiol.*, 6:547.

Farley, J. M., Narahashi, T., and Holan, G., 1979, The mechanism of action of a DDT analog on the crayfish neuromuscular junction, *Neurotoxicology*, 1:191.

Hall, L. M., Gitschier, J., and Strichartz, G. R., 1979, Saxitoxin binding to sodium channels from wild-type and mutant *Drosophila melanogaster*, *9th Ann. Mtg. Soc. Neurosci. Abstr.*, p. 247.

Hille, B., 1968, Pharmacological modifications of the sodium channels of frog nerve, *J. Gen. Physiol.*, 51:199.

Hille, B., 1975, The receptor for tetrodotoxin and saxitoxin: A structural hypotehsis, *Biophys. J.*, 15:615.

Hille, B., 1977, Local anesthetics: Hydrophilic and hydrophobic pathways for the drug receptor interaction, *J. Gen. Physiol.*, 69:497.

Holan, G., 1971, Rational design of degradable insecticides, *Nature*, 232:644.

Kao, C. Y., and Nishiyama, A., 1965, Actions of saxitoxin on peripheral neuromuscular systems, *J. Physiol.*, 180:50.

Lund, A. E., and Narahashi, T., 1979, The effect of the insecticide tetramethrin on the sodium channel of crayfish giant axons, *9th Ann. Mtg. Soc. Neurosci. Abstr.*, p. 293.

Matsumura, F., and Hayashi, M., 1966, Dieldrin: Interaction with nerve components of cockroaches, *Science*, 153:757.

Matsumura, F., and Hayashi, M., 1969, Dieldrin resistance. Biochemical mechanisms in the German cockroach, *J. Agr. Food Chem.*, 17:231.

Miyake, S. S., Kearns, C. W., and Lipke, H., 1957, Distribution of DDT-dehydrochlorinase in various tissues of DDT-resistant house flies, *J. Econ. Entomol.*, 50:359.

Moore, J. W., Narahashi, T., and Shaw, T. I., 1967, An upper limit to the number of sodium channels in nerve membrane? *J. Physiol.*, 188:99.

Narahashi, T., 1962a, Effect of the insecticide allethrin on membrane potentials of cockroach giant axons, *J. Cell. Comp. Physiol.*, 59:61.

Narahashi, T., 1962b, Nature of the negative after-potential increased by the insecticide allethrin in cockroach giant axons, *J. Cell. Comp. Physiol.*, 59:67.

Narahashi, T., 1964, Insecticide resistance and nerve sensitivity, *Japan. J. Med. Sci. Biol.*, 17:46.

Narahashi, T., 1971, Effects of insecticides on excitable tissues, *in:* "Advances in Insect Physiology," J. W. L. Beament, J. E. Treherne and V. B. Wigglesworth, eds., Vol. 8, pp. 1-93, Academic Press, London and New York.

Narahashi, T., 1976, Effects of insecticides on nervous conduction and synaptic transmission, *in:* "Insecticide Biochemistry and Physiology," C. F. Wilkinson, ed., pp. 327-352, Plenum Press, New York.

Narahashi, T., and Anderson, N. C., 1967, Mechanism of excitation block by the insecticide allethrin applied externally and internally to squid giant axons, *Toxicol. Appl. Pharmacol.*, 10:529.

Narahashi, T., and Haas, H. G., 1967, DDT: Interaction with nerve membrane conductance changes, *Science*, 157:1438.

Narahashi, T., and Haas, H. G., 1968, Interaction of DDT with the components of lobster nerve membrane conductance, *J. Gen. Physiol.*, 51:177.

Narahashi, T., and Lund, A. E., 1980, Giant axons as models for the study of the mechanism of action of insecticides, *in:* "Insect Neurobiology and Pesticide Action (Neurotox 79)," Proc. Soc.

Chem. Ind. Symp., Univ. York, Sept. 3-7, 1979, pp. 497-505, Soc. Chem. Ind., London.

Narahashi, T., and Yamasaki, T., 1960a, Mechanism of the after-potential production in the giant axons of the cockroach, *J. Physiol.*, 151:75.

Narahashi, T., and Yamasaki, T., 1960b, Mechanism of increase in negative after-potential by dicophanum (DDT) in the giant axons of the cockroach, *J. Physiol.*, 152:122.

Narahashi, T., and Yamasaki, T., 1960c, Behaviors of membrane potentials in the cockroach giant axons poisoned by DDT, *J. Cell. Comp. Physiol.*, 55:131.

Narahashi, T., Moore, J. W., and Scott, R., 1964, Tetrodotoxin blockage of sodium conductance increase in lobster giant axons, *J. Gen. Physiol.*, 47:965.

Narahashi, T., Anderson, N. C., and Moore, J. W., 1967a, Comparison of tetrodotoxin and procaine in internally perfused squid giant axons, *J. Gen. Physiol.*, 50:1413.

Narahashi, T., Haas, H. G., and Therrien, E. F., 1967b, Saxitoxin and tetrodotoxin: Comparison of nerve blocking mechanism, *Science*, 157:1441.

Narahashi, T., Frazier, D. T., and Yamada, M., 1970, The site of action and active form of local anesthetics, I, Theory and pH experiments with tertiary compounds, *J. Pharmacol. Exp. Therap.*, 171:32.

Neumcke, B., Nonner, W., and Stämpfli, R., 1978, Gating currents in excitable membranes, *in:* "Internat. Rev. Biochem., Biochemistry of Cell Walls and Membranes II," Vol. 19, J. C. Metcalfe, ed., pp. 129-155, Univ. Park Press, Baltimore.

Perry, A. S., and Hoskins, W. M., 1951, Detoxification of DDT as a factor in the resistance of house flies, *J. Econ. Entomol.*, 44:850.

Pichon, Y., 1969a, Effets du D.D.T. sur la fibre nerveuse isolée d'insecte. Étude en courant et an voltage imposés, *J. Physiol.*, Paris, 61 (Suppl. 1):162.

Pichon, Y., 1969b, Aspects Électriques et Ioniques du Fonctionnement Nerveux chez les Insectes, Cas Particulier de la Chaine Nerveuse Abdominale d'une Blatte *Periplaneta americana* L., These, Univ. Rennes.

Pratt, J. J., Jr., and Babers, F. H., 1953, Sensitivity to DDT of nerve ganglia of susceptible and resistant house flies, *J. Econ. Entomol.*, 46:700.

Ritchie, J. M., 1979, A pharmacological approach to the structure of sodium channels in myelinated axons, *Ann. Rev. Neurosci.*, 2:341.

Ritchie, J. M., Rogart, R. B., and Strichartz, G. R., 1976, A new method for labelling saxitoxin and its binding to non-myelinated fibres of the rabbit vagus, lobster walking leg, and garfish olfactory nerves, *J. Physiol.*, 261:477.

Smyth, T., Jr., and Roys, C. C., 1955, Chemoreception in insects and the action of DDT, *Biol. Bull.*, 108:66.

Steinburg, J., Kearns, C. W., and Bruce, W. N., 1950, Absorption and metabolism of DDT by resistant and susceptible house flies, *J. Econ. Entomol.*, 43:214.

Tahori, A. S., and Hoskins, W. M., 1953, The absorption, distribution, and metabolism of DDT in DDT-resistant houseflies, *J. Econ. Entomol.*, 46:302, 829.

Takeno, K., Nishimura, K., Parmentier, J., and Narahashi, T., 1977, Insecticide screening with isolated nerve preparations for structure-activity relationships, *Pestic. Biochem. Physiol.*, 7:486.

Tsukamoto, M., and Suzuki, R., 1964, Genetic analysis of DDT-resistance in two strains of the house fly *Musca domestica* L., *Botyu-Kagaku*, 29:76.

Tsukamoto, M., Narahashi, T., and Yamasaki, T., 1965, Genetic control of low nerve sensitivity to DDT in insecticide-resistant houseflies, *Botyu-Kagaku*, 30:128.

van den Bercken, J., 1977, The action of allethrin on the peripheral nervous system of the frog, *Pestic. Sci.*, 8:692.

Wang, C. M., Narahashi, T., and Scuka, M., 1972, Mechanism of negative temperature coefficient of nerve blocking action of allethrin, *J. Pharmacol. Exp. Therap.*, 182:442.

Weiant, E. A., 1955, Electrophysiological and behavioral studies on DDT-sensitive and DDT-resistant house flies, *Ann. Entomol. Soc. Amer.*, 48:489.

Wouters, W., and van den Bercken, J., 1978, Action of pyrethroids, *Gen. Pharmacol.*, 9:387.

Wouters, W., van den Bercken, J., and van Ginneken, A., 1977, Presynaptic action of the pyrethroid insecticide allethrin in the frog motor end plate, *Europ. J. Pharmacol.*, 43:163.

Wu, C. H., van den Bercken, J., and Narahashi, T., 1975, The structure-activity relationship of DDT analogs in crayfish giant axons, *Pestic. Biochem. Physiol.*, 5:142.

Wu, C. H., Oxford, G. S., Narahashi, T., and Holan, G., 1980, Interaction of a DDT-analog with the sodium channel of lobster axon, *J. Pharmacol. Exp. Therap.*, 212:287.

Yamasaki, T., and Ishii (Narahashi), T., 1952, Studies on the mechanism of action of insecticides (V), The effects of DDT on the synaptic transmission of the cockroach, *Oyo-Kontyu* (J. Nippon Soc. Appl. Entomol.), 8:111.

Yamasaki, T., and Narahashi, T., 1957, Intracellular microelectrode recordings of resting and action potentials from the insect axon and the effects of DDT on the action potential, Studies on the mechanism of action of insecticides (XIV), *Boytu-Kagaku*, 22:305.

Yamasaki, T., and Narahashi, T., 1958, Resistance of house flies to insecticides and the susceptibility of nerve to insecticides, Studies on the mechanism of action of insecticides (XVII),

Botyu-Kagaku, 23:146.
Yamasaki, T., and Narahashi, T., 1962, Nerve sensitivity and resistance to DDT in houseflies, *Japan. J. Appl. Entomol. Zool.*, 6:293.

THE *KDR* FACTOR IN PYRETHROID RESISTANCE

T. A. Miller, V. L. Salgado and S. N. Irving*

Division of Toxicology and Physiology
Department of Entomology
University of California
Riverside, California 92521

INTRODUCTION

The pyrethroid and DDT insecticides share several properties: Both groups of chemicals are more toxic at lower temperatures and poison nerves in a superficially similar manner, and their *kdr*-resistance factors confer cross resistance to both pyrethroids and DDT. It is possible to describe the action of an insecticide as DDT-like or pyrethroid-like, but it is impossible to state categorically which structural features determine DDT-like or pyrethroid-like action.

In a well-known early work the *kdr* factor in house flies was shown to render the nervous system less sensitive to DDT (Tsukamoto et al., 1965), but despite a clear need, the neurotoxic actions of pyrethroids on strains of insect that possessed this *kdr* factor were not examined for years. Nevertheless, circumstantial evidence was strong enough that the term "site insensitivity" came into use to explain the *kdr* factor.

Site insensitivity can be inferred by showing that all other resistance mechanisms, including penetration and metabolism, are not present in sufficient amounts to explain toxicity. Several studies are now in press that vindicate the early suspicions of the biochemical geneticists.

---

*Present address: ICI, Jealott's Hill Research Station, Bracknell, Berks. RG12 6EY, England.

(a)

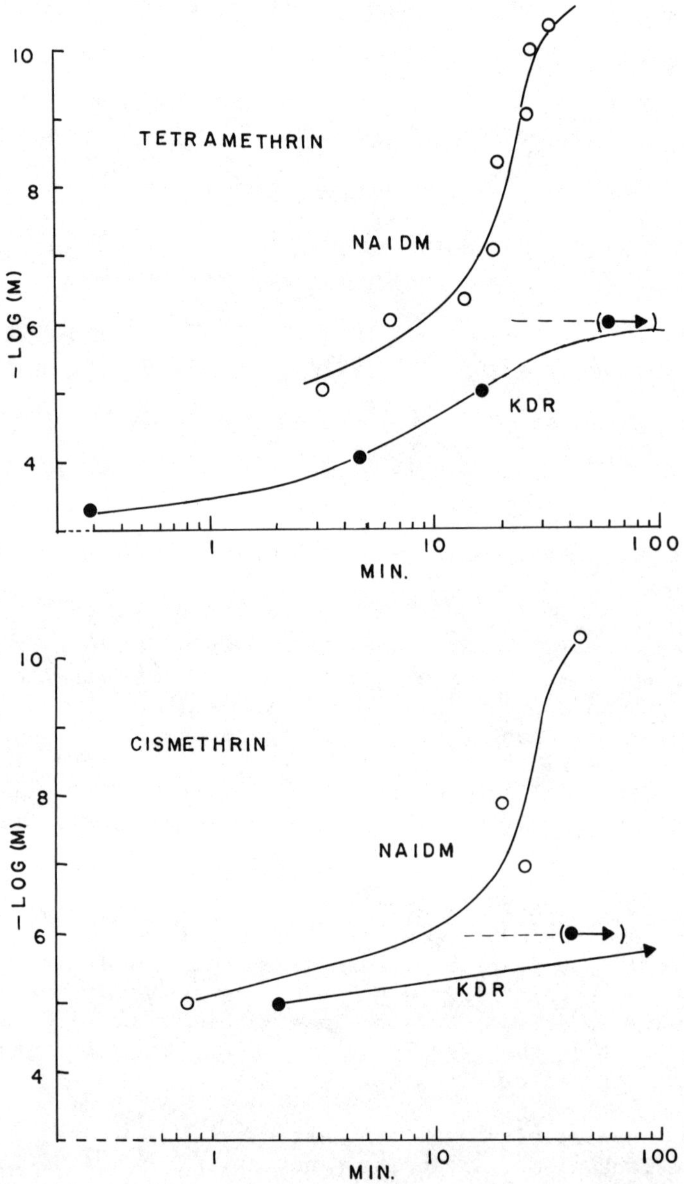

Figure 1. Latency (in min) to uncoupled convulsions monitored from flight motor units of adult S_{NAIDM} and *kdr* house flies after saline solutions of tetramethrin (a) or cismethrin (b) were perfused on the exposed thoracic ganglia. The dashed line indicates that the *kdr* house flies did not respond to that concentration up to 1 hr after application (Miller et al., 1979).

ACTIONS ON CENTRAL NERVOUS SYSTEM

The pyrethroids are toxic to the susceptible (NAIDM) and resistant (*kdr*) house fly strains with a resistance ratio of about 10, although the precise difference depends on the temperature. This resistance ratio is low compared to the large difference in threshold concentrations needed to show disruption in the sensory, motor or central nervous system (Figure 1a,b); however, the differences are smaller when comparing higher concentrations. Compare, for example, the concentrations producing convulsions in 10 min on the graphs in Figure 1.

The actions of pyrethroids when applied directly to the house fly thoracic ganglion have a negative temperature coefficient. Convulsions were recorded from flight motor units after a shorter latency at colder than at warmer temperatures. For example, tetramethrin produced uncoupled convulsions in 20 min at 21.0°C, but convulsions were recorded in 10 min when the same concentration (100 nM) was perfused on the ganglion at 17.0°C (Miller et al., 1979).

We were curious if the delay in response to pyrethroids (as shown in Figure 1) of the thoracic ganglion of *kdr* house flies could be explained by some change in the perineural sheath surrounding the ganglion. To test this, several bioassays were conducted with the sheath on the thoracic ganglion either intact (sheathed, Table 1) or removed (desheathed, Table 1). Although response times for a

Table 1. Mean Time to Uncoupled Convulsions of Exposed CNS to Two Strains of *Musca domestica* (NAIDM and *kdr*) with 10^{-5}M Bioneopynamin

Sheathed				Desheathed			
N	$\overline{X}$	SE	Range	N	$\overline{X}$	SE	Range
			NAIDM				
5	3.82	1.74	0.9-10.6	3	3.47	0.83	1.9-4.7
			kdr				
6	18.67	6.74	1.5-38.0 (plus 1 >60)	7	14.29	5.36	3.1-37.0

10 μM dose of bioneopynamin were longer for the *kdr* tissue, there was no significant difference between bioassays with the sheath intact and those with the sheath torn to allow faster access of the compound. Thus, the sheath was not considered to contribute to resistance in this insect in any way.

The bioassay data from Figure 1 show a clear difference between the susceptible and *kdr* strains. Whether sensory or motor nerves are examined in the peripheral nervous system, or whether the central nervous system is used for bioassay, the *kdr* tissues are always less suscpetible to pyrethroids.

These results, then, do not point out any particular part of the nervous system that might be more or less important in contributing to the *kdr* resistance. Instead, various parts of both the peripheral and central nervous systems in susceptible house flies appear to be sensitive to low concentrations (at or below nanomolar) of pyrethroids, while the nervous system of the *kdr* strain is uniformly resistant.

ACTION ON MOTOR AXONS

The lack of a more susceptible or more resistant part of the insect nervous system does not give a clue as to which site (or which tissue) might prove useful for structure-activity bioassays, and from the standpoint of studies on the mode of action of insecticides, it is still unclear as to which part of the nervous system is more important in pyrethroid poisoning. However, we can learn something by examining one site in detail, as long as we have evidence that the site chosen is involved in pyrethroid poisoning at reasonable concentrations. Since motor units are involved in such poisoning, work on the neuromuscular junction of the house fly maggot was undertaken in an effort to clarify the effect of pyrethroids at this site of action. Naturally, the house fly *kdr* strain (538ge) has been of enormous advantage in these studies.

Ten minutes after an LD_{50} dose of tetramethrin, house flies showed leg incoordination, slight tremors, and short bursts of flight activity (Adams and Miller, 1979). Concomitant with the onset of poisoning symptoms, the flight motor neurons showed repetitive discharges (i.e., short, high-frequency bursts of nerve impulses) in the flight motor units. These discharges either arose spontaneously in the motor nerve terminal or were triggered by single orthodromic action potentials, and, once initiated, the discharges were conducted both antidromically to the central nervous system, and orthodromically so that each repetitive impulse caused an excitatory postsynaptic potential (EPSP) in the muscle. Although repetitive discharges were not evoked below 19°C (which was termed 'transition temperature'), poisoning symptoms were more severe at lower temperatures. From this it was concluded that: "Motor unit discharges probably contribute to poisoning symptoms above the transition temperature, but their role in knockdown and toxicity at lower temperatures seems doubtful" (Adams and Miller, 1979).

Motor nerve discharges during pyrethroid poisoning have also been observed in locusts (Clements and May, 1977) and cockroaches (Gammon, 1978), but their origin and involvement in poisoning has never been fully investigated. Even though it seemed doubtful that repetitive discharges were important to poisoning, we decided to investigate the actions of pyrethroids on nerve terminals of the ventral longitudinal muscles of third instar larvae of two strains of house fly, *Musca domestica* (L.), because the nerve terminal regions of the flight motor axons were very sensitive to low concentrations of pyrethroids (Adams and Miller, 1979). The body wall muscles of house fly larvae have proven to be of great value in nerve-muscle studies (Irving and Miller, 1980). Each muscle fiber is innervated by one fast and one slow axon (Hardie, 1976; Irving and Miller, 1980), and the nerve-muscle synapses are all located superficially on the surface of the muscle with no glial barrier to the perfusion of chemicals.

After perfusion of the muscles with (1R)-*cis*-decamethrin at concentrations as low as 10^{-9}M, we found not only repetitive discharges in response to orthodromic stimuli, but complete disruption of normal synaptic function. Figure 2A1 shows normal unpoisoned slow, and composite fast plus slow EPSPs evoked by stimulation of the segmental nerve. The threshold of the slow axon was lower than that of the fast, in this case, so it could be selectively stimulated. Figure 2A2 shows a high-gain AC-coupled recording of miniature excitatory postsynaptic potentials (mEPSPs). Figure 2B was recorded 10 min after perfusion of 5 x 10^{-9}M (1R)-*trans*-permethrin. The stimulus evokes a composite EPSP, but the resulting repetitive discharge consists only of slow EPSPs (Figure 2B1). The repetitive EPSPs are known to be from the slow unit because of their size. At the same time, high-gain recordings show a great increase in resting mEPSP rate, and this is further increased following nerve activity, taking several seconds to return to the resting rate. Clements and May (1977) observed an increase in resting mEPSP rate of two- to three-fold after pyrethroid perfusion on locust leg muscles, while we have estimated the resting rate to be increased ten- to twenty-fold for a range of concentrations of pyrethroid.

Figure 2C shows that 1 hr after block of neuromuscular transmission, the mEPSP rate had declined to far below normal, and no EPSP was evoked by stimulation of either the fast or slow axons. At this point we stimulated both the fast and slow axon terminals with focal extra-cellular microelectrodes as described by Irving and Miller (1980). Although we could record the action potentials in

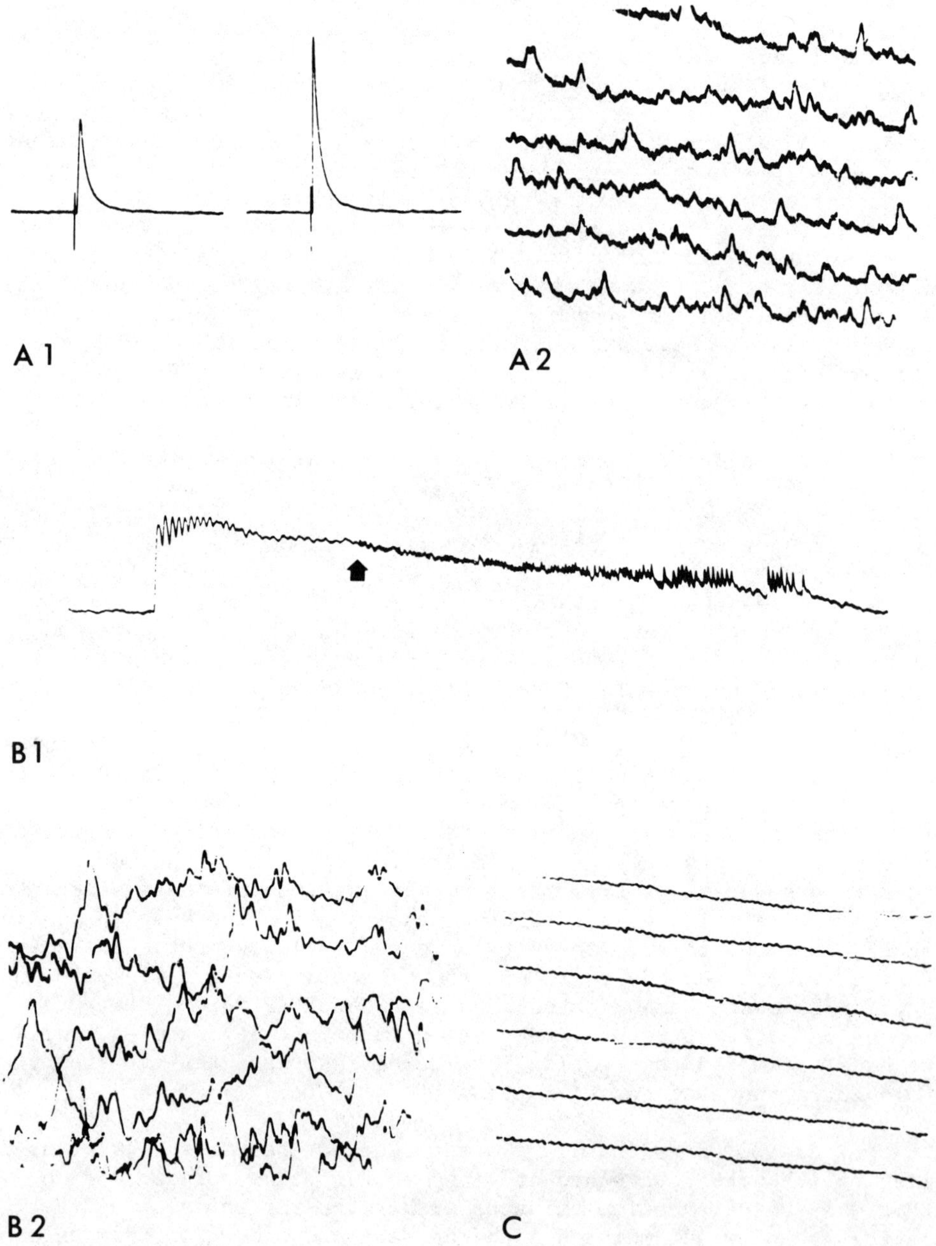
A1
A2
B1
B2
C

the nerve, no EPSP was evoked, indicating either that depolariaztion-release coupling was impaired, or presynaptic neurotransmitter stores were depleted. In either case, however, the functional result was neuromuscular block.

STRUCTURE-ACTIVITY CORRELATIONS WITH DECAMETHRIN

By using the mEPSP rate as an index, various pyrethroids were bioassayed on the larval body wall muscle preparation. An effective concentration (EC) was defined as the lowest concentration causing an increase in mEPSP rate within 30 min. Figure 3 shows a plot of mEPSP rate following perfusion of 5 x 10^{-8}M decamethrin.

As a general rule, esters of cyclopropane carboxylic acids with the (1R) configuration were more active than (1S), regardless of whether the substituent at position 3 was *cis* or *trans* (Elliott, 1977). This suggested a specific interaction of decamethrin with a chiral receptor.

The potency of synaptic action of pyrethroids was related to toxicity for four isomers of decamethrin (shown in Table 2). Threshold concentrations needed at 25°C to increase the mEPSP rate within 1 hr are also shown in Table 2 for (1R)-*trans*-permethrin and DDT. The (1S)-isomers of decamethrin are less active toxicologically than the (1R)-isomers, and were also less effective in increasing

←

Figure 2. Intracellular recordings from body wall muscle of *Musca domestica* larvae. A1: A neurally evoked "slow" excitatory postsynaptic potential (EPSP) shown on the left, and composite EPSP shown on the right in normal muscle. A2: Miniature excitatory postsynaptic potentials (mEPSP) from untreated muscle. Note gain is 100X compared to A1. Ten min after treatment of muscle with 5 x 10^{-9}M (1R)-*trans*-permethrin, a single shock to the nerve evoked a repetitive discharge response of "slow" EPSPs (B1). B2: The rate of mEPSPs was greatly increased. C: One hr after block of neuromuscular transmission, no mEPSPs. Amplitude calibration: A1, B1: 10 mV; A2, B2, C: 0.1 mV. Time calibration: all 100 msec, except B1 is compressed to 500 msec after the arrow.

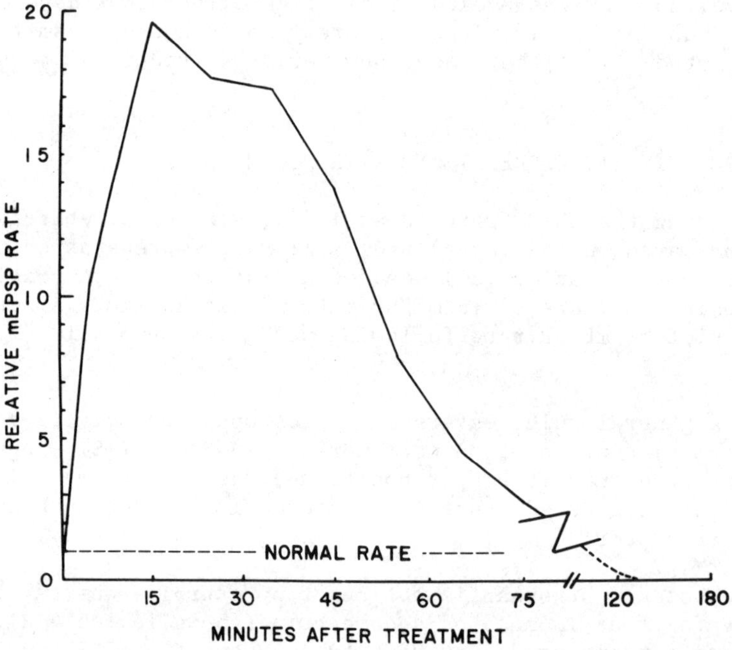

Figure 3. Plot of the relative mEPSP rate as recorded from a body wall muscle of a *Musca domestica* larvae. When 5 x 10^{-8}M decamethrin was perfused onto the preparation, the rate of mEPSPs (normal rate shown by the dashed line) increased to 20X greater than normal, then gradually declined. In this example, the mEPSP rate was still above normal when neuromuscular transmission failed; i.e., neurally evoked shocks to the nerve produced action potentials in the nerve but failed to produce postsynaptic potentials as recorded in the muscle. Then, after 2 hr, the mEPSP rate declined to zero. At neuromuscular transmission failure, the resting membrane potential of the treated muscle was the same as control, untreated muscles.

the mEPSP rate; it is interesting that the three isomers of decamethrin fall on a reasonably straight line when threshold concentration is plotted against house fly LD_{50} (Figure 4). DDT and permethrin both fall below this line, which may be due to differences in penetration, distribution and metabolism.

CORRELATION BETWEEN *kdr* RESISTANCE AND SYNAPTIC BIOASSAY

Another line of evidence correlating the synaptic site with the toxic site was obtained by studying the sensitivity of motor

Table 2. Adult House Fly Toxicity Data and Larval Nerve Terminal Activity by Pyrethroids

	Compound[b]	LD_{50} (ng/fly)	EC[a] (M)
I.	(1R)-*cis*-decamethrin	0.6	1×10^{-9}
II.	(1S)-*cis*-decamethrin	160.0	2×10^{-5}
III.	(1R)-*trans*-decamethrin	1.4	-
IV.	(1S)-*trans*-decamethrin	41.0	1×10^{-6}
V.	(1R)-*trans*-permethrin	11.0	1×10^{-9}
VI.	DDT	210.0	5×10^{-7}

[a]EC = lowest concentration effective in increasing mEPSP rate within 30 min.
[b]All decamethrin isomers are α-(R,S).

nerve terminals to decamethrin in susceptible strains of three species of insect. The best characterized site-insensitivity resistance factor is the *kdr* factor in house flies, *Musca domestica*. We used strain 538ge, which was obtained from A. W. Farnham, Rothamsted Experimental Station, Harpenden, England and bred in our laboratory since November, 1976. The pyrethroid resistance factor in 538ge is due to a single recessive gene on chromosome III (Farnham, 1977). The results are shown in Table 3 and indicate that the nerve terminals in the resistant strains are much less sensitive than those of the susceptible NAIDM strain.

We also tested pyrethroids on two mosquito species with laboratory-selected pyrethroid resistance believed to be at least partially due to site insensitivity of the *kdr* type: *Culex pipiens quinquefasciatus* (Priester and Georghiou, 1978, 1980) and *Anopheles stephensi* (Omer et al., 1980). These data are also shown in Table 3 and demonstrate that strains believed to possess a site-insensitivity resistance component consistently show nerve terminal site insensitivity.

POISONING SYMPTOMS CORRELATED WITH SYNAPTIC IMPAIRMENT

The motor nerve terminals of the ventral longitudinal muscles of mosquito larvae (*Anopheles stephensi*) are severely affected by pyrethroid poisoning. Two hours after treatment with an LC_{95} of (1R)-*trans*-permethrin, mosquito larvae responded in one of three

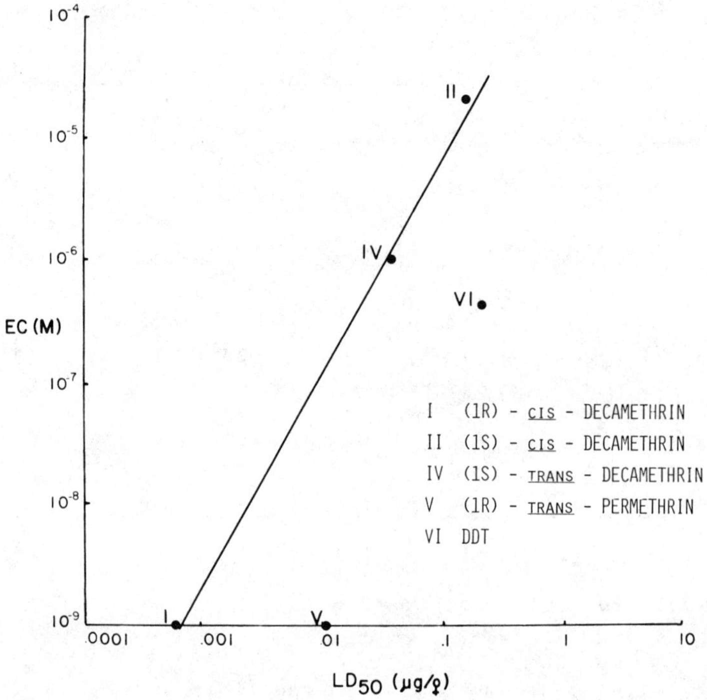

Figure 4. Relation between toxicity and effectiveness on nerve terminals for four pyrethroids and DDT. I, (1R)-*cis*-decamethrin; II, (1S)-*cis*-decamethrin; IV, (1S)-*trans*-decamethrin; V, (1R)-*trans*-permethrin; IV, DDT.

ways when touched with an artist's brush: (1) swam away from the stimulus with vigorous movements no different from untreated larvae, (2) made small sluggish ineffective movements, or (3) did not respond.

Ten larvae showing each type of response were dissected and the mEPSP rates in their ventral longitudinal muscles examined. Those that swam away vigorously had normal mEPSP rates; those that swam sluggishly had greatly increased mEPSP rates, and those that made no response had lower mEPSP rates than normal. Furthermore, those larvae that showed increased mEPSP rates initially, showed subnormal rates within one hour. Subnormal mEPSP rates were correlated with neuromuscular block.

DDT-like compounds appear to act in a manner similar to pyrethroids since both types of compounds have a negative temperature coefficient of toxicity (see Narahashi, 1971, for references), and site insensitivity resistance to one type invariably confers cross

Table 3. Comparison of Toxicity and Nerve Terminal Activity in Three Insect Species with a Site Insensitivity Factor

Insect	Compound[a]	LC_{50}[b] or LD_{50}	(R/S ratio)[c]	EC[d] (M)	(R/S ratio)[c]
Musca S[e]	*cis*-decamethrin	0.6	(13)	1×10^{-9}	(100)
Musca R[f]	*cis*-decamethrin	8.0	(13)	1×10^{-7}	(100)
Culex S[g]	*trans*-permethrin	0.0021	(4143)	5×10^{-10}	(2000)
Culex R[g]	*trans*-permethrin	8.7	(4143)	1×10^{-6}	(2000)
Anopheles S[h]	*cis*-permethrin	0.0035	(23)	5×10^{-11}	(20)
Anopheles R[h]	*cis*-permethrin	0.082	(23)	1×10^{-9}	(20)

[a]All compounds (1R) and (R,S)- α-cyano.
[b]Toxicity is in ng/♀ or ppm.
[c]The ratios of the immediately preceding values for a species are given in parentheses for comparison.
[d]EC = lowest concentration effective in increasing mEPSP rate within 30 min.
[e]NAIDM susceptible strain.
[f]538ge strain
[g]*Culex quinquefasciatus* susceptible (S) or resistant (R) strains.
[h]*Anopheles stephensi* susceptible (S) or resistant (R) strains.

resistance to the other (Omer et al., 1980; Priester and Georghiou, 1980; Farnham, 1977). In addition, both types of compounds produce repetitive discharge and block in nervous systems.

Insecticides of the pyrethroid or DDT group have a negative temperature coefficient of toxicity, and in the early stages, poisoning symptoms are readily reversed by raising the ambient temperature. This indicates that the negative temperature coefficient is due to increased sensitivity of the primary site of action at lower temperatures (cf. Narahashi, 1971). In agreement with this, we have found that the synaptic site is more sensitive at lower temperatures. A house fly larval preparation was treated with (1R) *cis*-decamethrin at 5×10^{-10}M for 1 hr at 25°C, and then the temeprature was lowered at 5° per minute. A temperature was reached at which the mEPSP rate suddenly increased and the slow axon became repetitive in response to stimulation. In untreated controls, the mEPSP rate decreased as temperature was lowered, as seen at other synapses (Mann and Joyner, 1978; Barrett et al., 1978).

When the temperature was raised, the mEPSP rate returned to normal. The symptoms could be reinduced by lowering the temperature again. There was not one characteristic temperature at which mEPSP activity reverted to normal rates; it occurred anywhere between 10°C and 21°C, and although we have not studied dose-response relationships in detail, the actions of pyrethroids on a particular synapse probably depend on temperature, concentration, time and individual variation, as well as other factors.

SUMMARY AND COMMENTS

We have provided several lines of evidence associating the synaptic action of pyrethroids with their toxicity. In summary, these are: (1) in the case of *Anopheles stephensi* larvae and *Musca domestica* adults (Adams and Miller, 1979), motor nerve terminals are affected during poisoning: (2) the synaptic site of action has stereochemical requirements similar to the toxic site; (3) the synaptic site is less sensitive in three strains believed to possess a site insensitivity resistance factor; (4) the synaptic site is also affected by DDT, which is believed to act at the same site as the pyrethroids, and (5) the synaptic site and the toxic site both show a negative temperature coefficient of action and temperature reversibility.

While repetitive firing and blocking in the neuromuscular system can be produced by selective action on the nerve terminals, it is also possible that a similar selective action on sensory neurons might be responsible for the fact that specific sites in the terminal extremeties of sensory or motor neurons appear to be much more sensitive to DDT and pyrethroids than are their axons (Roeder and Weiant, 1951; Clements and May, 1977; Wouters and van den Bercken, 1978). The great sensitivity of both motor nerve terminals and sensory structures to pyrethroids and DDT may be due to characteristics of the membrane ion channels in these two types of structures. Motor nerve terminal effects, however, are probably more important in the visible manifestation of poisoning. Gammon (1978) has found that in a poisoned, free-walking cockroach, sensory discharges appear first and persist for a period in which the only symptom is restlessness. Motor nerve discharges appear next, and these are correlated with symptoms of tremor and incoordination.

No clear correlation between neurotoxic actions and the toxicity of pyrethroids has been found in the past (Wouters and van den Bercken, 1978), but our evidence indicates that synaptic actions are more important than previously thought. Furthermore, although our results indicate that neuromuscular block is involved in poisoning, pyrethroids also act on central nervous sites, as ganglionic effects are often associated with poisoning symptoms (Miller and Adams, 1977; Gammon, 1978; Burt and Goodchild, 1971).

REFERENCES

Adams, M. E., and Miller, T. A., 1979, Site of action of pyrethroids: Repetitive 'backfiring' in flight motor units of house fly, *Pestic. Biochem. Physiol.*, 11:218.

Barrett, E. F., Barrett, J. N., Botz, D., Chang, D. B., and Mahaffey, D., 1978, Temperature-sensitive aspects of evoked and spontaneous release at frog neuromuscular junction, *J. Physiol. (London)*, 279:253.

Burt, P. E., and Goodchild, R. E., 1971, The site of action of pyrethrin I in the nervous system of the cockroach *Periplaneta americana*, *Entomologia Exp. Appl.*, 14:159.

Clements, A. N., and May, T. E., 1977, The actions of pyrethroids upon the peripheral nervous system and associated organs in the locust, *Pestic. Sci.*, 8:661.

Elliott, M., 1977, "Synthetic Pyrethroids," *ACS Symp. Ser.*, 42:1.

Farnham, A. W., 1977, Genetics of resistance of houseflies (*Musca domestica* L.) to pyrethroids. I. Knock-down resistance. *Pestic. Sci.*, 8:631.

Gammon, D. W., 1978, Neural effects of allethrin on the free walking cockroach *Periplaneta americana:* An investigation using defined doses at 15° and 32°C, *Pestic. Sci.*, 9:79.

Hardie, J., 1976, Motor innervation of the supercontracting longitudinal ventro-lateral muscles of the blowfly larva, *J. Insect Physiol.*, 22:661.

Irving, S. N., and Miller, T. A., 1980, Aspastate and glutamate as possible transmitters at the 'slow' and 'fast' neuromuscular juntions of the body wall muscles of *Musca* larvae, *J. Comp. Physiol.*, in press.

Mann. D. W., and Joyner, R. W., 1978, Miniature synaptic potentials at the squid giant synapse, *J. Neurobiol.*, 9:329.

Miller, T. A., and Adams, M. E., 1977, Central vs. peripheral action of pyrethroids on the housefly nervous system, *ACS Symp. Ser.*, 42:98.

Miller, T. A., Kennedy, J. M., and Collins, C., 1979, CNS insensitivity to pyrethroids in the resistant *kdr* strain of houseflies. *Pestic. Biochem. Physiol.*, 12:224.

Narahashi, T., 1971, Effects of insecticides on excitable tissues, *Adv. Insect Physiol.*, 8:1.

Omer, S. M., Georghiou, G. P., and Irving, S. N., 1980, DDT/pyrethroid resistance inter-relationships in *Anopheles stephensi*, *Mosq. News*, 20:200.

Priester, T. M., and Georghiou, G. P., 1978, Induction of high resistance to permethrin in *Culex pipiens quinquefasciatus*,

J. Econ. Entomol., 71:197.
Priester, T. M., Georghiou, G. P., 1980, Cross-resistance spectrum in pyrethroid-resistant *Culex quinquefasciatus*, *Pestic. Sci.*, 11:617.
Roeder, K. D., and Weiant, E. A., 1951, The effect of concentration, temperature, and washing on the time of appearance of DDT-induced trains in sensory fibers of the cockroach, *Ann. Ent. Soc. Am.*, 44:374.
Tsukamoto, M., Narahashi, T., and Yamasaki, I., 1965, Genetic control of low nerve sensitivity to DDT in insecticide-resistant houseflies, *Botyu-Kagaku*, 30:128.
Wouters, W., and van der Bercken, J., 1978, Action of pyrethroids, *Gen. Pharmacol.*, 9:387.

PENETRATION, BINDING AND TARGET INSENSITIVITY AS CAUSES OF RESISTANCE TO CHLORINATED HYDROCARBON INSECTICIDES

Fumio Matsumura

Pesticide Research Center
Michigan State University
East Lansing, Michigan 48824

INTRODUCTION

In the past 20 years scientists have worked hard to understand the biochemical basis of insecticide resistance in insect and acarine species, and in many instances, enough evidence has accumulated to offer some rational judgments on the cause of such resistance.

The most studied aspect of resistance mechanisms is the metabolic one, and with reason: in the overall biochemical defense system for developing resistance in insects, metabolic factors generally do play the most significant role (e.g., see Brown, 1971; Oppenoorth and Welling, 1976). Metabolic alterations of insecticidal chemicals are clearly recognizable, and the enzyme systems responsible for these changes can now be identified and characterized.

As the knowledge of resistance to insecticides has increased, however, it has become evident that there are mechanisms operating other than purely metabolic ones. Some of these factors may not be significant by themselves, but when they are combined with metabolic processes, they could cause high levels of resistance. Resistance can be developed by any means that reduces the amount of the actual toxic principle reaching the vulnerable site (i.e., the "target"), that reduces the sensitivity of the "target" towards the toxic principle, or by any combination of the above reduction processes that protects the vulnerable site. The contribution of these non-metabolic resistance factors to the final expression of resistance must be considered along with the metabolic fate of the insecticidal chemicals.

The role of reduced penetration in the final expression of resistance has been debated for some time (see Plapp and Hoyer, 1968; Sawicki and Farnham, 1967; Oppenoorth and Welling, 1976). Certainly there is a general agreement that reduced penetration, particularly when it is coupled with other resistance factors (Georghiou, 1971), can be an important factor in resistance. However, there are many questions that remain unanswered: e.g., the lack of specificity of resistance conferred by this method, the mechanism of reduction of penetration, and the question of the effectiveness, if any, of such a defense mechanism to stable insecticides. As for target insensitivity as a mechanism of resistance, there are now a number of good examples suggesting that such mechanisms do indeed play important roles in resistance by insect and acarine species. In the case of organophosphate and carbamate resistance, "target insensitivity" may evolve as the result of modifications of cholinesterase, the target enzyme of these insecticides (see paper by Hama in this volume). Cyclodiene resistance and many cases of DDT, BHC and pyrethroid resistance also involve target insensitivities. One aspect often forgotten by many scientists is that there are also many possible ways to develop target insensitivity, including reduced penetration and reduced binding at the target itself. Thus, in this presentation, I shall mainly describe those resistance mechanisms that result in the reduction of the toxic principles reaching the target or those involving chemical interactions with the target systems themselves. I shall rely upon the data we have obtained in our laboratory as well as pertinent data generated by other groups.

MATERIALS

We have chosen the German cockroach as the main study material because of the feasibility of both electrophysiological and genetic approaches. The strains used and their susceptibilities toward DDT and dieldrin have been described (Matsumura, 1971). The history of each strain has been described in detail by Kadous (1978). Both LP and VT strains were purified for four generations by a series of backcrossing schemes analogous to the one used by Telford and Matsumura (1970).

In addition, larvae of two strains of *Aedes aegypti* mosquitoes were used: a CSMA susceptible strain was obtained from the Wisconsin Alumni Research Foundation, and the resistant Isla Verde strain was kindly supplied by Dr. A. W. A. Brown, formerly at the University of Western Ontario, and was the original parental stock.

PENETRATION

There is no question about the role of penetration barriers as a viable mechanism of resistance (Forgash et al., 1962). Plapp and Hoyer (1968) studied the reduced penetration of dieldrin and DDT in a strain of house fly, and named the gene *organotin*-R with the mutant gene symbol of "*tin*." Apparently this gene introduces a general penetration barrier to many chemicals as judged by its cross resistance characteristics. Though the precise mechanisms by which it reduces penetration are not known, it is apparent that the system is more effective when the chemical is given in the form of residues (i.e., contact method) than in the cases where topical application techniques are employed (Plapp and Hoyer, 1968). These workers suggested that the gene works as an intensifier for the already existing resistance factors. Indeed, when a reduced penetration factor was genetically combined with other resistance factors (Plapp and Hoyer, 1968; Sawicki and Farnham, 1967; Georghiou, 1971), the degree of resistance increased much beyond the level expected by a simple additive effect. The tendency was particularly pronounced when the reduced penetration factor was combined with metabolic resistance factors (Georghiou, 1971). This is understandable because the reduction in the rate of cuticular penetration should give the metabolic systems the opportunity to degrade the insecticide to innocuous materials. Thus, it may be expected that a reduced penetration factor would be more significant in resistance to the readily metabolized insecticides such as malathion (Benezet and Forgash, 1972; Matsumura and Brown, 1963) than to the more stable chemicals such as dieldrin. Indeed, Plapp and Hoyer (1968) could not find any difference in LC_{50} between susceptible Orlando Regular and "tin" resistant *tin;stw* strains against dieldrin despite the fact that the latter strain showed a reduced uptake of ^{14}C-dieldrin and a high level of resistance to tributyltin chloride, a readily degradable pesticide. Reduction in penetration may be viewed as a reduced penetration constant P as shown below:

$$C_i = C_o(1-e^{-PAt})$$

where C_i is the concentration of insecticide inside the insect, C_o is the original concentration outside, P is the penetration constant, A is the area of contact, and t is the time. In such a relationship reduced P merely indicates the slowdown of penetration, and in due course the same amount of pesticide may enter into the insect body.

In the case of the dieldrin-resistant cockroach strains, Matsumura and Hayashi (1969) were not able to find any difference in the rate of cuticular penetration of dieldrin, nor were they able to show reduced dieldrin pickup in the dieldrin-resistant Isla Verde strain of *Aedes aegypti* larvae (Matsumura and Hayashi, 1966).

On the other hand, there appears to be a penetration barrier for dieldrin at the level of the nervous system. The first evidence of this was presented by Matsumura and Hayashi (1969) who isolated the abdominal nerve cords from the resistant (R) and susceptible (S) cockroaches treated in vivo with ^{14}C-dieldrin (Table 1). In this

Table 1. Dieldrin Takeup by Nerve Cords of German Cockroach Males in vivo[a]

Applied amount of dieldrin, µg/male adult	Time after application (hr)	Strains London (R)	CSMA (S)
0.381	5	22 (alive)	26 (alive)
	24	84.5 ± 16.7 (alive)	175.5 ± 24.4 (LT_{90})
	120	98.0 ± 14.7 (LT_{50})	
3.81	5	163.4 ± 45.5 (alive)	186.8 ± 69.1 (LT_{50})
	120	679.0 ± 80.0 (LT_{90})	

[a]Data expressed in C^{14} counts/3 minutes (1 µg dieldrin = 63,000 counts/3 minutes).
Average of two to four experiments ± standard error.

experiment, cockroaches were topically treated with 0.381 or 3.81 µg per insect of C^{14}-dieldrin with 1 µl of acetone. After various time intervals the cockroaches were successively surface-washed with saline solution and acetone, and dissected to isolate the abdominal nerve cords. Each nerve cord was homogenized in the counting solution with 0.5 ml of cockroach saline to assay the amount of dieldrin taken up by the nerve cords. The comparison of the amounts of dieldrin taken up by the R and the S cockroaches at identical poisoning time periods (Table 1) indicates that S nerve cords generally accumulate more dieldrin than the R counterparts. That the London individuals could tolerate much higher amounts of dieldrin in the nerve cords than the S insects became evident, however, when the amounts of dieldrin accumulated in the nerve cords from these strains were compared at the LT_{90}s for each strain.

Additional evidence was found by Telford and Matsumura (1970, 1971) who examined the ^{14}C-dieldrin exposed nervous system at an electron microscopic level by using an autoradiographic technique. By counting the silver grains developed in the electron micrographs (see Figure 1), these workers concluded that the only significant difference between the R strain (London) and the S strain was in the total number of grains found per tissue slide (Table 2). Although earlier Ray (1963) found no interstrain difference between the R and S German cockroach strains in the level of dieldrin pickup by the central nervous system, the R strains involved were different

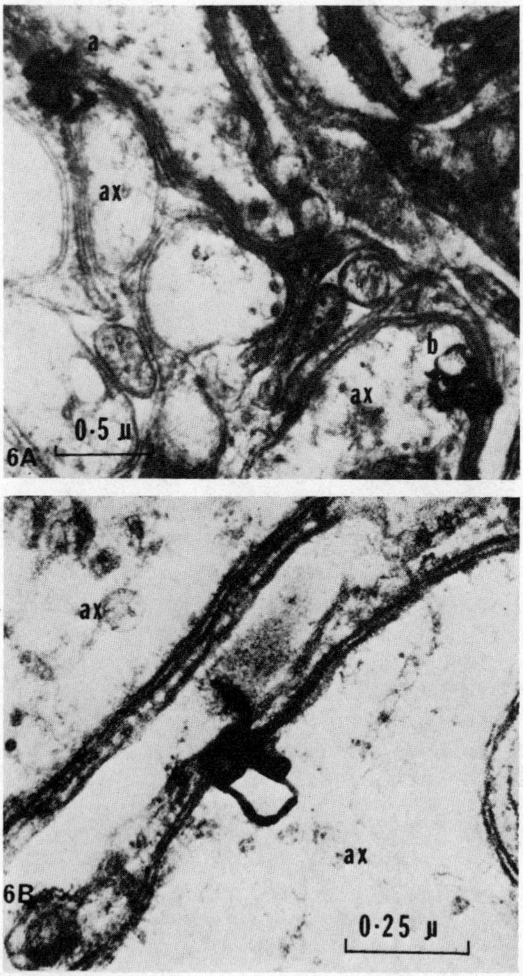

Figure 1. A transverse section of abdominal nerve cord (interganglionic axonic portion) in male CSMA susceptible cockroaches. Silver grains are observed situated over glial tissue (s) and over axonic tissue (b). B, transverse section of abdominal nerve cord in male CSMA susceptible cockroach. Silver grain is observed situated over axonic membrane. Axon (ax).

from the ones used in this study. Thus, one may conclude that at least in the case of the London strain, the reduced nerve penetration must play a role in the final expression of dieldrin resistance. The above observation supports the view that there are two different

Table 2. Distribution of Silver Grains over the Cellular Components of the Interganglionic Connectives in German Cockroach Strains

Strain	Silver grains/section[b]	Number of silver grains[a]					
		On axonic membrane	On glial membrane	In axon	In nerve sheath	In extra-cellular spaces	On other sites
CSMA	123.5	47.6	14.3	32.6	3.3	1.3	0.9
London pure	61.0	42.2	11.8	33.6	4.9	5.9	1.6

[a]Number expressed as percent of the total grain count over the sections.
[b]Average number of silver grains per section.

types of resistance cases based upon reduced penetration: one at the cuticular level and the other at the nervous system.

Although there are no detailed studies on differences between the mechanism of penetration through the insect cuticle and the nerve sheath, certain basic differences can be pointed out. First, in the former case, reduced penetration does not result in the decrease in the total amount of insecticide eventually getting into the body so long as the amount of the insecticide applied per insect is constant (e.g., the case with topical application). On the other hand, the amount of the insecticide reaching the target (i.e., the nervous system) is not constant. As shown in Figure 2, the amount of insecticide reaching the target is determined by both distribution and elimination kinetics. One must also note that the kinetics of insecticide pickup by the target are based upon the absorption-desorption equilibrium as well as the competition between the target and other non-target tissues for available insecticide.

In the case of a highly stable insecticide, therefore, development of reduced uptake by the target tissues may offer more advantages as a defense mechanism than the reduced cuticular penetration.

DISTRIBUTION

In the foregoing discussion the idea of competition for the available insecticide by different body compartments was introduced. The reduced uptake by the target organ could come from either a reduction in affinity of the target for the compound, the formation

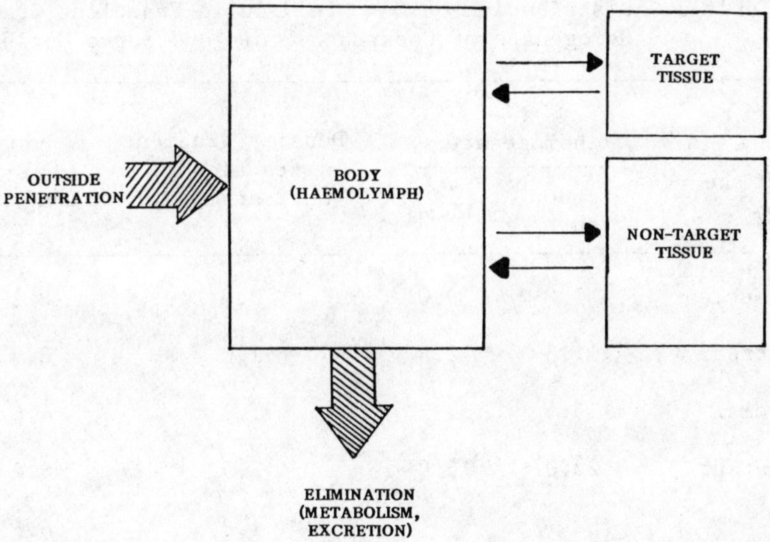

Figure 2. Schematic diagram of penetration routes of insecticides.

of a penetration barrier, or an increase in uptake by other tissues, such as fatbody.

In the experiment shown in Table 3, the heads and nerve cords of male cockroaches of the S (CSMA) strain and the R strain (London pure), which was a genetically purified London strain developed by backcrossing to CSMA for four generations, were homogenized in 0.25 M sucrose solution, and the large tissue fragments were removed by filtration through biosilicate glass wool. The filtrates were incubated with 10^{-5}M ^{14}C dieldrin for 1 hr at 24°C (Matsumura and Hayashi, 1969). Subcellular components of the homogenate were then separated by centrifugation and the radioactivity in each centrifugal fraction was determined. The results indicate that the level of dieldrin is higher in all heavy particulates of the CSMA strain than in those of the R strain. Conversely, the dieldrin levels in the supernatant and the microsomal fractions were higher in the R strain than in the S strain. Clearly, then, there is a difference in the pattern of distribution of dieldrin within the nervous system. Earlier we showed that the heavy particulates in these centrifugal fractions consisted of intact nerve cells, pinched off nerve ending particles, and nerve sheath materials (Telford and Matsumura, 1971).

A similar observation was made earlier with the head homogenate of the R (Isla Verde) and the S (CSMA) strains of *Aedes aegypti* larvae (Matsumura and Hayashi, 1966a).

Table 3. Distribution of C^{14} Dieldrin in Fractions from Homogenate of Cockroach Brain and Nerve Cord[a]

Centrifugal fractions	Brain homogenate		Density gradient fractions (Sucrose molar)	Brain homogenate	
	London pure[b]	CSMA		London pure	CSMA
Crude Nuclear	51.2	59.0	0.8	8.3	20.5
Mitochondrial	11.3	11.3	1.0	8.9	12.2
Microsomal	15.6	13.9	1.2	4.6	8.9
Supernatant	22.0	15.8	1.5	6.2	5.5
			1.8	6.6	6.1
			Supernatant	65.4	46.9

[a]Results expressed in percentages of given dieldrin (10 nanomoles per ml.) recovered in each fraction.
[b]Genetically purified from London strain.

To find whether such distribution differences exist in the axonic or ganglionic portion of the nervous system, 15 nerve cords were collected from each strain and the ganglionic portions were separated out from the axonic (connective) portion by dissection. The homogenate was prepared from each preparation and was incubated with 10^{-5}M dieldrin as before. The heavy particulates were collected by centrifugation at 20,000 g for 45 min at 0°C, suspended in 0.8 M sucrose, and then further separated by means of a discontinuous sucrose density centrifugation technique. Each particulate fraction was collected, and the level of radioactivity was determined. The results shown in Figure 3 clearly indicate that dieldrin levels were uniformly low in the R strain. This must mean that the "factor" that promotes reduced uptake of dieldrin in the R strain is widely distributed among nerve components.

BINDING TO NERVE COMPONENTS

Binding characteristics of dieldrin to the nerve particulates of the R and S cockroaches were then studied (Matsumura and Hayashi, 1966). The previous findings had been that (1) the crude nucleus fraction from the S strain absorbed more dieldrin than those from R strains, and (2) the rates of dieldrin recovery from the R

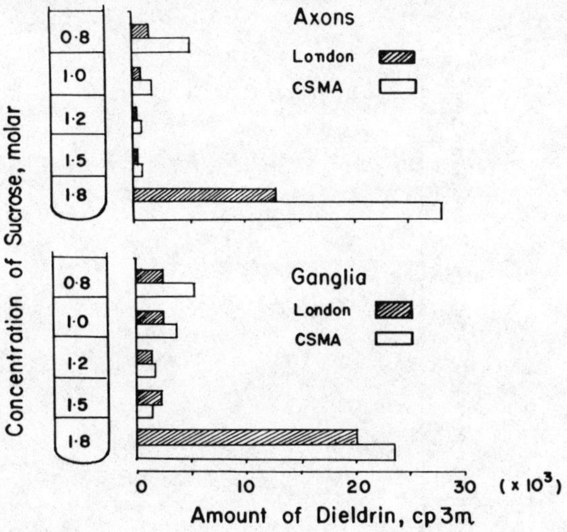

Figure 3. Interstrain comparison of binding patterns of C^{14}-dieldrin among axonic and ganglionic fractions of nerve cord separated by sucrose density gradient centrifugation technique.

supernatant were much higher than those from the S counterparts. The supernatant fraction may contain unknown amounts of free dieldrin molecules that should be extractable by any organic solvent upon partitioning. The aqueous phases of the R strains had higher radioactivity than those of the S strain. Partitioning the supernatants with n-pentane transferred 69.3 ± 5.3 percent (for CSMA) of radioactivity into the solvent phase; similar treatment with n-butanol extracted 90.0 ± 0.3 (London), 88.2 ± 1.2 (Fort Rucker), and 86.1 ± 3.8 percent (CSMA) of the total radioactivity.

Experiments with chloroform and benzene confirmed this tendency for the R supernatant to have slightly more solvent-extractable dieldrin than its S counterpart, but this tendency does not necessarily indicate that absorption by the soluble components of the R strains is significantly higher than that of the S strain. The radioactivity recovered from the R solvent phases was also high; the interstrain difference could be caused by the difference in the true substrate concentration, which is secondarily controlled by the rate of absorption by other particulate fractions.

Previously, the most conspicuous interstrain difference was in the crude-nucleus fraction. The rate of absorption of dieldrin by this fraction was investigated at various substrate concentrations by first incubating C^{14}-dieldrin for 1 hr with washed nucleus fractions, and then collecting and rewashing the fractions twice with

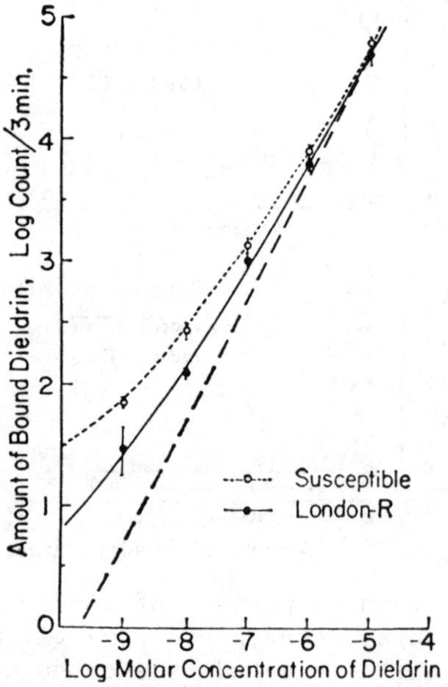

Figure 4. Binding of C^{14}-dieldrin with the nerve components of susceptible and resistant German cockroaches. The straight line indicates the rate of a theoretical binding estimated by nonspecific absorption of dieldrin at high concentrations. Vertical lines represent standard errors of the experimental data.

fresh sucrose solution (Figure 4). For each strain there is a saturation plateau that deviates from the theoretical absorption line at low concentrations (the components causing the plateau and the linear absorption are designated α and β, respectively); the R (London) nucleus shows a low plateau, α. The interstrain difference in terms of t-value at 10^{-8}M, for instance, was 3.21; it is highly significant at the 95% confidence level. The absorption constant for the S component α, as estimated by the median-saturation substrate concentration at equilibrium, was 9.1×10^{-7}M; that for the R component was 1.25×10^{-6}M.

A similar experiment with different incubation periods, at a dieldrin concentration of 1×10^{-6}M (Figure 5), indicated that the S component (possibly α) had a higher binding speed (bimolecular constant, 219×10^{5} liter mole^{-1} min^{-1}) than the R counterpart (bimolecular constant, 1.4×10^{5} liter mole^{-1} min^{-1}). The constants for the slow-binding components (possibly β) for each strain, on the other hand, scarcely differ from each other, the values being 4.9 and 5.3×10^{3} liter mole^{-1} min^{-1} for the S and R components, respectively.

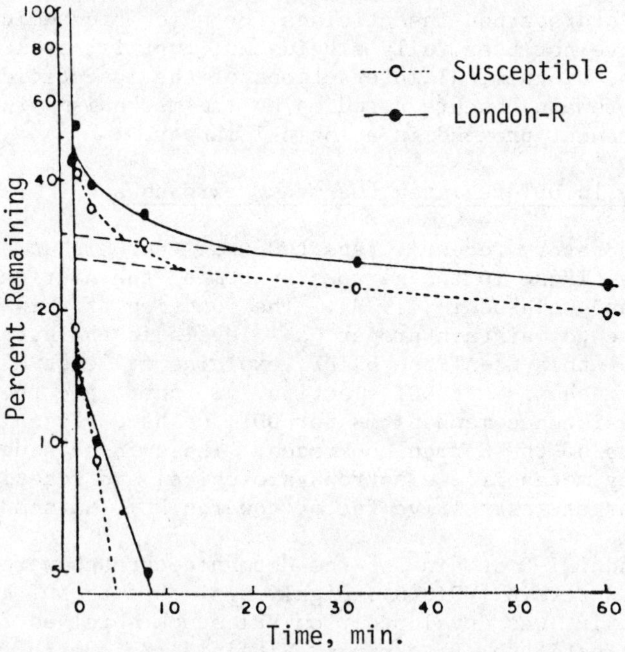

Figure 5. Effect of time on the rate of binding of C^{14}-dieldrin with the nerve components of susceptible and resistant cockroaches. Percent remaining (calculated) is the free dieldrin remaining at the end of the reaction as a percentage of the total dieldrin added initially. The fast phase of the main curve (bottom left) was obtained by subtraction of the straight line extrapolated to zero time.

CONSIDERATION OF TARGET SENSITIVITY

It has been known for sometime that some insects and acarine species are capable of developing resistance to organophosphates and carbamates by changing the properties of acetylcholinesterase (see paper by Hama in this volume). Likewise, the presence of R strains with target insensitivity against DDT has been well documented. Even in the case of house flies, which are known to develop resistance largely through increased detoxification processes through DDT-dehydrochlorinase, a mechanism known as Kdr (for knockdown resistance) contains the ability to withstand the action of DDT at the nervous system (see paper by Miller et al. in this volume).

While the involvement of cholinesterases in the development of target insensitivities toward organophosphate and carbamate insecticides is a logical expectation, the same cannot be said about chlorinated hydrocarbon insecticides for which biochemical mechanisms of action have not been fully elucidated; that is, as long as the nature of the biochemical interactions of the insecticides with the target site remain obscure, studies on the mechanism of this type of resistance cannot proceed in a logical manner.

A Case Study in DDT-resistant German Cockroaches

Our laboratory recently reported that there is an extremely DDT-sensitive Ca-ATPase in the nervous system of the American lobster (Matsumura and Ghiasuddin, 1979). The function of this Ca-ATPase appears to be to maintain proper Ca^{2+} levels in the nerve membrane. Inhibition of this Ca-ATPase by DDT explains at least a part of the biochemical mechanism of DDT's action. To correlate the above finding to the resistance mechanisms for DDT, we have chosen several DDT-R strains of the German cockroach. These insects have already been shown by metabolic, electrophysiological and genetical studies to have a target insensitive factor towards DDT (Matsumura, 1971).

CSMA and DDT-R strains of the German cockroach were used in this study. One R strain (VPIDLS), highly resistant to DDT but susceptible to dieldrin, was developed from VPI stock obtained from Virginia Polytechnic Institute, Blacksburg, Virginia (Matsumura, 1971; Telford and Matsumura, 1970). Another DDT-R strain (VT) was developed from VPIDLS and genetically purified by crossing and backcrossing with the susceptible CSMA strain for eight generations.

For preparation of the enzyme source, cockroaches were decapitated and isolated heads were homogenized for three minutes at $0^{o}C$ in 0.8 M sucrose + 1 mM EDTA solution using a Potter-Elvehjem, teflon-glass tissue homogenizer (clearance 0.1 mm). The brain homogenate was first centrifuged at 1000 g at $4^{o}C$ for 10 min. to separate the cuticular materials and the cell debris. The partially clear supernatant was then filtered through glass wool to remove heavy and light mitochondria. The clear supernatant was once more filtered through glass wool and diluted to 0.25 M sucrose using chilled deionized H_2O and centrifuged at 90,000 g (R max) for 120 min. The 90,000 g pellet thus obtained was resuspended in the original sucrose-EDTA solution. The enzyme solution thus prepared was divided into small portions and frozen until use.

The ATPase assay was conducted in small culture test tubes (12 x 100 mm). Each tube contained 0.1 ml of enzyme and 0.9 ml of assay buffer. The enzyme solutions were adjusted in such a way that each 0.1 ml contained 10 μg of protein. The ionic composition of assay buffer and the method of ATPase assay were identical to those used earlier (Ghiasuddin and Matsumura, 1979). Ouabain (0.1 mM) and KCN (2 mM, both expressed as final concentrations) were always added to each assay tube as aqueous solution in 10 μl volume.

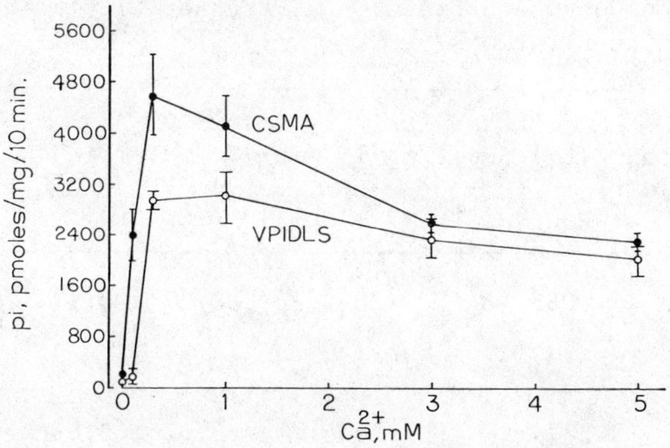

Figure 6. Effect of Ca^{2+} on ATPase activity of the membrane fraction from DDT-susceptible (CSMA) and DDT-resistant strains (VPIDLS) of German cockroach under standard assay conditions at 30 °C as given in materials. Ten µg equivalent protein per assay tube was employed. The assay buffer contained 60 mM Na^+, 60 mM K^+, pH 7.4 previously adjusted with 30 mM Tris-HCl. The vertical lines indicate standard error for each point. Averages of three independent experiments.

DDT was added in an aliquot of 10 µl ethanol while an equivalent amount of ethanol was added to the control. These agents and inhibitors were incubated for 10 min at room temperature before the reaction was started by adding 1 µl of γ-^{32}P ATP (8 x 10^{-8}M final concentration) and incubated at 30°C for an additional 10 min. The reaction was stopped by adding 0.2 ml with 10% TCA. The amount of inorganic phosphate (Pi) released was measured using a charcoal method (Ghiasuddin and Matsumura, 1979) and was used as the measure of ATPase activity.

Each experiment was repeated at least three times. To estimate the variation and difference in percent inhibition of Ca-APTase between the DDT-S and DDT-R strains, standard errors were calculated and the data further subjected to "t" tests.

An effort was first made to clearly demonstrate the presence of a Ca-APTase in the 90,000 g fraction of the cockroach brain similar to the one found in the nervous system of the American lobster (Ghiasuddin and Matsumura, 1979). The results (Figure 6) indicate that there is an ATPase that is stimulated by 0.3 to 1 mM of Ca^{2+} in the cockroach brain. It must be mentioned here that the reaction mixture contained ouabain (0.1 mM) and CN^- (2 mM) to inhibit the Na-K ATPase and mitochondrial ATPases. The degree of Ca^{2+} stimulation was more pronounced in the S strain than in the R strains.

Table 4. German Cockroach Brain Ca-ATPase Activity at Different ATP Concentrations[a]

ATP Conc.	Roach Strain[b]	ATPase Activity (Pi) pmole/mg/10 min		% Inhibition by DDT S.E.	(Number of experiments)
		Control	+ DDT		
$8x10^{-8}M$	CSMA	4086.45	2833.14	30.67 ± 3.11	(5)
	VPIDLS	3121.36	2487.42	20.31 ± 3.8[c]	(5)
	VT	3374.94	2604.74	22.82 ± 2.85	(5)
$8x10^{-9}M$	CSMA	387.24	240.19	37.97 ± 2.34	(4)
	VPIDLS	258.92	184.05	28.92 ± 2.51	(4)
	VT	336.77	239,44	28.90 ± 2.83[c]	(4)
$8x10^{-10}M$	CSMA	47.25	24.69	47.75 ± 2.83	(4)
	VPIDLS	30.98	18.28	40.99 ± 2.18[c]	(4)
	VT	34.44	21.43	37.78 ± 2.67[d]	(4)

[a]For standard assay conditions see legends for Figures 1 and 2.
[b]CSMA = DDT-susceptible, VPIDLS and VT are DDT-resistant strains.
[c,d]Significant by "t" test at $P \leq 0.05$ and $P \leq 0.025$, respectively.

It was previously shown (Matsumura and Ghiasuddin, 1979) that the level of DDT inhibition of this enzyme varies according to the changes in the ATP concentration and the total enzyme added to the incubation system. When the concentration of ATP was varied from $8 \times 10^{-8}M$ to $8 \times 10^{-10}M$, the degree of DDT inhibition increased (Table 4). It must be noted that in every case the Ca-ATPase from the R (VPIDLS) strain showed less sensitivity toward DDT than did the S counterpart.

The effect of DDT concentration changes on the Ca-ATPase activity was then examined. The results (Figure 7) clearly show that the enzyme from the CSMA strain is more sensitive to DDT than that from VPIDLS. The enzyme from VT generally behaved in the same manner as that of VPIDLS.

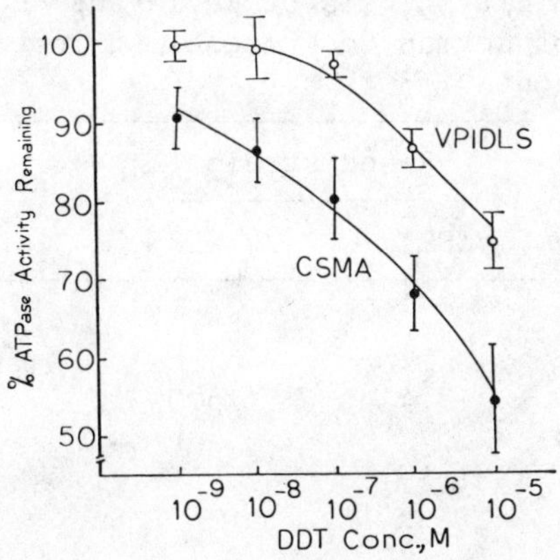

Figure 7. DDT inhibition of roach brain Ca-ATPase activity of DDT-susceptible (CSMA) and DDT-resistant (VPIDLS) strains of German cockroach at 30°C using the standard protein concentration (i.e., 10 µg/assay tube). The ionic composition was identical to that shown in Figure 6 except for Ca^{2+} which was maintained at 1 mM level for both strains. The vertical lines indicate standard error for each point. Averages of six independent experiments.

To examine the possibility that such an ATPase is involved in the resistance mechanism, various ATPase inhibitors and neuroactive agents known to affect Ca^{2+} levels were tested for their in vivo toxicities toward the S and the R cockroaches. The results (Table 5) show that the R strain is more tolerant to various inhibitors such as NaF, iodoacetate, ruthenium red, and lanthanum. It must be mentioned here that these inhibitors attack other enzymes and organs than the nervous system, and as such, in vivo poisoning symptoms induced by them do not necessarily resemble those caused by DDT. Nevertheless, if the biochemical change involves an enzyme or protein that has various functions in many parts of the body, the effects of the inhibitor of the enzyme should express themselves as an interstrain difference in toxicity in vivo. For instance, lanthanum is known to be a specific inhibitor of Ca^{2+} requiring ATPases, which are known to be present in many tissues carrying out various functions. NaF is also known to affect Ca^{2+} uptake by interacting with proteins (Hewitt and Nicholas, 1963). On the other hand, the R strain was equally sensitive or more sensitive than the S strain toward several -SH inhibitors such as N-ethylmalimide, mersalyl acid, 5,5-dithio *bis*-2-nitrobenzoic acid (DTNB), p-chloromercuriphenyl sulfonic acid (PCMPS), and phenylthiourea, all of which also work as inhibitors for several ATPases with active -SH moieties (White et al., 1968; Carsten and Mommaerts, 1964; Ruegg,

Table 5. Susceptibility Levels of Resistant and Susceptible Strains of German Cockroaches Against Various Neuroactive Agents[a]

Neuroactive agent	Roach strain		Relative resistance factor[b]
	CSMA	VPIDLS	
Insecticides			
DDT	7.4	115.0	15.54
Dieldrin	4.7	5.3	1.13
ATPase inhibitors			
(a) Agents affecting Ca^{2+} availability			
Lanthanum	4.8	30.0	6.25
Ruthenium red	10.0	24.0	2.40
NaF	9.46	25.65	2.71
EDTA	150.0	92.0	0.61
(b) -SH inhibitors			
Iodoacetate	14.0	32.0	2.29
Mersalyl acid	120.0	105.0	0.88
DTNB[c]	115.0	98.0	0.85
PCMPS[c]	125.0	98.0	0.78
Phenylthiourea	230.0	180.0	0.78
Diphenylthiourea	115.0	100.0	0.87
N-ethylmalimide	28.0	13.0	0.46
(c) Others			
Oligomycin	12.0	12.0	1.00

[a]Data are expressed in LT_{50} in terms of hours needed to kill 50% of population.

[b]Relative resistance between resistant (VPIDLS) against susceptible (CSMA) strain. The values less than 1.0 indicate negatively correlated resistance (i.e., the susceptible strain is more resistant).

[c]DTNB = 5,5-dithio *bis*-2-nitrobenzoic acid; PCMPS = p-chloromercuriphenyl sulfonic acid.

1961). The fact suggests that the modification in the R strain, if one accepts the above hypothesis, does not involve overall changes in all -SH enzymes. In addition, the VPIDLS strain was found to be resistant to theophylline (Kadous, 1978), which, like caffeine, is known to promote release of Ca^{2+} from the membrane (Duncan, 1978). This phenomenon is specific to the DDT-R strain. Oligomycin, known to affect mitochondrial ATPases, which are quite sensitive to chlorinated insecticides (e.g., Cutkomp, 1971; Dessaiah and Koch, 1975), is equally affected by both strains.

The most important aspect of this study is that the Ca-ATPase, which has been proposed as the site of DDT attack, has been found to be altered in the R cockroach so that it is not as sensitive as that of the S atrain. The fact that such an interstrain difference can be observed in an R strain (VT) that has been genetically purified by backcrossing for eight generations to the S strain (CSMA) suggests that it is not due to a simple geographical variation in ATPase activities. The finding also serves as evidence for the involvement of the Ca-ATPase in the course of DDT-induced nerve excitation. This is an analogous situation to the case of the strain of seizure-prone mice, which have been shown to have abnormally low levels of the Ca-APTase in the brain (Rosenblatt et al., 1976). These two sets of examples suggest that lowering of the Ca-ATPase, either genetically or from inhibition by DDT, induces nerve instability and leads to seizure and hyperexcitation in vivo.

Because of the cross resistance of this strain to various inhibitors of this enzyme in vivo, and because of the suggestion that this enzyme serves as a target for DDT in the lobster nerve, it has been concluded that the modification of this Ca-ATPase may serve as at least a part of the resistance mechanism for this strain of the German cockroach.

CONCLUSION

There is no question about the role of reduced penetration in inducing insecticide resistance in many cases. The reduced cuticular penetration is effective when it is coupled with metabolic defense mechanisms of the insects. For the same reason this mechanism appears more effective against insecticides that are readily degradable. The existence of a penetration barrier at the target level (i.e., the nervous system) in the R insects has been established in the case of dieldrin-R German cockroaches. Since the kinetics of penetration into the nervous system are different from those of cuticular penetration, the former mechanism appears to be more suited for stable insecticides. The amount of active insecticide that finally ends up in the nervous system is also determined by competitive equilibrium of the chemical between the nervous system and other tissues. For stable insecticides such as dieldrin, this method for reduction of the insecticide level in the nervous system may play an important role as one of the determinants of the expression of resistance.

The most effective resistance mechanism is "target insensitivity." In the case of the DDT-R German cockroach we were able to demonstrate the alteration of a Ca-ATPase, a proposed target enzyme, in the nervous system. These mechanisms would operate efficiently in union to protect the sensitive systems long enough for the poisons present in the body to be eliminated through metabolism, excretion or storage in nonvital tissues.

Acknowledgment: Supported by a research grant 71-6788 from the Michigan Agricultural Experiment Station (Journal Article No. 9436). This paper summarizes the work of Drs. M. Hayashi, A. A. Kadous and S. M. Ghiasuddin of this laboratory.

REFERENCES

Benezet, H. J., and A. J. Forgash, 1972, Reduction of malathion penetration in houseflies pretreated with silicic acid, *J. Econ. Entomol.*, 65:895.

Brown, A. W. A., 1971, Pest resistance to pesticides, *in:* "Pesticides in the Environment," Vol. I (Part II), R. White-Stevens, ed., pp. 457-533, Marcel Dekker, New York.

Carsten, M. E., and Mommaerts, W. F. H. M., 1964, The accumulation of calcium ions by sarcotubular vesicles, *J. Gen. Physiol.*, 48:183.

Cutkomp, L. K., Yap, H. H., Vea, E. V., and Koch, R. B., 1971, Inhibition of oligomycin-sensitive mitochondrial Mg^{2+} ATPase by DDT and selected analogs in fish and insect tissue, *Life Sci.*, 10:120.

Desaiah, D., and Koch, R. B., 1975, Inhibition of fish brain ATPases by aldrin-transdiol, aldrin, dieldrin and photodieldrin, *Biochem. Biophys. Res. Commun.*, 64:13.

Duncan, C. J., 1978, Role of intracellular calcium in promoting muscle damage: A stragety for controlling the dystrophic condition, *Experientia*, 34:1531.

Forgash, A. J., Cook, B. J., and Riley, R. C., 1962, Mechanisms of resistance in diazinon-selected multiresistant *Musca domestica*, *J. Econ. Entomol.*, 55:544.

Georghiou, G. P., 1971, Isolation, characterization and resynthesis of insecticide resistance factors in the housefly, *Musca domestica*, *in:* "Proceedings of the 2nd International IUPAC Congress of Pesticide Chemistry," Vol. II, A. S. Tahori, ed., pp. 77-94, Gordon and Breach, New York.

Ghiasuddin, S. M., and Matsumura, F., 1971, DDT inhibition of Ca-ATPase of the peripheral nerves of the American lobster, *Pestic. Biochem. Physiol.*, 10:151.

Hewitt, E. J., and Nicholas, D. J. D., 1963, *in:* "Metabolic Inhibitors," R. M. Hochster and J. H. Quastel, eds., Academic Press, New York.

Hoyer, R. F., and Plapp, F. W., Jr., 1966, A gross genetic analysis of two DDT-resistant house fly strains, *J. Econ. Entomol.*, 59:495.

Iwata, T., and Hama, H., 1972, Insensitivity of cholinesterase in *Nephotettix cincticpes* resistant to carbamate and organophosphorus insecticides, *J. Econ. Entomol.*, 65:643.

Kadous, A. A., 1978, Chlorinated hydrocarbon insecticides: Mode of action, resistance mechanisms and metabolism, Ph. D. Thesis, University of Wisconsin, Madison, 219 pp.

Matsumura, F., 1971, Studies on the biochemical mechanisms of resistance in strains of the German cockroach, *Proc. 2nd Intern. Congr. Pestic. Chem.*, 2:95.

Matsumura, F., and Brown, A. W. A., 1963, Studies on organophosphorus tolerance in *Aedes aegypti*, *Mosquito News*, 23:26.

Matsumura, F., and Ghiasuddin, S. M., 1979, Characteristics of DDT-sensitive Ca-ATPase in the axonic membrane, *in:* "Neurotoxicity of Insecticides and Pheromones," T. Narahashi, ed., pp. 245-247, Plenum Press, New York.

Matsumura, F., and Hayashi, M., 1966a, Interaction of dieldrin with the subcellular components of both resistant and susceptible strains of *Aedes aegypti* L., *Mosquito News*, 26:190.

Matsumura, F., and Hayashi, M., 1966b, Dieldrin: Interaction with nerve components of cockroach, *Science*, 153:757.

Matsumura, F., and Hayashi, M., 1969, Dieldrin resistance. Biochemical mechanisms in the German cockroach, *J. Agr. Food Chem.*, 17:231.

Oppenoorth, F. J., and Welling, W., 1976, Biochemistry and physiology of resistance, *in:* "Insecticide Biochemistry and Physiology," C. F. WIlkinson, ed., pp. 507-551, Plenum Press, New York.

Plapp, F. W., and Hoyer, R. F., 1968, Insecticide resistance in house fly: Decreased rate of absorption as the mechanism of action of a gene that acts as an intensifier of resistance, *J. Econ. Entomol.*, 61:1298.

Ray, J. W., 1963, Insecticide absorbed by the central nervous system of susceptible and resistant cockroaches exposed to dieldrin, *Nature*, 197:1226.

Rosenblatt, D. E., Lauter, C. J., and Trams, E. G., 1976, Deficiency of a Ca^{2+}-ATPase in brains of seizure-prone mice, *J. Neurochem.*, 27:1299.

Ruegg, J. C., 1961, On the effect of inhibiting the actin-myosin interaction on the viscous tone of a lamellibranch catch muscle, *Biochem. Biophys. Res. Commun.*, 6:24.

Sawicki, R. M., and Farnham, A. W., 1967, Examination of the isolated autosomes of the SKA strain of houseflies for resistance to several insecticides with and without pretreatment with sesamex and TBTP, *Bull. Entomol. Res.*, 59:409.

Telford, J. N., and Matsumura, F., 1970, Dieldrin binding in subcellular nerve components of cockroaches. An electron microscopic and autoradiographic study, *J. Econ. Entomol.*, 63:795.
Telford, J. N., and Matsumura, F., 1971, Electron microscopic and autoradiographic studies on distribution of dieldrin in the intact nerve tissues of German cockroaches, *J. Econ. Entomol.*, 64:230.
White, A. P., Handler, A. P., and Smith, E. L., 1968, *in:* "Principles of Biochemistry," McGraw-Hill Book Company, New York.

MECHANISMS OF PESTICIDE RESISTANCE IN NON-TARGET ORGANISMS

G. M. Booth, D. J. Weber, L. M. Ross, S. D. Burton,
W. S. Bradshaw, W. M. Hess and J. R. Larsen*

Departments of Botany, Microbiology and Zoology
Brigham Young University
Provo, Utah 84602

INTRODUCTION

Published literature on pesticide resistance is voluminous. In preparation of this paper, a computerized literature search of over 50,000 journals resulted in off-line printouts of about 2,000 resistance articles from the data banks of Chemical Abstracts, Biological Abstracts, Enviroline, Toxline, and Agricola. These references spanned approximately the last ten years.

However, excluding laboratory mammals, less than 200 of these articles dealt with non-target resistance, and fewer than 50 of these addressed detailed mechanisms of resistance. If these figures are reasonably accurate, it appears that less than 3% of the published literature relates to mechanisms of resistance in non-target organisms. In other words, much of the information on these organisms is simply a catalog of LD_{50}s, organisms displaying resistance, the presence or absence of enzyme systems, and conservative guesses as to mechanisms but without supporting data. Of course, this kind of data is needed to document the occurrence and distribution of resistance, but if we are to understand the impact of this phenomenon in the total ecosystem, comparative details on mechanisms are essential.

The intent of this paper is to review mechanisms of resistance in non-target organisms with emphasis on microorganisms but inclusive of information on aquatic organisms and terrestrial vertebrates and

* Department of Entomology, University of Illinois, Urbana, Illinois, 61821.

invertebrates. Most of the resistance data on arthropod predators and parasites are intentionally omitted since this topic is surveyed in another paper (see Croft and Strickler, this volume). Information on laboratory mammals will be referred to only as the data clarify mechanisms of resistance in other non-target species. Several comparative examples are included from our own laboratories to add a current perspective and balance to the total review.

NON-TARGET MECHANISMS OF RESISTANCE

Microorganisms

Microbial activities are clearly some of the most important in eliminating pesticidal chemicals from the environment. These organisms are also just as capable of developing resistance as are higher animals and plants (Matsumura, 1974) due in part to the same types of detoxifying enzymes (Hodgson, 1974, 1976; Shishido, 1978).

Some microorganisms are rather unique in that they can utilize certain pesticides as their sole carbon source, something that eucaryotic organisms cannot do. *Trichoderma viride*, for example, can degrade dieldrin to CO_2 without additional carbon sources (Bixby et al., 1971). *Pseudomonas* sp. can utilize diflubenzuron as its only carbon source, incorporating as high as about 10,000 ppm diflubenzuron, but with very limited metabolism of the parent compound (Booth and Ferrell, 1977; Metcalf et al., 1975). The resistance of *Pseudomonas* to diflubenzuron apparently is related to its ability to absorb or adsorb the chemical without metabolism and without lethal effects; that is, *Pseudomonas* sp. may be naturally tolerant to this chemical just as snails are to DDT analogs (Hansen et al., 1971). However, *Plectonema boryanum*, a blue-green alga, is highly resistant to diflubenzuron mainly because of metabolism. In just one hour, 5 mg of algal cells can metabolize 80% of a 1 ppm solution of diflubenzuron primarily to *p*-chlorophenyl urea (Booth and Ferrell, 1977). Diflubenzuron is a reversible inhibitor of chitin synthesis in insects, probably by blocking chitin synthetase (Verloop and Ferrell, 1977). This novel mode of action could explain why most non-target organisms, including microorganisms, are resistant to this chemical (Booth, 1978).

Depending on the chemical, the majority of the papers on pesticide resistance in microorganisms relates to detoxification processes via hydrolysis, oxidation and reduction reactions (Williams, 1977; Matsumura, 1974). Chemical structure often determines which of these processes predominates. Subba-Rao and Alexander (1977) studied 37 analogs of DDT and found that those compounds that were unsubstituted and those with a single hydroxyl, amino or methoxy group in the para position were readily metabolized by soil organisms, whereas chemicals with hydroxyl, chlorine or methoxy groups in both para positions were resistant to biodegradation.

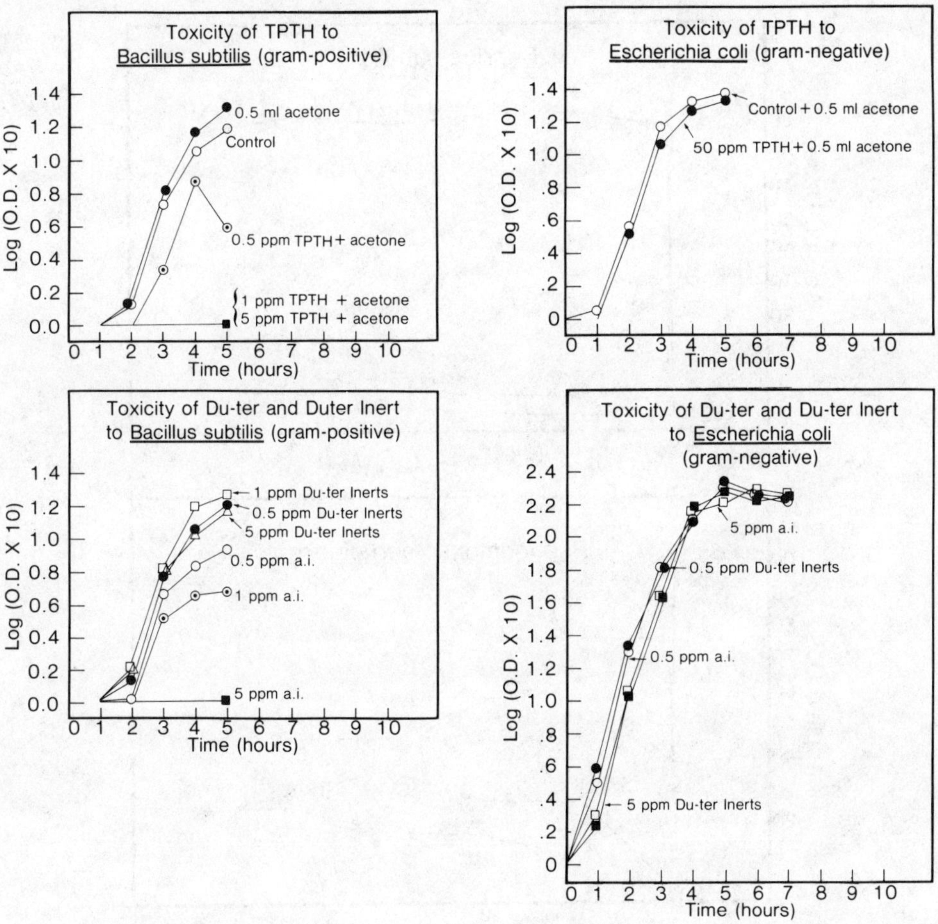

Figure 1. Toxicity of triphenyltin hydroxide (TPTH) and Du-ter to *Bacillus subtilis* and *Escherichia coli*.

What in some cases may appear to be a possible buildup of resistance by target organisms in nature may be a result of rapid microbial degradation of the compound to non-toxic levels, as seems to be the case for carbofuran (Williams et al., 1976). In addition, mixed microbial communities degrade organic chemicals more rapidly than freshly isolated members of the group. Known community resistance stratagems include co-metabolism, enzyme induction, transfer of metabolic plasmids, mutations generating new enzyme activities, and microbial interactions (Evans, 1977; Pemberton and Fisher, 1977).

Lindane apparently inhibits eucaryotic-microorganism growth, but bacteria are almost totally resistant to it, possibly due to

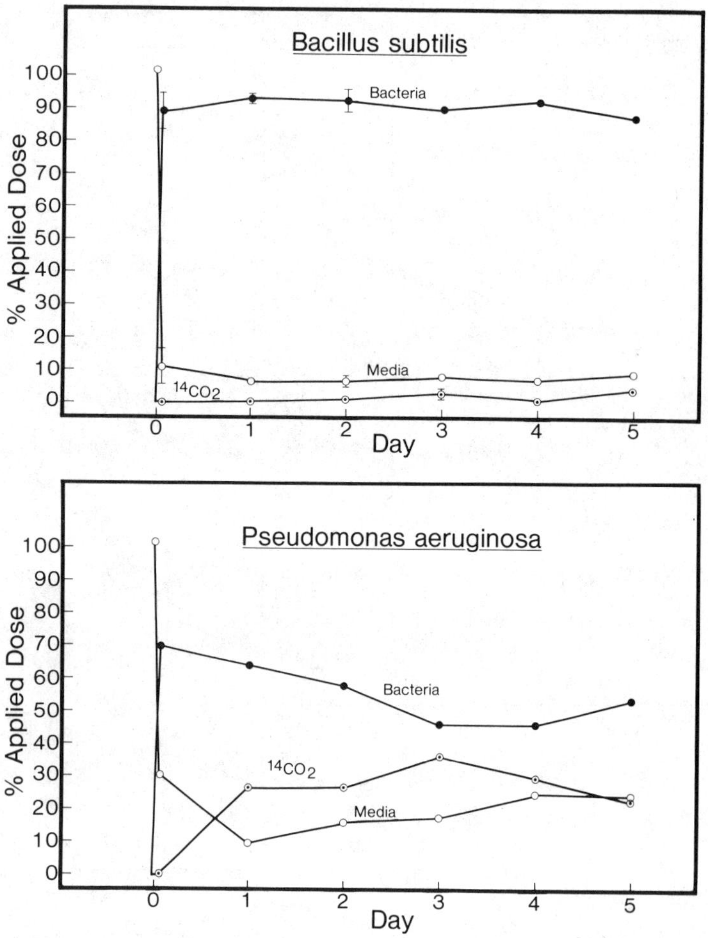

Figure 2. Uptake and elimination of ^{14}C-triphenyltin hydroxide and associated radiocarbon measured as percent applied dose in *Bacillus subtilis* and *Pseudomonas aeroginosa*.

significant differences in cellular membranes (Puiseux-Dao et al., 1977). The resistance of *Escherichia coli* to the fungicide dexon (p-dimethylaminobenzenediazosulfonate, Na-salt) is clearly related to an insensitive enzyme system: dexon acts directly on the flavin of NADH dehydrogenase and inhibits soluble NADH-cytochrome c-oxidoreductase and rotenone-insensitive NADH ubiquinone reductase, although the soluble NADH-oxidase from *E. coli* is not inhibited (Schewe et al., 1975).

Although *Bacillus subtilis* has been found to convert dinitrothion and EPN to amino derivatives (Miyamoto et al., 1966), this

Table 1. The Percent of Total ^{14}C-Triphenyltin Hydroxide in the Media and Cells of Two Species of Bacteria, *Bacillus subtilis* and *Pseudomonas aeruginosa*, Over a Five-day Incubation Period

Day	*B. subtilis*		*P. aeruginosa*	
	Media[a]	Bacterial cells	Media	Bacterial cells
0	8.80	3.30	26.00	23.67
1	3.75	21.00	38.75	6.10
2	21.75	18.03	1.00	5.25
3	14.25	38.10	0	1.60
4	21.33	29.08	0.70	9.00
5	21.75	29.13	2.10	8.00

[a]Values are mean percents of the total radiocarbon in the experiment.

bacterium is totally susceptible to other compounds. In fact, our laboratories became interested in studying mechanisms of microbial resistance to the fungicide triphenyltin hydroxide (TPTH) when we observed a remarkable resistance of gram-negative bacteria to TPTH and, conversely, a significant sensitivity of gram-positive bacteria such as *B. subtilis* (Figure 1). Note that Figure 1 shows that at least 1 ppm of TPTH or at least 5 ppm of its formulated product Du-ter totally inhibited the growth of *B. subtilis*, but concentrations of 50 ppm did not affect the growth of *E. coli*. This pattern of bacterial resistance and susceptibility to triaryltin compounds was consistent with the sensitivity of other gram-positive and gram-negative bacteria tested here and in other laboratories (Gruen, 1966).

Figure 2 suggests that TPTH resistance in gram-negative bacteria (in this case, *P. aeruginosa*) is due in part to increased metabolism. This bacterium absorbs approximately 70% of the residues, which drops to 54% by the fifth day, and concomitantly, it eliminates approximately 22% of the radiocarbon as $^{14}CO_2$. However, *B. subtilis* (gram-positive) immediately absorbs 90% of the total residues that remain with the organism over the five-day period with very little $^{14}CO_2$ given off. With the aid of recently developed analytical methodologies for TPTH (Wright et al., 1979), we have found very limited amounts of the parent compound in cells of *P. aeruginosa* but greater

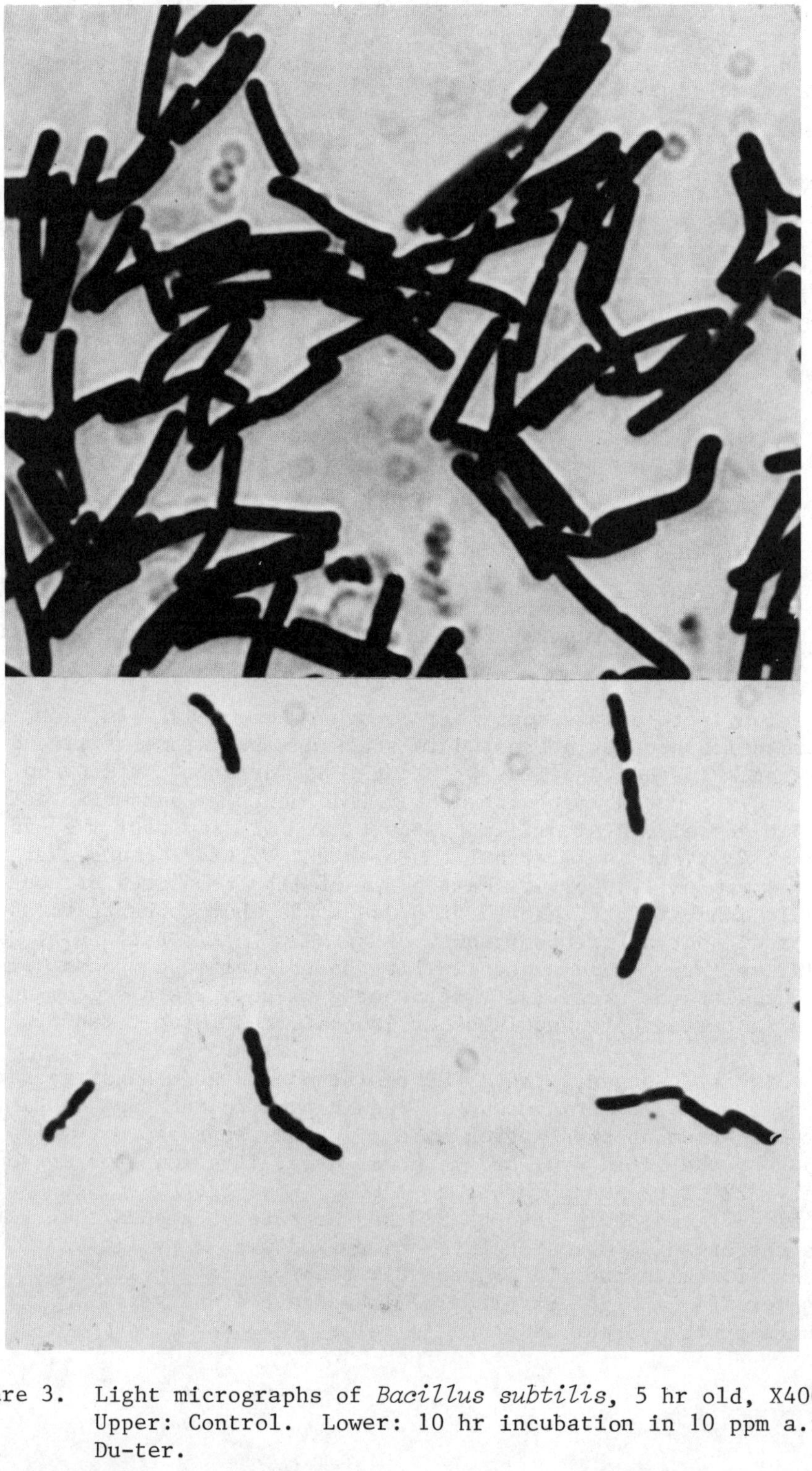

Figure 3. Light micrographs of *Bacillus subtilis*, 5 hr old, X400. Upper: Control. Lower: 10 hr incubation in 10 ppm a.i., Du-ter.

than 3x the amount of parent material in *B. subtilis* cells by day 5 (Table 1).

But are there morphological changes that also occur in these two TPTH-treated bacteria? At the light-microscope level, *B. subtilis* was clearly reduced in numbers, and the treatment resulted in some individuals with "wrinkled" cell walls (Figure 3, lower). No such changes were observed with the gram-negative species. Figure 4 (upper) shows transmission electron micrographs (TEM) of control *B. subtilis;* the triangular arrows indicate examples of normal beginning and final stages of fission in this species. The inset shows the cell wall and membrane typical of gram-positive bacteria. Figure 4 (lower) at 1 hr post-treatment with 10 ppm indicates the bacterium undergoes normal division (triangular arrows), although some individuals were obviously killed (thin arrow). The inset again shows the cell wall and membrane, which did not appear to be affected at this time interval of treatment, although several hours later there was a significant disruption of fission with the bacteria forming pleiomorphic bodies (triangular arrows) in conjunction with a thickening of the cell-wall cell-membrane area from about 35 nm (control) to 70 nm (Figure 5, inset). The apparent addition of cellular layers may be an unsuccessful survival adaptation of this species.

In contrast, TEM of the resistant bacterium *P. aeruginosa* clearly showed many individuals with disrupted membranes, but others were dividing normally and without the presence of pleiomorphic bodies (Figure 6). It appears that this species can sustain limited damage without changing the total population growth or mitotic divisions. This observation is often missed in resistance studies without the appropriate use of electron microscopic techniques.

TPTH has also been tested on numerous target and non-target fungi. Figure 7 summarizes the effect of TPTH on four species of fungi in liquid cultures. *Fusarium* was notably resistant up to 0.25 ppm TPTH, whereas in agar cultures this organism was clearly susceptible. *Fusarium* is one of the dimorphic fungi that grows with a mycelial form in agar but yeast-like form in liquid culture. Other authors have found that a closely related analog, triphenyltin acetate, was also less toxic to microorganisms in submerged liquid cultures (Barnes et al., 1971). Microorganisms of many kinds show 1-50x more resistance to dinitrocresol (DNOC) when in liquid culture, which is related undoubtedly to the mode of action of DNOC as an oxidative uncoupler (Tarkov et al., 1971). It is clear that information on the mode of action of TPTH and other organotin compounds is essential to understanding this type of resistance.

We know from previous work that Du-ter causes leaky membranes and depressed respiration in *Aspergillus niger* (Figure 8), and Figure 9 (top left) summarizes the inhibition kinetics of TPTH on

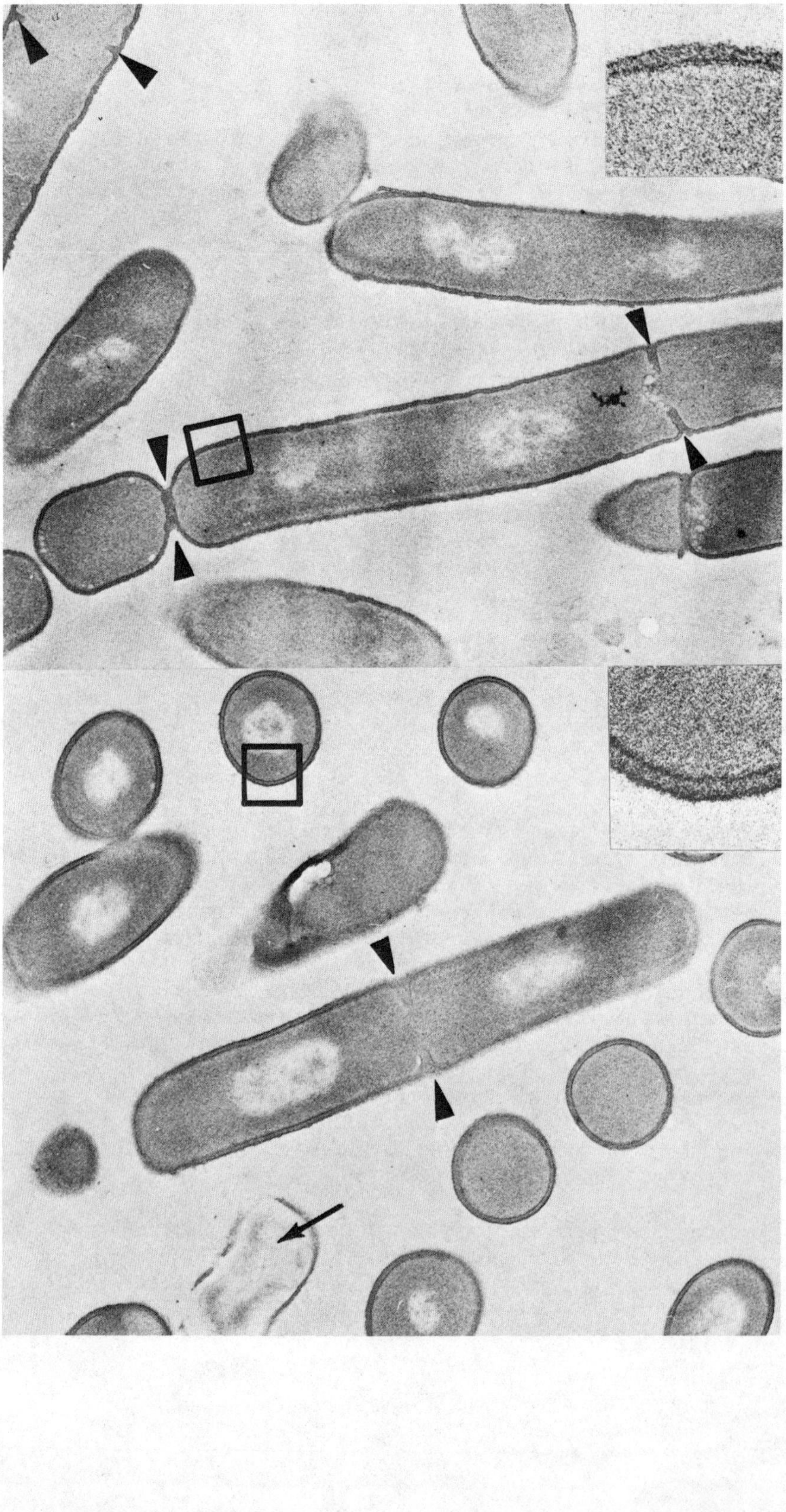

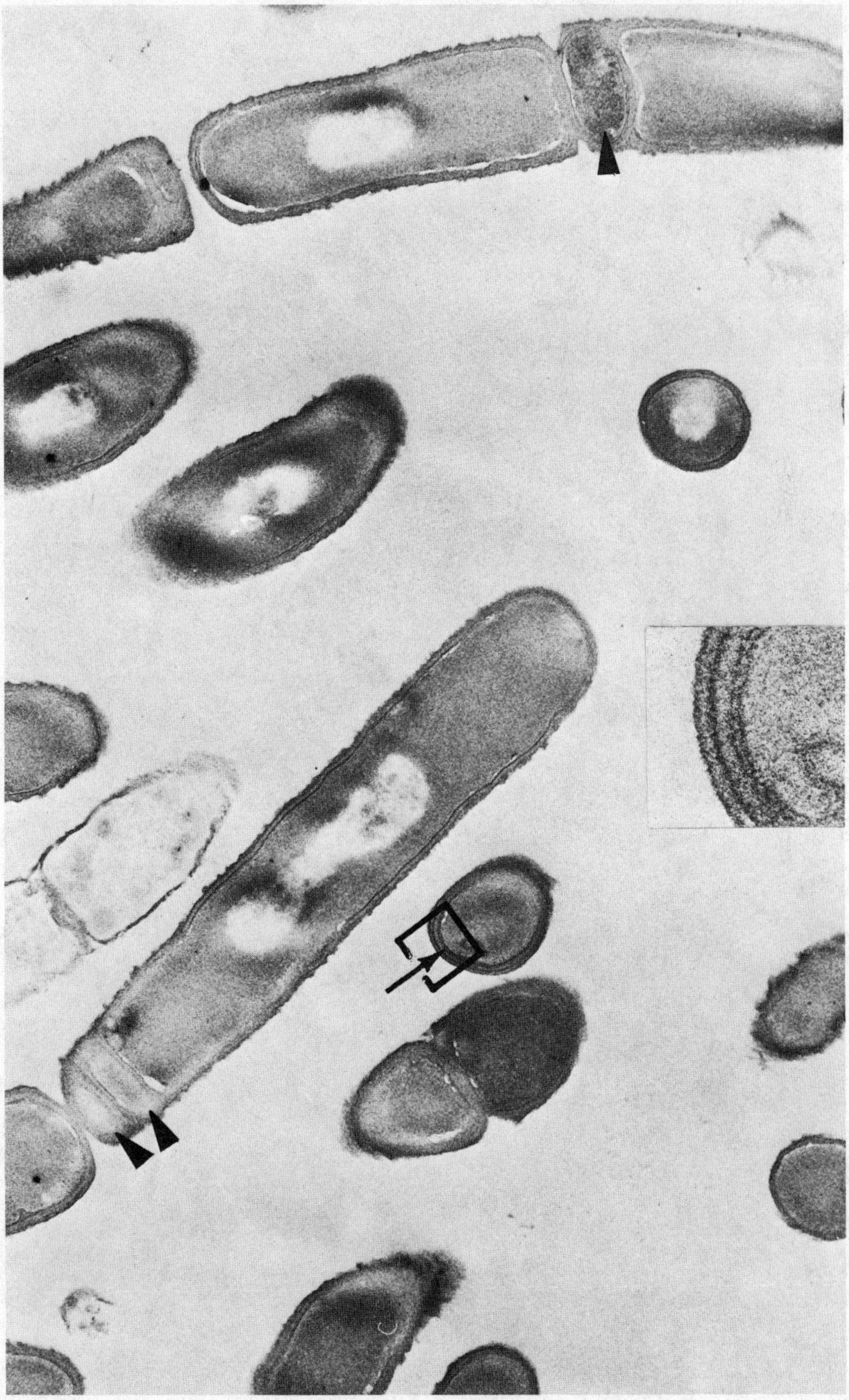

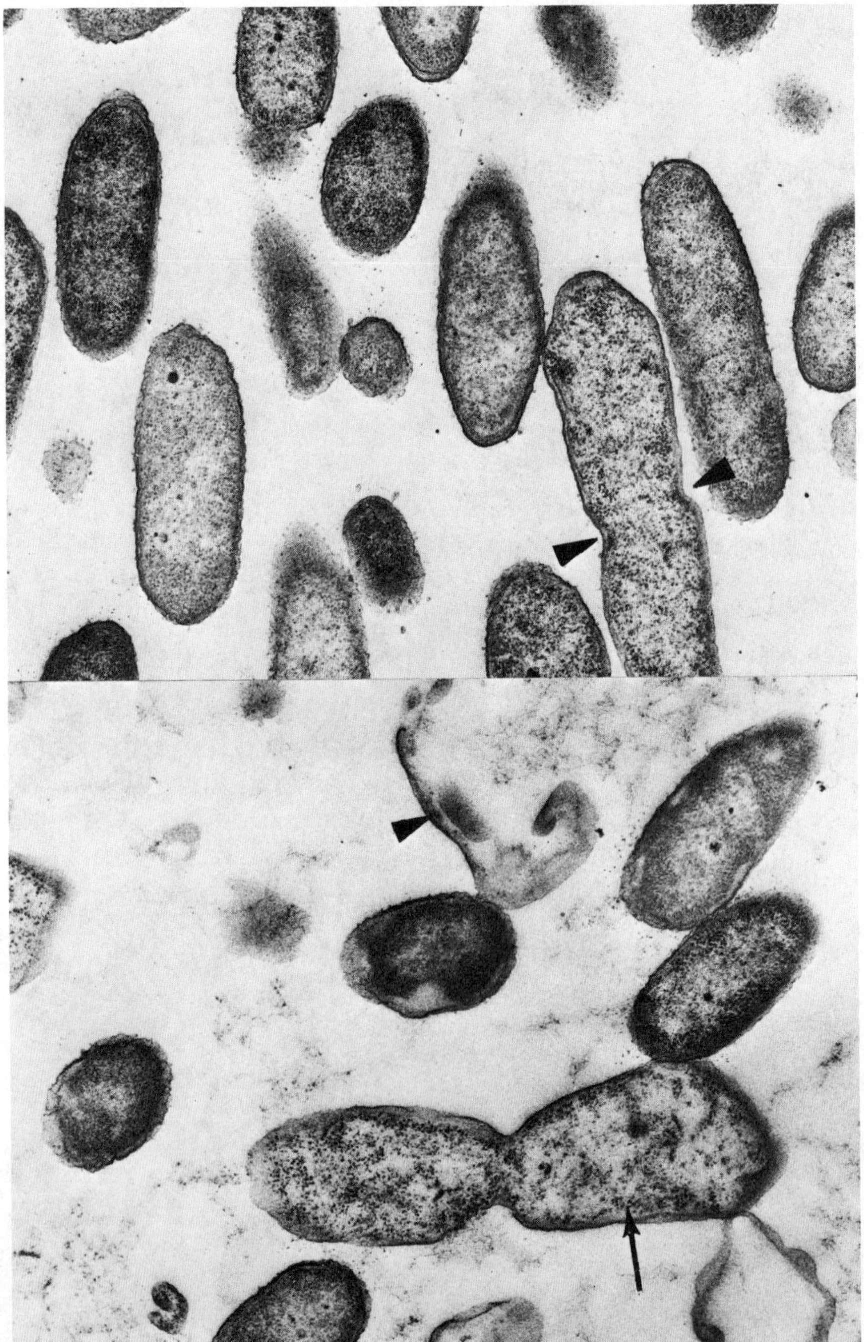

Figure 4. Page 394 — Transmission electron micrographs of *Bacillus subtilis*, X22,500. Upper: Control. Inset: Cell wall and membrane, X100,000. Lower: 1 hr incubation in 10 ppm triphenytin hydroxide. Inset: Cell wall and membranes, X100,000.

Figure 5. Page 395 — Transmission electron micrograph of *Bacillus subtilis* following a 10-hr incubation in 10 ppm a.i. Du-ter, X22,500. Inset: Cell wall and membranes, X100,000.

Figure 6. Page 396 — Transmission electron micrographs of *Pseudomonas aeruginosa*, X21,600. Upper: Control. Lower: 10 hr incubation in 10 ppm a.i. Du-ter.

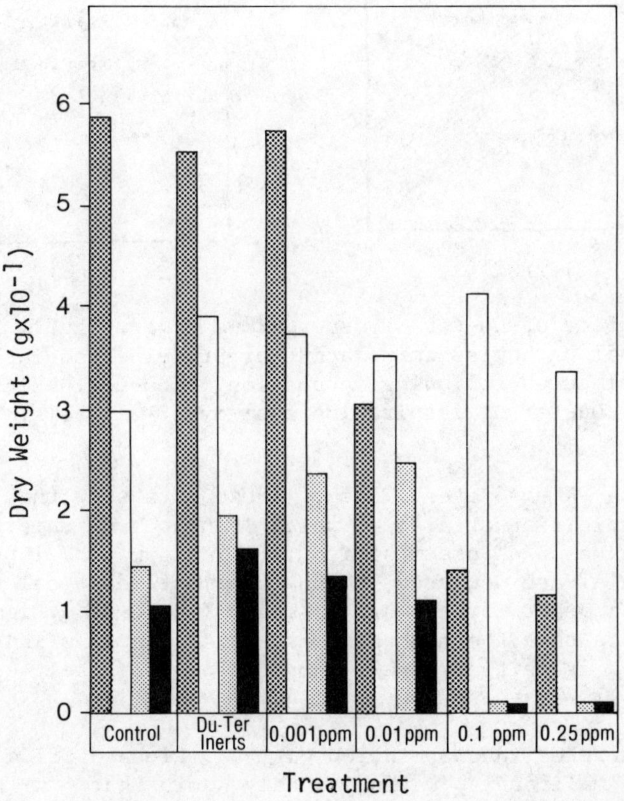

Figure 7. Toxicity of 0.001-0.25 ppm a.i. Du-ter and Du-ter inerts (10 ppm) to four species of fungi. Cross-hatched bars: *Aspergillus niger*. Open bars: *Fusarium oxysporum*. Striped bars: *Pythium debaryanum*. Black bars: *Trichoderma viride*.

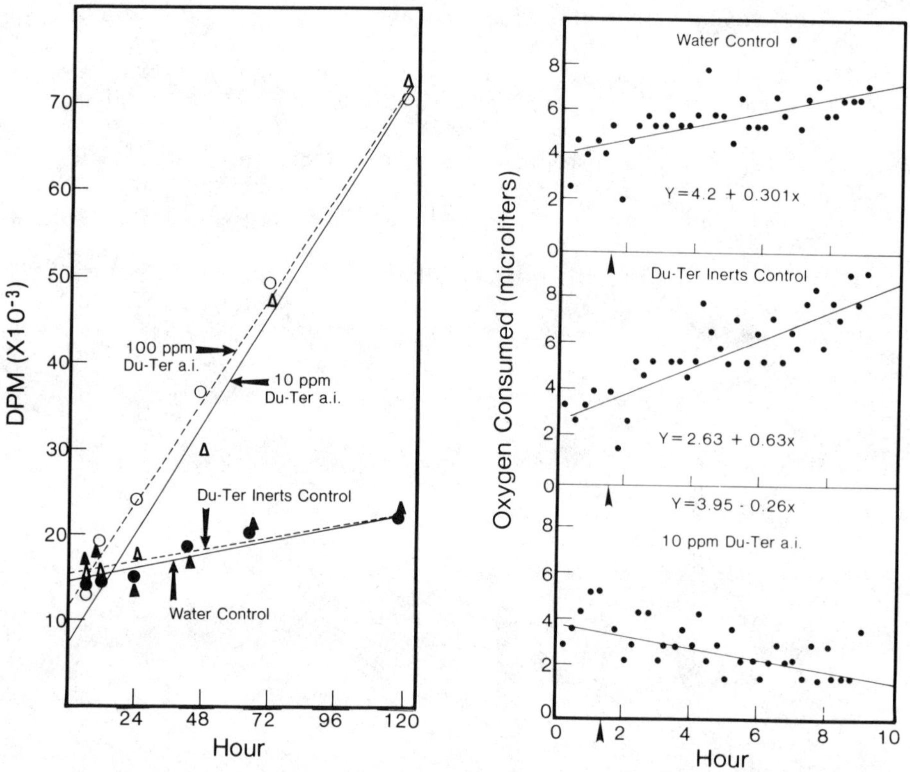

Figure 8. Effect of Du-ter on *Aspergillus niger*. Left: loss of radiocarbon as amino acids or protein-associated macromolecules following incubation in Du-ter. Right: effect of Du-ter on respiration from mycelial balls.

a mammalian Na^+-K^+-mediated ATPase. These data show that TPTH causes 67% inhibition at 14 ppm, which is near the inhibitory range of ouabain, a known ATPase inhibitor. At 30 ppm, 78% inhibition of the mammalian ATPase occurred, which is almost identical to the inhibition seen in yeast mitochondrial ATPase (Table 2). Figure 9 (top right) suggests that the mechanism of inhibition is mixed; that is, the inhibition is neither purely competitive nor purely non-competitive.

Data collected in cooperation with Dr. Richard Criddle at the University of California, Davis, clearly shows that ^{14}C-TPTH binds preferentially to the ninth subunit of the purified yeast ATPase cited in Table 2 (Figure 9, bottom). This subunit has a molecular weight of 7500 and appears to be the mitochondrial regulatory enzyme of this species.

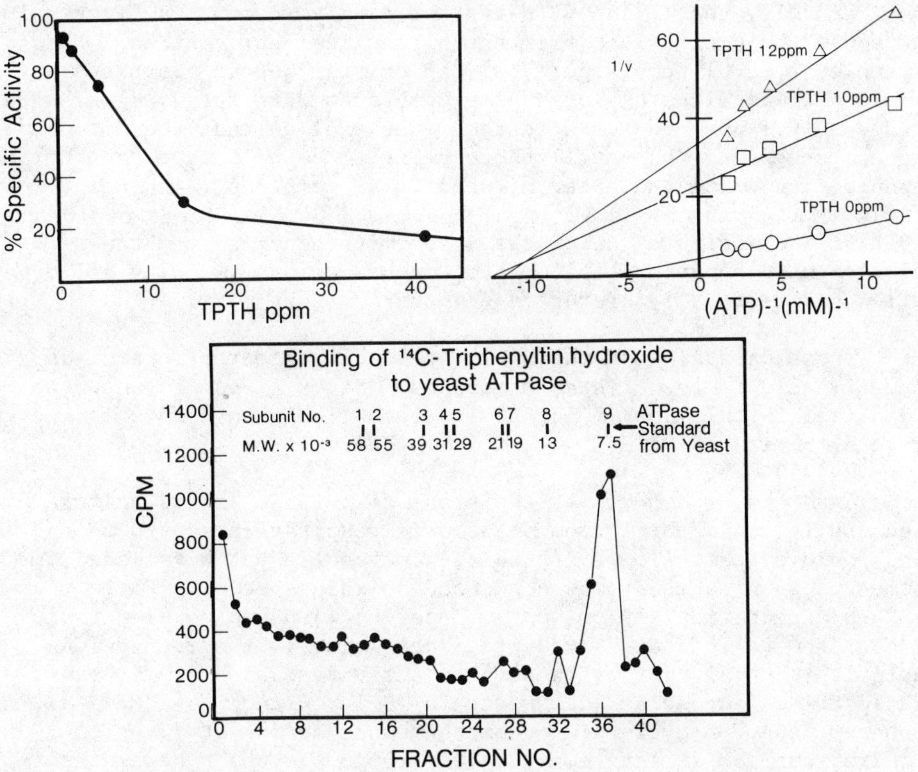

Figure 9. Effect of triphenyltin hydroxide (TPTH) on mammalian and yeast APTase. Upper figures: inhibition data on mammalian Na^{+}-K^{+}-mediated APTase. Lower: binding of ^{14}C-TPTH to yeast ATPase.

Table 2. Inhibition of Yeast Mitochondrial ATPase Activity by Triphenyltin Hydroxide (TPTH)

Treatment	Conc. (ppm)	ΔA_{340}/min	% Inhibition
control	--	0.68	--
ethanol	--	0.67	--
oligomycin	20	0.11	84
TPTH	30	0.16	76

Finally, the TEM of TPTH-treated *A. niger* shown in Figures 10-12 supports the conclusions from the biochemical information. For example, at 0.01 ppm (Figure 10) the cristae become distorted and swollen while all other organelle membranes were not visually affected. At 0.1 ppm, the cristae become so swollen that the mitochondria appear spherical, but still there seems to be no damage to other membranes, including the outer mitochondrial membrane (Figure 11, top). Figures 11 (bottom) and 12 (top) show a range of effects of 10 ppm TPTH from disintegration of the inner mitochondrial membranes to swollen mitochondria. At 100 ppm, membranes of virtually all organelles become totally disrupted (Figure 12, bottom).

These data suggest that TPTH at lower concentrations binds specifically to the cristae membranes. Since ATPase is known to be localized in the inner membrane it seems likely that this is the site of action of TPTH.

Fusarium oxysporum was resistant to TPTH in liquid culture because it could shift from an aerobic respiration system to an anaerobic system. To verify this, a metabolic pathways assay, summarized in Table 3, was conducted on this species. Following a 24-hr incubation, *F. oxysporum* apparently showed no detectable $^{14}CO_2$ from radiolabeled pyruvate, suggesting little if any TCA activity; but significant anaerobic activity was observed using radioactive glucose as the substrate (G-1-^{14}C and G-6-^{14}C data). However, the TCA cycle did become somewhat active at 72 hr of growth, but not as active as the anaerobic system. The agar culture appeared to have much greater TCA activity. When the 72 hr liquid culture was treated with 10 ppm TPTH, the activity of the TCA pathway decreased by 11X (e.g., 0.196 to 0.018), and the G-6/G-1 ratio dropped from 0.341 to 0.116 nmoles $^{14}CO_2$/min/mg protein, thereby modifying anaerobic energy production probably through stimulation of the pentose phosphate shunt. The following three possible survival mechanisms support these findings:

1. The reducing power of NADPH, which comes from the pentose shunt, is needed for the biosynthesis of cell wall constituents that repair membranes.

2. NADPH and NAD $\xrightleftharpoons{\text{PNT}}$ NADP and NADH $\rightarrow$ ETS. This reaction could still contribute protons to the ETS, however damaged it might be.

3. ATP could be produced by substrate phosphorylation via glycolysis and NADP could be used to stimulate the Entner-Duodoroff pathway for glucose degradation.

It is not known if the resistance mechanisms cited above for gram-negative bacteria and *Fusarium oxysporum* are natural or induced. Additional work is needed to clarify this matter.

Table 3. Metabolic Pathways Summary for the Utilization of Pyruvate-2-^{14}C (Pyr-2-^{14}C), Glucose-1-^{14}C (G-1-^{14}C) and Glucose-6-^{14}C (G-6-^{14}C) by *Fusarium oxysporum* in Agar or Liquid Matrix

Treatment	Matrix	Time	nmoles $^{14}CO_2$/min/mg protein ± S.D.			
			Pyr-2-^{14}C	G-1-^{14}C	G-6-^{14}C	G-6/G-1
0	Liquid	24 hr	0	3.480 ± 0.049	0.452 ± 0.038	0.130
0	Liquid	72 hr	0.196 ± 0.054	6.636 ± 0.599	2.261 ± 0.127	0.341
10 ppm TPTH	Liquid	72 hr	0.018 ± 0.020	2.740 ± 0.380	0.320 ± 0.034	0.116
0	Liquid	96 hr	0.176 ± 0.072	7.307 ± 0.358	1.919 ± 0.162	0.263
0	Agar	9 days	0.815 ± 0.084	1.338 ± 0.316	0.238 ± 0.042	0.178

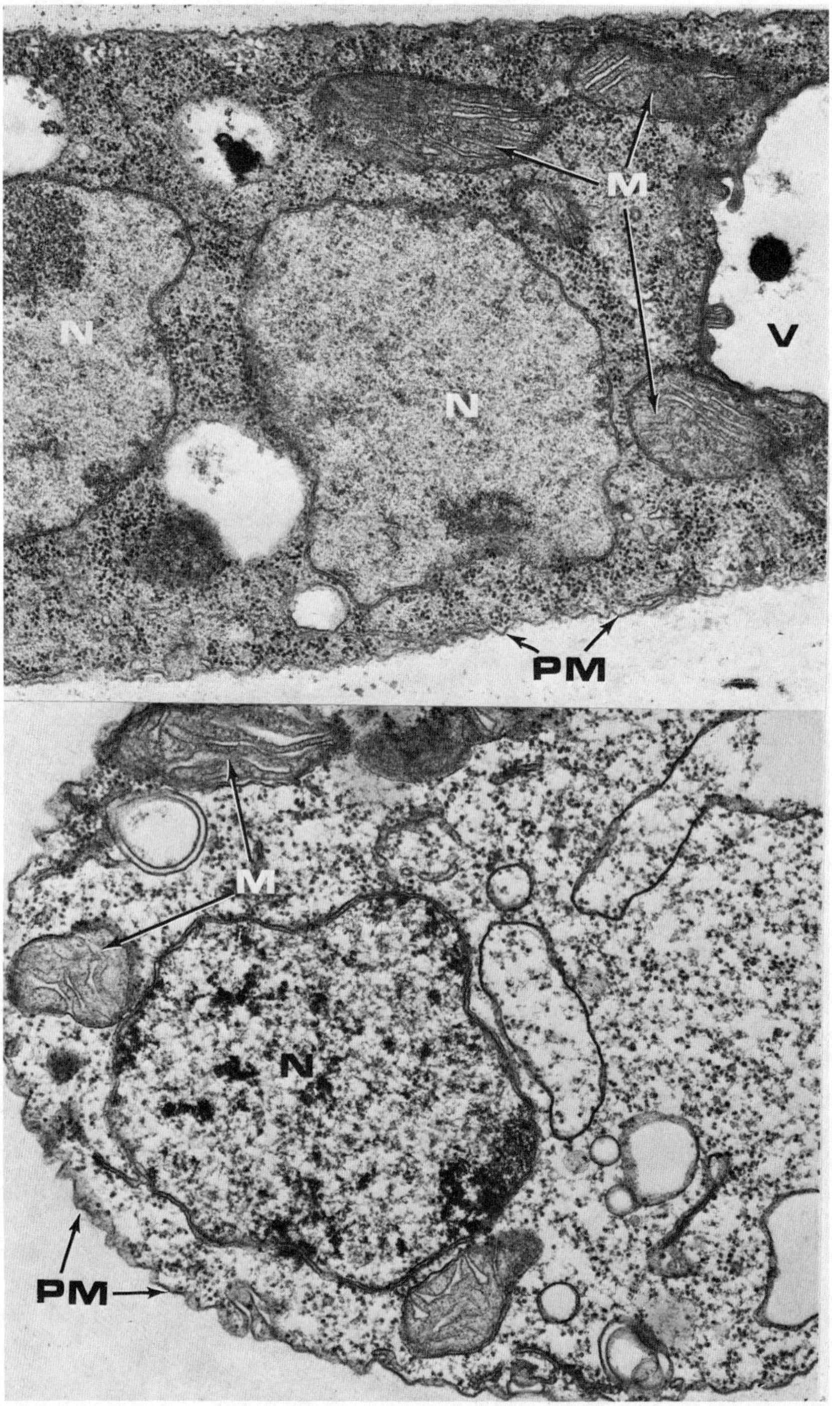
M
N
N
V
PM
M
N
PM

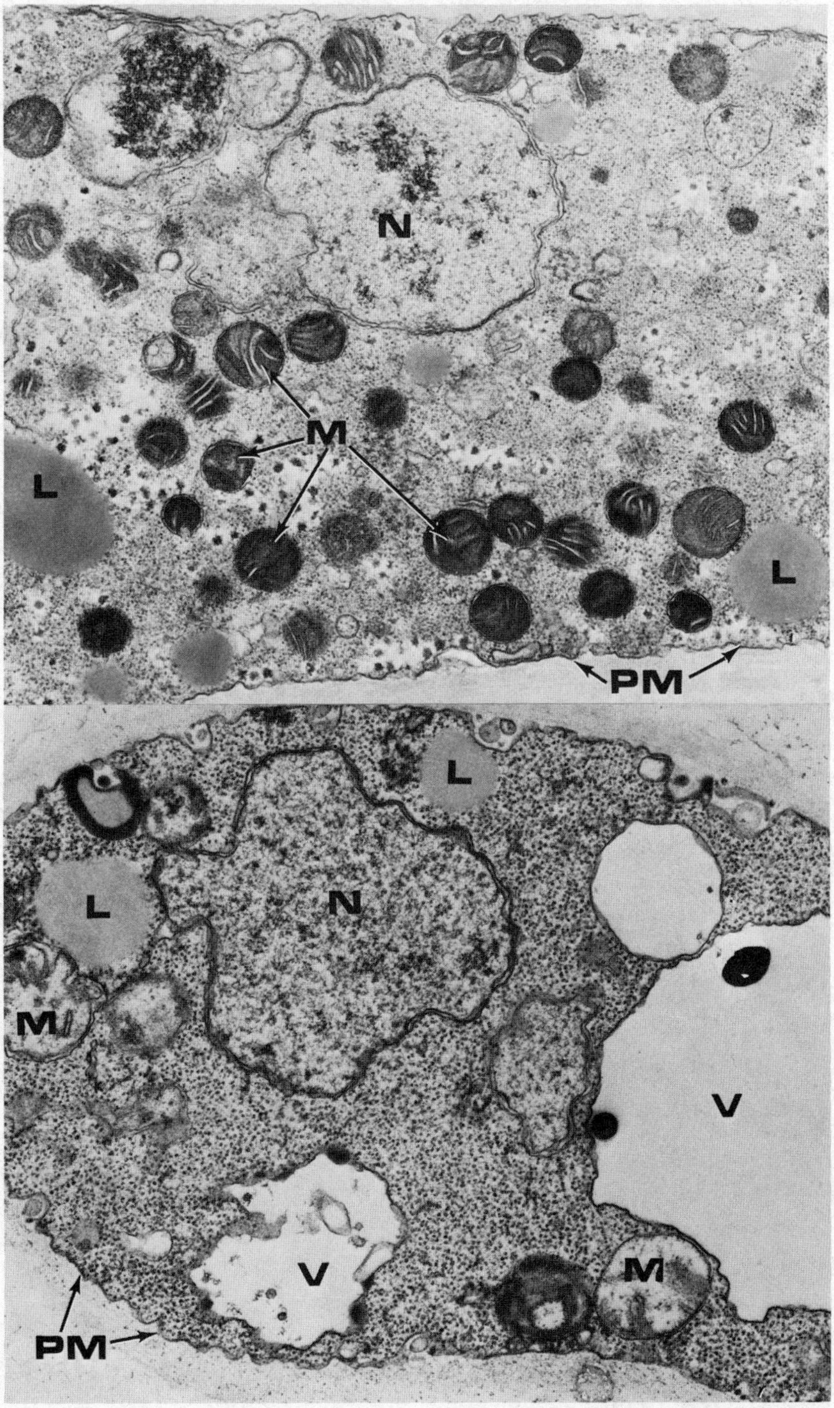
N
M
L
L
PM
L
L
N
M
V
V
M
PM

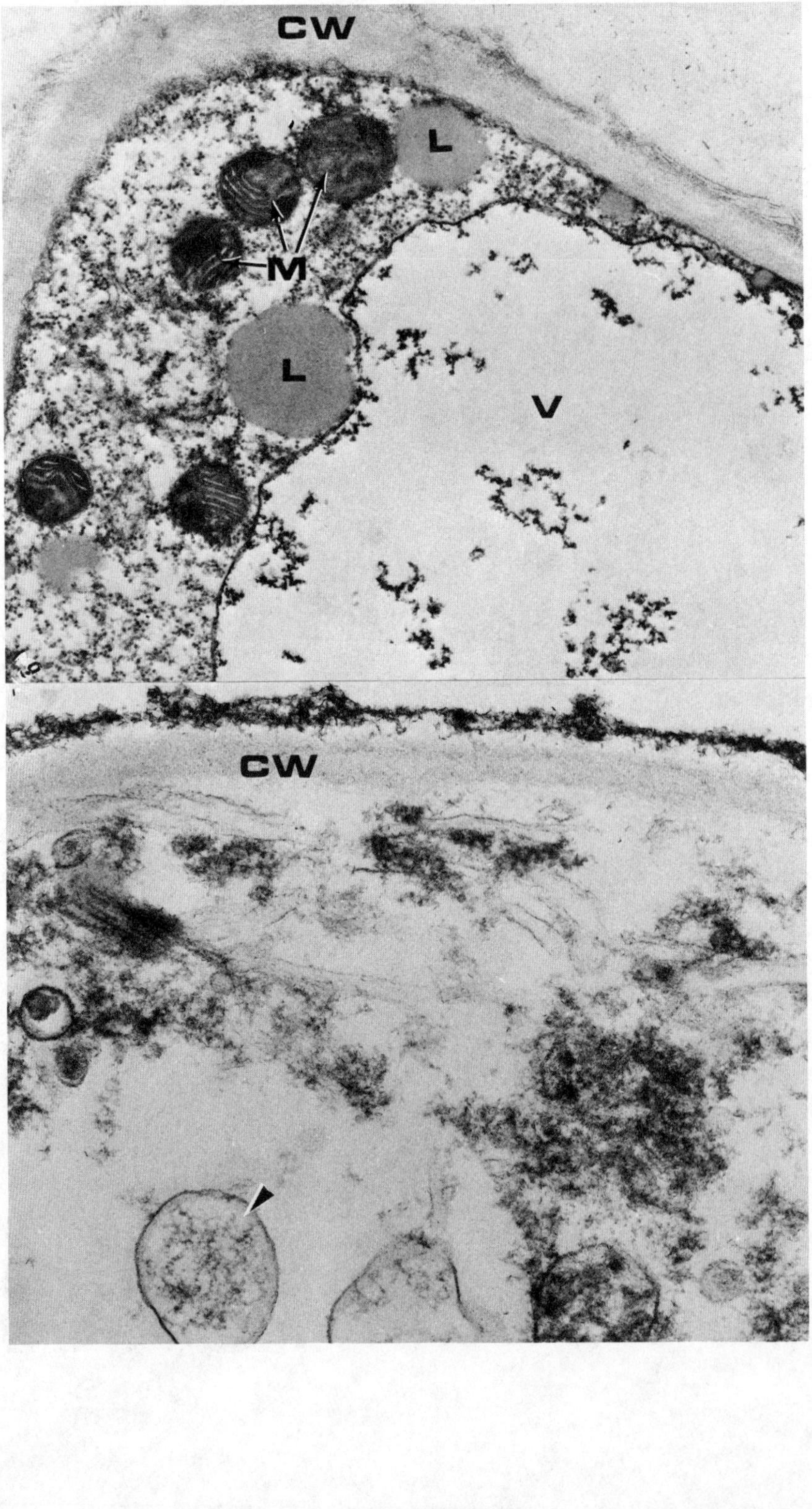
CW
L
M
L
V
CW

Aquatic organisms

Vertebrates. The most widely documented resistant aquatic vertebrate is the mosquito fish, *Gambusia affinis*. The environmental behavior of a variety of chlorinated hydrocarbon insecticides has been studied in this species, and the data generally show multiple resistance factors, including avoidance behavior (Kynard, 1974), less retention of parent compounds, membrane barriers, blood-brain barriers, increased tissue and body fat, structural differences in myelin (Yarbrough, 1974), and, related to the first factor, greater *O*-dealkylation of DDT-type compounds (Hansen et al., 1971).

The resistance mechanisms of mosquito fish to 2,4-D seem less clear, but decreased activation is a contributing factor (Chambers et al., 1977). Environmentally induced tolerance of *Gambusia* to parathion is the result of increased levels of mixed-function oxidases (MFO) that dearylate parathion (Chambers and Yarbrough, 1973).

It becomes increasingly clear in the literature that other than for *Gambusia*, most fish have only 11-40% of the mammalian hydrolytic capability used to metabolize organophosphates (Khan et al., 1977). Therefore, numerous examples of resistance in aquatic vertebrates are not likely to be found, but compounds with unique modes of action such as chitin-synthetase inhibitors generally are non-toxic to fish (Booth and Ferrell, 1977).

Figure 10. Page 402 — Transmission electron micrographs of hyphal cells from *Aspergillus niger*, X28,000. Upper: control. Lower: hyphal cell following 6 hr incubation in 0.01 ppm a.i. Du-ter. (M = mitochondria; V = vacuole; N = nucleus; PM = plasma membrane).

Figure 11. Page 403 — Transmission electron micrographs of hyphal cells from *Aspergillus niger* previously incubated with Du-ter. Upper: thin section following a 6-hr incubation with 0.1 ppm a.i. Du-ter, X16,800. Lower: thin section following a 6-hr incubation with 10 ppm a.i. Du-ter, X21,600. (N = nucleus; L = lipid body; M = mitochondria; V = vacuole; PM = plasma membrane).

Figure 12. Page 404 — Transmission electron micrographs of hyphal cells from *Aspergillus niger* previously incubated with Du-ter. Upper: thin section following a 6-hr incubation with 10 ppm a.i. Du-ter, X21,600. Lower: thin section following a 6-hr incubation with 100 ppm a.i. Du-ter, X36,480. Arrow: unidentified organelle. (CW = cell wall; M = mitochondria; L = lipid body; V = vacuole).

Invertebrates. Johnson et al. (1971) studied the in vivo metabolism of DDT in six non-target freshwater species and found the mayfly, *Hexagenia bilineate*, able to detoxify to DDE 85% of a given dose of DDT, which was 20% higher than any other organism examined. Apparently DDT dehydrochlorinase contributes to DDT resistance in mayflies.

In general, marine fish possess significantly lower levels of MFO than freshwater fish. Perhaps related to the low MFO levels is the greater susceptibility of marine fish to organophosphorus pesticides compared to freshwater fish (Kagoshima, 1977). Also, although MFO activity has been found in most aquatic animals, the levels are generally from 1.4x (blue gill fry) to 335x (lobster) lower than the mouse (Khan et al., 1977).

As previously mentioned, the snail *Physa* seems to be naturally tolerant to DDT-type analogs (Hansen et al., 1971). In the snail, uptake and metabolism of DDT and associated analogs proceed relatively slowly with no observable harm to the organism.

Birger and Malyarevskaya (1977) showed a direct correlation between the resistance of some mollusks and crustaceans to DDT and their ability to change their metabolism from aerobic to anaerobic. However, this was reported only under reduced oxygen tension.

A review of the literature leaves little doubt that resistance mechanisms in non-target aquatic vertebrates and invertebrates are poorly documented; more work in this area is clearly needed.

Terrestrial Organisms

Vertebrates. Most of the published resistance literature deals either with pests or laboratory animals, or comparative studies of both. Other than laboratory mice and rats, very little resistance mechanism data are available on vertebrates.

However, the wild pine vole (*Microtus pinetorum*) is considered by some investigators to be non-target, and it has been shown to be resistant to endrin. A marked species difference in microsomal MFO activity was noted between pine voles and white laboratory mice. Microsomal activity of endrin-treated white mice could be greatly induced relative to basal levels, but in pine voles the induction seemed to be a function of the substrate used (Hartgrove et al., 1977). The data caution investigators against extrapolating from laboratory animals to wild mammals.

Invertebrates. Non-target insects seem to possess many of the MFO enzymes that the pest species have, although it is questionable whether or not these enzymes contribute totally to insecticide resistance (Brattsten and Metcalf, 1970).

Leger and Millette (1977) reported resistance of two species of earthworms to as high as 800 ppm of captan. Since no residues were found in the worms, it might be inferred that these organisms metabolize and eliminate the compound; however, details on the mechanism of resistance were not included.

One final note on non-target resistance: Some investigators believe that resistance will probably develop up through the trophic levels. In other words, mosquitoes develop resistance, then mosquito fish, etc. However, empirical resistance data are generally lacking for most food chains. This offers a fruitful area of research in the future.

REFERENCES

Barnes, R. D., Jr., Bull, A. T., Jr., and Poller, R. C., Jr., 1971, Behavior of triphenyltin acetate in soil, *Chem. Ind. (London)*, 7:204.

Birger, T. I., and Malyarevskaya, A. Ya., 1977, Some biochemical mechanisms of resistance of aquatic invertebrates to toxicants, *Gidrobiol. Zh.*, 13(6):69.

Bixby, M., Boush, G. M., and Matsumura, F., 1971, Degradation of dieldrin to carbon dioxide by the soil fungus, *Trichoderma koningi*, *Bull. Environ. Contam. Toxicol.*, 6:491.

Booth, G. M., 1978, Dimilin and the environment, *in:* "Dimilin: Breakthrough in Pest Control," p. 5, Agri-fieldman and Consultant.

Booth, G. M., and Ferrell, D., 1977, Degradation of dimilin by aquatic foodwebs, *in:* "Pesticides in Aquatic Environments," M. A. Q. Khan, ed., pp. 221-243, Plenum Press, New York.

Brattsten, L. B., and Metcalf, R. L., 1970, The synergistic ratio of carbaryl with piperonyl butoxide as an indicator of the distribution of multi-function oxidases in the Insecta, *J. Econ. Entomol.*, 36:101.

Chambers, H., Dziuk, L. J., and Watkins, J., 1977, Hydrolytic activation and detoxication of 2,4-D acid esters in mosquito fish, *Pestic. Biochem. Physiol.*, 7(3):297.

Chambers, J. E., and Yarbrough, J. D., 1973, Organophosphate degradation by insecticide-resistant and susceptible populations of mosquito fish (*Gambusia affinia*), *Pestic. Biochem. Physiol.*, 3(3):312.

Evans, W. C., 1977, Biochemistry of the bacterial catabolism of aromatic compounds in anaerobic environments, *Nature*, 270(5632): 312.

Gruen, L., 1966, Trialkyl (aryl) tin toxicity to bacteria, *Gesundheitsw. Disinfex.*, 58:81 (Chem. Abs., 1965, 66, 610n).

Hansen, L. G., Kapoor, I. P., and Metcalf, R. L., 1971, Biochemistry of selective toxicity and biodegradability: Comparative *O*-dealkylation by aquatic organisms, *Comp. Pharm. Toxicol.*, 3(11):339.

Hartgrove, R. W., Jr., Hundley, S. G., and Webb, R. E., 1977, Characterization of the hepatic mixed-function oxidase system in endrin-resistant and susceptible pine voles, *Pestic. Biochem. Physiol.*, 7(2):146.

Hodgson, E., 1974, Comparative studies of cytochrome P-450 and its interaction with pesticides, *in:* "Survival in Toxic Environments," M. A. Q. Khan and J. P. Bederka, eds., pp. 213-260, Academic Press, Inc., New York.

Hodgson, E., 1976, Comparative toxicology: Cytochrome P-450 and mixed-function oxidase activity in target and non-target organisms, *in:* "Essays in Toxicology Volume 7," W. J. Hayes, Jr., ed., pp. 73-97, Academic Press, Inc., New York.

Johnson, B. T., Saunders, and Saunders, H. O., 1971, Metabolism of DDT by fresh water invertegrates, *J. Fish. Res. Bd. Can.*, 28:705.

Kagoshima, J., 1977, Effect of river and sea pollution from organophosphorus pesticides on marine animals, *Nippon Suisan Hogo Kyoka; Geppo.*, 151:4.

Khan, M. A. Q., Korte, F., and Payne, J. F., 1977, Metabolism of pesticides by aquatic animals, *in:* "Pesticides in the Aquatic Environments," M. A. Q. Khan, ed., Plenum Press, New York.

Kynard, B., 1974, Avoidance behavior of insecticide susceptible and resistant populations of mosquito fish to four insecticides, *Trans. Am. Fish. Soc.*, 103(3):557.

Leger, R. G., and Millette, G. J. F., 1977, Resistance of earthworms *Lumbricus terrestris* and *Allolobophora turgida* to captan 50 W. P., *Rev. Can. Biol.*, 36(4):351.

Matsumura, F., 1974, Microbial degradation of pesticides, *in:* "Survival in Toxic Environments," M. A. Q. Khan and J. P. Bederka, Jr., eds., pp. 129-154, Academic Press, Inc., New York.

Metcalf, R. L., Po-Yung, L., and S. Bowlus, 1975, Degradation and environmental fate of 1-(2,6-difluorobenzoyl)-3-(4-chlorophenyl) urea, *J. Agric. Fd. Chem.*, 23(3):359.

Miyamoto, J., Kitagawa, K., and Sato, Y., 1966, Reductive metabolism of organophosphorus compounds by microbial organisms, *Jap. J. Expt. Med.*, 36:211.

Pemberton, J. M., and Fisher, P. R., 1977, 2,4-D plasmids and persistence, *Nature*, 268(5622):72.

Puiseux-Dao, S., Jeanne-Levain, N., Roux, F., Ribier, J., Borghi, H., and Brun, C., 1977, Analysis of the effects of the organochlorine insecticide lindane at the cellular level, *Protoplasma*, 91(3):325.

Schewe, T., Hiebsch, C., and Halangk, W., 1975, The action of the systemic fungicide dexon, on several NADH dehydrogenases, *Acta Biol. Med. Ger.*, 34(11-12):1767.

Shishido, T., 1978, The role of glutathione S-transferases in pesticide metabolism, *Nippon Noyaku Gakkaish (J. Pestic. Sci.)*, 3(suppl.):465.

Subba, Rao, R. V., and Alexander, M., 1977, Effect of chemical structure on the biodegradability of 1,1,1-trichloro-2,2-bis (p-chlorophenyl) ethane (DDT), *J. Agric. Fd. Chem.*, 25(2):327.

Tarkov, M. I., Merenyu, G. V., and Timchenko, L. A., 1971, The action of dinitro-ortho-cresol on the growth of certain saprophytic and pathogenic microorganisms, *Gig. Sanit.*, 36(3):57.

Verloop, A., and Ferrell, C. D., 1977, Benzoylphenyl ureas -- A new group of larvicides interfering with chitin deposition, *in:* "ACS Symposium Series, No 37, Pesticide Chemistry in the 20th Century," J. R. Plimmer, ed., pp. 237-270, American Chemical Society, Washington, D.C.

Williams, I. H., Pepin, H. S., and Brown, M. J., 1976, Degradation of carbofuran by soil microorganisms, *Bull. Environ. Contam. Toxicol.*, 15(2):244.

Williams, P. P., 1977, Metabolism of synthetic organic pesticides by anaerobic microorganisms, *Res. Rev.*, 66:63.

Wright, B. W., Lee, M. L., and Booth, G. M., 1979, Determination of triphenyltin hydroxide derivatives by capillary GC and tin-selective FPD, *J. High Resol. Chrom. and Chrom. Comm.*, 2(4):189.

Yarbrough, J. D., 1974, Insecticide resistance in vertebrates, *in:* "Survival in Toxic Environments," M. A. Q. Khan and J. P. Bederka, Jr., eds., pp. 373-398, Academic Press, Inc., New York.

PATTERNS OF CROSS RESISTANCE TO INSECTICIDES IN THE HOUSE FLY IN JAPAN

Akio Kudamatsu,* Akifumi Hayashi and Rokuro Kano

Department of Medical Zoology
Tokyo Medical and Dental University
Bunkyo-ku, Tokyo, Japan

INTRODUCTION

In Japan, resistance to organic insecticides in the house fly, *Musca domestica*, was first reported by Yasutomi (1956) and Tsukamoto et al. (1957) who confirmed that house flies in Hikone were not sufficiently controlled with DDT in 1954. Soon afterward, house flies resistant to lindane and dieldrin were collected from several areas where the chemicals had been sprayed often (Suzuki et al., 1958), and in 1960, diazinon-resistant house flies were found in some areas near Tokyo, but the cross resistance to dichlorvos and malathion was indistinct (Suzuki et al., 1961). Similar occurrences were observed by Oshima et al. (1963) and Yokohama and Ohtaki (1965) in Saitama Prefecture.

In 1972, surprisingly high resistance to all tested organophosphorus insecticides developed in house flies in Kanagawa Prefecture (Hayashi et al., 1973). Since then, the serious problem of resistance to organophosphorus insecticides has rapidly become widespread.

The development of resistance in the house fly under fenitrothion, fenthion and dichlorvos pressure at the dumping island (Yumenoshima) of Tokyo Bay for 15 years has been successively reported by Yasutomi (1966, 1973, 1975) and Yasutomi and Shudo (1978) as a typical case. The cross resistance to various insecticides in the Yumenoshima III strain was investigated and some of the results have been reported by Kudamatsu et al. (1977, 1979). Further

*Also of the Agricultural Chemicals Research Institute, Nihon Tokushu Noyaku Seizo Co., Ltd., Hino-shi, Tokyo.

information is presented here, including a review of results reported in the past.

MATERIALS AND METHODS

Susceptible (S) house flies of the Takatsuki and Indonesian strain were obtained from a culture originally supplied by the National Institute of Health and by Tokyo Medical and Dental University, respectively. In 1975, the multi-resistant house flies "Yumenoshima III" were collected at the dumping island in Tokyo Bay where full coverage with organophosphorus insecticide sprays was being performed almost daily. They have been maintained at the Institute of Nihon Tokushu Noyaku Seizo Co., Ltd.

The average body weight of a female house fly in the Takatsuki strain was 20 mg, and 24 mg in the Yumenoshima strain. LD_{50} determinations on 3- to 5-day-old female house flies were based on mortality counts at 28°C 24 hr following topical application of the insecticidal solutions in 1 µl of acetone to individual flies under carbon dioxide anesthesia.

RESULTS AND DISCUSSION

Organochlorine Insecticides

High resistance to DDT and γ-BHC in the Yumenoshima III strain was found (Table 1), as reported by Yasutomi (1973); however, it is of great interest that plifenate (MEB 6046) was effective against both susceptible (S) and resistant (R) house flies. Furthermore, the addition of synergists such as chlorfenethol (DMC) and S-421 apparently increased the toxicity to house flies.

Organophosphorus Insecticides

The chemical structure and cross resistance relationships of organophosphorus insecticides in house flies have been studied since 1976. As shown in Table 2, the Yumenoshima III strain had a remarkably wide range of cross resistance to most of the organophsophorus insecticides. However, there were some interesting compounds showing resistance ratios of less than five-fold. These included a cyclic phosphorothionate (salithion), an ethylphosphonothionate (trichloronate), and phosphorothiolothionates and thiolates (prothiophos, sulprophos, profenofos and methidathion). The reasons for this lack of cross resistance to these insecticides have not been clarified.

Table 1. LD_{50} Values of Organochlorine Insecticides in Multi-resistant and Susceptible House Flies

Insecticide	LD_{50} µg/♀ fly		
	Susceptible (I)	Susceptible (S)	Resistant (R)
p,p'-DDT	0.43[a]	15.8	> 100
γ-BHC	0.01[a]	1.4	> 100
plifenate[b]	1.3	2.7	6.3
plifenate + DMC (1:1)		1.5	3.5
plifenate + S-421 (1:1)		1.9	2.8

(I), Indonesian strain; (S), Takatsuki strain; (R), Yumenoshima III strain.
[a]Data were cited from Hayashi et al. (1974).
[b]2,2,2-trichloro-1-(3,4-dichlorophenyl)-ethanol acetate.

Carbamate Insecticides

In Japan, carbamate insecticides were not introduced for house fly control but for ectoparasite control in domestic animals because the effectiveness of most carbamate insecticides against house flies was weak compared to that of several marketed organophosphorus compounds.

As shown in Table 3, the Yumenoshima III strain was cross resistant to all carbamate insecticides except methomyl, an oxime-carbamate insecticide. Piperonyl butoxide, safroxan and S-421, well known as synergists with pyrethroids, have proved that they are still impressively synergistic with carbamate insecticides for the R house fly, and Matsubara et al. (1964) previously reported that carbaryl synergized with safroxan or piperonyl butoxide has an excellent effect on diazinon-R house flies.

Pyrethroids

House flies that have developed high resistance to organochlorine, organophosphorus and carbamate insecticides are still susceptible to synthetic pyrethroids, as shown in Table 4. Additionally, the cross resistance to synthetic pyrethroids in the house fly has been studied on several occasions, and significant differences in

Table 2. LD_{50} Values and Resistance Ratios of Various Organophosphorus Insecticides in the Multi-resistant House Fly (Yumenoshima III Strain)

Common name	LD_{50} µg/♀ fly Susceptible (S)	LD_{50} µg/♀ fly Resistant (R)	Resistance ratio (R/S)
phosphorothionates			
bromophos	0.12	48.0	400
chlorpyrifos methyl	0.075	9.8	131
chlorpyrifos	0.095	5.5	58
coumaphos	0.11	> 100	> 909
cyanophos	0.08	101	1263
diazinon	0.15	15.9	106
fenchlorphos	0.18	16.5	92
fenitrothion	0.05	62.4	1248
fenthion	0.04	9.7	243
methyl parathion	0.03	35.0	1166
parathion	0.04	14.0	350
phoxim	0.034	25.0	735
salithion	0.07	0.24	3.3
sulfotepp	0.17	18.5	109
phosphonothionates and phosphonates			
cyanofenphos	0.036	170	4772
EPN	0.056	8.0	143
leptophos	0.26	> 100	> 417

Common name	LD_{50} µg/♀ fly		Resistance ratio (R/S)
	Susceptible (S)	Resistant (R)	
trichlornate	0.62	1.43	2.4
trichlorfon	0.32	> 100	> 313
phosphoro-thiolothionates and thiolates			
dimethoate	0.021	0.31	15
disulfoton	0.8	8.1	10
ethion	0.74	55.0	74
malathion	0.56	> 200	> 358
mecarbam	0.11	1.40	13
methidathion	0.35	1.6	4.6
phenthoate	0.067	> 100	> 1493
phosalone	0.58	49.0	85
phosmet	0.23	2.1	9.1
prothiophos	0.36	0.66	1.8
sulprophos	0.38	0.84	2.2
demeton	1.06	15.3	14
methyl demeton	1.1	30.5	29
profenofos	0.31	0.52	1.7
phosphoroamido-thionates and thiolates			
isophenphos	0.22	2.9	13
Zytron	0.12	4.3	36

(continued)

Table 2. (Continued)

Common name	LD_{50} μg/♀ fly		Resistance ratio (R/S)
	Susceptible (S)	Resistant (R)	
acephate	0.030	0.57	19
methamidophos	0.026	0.74	28
phosphates			
chlorfenvinfos	0.039	1.25	32
diazoxon	0.043	1.70	41
dichlorvos	0.01	0.98	98
fenitro-oxon	0.32	115	359
naled	0.0098	0.66	67
naphthalophos	0.32	> 40	> 125
paraoxon	0.0075	10.7	1427
propaphos	0.26	35.6	137
tetrachlorvinphos	0.056	> 40	> 714

the susceptibility levels have not been found (Yasutomi, 1973, 1975; Yasutomi and Shudo, 1978; Hayashi et al., 1973, 1977). Actually, pyrethroids have long been used against household pests, but their instability has severely restricted their use for house fly control with residual treatments. Permethrin, a novel synthetic pyrethroid with photo-stability, has been used for the control of multi-resistant house flies in Japan since 1978.

Table 3. Toxicity of Carbamate Insecticides and Their Synergism with Appropriate Synergists Against House Flies

Insecticide	LD$_{50}$ µg/♀ fly							
	Susceptible (S)				Resistant (R)			
	alone	+Saf.	+PB	+S-421	alone	+Saf.	+PB	+S-421
propoxur	0.92	0.30	0.36	0.72	>200	0.35	0.98	1.26
BPMC, Bassa	5.2	0.50	-	-	>200	2.8	-	-
carbaryl	106	0.86	-	-	>200	3.5	-	-
methomyl	0.18	-	-	-	0.52	-	-	-

Saf., safroxan; PB, piperonyl butoxide.

Table 4. Comparative Toxicity of Synthetic Pyrethroids in Two Strains of House Flies

Insecticide	LD$_{50}$ µg/♀ fly	
	Takatsuki	Yumenoshima III
allethrin, Pynamin	1.2	2.7
phthalthrin, Neopynamin	0.61	0.63
Esbiol	0.24	0.27
Kadethrin	0.074	0.098
fenvalerate, Sumicidin	0.039	0.048
phenothrin, Sumithrin	0.032	0.039
resmethrin, Chrysron	0.010	0.023
permethrin	0.009	0.011

REFERENCES

Hayashi, A., Hatsukade, M., and Moriya, K., 1973, Sur la sensibilité aux insecticides chez la mouche domestique à la préfecture Kanagawa, *Botyu-Kagaku,* 38:35.

Hayashi, A., Hatsukade, M., Shinonaga, S., Kano, R., Saroso, J. S., and Koiman, I., 1974, The resistant level of the house flies to several synthetic insecticides in Indonesia, *Botyu-Kagaku,* 39:88.

Hayashi, A., Funaki, E., Fujimagari, M., Kano, R., and Nomura, K., 1977, The resistant level of the house fly to several synthetic insecticides in west of Kanto, and Kyushu, Japan, *Botyu-Kagaku,* 42:198.

Kudamatsu, A., Hayashi, A., and Kano, R., 1977, The toxicity of prothiophos to organic phosphate resistant house flies collected from various localities in Japan, *Jap. J. Sanit. Zool.,* 28:285.

Kudamatsu, A., Sato, T., Hayashi, A., and Kano, R., 1979, Cross resistance to various organophosphorus insecticides in the third Yumenoshima strain of the house fly, *Musca domestica* L., *Jap. J. Sanit. Zool.,* 30:255.

Matsubara, H., Ito, H., Kawasaki, M., Kai, K., and Kanamori, S., 1964, On the synergism of barthrin, dimethrin and 1-naphthyl N-methyl-carbamate with pyrethrum synergists against diazinon resistant and susceptible house flies, *Botyu-Kagaku,* 29:1.

Ohtaki, T., 1965, The resistant level of the house fly to several synthetic insecticides in Saitama Prefecture, Japan, *Jap. J. Sanit. Zool.,* 16:253.

Oshima, S., Sugita, K., and Fuse, Y., 1963, The insecticide resistances in the house fly, *Musca domestica vicina* M. from Yokohama area, *Jap. J. Sanit. Zool.,* 14:109.

Suzuki, T., Ikeshoji, T., and Shirai, M., 1958, Insecticide resistance in several strains of house flies in Japan, *Japan. J. Exp. Med.,* 28:395.

Suzuki, T., Hirakoso, S., and Matsunaga, H., 1961, Diazinon resistance in house flies of Japan, *Japan. J. Exp. Med.,* 31:351.

Tsukamoto, M., Ogaki, M., and Kobayashi, H., 1957, Malaria control and the development of DDT resistant insects in Hikone City, Japan, *Jap. J. Sanit. Zool.,* 8:118.

Yasutomi, K., 1956, Studies on the insect-resistance to insecticides, IV, Relative toxicity of p,p'-DDT and related materials, *Jap. J. Sanit. Zool.,* 7:87.

Yasutomi, K., 1966, Insecticide resistance of house flies outbroken at the dumping site, Yumenoshima-island, Tokyo, *Jap. J. Sanit. Zool.,* 17:71.

Yasutomi, K., 1973, Insecticide resistance in the house flies of new Yumenoshima, the 15th dumping-island of Tokyo, *Jap. J. Sanit. Zool.,* 23:255.

Yasutomi, K., 1975, Insecticide resistance in the house flies of the third Yumenoshima, a new dumping-island of Tokyo, *Jap. J.*

Sanit. Zool., 26:257.

Yasutomi, K., and Shudo, C., 1978, Insecticide resistance in the house flies of the third Yumenoshima, a new dumping-island of Tokyo (II), *Jap. J. Sanit. Zool.*, 29:205.

EFFECT OF A RICE BLAST CONTROLLING AGENT, ISOPROTHIOLANE, ON *NILAPARVATA LUGENS* STAL WITH DIFFERENT LEVELS OF SUSCEPTIBILITY TO DIAZINON

Matazaemon Uchida and Minoru Fukada

Biological Research Center, Nihon Nohyaku Co., Ltd.
4-31 Honda-Cho, Kawachi-Nagano, Osaka, Japan 586

INTRODUCTION

Isoprothiolane (Fuji-one®, diisopropyl 1,3-dithiolan-2-ylidene-malonate, Figure 1) has been developed as a fungicide showing eradicant as well as protectant activities against the most important fungal disease of rice plant caused by *Pyricularia oryzae* Cav. (Taninaka et al., 1976). It acts on the fungus at the penetration and growth stages of the infecting hyphae rather than at those of conidial germination and appressorium formation (Araki and Miyagi, 1977). Besides being applied as a foliar spray, a submerging technique has been developed in which isoprothiolane is applied to the irrigation water as granules containing 12% active ingredient.

In the course of its development, isoprothiolane was also found to be effective in controlling two major planthopper pests on rice fields in Japan: the brown planthopper (*Nilaparvata lugens* Stal) and the white-backed planthopper (*Sogatella furcifera* Horváth) (Fukada and Miyake, 1978; Moriya et al., 1977). In the paddy fields treated with the isoprothiolane granules, the population density of these planthoppers was remarkably suppressed for more than one month. In view of the fact that isoprothiolane did not exert acute insecticidal activity even at high doses, its ability to control planthoppers was quite surprising.

Although these species represent two major planthoppers in Japan, their source of infestation is not yet fully understood. After invading rice fields in late June or early July, the brown planthopper goes through two or three generations, whereas the white-backed plant-

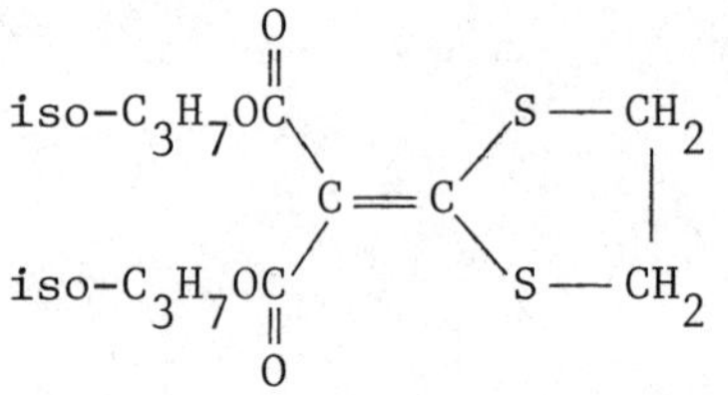

Figure 1. Structure of isoprothiolane

hopper passes through only one or two generations. On early cultivated rice plants the brown planthopper sometimes produces patched hopperburns during late August, although it usually causes them in September or October. Hopperburns near harvest periods result in a great loss of yield. Japanese archives indicate that from as early as the 6th century, great famines were sometimes induced by the brown planthopper (Suenaga, 1954). In South East Asia, however, it was not an important pest of rice plants until quite recently (Mochida et al., 1977).

From a comparison of LD_{50} values obtained from 1967 to 1976, it can be confirmed that the brown planthopper has developed resistance to organophosphorus insecticides, whereas the white-backed planthopper has remained susceptible (Nagata et al., 1979). In the southern parts of Japan, the brown planthopper shows a widespread trend toward the development of resistance to organophosphorus insecticides, although significant variations in the resistance levels have been observed in the various regions (Ozaki, 1978). Miyake (1978) has reported regional differences in levels of resistance toward diazinon and malathion and has observed a relationship between the development of resistance and the cumulative amounts of diazinon and malathion used. The effectiveness of brown planthopper control by diazinon, for example, has been decreasing steadily, probably because of developing resistance. This suggests a serious problem for the control of this planthopper in the near future.

Because of differences in chemical structure and mode of action, isoprothiolane was anticipated to be effective against organophosphorus- and/or carbamate-resistant planthoppers. The following discussion will be concerned with the effect of isoprothiolane on brown planthoppers exhibiting different levels of susceptibility to diazinon and the mode of action of isoprothiolane itself.

Table 1. Effect of Isoprothiolane on the Population Density of Two Major Planthoppers in Paddy Fields

	Population Density (insects/hill) of			
	S. furcifera		*N. lugens*	
Field	Adult	Nymph	Adult	Nymph
Isoprothiolane-treated	0.8	9.0	1.5	0.5
Untreated	13.3	144.0	17.5	39.0

Isoprothiolane granules were applied on July 28 at the rate of 360 g active ingredient per 10 ares, and the population density of the planthoppers was investigated on August 28. (Fukada and Miyake, 1978.)

EFFECT OF ISOPROTHIOLANE ON THE BROWN PLANTHOPPER IN RICE PADDY FIELDS

In the course of developing isoprothiolane as a fungicide for rice blast, it was found in 1973 to exert an effect on the brown and white-backed planthoppers (Fukada and Miyake, 1978), although it was not insecticidal against adult insects even at concentrations as high as 1,000 ppm. As shown in Table 1, the population densities of these two planthopper species were remarkably suppressed on the rice paddy fields where isoprothiolane had been applied to the irrigation water in the form of granules containing 12% active ingredient at the rate of 3 kg/10 ares. Our interest in developing isoprothiolane as an insect control agent led to a reinvestigation of the population densities of the planthoppers on rice fields in the following year. This confirmed the ability of isoprothiolane to decrease the population density of the planthoppers. On untreated fields, populations of both species of planthoppers increased progressively, and as a result, they caused severe damage to the rice plants. In the treated fields, however, populations were suppressed even 50 days after treatment with isoprothiolane granules.

Since the suppression of population growth became significant after about one month and was most outstanding during early instar stages rather than late instar or adult stages (Figure 2), the effect of isoprothiolane on these planthoppers may be indicated mostly by the decreased number in the subsequent generation. It also suggested that the mode of action of isoprothiolane is quite different from that of organophosphorus or carbamate insecticides.

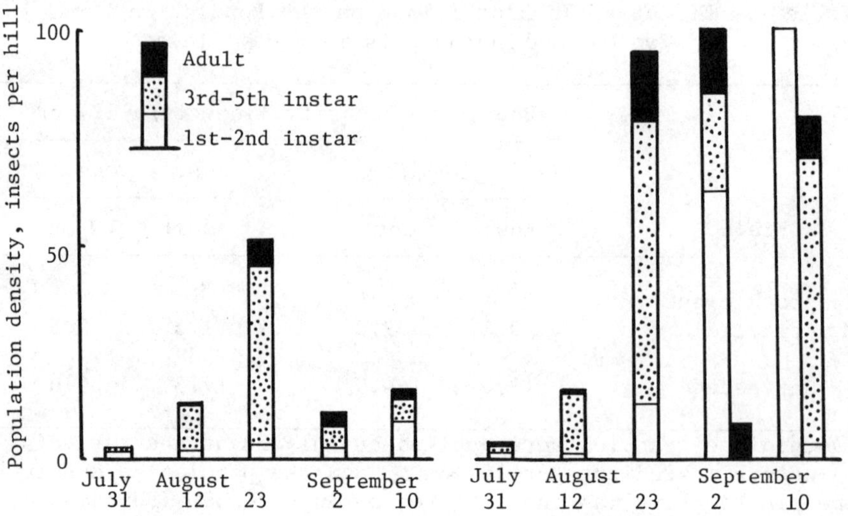

Figure 2. Population density of *Nilaparvata lugens* in the untreated and isoprothiolane-treated fields. Isoprothiolane granules were applied on August 1 at the rate of 400 g active ingredient per 10 ares.

Further experiments showed promise for control of the brown and white-backed planthoppers not only by application to paddy water but also by use of nursery box applications (Fukada et al., unpublished). Using both types of application, rice plants were well protected against these insect pests for more than 50 days without any phytotoxicity. Rates of application were 480 g of active ingredient per 10 ares and 120 to 180 g per 20 nursery boxes per 10 ares, respectively. Perhaps because of the long period of inhabitation and the rapid growth of population, it was found that the planthoppers sometimes sufficiently recovered their population level in September or October to again harm rice plants. Consequently, another granular application applied in early August was required for controlling the planthoppers throughout the rice-planting period.

Since diazinon is effective against several insect pests such as the rice stem borer (*Chilo suppressalis* Walker), the brown planthopper, the white-backed planthopper and the green rice leafhopper (*Nephotettix cincticeps* Uhler), it has been widely used in Japan in the form of foliar and/or submerged applications. However, it has been pointed out that LD_{50} values for diazinon toward the brown planthopper have increased by a factor of 2 to 5 (Tsurumachi, 1978) and that the effectiveness of control of this hopper by organophosphorus insecticides such as diazinon, fenitrothion and malathion has been lowering steadily to the point where failures have become appreciable in some regions since 1973. In 1978, Miyake pointed out

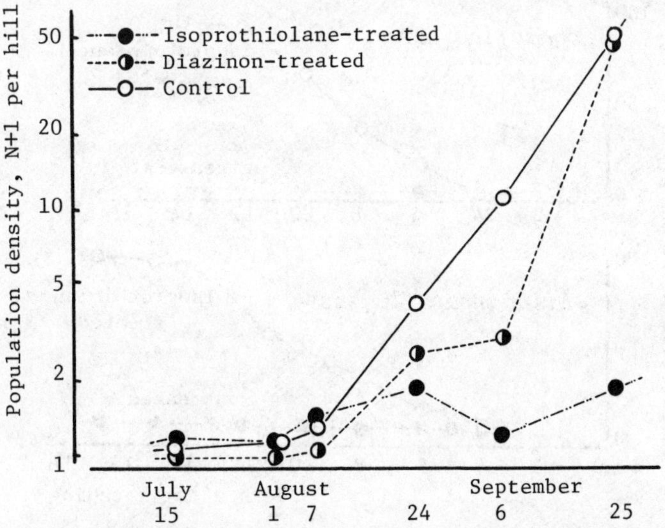

Figure 3. Change of the population density of *Nilaparvata lugens* in the paddy fields treated with isoprothiolane and diazinon as granules on July 11 and August 3.

the elevated LD_{50} values for diazinon against the brown planthopper (25 µg/g of insect at Hiroshima, 22 µg/g at Ehime, and 7 µg/g at Kagawa), and he has shown that there is a relationship between the extent of resistance and the amount of cumulative usage of diazinon in the various regions.

Figure 3 shows the population growth of the brown planthopper on paddy fields of Hiroshima Prefecture in 1974 (Nakasawa and Hayashi, 1974). On the diazinon-treated field, the population of the brown planthopper became progressively greater and control measures resulted in failure even though the insect had been temporarily suppressed after two applications. According to these results, brown planthoppers in Hiroshima Prefecture were less susceptible to diazinon as early as 1974. The data in Figure 3 show that against this planthopper, isoprothiolane exerted no obvious effect until August 7. Thereafter, however, growth of the planthopper population was so slow that its population density was suppressed for a period of about two months. Isoprothiolane thus proved effective against diazinon-resistant strains of the brown planthopper.

MODE OF ACTION OF ISOPROTHIOLANE ON THE BROWN PLANTHOPPER

As mentioned earlier, the mode of action of isoprothiolane against the brown and the white-backed planthoppers appears to be quite different from that of organophosphorus insecticides. Since

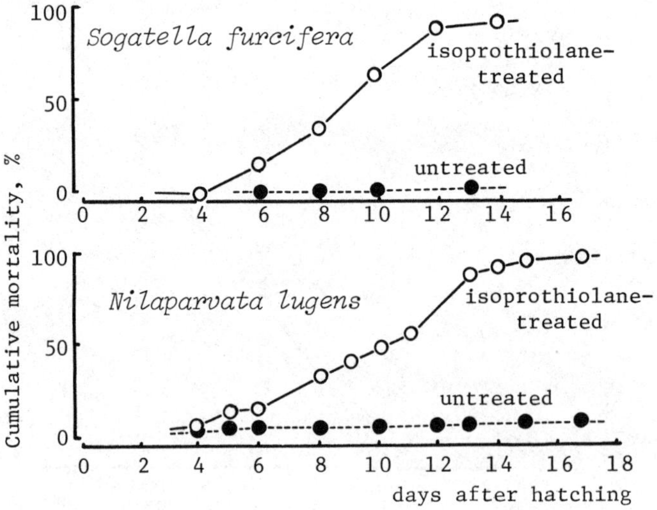

Figure 4. Cumulative mortality of *Nilaparvata lugens* and *Sogatella furcifera* reared on rice plants untreated and treated with isoprothiolane.

it exerted no fast activity against these planthoppers, the observation of symptoms occurring throughout the lifetime of the insect seemed to be neccessary for the elucidation of its unique mode of action.

When first instar nymphs of the brown planthopper were placed on rice leaf blades in test tubes, they molted about every three days at 25°C and emerged after five molts. More than 80% of the nymphs were able to develop into adults on untreated leaves, as shown in Figure 4. On the leaf blades previously treated with isoprothiolane granules at a rate of 480 g of active ingredient per 10 ares, however, the planthopper nymphs began to die on the fifth day after starting to ingest the treated plants, and the number of individuals dying gradually increased until mortality was greater than 95%. Although a few individuals were able to emerge, their longevity was observed to be less than three days. Examination of affected planthoppers showed abnormalities in their ability to shed their exuviae. Figure 4 shows a similar effect of isoprothiolane against the white-backed planthopper.

As described earlier, isoprothiolane apparently exerts a delayed effect against the brown and the white-backed planthoppers. Perhaps because a period of time is needed for the insecticide to show its effect, the susceptibility of later instar nymphs was much lower than that of earlier ones. Additional effects such as a shortening of longevity after emergence and a suppression of the oviposition rate were also observed in subsequent studies (Fukada and Miyake, 1978;

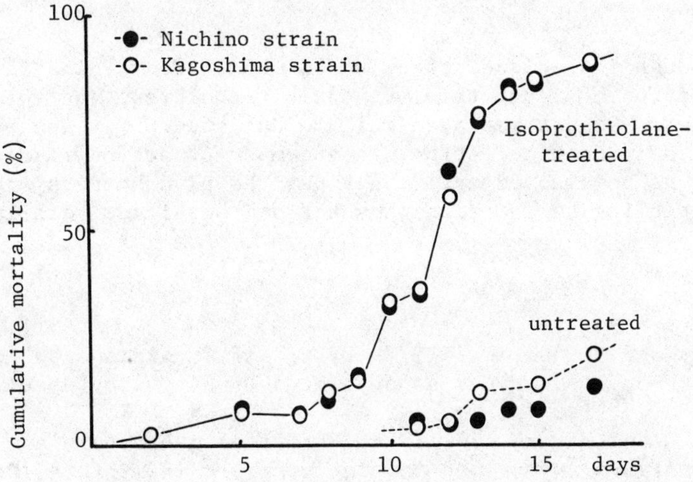

Figure 5. Cumulative mortality of the Nichino and the Kagoshima strains of *Nilaparvata lugens* reared on rice plants untreated or treated with isoprothiolane.

Moriya et al., 1977). These effects seemed to minimize the population size of the next generation of brown and white-backed planthoppers, as observed on the isoprothiolane-treated fields.

The hatchability of eggs laid by the female brown planthoppers treated with 40 ppm of isoprothiolane is slightly lower than in untreated insects (Moriya et al., 1977), but the hatched nymphs showed normal growth. There exists no ovicidal activity of isoprothiolane against the brown planthopper.

A strain of brown planthoppers collected from paddy fields of Kagoshima Prefecture in August of 1976 has been reared in our laboratory. This strain, named "Kagoshima strain," was 30 times less susceptible to topically applied diazinon than the Nichino strain, which was used in the study described above.

In spite of their different levels of susceptibility to diazinon, isoprothiolane was equally effective against both the Kagoshima and the Nichino strains of brown planthoppers. When these two strains of planthoppers were reared on rice seedlings with 0 or 10 ppm of isoprothiolane aqueous solution (1 ml/test tube), the cumulative mortality curves of the two strains agreed very closely with each other, as shown in Figure 5. This seemed to indicate not only that there was no cross resistance between isoprothiolane and diazinon, but also that each had a different mode of action.

CONCLUSION

Although its unique effectiveness has been fascinating, the mode of action of isoprothiolane against the brown and white-backed planthoppers has not yet been fully elucidated. Perhaps because of differences in chemical structure and mode of action, isoprothiolane was shown to be also effective against the planthoppers whose level of susceptiblity to an organophosphorus insecticide, diazinon, was low.

REFERENCES

Araki, F., and Miyagi, Y., 1977, Effect of fungicides on penetration by *Pyricularia oryzae* as evaluated by an improved cellophane method, *J. Pestic. Sci.*, 2:457.

Fukada, M., and Miyake, T., 1978, Effect of isoprothiolane on *Nilaparvata lugens* and *Sogata furcifera* (Hemiptera, Delphacidae), *Jap. J. appl. Ent. Zool.*, 22:191.

Fukada, M., Asai, T., and Yonekubo, T., unpublished.

Miyake, T., 1978, Decrease of insecticide susceptibility on the brown planthopper, *Jap. J. appl. Ent. Zool.*, 22:Supplement (in Japanese).

Mochida, O., Tatang, S., and Ayuk, W., 1977, Recent outbreaks of the brown planthopper in South East Asia, *ASPAC* (Tokyo), pp. 170-191.

Moriya, S., Maeda, Y., Yonekubo, T., and Asakawa, K., 1977, Effect of isoprothiolane on the reproduction of the brown planthopper, *Nilaparvata lugens* Stål, *Jap. J. appl. Ent. Zool.*, 21:220.

Nagata, T., Maeda, E., and Moriya, S., 1979, Development of insecticide resistance in the brown planthopper, *Nilaparvata lugens* Stål (Hemiptera, Delphacidae), *Appl. Ent. Zool.*, 14:264.

Nakasawa, D., and Hayashi, H., 1974, Report on field evaluation of insecticides in paddies in Hiroshima Prefecture, *Itaku Shiken Seiseki*, 19:305 (in Japanese).

Ozaki, K., 1978, Development of insecticide resistance in the brown planthopper, *Jap. J. appl. Ent. Zool.*, 22:Supplement (in Japanese).

Suenaga, H., 1954, Chronological records of outbreak of rice planthoppers in Japan, *Oyo Kontyu*, 10:85.

Taninaka, K., Kurono, H., Hara, T., and Murata, K., 1976, Rice-blast controlling activities of bis-(alkoxycarbonyl)-ketenedithioacetals and their related compounds, *J. Pestic. Sci.*, 1:115.

Tsurumachi, M., 1978, Annual fluctuation of insecticide susceptibility to the brown planthopper, *Jap. J. appl. Ent. Zool.*, 22:Supplement (in Japanese).

MECHANISMS OF ACARICIDE RESISTANCE

WITH EMPHASIS ON DICOFOL

Tetsuo Saito, Katsuhiro Tabata* and Satoshi Kohno†

Laboratory of Applied Entomology and Nematology
Faculty of Agriculture, Nagoya University
Furo-cho, Chikusa-ku, Nagoya, 464 Japan

INTRODUCTION

Resistance in mites to acaricides was first observed by Compton and Kearns in the two-spotted spider mite when it developed resistance to ammonium potassium selenosulfide (Selecide®) in 1937. The introduction of organophosphorus acaricides in 1947, first TEPP and later parathion, resulted in the virtual elimination of mites in greenhouses. Resistance to parathion and TEPP became evident in 1949, and by 1950 resistant mites were present in a large percentage of rose houses in the eastern United States (Jeppson et al., 1975).

In Europe, resistance to organophsophorus acaricides occurred in several countries by 1950. Schradan resistance in the citrus red mite, *Panonychus citri*, was found by Seki in 1958 and was the first case of resistance in mites reported in Japan. In 1959, resistance in the European red mite, *Panonychus ulmi*, to organophosphorus acaricides was observed in Hokkaido and Aomori, and to tetradifon (Tedion®) in Akita (Nomura, 1973; Asada, 1978). On ornamental plants resistance in the carmine mite, *Tetranychus cinnabarimus*, to demeton-S-methyl (Metasystox®) was demonstrated by Nomura and Nakagaki in 1959, and on tea plants, resistance to phencapton in the Kanzawa spider mite, *Tetranychus kanzawai*, had developed in Shizuoka by 1961 (Osakabe, 1971).

*Forestry and Forest Products Research Institute, P.O. Box 2, Ushiku, Ibaraki, 300-12 Japan
†Hyogo Prefectural Agricultural Center for Experiment, Extention and Education, Kitaouji, Akashi, 673 Japan

In the 1960s the development of resistance in mites to insecticides and acaricides, except petroleum oil, had become a worldwide problem.

Mites are capable of rapidly developing extensive resistance to many acaricides. A very high level of resistance to many compounds can develop after only one to four years of use and often induces a high degree of cross resistance. The factors contributing to this are great egg-laying potential, very short life cycle, cross fertilization, and high mutation rates.

On the basis of genetic studies, most species of the Tetranychidae are known to be bisexual and to reproduce by means of arrhenotokous parthenogenesis. Haplo-diploid sex determination has particular implications in genetic variability. Because of male haploidy, the peculiar situation occurs that mutations are immediately expressed in the male without respect to whether these mutations are dominant or recessive. This implies that there is an immediate interaction between the process of mutation and selection (Helle and Overmeer, 1973; Dittrich, 1969; Brader, 1977; Croft, 1977).

Studies on mechanisms of resistance to acaricides in mites have lagged behind comparable works in insects because of the difficulty in handling presented by their minute size. Physiological studies have concerned the mode of action of organophosphorus compounds in mites, and Casida (1955) was the first to report the presence of cholinesterase in the mite *Acarus siro*. Voss (1959, 1960) and Dauterman and Mehrotra (1963) demonstrated the presence of cholinesterase in the two-spotted spider mite. Sakai (1967) and Motoyama and Saito (1968) examined the substrate specificity of mite cholinesterases, and confirmed that insect cholinesterases are specific while those in mites are nonspecific.

Smissaert (1964) reported that a strain of two-spotted spider mite resistant (R) to organophosphorus compounds had developed a diazoxon-insensitive cholinesterase. This finding was subsequently confirmed by Voss and Matsumura (1964, 1965) and by Smissaert et al. (1970).

Matsumura and Voss (1964) revealed that the R two-spotted spider mite (Blauvelt strain) had a superior ability to detoxify malathion, parathion and malaoxon than the susceptible (S) Niagara strain. In the case of malathion metabolism, the greatest interstrain difference was found in the amount of carboxyesterase products, but a substantial difference in the phosphatase activity was also observed. The amount of cuticular absorption of ^{14}C-malathion and ^{3}H-parathion by the Niagara strain was slightly greater than that of the Bauvelt strain, and no significant interstrain differences were observed in the amount of total insecticides absorbed into the mite body.

The synergistic action of 46 compounds with malathion and dimethoate in several organophosphorus-R citrus red mite strains and an S strain was evaluated by Takahashi et al. (1972, 1973). Triphenyl phosphate, tri-*o*-cresyl phosphate, EPN, Kitazin-P® (*O,O*-bis-(1-methylethyl) *S*-(phenyl methyl) phosphorothioate) and saligenin cyclic phosphates were shown to have synergistic action with malathion in both S and R strains. K-1 (2-phenyl 4H-1,3,2-benzodioxaphosphorin-2-oxide) had the most effective synergistic activity with malathion in both strains. It was supposed that the mechanism of synergistic action of K-1 with malathion was the inhibition of the carboxyesterase by K-1, one of the metabolic enzymes of malathion. There are no effective synergists for dimethoate, and methylene dioxyphenyl compounds and dichlorvos show only an antagonistic action to malathion.

Kuwahara (1977) selected Kanzawa spider mites with Metasystox-S® for 20 generations and this selected strain became cross-resistant to organophosphates but not to carbamates. K-1 showed remarkable synergism with malathion but only little synergism with phenthoate, suggesting that K-1 may be a selective malathion carboxyesterase inhibitor in the Kanzawa spider mite.

Henneberry et al. (1964) reported that the cuticle of the organophosphate-R two-spotted spider mites was thicker than that of the S strain. Hirai et al. (1972, 1973) suggested that the difference in penetration of ^{32}P-dimethoate is one of the factors involved in citrus red mite resistance.

MECHANISM OF DICOFOL RESISTANCE IN THE CITRUS RED MITE

Dicofol (Kelthane®) has excellent toxicity to all mite stages and a high residual effect against not only the Tetranychidae, but also the Tenuipalpidae and the genera *Steneotarsonemus* and *Polyphagotarsonemus* of the Tarsonemidae (Jeppson et al., 1975). Inoue and Saito (1972) reported that the larvae are most susceptible to dicofol and that the egg is less susceptible than the nymph and adult in both dicofol-R and S citrus red mites.

Resistance to dicofol by the McDaniel mite, *Tetranychus mcdanieli*, was first noted in 1958 (Hoyt and Harries, 1961); thereafter this problem occurred in many apple orchards throughout north central Washington. Laboratory studies confirmed the presence of dicofol-R strains of the McDaniel mite at several locations. One strain from Orndo, where control had been especially difficult, showed a 200-fold increase in the LD_{50} over an S strain. Several other chlorinated hydrocarbon acaricides also failed to control the mites in these areas. Laboratory studies with strains of the two-spotted spider mite from different areas showed that they were about equally resistant to dicofol (Hoyt and Harries, 1961).

Under field conditions citrus red mite populations became resistant to dicofol after seven to twelve treatments (Jeppson et al., 1962). Eight local strains of two species of spider mites, *Tetranychus arabicus* and *T. cucurbitacearum*, occurring on cotton in Egypt, were monitored for their resistance levels for three years (Dittrich and Ghobrial, 1974), and on the basis of LC_{50} determinations with the slide-dip method, the mites showed increasing resistance to dicofol.

Osakabe (1973) observed that resistance to dicofol had developed in the Kanzawa spider mite in Shizuoka, and Inoue (1979) collected dicofol-R citrus red mites from Fukuoka. It is interesting that dicofol resistance was manifested as a recessive character in both species.

Dicofol-R citrus red mites (Jeppson, 1963) and two-spotted spider mites (Hansen et al., 1963) both display cross resistance to organophosphorus compounds. Matsumoto and Shinkaji (1974) studied the cross resistance in dicofol-R citrus red mites and found cross resistance to chlorobenzilate, Kilacar® [di-(*p*-chlorophenyl)-cyclopropyl carbinol], Akrol® (isopropyl 4,4'-dibromobenzilate), tetradifon, dialifor, dimethoate, and phencapton.

In the present study, the authors made an attempt to investigate the mechanisms of dicofol resistance in citrus red mites by comparing the differences in penetration and metabolism of the chemical in dicofol-R and S strains.

Topical Toxicity of Dicofol to Susceptible and Resistant Mites

The accurate toxicity of an insecticide to an insect is obtained when precisely known doses are applied to selected parts of an individual and an extremely high rate of recovery of the insecticide is obtained. Many devices have been used for topical application of insecticides to insects; however, topical treatment on smaller organisms such as mites is difficult. Dicofol was applied to the mite by means of a newly devised applicator that consists of a screw micrometer and an ultra microsyringe that is used for injecting the samples in gas chromatographic analysis (Tabata and Saito, 1970). This device enables the easy application of 2 nl of a solution in furfuryl alcohol on the idosoma of the female adult mite under a binocular microscope. Twenty mites of each strain were collected on a round coverglass (22 mm in diameter), placed on water-flooded filter paper. After orientation, each mite was topically treated with dicofol and held at 25°C, 74% R.H. Topical toxicities of dicofol to susceptible (Wakayama-S) and resistant (Wakayama-R) strains are shown in Tables 1 and 2. The Wakayama-R strain is very highly resistant, while the Nagoya-R and Saga-R strains show little resistance.

Table 1. Topical Toxicities of Dicofol to Adult Female Citrus Red Mites

Strain	Time (hrs.)	LD_{50} (ng/mite)	Resistance ratio
Wakayama-S	24	3.57	1
	48	1.36	1
	72	1.11	1
Wakayama-R	48	366	269
	72	268	241

Table 2. Topical Toxicities of ^{3}H-Dicofol (20 ng/2 nl/mite) to Resistant and Susceptible Female Citrus Red Mites

Strain	Mortalities (%)					
	1 hr.	2	4	8	16	32
Wakayama-S	0.0	0.0	0.0	10.0	55.9	98.0
Saga-R	0.0	0.0	0.0	16.2	61.0	57.1
Nagoya-R	0.0	0.0	0.0	14.7	16.7	23.4
Wakayama-R	0.0	0.0	0.0	0.0	0.0	0.0

In Vivo Metabolism of ^{3}H-Dicofol in Susceptible and Resistant Mites

^{3}H-ring labeled dicofol (0.96 mCi/mg) in furfuryl alcohol was applied to the idosome of adult female mites, and the isolation and determination of radioactive metabolites was performed as shown in Figure 1. The results of cuticle permeability, metabolism and excretion are shown in Figures 2, 3 and 4, respectively.

In all strains, about 40% to 50% of the dicofol had penetrated the mite 32 hours after topical application. No significant difference between S and R strains was found. The radioactivity of the water-soluble dicofol metabolites in mites of the Wakayama-R strain

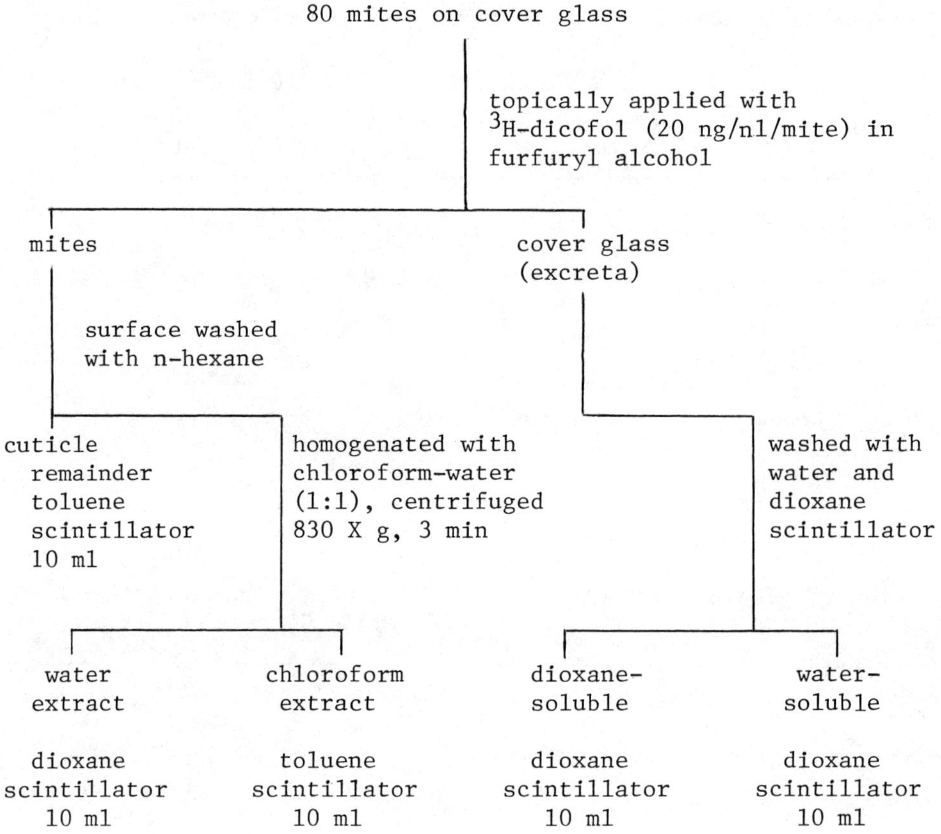

Figure 1. Isolation and determination of radioactive metabolites from citrus red mite adult females topically applied with ^{3}H-dicofol.

was found. The radioactivity of the water-soluble dicofol metabolites in mites of the Wakayama-R strain was greater than that of the other strains. The ratio of the water-soluble materials to the dioxane-soluble materials in excreta of the Wakayama-R and Nagoya-R strains increased gradually 8 hr after topical application, but those of the Saga-R and Wakayama-S strains remained small up to 32 hr after topical application. It has been suggested that dicofol-R mites have a higher ability to metabolize dicofol to the water-soluble materials than do the S mites (Tabata and Saito, 1971).

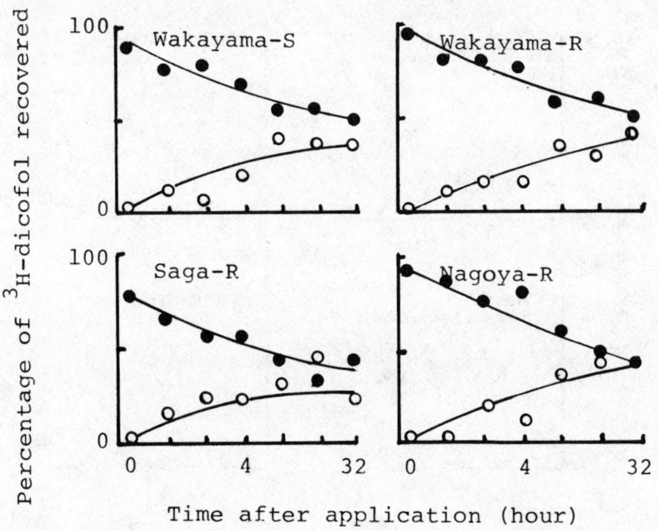

Figure 2. The change of radioactivity of cuticle remainder-^{3}H and penetrated-^{3}H of citrus red mite adult female treated topically with ^{3}H-dicofol (Tabata and Saito, 1971). ●-●, cuticle remainder-^{3}H; O-O, penetrated-^{3}H.

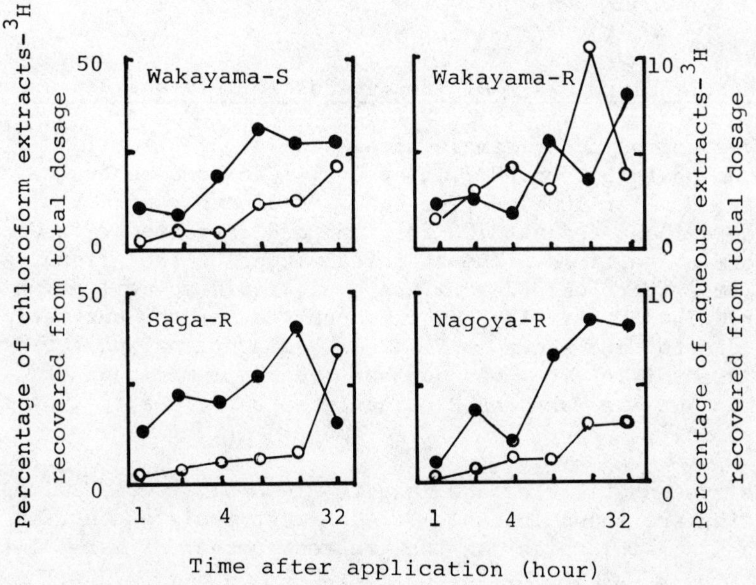

Figure 3. The change of radioactivity of chloroform and aqueous extracts-^{3}H in the body of topically applied citrus red mite adult female (Tabata and Saito, 1971). ●-●, chloroform extracts-^{3}H; O-O, aqueous extracts-^{3}H.

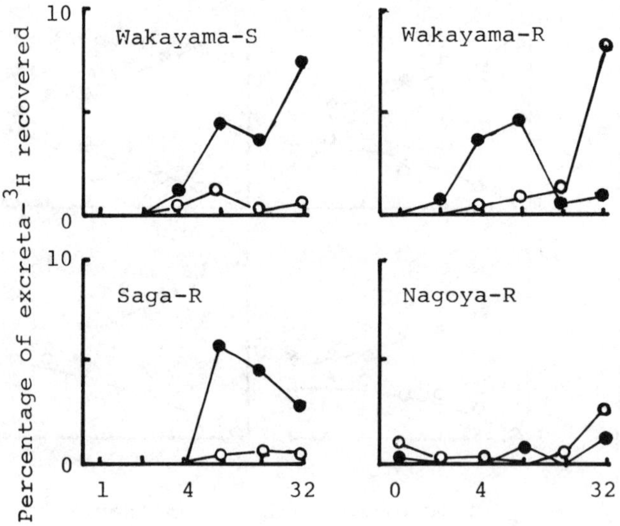

Figure 4. The change of radioactivity of dioxane-soluble-^{3}H and water-soluble-^{3}H in excreta of citrus red mite adult female treated topically with ^{3}H-dicofol (Tabata and Saito, 1971). ●-●, dioxane-soluble-^{3}H; O-O, water-soluble-^{3}H.

In Vitro Metabolism of ^{3}H-Dicofol in Susceptible and Resistant Mites

Five hundred adult female mites were homogenized in 0.5 ml of 1/15 M phosphate buffer pH 7.13 at 0°C. The homogenate was centrifuged at 800 g for 10 min. Two μg of ^{3}H-dicofol and 0.1 ml of supernatant were incubated at 37°C for 1 hr and extracted with 0.2 ml of chloroform three times. The radioactivities of the aliquot and chloroform extract were determined by a liquid scintillation spectrometer. The pH-activity and time courses for the enzymatic metabolism of dicofol are shown in Figure 5. The optimal pH for degradation of dicofol in mites was 7.3 and the maximum amount of dicofol degradation in the Wakayama-R strain was four times that of the Ogi-S strain.

The susceptibilities and in vitro metabolism of ^{3}H-dicofol in the strains are shown in Table 3. The most resistant strain is the Wakayama-R strain, which has the greatest amount of metabolites.

The effects of cofactors on the in vitro metabolism of dicofol are shown in Table 4. The addition of NADPH, GSH and Mg were not increased in the water-soluble metabolites by the mite enzyme.

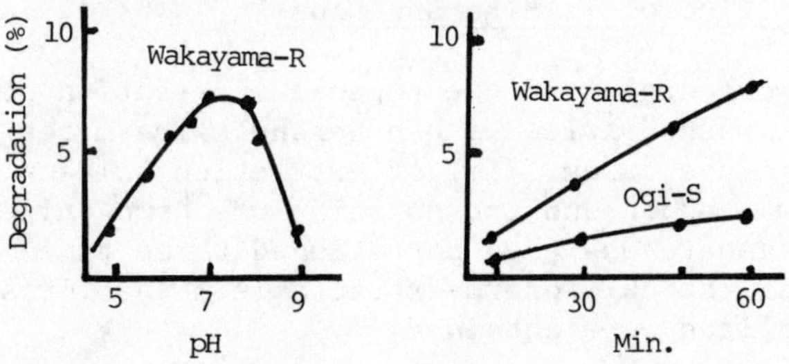

Figure 5. In vitro metabolism of ^{3}H-dicofol by citrus red mites. 500 mites + 0.5 ml 1/15 M phosphate buffer ⟶ homogenate ⟶ 800 g, 10 min ⟶ supernatant 0.1 ml + 2 μg ^{3}H-dicofol ⟶ 37°C, 1 hr ⟶ + 0.2 ml chloroform extractions 3 times ⟶ chloroform extract, water extract ⟶ scintillation counting.

Table 3. In Vitro Metabolism of ^{3}H-dicofol by Citrus Red Mites

Strain	Mortality (20 ng/mite) 24 hrs	48 hrs	In vitro metabolism (ng/100 mites/hr)
Ogi-S	98.8%	100 %	17.2
Gotanda-R	43.2	56.4	0.0
Okaguchi-R	32.9	46.7	3.5
Tarumo-R	17.3	35.6	51.4
Wakayama-R	0.0	10.3	120.4

Table 4. Effects of Cofactors on the Metabolism of Dicofol by the Wakayama-R Strain

Cofactor	Metabolism (ng/100 mites/hr)
None	113.7
2.4 μM NADPH	68.9
4.0 μM GSH	104.8
10 μM Mg^{++}	72.3
2.4 μM NADPH + 4.0 μM GSH	64.1

Metabolites of Dicofol in Mite and Mouse

Thirty-two hours after the topical application of ^{3}H-dicofol, 300 mites were washed twice with n-hexane. The mites were transferred to 10 ml centrifuge tubes, 2 ml of extraction solvent (chloroform-water = 1:1) was added and the solution was homogenized by a glass bar. The homogenate was then partitioned three times with 1 ml of chloroform, and the chloroform-extractable metabolites and water-soluble metabolites were obtained.

Adult male mice (ICR-JCL), 25 to 28 g, orally received 100 μg of ^{14}C-dicofol (ring labeled 01745 mCi/g) or ^{3}H-dicofol and non-labeled dicofol dissolved in 0.1 ml propylene glycol containing 10% acetone. Urine samples were collected 72 hr after administration and were extracted with chloroform as mentioned above. Both extracts were concentrated *in vacuo* at 40°C. The concentrates were spotted

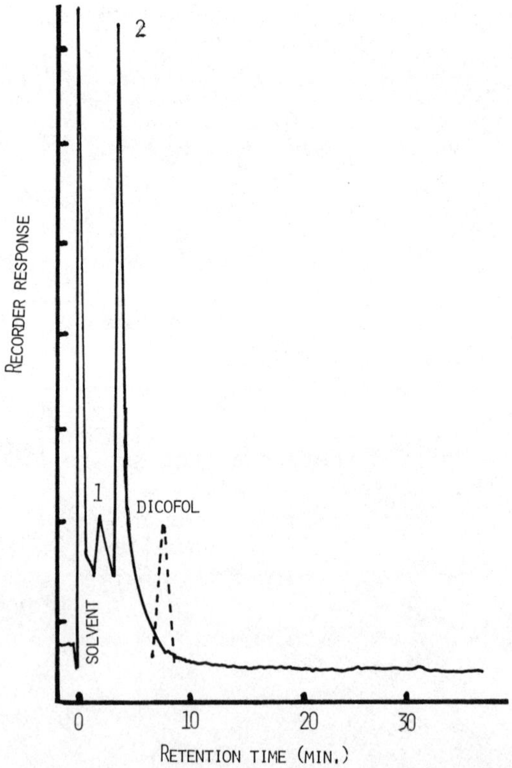

Figure 6. Gas chromatogram of the water-soluble metabolites in the dicofol-resistant mite. The water-soluble metabolites are peaks 1 and 2 (Tabata et al., 1979).

Table 5. The Cleavage of the Water-soluble Radioactive Metabolites from Mouse Urine and Mites with β-Glucuronidase, β-Glucosidase and Under Acidic Conditions

	Radioactivity dpm (%)			
	Mouse urine		Citrus red mite	
	Aqueous	Chloroform	Aqueous	Chloroform
β-glucuronidase[a]	2998(96.9)	98(3.1)	653(94.0)	42(6.0)
β-glucosidase[a]	3817(95.7)	173(4.3)	846(97.8)	19(2.2)
1 N-HCl[b]	3622(77.2)	1070(22.8)	607(93.1)	45(6.9)
None	3639(97.0)	132(3.5)	651(97.0)	20(3.0)

[a]Incubated 300 μl (mite) or 50 μl (mouse) of the aqueous fractions with 20 mg β-glucuronidase (15 units per gram, beef liver, Wako Chemical Ind., Ltd.) or β-glucosidase (800 units per mg, sweet almond, Chemical Ind., Ltd.) at 37°C for 20 hr.
[b]Incubated 200 μl (mite) or 50 μl (mouse) of the aqueous fractions with 0.3 ml (mite) or 0.1 ml (mouse) of 1 N-HCl at 60°C for 10 hr.

on a precoated silica gel thin-layer plate using a mixture of n-butanol:acetic acid:water (4:1:2) as a developing solvent. Detection of spots was made under a UV lamp (254 nm) and by scraping the silica gel and measuring the radioactivity in the liquid scintillation spectrometer (Tabata and Saito, 1973). The water-soluble metabolites were investigated by the ECD-gas chromatograph (Tabata et al., 1979).

Thin-layer radiochromatograms of the water extract of the mouse urine revealed major metabolites with Rf values of 0.6 to 0.7 and 0.46. In mites of the dicofol-R strain, the compound of Rf 0.6 to 0.7 was observed (Tabata and Saito, 1973). The water-soluble metabolites for the mice and mites were analyzed by gas chromatography. Two peaks for the R mite (Figure 6) and four peaks for the mice were detected (Tabata et al., 1979). The water-soluble metabolites were not cleaved by the addition of β-glucuronidase, β-glucosidase or hydrochloric acid (Table 5).

The results of joint action between dicofol and the related compounds in the dicofol-R citrus red mite (the Tarumon-R strain) are shown in Table 6. P,p'-DDT and p,p'-DDE showed slight synergistic activities as compared to dicofol alone, and the in vitro metabolism of ^{3}H-dicofol was slightly decreased. DBP (4,4'-dichlorobenzophenone)

Table 6. Joint Action Between Dicofol and Selected Compounds in the Dicofol-resistant Citrus Red Mite, Tarumon-R

	Mortalities (%) 24 hrs	48 hrs	In vitro metabolism (ng/hr/2 μg)
20 ng dicofol	25.5	33.1	41
20 ng dicofol + 20 ng p,p'DDT	15.0	50.8	26
20 ng dicofol + 20 ng p,p'-DDE	28.9	41.3	34
20 ng dicofol + 20 ng DBP	0.0	0.0	48
20 ng dicofol + 20 ng DBH	24.4	29.4	43

showed a strong antagonistic activity and increased the in vitro metabolism of ^{3}H-dicofol.

In the present experiment, it is suggested that the mechanisms of dicofol resistance in the citrus red mite may be correlated with the increase in detoxification activity but not the decrease in the cuticle permeability to dicofol.

The metabolites of detoxification were water-soluble materials, and these may not conjugate with glucuronide or glucoside. The water-soluble metabolites were analyzed by ECD-gas chromatography. Two peaks in the R mite metabolites and four peaks in the mouse urine were observed. The retention times of the R mite metabolites were the same as in two of four of the metabolites observed in mouse urine. Therefore, the mouse also has the ability to detoxify dicofol to water-soluble metabolites.

Further experiments are now in progress to identify these water-soluble metabolites of dicofol.

DDT and DDE showed slight synergistic activities, but DBP showed a strong antagonistic activity with dicofol in R mites. Subsequently it was found that the joint action of these compounds resulted from a decrease or increase in the detoxification activities of dicofol in the citrus red mite.

MECHANISM OF DICOFOL RESISTANCE IN THE TWO-SPOTTED SPIDER MITE

To determine the possible role of detoxication in dicofol-R two-spotted spider mites, the rate of dicofol detoxication was

Table 7. Metabolism of ^{3}H-Dicofol by Adult Two-spotted Spider Mites

Strain	LC_{50} (ppm)	Water metabolites (%)		
		1.5 hrs	6 hrs	24 hrs
Susceptible	79.4	5.5	8.2	28.3
Resistant	1828	7.5	13.3	45.2

R/S = 23.0

measured directly in vitro or in vivo as was done for the citrus red mite.

Two hundred adult mites were transferred to a test tube, 20 mm in diameter, to which 15 µl of 25 ppm ^{3}H-dicofol (3.82 µCi/mg) in acetone and water (1:2) was added. The tube was then rotated for several minutes to ensure a homogeneous application of acaricide to all individual mites. The mites were then allowed to stand for 1.5, 6 and 24 hr at 28°C. One milliliter of water and 1 ml of chloroform were added to the tube and homogenized by glass rod. After centrifugations at 3,000 rpm for 3 min, the chloroform layer was separated. Two additional chloroform extractions were conducted. The water extract, chloroform extract and precipitate were analyzed for radioactivity by liquid scintillation spectrometry, and the results are shown in Table 7. The percentage of water-soluble metabolites in the R strains increased with time more than did that of the S strain. Further investigations to identify these water-soluble metabolites of dicofol are now in progress.

SUMMARY

It is known that a number of mites can develop a high degree of resistance against many acaricides, but except for malathion, the mechanisms of this resistance have not been determined. Comparative studies on the penetration and metabolism of dicofol in R and S strains of the citrus red mite and the two-spotted spider mite have revealed that R strains have a superior ability to detoxify dicofol to water-soluble materials.

Acknowledgment: This research was supported in part by a grant from The Ministry of Education, Japan. Thanks are due to Rohm and Haas Co., U.S.A., and Sanyo Trading Co., Ltd., Japan, for supplying the radioactive dicofol and authentic dicofol.

REFERENCES

Asada, M., 1978, Genetics and biochemical mechanisms of acaricide resistance in phytophagous mites, *J. Pesti. Sci.*, 3:61.

Brader, L., 1977, Resistance in mites and insects affecting orchard crops, *in:* "Pesticide Management and Insecticide Resistance," D. L. Watson and A. W. A. Brown, eds., Academic Press, New York, pp. 353-376.

Casida, J. E., 1955, Comparative enzymology of certain insect acetylesterases in relation to poisoning by organophosphorus insecticides, *Biochem. J.*, 60:487.

Compton, C. C., and Kearns, W. W., 1937, Improved control of red spider on greenhouse crops with sulfur and cyclohexylamine derivatives, *J. Econ. Entomol.*, 30:512.

Croft, B. A., 1977, Resistance in arthropod predators and parasites, *in:* "Pesticide Management and Insecticide Resistance," D. L. Watson and A. W. A. Brown, eds., Academic Press, New York, pp. 377-393.

Dauterman, W. C., and Mehrotra, K. N., 1963, The N-alkyl group specificity of cholinesterase from the house fly, *Musca domestica* L., and the two-spotted spider mite, *Tetranychus telarius* L., *J. Insect Physiol.*, 9:257.

Dittrich, V., 1969, Chlorphenamidine negatively correlated with OP resistance in a strain of two-spotted spider mite, *J. Econ. Entomol.*, 62:44.

Dittrich, V., and Chobrial, A., 1974, Dynamics of resistance to acaricides in two mite species, *Tetranychus arabicus* Attiah, and *T. cucurbitacearum* Sayed, occurring on Egyptian cotton, *A. ang. Ent.*, 76:418.

Hansen, C. O., Naegele, J. A., and Everett, H. E., 1963, Cross-resistance patterns in the two-spotted spider mite, *in:* "Advance in Acarology I," J. G. Matthysse, W. D. McEnroe, B. V. Travis, K. N. Mehrotra and V. Dittrich, eds., Comstock Publ Co., New York, pp. 257-275.

Helle, W., and Ovemeer, W. P. J., 1973, Variability in tetranychid mites, *Ann. Rev. Ent.*, 18:97.

Henneberry, T. J., Adams, J. R., and Cantwell, G. E., 1964, Comparative electron microscopy of the integument of organophosphate resistant and non-resistant two-spotted spider mties (*Tetranychus telarius* L.), *Acalorogia*, 6:414.

Hirai, K., Miyata, T., and Saito, T., 1972, A comparison of the pesticide susceptibility of citrus red mite, *Panonychus citri* McGregor, treated by micro syringe application method and spraying method, *Jap. J. appl. Ent. Zool.*, 16:215.

Hirai, K., Miyata, T., and Saito, T., 1974, Penetration of ^{32}P-dimethoate into organophosphate-resistant and susceptible citrus red mite, *Panonychus citri* McGregor (Acarina: Tetranychidae), *Appl. Ent. Zool.*, 8:183.

Hoyt, S. C., and Harries, F. H., 1961, Laboratory and field studies on orchard-mite resistance to Kelthane, *J. Econ. Entomol.*, 54:12.

Inoue, K., 1979, The change of susceptibility of mite population to dicofol and genetic analysis of dicofol resistance in the citrus red mite, *Panonychus citri* (McG.), *J. Pesti. Sci.*, 4:337.

Inoue, T., and Saito, T., 1972, The susceptibilities of various stages of dicofol resistant and susceptible citrus red mite, *Panonychus citri* McGregor, against dicofol, *Jap. J. appl. Ent. Zool.*, 16:152.

Jeppson, L. R., 1963, Cross resistance in Acarina, *in:* "Advances in Acarology I," J. G. Matthysse, W. D. McEnroe, B. V. Travis, K. N. Mehrotra, and V. Dittrich, eds., Comstock Publ. Co., New York, pp. 276-282.

Jeppson, L. R., Complin, J. O., and Jesser, M. J., 1962, Effects of application programs on citrus red mite control and development of resistance to acaricides, *J. Econ. Entomol.*, 55:17.

Jeppson, L. R., Keifer, H. H., and Baker, E. W., 1975, "Mites Injurious to Economic Plants," University of California Press, Berkeley, pp. 614.

Kuwahara, M., 1977, Joint action of organophosphates, carbamates and synthetic synergists against ESP-selected and ESP-reversely-selected strains of Kanzawa spider mite, *Tetranychus kanzawai* Kishida, *Jap. J. appl. Ent. Zool.*, 21:94.

Matsumura, F., and Voss, G., 1964, Mechanism of malathion and parathion resistance in the two-spotted spider mite, *Tetranychus urticae*, *J. Econ. Entomol.*, 57:911.

Matsumoto, K., and Shinkaji, J., 1974, Difference of susceptibility against various acaricides between dicofol-resistant strain and susceptible strains of the citrus red mite, *Panonychus citri* (McGregor), *Jap. J. appl. Ent. Zool.*, 18:147.

Motoyama, N., and Saito, T., 1968, Substrate specificity of cholinesterase in mites, *Botyu-Kagaku*, 33:77.

Nomura, K., 1973, Review of acaricide resistance in red spider mites in Japan, *Rev. Plant Protec. Res.*, 6:44.

Nomura, K., and Nakagaki, S., 1959, On resistance of red spider mite, *Tetranychus cinnabarinus*, to methyl demeton (Metasystox), *Tech. Bull. Fac. Hort. Chiba Univ.*, 7:39.

Osakabe, M., 1971, Studies on insecticide resistance of the Kanzawa spider mite, *Tetranychus kanzawai* Kishida, I-III, *Rev. Plant Protect. Res.*, 4:132.

Osakabe, M., 1973, Studies on acaricide resistance of the Kanzawa spider mite, *Tetranychus knazawai* Kishida, parasitic on tea plant, *Bull. Nat. Res. Inst. Tea*, 8:1.

Sakai, M., 1967, Hydrolysis of acetylthiocholine and butyrylthiocholine by cholinesterases of insects and a mite, *Appl. Ent. Zool.*, 2:111.

Seki, M., 1958, Control of the citrus red mite, *Panonychus citri*, *Record of II Symposium Jap. Soc. Appl. Ent. Zool.*, 59:62.

Smissaert, H. R., 1964, Cholinesterase inhibition in spider mites

susceptible and resistant to organophosphates, *Science*, 143:129.
Smissaert, H. R., Voerman, S., Oostenbrugge, L., and Renooy, J., 1970, Acetylcholinesterases of organophosphate-susceptible and resistant spider mites, *J. Agr. Food Chem.*, 18:66.
Tabata, K., and Saito, T., 1970, Topical application of insecticide solutions to citrus red mite, *Panonychus citri* McGregor, *Jap. J. appl. Ent. Zool.*, 14:218.
Tabata, K., and Saito, T., 1971, Mechanism of dicofol resistance in spider mites I: Fate of topically applied ^{3}H-dicofol in citrus red mite, *Panonychus citri* McGregor, *Botyu-Kagaku*, 36:169.
Tabata, K., and Saito, T., 1973, Mechanism of dicofol resistance in spider mites II: Thin layer chromatographic identification of dicofol metabolites in citrus red mite, *Panonychus citri* McGregor, *Botyu-Kagaku*, 38:151.
Tabata, K., Miyata, T., and Saito, T., 1979, Water soluble metabolites of dicofol in mouse urine, *Appl. Ent. Zool.*, 14:490.
Takahashi, Y., Saito, T., Iyatomi, K., and Eto, M., 1972, Joint toxic action of organophosphorus compounds and various compounds to resistant citrus red mite I: Joint toxic action of various compounds with malathion and dimethoate to organophosphate-resistant citrus red mite, *Botyu-Kagaku*, 37:13.
Takahashi, Y., Saito, T., Iyatomi, K., and Eto, M., 1973, Joint toxic action of organophosphorus compounds and various compounds to resistant citrus red mite II: Mechanism of synergistic action between malathion and K-1 (2-phenyl-4H-1,3,2-benzodioxaphosphorin-2-oxide) in organophosphate-resistant citrus red mites, *Botyu-Kagaku*, 38:13.
Voss, G., 1959, Esterasen bei der Spinnmilbe, *Tetranychus urticae* Koch (Acari, *Trombidiformes*, Tetranychidae), *Naturwissenschaften*, 46:652.
Voss, G., 1960, Esterasen bie der Spinnmible, *Tetranychus urticae* Koch (Acari, *Trombidiformes*, Tetranychidae), *Naturwissenschaften*, 47:400.
Voss, G., and Matsumura, F., 1964, Resistance to organophosphorus compounds in the two-spotted spider mite: Two different mechanisms of resistance, *Nature*, 202:319.
Voss, G., and Matsumura, F., 1965, Biochemical studies on a modified and normal cholinesterase found in the Leverkusen strains of the two-spotted spider mite, *Tetranychus urticae*, *Can. J. Biochem.*, 43:63.

RESISTANCE TO BENZOMATE* IN MITES

Tomia Yamada, Hiromi Yoneda and Mitsuo Asada

Laboratory of Applied Entomology
Nisso Institute for Life Science
Nippon Soda Co., Ltd.
Oiso-machi, Kanagawa, Japan

INTRODUCTION

Benzomate (ethyl O-benzoyl 3-chloro-2,6-dimethoxybenzohydroximate) possesses high acaricidal activity against *Panonychus* mites (Asada and Yoneda, 1971; Sato, 1976) and was first used on citrus in Japan in 1971 for the control of citrus red mites, *Panonychus citri* (McGregor). However, it was found that resistance of mites to benzomate had developed in a few of these citrus areas in the period of 1973 to 1974 (Yoneda, 1973, unpublished; Mori and Asano, 1975). Mori et al. (1977) and Nishino (1977) reported that citrus red mite populations became resistant to benzomate after three applications of this acaricide in selection tests in the field.

In the present study, the cross resistance of benzomate-resistant citrus red mites to derivatives of benzomate and other acaricides was examined. Effects of synergists on the toxicity of benzomate to these mites were also evaluated.

MATERIALS AND METHODS

A strain of the citrus red mite, *Panonychus citri* (McGregor), resistant to benzomate was collected from an orchard at Arita City in Wakayama Prefecture in 1973 where benzomate had not shown satisfactory efficacy. Since then, the strain has been selected with benzomate two to three times a year in our laboratory. A susceptible

*Another common name: benzoxamate (ISO).

strain was obtained from citrus plants in our institute. Both resistant and susceptible strains were fed on citrus plants grown in pots in separate rearing rooms.

For the toxicological test, a detached citrus leaf placed on a petri dish was used. The surface of the leaf was surrounded with Tanglefood to prevent mites from escaping. A piece of wet cotton was placed on the tip of the leaf to provide moisture for the leaf. In the adulticidal test, 30 adult females inoculated on a detached leaf, and in the ovicidal test, about 100 eggs zero to three days old oviposited on a detached leaf, were sprayed with 3 ml of chemical solution by a rotary spray tower and then kept in a room at 25°C. Mortality was assessed in the adulticidal test at 72 hr and in the ovicidal test at seven days after treatment. For the oral test, a polyehtylene tube 0.017 mm thick was filled with chemical solution. The surface of the tube was surrounded with Tanglefood. Thirty adult females were inoculated and allowed to suck the toxicant through the tube. The tube was placed in a desicator that was regulated at 95% relative humidity with sodium phosphate and was kept in a room at 25°C. Mortality was assessed 48 hr after treatment, and LC_{50} was calculated by the Guddum method. The formula of acaricides and chemicals used in the present study were:

Benzomate	20% emulsifiable concentrate (EC)
Chlorfenethol (25%) - Chlorfenson (25%)	50% wettable powder (WP)
Chloropropylate	22% EC
Proclonol	40% WP
Chinomethionat	25% WP
Chlordimeform	60% soluble powder
Dialifor	40% EC

Compound 1: ethyl 3-chloro-2,6-dimethoxybenzohydroxamate
Compound 2: methyl 3,6-dichloro-2-methoxybenzohydroxamate
Compound 3: ethyl 3,6-dichloro-2-methoxybenzohydroxamate
Compound 4: n-propyl 3,6-dichloro-2-methoxybenzohydroxamate
Compound 5: allyl 3,6-dichloro-2-methoxybenzohydroxamate
Compound 6: ethyl 3-chloro-2,6-dimethoxybenzamide
Compound 7: allyl N-diethyl acetyl-2-methoxy-1-naphthohydroxamate
Piperonyl butoxide
Sesamex

Piperonyl butoxide, sesamex and Compounds 1 through 7 were dissolved in dimethylformamide containing 1.5% Tween 20.

RESULTS AND DISCUSSION

The susceptibilities of the benzomate-resistant and susceptible strains to several acaricides are shown in Table 1. The resistant strain was not cross-resistant to the mixture of chlorfenethol and

Table 1. Susceptibility of Benzomate-resistant Citrus Red Mites to Some Acaricides in the Adulticidal Test

Acaricide	LC_{50} (ppm)		R/S
	Susceptible Strain	Resistant Strain	
benzomate	6.9	374.6	54.3
dicofol	36.1	107.3	3.0
chlorfenethol-chlorfenson	324.6	252.6	0.8
chloropropylate	44.0	85.7	1.9
proclonol	82.9	364.3	4.4
chinomethionat	42.8	48.0	1.1
chlordimeform	355.0	446.0	1.3
dialifor	11.6	30.8	2.7

chlorfenson or to chloropropylate, chinomethionat or chlordimeform, but it was slightly tolerant to dicofol, proclonol and dialifor. Mori and Asano (1975) reported that the Kawachi strain of citrus red mites resistant to benzomate was not cross resistant to dicofol, proclonol, chlorobenzilate, chinomethionat or chlordimeform. Nomura et al. (1978) suggested that in citrus red mites resistant to both benzomate and dicofol, there was no cross resistance relationship between benzomate and dicofol, but there was multiple resistance to these two acaricides. Therefore, it should be investigated whether the different susceptibility of the Arita strain to dicofol, proclonol and dialifor was due to the cross resistance or to the difference in susceptibility of the original population. Table 2 presents the cross resistance relationships between benzomate and its derivatives in eggs of the resistant strains.

Citrus red mites resistant to benzomate were highly cross resistant to Compound 1, which was a main metabolite of benzomate in the citrus plant (Sato, 1976) and which possessed acaricidal activity against susceptible strains (LC_{50} of Compound 1: 17 ppm for adult females and 57 ppm for eggs). Mites were also resistant to Compounds 2, 3, 4 and 5, in which a methoxy group at the sixth position of benzomate was substituted with chlorine, and to Compound 7, which had a naphthyl ring instead of a phenyl ring of benzomate. Overmeer (1967) reported that two-spotted spider mites resistant to tetradifon were highly cross resistant to tetrasul and CPAS, which are related compounds. Citrus red mites resistant to benzomate were also highly cross resistant to all derivatives of benzomate.

Table 2. Cross Resistance of Benzomate-resistant Citrus Red Mites to Derivatives of Benzomate in the Ovicidal Test

No. of Compound	Chemical Structure	Mortality (%)		
		Susceptible Strain		Resistant Strain
		100 ppm	500 ppm	1000 ppm
benzomate	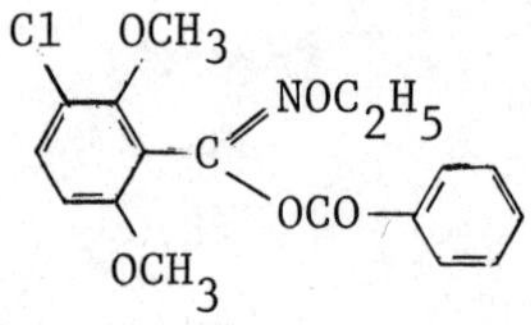	100	100	0
1	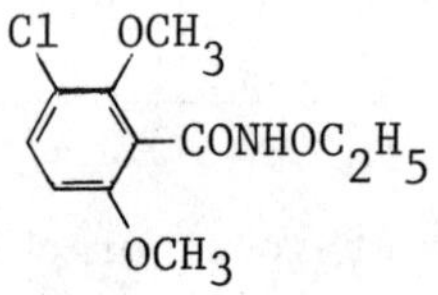	89	100	11
2	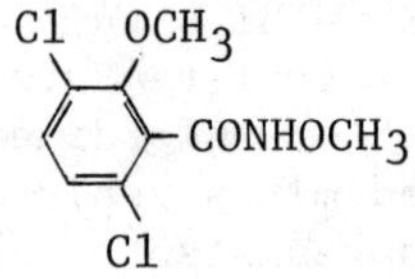	20	100	0
3	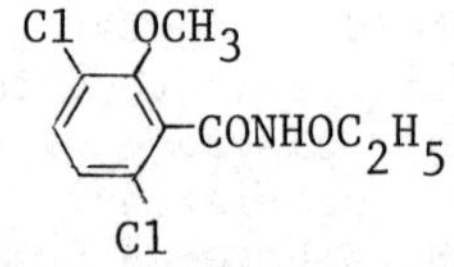	87	100	0
4	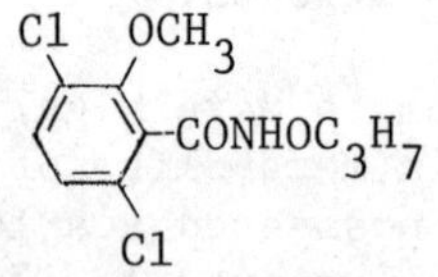	74	100	0
5	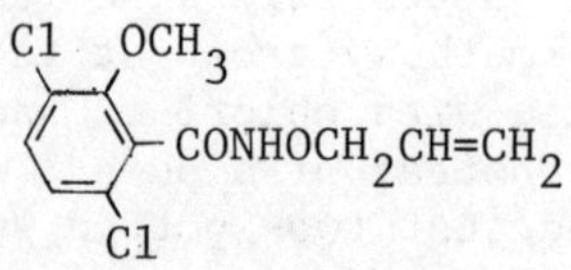	82	100	0

No. of Compound	Chemical Structure	Mortality (%)		
		Susceptible Strain		Resistant Strain
		100 ppm	500 ppm	1000 ppm
6	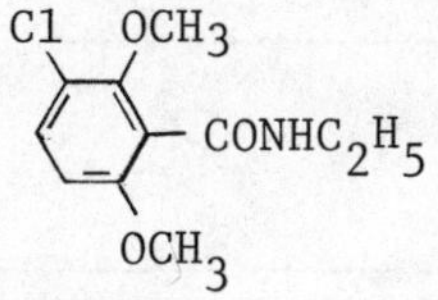		60	0
7	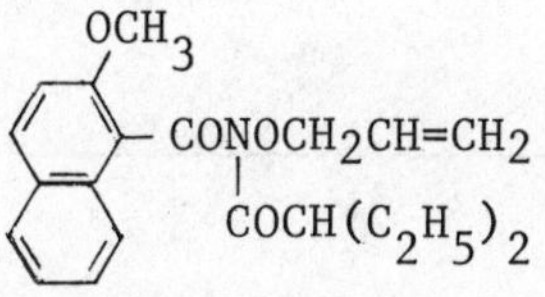		100	0

Benzomate possessed a differential toxicity to the citrus red mites and two-spotted spider mites (Asada and Yoneda, 1971; Sato, 1976), and a mixture of benzomate with piperonyl butoxide showed a synergistic action against two-spotted spider mites that were not susceptible to benzomate (Asado, 1976). Therefore, effects of piperonyl butoxide and sesamex on the toxicity of benzomate in the Arita strain were examined. The results are summarized in Table 3. It was found that these two synergists showed no effect on the toxicity of benzomate to the resistant strain by either spray or oral methods. Mori and Asano (1975) reported that a mixture of benzomate with piperonyl butoxide, safroxane, MGK-264 or synepirin-500 showed no synergistic action against the Kawachi strain by the spray method. In Table 3, it is also shown that the Arita strain was resistant to benzomate even when administered by the oral method.

SUMMARY

The citrus red mite, *Panonychus citri,* resistant to benzomate (ethyl O-benzoyl 3-chloro-2,6-dimethoxybenzohydroximate) was not cross-resistant to chloropropylate, chinomethionat, chlordimeform or the mixture of chlorfenethol and chlorfenson, but was highly cross-resistant to all derivatives of benzomate. Piperonyl butoxide and sesamex showed no effect on the toxicity of benzomate to the resistant strain.

Table 3. Effects of Synergists on the Toxicity of Benzomate to Benzomate-resistant Citrus Red Mites in Adulticidal Test

Method of Treatment	Acaricide or Synergist	Concn. (ppm)	Mortality (%)			
			Resistant Strain			Susceptible Strain
			Single Compound	P.B.[a] (1:1)[b]	Sesamex (1:1)	Single Compound
Spray	piperonyl butoxide	500	43			
		250	20			
	sesamex	500	23			
		250	20			
	benzomate	500	38	44	45	
		250	15	14	19	
Oral	piperonyl butoxide	100	0			
	sesamex	100	0			
	benzomate	100	8	0	0	
		2				100

[a]piperonyl butoxide.
[b]mixed ratio.

REFERENCES

Asada, M., 1979, Acaricide resistance in mites, *in:* "Pesticide Design," pp. 693-717, Soft Science Company, Tokyo.

Asada, M., and Yoneda, H., 1971, New acaricide, Benzomate, *AGCHE AGE* (Nippon Soda Co., Ltd.), 103:1.

Mori, S., Ogihara, H., and Ohmasa, Y., 1977, Study on acaricide resistance in citrus red mite, *in:* "Results of Tests on Insecticide and Acaricide Resistance," pp. 122-124, Japan Plant Protection Association.

Mori, Y., and Asano, K., 1975, Test of citrus red mite resistant to benzomate, Abstract of the 19th annual meeting of the Japanese

Society of Applied Entomology and Zoology at Tokyo, p. 537.
Nishino, M., 1977, Study on acaricide resistance in citrus red mite, *in:* "Results of Tests on Insecticide and Acaricide Resistance," pp. 109-116, Japan Plant Protection Association.
Nomura, K., Tokumura, J., and Uchida, T., 1978, Study on acaricide resistance in citrus red mite, *in:* "Results of Tests on Insecticide and Acaricide Resistance," pp. 92-100, Japan Plant Protection Association.
Overmeer, W.P.J., 1967, Genetics of resistance to Tedion in *Tetranychus urticae* C.L.Koch, *Arch. Neerl. Zool.,* 17:295.
Sato, N., 1976, Benzomate, *J. Pest. Sci.,* 3:123.

HERBICIDE RESISTANCE IN HIGHER PLANTS

Steven R. Radosevich

Department of Botany
University of California
Davis, California 95616

EXTENT OF HERBICIDE RESISTANCE

Many agricultural systems still rely heavily on mechanical methods (tillage) for weed control. Such control methods are highly effective in reducing the density of undesirable plants but are not believed to appreciably influence weed species composition. However, weed species have evolved in response to other agricultural practices (Baker, 1974; Young and Evans, 1976), and it is likely that they have responded to tillage, as well. Recent studies by Price et al. (1980) suggest that wild oat (*Avena fatua*) can be genetically manipulated by cultivation practices.

Herbicides also impose selection pressure on weed communities (Way and Chancellor, 1976). Differential susceptibility of various plant species to herbicides is well documented (Ashton and Crafts, 1973), and the continued use of a particular herbicide often causes a shift in plant populations from susceptible to more tolerant species. For example, a species shift that favors grass weeds is known to occur from yearly applications of 2,4-D [(2,4-dichlorophenoxy)acetic acid] for broadleaved weed control in cereals (Fryer and Chancellor, 1979; Hay, 1968). In spite of the widespread occurrence of pesticide resistance in other organisms, there are few examples of formerly susceptible weed species that have developed resistance. It has been suggested that the apparent lack of extensive herbicide resistance in plants is due to a combination of factors: the low selection pressure of most herbicides, lower fitness of resistant weed strains than susceptible ones, the ability of herbicide-thinned stands of susceptible weeds to produce high amounts of seed, and the large soil reservoir of susceptible weed seeds

Table 1. Appearance of Genetic Tolerance and Resistance to Herbicides[a]

Herbicide	Species	Type of Tolerance	Notes
Phenols		None reported	
Benzonitriles			
ioxynil	*Tripleurospermum inodorum*	diff. tolerance	Found with natural variations
Thiocarbonyls		None reported	
Quarternary ammoniums			
paraquat	*Lolium perenne*	diff. tolerance	Selected from strain differences
diquat	*Zea mays*	RESISTANT	Artificially selected albino
Phenoxy acids			
MCPA	*Tripleurospermum inodorum*	diff. tolerance	Found with natural variation
	Linum usitatissiumum	diff. tolerance	Varietal differences
2,4-D	*Cardaria chalapensis*	diff. tolerance	Strain differences in nature

Herbicide	Species	Type of Tolerance	Notes
Phenoxy acids (cont.)			
2,4-D (cont.)	*Citrus sinensis*	diff. tolerance	Selected in callus cultures
	Cirsium arvense	diff. tolerance	Clonal ecotypes
	Convolvulus arvensis	diff. tolerance	Clonal differences
	Cyperus esculentus	diff. tolerance	Varietal differences
	Daucus carota (wild)	diff. tolerance	Biotype variations
	Daucus carota (cultivated)	RESISTANT	Selected in cell cultures
	Lotus corniculatus	diff. tolerance	Repeated field selection
	Nicotiana sylvestris	RESISTANT	Selection in haploid tissue cultures
	Saccharum L.	diff. tolerance	Clonal differences
	Trifolium repens	diff. tolerance	Selected in cell cultures, resistant to other phenotypes
Benzoic acids			
TBA	*Cardaria chalepensis*	diff. tolerance	Strain difference in nature
dicamba	*Cirsium arvense*	diff. tolerance	Clone ecotypes

(continued)

Table 1. (continued)

Herbicide	Species	Type of Tolerance	Notes
Halogenated aliphatics			
dalapon	*Cynodon dactylon*	diff. tolerance	Biotype variation
	Echinochloa crusgalli	diff. tolerance	Biotype differences
	Saccharum L.	diff. tolerance	Clonal variations
	Setaria	diff. tolerance	-
	Sorghum halapense	diff. tolerance	Ecotype variations
	Lolium perrene	diff. tolerance	Cultivar variations
TCA	*Cynodon dactylon*	diff. tolerance	Biotype differences
Carbamates and Thiocarbamates			
propham	*Nicotiana tabacum*	diff. tolerance	Selected in isolated protoplasts
Amides			
propachlor	*Sorghum bicolor*	diff. tolerance	Found among 40 varieties
alachlor	*Zea mays*	diff. tolerance	Found in inbred lines and hybrids
	Daucus carota (wild)	diff. tolerance	Biotype variations

Herbicide	Species	Type of Tolerance	Notes
Ureas			
diuron	*Saccharum* L. cvs.	diff. tolerance	Clonal types
linuron and norea	*Sorghum bicolor*	diff. tolerance	Found among 40 varieties
siduron	*Hordeum jubatum*	diff. tolerance	Biotype differences controlled by three dominant genes
metoxuron	*Poa annua*	diff. tolerance	Artificially selected
	Saccharum L.	diff. tolerance	Clonal differences
Diazines			
bentazon	*Glycine max*	diff. tolerance	
Triazines[b]			
simazine	*Brassica napa*	diff. tolerance	Varietal responses
	Capsella bursa pastoris	diff. tolerance	Response related to number of repeated treatments
	Chenopodium album	diff. tolerance	Response related to number of repeated treatments
	Senecio vulgaris	diff. tolerance	Response related to number of repeated treatments

(continued)

Table 1. (continued)

Herbicide	Species	Type of Tolerance	Notes
Triazines (cont.)			
simazine (cont.)	*Sinapis alba*	diff. tolerance	Varietal difference
	Tripleurospermum inodorum	diff. tolerance	Wide natural variation
	Triticum aestivum	diff. tolerance	Varietal difference
atrazine	*Amaranthus retroflexus*	RESISTANT	Field strains
	Ambrosia artemisiifolia	RESISTANT	Field strains
	Brassica campestris	RESISTANT	Field strains
	Chenopodium album	RESISTANT	10 years repeated treatment in maize
	Cyperus esculentus	diff. tolerance	Varietal differences
	Echinochloa crusgalli	diff. tolerance	5 repeated treatments
	Glycine max	RESISTANT	Selected in cell cultures
	Linum usitatissimum	diff. tolerance	Quantitatively inherited
	Senecio vulgaris	RESISTANT	10 years repeated treatment in nursery
	Setaria sp.	diff. tolerance	In repeated sprayed vineyard

Herbicide	Species	Type of Tolerance	Notes
Triazines (cont.)			
propazine	*Sorghum bicolor*	diff. tolerance	Among 40 varieties
Miscellaneous			
amitrole	*Cirsium arvense*	diff.tolerance	Clonal ecotypes
picloram	*Nicotiana tabacum*	RESISTANT	Selected in cell cultures

[a]Adapted from Gressel, 1979.
[b]Species tolerant to one *s*-triazine are usually tolerant to others.

(Conard and Radosevich, 1979; Gressel and Segal, 1978; Holiday et al., 1976).

Resistant strains of several plant species have been laboratory selected or developed through cell culture. Miles (1976) has artificially selected albino seedlings of corn (*Zea mays* L.) that are physiologically resistant to diquat (1,1'-ethylene-2-2'-dipyridylium bromide). Other workers have developed 2,4-D-resistant plants of carrot (*Daucus carota*) and picloram (4-amino-3,5,6-trichloropiclonic acid) resistant strains of tobacco (*Nicotiniana tabacum*) by cell culture (Chaleff and Parsons, 1978; Gressel, 1979). Other studies have demonstrated a genetic basis for differential herbicide tolerance within a plant species (Schooler et al., 1972). Such studies indicate a potential for herbicide resistance to be developed by various plant species if proper conditions exist.

Some natural weed populations have developed resistance to one or more herbicides (Table 1). In every case, resistance has occurred in the field after approximately ten years of successive atrazine [2-chloro-4-(ethylamino)-6-(isopropylamino)-*s*-triazine] or other *s*-triazine treatment. Biotypes resistant to atrazine are also resistant to other *s*-triazines, but they are susceptible to non-triazine herbicides (Machado et al., 1978; Radosevich and Appleby, 1973a; Ryan, 1970).

MECHANISM FOR *s*-TRIAZINE RESISTANCE IN PLANTS

Atrazine and other triazine herbicides are usually applied to the soil, absorbed by roots, and translocated in the xylem to the leaves where they inhibit photosynthesis (Ashton and Crafts, 1973; Gunther, 1976). Differential susceptibility among plant species to atrazine has been well documented (Ashton and Crafts, 1973), and soil placement of the herbicide in relation to germination or rooting depth, differential uptake, or metabolism to nontoxic herbicide metabolites are cited most to account for the different responses of plant species to atrazine (Ashton and Crafts, 1973).

Radosevich and Appleby (1973a) grew two biotypes of common groundsel (*Senecio vulgaris* L.), reported to be susceptible and resistant to atrazine (Ryan, 1970), in aqueous nutrient solutions containing various concentrations of several *s*-triazine herbicides. Plants of the susceptible (S) biotype soon became chlorotic and died, but plants of the resistant (R) biotype did not (Table 2, Figure 1). Experiments conducted in this way excluded germination and rooting characteristics as factors affecting herbicide tolerance. These data (Radosevich and Appleby, 1973a) and those of Ryan (1970), in which both biotypes were killed by non-triazine herbicides suggested that the mechanism of resistance was physiological in nature and restricted to the triazine structure.

Table 2. Dosages of Six *s*-Triazine Herbicides Required to Cause Complete Necrosis (LD_{100}) in Two Biotypes of *Senecio vulgaris* Grown in Greenhouse Nutrient Cultures[a]

Herbicide	LD_{100}[b]	
	Susceptible Biotype	Resistant Biotype
simazine	0.1 - 0.5	> 4[c]
atrazine	0.1 - 0.5	>30
GS-14254	0.5 - 1.0	>30
prometone	0.1 - 1.0	>30
terbutryn	>30	>30
prometryn	0.0 - 4.0	>30

[a]From Radosevich and Appleby 1973a.
[b]LD_{100} estimated visually in four replications in each of two experiments. Exposure times and experimental conditions were identical for both biotypes.
[c]The maximum concentration of simazine used was 4 ppm because of its low water solubility.

Uptake and Translocation

Most plants are capable of absorbing atrazine, simazine [2-chloro-4,6-bis(ethylamine)-*s*-triazine] and other triazine herbicides by their roots (Warwick, 1976). Uptake studies with S and R biotypes of common groundsel (Radosevich and Appleby, 1973b) revealed no difference in the ability of either biotype to absorb simazine-^{14}C (Figure 2). A later study by Radosevich and DeVilliers (1977) demonstrated a similar pattern of atrazine-^{14}C translocation by the two biotypes. Other absorption studies with atrazine-susceptible and -resistant biotypes of common lambsquarters (*Chenoposium album* L.) and redroot pigweed (*Amaranthus retroflexus* L.) confirmed these observations (Jensen et al., 1977; Radosevich, 1977).

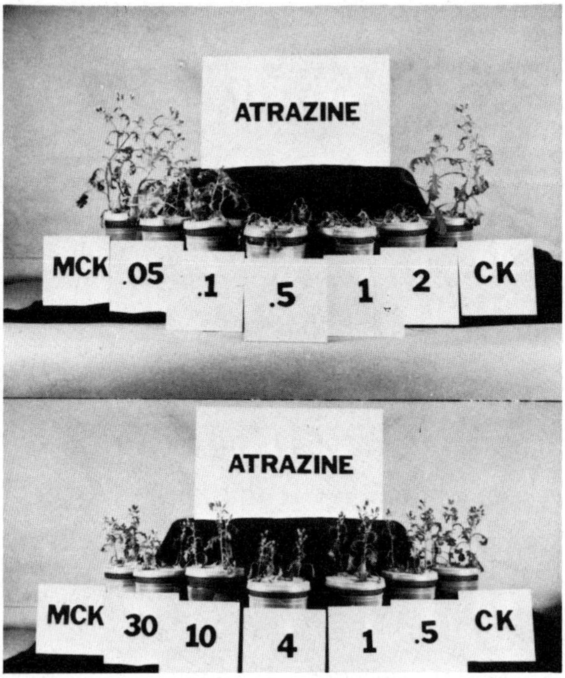

Figure 1. Effect of atrazine on a susceptible (top) and a resistant (bottom) biotype of *Senecio vulgaris*. Rates are in ppm (w/v) in nutrient solution. CK refers to plants grown in untreated nutrient solution. MCK (methanol check) refers to plants subjected to the amount of methanol required in the highest atrazine rates (from Radosevich and Appleby, 1973a).

Herbicide Metabolism

Extensive research (Hamilton and Moreland, 1952; Castelfranco et al., 1961; Shimabukuro et al., 1966, 1970, 1971) has revealed three metabolic pathways for triazine (atrazine) metabolism in plants (Figure 3). These include: dealkylation of aminoalkyl side chains (forming chloroform-soluble metabolites), hydroxylation at the 2-position of the triazine ring, and conjugation with glutathione. Both hydroxylation and glutathione conjugation form water-soluble degradation products (Figure 3). Corn (*Zea mays* L.), which is highly tolerant to atrazine, possesses each of these mechanisms. They are so effective in that species that atrazine does not accumulate to toxic levels.

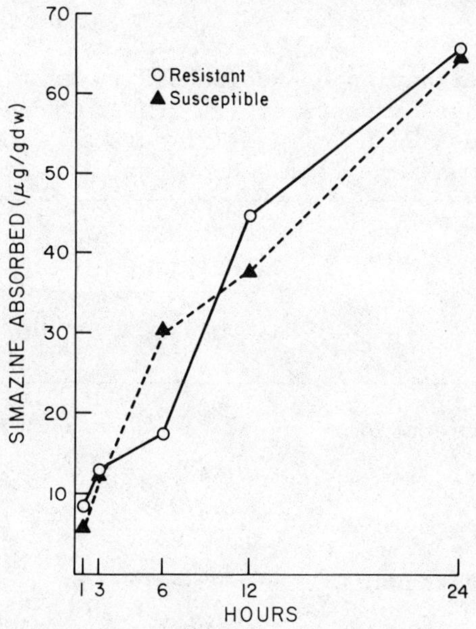

Figure 2. Simazine-^{14}C absorption by two biotypes of *Senecio vulgaris* (from Radosevich and Appleby, 1973b).

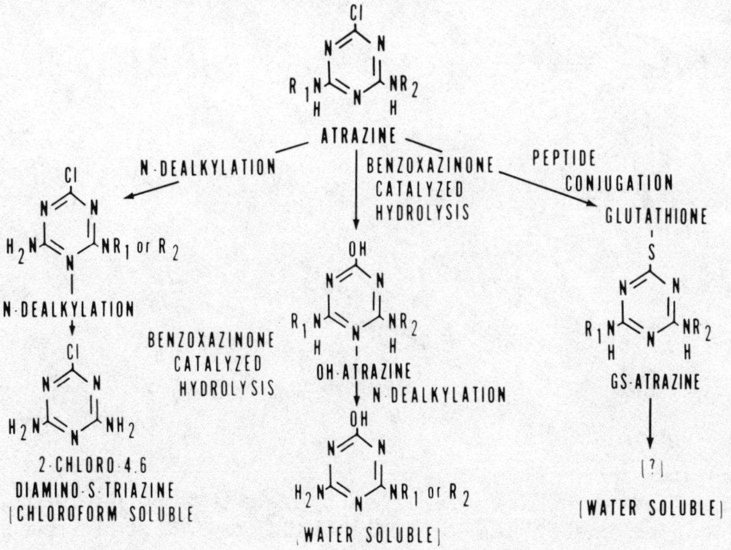

Figure 3. Metabolic scheme for atrazine metabolism in higher plants (adapted from Shimabukuro, 1966, 1970, 1971).

Table 3. The Distribution of Radioactivity from Applied Simazine-^{14}C in Foliage Extracts of Two *Senecio vulgaris* Biotypes and Atrazine-^{14}C in Foliage Extracts of *Chenopodium album* and *Amaranthus retroflexus* Biotypes[a]

Species	Biotype	Exposure Time	^{14}C Recovered from Foliage Extraction	
		(hr)	Chloroform (%)	Water (%)
S. vulgaris	Susceptible	24	79	9
		48	81	14
		96	83	14
S. vulgaris	Resistant	24	72	12
		48	82	13
		96	83	13
C. album	Susceptible	12	95	5
		36	88	12
		60	85	12
C. album	Resistant	12	93	7
		36	91	9
		60	81	19
A. retroflexus	Susceptible	12	82	18
		36	64	36
		60	56	44
A. retroflexus	Resistant	12	84	16
		36	55	45
		60	50	50

[a]Differences between biotypes of each species were not statistically significant at the 5% level of probability. Adapted from Radosevich 1977 and Radosevich and Appleby 1973b.

Table 4. Rf Values of Simazine-^{14}C or Atrazine-^{14}C Recovered from Chloroform Foliage Extracts of Two Biotypes of *Senecio vulgaris*, *Chenopodium album* and *Amaranthus retroflexus*[a]

Simzaine	Atrazine	*Senecio vulgaris*		*Chenopodium album*		*Amaranthus retroflexus*	
		R	S	R	S	R	S
.35	.57	.34	.33	.57	.55	.57	.58

[a]Averages of six replicates for common groundsel and nine prelicates for *C. album* and *A. retroflexus*. Differences in Rf values between simazine and *S. vulgaris* and atrazine and *C. album* and *A. retroflexus* are not significantly different at 5% level of probability. Adapted from Radosevich 1977 and Radosevich and Appleby 1973b.
[b]R = resistant; S = susceptible.

When metabolism studies were conducted with R and S biotypes of common groundsel (Radosevich and Appleby, 1973b), common lambsquarters (Jensen et al., 1977; Radosevich, 1977) and redroot pigweed (Radosevich, 1977), the occurrence of water-soluble herbicide metabolites varied among species, but the greatest concentration of ^{14}C (from either applied simazine-^{14}C or atrazine-^{14}C) occurred in chloroform partitions of foliage extracts of both biotypes (Table 3). Little ^{14}C was observed in root extracts. The radioactivity in the chloroform fraction of foliage extract was determined by thin layer chromatography to be intact simazine-^{14}C or atrazine-^{14}C (Table 4). Some water-soluble atrazine or simazine metabolites were always observed, but time course studies revealed no differences between biotypes in either the amount or rate of accumulation of these metabolites (Table 3). These data are in contrast to the abundance of literature that implicates herbicide metabolism to nontoxic metabolites as the predominant mechanism for differential tolerance among species (Davis et al., 1959; Hamilton, 1964; Shimabukuro et al., 1966, 1970, 1971; Thompson et al., 1971). It must be concluded, therefore, that differential degradation of the herbicide does not play a major role in accounting for the extreme differences in atrazine tolerance observed between S and R weed biotypes.

Photosynthetic Responses of S and R Weed Biotypes

Atrazine and simazine are well known as potent inhibitors of photosynthesis. Radosevich and Appleby (1973b) measured the net photosynthetic response of S and R common groundsel biotypes in the presence and absence of simazine. A similar study, involving the

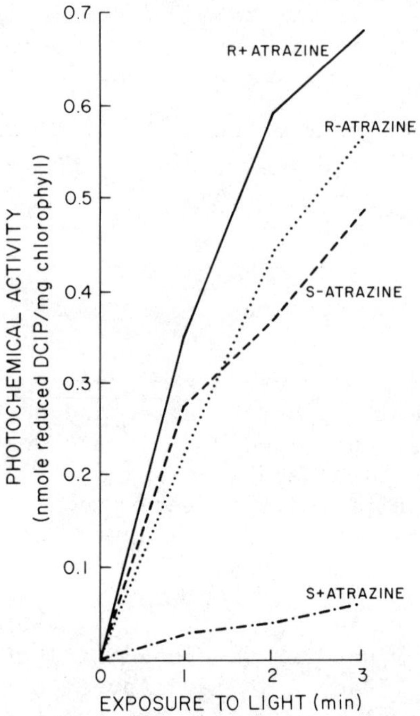

Figure 4. Photochemical activity of isolated chloroplasts of resistant (R) and susceptible (S) biotypes of *Senecio vulgaris*, each exposed to 5 μM atrazine (+ atrazine) or untreated (- atrazine). Averages of six isolations for each biotype. LSD (0.05) were 0.20, 0.29 and 0.17, respectively (from Radosevich and DeVilliers, 1976).

differential gas exchange between S and R biotypes of redroot pigweed, has been conducted by West et al. (1976). Rapid inhibition of net CO_2 fixation occurred when the S biotype of either species was subjected to atrazine or simazine. Net photosynthesis of the R biotypes was not altered by the presence of either herbicide.

Interference with photochemically induced electron transport and oxygen evolution (photosystem II) has been demonstrated for a large number of structurally divergent chemicals (Moreland, 1969). These inhibitors include many members of the triazine, urea and uracil chemical families. Simazine and atrazine are known to rapidly inhibit the oxygen evolution (Hill reaction) of isolated chloroplasts of barley (*Hordeum vulgare* L.), turnip (*Brassica* sp.), soybean (*Glycine max* L.) and corn; however, the photochemical activity of isolated chloroplasts of R common groundsel, redroot pigweed and common lambsquarters was not inhibited by atrazine (Table 5 and

Table 5. Photochemical Activity of Isolated Chloroplasts of Two Biotypes of Common Lambsquarters and Redroot Pigweed[a]

		Atrazine					
		Present			Absent		
		min light exposure			min light exposure		
Species	Biotype	1	2 (%)	3	1	2 (%)	3
lambs-quarters	resistant	17	32	44	21	37	52
	susceptible	8	21	31	40	73	86
	$LSD_{.05}$	6	12	13	6	12	13
pigweed	resistant	20	38	54	22	42	54
	susceptible	7	13	19	15	37	58
	$LSD_{.05}$	5	10	14	5	10	14

[a]Isolated chloroplasts of each biotype exposed to 5 μM atrazine or untreated. Data are nmole reduced DCIP/mg chlorophyll expressed as percent of dark control. Averages of six isolations for each biotype (from Radosevich, 1977).

Figure 4). Chloroplasts of S biotypes of those species were severely inhibited by the herbicide (Radosevich and DeVilliers, 1976; Radosevich, 1977). More extensive studies (Arntzen et al., 1979; Machado et al., 1978; Radosevich et al., 1979) indicated that differences in triazine sensitivity between the S and R weed biotypes was related to inherent differences in the thylakoid membranes of the biotypes and not to selective herbicide exclusion by R-type chloroplasts. Stroma-free thylakoid membrane suspensions isolated from R weed biotypes were 60 to 3200 times more tolerant to triazines and other photosynthetic inhibitors than thylakoids from S biotypes (Figure 5).

Structure and the Activity of Hill Reaction Inhibitors in S and R Weed Biotypes

Several studies have attempted to relate chemical structure to the activity of various Hill reaction (PS II) inhibitors (van Assche

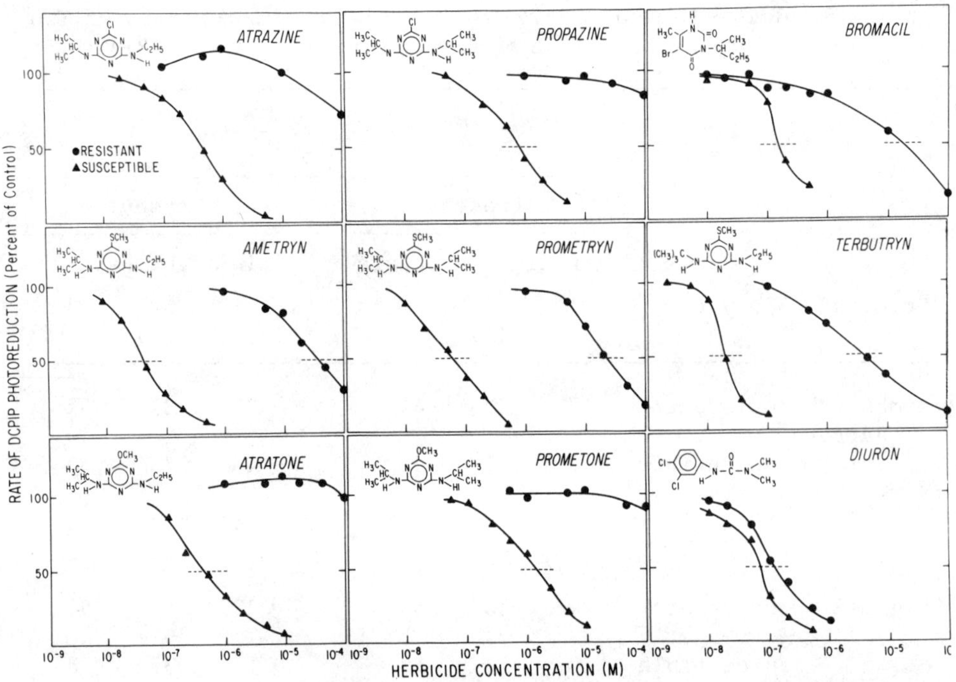

Figure 5. Rate of reduction of DCPIP by isolated chloroplasts of atrazine-resistant (●) and -susceptible (▲) biotypes of *Senecio vulgaris* with various concentrations of chloro-s-triazines (atrazine, propazine), methylthio-*s*-triazines (ametryn, prometryn, terbutryn), methoxy-*s*-triazines (atratone, prometone), bromacil, and diuron. Data points represent means of replicated observations. Standard errors of the mean are no larger than the plotted points (from Radosevich et al., 1979).

and Ebert, 1976; Hansch, 1969; Moreland, 1969). Thylakoid suspensions of S and R biotypes of common groundsel and common lambsquarters were most affected by the methylthio-*s*-triazines and least affected by the methoxy-*s*-triazines (Figure 5). Chloro-*s*-triazines were intermediate in effect (Figure 5). These observations agree with those of other workers concerning the activity of numerous substituted *s*-triazines (van Assche and Ebert, 1976). Hill reaction inhibition of S-type common groundsel thylakoids was favored by asymmetric alkylamino substitution at the 4 and 6 positions of the triazine ring (Figure 5). These data support the observations of Moreland (1969) and indicate that the orientation of the triazine molecule at the active site may be controlled by alkylamino substitutes of ring positions 4 and 6. This relationship was less apparent with the R type because of its greater tolerance.

Resistant-type thylakoid membranes of common groundsel were also more tolerant of diuron [3-(3,4-dichlorophenyl)-1,1-diethylurea] and bromacil (5-bromo-3-*sec*-butyl-6-methyluracil) than S-type thylakoids. Diuron was the most effective inhibitor of the R-type, and bromacil was intermediate in activity. All photosystem II inhibitors (e.g., ureas, uracils and triazines) have a -C-NH group as a basic structural element, and the lipophilic substituent at the nitrogen is usually an aromatic ring (Hansch, 1969). The data indicate that the affinity of these inhibitors for the reactive site may be altered in R thylakoids but not in S thylakoids.

Polypeptide Alteration of R-Type Thylakoid Membranes

Arntzen et al. (1979) reported alterations in the electrophoretic mobility of two thylakoid membrane polypeptides in the R biotype of redroot pigweed (Figure 6). However, electrophoretic analysis of thylakoid membrane polypeptides of S and R biotypes of common groundsel revealed no detectable changes in molecular weight of any of the membrane polypeptides of the R biotype when compared to the S biotype (Radosevich et al., 1979). This observation does not rule out the possibility of changes in primary structure of membrane polypeptides that do not result in detectable changes in molecular weight as determined by electrophoretic mobility. The mechanism for herbicide resistance appears to be due to specific structure or conformational changes in one or more of the thylakoid membrane constituents and such alterations can apparently reduce the affinity of the triazine herbicides for the active site itself.

Inhibitor Binding to Thylakoid Membranes of S and R Common Groundsel Biotypes

Diuron is among the most extensively studied inhibitors of photosynthesis. It is commonly believed that this compound acts as a specific photosystem II inhibitor between a primary electron acceptor (Q) and the plastoquinone pool of the photosynthetic electron transport chain (Isawa, 1977). Recent observations suggest that diuron interacts directly with the second carrier on the reducing side of photosystem II, thereby altering its redox properties (Velthuys, 1976). Atrazine, simazine and diuron are generally concluded to posses a similar site of action (Isawa, 1977). Additional evidence indicates that inhibition of photosynthetic electron transport by diuron and atrazine is related to high affinity binding of the inhibitor to electron carrier molecules associated with the thylakoid membrane (Tischer and Strotman, 1977). Binding studies with diuron-^{14}C and atrazine-^{14}C were conducted by Pfister et al. (1979) to directly characterize the amount of bound inhibitor, affinity and number of binding sites for S and R common groundsel chloroplasts.

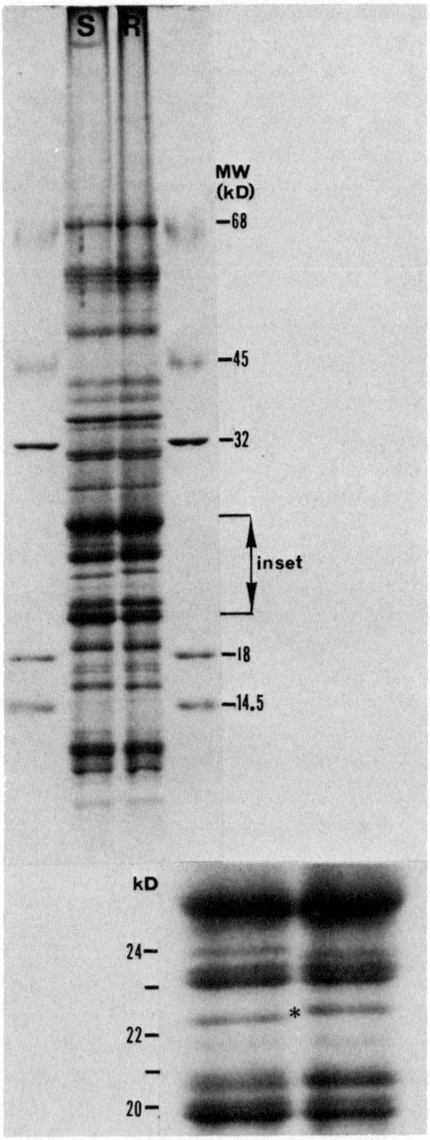

Figure 6. Polypeptide complement of chloroplast thylakoids isolated from triazine-susceptible (S) and triazine-resistant (R) biotypes of *Amaranthus retroflexus*. Thylakoids were washed twice with 0.75 mM EDTA to remove extrinsic and stromal proteins. Proteins of known molecular weight were electrophoresed in slots on either side of the membrane samples; the sizes (in K daltons) of these standards are indicated. A protein with molecular weight near 22Kd (indicated by * in the insert) was found to differ in

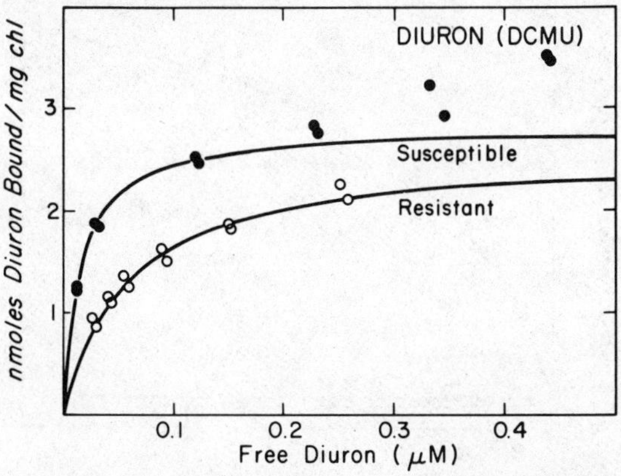

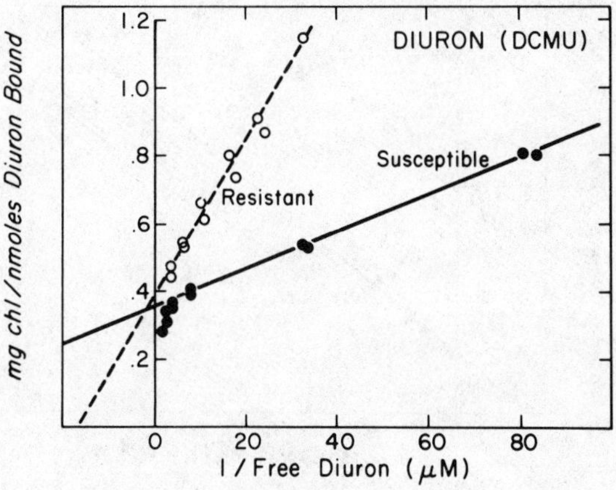

Figure 7. (A) Binding of ^{14}C-diuron to susceptible and resistant chloroplast membranes (from Pfister et al., 1979). (B) Double-reciprocal plot of the data from Figure 7A (from Pfister et al., 1979).

The amount of diuron binding to chloroplast membranes increased as the concentration of free, unbound diuron was increased (Figure 7A). From these plots (bound vs. free diuron) it is apparent that R common groundsel chloroplasts have slightly less affinity for diuron than the S chloroplasts. A more quantitative analysis of the binding data is possible in a double reciprocal plot (Figure 7B), in which the data are transformed to linear relationships. The intercept on the ordinate is a measure of the maximum number of available binding sites (chlorophyll basis) and thus allows the determination of the amount of bound inhibitor per "photosynthetic

---

apparent molecular weight between the two membrane samples (from Arntzen et al., 1979).

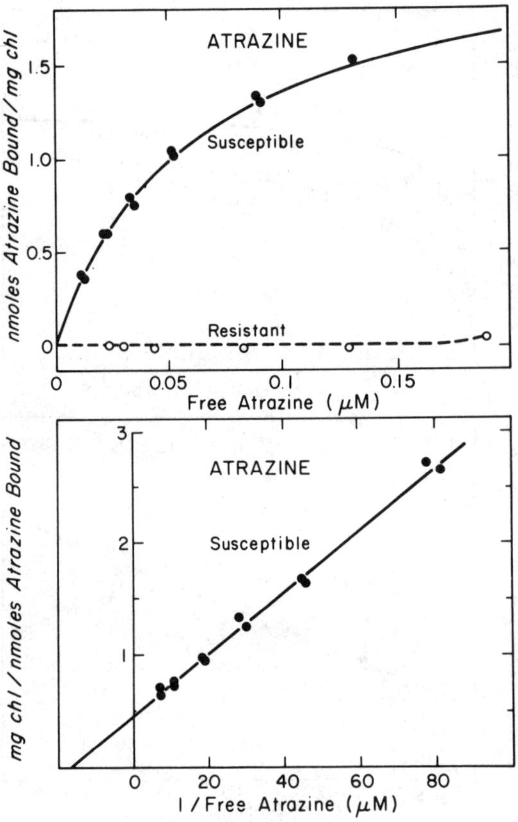

Figure 8. (A) Binding of ^{14}C-atrazine to susceptible and resistant chloroplast membranes. Note that no atrazine binding to resistant chloroplasts was observed at these inhibitor concentrations (from Pfister et al., 1979).

unit." The double reciprocal plot for diuron in the S chloroplasts clearly shows two different absorption processes: one occurring at low inhibitor concentrations with high affinity and another at higher inhibitor concentrations with lower affinity. Tischer and Strotmann (1977) have called the second process "unspecific absorption" because this inhibitor binding was not correlated with the inhibition of electron transport.

Atrazine binding to S- and R-type chloroplasts of common groundsel (Figure 8) was analyzed and the data are presented in the same fashion as was described for diuron. The calculated values for the binding constants and the number of binding sites for both inhibitors are presented in Table 6. Over the same atrazine concentration range, a completely different binding pattern was observed

Table 6. Calculated Binding Constants (K) and Number of Binding Sites (X_I; Chlorophylls per One Bound Inhibitor Molecule) for Diuron and Atrazine[a]

Inhibitor	Susceptible Chloroplasts		Resistant Chloroplasts	
	K	X_I	K	X_I
Diuron	1.4×10^{-8} M	420 chl/inh.	5×10^{-8} M	50 chl/inh.
Atrazine	4.0×10^{-8} M	450 chl/inh.	- no binding detected -	

[a]The values were determined by regression analysis of reciprocal plots similar to those shown in Figures 7B and 8B (correlation coefficients were between 0.997 and 0.970). The data are averages of six experiments (from Pfister et al., 1979).

between the two chloroplast samples. Susceptible chloroplasts bound atrazine, but no binding was detected in the R chloroplast sample.

The fact that diuron and atrazine appear to inhibit the same step in electron transport does not necessarily indicate that the two inhibitors act at the same binding site. This information can only be obtained by direct measurements of binding and anlaysis of competition between inhibitors. Tischer and Strotmann (1977) have analyzed competition between these herbicides in spinach chloroplasts and have indicated that both inhibitors act at the same site. This conclusion has now been verified for chloroplasts isolated from the triazine-S biotype of common groundsel (Pfister et al., 1979). In contrast, very weak competition between diuron and atrazine was observed in the R chloroplasts. This latter observation is directly related to the fact that a high affinity binding site for diuron but not atrazine was detected in the R chloroplasts.

Herbicide Binding Model for Resistance. The relationship between structural specificity of atrazine and diuron and the activity of these inhibitors in S and R common groundsel chloroplasts is summarized in Figure 9 (Pfister et al., 1979). In this diagram, the general asymmetric organization of electron transport carriers in the chloroplast thylakoid is indicated and two different regions within the herbicide binding component (protein) of the thylakoid are apparent. There is an "essential region," which interacts with a common structural element of different herbicides and thus interrupts electron transport. Another domain is also necessary for specific attachment or orientation of the inhibitors to the binding

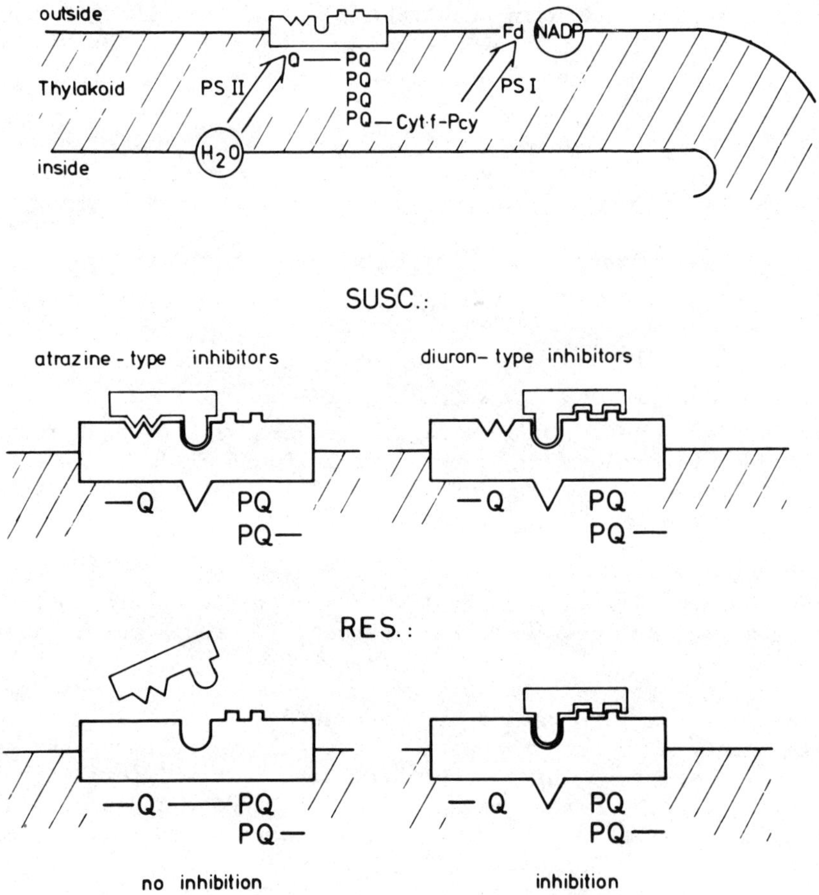

Figure 9. Model of inhibitor binding to chloroplast membranes (from Pfister et al., 1979).

component. Modification of one substructure of the binding component in the R chloroplast causes loss of atrazine binding, whereas diuron binding to the same constituent is not substantially affected.

COMPETITION AND ECOLOGICAL FITNESS

Field observation of weed populations after atrazine application always reveals very few resistant (surviving) plants. Furthermore, under noncompetitive conditions atrazine-susceptible (S) plants always produce more dry matter than resistant (R) ones (Conard and Radosevich, 1979). These observations suggest that R biotypes are less fit than S ones when grown in the absence of atrazine. Under competitive conditions S biotypes of both common

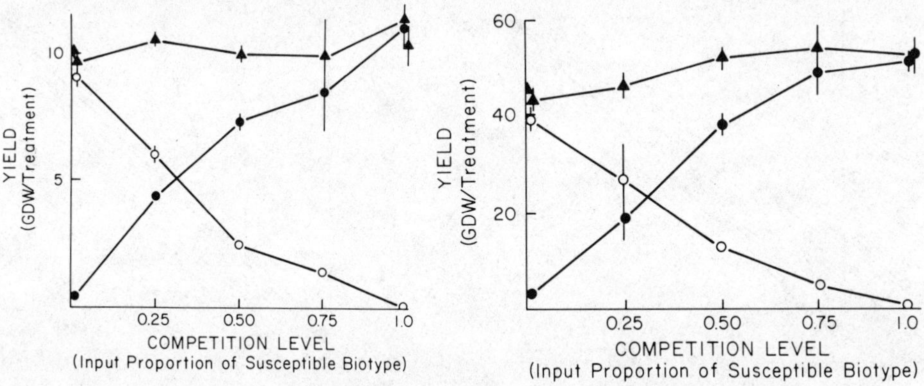

Figure 10. The dry weight of biotypes of common groundsel (left) and redroot pigweed (right) that are susceptible (•) or resistant (o) to atrazine. The biotypes were grown together in various proportions but at constant overall density (de Witt, 1960). The yields of the mixture of susceptible and resistant biotypes (▲) are also shown. Vertical lines represent 95 percent confidence limits (from Conard and Radosevich, 1979).

groundsel and redroot pigweed also produce more biomass and seed yield than R plants (Figures 10 and 11). The changeover of the population from 98 percent R to 98 percent S (Figure 11) resulting from this competitive advantage would require about 10 generations for both weed species (Conard and Radosevich, 1979). Selective pressure of such magnitude is sufficient to explain the low incidence of R biotypes when no herbicide is applied.

Light-dependent CO_2 fixation (intact leaves) and O_2 evolution (chloroplasts) revealed striking differences between S and R biotypes of common groundsel (Holt et al., 1981). At similar conditions of illumination and CO_2 concentration the photosynthetic capacity of S plants is markedly higher than that of R plants. Higher levels of oxygen evolution are also evident in S common groundsel chloroplasts than in those from the R biotype. These studies reveal that the R biotype of common groundsel is much less efficient in the light reactions of photosynthesis and carbon fixation than the S biotype. Given the poor photosynthetic performance, lowered vigor and reduced competitive fitness of the resistant biotype, triazine resistance is apparently only of benefit under conditions where those herbicides are repeatedly used.

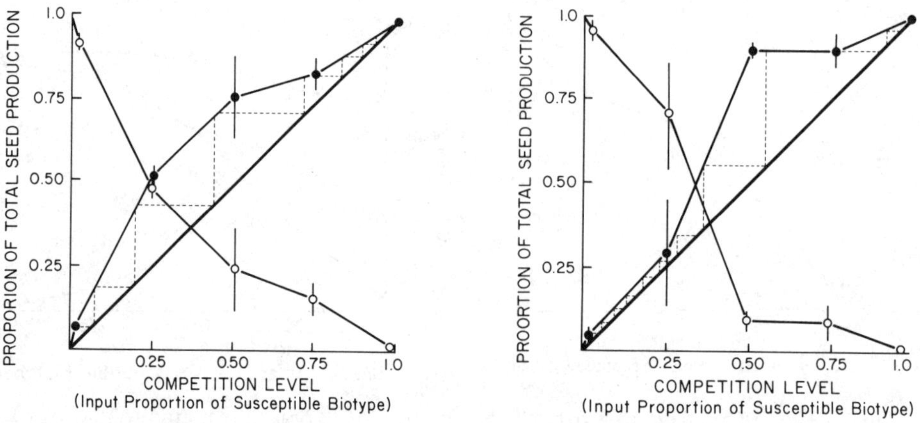

Figure 11. The proportion of total seed production (dry weight) of biotypes of common groundsel (left) and redroot pigweed (right) that are susceptible (•) and resistant (o) to atrazine. The biotypes were grown together at varying proportions but at constant overall density (de Witt, 1960). The predicted number of generations for the population to change from 98 percent resistant to 98 percent susceptible biotype is indicated by the dashed line. Theoretical yield of the biotypes, if they had identical competitive abilities, is indicated by the solid diagonal line. Vertical lines represent 95 percent confidence limits (from Conard and Radosevich, 1979).

REFERENCES

Arntzen, C. J., Ditto, C., Brewer, P., 1979, Chloroplast membrane alterations in triazine-resistant *Amaranthus* biotypes, *Proc. Natl. Acad. Sci. U.S.*, 76:278.

Ashton, F. M., and Crafts, A. S., 1973, "Mode of Action of Herbicides," Wiley and Sons, New York, 504 p.

Baker, H. G., 1974, The evolution of weeds, *Ann. Rev. Ecol. and System.*, 5:1.

Castelfranco, P. C., Foy, C. L., and Deutch, D. B., 1961, Nonenzymatic detoxification of 2-chloro-4,6-bis(ethylamino)-*s*-triazine (simazine) by extracts of *Zea mays*, *Weeds*, 9:580.

Chaleff, R. S., and Parsons, M. F., 1978, Direct selection in vitro for herbicide resistant mutants of *Nicotiana tabacum*. *Proc. Natl. Acad. Sci. U.S.*, 75:5108.

Conard, S. G., and Radosevich, S. R., 1979, Ecological fitness of *Senecio vulgaris* and *Amaranthus retroflexus* biotypes susceptible or resistant to atrazine, *J. Appl. Ecol.*, 16:171.

Davis, D. E., Funderburk, H. H., and Sansing, N. G., 1959, The absorption and translocation of ^{14}C-labeled simazine by corn, cotton, and cucumber, *Weeds*, 7:300.

Fryer, J. D., and Chancellor, R. J., 1979, Herbicides and our changing weeds, *in:* "The Flora of a Changing Britain," F. Perring, ed., pp. 105-117, Classey, Hampton, U.K.

Gressel, J., 1979, Genetic herbicide resistance: Projections on appearance in weeds and breeding for it in crops, *in:* "Plant Regulation and World Agriculture," T. K. Scott, ed., Plenum Press, New York.

Gressel, J., and Segal, S. D., 1978, The paucity of plants evolving resistance to herbicides: Possible reasons and implications, *J. Theo. Biol.*, 75:349.

Gunther, F. A., 1976, Effects of triazine herbicides on the physiology of plants, *Res. Rev.*, 65:1.

Hamilton, R. H., 1964, Tolerance of several grass species to 2-chloro-s-triazine herbicides in relation to degradation and content of benzoxazinone derivatives, *J. Agr. Food Chem.*, 12:14.

Hamilton, H. R., and Moreland, D. E., 1962, Simazine degradation by corn seedlings, *Science*, 135:373.

Hansch, C., 1969, Theoretical considerations of the structure-activity relationship in photosynthesis inhibitors, *Prog. in Photosynth. Res.*, 3:1685.

Hay, J. R., 1968, The changing weed problem in the prairies, *Agr. Inst. Rev.*, 23:17.

Holliday, R. J., Putwain, P. D., and Dafni, A., 1976, The evolution of herbicide resistance in weeds and its implications for the farmer, *Proc. 1976 Brit. Crop Prot. Conf. Weeds*, p. 937.

Holt, J. S., Stemler, A. J., and Radosevich, S. R., 1981, Differential light responses of photosynthesis by triazine resistant and susceptible *Senecio vulgaris* biotypes, *Plant Physiol.*, 67:744.

Isawa, S., 1977, Inhibitors of electron transport, *in:* "Encyclopedia of Plant Physiology, New Series Vol. 5, Photosynthesis I," A. Trebst and M. Avron, eds., pp. 266-282, Springer Publ. Co., New York.

Jensen, K. I. N., Bandeen, J. D., and Souze-Machado, V., 1977, Studies on the differential tolerance of two lambsquarters selections to *s*-triazine herbicides, *Can. J. Plant Sci.*, 57:1169.

Machado, V. S., Arntzen, C. J., Bandeen, J. D., and Stephenson, G. R., 1978, Comparative triazine effects upon system II photochemistry in chloroplasts of two common lambsquarters (*Chenopodium album*) biotypes, *Weed Sci.*, 26:318.

Miles, C. D., 1976, Selection of diquat resistant photosynthesis mutants from maize, *Plant Physiol.*, 57:284.

Moreland, D. E., 1969, Inhibitors of chloroplast electron transport: Structure-activity relations, *Prog. Photosyn. Res.*, 3:1693.

Pfister, L. K., Arntzen, C. J., and Radosevich, S. R., 1979, Modification of herbicide binding to photosystem II in two biotypes of *Senecio vulgaris* L., *Plant Physiol.*, 64:995.

Price, S. C., Hill, J. E., Naylor, J., and Allard, R. W., 1980, Genetic response of *Aveno fatua* to cultivation, Abstracts, Weed Science Society of America.

Radosevich, S. R., 1977, Mechanisms of atrazine resistance in lambsquarters and pigweed, *Weed Sci.*, 25:316.

Radosevich, S. R., and Appleby, A. P., 1973a, Relative susceptibility of two common groundsel (*Senecio Vulgaris* L.) biotypes to six s-triazines, *Agron. Jour.*, 65:553.

Radosevich, S. R., and Appleby, A. P., 1973b, Studies on the mechanism of resistance to simazine in common groundsel, *Weed Sci.*, 21:497.

Radosevich, S. R., and DeVilliers, O. T., 1976, Studies on the mechanism of *s*-triazine resistance in common groundsel, *Weed Sci.*, 24:229.

Radosevich, S. R., Steinback, K. E., and Arntzen, C. J., 1979, Effect of photosystem II inhibitors on thylakoid membranes of two common groundsel (*Senecio vulgaris*) biotypes, *Weed Sci.*, 27:216.

Ryan, G. E., 1970, Resistance of common groundsel to simazine and atrazine, *Weed Sci.*, 18:614.

Schooler, A. B., Bell, A. R., and Nalewaja, J. D., 1972, Inheritance of siduron tolerance in foxtail barley, *Weed Sci.*, 20:167.

Shimabukuro, R. H., Kadunce, R. E., and Frear, D. S., 1966, Dealkylation of atrazine in mature pea plants, *J. Agr. Food Chem.*, 14:392.

Shimabukuro, R. H., Swanson, H. R., and Walsh, W. L., 1970, Glutathione conjugation atrazine detoxification mechanism in corn, *Plant Physiol.*, 46:104.

Shimabukuro, R. H., Frear, D. S., Swanson, H. R., and Walsh, W. C., 1971, Glutathione conjugation, an enzymatic basis for atrazine resistance in corn, *Plant Physiol.*, 47:10.

Tischer, W., and Strotmann, H., 1977, Relationship between inhibitor binding by chloroplasts and inhibition of photosynthetic electron transport, *Biochem. Biophys. Acta*, 460:113.

Thompson, L., Jr., Houghton, J. M., Slife, F. W., and Butler, H. S., 1971, Metabolism of atrazine by fall panicum and large crabgrass, *Weed Sci.*, 19:409.

Van Assche, C. J., and Ebert, E., 1976, Photosynthesis, *Res. Rev.*, 65:2.

Velthuys, B. R., 1976, Charge accumulation and recombination in system 2 of photosynthesis. Thesis, University of Leiden, The Netherlands.

Warwick, D. D., 1976, Plant variability in triazine resistance, *Res. Rev.*, 65:56.

Way, J. M., and Chancellor, R. J., 1976, Herbicides and higher plant ecology, *in:* "Herbicides: Physiology, Biochemistry, Ecology,"

L. J. Audus, ed., Academic Press, New York.
West, L. D., Muzik, T. J., and Witters, R. E., 1976, Differential gas exchange responses to two biotypes of redroot pigweed to atrazine, *Weed Sci.*, 24:68.
Witt, C. T., de, 1960, On Competition. Inst. Voor. Biol. en Schemkundig Onderzoek van Landbouwg. Wagening en Netherlands.
Young, J. A., and Evans, R., 1976, Response of weed populations to human manipulations of the natural environment, *Weed Sci.*, 24:186.

MECHANISMS OF FUNGICIDE RESISTANCE —

WITH SPECIAL REFERENCE TO ORGANOPHOSPHORUS FUNGICIDES

Yasuhiko Uesugi

National Institute of Agricultural Sciences

Yatabe-machi, Tsukuba, Ibaraki 305, Japan

INTRODUCTION

Resistance of plant pathogenic fungi to agricultural fungicides was not important in field practice until about 1970, but it is now an urgent problem in many countries. The resistance problem is mainly due to the recent development and widespread use of new types of fungicides that act more specifically on target fungi and are generally less toxic to mammals and higher plants than conventional fungicides. While greater specificity is desirable, it is not altogether beneficial because it increases the possibility that toxicity will be overcome by fungal resistance. In some cases, a single mutation in a fungal pathogen may lead to resistance problems.

Resistant (R) mutants of fungi can be obtained by in vitro selections made in the laboratory, and several kinds of R mutants that differ genetically and biochemically from each other may be obtained from a mother strain of a fungus. R mutants fit for survival in natural environments may emerge in the field when fungicides are used, but in some cases, no emergence of R strains is found in the field even though they are easily obtained in the laboratory. Sometimes laboratory-derived mutants differ from field isolates in resistance mechanisms and other genetic factors.

In this paper, the mechanisms of resistance to fungicides are reviewed and a study is presented on the mechanisms of resistance to organophosphorus fungicides in laboratory-derived mutants and field isolates of *Pyricularia oryzae*.

GENERAL REVIEW OF MECHANISMS OF RESISTANCE TO FUNGICIDES

The biochemical bases of resistance may be summarized as follows: (1) modification at the site of action of the fungicide; (2) decreased permeability of the fungal cell membranes to the fungicide, and (3) altered metabolism of the fungicide within the fungal cells, i.e., increased detoxification or decreased activation of the fungicide.

These are basically the same as the mechanisms involved in resistance to bactericides and insecticides.

Modification at the Site of Action

In this category, many mechanisms are possible, including decreased affinity of the binding site for the fungicide; increased production of fungal cell components (i.e., enzymes, their substrates, etc.) that compensate for action of the fungicide; decreased requirements for the physiological activity of the target site, and development of a bypass for the inhibited pathway are examples of several such resistance mechanisms.

Fungicidal action of carbendazim, a benzimidazole compound that is the active principle of benomyl and thiophanate-methyl, is attributed to the inhibition of chromosome separation at mitosis. Carbendazim inhibits mitosis by binding to the protein subunits that are to constitute the spindles required for the process of mitosis. Decreased affinity of the site for carbendazim was proposed as a resistance mechanism in mutants of *Aspergillus nidulans* by Davidse and Flach (1977), and a close correlation was found between the affinity and the resistance level. Since resistance in the strains tested involved mutations in one chromosome locus, the results seem reasonable. In *A. nidulans* three loci have been found for resistance to carbendazim by van Tuyl (1977). Resistance mechanisms for the other two loci have not yet been identified.

The antibiotic kasugamycin binds to the ribosomes of sensitive (S) fungi resulting in an inhibition of protein synthesis. It was proposed that the resistance mechanism in a mutant obtained by in vitro selection was the decreased affinity of the site of action, i.e., ribosomes, for this antibiotic (Misato and Ko, 1975).

Carboxin is an oxathiin carboxamide fungicide that inhibits electron transfer in fungal respiration by affecting the succinic dehydrogenase complex. Mutants resistant to the fungicide due to modification of the target site have been reported in *Ustilago maydis* (Georgopoulos et al., 1972, 1975; Georgopoulos and Ziogas, 1977) and in *Aspergillus nidulans*. Recently, a number of oxathiin carboxamide compounds were tested for toxicity to fungal growth and inhibition of succinic dehydrogenase activity in cell-free preparations of R mutants of the two fungal species (White et al., 1978). With few exceptions, growth inhibition closely paralleled enzyme inhibition. In these tests a remarkable negative correlation in cross resistance was found between carboxin and 4'-phenylcarboxin, the latter being far more toxic to a category of the mutants resistant to carboxin than to the wild-type strain of *U. maydis*. Another mutant of *U. maydis* that is moderately resistant to carboxin and, at the same time, more sensitive to antimycin A than the wild-type strain, has been reported (Georgopoulos and Sisler, 1970). In this fungus, an alternate electron transport pathway insensitive to antimycin A is present in wild-type cells. The alternate pathway, which may be somehow related to the site of action of the fungicide, was genetically blocked in the R mutant.

Field emergence of resistance to oxathiin carboxamide fungicides has been reported only in Japan for *Puccinia horiana* against oxycarboxin, where kasugamycin resistance in *Pyricularia oryzae* has also been found. The resistance mechanisms in these field isolates have not been studied, although they are well worth investigation.

Decreased Permeability of Fungal Cell Membranes to Fungicides

An R mutant of *Pyricularia oryzae* obtained by successive transfers on agar medium supplemented with increasing concentrations of blasticidin S was investigated for the mechanism of the resistance to the antibiotic by Huang et al. (1964). The antibiotic is an inhibitor of protein biosynthesis; therefore, incorporation of ^{14}C-labeled amino acids in the protein fraction was investigated. In intact cells, inhibition of incorporation by the antibiotic was far less in the R than in the S cells, but in cell-free extracts there was equal inhibition in both extracts from R and S cells. The resistance was therefore attributed to decreased permeability of the protoplasmic membrane.

A similar investigation was carried out with field isolates of *Alternaria kikuchiana* resistant to polyoxins. The inhibition of chitin synthetases is the mode of action of these antibiotics and the effect of polyoxin B on chitin biosynthesis was compared in S and R strains. Differences were found with intact cells but not with cell-free enzymes (Hori et al., 1974). Uptake of ^{14}C-labeled polyoxin A and B by the fungus was investigated and far less of the antibiotics was found to be taken up by the R than by the S strains (Hori et al., 1976). These results indicate that resistance is due to reduced permeation of the antibiotics.

Mutants of *Sporobolomyces roseus* resistant to carbendazim because of decreased permeability were reported by Nachmias and Barash (1976).

Among 12 strains of *Verticillium malthousei* having various levels of tolerance to benomyl, it was found that tolerance was associated with the ability to produce acids. The probable mechanism of tolerance was presumed to be a pH-regulated reduction in uptake or binding of the fungicide (Lambert and Wuest, 1976).

Recently, de Waard and van Nistelrooy (1979) reported that uptake of labeled fenarimol by *Aspergillus nidulans* in a short time period (10 min) was significantly lower in all 20 mutants resistant to the fungicide than in the wild-type strain. The resistance mechanism was assumed to be a decreased uptake of this fungicide. Fenarimol, triarimol, triforine, buthiobate (S-1358), triadimefon, triadimenol and imazalil have a common mode of action as inhibitors of ergosterol biosynthesis even though the chemical structures in some cases are rather different. It seems reasonable that cross resistance among these compounds has often been reported in mutants of several fungi and that the resistance mechanism in these cases may be site modification. Among the mutants of *A. nidulans* tested by de Waard and van Nistelrooy, some may display cross resistance to other inhibitors of ergosterol biosynthesis. The resistance mechanism was, nevertheless, concluded to be decreased uptake. A possible explanation may be that the amount of fungicide taken up by the fungal cells is related to its toxicity to the fungus. In other words, affinity of the fungicide for the site of action may be reflected in the uptake by the fungal cells.

Altered Metabolism of Fungicides

Increased detoxification of insecticides and antibiotics is often the mechanism of resistance to these agents in insects and bacteria. Various types of insecticide metabolism related to insect resistance to insecticides have been elucidated, and the enzymes involved include hydrolases such as carboxylesterases and phosphatases, transferases such as glutathion-*S*-transferases, oxidases such as mixed-function oxidases, and lyases such as dehydrochlorinases. In most cases, these enzymes detoxify by splitting the insecticide molecules.

On the other hand, detoxification through conjugation, such as acetylation, adenylation and phosphorylation, is a resistance mechanism in many cases of bacterial resistance to antibiotics. Such biochemical reactions are mostly mediated by transferases such as acetyltransferases, adenyltransferases and phosphotransferases. Cleavage by hydrolases such as β-lactamases is important in bacteria as a resistance mechanism to penicillin and other β-lactam antibiotics. Interestingly, such a mechanism is often found in clinical isolates but is infrequently found in R mutants obtained in the laboratory, among which modification of some cellular component, such as the site of action of the antibiotics, is more frequently seen as a resistance mechanism (Benveniste and Davies, 1973; Davies and Smith, 1978). Although there are many possible explanations for this phenomenon based on genetic, biochemical and other viewpoints, it can be deduced that R strains found in practice are more fitted for survival in natural environments than mutants obtained under laboratory conditions.

As compared with insecticides and bactericides, metabolism of fungicides has been less studied in relation to resistance. Inactivation of organomercurials through binding by either red anthraquinone pigments (Greenaway, 1971, 1972) or thiol compounds (Ross and Old, 1973; Ross, 1974) in the fungal cells has been proposed as a resistance mechanism in some strains of *Pyrenophora avenae*, but no clear-cut conclusion has been obtained.

Fungal metabolism of quintozene (pentachloronitrobenzene) to pentachloroaniline and pentachlorothioanisole was reported by Nakanishi and Oku (1969) as a detoxification mechanism in *Fusarium oxysporum* and *Rhizoctonia solani*. The differential rate of the metabolism was regarded as the basis for the difference in sensitivity to quintozene of the R fungus, *F. oxysporum*, and the S fungus, *R. solani*, and of an R strain of *Botrytis cinerea* and an S strain of the same fungus (Nakanishi and Oku, 1970). However, fungal metabolism of quintozene to pentachloroaniline was regarded by Ko (1968) as not necessarily the resistance mechanism in strains of *R. solani* with different sensitivities to quintozene. Although the conversion was higher in R strains, this was assumed to be due to better growth than to more efficient conversion. Furthermore, pentachloroaniline was not always less toxic than quintozene to this fungus.

Conversion of dodine (dodecylguanidine acetate) into less toxic compounds was assumed to be the basis for resistance of *Fusarium solani* f. sp. *phaseoli* to this fungicide (Bartz and Mitchell, 1970a, 1970b). A similar metabolic conversion may also be the mechanism of resistance of a mutant of *Nectria haematococca* to dodine (Kappas and Georgopoulos, 1970). The chemical structure of the inactive metabolites and the nature of the conversion reaction have not been elucidated, however.

Development of resistance by decreased conversion of an inactive precursor to an active compound is not often seen in insects, bacteria or fungi. A good example of decreased activation was demonstrated by Dekker (1968) in a UV-induced mutant of *Cladosporium cucumerinum* resistant to 6-azauracil, an experimental fungicide. Azauracil was converted in the fungal cells via 6-azauridine to 6-azauridine-5'-phosphate, which inhibits orotic acid metabolism presumably by inhibiting the enzyme orotidine-5'-phosphate decarboxylase. In the R mutant, the ability to metabolize azauracil to azauridine is lacking, and therefore, the mutant is resistant to azauracil but not to azauridine and azauridinephosphate.

MECHANISM OF RESISTANCE TO ORGANOPHOSPHORUS FUNGICIDES IN *Pyricularia oryzae*

In *Pyricularia oryzae*, which causes rice blast diseases, the development of resistance to organophosphorus fungicides in paddy fields was recently reported (Yaoita et al., 1977; Katagiri et al., 1978). Organophosphorus fungicides have been used in Japan for 14 years, and mutants selected for resistance to these chemicals under laboratory conditions have also been reported (Uesugi et al., 1969; Katagiri and Uesugi, 1978). Studies on fungal metabolism of organophosphorus compounds will be summarized and discussed in this paper in relation to resistance of laboratory mutants and field isolates of the fungus.

Development of Resistance

Mutants resistant to organophosphorus thiolate fungicides. Organophosphorus fungicides developed to control rice blast are primarily phosphorothiolate and phosphonothiolate compounds, except for a few phosphates having no sulfur atom, such as phosdiphen [bis(2,4-dichlorophenyl) ethyl phosphate] and H0034 (ethyl phenyl 2,4,5-trichlorophenyl phosphate). Organophosphorus thiolate (PTL) fungicides have been widely used in Japan. Compounds in this group include EBP (*S*-benzyl diethyl phosphorothiolate), IBP (*S*-benzyl diisopropyl phosphorothiolate), ESBP (*S*-benzyl ethyl phenylphosphonothiolate), ESTP (*S*-benzyl ethyl tolylphosphonothiolate), BEBP (*S*-benzyl *S*-ethyl butyl phosphorodithiolate), MHCP (*S*-*p*-chlorophenyl cyclohexyl methyl phosphorothiolate) and edifenphos (EDDP, ethyl *S*,*S*-diphenyl phosphorodithiolate). Among these PTL fungicides, only IBP and edifenphos are being used in Japan at the present time.

Mutants of *Pyricularia oryzae* resistant to PTL fungicides were easily obtained by selection from a large number of conidia plated on an agar medium containing a PTL fungicide. EBP-resistant mutants thus selected were cross-resistant to IBP, ESBP and edifenphos, and an edifenphos-R mutant was cross-resistant to IBP (Uesugi et al., 1969). A PTL-R mutant was found to be cross-resistant to isoprothiolane (diisopropyl 1,3-dithiolan-2-ylidenemalonate), which is a fungicide having no phosphorus atom. A mutant selected for resistance to isoprothiolane by a similar method was also found to be cross-resistant to IBP and to edifenphos (Katagiri and Uesugi, 1977). Furthermore, almost equal frequency of emergence of R mutants was observed in selection of conidia on agar media containing 200 μM IBP, 30 μM edifenphos or 140 μM isoprothiolane, and the mutants so obtained were cross-resistant to IBP and collaterally sensitive to a phosphoramidate as stated below (Katagiri and Uesugi, 1978).

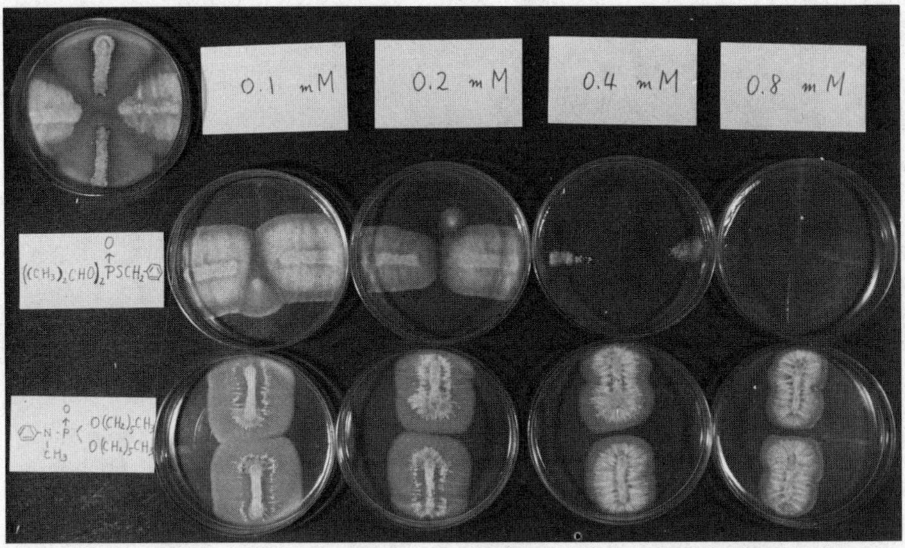

Figure 1. Negative correlation in cross resistance between a PTL fungicide, IBP, and a PA compound, HPA. Potato sucrose agar plate — untreated (top left), amended with IBP (middle row), and amended with HPA (bottom row) were seeded in a vertical direction with a normal wild-type strain, Hoku 373, and laterally with a mutant resistant to PTL fungicides.

In the course of the study on cross resistance of PTL-R mutants to various organophosphorus compounds, a certain range of compounds belonging to *N*-phenylphosphoramidate (PA) were found to be more toxic to the PTL-R mutants than to the normal wild-type strains (Uesugi et al., 1974). This negatively correlated cross resistance (i.e., the collateral sensitivity) was most clearly shown with dihexyl *N*-methyl-*N*-phenylphosphoramidate (HPA) among the PA compounds tested (Figure 1).

To investigate the relationship in fungitoxicity between PTL and PA compounds, their joint action was tested (Uesugi et al., 1974). Two strips of filter paper impregnated with a PTL and a PA compound, respectively, were crossed at right angles on an agar plate seeded uniformly with conidia of a test strains of *P. oryzae*. If the two test fungicides were entirely different in mode of action, the growth-inhibitory zones around the paper strips would be formed independently with no interaction at the intersection of the paper strips (Figure 2, independent); two test fungicides having a similar mechanism of action should result in an additive zone of growth inhibition at the intersection of the paper strips (Figure 2,

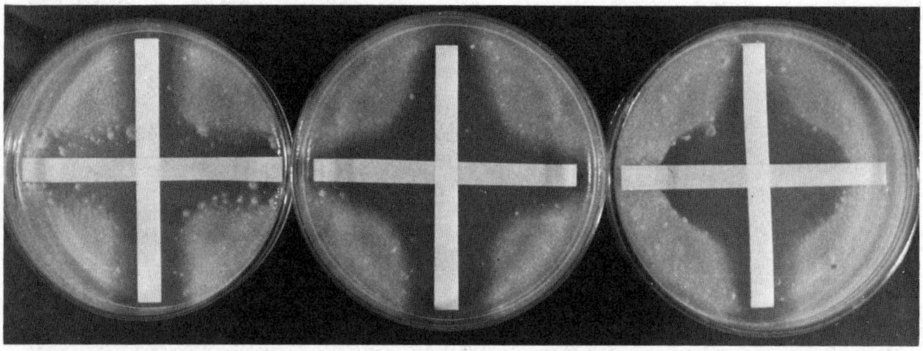

Figure 2. Joint actions as tested by the crossed paper technique. The compounds impregnated in the filter paper strips were, from left to right, IBP (vertical) x phenylmercury acetate (lateral), IBP x IBP, and IBP x BPA.

additive). The synergistic action of PTL and PA compounds on wild-type strains is shown by expansion of the inhibitory zones at the intersection of the paper strips impregnated with the compounds (Figure 2, synergistic). This synergism was observed with normal wild-type strains but not with PTL-R mutants, and was revealed with remarkable clarity when dibutyl *N*-methyl-*N*-phenylphosphoramidate (BPA) and IBP were used as the PA and the PTL compounds, respectively.

Recently, other mutants moderately resistant to PTL fungicides (PTL-MR) were found in a selection with IBP at a lower concentration (100 μM) (Katagiri and Uesugi, 1978). The PTL-MR mutants had lower resistance to PTL fungicides and isoprothiolane and lower sensitivity to PA compounds than the PTL-R mutants (Table 1). The negatively correlated cross resistance between PTL and PA was also observed in the PTL-MR mutants.

Field isolates resistant to phosphorothiolate fungicides. Although PTL-R mutants were thus easily obtained in the laboratory, emergence of R strains in the field was not recognized until recently. Only one exceptional case was observed in 1973 in an experimental greenhouse in which IBP had been used intensively as a standard in evaluations of fungicides. The preventive effects of IBP against blast on rice seedling were remarkably decreased, but

Table 1. Sensitivity of Field Isolates and Mutants of *Pyricularia oryzae* to PTL and Related Fungicides

Fungicide and concentration (μM)	Isolates			Mutants[a]	
	Tn 14-1	Fm 3-1	Hoku 373	PTL-MR	PTL-R
	Fungal growth[b]				
IBP					
100	++	+	-	+	++
200	++	-	-	-	+
Ediphenphos					
20	++	+	-	++	++
40	+	-	-	-	+
Isoprothiolane					
70	++	+	-	+	++
140	+	-	-	-	+
HPA					
100	-	++	++	++	-
200	-	++	++	-	-
BPA					
200	-	++	++	+	-
	Synergistic fungitoxicity[c]				
IBP and BPA	-	±	++	±	-

[a]Mutants were selected from conidia of a normal wild-type strain Hoku 373. PTL-MR is moderately resistant, and PTL-R is resistant to PTL fungicides.

[b]Fungal growth on potato sucrose agar medium containing fungicides was observed 4 to 7 days after inoculation with a suspension of the conidia and compared with that on untreated medium. ++, normal

Table 2. Result of Monitoring for Resistance of Field Isolates of *Pyricularia oryzae* to a PTL Fungicide

Minimal growth-inhibitory concentration of IBP (μM)	Number of isolates[a]			
	Before 1964	1974	1976	1977
≧ 200	0	0	8	4
100	3	} 165	} 165	28
≦ 50	27			46

[a]Source of isolates: stock strains isolated before 1964, 1974 isolates from Ibaraki and Yamagata Prefectures, and 1976 and 1977 isolates from Toyama Prefecture.

a PA compound, *p*-chlorophenyl ethyl *N*-methyl-*N*-phenylphosphoramidate, which is not fungicidal to normal strains but is specifically toxic to PTL-R mutants of *P. oryzae*, was quite effective. Since detailed investigation on the fungus has not been done, a definite conclusion has not been obtained, but the case may be an example of the emergence of R fungi similar to laboratory-derived PTL-R mutants on the host plants.

Monitoring for resistance to PTL fungicides was therefore conducted in paddy fields of Ibaraki and Yamagata Prefectures in 1974, but no isolate was found that grew on an agar medium containing IBP at 100 μM. In 1976, eight out of 173 isolates from Toyama Prefecture were found to be resistant to 100 μM IBP (Table 2). Four out of the eight R isolates were tested further for their sensitivity to PTL and related fungicides, and one isolate, coded as Tn 14-1, showed a level of resistance to PTL and a sensitivity to PA similar to the PTL-R mutants obtained in vitro. A synergistic fungicidal action of PTL and PA was not observed with Tn 14-1, as had been the case for the PTL-R mutants. The other three field isolates were resistant to IBP at 100 μM but sensitive at 200 μM, as were PTL-MR mutants obtained in vitro. These PTL-MR isolates, however, were different from the laboratory-derived PTL-MR mutants in sensitivity to PA compounds. The former were more resistant to PA than the latter,

growth; +, growth but slightly inhibited; ±, faint growth; -, entirely inhibited.

[c]Synergism was observed by the crossed paper technique described in the text. ++, remarkably synergistic; +, synergistic; ±, slightly synergistic; -, not synergistic.

so the negative correlation of cross resistance between PTL and PA was lacking in the former. Synergism of PTL and PA in the PTL-MR field isolates was not as evident as in the normal wild-type strains.

The sensitivity to fungicides of the field isolates as represented by Tn 14-1 and a PTL-MR isolate Fm 3-1 is compared with a normal wild-type strain coded as Hoku 373 and its PTL-MR and PTL-R mutants in Table 1. The five strains shown may be typical of *P. oryzae* types known to date in terms of their sensitivity to PTL and related fungicides.

Since the monitoring of resistance in 1976 revealed the emergence of PTL-R and PTL-MR strains in the field, the sensitivity to IBP of field isolates collected from Toyama Prefecture in 1977 was compared with that of stock strains isolated from lesions of rice leaves and panicles from various areas of Japan before 1964 when no organophosphorus fungicides were used. Four out of 78 of the 1977 isolates were found to be resistant to IBP at 100 μM, although no isolates obtained before 1964 were at the same resistance level. Furthermore, only three out of 30 of the latter isolates were resistant to IBP at 50 μM, whereas 32 out of 78 of the 1977 isolates were resistant to this concentration (Katagiri et al., 1979). These results suggest that the development of resistance to PTL in the field was occurring widely, although the resistance level was so low that it was not readily detected. A similar development of PTL resistance in the field is suggested by the observation made in 1977 in Niigata Prefecture that the frequency distribution of IBP sensitivity of field isolates of *P. oryzae* differed with locality in the prefecture (Yaoita et al., 1978).

As stated above, synergistic fungicidal action of PTL and PA was prominent only with normal wild-type strains (Table 1). The synergistic action between IBP and BPA was tested with all 1977 isolates and also with those collected before 1964 (Katagiri et al., 1979). For 1977 isolates, distribution of the synergism as expressed by the width of the expanded growth-inhibitory zone clearly showed two peaks (Figure 3). The peak representing the greater synergistic response (the right peak in Figure 3) consists of all isolates sensitive to 50 μM IBP and a small number of those isolates that were sensitive to 100 μM IBP but resistant to 50 μM. The peak representing a lesser synergistic response (the left peak in Figure 3) consists of all isolates resistant to 100 μM IBP and most of those resistant to 50 μM. The latter peak, consisting of strains resistant to this type of synergism, was not found for isolates collected before 1964. Therefore, strains belonging to this group were assumed to have emerged in the field after the application of PTL fungicides. Since field isolates resistant to 200 μM IBP were few, most strains belonging to the group more resistant to synergistic fungitoxicity (Figure 3) appear to be basically the same as the PTL-MR strains,

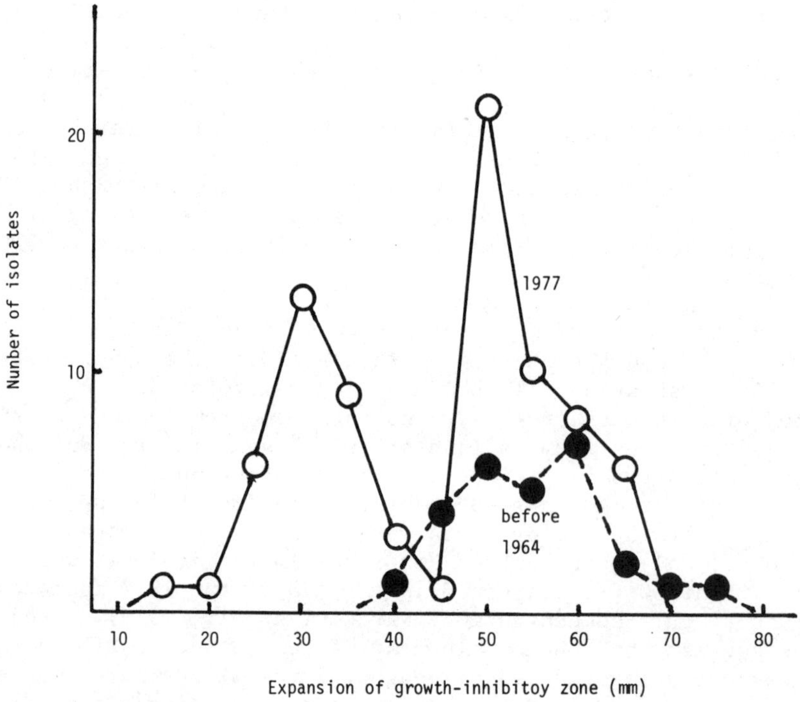

Figure 3. Distribution of response of field isolates to synergistic fungicidal action of IBP and BPA as tested by the crossed paper technique. The test strains were stock strains isolated before 1964 (solid circle and dashed line) and strains isolated from fields in Toyama Prefecture in 1977 (circle and solid line). The growth-inhibitory zone measured at the cross section of the paper strips includes the width of strips (7 mm) used in the test.

such as Fm 3-1, found in 1976. Discrimination between these MR strains and normal wild-type strains may be possible by observation of synergistic fungitoxicity of IBP and BPA, even when such discrimination by a test for sensitivity to a PTL fungicide is difficult.

Fungal Metabolism of Organophosphorus Fungicides in Relation to Resistance Mechanism

Metabolism of a phosphoramidate. Metabolites of BPA produced by mycelial cells of *P. oryzae* have been investigated by gas-liquid and thin-layer chromatography (Uesugi and Katagiri, 1977; Uesugi and Sisler, 1978). Since cleavage of ester linkages is often an important metabolic pathway of this type of compound, a search was made

Table 3. Metabolites of PTL and PA Fungicides Produced by Incubation with Mycelial Cells of *Pyricularia oryzae*

Fungicide	Main metabolite	Other metabolites
$[(CH_3)_2CHO]_2P(O)SCH_2C_6H_5$		
(IBP)	$[(CH_3)_2CHO]_2POSH$	$[(CH_3)_2CHO]_2P(O)SCH_2C_6H_4OH$ (*m*)
		$[(CH_3)_2CHO]_2P(O)OH$
		$(CH_3)_2CHOP(O)(SCH_2C_6H_5)OH$
		$(CH_3)CHOP(O)(OH)_2$
		non-ionic water-soluble metabolite
$C_2H_5OP(O)(SC_6H_5)_2$		
(edifenphos)	$C_2H_5OP(O)(SC_6H_5)OH$	$C_2H_5OP(O)(SC_6H_5)SC_6H_4OH$ (*p*)
		$(C_6H_5S)_2P(O)OH$
		$C_2H_5OP(O)(OH)_2$
		non-ionic water-soluble metabolite
$C_6H_5P(O)(OC_2H_5)SCH_2C_6H_5$		
(ESBP)	$C_6H_5P(OC_2H_5)OSH$	$C_6H_5P(O)(OC_2H_5)SCH_2C_6H_4OH$ (*m*)
	$C_6H_5P(O)(OC_2H_5)OH$	$C_6H_5P(O)(OH)_2$
$C_6H_5N(CH_3)P(O)(OCH_2CH_2CH_2CH_3)_2$		
(BPA)	$C_6H_5N(CH_3)P(O)(OCH_2CH_2CH_2CH_3)OCH_2CH_2CH(OH)CH_3$	
		$C_6H_5NHP(O)(OCH_2CH_2CH_2CH_3)_2$

Table 4. Fungitoxicity of PTL and PA Fungicides and their Metabolites Evaluated by Conidial Germination Test

Fungicide or metabolite[a]	LD_{50} (μM)				
	Field isolates			Mutants	
	Tn 14-1	Fm 3-1	Hoku 373	PTL-MR	PTL-R
IBP	530	106	95	140	420
$[(CH_3)_2CHO]_2POSH$	> 1000	> 1000	> 1000	> 1000	> 1000
$[(CH_3)_2CHO]_2POOH$	> 1000	> 1000	> 1000	> 1000	> 1000
HOIBP	n.t.[b]	n.t.	760	n.t.	n.t.
BPA	300	350	330	340	300
HOBPA	> 600	> 600	> 600	> 600	> 600
deMeBPA	440	480	480	400	340

[a]HIOPB, *m*-hydroxyIBP; HOBPA, butyl 3-hydroxybutyl *N*-methyl-*N*-phenylphosphoramidate; deMeBPA, dibutyl *N*-phenylphosphoramidate.
[b]Not tested.

for metabolites expected to result from such reactions, but results indicated no appreciable amounts of such metabolites or their methylation products identical in Rf values or retention times with authentic samples. Instead, two water-insoluble metabolites were found, each slightly more polar than the original compound. One was identified as the *N*-demethylation product of BPA on the basis of its mass spectrum and methylation with diazomethane to yield the original BPA. The other metabolite had a molecular ion peak of 16 greater than that of BPA, and was therefore assumed to be a hydroxylated product. Later work revealed that it was a BPA derivative hydroxylated at the 3-position of one of the two butyl radicals (Table 3). Usually the hydroxylated metabolite was found in greater abundance than the *N*-demethylated product.

The antifungal activity of the hydroxylated and *N*-demethylated metabolites for field isolates and mutants of *P. oryzae* was determined by conidial germination tests, the results of which are summarized in Table 4.

In regard to the method of evaluation of antifungal activity, mycelial growth might be preferred over conidial germination as an indicator of toxicity since metabolism was determined with mycelial cells. Since only limited quantities of the hydroxylated metabolite were available, it was necessary to use the conidial germination test, which required less material than the mycelial growth test.

Some discrepancies between the results of conidial germination tests (Table 4) and mycelial growth tests (Table 1) were observed with BPA. However, as far as inhibition of conidial germination is concerned, hydroxylation of BPA clearly decreased antifungal activity although no remarkable decrease in the activity was observed in *N*-demethylation. Hydroxylation is the main pathway of BPA metabolism in *P. oryzae* and may be regarded as a mechanism of detoxification of this fungicide.

The rate of fungal metabolism of BPA varies widely among field isolates and mutants, as shown in Table 5. BPA metabolism was hardly observed in a PTL-R isolate (Tn 14-1) or in a PTL-R mutant derived from Hoku 373, although it was clearly evident in a PTL-MR isolate (Fm 3-1) and in a normal wild-type strain (Hoku 373). These results are in good accordance with the fact that PA fungicides are specifically fungicidal to the PTL-R mutant and the PTL-R isolate but not to the PTL-MR isolate and the normal wild-type strain. The PTL-MR mutant metabolized BPA moderately, and it was moderately resistant to BPA. Therefore, the sensitivities of all the field isolates and mutants tested on BPA were related to the rate of the detoxifying metabolism.

BPA metabolism in normal wild-type strains has been reported to be inhibited by IBP and isoprothiolane (Uesugi and Sisler, 1978), and also by inhibitors of mixed-function oxidases such as piperonyl butoxide (Uesugi et al., 1978). Synergism of the fungicidal action of BPA by IBP on wild-type strains may well be explained by this phenomenon. In the case of PTL-R mutants and PTL-R isolates, BPA metabolism is low originally, and synergism by IBP may not be very remarkable.

<u>Metabolism of organophosphorus thiolate fungicides</u>. Metabolism of PTL fungicides has been studied using normal wild-type strains of *P. oryzae* and their PTL-R mutants as test fungi, and edifenphos (Uesugi and Tomizawa, 1971), IBP (Tomizawa and Uesugi, 1972) and ESBP (Uesugi and Tomizawa, 1972) as test fungicides. Metabolites identified in these studies are sumamrized in Table 3.

The main metabolic degradation pathway for these fungicides was determined to be cleavage of the thiolic ester bond, P-S-C. In metabolism of edifenphos, cleavage of the P-S bond in the thiolic ester was the main reaction, whereas cleavage of the S-C bond was

Table 5. Amount of PTL and PA Fungicides Remaining and Metabolites Produced in the Incubation Mixture with Mycelial Cells of *Pyricularia oryzae*

Fungicide or metabolite	Amount found[a] (μM)				
	Field isolates			Mutants	
	Tn 14-1	Fm 3-1	Hoku 373	PTL-MR	PTL-R
IBP	194	0	93	172	218
$[(CH_3)_2CHO]_2POSH$	5	194	87	17	3
BPA	195	21	19	156	185
HOBPA	0	175	125	29	0
deMeBPA	0	5	35	3	0

[a]The culture consisted of 0.2 g fresh wt. mycelial cells in 15 ml M/30 phosphate buffer (pH 7.0) containing 200 μM test fungicides and 0.2% glucose, and was incubated at 26°C for 4 hr. Unmetabolized fungicides and metabolites were extracted and determined by gas-liquid chromatography.

the main reaction in IBP metabolism. Cleavage of both the P-S and S-C bond was observed in ESBP metabolism.

Using thin-layer chromatography to separate toluene-soluble metabolites of labeled fungicides, a metabolite slightly more polar than the original fungicide was found for each of the three fungicides. The metabolite was identified as a product *p*-hydroxylated on one of the two phenyl radicals in the case of edifenphos, and a product *m*-hydroxylated on a benzyl radical in the case of IBP and ESBP. Identification was based on gas-liquid chromatograms of the intact metabolites and of their methylation and acetylation products compared with their respective authentic compounds.

Differences in the rate of metabolism of IBP, edifenphos and ESBP were not found among the strains when a large amount of mycelial cells (0.15 g fresh wt./ml) was incubated with the fungicides at a low concentration (about 10 μM) (Uesugi and Tomizawa, 1971, 1972; Tomizawa and Uesugi, 1972); however, differences were found when cells were incubated at a higher concentration of IBP (about 200 μM) (Tomizawa and Uesugi, 1972; Uesugi and Katagiri, 1977). The differences were more remarkable when the amount of mycelial cells was smaller (0.2 g fresh wt./ 15 ml); therefore, IBP metabolism was compared among strains under these appropriate conditions, as shown in Table 5.

IBP metabolism, i.e., detoxification, was less in the PTL-R mutant than in the normal wild-type strain, Hoku 373. The resistance mechanism in this mutant, therefore, is not increased detoxification. Tn 14-1, a PTL-R isolate, and a PTL-MR mutant both resembled the PTL-R mutant in their capacity to metabolize IBP; however, the PTL-MR mutant was slightly more active. On the other hand, IBP was metabolized remarkably well in Fm 3-1, a PTL-MR isolate. The increased metabolism, i.e., detoxification, of IBP may be, at least in part, the mechanism of resistance to the fungicide in this isolate. The increased metabolism of IBP may also be the basis for lowered synergism of BPA toxicity by IBP in this strain.

Mechanism of resistance to a phosphorothiolate fungicide. As stated above, the mechanism of resistance to IBP was explained by increased detoxification only in the case of an isolate moderately resistant to PTL fungicides. Since IBP metabolism was inhibited by inhibitors of mixed-function oxidases, and since IBP seemed to be a competitor of oxidative metabolism of BPA (Uesugi et al., 1978), the IBP metabolism may be oxidative rather than hydrolytic. The relation of BPA metabolism and IBP metabolism was also suggested by parallelism in rates among various strains. However, the details of the reaction have not yet been elucidated because of difficulties in obtaining the enzyme preparations.

The mechanism of resistance in a PTL-R isolate, which is rarely found in the field, and a PTL-R mutant obtained in vitro is definitely not increased detoxification. The detoxification in these strains is even decreased. The following evidence suggests that in these PTL-R strains, the lowered rate of metabolism of BPA is related to the mechanism of resistance to IBP. Selection from a PTL-R mutant, which is sensitive to PA fungicides, with a PA compound, HPA, gave rise to a PA-resistant mutant in which sensitivity to PTL fungicides and to the synergistic fungitoxicity of PTL plus PA were recovered at the same time. Thus, this PA-R mutant appeared to be a revertant that emerged by a back-mutation (Uesugi and Katagiri, 1977). In both the selections from normal wild-type strains to PTL-R mutants and from a PTL-R mutant to the revertant, no mutant was found that was resistant to both PTL and PA. These facts suggest that resistance to PTL is connected with increased sensitivity to PA in *P. oryzae*, and that a single genetic factor and a single biochemical mechanism cause resistance to PTL and a concomitant increase in sensitivity to PA. Since sensitivity to BPA in a PTL-R mutant was explained by a decreased metabolism of BPA, resistance to PTL may have some connection with this decreased metabolism.

If the decreased activity in metabolizing BPA is connected with the development of resistance to PTL in PTL-R mutants of *P. oryzae*, some enzyme metabolizing BPA in the normal strain may be a possible site of action for IBP, because IBP is an inhibitor of BPA metabolism and a change in the enzyme leads to a PTL-R mutant. The resistance mechanism in this case apparently involves a modification of the site of action of the PTL fungicide. This may also be the case with a PTL-R isolate, Tn 14-1, which resembles the PTL-R mutant in PTL sensitivity and its ability to metabolize this and other related fungicides. An important point that remains to be established in this hypothesis is whether or not the enzyme originally metabolizing BPA is vital to the physiology of the fungus.

Another possible explanation for the connection of the decreased metabolism of BPA with the resistance to IBP in the PTL-R strains may be that the enzyme metabolizing BPA activates IBP before it inhibits some physiologically important site in the fungal cells. In this case, the resistance mechanism is a decreased activation. However, the main metabolites of IBP detected were inactive, as stated above; therefore, the activated metabolite of IBP must be an unstable or minor one, and it has not been identified.

Maeda et al. (1970) investigated the incorporation of labeled glucosamine into the cell wall fraction of *P. oryzae* and found that treatment with IBP resulted in the accumulation of UDP-*N*-acetylglucosamine, an intermediate of chitin biosynthesis. They assumed that this apparent inhibition of chitin biosynthesis in intact cells resulted from an influence of the fungicide on the cytoplasmic membrane rather than from an inhibition of polymerization of the intermediate. A similar observation with edifenphos and IBP was made by de Waard (1972, 1974). Recently, Akatsuka et al. (1977) and Kodama et al. (1979) proposed the mode of action of IBP to be an inhibition of biosynthesis of phosphatidylcholine through the Greenberg pathway, i.e., through inhibition of transmethylation from *S*-adenosylmethionine to phosphatidylethanolamine. Phosphatidylcholine is presumably an important component of the cytoplasmic membrane of the fungal cells.

The relationship between phospholipid biosynthesis and the enzymes related to BPA metabolism has not yet been elucidated, nor has the relationship between phospholipid biosynthesis in normal strains and that in the PTL-R mutant been determined. Final clarification of the resistance mechanism in PTL-R mutants, therefore, has not yet occurred. Of course, it is possible that the mechanism in PTL-R strains is a decreased permeability of fungal cell membranes to PTL fungicides. If this is so, and in view of the fact that PA fungicides can penetrate the membrane of both the resistant and the normal strains and act or be metabolized, the resistance to PTL and the increased sensitivity to PA should probably be explained by quite different mechanisms in the PTL-R mutant. This seems unlikely from the results of the back-mutation experiment, as stated above.

The resistance mechanism of the PTL-MR mutant obtained in vitro is unknown, but it may be similar to that of the PTL-R mutant, since these two mutants differ only in their relative sensitivities to the fungicides and their ability to metabolize them.

In any case, results of the present study indicate that the important mechanism of resistance to PTL fungicides in the field is increased detoxification, although the resistance level is not remarkably high, whereas the mechanism occurring in mutants obtained in vitro is probably some modification of the fungal cell components related to the action of the fungicides. The results are similar to those obtained with antibiotics and bacteria, which show that the resistance mechanism in clinical isolates is often increased detoxification, although in most laboratory-derived mutants the mechanism is not increased detoxification, but either a modification of the site of action or other mechanism. More work on fungicide resistance mechanisms in laboratory-derived mutants and in field isolates of phytopathogenic fungi must be done, however, before broad generalizations can be made.

SUMMARY

From a toxicological point of view, possible mechanisms of resistance to a fungicide are: modification at the site of action of the fungicide, decreased permeability of the fungal cell membrane to the fungicide, and altered metabolism of the fungicide within the fungal cells.

Increased metabolism of the toxicant, mostly by cleavage reactions, is often the basis of insect resistance to insecticides. A similar mechanism of resistance to antibiotics, not only by cleavage but also by conjugation, is often seen in clinical isolates of bacteria, but is seldom seen in mutants obtained in vitro. However, there has been a rather small number of examples of fungal resistance to fungicides due to altered metabolism.

Fungal metabolism of organophosphorus thiolate (PTL) and related fungicides was reviewed in relation to resistance to the fungicides in strains of *Pyricularia oryzae* of various sensitivities. In one of the moderately resistant isolates, which are frequently found in the field, increased detoxification by cleavage of the thiolate bond was seemingly a resistance mechanism to the PTL fungicide IBP. On the other hand, in a mutant selected in vitro for resistance to PTL fungicides, the metabolic detoxification of IBP was not increased but was even decreased. A strain having characteristics similar to this R mutant obtained under laboratory conditions has also been isolated from the field, although quite rarely. In this PTL-R mutant, *N*-phenylphosphoramidate (PA) compounds are specifically fungicidal, possibly because of decreased detoxification of the PA compounds through hydroxylation. An example was demonstrated with dibutyl *N*-methyl-*N*-phenylphosphoramidate (BPA). The BPA metabolism in this fungus was inhibited by IBP, and the development of resistance to PTL and the increase in sensitivity to PA in the mutant seemed to be caused by a single genetic factor and, therefore, by a single biochemical mechanism. A possible site of action of IBP may be an enzyme involved in BPA metabolism, and a change in its capacity to metabolize BPA may be connected with PTL resistance. However, a conclusion could not be reached because it was not determined whether or not this site of action of IBP is vital to the fungus.

There is another possibility that IBP is activated by some enzyme involved in BPA metabolism before it acts on some physiologically important site in the fungal cells. Decreased activation of IBP caused by decreased activity of the enzyme related to BPA metabolism may be the mechanism of resistance to IBP in this case. However, the activated metabolite of IBP has not yet been identified.

A proposal was recently made that a mode of action of IBP may be the inhibition of phosphatidylcholine biosynthesis through the Greenberg pathway, but no work has been done to relate this inhibition to the resistance mechanism.

In this study, the mechanism of resistance to IBP in most of the field isolates moderately resistant to PTL fungicides is probably an increased metabolism detoxifying the fungicides, and in the mutants selected in vitro, it is probably a modification of some cellular component that is related to the action of PTL fungicides.

REFERENCES

Akatsuka, T., Kodama, O., and Yamada, H., 1977, A novel mode of action of Kitazin P in *Pyricularia oryzae*, *Agric. Biol. Chem.*, 41:2111.

Bartz, J. A., and Mitchell, J. E., 1970a, Comparative interaction of *N*-dodecylguanidine acetate with four plant pathogenic fungi, *Phytopathology*, 60:345.

Bartz, J. A., and Mitchell, J. E., 1970b, Evidence for the metabolic detoxification of *N*-dodecylguanidine acetate by ungerminated macroconidia of *Fusarium solani* f. sp. *phaseoli*, *Phytopathology*, 60:350.

Benveniste, R., and Davies, J., 1973, Mechanism of antibiotic resistance in bacteria, *Ann. Rev. Biochem.*, 42:471.

Davidse, L. C., and Flach, W., 1977, Differential binding of methyl benzimidazol-2-ylcarbamate to fungal tubulin as a mechanism of resistance to this antimitotic agent in mutant strains of *Aspergillus nidulans*, *J. Cell Biol.*, 72:174.

Davies, J., and Smith, D. I., 1978, Plasmid-determined resistance to antimicrobial agents, *Ann. Rev. Microbiol.*, 32:469.

Dekker, J., 1968, The development of resistance in *Cladosporium cucumerinum* against 6-azauracil, a chemotherapeutant of cucumber scab, and its relation to biosynthesis of RNA-precursors, *Neth. J. Plant Pathol.*, 74 Suppl. 1:127.

de Waard, M. A., 1972, On the mode of action of the organophosphorus fungicide Hinosan, *Neth. J. Plant Pathol.*, 78:186.

de Waard, M. A., 1974, Mechanism of action of the organophosphorus fungicide pyrazophos, *Mededelingen Landbouwhogeschool Wageningen*, 74-14:45.

de Waard, M. A., and van Nistelrooy, J. G. M., 1979, Mechanism of resistance to fenarimol in *Aspergillus nidulans*, *Pestic. Biochem. Physiol.*, 10:219.

Georgopoulos, S. G., and Sisler, H. D., 1970, Gene mutation eliminating antimycin A-tolerant electron transport in *Ustilago maydis*, *J. Bacteriol.*, 103:745.

Georgopoulos, S. G., and Ziogas, B. N., 1977, A new class of carboxin-resistant mutants of *Ustilago maydis*, *Neth. J. Plant Pathol.*, 83 Suppl. 1:235.

Georgopoulos, S. G., Alexandri, E., and Chrysayi, M., 1972, Genetic evidence for the action of oxathiin and thiazole derivatives on the succinic dehydrogenase system of *Ustilago maydis* mitochondria, *J. Bacteriol.*, 110:809.

Georgopoulos, S. G., Chrysayi, M., and White, G. A., 1975, Carboxin resistance in the haploid, the heterozygous diploid, and the plant parasitic dicaryotic phase of *Ustilago maydis*, *Pestic. Biochem. Physiol.*, 5:543.

Greenaway, W., 1971, Relation between mercury resistance and pigment production in *Pyrenophora avenae*, *Trans. Brit. Mycol. Soc.*, 56:37.

Greenaway, W., 1972, Permeability of phenyl-Hg^+-susceptible isolates of *Pyrenophora avenae* to the phenyl-Hg^+ ion, *J. Gen. Microbiol.*, 73:251.

Hori, M., Eguchi, J., Kakiki, K., and Misato, T., 1974, Study on the mode of action of polyoxins. VI. Effect of polyoxin B on chitin synthesis in polyoxin-sensitive and resistant strains of *Alternaria kikuchiana*, *J. Antibiotics*, 27:260.

Hori, M., Kakiki, K., and Misato, T., 1976, Mechanism of polyoxin-resistance in *Alternaria kikuchiana*, *J. Pestic. Sci.*, 1:31.

Huang, K. T., Misato, T., and Asuyama, H., 1964, Selective toxicity of blasticidin S to *Piricularia oryzae* and *Pellicularia sasakii*, *J. Antibiotics*, A 17:71.

Kappas, A., and Georgopoulos, S. G., 1970, Genetic analysis of

dodine resistance in *Nectria haematococca* (syn. *Hypomyces solani*), *Genetics*, 66:617.
Katagiri, M., and Uesugi, Y., 1977, Similarities between the fungicidal action of isoprothiolane and organophosphorus thiolate fungicides, *Phytopathology*, 67:1415.
Katagiri, M., and Uesugi, Y., 1978, In vitro selection of mutants of *Pyricularia oryzae* resistant to fungicides, *Nippon Shokubutsu Byori Gakkaiho*, 44:218.
Katagiri, M., Uesugi, Y., and Umehara, Y., 1978, Field emergence of resistance to organophosphorus fungicides in *Pyricularia oryzae* (Abstract), *Nippon Shokubutsu Byori Gakkaiho*, 44:401.
Katagiri, M., Uesugi, Y., and Umehara, Y., 1979, Some characteristics of field isolates resistant to organophosphorus fungicides in sensitivity to fungicides (Abstract), *Nippon Shokubutsu Byori Gakkaiho*, 45:548.
Ko, W. H., 1968, Evaluation of two suggested factors determining the specificity of pentachloronitrobenzene, *Phytopathology*, 58:1715.
Kodama, O., Yamada, H., and Akatsuka, T., 1979, Kitazin P, inhibitor of phosphatidylcholine biosynthesis in *Pyricularia oryzae*, *Agric. Biol. Chem.*, 43:1719.
Lambert, D. H., and Wuest, P. J., 1976, Acid production, a possible basis for benomyl tolerance in *Verticillium malthousei*, *Phytopathology*, 66:1144.
Maeda, T., Abe, H., Kakiki, K., and Misato, T., 1970, Studies on the mode of action of organophosphorus fungicide, Kitazin. Part II. Accumulation of an amino sugar derivative on Kitazin-treated mycelia of *Pyricularia oryzae*, *Agric. Biol. Chem.*, 34:70.
Misato, T., and Ko, K., 1975, The development of resistance to agricultural antibiotics, *in:* "Environmental Quality and Safety Suppl. 3, Pesticides," F. Coulston and F. Korte, eds., pp. 437-440, George Thieme Publishers, Stuttgart.
Nachmias, A., and Barash, I., 1976, Decreased permeability as a mechanism of resistance to methyl benzimidazol-2-ylcarbamate (MBC) in *Sporobolomyces roseus*, *J. Gen. Microbiol.*, 94:167.
Nakanishi, T., and Oku, H., 1969, Metabolism and accumulation of pentachloronitrobenzene by phytopathogenic fungi in relation to selective toxicity, *Phytopathology*, 59:1761.
Nakanishi, T., and Oku, H., 1970, Mechanism of selective toxicity of fungicide: Absorption, metabolism and accumulation of pentachloronitrobenzene by phytopathogenic fungi, *Nippon Shokubutsu Byori Gakkaiho*, 36:67.
Ross, I. S., 1974, Non-protein thiols and mercury resistance of *Pyrenophora avenae*, *Trans. Brit. Mycol. Soc.*, 63:77.
Ross, I. S., and Old, K. M., 1973, Thiol compounds and resistance of *Pyrenophora avenae* to mercury, *Trans. Brit. Mycol. Soc.*, 60:301.
Tomizawa, C., and Uesugi, Y., 1972, Metabolism of *S*-benzyl *O,O*-diisopropyl phosphorothiolate (Kitazin P) by mycelial cells of *Pyricularia oryzae*, *Agric. Biol. Chem.*, 36:294.

Uesugi, Y., and Katagiri, M., 1977, Back-mutation of an organophosphorus thiolate-resistant mutant of *Pyricularia oryzae*, *Neth. J. Plant Pathol.*, 83 Suppl. 1:243.

Uesugi, Y., and Sisler, H. D., 1978, Metabolism of a phosphoramidate by *Pyricularia oryzae* in relation to tolerance and synergism by a phosphorothiolate and isoprothiolane, *Pestic. Biochem. Physiol.*, 9:247.

Uesugi, Y., and Tomizawa, C., 1971, Metabolism of *O*-ethyl *S,S*-diphenyl phosphorodithiolate (Hinosan) by mycelial cells of *Pyricularia orzyae*, *Agric. Biol. Chem.*, 35:941.

Uesugi, Y., and Tomizawa, C., 1972, Metabolism of *S*-benzyl *O*-ethyl phenylphosphonothiolate (Inezin) by mycelial cells of *Pyricularia orzyae*, *Agric. Biol. Chem.*, 36:313.

Uesugi, Y., Katagiri, M., and Fukunaga, K., 1969, Resistance in *Pyricularia oryzae* to antibiotics and organophosphorus fungicides, *Nogyo Gijutsu Kenkyusho Hokoku*, 23:93.

Uesugi, Y., Katagiri, M., and Noda, O., 1974, Negatively correlated cross-resistance and synergism between phosphoramidates and phosphorothiolates in their fungicidal actions on rice blast fungi, *Agric. Biol. Chem.*, 38:907.

Uesugi, Y., Kodama, O., and Akatsuka, T., 1978, Inhibition of fungal metabolism of some organophosphorus compounds by piperonyl butoxide, *Agric. Biol. Chem.*, 42:2181.

van Tuyl, J. M., 1977, Genetics of fungal resistance to systemic fungicides, *Mededelingen Landbouwhogeschool Wageningen*, 77-2:1.

White, G. A., Thorn, G. D., and Georgopoulos, S. G., 1978, Oxathiin carboxamides highly active against carboxin-resistant succinic dehydrogenase complexes from carboxin-selected mutants of *Ustilago maydis* and *Aspergillus nidulans*, *Pestic. Biochem. Physiol.*, 9:165.

Yaoita, T., Go, N., Aoyagi, K., and Sakurai, H., 1977, Epidemiological trend of strains of rice blast fungus resistant to fungicides in Niigata Prefecture (Abstract), *Nippon Shokubutsu Byori Gakkaiho*, 43:357.

Yaoita, T., Go, N., Aoyagi, K., and Sakurai, H., 1978, Distribution of sensitivity to organophosphorus fungicides among isolates of rice blast fungus from Niigata Prefecture (Abstract), *Nippon Shokubutsu Byori Gakkaiho*, 44:401.

NATURE OF PROCYMIDONE-TOLERANT *BOTRYTIS CINEREA* STRAINS OBTAINED IN VITRO

Toshiro Kato, Yoshio Hisada and Yasuo Kawase

Research Department, Pesticides Division
Sumitomo Chemical Co., Ltd.
4-2-1 Takatsukasa, Takarazuka, Hyogo 665, Japan

INTRODUCTION

Tolerance of plant pathogens to fungicides was not a serious problem when mainly non-systemic conventional fungicides were used. These fungicides have been called "multi-site inhibitors" because they non-selectively inhibit several biologically important functions of living fungal cells. The nature of their fungitoxic mechanisms implies that there is little possibility for the development of tolerant mutants because mutation of a single gene cannot overcome the lethal effect of their multisite activity. On the other hand, we have seen emergence of tolerant pathogens with the recently introduced systemic fungicides. Since these systemic fungicides can easily penetrate into plant tissues, they must be selectively toxic to pathogens at concentrations that do not produce phytotoxic effects. Therefore, systemic fungicides have been understood to be "specific-site inhibitors" as they affect biochemically restricted regions of fungal cells. This intrinsic nature of systemic fungicides accelerated the selection of the fungicide-tolerant pathogens that appeared through mutation of a single gene.

However, it recently became apparent that not all of the systemic fungicides possessing a specific site of action cause tolerance problems in the field. For example, the systemic fungicide triforine, which is known to interfere with ergosterol biosynthesis (Sherald and Sisler, 1975), has been used for several years to control powdery mildews and rusts of agricultural crops, and has never been associated with the development of tolerance problems in the field. Fuchs et al. (1977) found triforine-tolerant strains of *Cladosporium cucumerinum* on a nutrient agar medium, but further investigation indicated that the tolerance was associated with decreased fungal growth rates, sporulation and pathogenicity, which might possibly explain why such tolerance did not develop in practice. Similarly, tolerance of *C. cucumerinum* and *Fusarium oxysporum* to the antibiotic fungicide pimaricin was shown to be associated with decreased pathogenicity and viability (Dekker and Gielink, 1979). As suggested by Wolfe (1971), the above examples imply that the occurrence of tolerant mutants in vitro does not necessarily mean that there is a potential for the development of such mutants under natural conditions. Dekker (1977) indicated that resistance problems may be affected not only by the type of disease involved and the selection pressure exerted by the fungicide, but also by the characteristics of the resistant mutants. Thus, laboratory examination of in vitro strains tolerant to a certain fungicide might give us an indication of the possibility of tolerance developing in the field. This procedure, especially on a newly developed fungicide, should be used before the compound is marketed since fungicide tolerance causes serious consequences for both the farmers and the pesticide producers.

Procymidone [*N*-(3',5'-dichlorophenyl)-1,2-dimethylcyclopropane-1,2-dicarboximide] is a new systemic fungicide that is especially effective in controlling botrytis diseases of vegetables and grape vines, sclerotinia rot of beans, monilinia diseases of deciduous fruit crops, and helminthosporium diseases of cereals (Hisada et al., 1976). Although the mechanism of fungitoxic action of procymidone is not clearly understood, it has been suggested that the fungicide affects a certain function of fungal membranes (Hisada et al., 1978), indicating that it is a specific-site inhibitor. Leroux et al. (1977) found in vitro some *Botrytis cinerea* strains tolerant to both iprodione [promidione or glycophene, 3-(3',5'-dichlorophenyl)-1-isopropylcarbamoylimidazolidine-2,4-dione] and vinclozolin [3-(3',5'-dichlorophenyl)-5-methyl-5-vinyloxazolidine-2,4-dione]. These fungicides are similar to procymidone in chemical structure and can be termed cyclic *N*-(3,5-dichlorophenyl)imides. Moreover, the iprodione-tolerant *B. cinerea* strains were also tolerant to procymidone and another cyclic *N*-(3,5-dichlorophenyl)imide, dichlozoline [3-(3',5'-dichlorophenyl)-5,5-dimethyloxazolidine-2,4-dione] (Fritz et al., 1977). Most of these strains exhibited cross tolerance to the aromatic hydrocarbon fungicides, DCNA (dicloran, 2,6-dichloro-4-nitroaniline), PCNB (quintozene, pentachloronitrobenzene), chloroneb (1,4-dichloro-2,5-dimethoxybenzene) and diphenyl. It is therefore of great importance that the possibility of emergence in the field of *B. cinerea* strains tolerant to procymidone and its related fungicides be estimated before they are used extensively. Thus, the present study was undertaken to investigate this possibility.

MATERIALS AND METHODS

Culture of Fungi

Twelve field isolates of *B. cinerea* were employed in the present experiments. The strains Bc-1, 2, 4 and 5, respectively, were isolated from strawberry, mandarin orange, eggplant and grape. Seven isolates from cucumber (Bc-9, 11, 12, 15, 16, 17, 18) and one isolate from tomato (Bc-14) were kindly supplied by Dr. Kiso (Vegetable and Ornamental Crops Research Station, Kurume Branch, Japan). Bc-5, 11, 12, 14, 15 and 18 were benomyl-tolerant, and the other isolates were benomyl-sensitive. *Cochliobolus miyabeanus, Sclerotinia sclerotiorum, Alternaria kikuchiana* and *Rhizoctonia solani* were the stock cultures of this laboratory, and all cultures were maintained on a potato sucrose agar medium. Production of conidia was made by culturing the respective strains on Hislop's medium (Hislop, 1967) under fluorescent lamps. The incubation temperature for fungus culture was 25 °C.

Fungicide

All of the fungicides employed throughout the experiments were commercially available products.

Pathogenicity Test

Cucumber seedlings (cv. Sagamihanjiro) at the primary leaf stage were used to determine pathogenicity of the procymidone-tolerant strains. Half of the primary leaf was inoculated with a mycelial disc (5mm in diameter) of the tolerant strain and the other half with a disc of the parent strain (Bc-2). Disease severity was assessed after incubating the test plants in a humid chamber at 20°C for three days. Pathogenicity was expressed as the percentage of average diameter of the lesion produced by the tolerant strain compared with that produced by the parent strain.

Competition Between the Tolerant and Sensitive Strains

The procymidone-tolerant strain, R-22 originating from Bc-2, was employed as a representative of the tolerant strains because of its high virulence and abundant spore production. Detached rose flowers (cv. Peace) were inoculated in the beginning of the experiment with a mixed suspension containing the conidia of R-22 and of the parent strain Bc-2 at a ratio of 20 to 1 in cell density. The flowers are infected in a humid chamber at 25°C. At intervals of 7 to 10 days, five successive transfers of spores produced on these diseased flowers were made onto fresh flowers.

Field Trial for Development of Fungicide Tolerance

Flowers of rose plants (cv. Peace, 3 to 6 years old) grown in two separate greenhouses were inoculated with conidia of *B. cinerea* (Bc-2) on May 10, 1976. Procymidone and benomyl were separately sprayed on half of the plants (30 in total) in each greenhouse 19 times during the three years of the experimental period. The dates for spraying were May 26, June 2, Oct. 20, Nov. 16 and 24 in 1976, Apr. 11, 18 and 25, May 5, June 20 and 27, July 7 and 14, Oct. 3, 11 and 17, Nov. 7 and 21 in 1977, and July 4 in 1978. The fungi were isolated from the plants at appropriate times, and the sensitivity of the single spore isolates to the fungicides was determined.

RESULTS

Development of Tolerant Strains In Vitro

It is known that fungi may increase to some extent their tolerance to a specific fungicide after continuous exposure to sublethal concentrations, although such tolerance is not inherited (Georgopoulos and Zaracovitis, 1967). In this experiment, *B. cinerea* mycelia were examined for their ability to acquire this type of tolerance to procymidone by successive transfers to a Hislop's medium containing a 50% growth-inhibitory concentration of the fungicide. The incubation period prior to the next transfer was five to ten days. Even after 30 transfers, a significant change in the sensitivity of the fungi to the toxicant was not detected (Table 1), indicating that this type of adaptation is not likely.

On the contrary, *B. cinerea* strains tolerant to procymidone were easily obtained by culturing a large number of conidia on an agar plate of a Hislop's medium containing the fungicide at lethal concentrations (Table 2). The frequency of occurrence of the tolerant strains was not markedly affected by the concentration of the toxicant, indicating that most of the strains identified possess a high degree of tolerance. If strains having a low degree of tolerance were present, the frequency of appearance of tolerant colonies would be higher at lower concentrations. Although the number of tolerant colonies was usually counted after five days of incubation at 25°C, it tended to increase with longer incubation periods. In the same series of experiments, several isolates of *B. cinerea* were examined for the possibility of tolerant colonies from a mass of conidia using the other cyclic *N*-(3,5-dichlorophenyl)imides, dichlozoline, iprodione and vinclozolin, and the aromatic hydrocarbon fungicide, DCNA. The results indicated that the tolerant strains could be similarly obtained by incubating conidia in a medium containing one of the test fungicides (Table 3). The frequency of appearance of tolerant colonies varied depending on the fungicide and the *B. cinerea* isolate. The benomyl-tolerant isolates produced

Table 1. Changes in the Sensitivity of *B. cinerea* to Procymidone During Successive Transfers to the Fungicide-containing Medium

Times of Transfer	ED_{50} values (μg/ml)	
	Bc-2	Bc-3
0	0.30	0.25
10	0.26	0.24
20	0.36	0.27
30	0.37	0.21

tolerant colonies at a higher rate than did the sensitive ones, although the reason was not apparent.

The procymidone-tolerant strains were also obtained from the mycelium as readily as from a mass of the conidia. In the absence of the toxicant, the mycelial disc (5mm in diameter) grew over the nutrient agar plate (90 mm in diameter) three days after inoculation, whereas on the fungicide-containing medium development of tolerant sectors usually required a minimum of five days. The rate of appearance of the tolerant sectors depended on the isolates, but the concentration of procymidone did not markedly affect the frequency of development (Tables 4 and 5). These strains had a very high degree of tolerance to procymidone as demonstrated by their growth on a

Table 2. Frequency of Occurrence of Procymidone-tolerant Strains from the Conidia of *B. cinerea* (Bc-2)

Concentration of the fungicide in medium (μg/ml)	Frequency of occurrence of tolerant strains ($x10^{-6}$)	
	Exp. 1	Exp. 2
500	2.1	1.4
100	2.2	1.9
20	1.7	1.8

Table 3. Frequency of Occurrence of *B. cinerea* Strains Tolerant to Procymidone, Dichlozoline, Iprodione, Vinclozolin or DCNA

Fungicide (500 μg/ml)	Frequency of Occurrence of Tolerant Strains from Conidia ($x10^{-7}$)							
	benomyl-sensitive			benomyl-tolerant				
	Bc-2	Bc-16	Bc-17	Bc-11	Bc-12	Bc-14	Bc-15	Bc-18
Procymidone	3.0	1.8	2.8	11.8	10.8	27.2	24.5	15.5
Dichlozoline	1.6	4.0	4.7	12.8	17.9	18.1	50.6	10.0
Iprodione	0.2	2.2	1.7	4.9	6.4	5.0	32.7	2.8
Vinclozolin	0.5	8.1	8.9	8.9	0.4	12.2	28.0	5.3
DCNA	0.6	1.1	2.5	9.8	0.6	10.2	6.8	1.3

Table 4. Development of Procymidone-tolerant Sectors from *B. cinerea* Mycelia

B. cinerea isolates	Number of mycelial discs developing tolerant sectors per 20 inoculated discs	
	Exp. 1	Exp. 2
Bc-1	3	4
Bc-2	4	2
Bc-4	0	1
Bc-5	9	9
Bc-9	1	0

Note: Mycelial discs (5mm in diameter) were placed on the PDA medium containing 500 μg/ml of the fungicide. Observations of tolerant sectors were made after eight days of incubation at 25°C.

potato dextrose agar (PDA) medium containing 1600 μg/ml of the fungicide, whereas the minimum inhibitory concentration to the parent strain (Bc-2) was 1.56 μg/ml (Table 6). It was interesting to note that some of the strains grew better in the presence of the fungicide.

The development of tolerance in vitro was not restricted to *B. cinerea*: *C. miyabeanus*, *S. sclerotiorum*, *A. kikuchiana* and *R. solani* were found to form tolerant sectors from mycelial discs grown on a PDA plate containing the toxicant (Table 7).

Cross Tolerance to Structurally Related Fungicides and the Aromatic Hydrocarbon Fungicides

The procymidone-tolerant *B. cinerea* strains exhibited cross tolerance to the cyclic *N*-(3,5-dichlorophenyl)imides, dichlozoline, iprodione and vinclozolin, and the aromatic hydrocarbon fungicides, PCNB, DCNA and chloroneb (Figure 1). The mycelial disc of the tolerant strain (R-22) or the sensitive parent strain (Bc-2) was placed on the center of a PDA plate, and the filter paper discs (8 mm in diameter) containing the test chemical were placed at the margin of the plate. After incubation, growth inhibition by the fungicides was apparent in the plate inoculated with the sensitive strain but

Table 5. Effect of Concentration of Procymidone on Development of Tolerant Sectors

Concentration (μg/ml)	Percentage of mycelial discs developing tolerant sectors
500	16
50	11
5	18

Note: The experiment was conducted in the manner presented in Table 4 except for the fungicide concentration. The test fungus was *B. cinerea* (Bc-2).

with the exception of benomyl, the chemicals did not inhibit the mycelial growth of the resistant strain.

<u>Growth Rate, Sporulation, Sclerotium Formation, Pathogenicity and Morphology of the Procymidone-tolerant Strains</u>

Fifty-six strains tolerant to procymidone were obtained from the conidia of *B. cinerea* (Bc-2) by selection on the medium containing 500 μg/ml of the fungicide. The rates of hyphal growth, sporulation and sclerotium formation of the strains were examined on the Hislop's medium in the presence of 100 μg/ml of the fungicide. The characteristics of the parent Bc-2 were also examined for comparison, using the same medium but without the fungicide. The growth rate on the medium varied depending on the strains, and some of the strains extended their hypha more rapidly than did the parent strain. In general, the tested strains were inferior to the parent strain in respect to the rates of sporulation and sclerotium formation, and to pathogenicity. Among the 56 strains, 20 lost their ability to form conidia on the medium and all of the rest sporulated less abundantly than did the parent. Moreover, the ability to form sclerotia was lost in 12 tolerant strains and the pathogenicity to cucumber seedlings was lost in 20 strains. All of the tested tolerant strains exhibited lower pathogenicity than did the parent strain. The characteristics of the tolerant strains possessing relatively high virulence are summarized in Table 8.

Microscopic observations of the tolerant strains further indicated that their mycelial morphology was generally abnormal. The growing hyphae on the agar plate were often distorted and were sometimes excessively branched, even in the absence of the fungicide. Each tolerant strain appeared distinct from the others in its hyphal morphology.

Table 6. Effect of Procymidone on the Mycelial Growth of the Tolerant Strains of *B. cinerea*

Concentration (μg/ml)	Mycelial growth (mm)		
	R-20	R-22	R-37
1600	20	35	28
400	25	31	34
100	32	35	36
25	34	36	38
0	33	33	18

Note: The tolerant strains employed in the experiment were originally isolated from Bc-2. Measurements of mycelial growth were made after four, three and seven days of incubation for R-20, 22 and 37, respectively.

Competition of the Tolerant Strain with the Parent Strain on Host Plants

Competition between fungicide-tolerant strains and sensitive strains on host plants may be one of the important factors in determining whether or not tolerant strains become dominant in the field. The following experiment was therefore conducted by using a mixed inoculation of the procymidone-tolerant strain (R-22) and the parent strain (Bc-2) onto rose flowers. After five successive transfers of spores produced on the flowers, the tolerant strain could not be detected by a single spore isolation from the infected flowers (Table 9). Even if the tolerant strains have pathogenicity, the above result implies that they possess little ability to retain their chemical tolerance and may disappear under natural conditions.

Greenhouse Trial for the Possibility of Emergence of the Tolerant Strain

In order to examine whether repeated use of procymidone causes a build-up of the tolerant strains in the field, the fungicide was sprayed on rose flowers that had been grown in greenhouses under suitable conditions for *B. cinerea* infection. After 19 applications of procymidone over three years, no tolerant strains were found, although a large increase in the population of the benomyl-tolerant

Table 7. Development of Procymidone-tolerant Sectors from the Mycelia of Several Fungal Species

Fungus	Number of mycelial discs developing tolerant sectors per 20 inoculated discs	
	Exp. 1	Exp. 2
C. miyabeanus	1	0
S. sclerotiorum	4	1
A. kikuchiana	2	1
R. solani	1	0

Note: The experiment was conducted in the manner presented in Table 4.

strains was recognized in the benomyl-treated plot. After benomyl had been sprayed five times, about 80% of the 139 isolates were highly tolerant to benomyl and all isolates were tolerant after 18 sprayings. The results, therefore, indicate that the development of tolerance to procymidone does not occur in the field as easily as in vitro.

DISCUSSION

This study demonstrates that tolerant strains of *B. cinerea* could be obtained easily under laboratory conditions without UV irradiation or other mutagenic treatments. They developed after either dense inoculation on fungicide-containing agar plates with the fungal spores or prolonged incubation of the mycelial discs on the fungicide-containing medium. The present experiments also indicate that tolerance of *B. cinerea* to procymidone resulted in tolerance to the other cyclic *N*-(3,5-dichlorophenyl)imides, dichlozoline, iprodione and vinclozolin, and the aromatic hydrocarbon fungicides, PCNB, DCNA and chloroneb. Leroux et al. (1977, 1978) and Fritz et al. (1977) obtained *B. cinerea* strains growing in the presence of iprodione, vinclozolin or DCNA in vitro and showed that most of the strains selected under one of the fungicides exhibited cross tolerance to the other cyclic *N*-(3,5-dichlorophenyl)imides, including procymidone and the aromatic hydrocarbon fungicides. Sztejnberg and Jones (1978) reported that tolerant strains of *Monilinia fructicola* were cross tolerant to iprodione, vinclozolin, procymidone and DCNA. These observations indicate that procymidone and the other cyclic *N*-(3,5-dichlorophenyl)imides can be classified

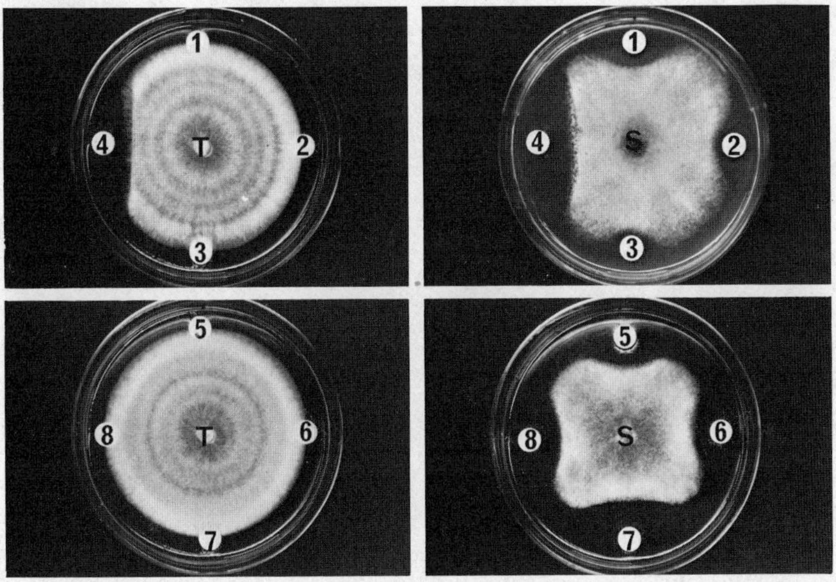

Figure 1. Cross tolerance of the procymidone-tolerant strain of *B. cinerea* to the *N*-(3,5-dichlorophenyl)imides and aromatic hydrocarbon fungicides. T, the tolerant strain (R-22); S, the sensitive strain (Bc-2); 1, DCNA (0.1 mg/disc); 2, PCNB (1 mg/disc); 3, chloroneb (1 mg/disc); 4, benomyl (0.1 mg/disc); 5, procymidone (0.1 mg/disc); 6, dichlozoline (0.1 mg/disc); 7, iprodione (0.1 mg/disc); 8, vinclozolin (0.1 mg/disc).

in the aromatic hydrocarbon group, as termed by Georgopoulos and Zaracovitis (1967). In addition, Georgopoulos et al. (1979) recently showed that iprodione, procymidone and vinclozolin markedly increased the frequency of somatic recombination in diploid colonies of *Aspergillus nidulans*, as was previously detected with several aromatic hydrocarbon fungicides, PCNB, TCNB (tetrachloronitrobenzene), DCNA, SOPP (sodium orthophenylphenate) and chloroneb (Georgopoulos et al., 1976). This supports the above classification. However, it should be noted that cross tolerance between the cyclic *N*-(3,5-dichlorophenyl)imides and the other members of the aromatic hydrocarbon group was not detected in some of the tolerant strains (Leroux et al., 1977; Fritz et al., 1977), suggesting slight differences in their mode of fungitoxic action.

Table 8. Characteristics of Procymidone-tolerant Strains of *B. cinerea*

Strain	Rate of mycelial growth (mm/day)	Sporulation	Sclerotium formation	Relative pathogenicity
R-9	14	sparse	+	76
R-12	19	moderate (1.0×10^6)	+	66
R-15	10	sparse	+	88
R-18	14	sparse	+	64
R-20	8	moderate (1.4×10^6)	+	66
R-21	8	sparse	+	71
R-22	12	moderate (5.8×10^5)	+	76
R-37	12	moderate (4.0×10^5)	-	62
Bc-2 (parent)	12	abundant (4.0×10^6)	+	100

Note: Figures in parentheses indicate conidial numbers formed per petri dish.

Table 9. Competition of the Procymidone-tolerant Strain (R-22) and the Parent Strain (Bc-2) on Rose Flowers

Times of transfer	Percentage of tolerant isolates	
	Exp. 1	Exp. 2
1	68	98
5	0	0

Note: Rose flowers were inoculated with a mixture of R-22 and Bc-2 at a ratio of 20 to 1. Conidia formed on the flowers were successively transferred to fresh flowers.

Tolerance of *B. cinerea* to procymidone was associated with decreased sporulation and pathogenicity to cucumber seedlings. Some of the tolerant strains lost their ability to form sclerotia, and microscopic observations indicated that in the absence of the fungicide, the tolerant isolates appeared different from the sensitive parent strain in the morphology of growing hyphae. The hyphae of the former strains were distorted and/or excessively branched, showing that tolerance was accompanied by morphological change. The most important defect of the tolerant strains was found to be their low ability to compete with the sensitive strain on host tissues, which was proved in the competition experiment on detached rose flowers. Furthermore, when the mixed inoculation of the tolerant and sensitive strains at a higher rate of the former was used with eggplants in a greenhouse, the population of the tolerant strain rapidly decreased (Hisada and Maeda, 1979), suggesting that the procymidone-tolerant strains like those obtained in vitro are not likely to become dominant in the field. In fact, continuous sprayings of procymidone on rose plants in a greenhouse did not result in the emergence of tolerant strains of *B. cinerea*, although tolerance to benomyl developed in the benomyl-treated plot. Hirota and Kato (1977) also showed that tolerant strains of *B. cinerea* did not appear even after 30 continuous applications of procymidone or iprodione on eggplants, while benomyl-tolerant strains occurred after 10 continuous sprayings of benomyl. A similar experiment was also conducted in the vineyards of West Germany. Lorenz and Eichhorn (1978) reported that adaptation of *B. cinerea* to the fungicide was not observed after 11 continuous sprayings of iprodione or vinclozolin under field conditions, although tolerant strains were obtained on the fungicide-containing medium. On the other hand, although dichlozoline and DCNA have been used for several years in Japan, all of the 74 isolates of *B. cinerea* from the various locations were almost equally sensitive to procymidone (Takaki et al., 1979). Schüepp and Küng (1978) also observed that tolerant strains developed on medium containing the cyclic *N*-(3,5-dichlorophenyl)imide, but they failed to isolate any tolerant strains from vineyards. It is therefore safe to say that adaptation of *B. cinerea* to the cyclic *N*-(3,5-dichlorophenyl)imide fungicides may not occur in natural conditions as easily as it does in the laboratory.

Occurrences of tolerant strains have been frequently reported for the aromatic hydrocarbon fungicides (Table 10). Most of the tolerant isolates were obtained under laboratory conditions from the fast-growing sectors of mycelial colonies on the fungicide-containing medium; however, some of the tolerant strains were isolated directly from fields. The tolerant strains of *Sclerotium cepivorum, Rhizoctonia solani* and *Penicillium digitatum* were isolated from rotted onion bulbs, diseased cotton seedlings and spores present in lemon packing houses, respectively (Locke, 1969; Shatla and Sinclair, 1963; Harding, 1962), but it is not known whether or not these strains had characteristics similar to those of the tolerant strains obtained in vitro. For the tolerant strains from the field, the degree of tolerance reported seemed to be rather small, while the procymidone-tolerant isolates obtained in vitro showed a high degree of tolerance. For instance, diphenyl-tolerant strains of *P. digitatum* were found to be only four times as tolerant to SOPP as the sensitive strains. Their growth was completely inhibited at 160 μg/ml of the fungicide (Harding, 1962), but the *B. cinerea* strains obtained in the present study were more than 1000 times as tolerant. Thus, the characteristics of the tolerant field isolates may possibly differ from those obtained in the laboratory.

The mechanism by which aromatic hydrocarbon tolerance is acquired by fungi is still obscure. Georgopoulos (1962) concluded that chlorinated nitrobenzene tolerance in *Hypomyces solani* f. *cucurbitae* originated from a gene mutation because crossing the tolerant and sensitive strains resulted in a Mendelian segregation in the progeny. Further investigations showed that tolerance was acquired by mutation at one of three loci in the genetic apparatus (Georgopoulos, 1963). By contrast, heterokaryosis has been suggested as a possible mechanism for inducing tolerance to the aromatic hydrocarbon fungicides (Esuruoso and Wood, 1971; Webster et al., 1968, 1970; Meyer and Parmeter, 1968). In *Rhizopus stolonifer,* the spores produced by the DCNA-tolerant strains were not always tolerant to the fungicide, indicating the heterokaryotic nature of the cells (Webster et al., 1968). Similarly, the *B. cinerea* strains tolerant to PCNB, TCNB or DCNA produced tolerant and sensitive spores at a rate of approximately 1 to 1 (Esuruoso and Wood, 1971), and in *Thanatephorus cucumeris,* the heterokaryon, which was synthesized from a certain combination of homokaryons sensitive to diphenyl,

Table 10. Laboratory and Field Occurrences of Fungal Strains Tolerant to the Aromatic Hydrocarbon Fungicides

Fungus	Fungicides	References
Aspergillus nidulans	dichlozoline, PCNB, TCNB	Dekker (1972), Threlfall (1968)
Botrytis allii	PCNB, TCNB	Priest and Wood (1961)
Botrytis cinerea	PCNB, TCNB, DCNA, chloroneb, iprodione, vinclozolin	Webster et al. (1970), Esuruoso and Wood (1971), Leroux et al. (1977, 1978), Lorenz and Eichhorn (1978)
Diplodia natalensis	diphenyl	Littauer and Gutter (1953)
Fusarium caeruleum	TCNB	McKee (1951)
Gilbertella persicaria	DCNA	Ogawa et al. (1963)
Hypomyces solani	PCNB, TCNB	Georgopoulos (1962)
Monilinia fructicola	iprodione, vinclozolin, procymidone	Sztejnberg and Jones (1978)

(continued)

Table 10. (Continued)

Fungus	Fungicides	References
Neurospora crassa	diphenyl	Georgopoulos et al. (1966)
Penicillium expansum	iprodione	Leroux et al. (1978)
Penicillium digitatum[a]	SOPP, diphenyl	Harding (1962)
Penicillium italicum	diphenyl	Harding (1959)
Rhizoctonia solani[a]	PCNB	Shatla and Sinclair (1963)
Rhizopus stolonifer	DCNA	Webster et al. (1968)
Sclerotium cepivorum[a]	DCNA	Locke (1969)
Sclerotium rolfsii	PCNB	Georgopoulos (1964)
Tiletia foetida[a]	HCB (hexachlorobenzene), PCNB	Kuiper (1965), Skorda (1977)
Ustilago maydis	iprodione, chloroneb	Leroux et al. (1978), Tillman and Sisler (1973)

[a]Field occurrence of tolerant strains.

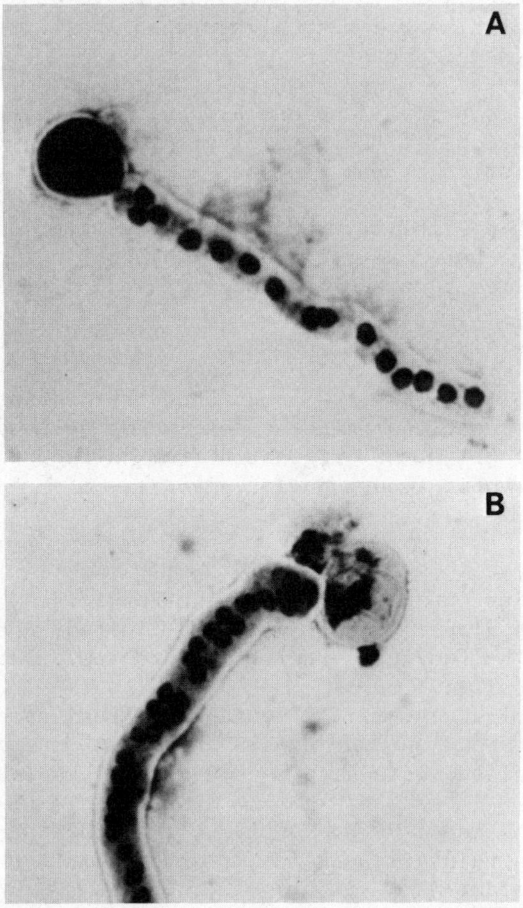

Figure 2. Nuclei of control and procymidone-treated germ tubes of *B. cinerea*. A, control; B, procymidone-treated (after 6-hr incubation with the fungicide).

was shown to be tolerant to both diphenyl and PCNB (Meyer and Parmeter, 1968).

Since cells of *B. cinerea* are multinucleate and heterokaryotic (Hansen and Smith, 1932), it is probable that procymidone tolerance occurred from certain heterokaryotic associations of nuclei, at least in *B. cinerea*. The following observations support this hypothesis:

1. The frequency of occurrence of the procymidone-tolerant strains was much higher than that expected from spontaneous mutation.

2. Tolerance to procymidone was always associated with many variations of the physiological and morphological characteristics of the fungus.

3. There is a possibility that procymidone itself induces the heterokaryotic association of nuclei in *B. cinerea* because the number of nuclei per cell would often increase in the presence of the fungicide (Figure 2).

However, the possibility that the occurrence of tolerant strains was due to genetic mutations cannot be excluded, because the TCNB-tolerant strains of *H. solani* originated from an ascospore isolate and must be homokaryotic (Georgopoulos, 1962).

REFERENCES

Dekker, J., 1972, Resistance, *in:* "Systemic Fungicides," R. W. Marsh, ed., pp. 156-174, Longman Group Limited, London.

Dekker, J., 1977, The fungicide-resistance problems, *Neth. J. Pl. Path.*, 83 (suppl. 1):159.

Dekker, J., and Gielink, A. J., 1979, Acquired resistance to pimaricin in *Cladosporium cucumerinum* and *Fusarium oxysporum* f. sp. *narcissi* associated with decreased virulence, *Neth. J. Pl. Path.*, 85:67.

Esuruoso, O. F., and Wood, R. K. S., 1971, The resistance of spores of resistant strains of *Botrytis cinerea* to quintozene, tecnazene and dicloran, *Ann. Appl. Biol.*, 68:271.

Fritz, R., Leroux, P., and Gredt, M., 1977, Mechanism of antifungal action of promidione (26019 RP or glycophene), vinclozolin and dicloran on *Botrytis cinerea* Pers., *Phytopath. Z.*, 90:152.

Fuchs, A., de Ruig, S. P., van Tuyl, J. M., and de Vries, F. W., 1977, Resistance to triforine: A nonexistent problem? *Neth. J. Pl. Path.*, 83 (Suppl. 1):189.

Georgopoulos, S. G., 1962, Genetic nature of tolerance of *Hypomyces solani* f. *cucurbitae* to penta- and tetra-chloronitrobenzene, *Nature*, 194:148.

Georgopoulos, S. G., 1963, Tolerance to chlorinated nitrobenzenes in *Hypomyces solani* f. *cucurbitae* and its mode of inheritance, *Phytopathology*, 53:1086.

Georgopoulos, S. G., 1964, Chlorinated-nitrobenzene tolerance in *Sclerotium rolfsii*, *Ann. Inst. Phytopath. Benaki N. S.*, 6:156.

Georgopoulos, S. G., and Zaracovitis, C., 1967, Tolerance of fungi to organic fungicides, *Ann. Rev. Phytopath.*, 5:109.

Georgopoulos, S. G., Kappas, A., and Macris, B., 1966, Gene-controlled resistance to aromatic hydrocarbons in *Neurospora crassa* and its relationship to the inhibition by L-sorbose, *Neurospora Newsletter*, 10:8.
Georgopoulos, S. G., Kappas, A., and Hastie, A. C., 1976, Induced sectoring in diploid *Aspergillus nidulans* as a criterion of fungitoxicity by interference with hereditary processes, *Phytopathology*, 66:217.
Georgopoulos, S. G., Sarris, M., and Ziogas, B. N., 1979, Mitotic instability in *Aspergillus nidulans* caused by the fungicides iprodione, procymidone and vinclozolin, *Pestic. Sci.*, 10:389.
Hansen, H. N., and Smith, R. E., 1932, The mechanism of variation in imperfect fungi: *Botrytis cinerea*, *Phytopathology*, 22:953.
Harding, P. R., Jr., 1959, Biphenyl-induced variations in citrus blue mold, *Pl. Dis. Reptr.*, 43:649.
Harding, P. R., Jr., 1962, Differential sensitivity to sodium orthophenylphenate by biphenyl-sensitive and biphenyl-resistant strains of *Penicillium digitatum*, *Pl. Dis. Reptr.*, 46:100.
Hirota, K., and Kato, K., 1977, On the development and decay of fungicide resistance in *Botrytis cinerea* to benomyl, *Res. Bull. Aichi Agric. Res. Centr.*, B9:48.
Hisada, Y., and Maeda, K., 1979, unpublished.
Hisada, Y., Maeda, K., Tottori, N., and Kawase, Y., 1976, Plant disease control by *N*-(3',5'-dichlorophenyl)-1,2-dimethylcyclopropane-1,2-dicarboximide, *J. Pestic. Sci.*, 1:145.
Hisada, Y., Kato, T., and Kawase, Y., 1978, Mechanism of antifungal action of procymidone in *Botrytis cinerea*, *Ann. Phytopath. Soc. Japan*, 44:509.
Hislop, E. C., 1967, Observations on the vapor phase activity of some foliage fungicides, *Ann. Appl. Biol.*, 60:265.
Kuiper, J., 1965, Failure of hexachlorobenzene to control common bunt of wheat, *Nature*, 206:1219.
Leroux, P., Fritz, R., and Gredt, M., 1977, Laboratory studies on strains of *Botrytis cinerea* Pers. tolerant to dichlozoline, dicloran, quintozene, vinclozolin and 26019 RP (or glycophene), *Phytopath. Z.*, 89:347.
Leroux, P., Gredt, M., and Fritz, R., 1978, Resistance to dichlozoline, dicyclidin, iprodione, vinclozolin and aromatic hydrocarbon fungicides in some phytopathogenic fungi, *Med. Fac. Landbouww. Rijksuniv. Gent*, 43:881.
Littauer, F., and Gutter, Y., 1953, Diphenyl-resistant strains of *Diplodia*, *Palestine J. Botany Rehovot Ser.*, 8:185.
Locke, S. B., 1969, Botran tolerance of *Sclerotium cepivorum* isolants from fields with different Botran-treatment histories, *Phytopathology*, 59:13.
Lorenz, D. H., and Eichhorn, K. W., 1978, Untersuchungen zur möglichen Resistenzbildung von *Botrytis cinerea* an Reben gegen die Wirkstoffe Vinclozolin und Iprodione, *Die Weinwissenschaft*, 33:251.

McKee, R. K., 1951, Mutations appearing in *Fusarium caeruleum* cultures treated with tetrachloronitrobenzene, *Nature*, 167:611.

Meyer, R. W., and Parmeter, J. R., Jr., 1968, Changes in chemical tolerance associated with heterokaryosis in *Thanatephorus cucumeris*, *Phytopathology*, 58:472.

Ogawa, J. M., Ramsey, R. H., and Moore, C. J., 1963, Behavior of variants of *Gilbertella persicaria* arising in medium containing 2,6-dichloro-4-nitroaniline, *Phytopathology*, 53:97.

Priest, D., and Wood, R. K. S., 1961, Strains of *Botrytis allii* resistant to chlorinated nitrobenzenes, *Ann. Appl. Biol.*, 49:445.

Schüepp, H., and Küng, M., 1978, Gegenüber Dicarboximid-Fungiziden tolerante Stämme von *Botrytis cinerea*, *Ber. Schweiz. Bot. Ges.*, 83:63.

Shatla, M. N., and Sinclair, J. B., 1963, Tolerance to pentachloronitrobenzene among cotton isolates of *Rhizoctonia solani*, *Phytopathology*, 53:1407.

Sherald, J. L., and Sisler, H. D., 1975, Antifungal mode of action of triforine, *Pestic. Biochem. Physiol.*, 5:477.

Skorda, E. A., 1977, Insensitivity of wheat bunt to hexachlorobenzene and quintozene (pentachloronitrobenzene) in Greece, *Proc. 9th Br. Insectic. Fungic. Conf.*, 1:67.

Sztejnberg, A., and Jones, A. L., 1978, Tolerance of the brown rot fungus *Monilinia fructicola* to iprodione, vinclozolin and procymidone fungicides, *Proc. Am. Phytopath. Soc.*, 5:187.

Takaki, H., Hisada, Y., Ozaki, T., and Kawase, Y., 1979, unpublished.

Threlfall, R. J., 1968, The genetics and biochemistry of mutants of *Aspergillus nidulans* resistant to chlorinated nitrobenzenes, *J. Gen. Microbiol.*, 52:35.

Tillman, R. W., and Sisler, H. D., 1973, Effect of chloroneb on the growth and metabolism of *Ustilago maydis*, *Phytopathology*, 63:219.

Webster, R. K., Ogawa, J. M., and Moore, C. J., 1968, The occurrence and behavior of variants of *Rhizopus stolonifer* tolerant to 2,6-dichloro-4-nitroaniline, *Phytopathology*, 58:997.

Webster, R. K., Ogawa, J. M., and Bose, E., 1970, Tolerance of *Botrytis cinerea* to 2,6-dichloro-4-nitroaniline, *Phytopathology*, 60:1489.

Wolfe, M. S., 1971, Fungicides and the fungus population problem, *Proc. 6th Br. Insectic. Fungic. Conf.*, 2:724.

PROBLEMS OF FUNGICIDE RESISTANCE IN PENICILLIUM ROT OF CITRUS FRUITS

Joseph W. Eckert and Brian L. Wild*

Department of Plant Pathology
University of California
Riverside, California 92521

PENICILLIUM MOLD OF CITRUS FRUITS: BIOLOGY AND CONTROL

Disease Etiology

Green and blue mold of citrus fruits are incited by the ubiquitous fungi *Penicillium digitatum* Sacc. and *P. italicum* Wehm., respectively. Green mold is the most important cause of postharvest decay of citrus fruits produced in areas with scant rainfall during the period of fruit maturation. Blue mold is of lesser overall importance, but may become the major problem under environmental conditions or fungicide treatments that selectively suppress the development of green mold. Penicillium molds are also important in humid production areas but tend to be overshadowed there by the stem-end rots. In nature, *P. digitatum* completes its life cycle only on citrus fruits, whereas *P. italicum* can infect an array of different fruits and vegetables. Spores of *P. digitatum* and *P. italicum* that form on diseased fruits on the ground in citrus groves and in packinghouses are transported by air currents to healthy fruits. The surface of virtually every citrus fruit is contaminated with these spores at harvest time, but they are unable to germinate and infect the fruit except at injury sites, where the rate of infection reaches a maximum at about 25°C under high humidity conditions. Infected fruit are decayed totally within 7 days and the fungus sporulates heavily on the fruit surface. The reproductive capacity of *Penicillium* is enormous. Potentially, a single spore inoculated into a

*Present address: New South Wales Department of Agriculture, Gosford Horticultural Postharvest Laboratory, New South Wales, Australia.

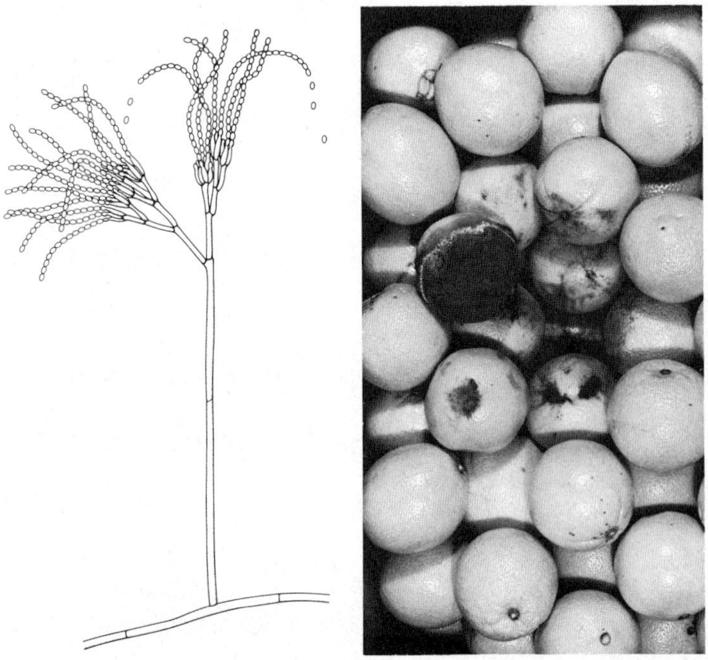

Figure 1. Left: Typical conidiophore and conidia (spores) of *Penicillium digitatum*. Right: Orange fruit infected by *P. digitatum* and several sound fruit contaminated superficially with spores.

susceptible fruit could give rise to a progeny of about 10^{10} spores in 7 days under optimum environmental conditions. These spores serve as inoculum for infection of other citrus fruits and they may visibly contaminate adjacent fruit in the same container, giving rise to a problem known as "soilage" (Figure 1), which seriously reduces the market value of the fruit.

Biology of *Penicillium digitatum* and *P. italicum*

The hyphal cells and spores of *P. digitatum* and *P. italicum* usually are haploid like other species of the Class Deuteromycotina (Fungi Imperfecti). The hyphal cells are multi-nucleate and homocaryotic, but the conidia have only one nucleus. No true sexual stage has been reported for either *P. digitatum* or *P. italicum*, but some isolates of *P. italicum*, *P. expansum* and other species may undergo a "parasexual cycle" in which the hyphae of two compatible haploid isolates fuse to form a heterocaryon. Some of the haploid nuclei fuse, giving rise to diploid colonies that are usually unstable in normal culture medium, and these diploids spontaneously

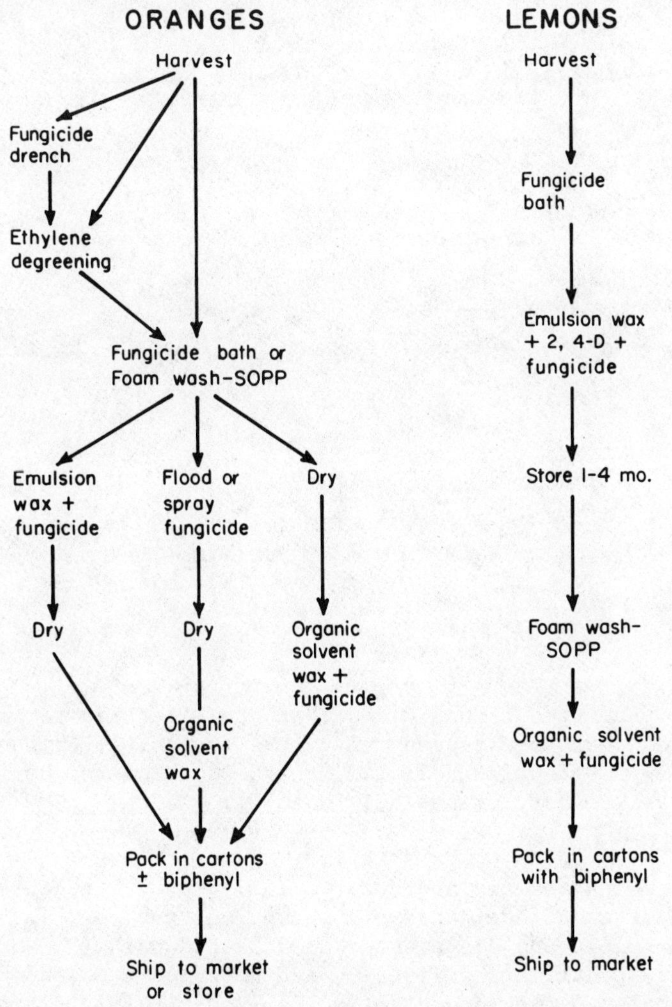

Figure 2. Flow diagram showing fungicides and other treatments applied to oranges and lemons after harvest.

revert to haploid segregants accompanied by a recombination of the chromosomes among the progeny (Strømnaes et al., 1964; Beraha and Garber, 1980). This parasexual cycle has never been observed in *P. digitatum*, despite repeated attempts utilizing techniques that had been successful with other *Penicillium* species.

Fungicide Control of Penicillium Mold

Figure 2 shows the sequence of fungicide treatments that are applied to California citrus fruits after harvest to control

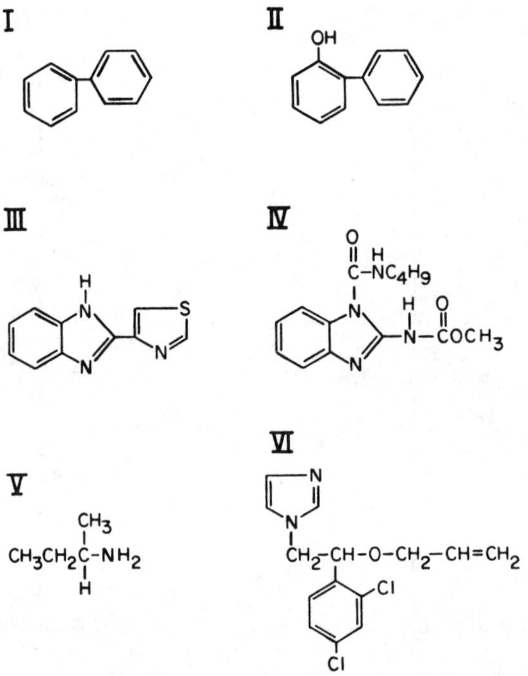

Figure 3. Structure of fungicides that are effective for control of postharvest decay of citrus fruits. I, biphenyl; II, *o*-phenylphenol; III, thiabendazole; IV, benomyl; V, *sec*-butylamine; VI, imazalil.

Penicillium molds (Dawson and Eckert, 1977). The structures of the recommended organic fungicides are shown in Figure 3. The harvested fruits are cleaned and disinfested in a heated bath (32-43°C) of one of the broad-spectrum fungicides -- sodium carbonate, sodium tetraborate or sodium *o*-phenylphenate (SOPP) (Eckert, 1977, 1978). The fruit are then rinsed with fresh water and are covered with a hydrophobic coating ("wax") containing a fungicide, e.g., thiabendazole (2 to 4 g/liter), benomyl (1 to 2 g/liter), SOPP (20 g/liter) or *sec*-butylamine (10 to 20 g/liter). The "wax"-fungicide treatment retards water loss and prevents Penicillium decay during storage and transport of the fruit to market. Finally, the fruits are placed in a fiberboard box with a paper sheet impregnated with the volatile fungistat, biphenyl. Commercial variations of this sequence can have a major influence upon the development of fungicide-resistant strains of *P. digitatum* and *P. italicum*.

Early and late-season oranges often require degreening with ethylene gas to attain the uniform orange color that is preferred by most consumers. The harvested fruits are held for several days,

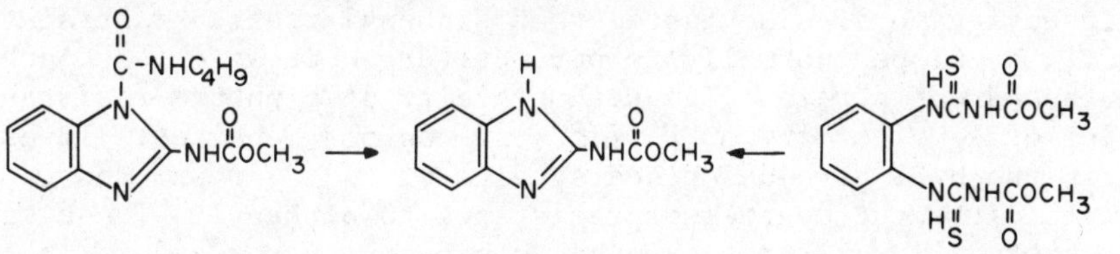

Figure 4. Transformation of benomyl and thiophanate-methyl to carbendazim.

before cleaning, in an atmosphere of 5-10 µl/liter ethylene gas at 20-25°C and 90% relative humidity. Since these environmental conditions are optimal for infection of the fruit by *Penicillium* species, it is advisable to drench the fruit with a solution of *sec*-butylamine (phosphate salt), thiabendazole or benomyl before degreening. Lemons produced in the coastal districts of California usually are stored for 1 to 4 months after harvest to improve fruit quality and to regulate supply and demand. Oranges and lemons may be stored for days or weeks after packaging for shipment, depending upon market demand for certain fruit sizes and grades. Since the residues of all of these fungicides are stable on fruit, the practice of storing fungicide-treated fruit creates intense selection pressure for the fungicide-resistance genes in the populations of *P. digitatum* and *P. italicum* in citrus packinghouses.

In Japan, satsuma mandarins usually are sprayed once with thiophanate-methyl or benomyl before harvest to prevent infection of the fruit during picking and storage (Kuramoto, 1976). Both of these compounds are progenitors of carbendazim (Figure 4), which is the actual fungitoxicant. No other fungicides are applied to the fruit either before or after storage.

RESISTANCE OF *Penicillium digitatum* AND RELATED SPECIES TO FUNGICIDES

Biphenyl and *o*-Phenylphenol

Several years after biphenyl was recognized to be an effective postharvest fungicide, Farkas and Aman (1940) in Israel reported that mutants with stable resistance to biphenyl arose spontaneously in pure cultures of *P. digitatum* that were exposed to biphenyl vapors. These variants were not observed in citrus groves nor packinghouses and, apparently, were not regarded by these investigators as a threat to the use of biphenyl as a citrus fruit fungicide. Littauer and Gutter (1953) observed that resistant mutants of *P. italicum* and *Diplodia natalensis*, as well as *P. digitatum*, arose in pure cultures of these fungi that were exposed for long periods to biphenyl vapors. They noted that biphenyl-resistant strains of *Diplodia* developed normally in pure culture, but were less pathogenic than the parent strain. The pathogenicity of biphenyl-resistant strains of *P. digitatum* or *P. italicum* was not reported. Investigators in the U.S. have described strains of *P. digitatum* and *P. italicum* with different degrees of resistance to biphenyl. These variants were observed in cultures exposed to biphenyl vapor and in isolates collected in citrus groves and packinghouses where neither biphenyl nor related fungicides had ever been used (Duran and Norman, 1961; Harding, 1964, 1965; Smoot and Winston, 1967). Harding (1965) distinguished two types of biphenyl-resistant mutants. Type A mutants grew faster in the presence of biphenyl and did not sporulate in the absence of this fungicide. Type B resistant mutants were variable in sporulation and growth in the presence of biphenyl and grew faster or slower than the parent strain in the absence of biphenyl. The pathogenicity of the resistant strains relative to the parent strains was not reported.

The existence of biphenyl-resistant mutants of *Penicillium* species was not regarded as a practical problem until the late 1960s when certain shipments of California lemons packed with biphenyl-impregnated papers arrived in distant markets with high levels of Penicillium decay and sporulation. Harding (1962, 1964) observed that many of these problem shipments originated in packinghouses where the lemons had been treated with SOPP several months prior to shipment. Harding (1962) found that strains of *P. digitatum* that were resistant to biphenyl were also resistant to SOPP, as would be anticipated from the structural similarity of the two fungicides (Figure 3, I and II). Harding (1962) further demonstrated that residues of *o*-phenylphenol on SOPP-treated lemons were sufficient to suppress the growth of biphenyl-sensitive strains, thereby allowing the biphenyl-resistant mutants in the population to proliferate on the fruit. In a survey of California citrus-growing areas, Harding (1964) observed biphenyl-resistant variants of *P. digitatum* at a frequency of about 0.5% of the isolates collected in citrus groves and packinghouses that had never been exposed to SOPP. Packinghouses that had a history of SOPP use showed a resistant strain frequency of 14-18% of the *P. digitatum* population. Sixty percent of the Penicillium isolates were resistant to biphenyl in certain packinghouses that had used SOPP continuously over a period of years. These observations explained the failure of the biphenyl treatment to control green mold on fruit shipped from these packinghouses. Biphenyl/SOPP-resistant isolates of *P. italicum* were rarely recovered from the atmosphere of California packinghouses, explaining why this species was not involved in the biphenyl resistance problem. Biphenyl-resistant strains of *P. digitatum* have been reported in Florida, but these isolates have not caused a significant commercial problem because SOPP-treated fruits are not stored routinely in the packinghouses there (Smoot and Winston, 1967; Houck, 1977). Thus, there is little opportunity for selection and dissemination of biphenyl/SOPP-resistant mutants in Florida.

SOPP inhibits several enzymes of the tricarboxylic acid cycle of *P. italicum*; i.e., isocitrate dehydrogenase, α-ketoglutarate dehydrogenase, succinate dehydrogenase, and, to a lesser degree, malate dehydrogenase (Rehm, 1969). *o*-Phenylphenol-resistant isolates of *P. italicum* apparently can utilize the glyoxylate cycle to bypass the inhibited enzyme reactions (Herbig and Rehm, 1967; Rehm, 1969). SOPP and other aromatic hydrocarbons increase somatic recombination in diploid strains of *Aspergillus nidulans* (Georgopoulos et al., 1976, 1979), and biphenyl influences the uptake of phosphate and potassium ions by conidia of *Fusarium solani* (Georgopoulos et al., 1967).

Beraha and Garber (1966, 1980) have investigated the genetics of SOPP resistance in *P. italicum* and *P. expansum* (an apple fruit pathogen) by means of their parasexual cycles. In *P. italicum*, SOPP resistance is based upon a single Mendelian gene that is dominant in heterozygous diploids. The pathogenicity of the SOPP-resistant mutants of *P. italicum*, produced by UV irradiation or N-methyl-N'-nitro-N-nitrosoguanidine treatment of conidia, was not reported. Earlier work (Beraha et al., 1964) had shown that approximately 44% of color and auxotropic mutants of *P. italicum* and 67% of the *P. digitatum* mutants were virulent when inoculated into orange fruit. SOPP resistance in *P. expansum* has been attributed to three unlinked genes that are recessive in heterozygous diploids. Three mutants resistant to 50 μg/ml SOPP were avirulent to apple fruit, whereas two mutants resistant to 30 μg/ml were virulent (Beraha and Garber, 1966).

Benzimidazole Fungicides

The benzimidazole fungicides have been used extensively over the past ten years to control postharvest decays of citrus and other fruits (Eckert, 1977, 1978). The best known compounds in this group of fungicides are benomyl, thiabendazole, carbendazim and thiophanate-methyl (Figures 3 and 4). Thiabendazole, benomyl or carbendazim, suspended in water or in a wax formulation, are applied to citrus fruits after harvest in all production areas except Japan. In Japan, benomyl or thiophanate-methyl is applied to satsuma mandarin fruit as a single spray before harvest. None of these fungicides are registered for postharvest application in Japan. Preharvest sprays of benomyl also have been used effectively in Florida (Brown, 1977), but this practice has been discontinued since it could select for benzimidazole-resistant fungi that would reduce the effectiveness of the postharvest benzimidazole treatment, which is necessary for successful marketing of the crop.

Thiabendazole is stable on citrus fruits, whereas both thiophanate-methyl and benomyl undergo transformation to carbendazim (Figure 4). Carbendazim appears to be the principal fungitoxicant present in plant tissue and in fungi that have been treated with benomyl or thiophanate-methyl (Vonk and Kaars Sijpesteijn, 1971; Hammerschlag and Sisler, 1973). Benomyl and carbendazim are about equally fungitoxic, but thiabendazole is significantly less inhibitory (Eckert and Rahm, 1979); however, some rare mutants are more sensitive to benomyl than to thiabendazole. Benomyl is more lipophilic than thiabendazole or carbendazim and, therefore, penetrates hydrophobic barriers (wax and cuticle) on the plant surface more readily than thiabendazole or carbendazim (Eckert et al., 1979). For this reason, benomyl is more effective than the other benzimidazole fungicides in situations that require penetration of the fungicide into the plant tissue, e.g., eradication of latent infections, internal protective action, and sporulation inhibition.

The benzimidazole fungicides have been highly effective against Penicillium decay of citrus fruits as well as the stem end rots incited by *Diplodia natalensis* and *Phomopsis citri*. They are also somewhat more effective against *P. digitatum* than *P. italicum* (Gutter, 1975) and are inactive against sour rot (*Geotrichum candidum*), Alternaria rot (*Alternaria citri*), and brown rot (*Phytophthora citrophthora*). These diseases may become more prominent in fruit lots that have been treated with benzimidazole fungicides and then stored for several weeks (Brown, 1977).

Benzimidazole resistance in *Penicillium* species was observed in California citrus packinghouses about 15 months after thiabendazole was incorporated in the wax formulation applied to lemons before storage (Houck, 1977). Harding (1972) reported that thiabendazole-resistant isolates of both *P. digitatum* and *P. italicum* occurred at a high frequency in the atmosphere of lemon packinghouses that used thiabendazole intensively as a pre-storage treatment for lemons. These isolates were not controlled by treating inoculated fruit with a high dosage (3 g/liter) of thiabendazole. Benzimidazole-resistant isolates of *P. digitatum* and *P. italicum* also have been isolated in Florida (Smoot and Brown, 1974), Israel, Australia (Muirhead, 1974; Wild and Rippon, 1975) and Japan (Kuramoto, 1976), and a survey of the frequency of benzimidazole-resistant isolates of *Penicillium* on decayed citrus fruit arriving in Rotterdam has revealed the magnitude of this problem (Table 1). Benzimidazole-resistant strains of *P. expansum* have been reported on harvested apples and pears in Oregon (Bertrand and Saulie-Carter, 1978), New York (Rosenberger and Meyer, 1979; Rosenberger et al., 1979) and Australia (Wicks, 1977; Koffmann et al., 1978). Most observations of benzimidazole-resistant strains of *Penicillium* species have followed intensive use of thiabendazole or benomyl over a period of some months, but Kuramoto (1976) reported a severe resistance problem on satsuma

Table 1. Benzimidazole-resistant Strains of *Penicillium digitatum* Isolated from Citrus Fruits Shipped to Rotterdam[a]

Fruit origin	No. isolates	% of isolates resistant to:	
		Thiabendazole	Benomyl
Algeria	27	37	30
Argentina	142	49	39
Australia	2	100	100
Brazil	86	33	19
California	388	55	50
Chile	13	15	8
Cuba	19	37	26
Cyprus	50	14	12
Egypt	22	27	14
Florida	65	28	15
Greece	57	16	9
Honduras	24	54	29
Israel	99	90	80
Italy	73	49	44
Morocco	56	57	43
South Africa	115	77	32
Spain	146	31	16
Texas	57	35	26
Turkey	7	29	14
Uruguay	43	58	51

[a]McDonald et al., 1979.

mandarins in Japan following application of a single spray of thiophanate-methyl immediately before harvest. Several investigators have recorded the isolation of benzimidazole-resistant strains from citrus groves and packinghouses where benzimidazole fungicides had never been used (Harding, 1972; Kuramoto, 1976; Wild, 1980).

Benzimidazole-resistant isolates of *Penicillium* species found in packinghouses are usually more sensitive to benomyl than to thiabendazole (Harding, 1972; Houck, 1977; Wild, 1980), although some isolates of *P. digitatum* and *P. italicum* are less sensitive to benomyl than to thiabendazole (Muirhead, 1974; Wild and Rippon, 1975). Furthermore, some thiabendazole-resistant *P. expansum* produced by mutagenic agents in vitro were more sensitive to benomyl than the wild parent strain (Van Tuyl 1975, 1977b). This phenomenon is known as "negatively correlated cross resistance." Most of the benzimidazole-resistant mutants that have been isolated in fruit packinghouses are sensitive to the other postharvest fungicides, *o*-phenylphenol, *sec*-butylamine and imazalil (Smoot and Brown, 1974; Wild and Rippon, 1975, Harding, 1976; Wicks, 1977). Less frequently, isolates of *P. digitatum* that are doubly resistant to benzimidazole fungicides and SOPP or *sec*-butylamine have been reported (Harding, 1976; Wild, 1980).

In general, benzimidazole-resistant strains that have been isolated from packinghouses have been as virulent as benzimidazole-sensitive isolates of the same species (Muirhead, 1974; Wild and Rippon, 1975; Wicks, 1977; Koffmann et al., 1978). Van Tuyl (1977b) reported that 15 mutants of *P. expansum* with resistance to three fungicides were as pathogenic as the wild strain; however, some benzimidazole-resistant strains collected from rotting apples and pears were less virulent than sensitive strains, especially when tested in certain varieties of apples (Bertrand and Saulie-Carter, 1978; Rosenberger and Meyer, 1979). Wild (1980) observed that 96% of the benzimidazole-resistant strains of *P. digitatum* that he isolated in California could be inhibited completely in vitro by 30 μg/ml carbendazim and were as pathogenic as the benzimidazole-sensitive isolates when pure cultures were inoculated into oranges. Four percent of the California isolates required 100 μg/ml carbendazim for complete growth inhibition and these highly resistant isolates appeared to be less pathogenic than benzimidazole-sensitive strains. In Japan, the most frequently isolated benzimidazole-resistant strains of *P. digitatum* had a minimum inhibitory concentration (MIC) of 400-800 μg/ml benomyl (263-527 μg/ml carbendazim) (Kuramoto, 1976), and these isolates appeared to be more virulent than isolates with intermediate resistance to benomyl (Kuramoto, personal communication).

Carbendazim binds to the protein tubulin, which is the principal component of the microtubules of the mitotic spindle. The defective spindle causes abnormal mitosis and growth inhibition in sensitive haploid fungi (Hammerschlag and Sisler, 1973; Davidse, 1973, 1976, 1977; Davidse and Flach, 1977) and sectoring (haploidization) in colonies of diploid fungi (Georgopoulos et al., 1976). Hastie and Georgopoulos (1971) reported two loci (ben-1 and ben-2) for carbendazim resistance in *Aspergillus nidulans* based upon the recombinants in the sexual cycle of this fungus. Van Tuyl (1975, 1977b) demonstrated that benomyl/carbendazim and thiabendazole resistance in *A. nidulans* was determined by one main gene, ben-A (same locus as ben-1), which confers a high level of resistance to benomyl or thiabendazole and by at least two other genes, ben-B and ben-C, which impart less pronounced resistance to the benzimidazole fungicides. A single gene was responsible for benzimidazole resistance in all mutants studied, and a mutation at one locus could produce a 43-fold increase in resistance over the wild-type parent. All three genes for resistance were in separate linkage groups, and a low degree of dominance was observed in diploid strains. Haploid recombinants of ben-A and ben-B or ben-C were intermediate to the parents in benomyl resistance; recombination of ben-B and ben-C in one strain resulted in only a slight increase in resistance over the sensitive parent. Van Tuyl (1975, 1977b) reported that cross resistance of benomyl-resistant strains to thiabendazole was the general rule in all ten species of fungi investigated. However, in *Aspergillus nidulans* and *A. niger* he found that about 1% of the benomyl-resistant strains were as sensitive to thiabendazole as the wild-type. Similarly, some thiabendazole-resistant mutants were more sensitive to benomyl than the wild-type.

Van Tuyl (1977b) treated conidia of *Penicillium expansum* with UV radiation and with N-methyl-N'-nitro-N-nitrosoguanidine and obtained mutants that were 3000 times more resistant to benomyl and 50 times more resistant to thiabendazole than the parent. Although most of these mutants were cross resistant to thiabendazole, approximately 15-20% of the mutants selected on thiabendazole-amended culture medium were more sensitive to benomyl than the wild-type. Beraha and Garber (1980) produced mutants of *P. italicum* by treatment of conidia with UV and nitrosoguanidine. They isolated mutants with low (L), intermediate (I) or high (H) resistance to thiabendazole, based upon the ability of the mutants to grow on medium amended with 0.5, 7.0 or 500 μg/ml thiabendazole, respectively. Thiabendazole resistance in this fungus involved two or three linked genes. Beraha and Garber (1980) found that a sensitive (wild) strain could not be mutated directly to an intermediate or high level of thiabendazole resistance; rather, it was obligatory that gene L first undergo mutation to give low-level mutants that could be mutagenized again to give strains with intermediate and high resistance. Diploids heterozygous for the H gene had a high level of resistance to thiabendazole; diploids heterozygous for L or I genes had low levels. The thiabendazole loci were not linked to the SOPP resistance locus. Beraha and Garber (1980) concluded that a two-loci model for thiabendazole resistance in *P. italicum* could better explain the behavior of the resistant strains than three alleles, each for a different level of resistance.

The biochemical basis for sensitivity or resistance to benzimidazole fungicides may involve: 1) the binding affinity of a specific tubulin for the benzimidazole structure, or 2) the relative permeability of the cells of resistant (R) and sensitive (S) isolates to benzimidazole fungicides. Davidse and Flach (1977) reported that a tubulin-containing fraction of benzimidazole-R species of fungi (i.e., *Pythium* and *Alternaria*) possessed little carbendazim-binding activity. The same fraction from three strains of *A. nidulans* (i.e., wild-type, benomyl-R and benomyl super-S), differing only in the ben-A gene, quantitatively bound carbendazim in direct relation to the sensitivity of the strain. This strongly suggested that the sensitivity of the three strains of *A. nidulans* was governed by the affinity of the tubulin receptor site for carbendazim, since no differences were observed in the uptake or detoxification of this fungicide by the three strains. The ben-A gene in *A. nidulans* may code for the primary structure of tubulin or for some regulatory function in tubulin synthesis (Davidse, 1977; Van Tuyl, 1977b).

The possibility that permeability phenomena are involved in the resistance of some strains of *A. nidulans* (e.g., ben-B or ben-C mutants) or in other species has not been completely discounted. Beraha and Garber (1980) discussed the possibility that an alteration of a thiabendazole transport system in *P. italicum* might explain the L mutation (low-level resistance) that is a prerequisite for a second mutation for a higher level of thiabendazole resistance. Nachmias and Barash (1976) described an active transport mechanism for the uptake of carbendazim by the fungus *Sporobolomyces roseus* and have demonstrated the reduced uptake of this fungicide by a resistant mutant. The report of Greenaway et al. (1978) suggests that there is no permeability barrier to thiabendazole in *P. digitatum;* rather, resistance to this fungicide is due to the ability of some strains to actively secrete thiabendazole from the hyphae.

sec-Butylamine

An aqueous solution of *sec*-butylamine (phosphate) pH 9 is applied to oranges and lemons after harvest to control *Penicillium* molds during storage. The treatment usually is applied to oranges before degreening with ethylene gas and to lemons before storage (Dawson and Eckert, 1977; Eckert, 1977, 1978). Both applications create a situation that is highly favorable for the selection and proliferation of *sec*-butylamine-R strains of *P. digitatum* and *P. italicum* in the natural population. Therefore, the appearance of a severe resistance problem within several years after the introduction of this fungicide was not unexpected. The *sec*-butylammonium cation may inhibit the growth of *P. digitatum* by two mechanisms: *sec*-butylamine (0.5 mM) inhibits pyruvate oxidation in *P. digitatum*, presumably by direct action upon pyruvic dehydrogenase (Yoshikawa and Eckert, 1976; Yoshikawa et al., 1976). *sec*-Butylamine (1.0 mM) also inhibits the active transport of amino acids into the hyphae (Bartz and Eckert, 1972). Since R strains of *P. digitatum* accumulate as much *sec*-butylamine as S strains, the mechanism of resistance does not seem to involve a permeability phenomenon (Eckert, unpublished data).

ECOLOGY OF BENZIMIDAZOLE-SENSITIVE AND -RESISTANT ISOLATES OF *P. digitatum*

Benzimidazole-R strains of *P. digitatum*, *P. italicum*, *P. expansum*, *P. corymbiferum* and *P. brevicompactum* have been isolated and, in most cases, several levels of resistance have been reported (Bollen, 1971; Harding, 1972; Kuramoto, 1976; Davidse, 1977; Van Tuyl, 1977b; Beraha and Garber, 1980). The isolation of fungicide-R strains, especially those obtained by treatment with mutagens, does not provide information on the potential prominence of the R strains in the natural population of the species. Even the demonstrated pathogenicity of most mutants of *P. italicum*, *P. digitatum* and *P. expansum* (Beraha and Garber, 1965, 1966; Beraha et al., 1964; Van Tuyl, 1977b) is not conclusive evidence that fungicide-R strains of these species will become a practical problem in disease control. Laboratory tests of pathogenicity cannot include many of the factors involved in "fitness" of a field population, and a mutation for fungicide resistance may have a pleiotropic effect upon the "fitness" of the fungicide-R individual in the natural environment, especially in pathogenic competition with the existing fully-adapted population. Wolfe (1971) has pointed out that although a fungicide treatment may select fungicide-tolerant strains, the whole population does not easily shift in this direction since there are many characteristics in the population that are held in a complex balance in the existing environmental situation. The population may tend to return to its original state of balanced adaptation if the fungicide selection pressure is relaxed.

Characteristics and Frequencies of Resistant Strains

Wild (1980) investigated the characteristics of the natural population of benzimidazole-R strains of *P. digitatum* in California. Approximately 130 carbendazim-R isolates of the species were collected in packinghouses and citrus groves and classified according to normal growth-limiting concentration, ED_{50}, minimum inhibitory concentration (MIC), and dosage required to inhibit sporulation (Table 2). Isolates resistant to carbendazim were cross resistant to thiabendazole at appropriate concentrations, but the relative

Table 2. Classification of *Penicillium digitatum* Isolates According to Tolerance to Carbendazim and Thiabendazole in Agar Medium[a] and in Inoculated Oranges[b]

Fungicide	Assigned resistance category	Fungicide concentration for: Upper limit for normal growth, μg/ml	50% growth reduction (ED_{50}), μg/ml	Sporulation inhibition, μg/ml	Complete growth inhibition (MIC), μg/ml	90% reduction in fruit infection, g/liter[c]
Carbendazim	0	0.005	0.03	0.03	0.05	0.03
	I	1.0	1.3	1.5	4.0	3.9
	II	6.0	5.5	8.0	14	>26[d]
	III	8.0	6.5	12	30	>26
	IV	50	40	80	100	>26
Thiabendazole	0	0.1	0.15	0.25	0.5	0.08
	I	40	21	50	90	>80[d]
	II	80	60	200	300	>80
	III	50	30	60	80	>80
	IV	25	30	40	50	>80

[a]PDA medium supplemented with yeast extract and neo-peptone, pH 6.0, 26°C.
[b]Each Valencia orange was injected with 5 μl of 10^6 spores/ml and incubated at 26°C for 18 hrs before treatment with benomyl.
[c]Carbendazim applied as benomyl (concn. carbendazim indicated x 1.519 = concn. benomyl applied).
[d]Highest concn. tested.
(Adapted from Wild, 1980.)

resistance of each strain to the two fungicides was not constant. Isolates that belonged to carbendazim resistance category II, of intermediate resistance, were the most resistant to thiabendazole. None of the isolates were "negatively cross resistant." Isolates in resistance categories II and III could not be separated solely on the basis of ED_{50} and normal growth-limiting concentrations of carbendazim; however, these categories were readily differentiated when the additional criteria of MIC and sporulation inhibition were applied to the classification. Approximately 5% of the R isolates obtained from packinghouses fell in category I, 33% into category II, 58% in category III and 4% in category IV. In a similar survey of packinghouses in Japan, Kuramoto (1976) found that the MIC for the great majority of R isolates was 400-800 µg/ml benomyl (263-527 µg/ml carbendazim), approximately 9 to 18 times greater than for the most abundant category of R strains in California. Wild (1980) also collected spores from citrus groves where benomyl and thiabendazole had never been applied in order to assess the range of carbendazim resistance in the wild population of *P. digitatum*. Approximately 3×10^8 spores from 30 decayed fruit were sampled from a grove that was within 6.4 km of a citrus packinghouse known to have a high frequency of R strains. Only five isolates developed on a culture medium containing 1 µg/ml carbendazim and these strains belonged to resistance category III. Two other groves, 7 and 35 km from commercial packinghouses, were sampled and 1×10^9 spores from 45 decayed fruit were plated on carbendazim-amended medium. No R strains were observed.

Isolates belonging to resistance category IV were less infectious than isolates belonging to categories 0, I, II, or III when spores (10^5/ml) from a pure culture of each strain were inoculated into untreated oranges. At a higher inoculum concentration, all isolates of *P. digitatum* gave the same level of infection. Although the benzimidazole-S strains of *P. digitatum* were well controlled by treatment of inoculated fruit with 0.03 g/liter carbendazim (as benomyl) or 0.08 g/liter thiabendazole, none of the benzimidazole-R strains, irrespective of resistance category, were controlled by 1-2 g/liter carbendazim (as benomyl) or 2-4 g/liter thiabendazole, dosages used in commercial packinghouses. Decay in fruit inoculated with a category I isolate was reduced about 50% by treatment with 2.6 g/liter carbendazim.

Interaction Between Sensitive and Resistant Isolates in Infected Fruits

The low frequency of carbendazim-R isolates in the natural population of *P. digitatum* in citrus groves prompted an investigation of the competition between carbendazim-S and -R isolates in infected oranges (Wild, 1980). The fruit were inoculated with a standardized mixture of two or more strains of *P. digitatum*, incubated at 26°C

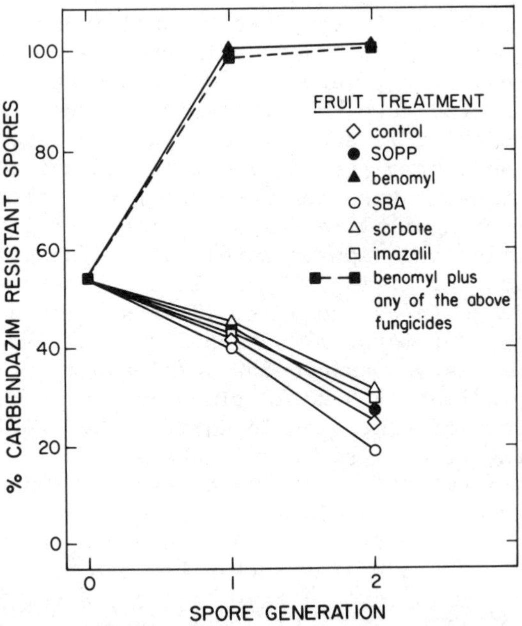

Figure 5. Effect of several fungicides, applied separately and in combination with benomyl, upon the selective multiplication of carbendazim-resistant and -sensitive isolates of *P. digitatum* (Wild, 1980).

for 5 days, and a representative sample of the spore population was collected from the surface of the diseased fruits. A suspension of these spores (first generation) was then inoculated into a second lot of healthy oranges. The relative abundance of fungicide-R and -S spores in the population of each spore generation was determined by plating a sample of the spores onto potato-dextrose-agar amended with carbendazim or another fungicide. In some experiments, the fruit were treated with a fungicide after each spore generation was inoculated into the fruit. Treatment of oranges with 0.5 g/liter benomyl (0.33 g/liter carbendazim), after inoculation with a mixture of carbendazim-R and -S spores, resulted in almost total selection for the R isolates in the first spore generation (Figure 5). The addition of a second unrelated fungicide to the benomyl treatment, as recommended by some authorities as a measure to control resistance development, had no measurable influence upon the selection and proliferation of the R strain.

Treatment of the inoculated fruit with a non-benzimidazole fungicide alone or no fungicide at all resulted in a rapid decline in the frequency of the R strain in subsequent spore generations (Figure 5). The rapid disappearance of the R strains from the

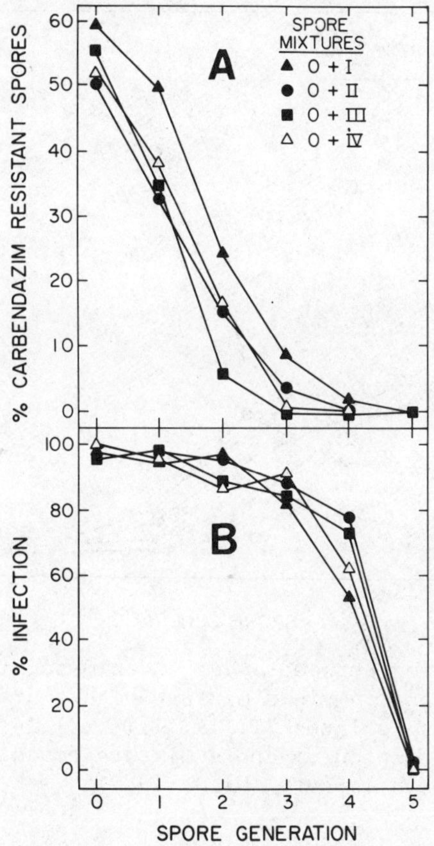

Figure 6. A. Suppression of several carbendazim-resistant isolates (I, II, III, IV) of *P. digitatum* in Valencia oranges by carbendazim-sensitive isolate M6R(0). B. Infection of benomyl-treated Valencia oranges by mixtures of spores of carbendazim-resistant and -sensitive strains collected at each spore generation in plot A (Wild, 1980).

spore population was unexpected because the R isolates appeared to be as virulent as the carbendazim-S isolates. The behavior of the R and the S strains in combination in untreated fruits was further investigated. Figure 6A shows that the frequency of *P. digitatum* isolates representative of resistance categories I, II, III and IV all declined in the spore population that appeared on diseased fruit following inoculation with an equal mixture of one R strain and the S strain M6R (category 0). The decrease in the frequency of the R strain in relation to the S strain continued for each spore generation. This experiment was repeated twice using other benzimidazole-R and -S isolates with similar results. The spores harvested after

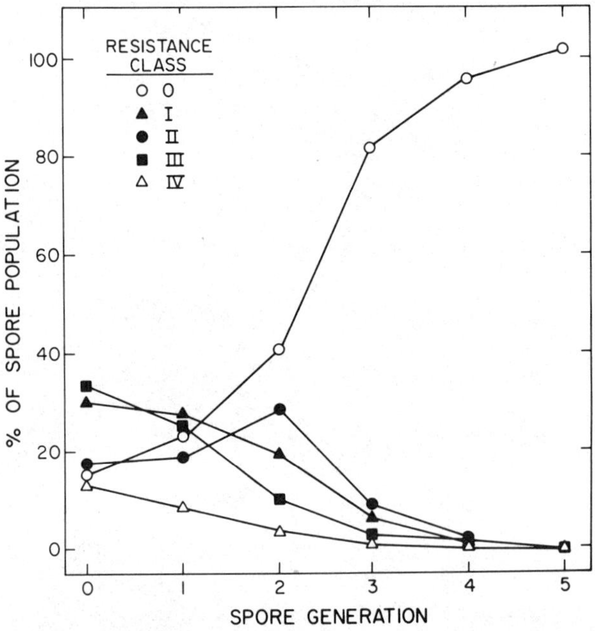

Figure 7. Competition between a carbendazim-sensitive isolate(O) of *P. digitatum* in a mixture with four carbendazim-resistant isolates (I, II, III, IV) in infected Valencia oranges. Inoculum spore mixture contained approximately 20% of each isolate (Wild, 1980).

each disease cycle (Figure 6A) were inoculated into five sound oranges (four inoculation sites per fruit), which, after 18 hrs incubation, were treated with 0.5 g/liter benomyl. The benomyl treatment did not significantly reduce decay until the spore population of the fourth generation (containing about 3% R spores) was used as inoculum (Figure 6B).

Figure 7 shows the competitive interaction between one S strain and five R isolates, representative of all four resistance categories, following injection of a mixture containing approximately 20% each isolate into an untreated orange. The frequency of the S strain increased from 16% in the original inoculum to 95% of the population by the fourth spore generation and the frequency of each R strain decreased proportionally. The dominance of benzimidazole-S isolates in competition with benzimidazole-R isolates in infected citrus fruits was evaluated further by pairing 31 R isolates with 18 S isolates in various combinations. Sixteen of the S strains dominated over the R strains in the combinations tested. These investigations clearly showed that R strains can be rapidly selected from a natural population of *P. digitatum* spores under the selection pressure of a benzimidazole fruit treatment. Furthermore, the weak competitive behavior of R strains in mixture with S strains in infected fruit (not treated with a benzimidazole fungicide) provided an explanation for the low frequency of benzimidazole-R strains in the natural spore population of citrus groves. Benzimidazole fungicides have never been used for control of field diseases of citrus trees in California, but enormous numbers of benzimidazole-R spores have been discharged from packinghouses into the atmosphere.

Despite the competitive advantage of benzimidazole-S strains over R strains in untreated fruit, it has been observed that benzimidazole-R strains do not disappear from the spore population of packinghouses after the benzimidazole fungicides have been replaced by SOPP or *sec*-butylamine. Wild (1980) found that many carbendazim-R isolates collected in these packinghouses were also R to *sec*-butylamine or SOPP. Oranges were inoculated with a mixture of a benzimidazole-S strain and a strain that was doubly resistant to both carbendazim and *sec*-butylamine. Treatment of fruit inoculated with this spore mixture with either benomyl or *sec*-butylamine resulted in an immediate increase in the frequency of the doubly resistant strain in the population of the first spore generation (Figure 8, top), whereas treatment with unrelated fungicides did not create selection pressure for either of the resistance genes (carbendazim or *sec*-butylamine). Treatment of oranges inoculated with a mixture of S and carbendazim/SOPP doubly resistant spores with either carbendezim or SOPP alone had an analogous influence on the buildup of SOPP/carbendazim doubly resistant strains (Figure 8, bottom).

Although doubly resistant carbendazim/*sec*-butylamine isolates were recognized earlier, their significance to the ecology and management of carbendazim-R strains was not fully recognized (Harding, 1976; Houck, 1977). Chastagner and Ogawa (1977) observed laboratory variants of *Botrytis* that were R to both benomyl (carbendazim) and DCNA (dicloran), but the doubly resistant strains were not considered to be of practical importance because they were weak in growth, sporulation and virulence.

MANAGEMENT OF FUNGICIDE-RESISTANT STRAINS

Factors Underlying the Severity of the Fungicide Resistance Problem in *P. digitatum* and *P. italicum*

The initial effectiveness of *sec*-butylamine, thiabendazole and benomyl against Penicillium decay of citrus fruits following the introduction of these treatments in the mid-1960s indicates that the level of resistance genes in the population of *P. digitatum* and *P. italicum* was very low at that time. Indeed, current surveys of the spore population in citrus groves indicate that the frequency of

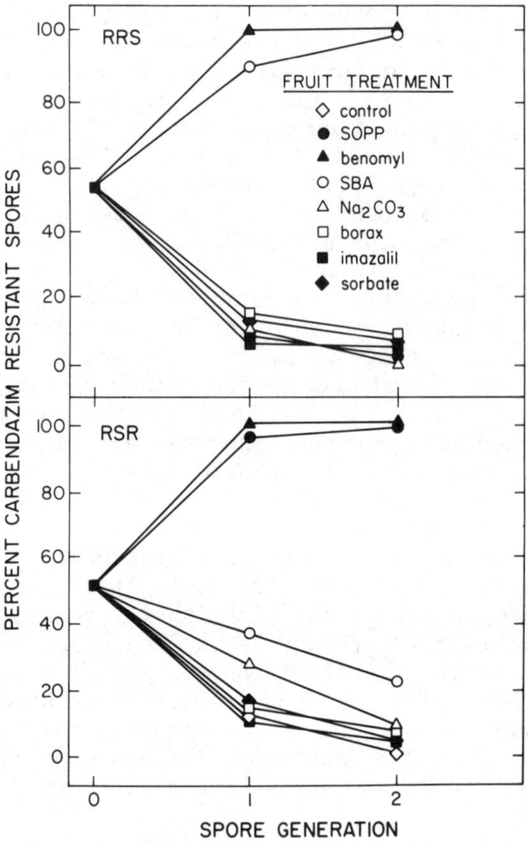

Figure 8. Effect of treatment of several lots of Valencia oranges with the same fungicide upon the frequency of carbendazim-resistant spores of *P. digitatum* in the population after two disease cycles (spore generations). The fruit were treated with the specified fungicides after infection was established by inoculation with a spore mixture containing: (upper plot) a fungicide-sensitive strain and a strain doubly resistant to carbendazim and *sec*-butylamine, but sensitive to SOPP (RRS), and (lower plot) a fungicide-sensitive strain and a strain doubly resistant to carbendazim and SOPP, but sensitive to *sec*-butylamine (RSR) (Wild, 1980).

fungicide-R strains is less than 1% of the natural population today.

The most important factors increasing the frequency of fungicide-R strains in haploid fungi such as *Penicillium* are mutation and selection. Since the perfect (sexual) stages of *P. digitatum* and *P. italicum* have never been observed, it is assumed that spontaneous mutation of one or more nuclear genes is the source of the various levels of resistance to the benzimidazoles, *sec*-butylamine and SOPP found in the population of these species. Although the benzimidazole fungicides and SOPP are known to increase somatic recombination in diploid colonies of *Aspergillus nidulans* (Georgopoulos et al., 1976, 1979), there is no substantial evidence that these fungicides cause genetic mutation in haploid fungi such as *Penicillium*. Several investigators have reported the occurrence of R strains in locations where it was highly improbable that the population was ever exposed to biphenyl, SOPP or a benzimidazole fungicide and most probably, mutations for resistance to these fungicides have been occurring continuously during the evolution of these species.

The degree of resistance of the mutants in a population depends upon the number and kinds of genes for resistance, their mutation frequency, the relationship of the resistance genes to virulence and "fitness" genes, and the evolutionary level of the R mutant (Wolfe, 1971, 1975; Georgopoulos, 1977). The interplay of these factors is indicated by the frequencies of the different levels of benzimidazole resistance in the populations of *P. digitatum* in Japan and California, each apparently of independent evolutionary origin. The rarity of benzimidazole-R strains of *P. digitatum* in California citrus groves suggests that mutations for resistance are accompanied by a pleiotropic reduction in ecological "fitness" at a level that is not measured by laboratory tests for vigor of growth and virulence (Wolfe, 1971). However, some less common mutants, such as those in resistance category IV (Wild, 1980) show a measurable reduction in virulence.

Although fungicide-R mutants of recent origin may be less "fit" initially, continuous selection pressure (fungicide, host and environment) could lead to the evolution of a fungicide-R strain that is well adapted ecologically. The existence of different levels of virulence in the population of *Penicillium* species (Kuramoto, 1976; Rosenberger and Meyer, 1979; Wild, 1980), and especially, changes in virulence of the population under selection pressure (Bertrand and Saulie-Carter, 1978) suggest that R strains are evolving continuously towards increased virulence and ecological fitness. Evolution of fungicide-R strains along these lines could be delayed or prevented by: 1) a pleiotropic interaction between the gene(s) for fungicide resistance and the genes for an essential physiological function, and 2) a conservative biochemical site of fungicide action that cannot be altered without a simultaneous reduction in functional efficiency. Fungicide-R strains of *Penicillium* and other fungi induced by mutagenic agents in the laboratory usually appear to be as pathogenic as the S parent strains (Beraha and Garber, 1966; Van Tuyl, 1975, 1977b). However, pathogenic R strains may not be sufficiently adapted to compete successfully with the wild strain when the fungicide pressure is relaxed (Dekker, 1977; Wild, 1980). Alternately, a fungicide-R mutant that is poorly adapted by other criteria may evolve for fitness in the competition vacuum created by the intensive use of the selective fungicide. Thus, the existence of laboratory-induced, fungicide-R mutants of low vigor should not be accepted as unequivocal evidence of the improbability of future development of practical resistance to that fungicide.

The development of a fungicide resistance problem requires that R strains exceed a certain critical frequency in the population and be self-sustaining at a level that causes economic damage (Wolfe, 1971, 1975). Postharvest fungicide treatments exert a heavy selection pressure for the emergence of resistance genes from a low frequency in the natural population of *Penicillium* in the citrus grove. A postharvest treatment results in complete coverage of every fruit with a persistant deposit of the fungicide, which prevents infection of the fruit by fungicide-S strains. In addition, biphenyl and the benzimidazole fungicides inhibit sporulation of all strains except those with a high degree of resistance, even though the interior of the fruit may be colonized with a mixture of R and S strains (Wild, 1980). Genes for biphenyl resistance are not uncommon in phenotypically S strains, since exposure of S field isolates of *Penicillium* species to biphenyl vapor, which is only fungistatic, results in the appearance of many sectors of biphenyl-R mutants in the colony (Farkas and Aman, 1940; Duran and Norman, 1961; Harding, 1965).

Resistance of *P. digitatum* and *P. italicum* to biphenyl was a practical problem before the introduction of the "systemic" fungicides, thiabendazole and benomyl. However, the importance of the benzimidazole fungicides to fruit marketing strategies and the alarming buildup of benzimidazole resistance have been responsible for several investigations as to the nature and control of the problem. The term "systemic" applied to postharvest fungicides identifies those compounds that can penetrate to a limited extent into the fruit where they inhibit the development of quiescent or incipient infections. This implies that the epidermal and outer cortical cells of the fruit are less sensitive to the "systemic" fungicide than are the cells of the pathogen. Several investigations have confirmed that the mitotic apparatus of the fungus cell is the primary site of action of the benzimidazole fungicides. Mitosis in plant cells seems to be considerably less sensitive to these compounds and, in addition, the surface cells of a mature fruit are not in a meristematic state. The selective action of systemic fungicides implies that they act against only one or a few biochemical sites in the pathogen, at least at the concentrations applied to fruit. In contrast, protective fungicides such as captan or chlorothalonil inhibit many sulfhydryl-dependent enzymes in both plants and fungi. Thus, the selectivity of these protective fungicides depends upon their exclusion from the host tissues. Davidse (1977) has provided abundant evidence that tubulin is the receptor for carbendazim and that R fungi and higher plants have a lower binding affinity for carbendazim than S species of fungi.

The development of a serious resistance problem is dependent upon the rapid multiplication and dissemination of the R strains to many other hosts treated with the same fungicide. The number of R propagules (spores) of *P. digitatum* can increase about 10^{10}-fold in 7 days and they are readily disseminated to other selective environments, since all fruits are treated uniformly with the same fungicide. The points in the handling sequence for citrus fruit that are most favorable to the development and proliferation of fungicide-R strains of *Penicillium* are shown in Figure 2. Early-season oranges are drenched with a solution of thiabendazole, benomyl, or *sec*-butylamine and then placed in an ethylene degreening room at 22-25°C. The fungicide residues suppress the S strains that are dominant in the field population of *Penicillium* species, thereby permitting the fungicide-R strains to colonize the fruit and sporulate after several days. The fruit are then dumped from the field boxes and the spores of the R strains are air-borne to the newly harvested fruit as they, in turn, are placed in the degreening room. After degreening, the fruit contaminated with benzimidazole-R spores receive a weak disinfectant treatment and then pass into the packinghouse where they receive a second treatment with a benzimidazole fungicide in "wax" before shipment to market. Inevitably, the first benzimidazole treatment selects for R strains and, therefore, the second treatment fails to control fruit decay. Figure 2 shows that lemon fruit are washed in a bath of SOPP or sodium carbonate, coated with a "wax" containing 2,4-D and thiabendazole, benomyl or *sec*-butylamine and then are stored at 15°C for several months. Benzimidazole-R strains that are selected during the storage period contaminate the sound fruit when they are removed from storage, leading to the failure of the second benzimidazole treatment (in wax). Also, the effectiveness of the biphenyl treatment applied in the shipping container may be greatly diminished if SOPP is applied to the lemons before storage. Another serious problem, and perhaps more difficult to solve, is the contamination of lemons entering storage with spores of R strains that have been selected by fungicide residues on the preceeding lot of fruit during storage. A final scenario for the development of a serious fungicide resistance problem unfolds when lemons or oranges are packed for market, but cannot be sold because of weak demand. These fruit remain in cold storage for several weeks before they are returned to the packinghouse for removal of decayed fruit from the boxes prior to shipment. Fungicide-R *Penicillium* species, selected during storage by the fungicide deposits on the fruit, contaminate all fruit passing through the packinghouse, resulting in a great reduction in effectiveness of the final fungicide treatments. Since the effective use of the benzimidazole fungicides is essential for success of current procedures for marketing citrus fruits, several strategies for the control of R strains of *Penicillium* are being investigated.

Strategies for the Management of Fungicide-resistant Strains of *Penicillium*

Sanitation. During periods of the year when fruit are not harvested, packinghouses can be disinfested with broad-spectrum antimicrobial agents (e.g., formaldehyde, quaternary ammonium compounds) that cannot be used when fruit are present. This practice assures a very low population of spores in the packinghouse at the beginning of the harvest period, resulting in high efficiency of all postharvest fungicide treatments. Within a month, however, the level of fungicide resistance in the *Penicillium* population begins to build up and is accompanied by a gradual deterioration in effectiveness of one or more fungicide treatments. Eventually, a point is reached where the decay control program is no longer cost-effective. This economic threshold can be delayed by excluding contaminated fruit from the packinghouse and by a rigorous sanitation program inside the packinghouse, consisting of a daily disinfection of the fruit-handling equipment with a broad-spectrum sanitizing agent. These practices do not reduce the frequency of R strains existing in the population nor do they influence the selection process per se, but sanitation does reduce the absolute number of R spores in the packinghouse to the same extent that it reduces the total spore population. A reduction in the number of R spores should delay the development of a fungicide-R epidemic because the probability of successful infection (by an R strain) depends upon the spore concentration (inoculum threshold) at a limited number of injured sites on the fruit surface. A reduction in total spore population also should delay the buildup of resistance since fungicide-S spores can increase the probability of infection of benzimidazole-treated fruit by a benzimidazole-R strain (Wild, 1980).

Non-selective Treatments. The elimination of selective fungicides and structurally related compounds from the postharvest fungicide program should produce a decrease in the frequency of the resistant strains due to the competition of the S strains (Figures 5 through 8). This strategy would be counterproductive, however, since heavy decay losses would arise through fruit infection by the uncontrolled fungicide-S strains of *Penicillium*. Furthermore, reduction of the fungicide-R population to a level that could be controlled by a benzimidazole treatment would require several disease cycles (Figure 6), and restoration of the benzimidazole treatment would cause an immediate increase in the frequency of the R strains (Figure 8).

Heated solutions of borax and sodium carbonate reduce fruit infection by *Penicillium* and have not been implicated in a fungicide resistance problem. The applications of borax are rather limited today because of the possibility of ground-water pollution, but sodium carbonate (3%, 43°C) is an effective treatment for disinfecting citrus fruit before they are brought into the packinghouse. None of the other well known multi-site fungicides (e.g., captan, maneb, chlorothalonil) are effective against *Penicillium* decay. Furthermore, the discovery of a non-selective protective fungicide with activity against *Penicillium* would not solve the benzimidazole resistance problem since an agent with these characteristics would not penetrate the fruit surface, as does benomyl, to inhibit incipient infections and fungus sporulation (Brown, 1977; Eckert, 1977, 1978; Eckert et al., 1979).

Mixtures of Selective and Non-selective Fungicides. This stratedy had been suggested by several authorities as a means of delaying the buildup of R strains of pathogens in the field, but unfortunately there is little published data supporting this contention. Mixtures of systemic fungicides (e.g., benomyl) and protective fungicides (e.g., captan) may prevent penetration of benzimidazole-R strains into treated fruits, thereby reducing the total pathogen population including the benzimidazole-R component (Koffman et al., 1978). However, this strategy has no advantage over the use of the protective fungicide alone when R strains are already present at significant levels in the population (Rosenberger et al., 1979). The combination of benzimidazole and protective fungicides does not appear to be a suitable strategy for control of R strains of *Penicillium* on citrus and other fruits that may be infected with a mixture of R and S strains before the fungicide treatment is applied. Non-systemic fungicides cannot eradicate deep-seated infections nor inhibit sporulation efficiently. Thus, the rapid buildup of benzimidazole-R strains following treatment of infected fruit with a mixture of a benzimidazole and a protective fungicide is predictable (Figure 5). The combination of two unrelated systemic fungicides, such as benomyl and imazalil (Figure 3, VI), each with a unique mechanism of action, appears to be a more promising, but expensive, strategy for the control of R strains (Laville et al., 1977; Koffmann et al., 1978; Rosenberger et al., 1979). The possibility of combining two synergistic compounds or two fungicides that exhibit "negatively correlated cross resistance" is an intriguing, but unproven, strategy for controlling fungicide-R plant pathogens (Dekker, 1977; Georgopoulos, 1977; Uesugi, 1980). Finally, combinations of certain fungicides, such as SOPP and *sec*-butylamine, may be beyond the genetic capabilities of the pathogen for resistance development (Wild, 1980).

Alternation of Two or More Unrelated Fungicides. This strategy involves the application of two or more unrelated fungicides in a sequence that, hopefully, will suppress strains of the pathogen that are resistant to the final fungicide in the series. This procedure is feasible for citrus fruits because they are treated customarily with several fungicides after harvest. For example, lemons may be treated with SOPP and/or *sec*-butylamine before storage and after storage with SOPP, a benzimidazole fungicide, and biphenyl (Dawson and Eckert, 1977; Eckert, 1978). This sequence is required because the benzimidazoles are the only approved fungicides that can control sporulation of *Penicillium* on diseased fruit. The success of this schedule for lemons depends upon an efficient sanitation program, since the buildup of a large population of SOPP- or *sec*-butylamine-R strains during storage could result in the selection of mutants that are doubly resistant to one of these compounds and the benzimidazole fungicides (Figure 8). A pre-storage sequence of SOPP followed by *sec*-butylamine might be effective in suppressing the *Penicillium* population since double resistance to these two fungicides has not been observed and may indicate that such doubly resistant (SOPP/*sec*-butylamine) mutants are poorly adapted as fruit pathogens (Wild, 1980). The multiple sites of action of both SOPP and *sec*-butylamine should further discourage the development of strains resistant to both of these fungicides. At low concentrations, *sec*-butylamine inhibits pyruvic dehydrogenase in *P. digitatum* (Yoshikawa and Eckert, 1976; Yoshikawa et al., 1976), while at higher concentrations, the transport of amino acids into the hyphae is inhibited (Bartz and Eckert, 1972). SOPP increases somatic recombination in diploid strains of *Aspergillus nidulans*, suggesting that this inhibitor may adversely affect hereditary processes in other fungi as well (Georgopoulos et al., 1976, 1979). SOPP also inhibits several TCA cycle enzymes in *P. italicum* (Rehm, 1969). The residue tolerance (U.S.) for *sec*-butylamine (30 mg/kg citrus fruit) would accommodate the application of higher dosages of this chemical than are now customary; however, the rate of application of SOPP could not be substantially increased without endangering the fruit or exceeding the residue tolerance (10 mg/kg).

Imazalil and Related Compounds: Immune to Resistant Strain Development? Tests conducted over the past several years have demonstrated that imazalil (Figure 3, VI) is highly effective in preventing infection of citrus fruits by *Penicillium* species and in suppressing sporulation of these fungi on diseased fruits (Harding, 1976; Laville et al., 1977). Imazalil-R strains of *Penicillium digitatum* were not observed following treatment of spores with mutagenic agents that produced mutants resistant to other fungicides (Van Hoorn and Eckert, unpublished data). Van Tuyl (1977b) found that the mutational frequency for resistance to imazalil in *P. expansum* was high (70×10^{-6}) compared to that for benomyl (1×10^{-6}), but that the increase in resistance level accompanying a single gene mutation was rather small. The increase in resistance level (ED_{50} mutant/ED_{50} parent) for a single mutation was 3300 for benomyl and

4 for imazalil. The pathogenicity of the imazalil-R mutants was not substantially different from the parent strain, but the effectiveness of imazalil treatments in controlling imazalil-R strains on inoculated apples was not reported. Van Tuyl (1977a, 1977b) found that imazalil resistance in *Aspergillus nidulans* was based on a multigenic system located at eight chromosome loci. All of the single gene mutants showed a relatively low level of imazalil resistance; the maximum MIC for the mutants was 12 μg/ml compared to 1.2 μg/ml for the wild-type strain. When two resistance genes were combined in one strain, the level of resistance was increased to an MIC of 50 μg/ml; three genes combined gave a strain with an MIC of 200 μg/ml imazalil. Mutations for resistance at different loci yielded strains that were pleiotropically resistant or hypersensitive to other fungicides. Imazalil inhibits ergosterol synthesis in *P. italicum* (Siegel and Ragsdale, 1978), *P. expansum* (Leroux and Gredt, 1978), and *Ustilago avenae* (Buchenauer, 1977), and the striking pleiotropic effects accompanying imazalil resistance in *A. nidulans* may be related to a site of action involving the cell membrane (Van Tuyl, 1977a). Other fungicides with a similar mechanism of action have not yet encountered a practical problem of pathogen resistance (Dekker, 1977).

Genetic and biochemical investigations on imazalil and related inhibitors in other fungi provide hope that the development of resistance in *Penicillium* species will be slow and will involve mutants with diminished virulence. Relatively high dosages of imazalil should be applied to fruit to suppress the development of low-level R mutants, which, if allowed to persist in the population, might acquire additional resistance genes, resulting in strains with a higher level of resistance. All strategies of imazalil use should include a strict sanitation program to minimize the resistance gene pool available for selection by imazalil. Trials are now under way in lemon packinghouses to evaluate several fungicide sequences including imazalil, benzimidazoles and other fungicides.

SUMMARY

Spores of *Penicillium digitatum* and *P. italicum* infect citrus fruits through harvest-related injuries and, under favorable conditions, may give rise to a progeny of 10^{10} spores on a single diseased fruit in 7 days. Several fungicides — sodium *o*-phenylphenate (SOPP), *sec*-butylamine, thiabendazole, benomyl and biphenyl — are applied to harvested fruits to prevent infection and sporulation of *Penicillium* on diseased fruits during storage and marketing. Spontaneous somatic mutations give rise to strains of *Penicillium* that are resistant to one or more of the fungicides. Resistant mutants are present at low frequencies in the natural population of *Penicillium* species in citrus groves, but under the selection pressure created by postharvest fungicide treatments, these strains may become the dominant component of the population in packinghouses. SOPP-R strains of *Penicillium* are also resistant to biphenyl; benzimidazole-R strains are not controlled by thiabendazole, benomyl or thiophanate methyl. Treatment of fruit with SOPP or benomyl before storage may cause an increase in the frequency of mutant strains that are not controlled by post-storage treatment with biphenyl or thiabendazole, respectively. If strains doubly resistant to the benzimidazole fungicides and SOPP or *sec*-butylamine are present in the population, a pre-storage treatment of SOPP or *sec*-butylamine may cause the proliferation of benomyl/thiabendazole-R strains during the storage period.

Although benzimidazole-R strains appear as virulent to citrus fruits as sensitive strains in laboratory inoculation tests, the development of the R strains is suppressed by the S strains when both are inoculated together into a fruit that has not been treated with a benzimidazole fungicide. The reduced fitness of benzimidazole-R strains, in the absence of selection pressure, may be responsible for the low frequency of benzimidazole-R strains in a benzimidazole-free environment.

The fungicide resistance problem can be managed by: 1) disinfestation of the packinghouse to minimize the number of fungicide-R spores available for fruit infection and fungicide selection, and 2) application of a fungicide combination or sequence that is unfavorable to the buildup of *Penicillium* strains resistant to the final fungicide treatment. Imazalil and several other fungicides that interfere with the function of the fungal cell membrane offer considerable promise for control of fungicide-R strains of *Penicillium*.

REFERENCES

Bartz, J. A., and Eckert, J. W., 1972, Studies on the mechanism of action of 2-aminobutane, *Phytopathology*, 62:239.

Beraha, L., and Garber, E. D., 1965, Genetics of phytopathogenic fungi. XI. A genetic study of avirulence due to auxotrophy in *Penicillium expansum*, *Am. J. Bot.*, 52:117.

Beraha, L., and Garber, E.D., 1966, Genetics of phytopathogenic fungi. XV. A genetic study of resistance to sodium orthophenylphenate and sodium dehydroacetate in *Penicillium expansum*, *Am. J. Bot.*, 53:1041.

Beraha, L., and Garber, E. D., 1980, A genetic study of resistance to thiabendazole and sodium *o*-phenylphenate in *Penicillium italicum* by the parasexual cycle, *Botan. Gaz.*, 141:204.

Beraha, L., Garber, E. D., and Strømnaes, Ø., 1964, Genetics of phytopathogenic fungi. X. Virulence of color and nutritionally deficient mutants of *Penicillium italicum* and *Penicillium digitatum*, *Can. J. Bot.*, 42:429.

Bertrand, P. F., and Saulie-Carter, J. L., 1978, The occurrence of benomyl-tolerant strains of *Penicillium expansum* and *Botrytis cinerea* in the mid-Columbia region of Oregon and Washington, *Plant Disease Reptr.*, 62:302.

Bollen, G. J., 1971, Resistance to benomyl and some chemically related compounds in strains of *Penicillium* species, *Neth. J. Pl. Pathol.*, 77:187.

Brown, G. E., 1977, Application of benzimidazole fungicides for citrus decay control, *Proc. Int. Soc. Citric.*, 1:273.

Buchenauer, H., 1977, Mechanism of action of the fungicide imazalil in *Ustilago avenae*, *Z. Pflanzenkrankr. u. Pflanzenschutz.*, 84:440.

Chastagner, G. A., and Ogawa, J. M., 1979, DCNA-benomyl multiple tolerance in strains of *Botrytis cinerea*, *Phytopathology*, 69:699.

Davidse, L. C., 1973, Antimitotic activity of methyl benzimidazole-2-yl carbamate (MBC) in *Aspergillus nidulans*, *Pestic. Biochem. Physiol.*, 3:317.

Davidse, L. C., 1976, The antimitotic properties of the benzimidazole fungicide carbendazim and a mechanism of resistance to this compound in *Aspergillus nidulans*, Doctoral Thesis, Agricultural University, Wageningen, Netherlands, 84 pp.

Davidse, L. C., 1977, Mode of action, selectivity and mutagenicity of benzimidazole compounds, *Neth. J. Pl. Pathol.*, 83(Suppl. 1): 135.

Davidse, L. C., and Flach, W., 1977, Differential binding of methyl benzimidazole-2-yl carbamate to fungal tubulin as a mechanism of resistance to this antimitotic agent in mutant strains of *Aspergillus nidulans*, *J. Cell. Biol.*, 72:174.

Dawson, A. J., and Eckert, J. W., 1977, Problems of decay control in marketing citrus fruits: Strategy and solutions, California and Arizona, *Proc. Int. Soc. Citric.*, 1:255.

Dekker, J., 1977, The fungicide-resistance problem, *Neth. J. Pl. Pathol.*, 83(Suppl. 1):159.

Duran, R., and Norman, S. M., 1961, Differential sensitivity to biphenyl among strains of *Penicillium digitatum* Sacc., *Plant Disease Reptr.*, 45:475.

Eckert, J. W., 1977, Control of postharvest diseases, *in:* "Antifungal Compounds," M. R. Siegel and H. D. Sisler, eds., Vol. 1, pp. 269-352, Marcel Dekker Inc., New York.

Eckert, J. W., 1978, Postharvest diseases of citrus fruits, *Outlook on Agric.*, 9:225.

Eckert, J. W., and Rahm, M. L., 1979, The antifungal activity of alkyl benzimidazole-2-yl carbamates and related compounds, *Pestic. Sci.*, 10:473.

Eckert, J. W., Kolbezen, M. J., Rahm, M. L., and Eckard, K. J., 1979, Influence of benomyl and methyl 2-benzimidazolecarbamate on development of *Penicillium digitatum* in the precarp of orange fruit, *Phytopathology*, 69:934.

Farkas, A., and Aman, J., 1940, The action of diphenyl on *Penicillium* and *Diplodia* moulds, *Palestine J. Bot. Jerusalem Ser.*, 2:38.

Georgopoulos, S. G., 1977, Development of fungal resistance to fungicides, *in:* "Antifungal Compounds," M. R. Siegel and H. D. Sisler, eds., Vol 2, pp. 439-495, Marcel Dekker Inc., New York.

Georgopoulos, S. G., Zafiratos, C., and Georgiadis, E., 1967, Membrane functions and tolerance to aromatic fungitoxicants in conidia of *Fusarium solani*, *Physiol. Plant.*, 20:373.

Georgopoulos, S. G., Kappas, A., and Hastie, A. C., 1976, Induced sectoring in diploid *Aspergillus nidulans* as a criterion of fungitoxicity by interference with hereditary processes, *Phytopathology*, 66:217.

Georgopoulos, S. G., Sarris, M., and Ziogas, B. M., 1979, Mitotic instability in *Aspergillus nidulans* caused by the fungicides iprodione, procymidone and vinclozolin, *Pestic. Sci.*, 10:389.

Greenaway, W., Ward, S., and Whatley, F. R., 1978, Uptake of fuberidazole and thiabendazole by *Penicillium digitatum*, *Cunninghamella echinulata* and potato slices, *New Phytol.*, 80:595.

Gutter, Y., 1975, Interrelationship of *Penicillium digitatum* and *P. italicum* in thiabendazole-treated oranges, *Phytopathology*, 65:498.

Hammerschlag, R. S., and Sisler, H. D., 1973, Benomyl and methyl-2-benzimidazolecarbamate (MBC): Biochemical, cytological and chemical aspects of toxicity to *Ustilago maydis* and *Saccharomyces cerevisiae*, *Pestic. Biochem. Physiol.*, 3:42

Harding, P. R., Jr., 1962, Differential sensitivity to sodium orthophenylphenate by biphenyl-sensitive and biphenyl-resistant strains of *Penicillium digitatum*, *Plant Disease Reptr.*, 46:100.

Harding, P. R., Jr., 1964, Assaying for biphenyl resistance in *Penicillium digitatum* in California lemon packing houses, *Plant Disease Reptr.*, 48:43.

Harding, P. R., Jr., 1965, The nature of biphenyl-resistant mutants of *Penicillium digitatum*, *Plant Disease Reptr.*, 49:965.

Harding, P. R., Jr., 1972, Differential sensitivity to thiabendazole by strains of *Penicillium italicum* and *P. digitatum*, *Plant Disease Reptr.*, 56:256.

Harding, P. R. Jr., 1976, R23979, a new imidazole derivative effective against postharvest decay of citrus by molds resistant to thiabendazole, benomyl and 2-aminobutane, *Plant Disease Reptr.*, 60:643.

Hastie, A. C., and Georgopoulos, S. G., 1971, Mutational resistance to fungitoxic benzimidazole derivatives in *Aspergillus nidulans*, *J. Gen. Microbiol.*, 67:371.

Herbig, G., and Rehm, J.-J., 1967, Stoffwechsel-physiologische Untersuchungen en diphenyl-und natrim-*o*-phenyl-phenolatresistenten Pilzen., *Naturiwiss.*, 54, 46.

Houck, L. G., 1977, Problems of resistance to citrus fungicides, *Proc. Int. Soc. Citric.*, 1:263.

Koffmann, W., Penrose, L. J., Menzies, A. R., Davis, K. C., and Kaldor, J., 1978, Control of benzimidazole tolerant *Penicillium expansum* in pome fruit, *Sci. Hortic.*, 9:31.

Kuramoto, T., 1976, Resistance to benomyl and thiophanate-methyl in strains of *Penicillium digitatum* and *P. italicum* in Japan, *Plant Disease Reptr.*, 60:168.

Laville, E. Y., Harding, P. R., Dagan, Y., Rahat, M., Kraght, A. J., and Rippon, L. E., 1977, Studies on imazalil as a potential treatment for control of citrus fruit decay, *Proc. Int. Soc. Citric.*, 1:269.

Leroux, P., and Gredt, M., 1978, Effets de quelques fongicides systemiques sur la biosynthèse de l' Ertostérol chez *Botrytis cinerea* Pers., *Penicillium expansum* Link. et *Ustilago maydis* (DC.) Cda. *Ann. Phytopathology*, 10:45.

Littauer, F., and Gutter, Y., 1953, Diphenyl-resistant strains of *Diplodia*, *Palestine J. Bot.*, *Rehovot Ser.*, 8:185.

McDonald, R. E., Risse, L. A., Hillebrand, B. M., 1979, Resistance to thiabendazole and benomyl of *Penicillium digitatum* and *P. italicum* isolated from citrus fruit from several countries, *J. Am. Soc. Hort. Sci.*, 104:333.

Muirhead, I. F., 1974, Resistance to benzimidazole fungicides in blue mould of citrus in Queensland, *Australian J. Exptl. Agr. Animal Husbandry*, 14:698.

Nachmias, A., and Barash, I., 1976, Decreased permeability as a mechanism of resistance to methyl benzimidazol-2-yl carbamate (MBC) in *Sporobolomyces roseus*, *J. Gen. Microbiol.*, 94:167.

Rehm, H.-J., 1969, Inhibiting action of sodium *o*-phenylphenate (SOPP) and biphenyl on specific reactions of the metabolism of microorganisms, *Proc. First Int. Citrus Symp.*, 3:1325.

Rosenberger, D. A., and Meyer, F. W., 1979, Benomyl-tolerant *Penicillium expansum* in apple packinghouses in Eastern New York, *Plant Disease Reptr.*, 63:37.

Rosenberger, D. A., Meyer, F. W., and Cecilia, C. V., 1979, Fungicide strategies for control of benomyl-tolerant *Penicillium expansum* in apple storages, *Plant Disease Reptr.*, 63:1033.

Siegel, M. R., and Ragsdale, N. N., 1978, Antifungal mode of action of imazalil, *Pestic. Biochem. Physiol.*, 9:48.

Smoot, J. J., and Brown, G. E., 1974, Occurrence of benzimidazole-resistant strains of *Penicillium digitatum* in Florida citrus packinghouses, *Plant Disease Reptr.*, 58:933.

Smoot, J. J., and Winston, J. R., 1967, Biphenyl-resistant citrus green mold reported in Florida, *Plant Disease Reptr.*, 51:700.

Strømnaes, Ø., Garber, E. D., and Beraha, L., 1964, Genetics of phytopathogenic fungi. IX. Heterocaryosis and the parasexual cycle in *Penicillium italicum* and *P. digitatum*, *Can. J. Bot.*, 42:423.

Sutton, T. B., 1978, Failure of combinations of benomyl and reduced rates of non-benzimidazole fungicides to control *Venturia inaequalis* resistant to benomyl and the spread of resistant

strains in North Carolina, *Plant Disease Reptr.*, 62:830.
Uesugi, Y., 1981, this volume.
Van Tuyl, J. M., 1975, Genetic aspects of acquired resistance to benomyl and thiabendazole in a number of fungi, *Med. Fac. Landbouw. Rijksuniv. Gent.*, 40:691.
Van Tuyl, J. M., 1977a, Genetic aspects of resistance to imazalil in *Aspergullus nidulans*, *Neth. J. Pl. Pathol.*, 83(Suppl. 1):169.
Van Tuyl, J. M., 1977b, Genetics of fungal resistance to systemic fungicides, Doctoral Thesis, Agricultural University, Wageningen, Netherlands, 137 pp.
Vonk, J. W., and Kaars Sijpesteijn, A., 1971, Methyl benzimidazole-2-yl carbamate, the fungitoxic principle of thiophanate-methyl, *Pestic. Sci.*, 2:160.
Wicks, T., 1977, Tolerance to benzimidazole fungicides in blue mold (*Penicillium expansum*) on pears, *Plant Disease Reptr.*, 61:447.
Wild, B. L., 1980, Resistance to citrus green mold *Penicillium digitatum* Sacc. to benzimidazole fungicides, Doctoral Thesis, University of California, Riverside, 89 pp.
Wild, B. L., and Rippon, L. E., 1975, Response of *Penicillium digitatum* strains to benomyl, thiabendazole and sodium *o*-phenylphenate, *Phytopathology*, 65:1176.
Wolfe, M. S., 1971, Fungicides and the fungus population problem, *Proc. 6th Brit. Insectic. Fungic. Conf.*, 3:724.
Wolfe, M. S., 1975, Pathogen response to fungicide use, *Proc. 8th Brit. Insectic. Fungic. Conf.*, 3:813.
Yoshikawa, M., and Eckert, J. W., 1976, The mechanism of fungistatic action of *sec*-butylamine. I. Effects of *sec*-butylamine on the metabolism of hyphae of *Penicillium digitatum*, *Pestic. Biochem. Physiol.*, 6:471.
Yoshikawa, M., Eckert, J. W., and Keen, N. T., 1976, The mechanism of fungistatic action of *sec*-butylamine. II. The effect of *sec*-butylamine on pyruvate oxidation by mitochondria of *Penicillium digitatum* and on the pyruvate dehydrogenase complex, *Pestic. Biochem. Physiol.*, 6:482.

SUPPRESSION OF METABOLIC RESISTANCE THROUGH CHEMICAL STRUCTURE MODIFICATION

T. Roy Fukuto and Narayana M. Mallipudi

Division of Toxicology and Physiology
Department of Entomology
University of California
Riverside, California 92521

INTRODUCTION

One of the major causes for the development of resistance to the toxic action of insecticides is the ability of the insect to modify or detoxify the insecticide at a rate fast enough to prevent critical buildup of the active material at the target site. This insecticide detoxification may occur by means of a variety of metabolic processes in which the parent material is converted into a nontoxic form or into a form that can be rapidly eliminated from the insect body. These processes are well described elsewhere in this volume.

Limited success has been achieved with the intentional modification of the chemical structure of organic insecticides for the purpose of counteracting insecticide resistance arising from metabolic alteration, but the idea still remains an attractive one. The absence of any notable achievement by this approach probably lies in the fact that insects are highly adaptable and resistance is conferred by a combination of different mechanisms, and structural modifications often result in compounds of lower insecticidal activity. Nevertheless, a number of attempts to counteract insecticide resistance by such structure modification have been made, and this chapter will briefly describe these efforts.

MODIFICATION OF DDT

More attempts have been made to modify the structure of DDT as a countermeasure for resistance than have been made with any other insecticide. Two pathways for the metabolic detoxification of DDT in resistant insects have been demonstrated: dehydrochlorination of DDT to DDE by DDT-dehydrochlorinase (Lipke and Kearns, 1960) and α-hydroxylation to α-hydroxy-DDT by a mixed-function oxidase (Tsukamoto, 1961). Of the two pathways for DDT resistance, dehydrochlorination is undoubtedly more significant, and a number of structural modifications have been made to counteract resistance by this process.

Deutero-DDT

Although the mechanism for DDT-dehydrochlorinase-catalyzed dehydrochlorination of DDT is not well understood, it is believed that the reaction takes place by an E_2-type elimination reaction in which attack of the α-hydrogen atom by the enzyme occurs with concomitant elimination of a chlorine atom (Metcalf and Fukuto, 1968).

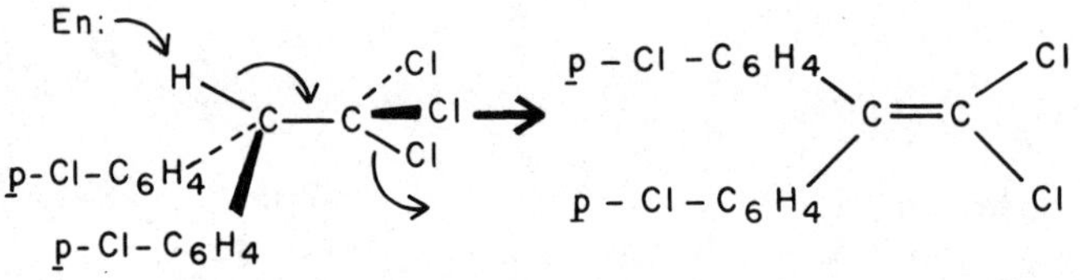

DDT DDE

In light of the greater stability of the carbon-deuterium bond (C-D) compared to the carbon-hydrogen bond (C-H), one of the earlier attempts in DDT modification was the substitution of deuterium for the α-hydrogen atom (Barker, 1960; Moorefield et al., 1962). Unfortunately, deutero-DDT proved to be only slightly more effective than DDT against either susceptible (S) or resistant (R) house flies, and no significant difference was observed in the rates of dehydrochlorination of the two compounds in vivo for R house flies. In marked contrast, however, deutero-DDT was highly toxic to the larvae of eleven different DDT-R strains of the mosquito *Aedes aegypti*; for example, deutero-DDT was 50- to 100-fold more toxic than DDT to the DDT-R strains (Pillai et al., 1963). Further, selection of an S strain (Trinidad S) with deutero-DDT for five generations resulted in only a 4.5-fold increase in resistance (LC_{50} from 0.008 ppm to 0.036 ppm), and no increase in resistance was observed with additional selection pressure. On the other hand, selection with DDT for five generations resulted in more than 1100-fold resistance (LC_{50} from 0.013 to 15.0). Deutero-DDT proved to be less readily dehydrochlorinated in R mosquito larvae than DDT either in vivo or in vitro, and the increased stability was attributed to the deuterium isotope effect.

The difference in the susceptibilities of DDT-R house flies and mosquitoes is curious, but it is not an unusual phenomenon. Evidently, house fly dehydrochlorinase is capable of overcoming the increased energy barrier for dehydrochlorination of deutero-DDT, but the mosquito enzyme is not. While the substitution of deuterium for the α-hydrogen atom in DDT has had no practical significance, presumably owing to the much greater cost of deutero-DDT compared to DDT, the approach was of substantive value in pointing out other ways of minimizing the role of dehydrochlorination in resistance.

Ortho-substituted DDT Analogs

Concurrent to the studies on the substitution of deuterium for the α-hydrogen atom in DDT, an interesting discovery was made which showed that the introduction of another substituent into one of the *ortho*-positions in the phenyl rings also served as a countermeasure for DDT resistance (Hennessy et al., 1961). For example, *o*-Cl-DDT [2-(4-chlorophenyl)-2-2(2,4-dichlorophenyl)-1,1,1-trichloroethane] was not only toxic to S house flies (4-fold less toxic than DDT), it was also very effective against those resistant to DDT. These compounds were synthesized to test a free radical hypothesis for the mode of action of DDT and not for the purpose of overcoming DDT resistance. In another study (Metcalf and Fukuto, 1968) the insecticidal activities of a series of *ortho*-substituted analogs of active *p,p'*-substituted DDT analogs were examined, and several were found to be effective against R house flies (Table 1). Although substitution in the *ortho* position of DDT resulted in a significant decrease in toxicity to the susceptible S-NAIDM strain as the *ortho* substituent increased in size, several of the analogs (3, 5 and 6) were moderately active against R house flies. Of the DDT series (compounds 1-4), *o*-Cl-DDT (3) was the only compound that showed effectiveness against the two DDT-R strains, while two of the three compounds in the deutero-DDT analogs (compounds 5-7) were considered effective.

The efficacy of *o*-Cl-DDT and the two deuterated analogs against R house flies may be attributed to their resistance to the action of DDT-dehydrochlorinase. For example, in several DDT-R strains of house flies, the conversion of *o*-Cl-DDT to *o*-Cl-DDE has been shown to occur to only 3.4 to 27%, while 93-99% conversion was observed with DDT. A substituent of significant size, e.g., chlorine, in the *ortho* substitution and deuterium substitution of the α-hydrogen atom resulted in slightly greater effectiveness against R house flies.

While *ortho* substitution in the DDT molecule resulted in greater effectiveness against DDT-R house flies, DDT-R *Aedes aegypti* larvae were remarkably tolerant to either *o*-Cl-DDT or deutero-*o*-Cl-DDT. This is quite surprising since resistant *A. aegypti*, as indicated earlier, are highly susceptible to deutero-DDT (Pillai et al., 1963).

Table 1. Toxicity of *Ortho*-substituted DDT Analogs to Susceptible and Resistant House Flies, *Musca domestica*[a]

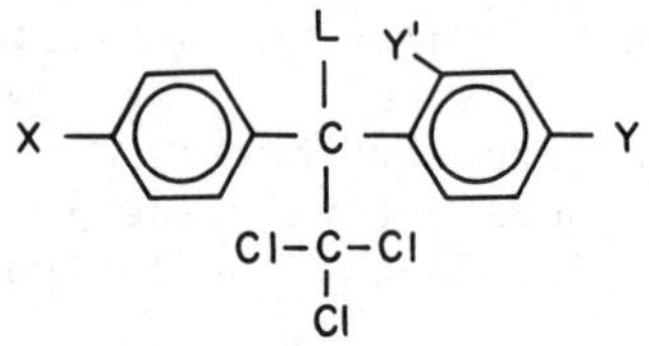

Number	X	Y	Y'	L	*Musca domestica*[b] (topical LD_{50}, µg/♀)		
					S-NAIDM	R_{SP}	R_{SC}
1 (DDT)	Cl	Cl	H	H	0.04	>100	>100
2	Cl	Cl	F	H	0.092	>10	>10
3	Cl	Cl	Cl	H	0.22	0.72	0.43
4	Cl	Cl	Br	H	0.51	>10	>1.0
5	Cl	Cl	F	D	0.086	0.66	0.80
6	Cl	Cl	Cl	D	0.19	0.44	0.40
7	Cl	Cl	Br	D	0.54	>10	>10

[a]Data from Metcalf and Fukuto (1968).
[b]The strains of house flies were: a susceptible S-NAIDM strain; a chlorinated hydrocarbon-resistant R_{SP} strain, under continuous selection pressure with DDT and lindane, and the R_{SC} strain, under continuous selection pressure with the organophosphate chlorthion.

These differences are difficult to explain and additional study of these compounds is required.

Cyclopropane Analogs of DDT

The design of deutero-DDT as a modified DDT structure to counteract DDT resistance was based on the greater stability of the carbon-deuterium bond and the expectation that enzymatic dehydrochlorination would take place at a slower rate. In contrast to this type of approach, the design of a number of highly active cyclopropane analogs of DDT finds its basis in a hypothetical model for the receptor or target site of DDT (Holan, 1969, 1971a, 1971b). Although the mode of action of DDT has not been clearly defined, DDT is believed to affect nerve membrane conductance by interacting directly with nerve membranes through physicochemical forces (Holan, 1969). Based on the insecticidal activity of DDT and lindane, an early proposal for the DDT receptor site was visualized as a cavity in the interspace between three membrane macromolecules (Mullins, 1956). The presence of DDT in this membrane cavity was believed to distort the membrane and cause leakage of sodium ions, leading to abnormal transmission of nervous activity.

Mullins' model for the DDT receptor provided a useful concept, but it was somewhat nebulous in terms of spatial and dimensional features. This model was subsequently modified (Holan, 1969) to explain the insecticidal activity of several DDT analogs that could not be readily explained by the Mullins model. Holan's model is shown in Figure 1, which provides projections of the fit of DDT and of an equally active cyclopropane analog in the receptor site. The broken line in part A shows the limit in the dimension of the receptor site for alkyl- or alkoxy-containing ring substituents, e.g., ethoxy, and the solid line denotes Van der Waal's limits of negative atom dipoles. The model, therefore, can accommodate a broader range of DDT analogs in the lipid-protein nerve membrane interface. The substituted phenyl rings bind to the protein layer in this interface,

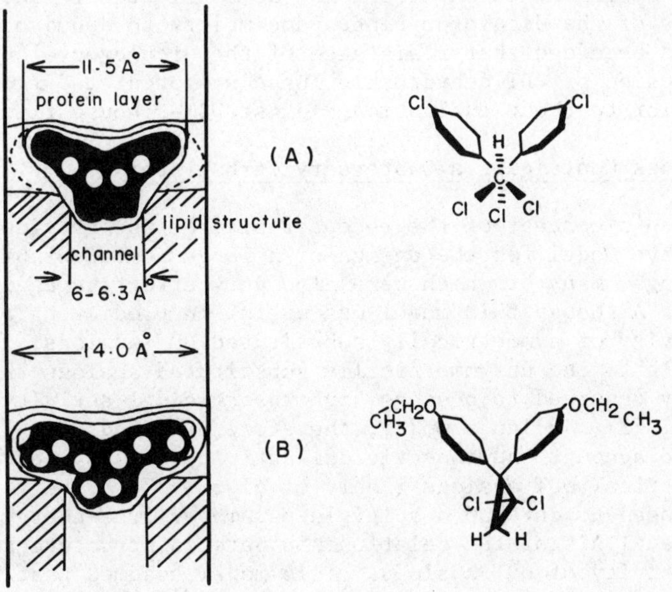

Figure 1. Fit of DDT (A) and 1,1-di-(*p*-ethoxyphenyl)-2,2-dichlorocyclopropane (B) in the Holan receptor.

perhaps through hydrophobic bonding and Van der Waal's forces, and the trichloromethyl or dichlorocyclopropane moieties fit within the channel in the membrane. In this position, the molecule in some manner keeps the channel open for sodium ion leakage.

Application of Holan's model for the DDT receptor resulted in the design of a number of insecticidally active DDT analogs (Holan, 1969, 1971b). Notable among these are the 2,2-dichlorocyclopropane analogs, whose structures are given below.

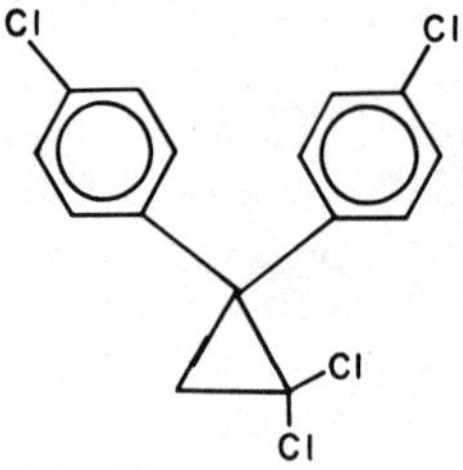

8

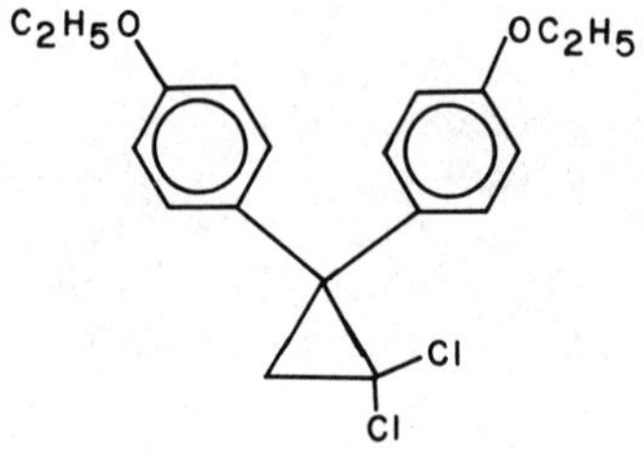

9

Both of these compounds were more effective than DDT against house flies and mosquito larvae, as shown by the data in Table 2. Compared to DDT and methoxychlor (12), compounds 8 and 9 were highly active against the R strain of house flies (Turramurra strain). For example, 9, the diethoxy-substituted compound, was 4000-fold more effective against this strain than methoxychlor. The outstanding effectiveness of 8 and 9 against the R strain may be attributed to the greater stability of the dichlorocyclopropane moiety to dehydrochlorination. It should be added that resistance of the dichlorocyclopropane analogs to the action of DDT dehydrochlorinase was predicted from their structures prior to their evaluation against DDT-R house flies.

DDT Analogs Containing a Quaternary Carbon Atom

Holan's concept of the receptor site of DDT provided an excellent qualitative model for the design of a large number of highly active DDT analogs, many of which were also very effective against DDT-R insects. Although this model was useful in predicting and explaining the activity of symmetrically substituted DDT analogs, it was not applicable to the unsymmetrically substituted analogs that were subsequently observed to possess high insecticidal activity (Metcalf et al., 1971). Holan's model, therefore, was modified in order to take into account the insecticidal activity of unsymmetrical as well as symmetrical DDT analogs (Fahmy et al., 1973). The revised model was proposed to develop a multiple-parameter free-energy approach for the analysis of the relationship between structure and insecticidal activity of DDT analogs. This model assumed that DDT or its analog fits into a receptor site in a nerve membrane similar to that suggested by Holan. The receptor site was visualized as a cavity or

Table 2. Insecticidal Properties of 2,2-Dichlorocyclopropane Analogs of DDT[a]

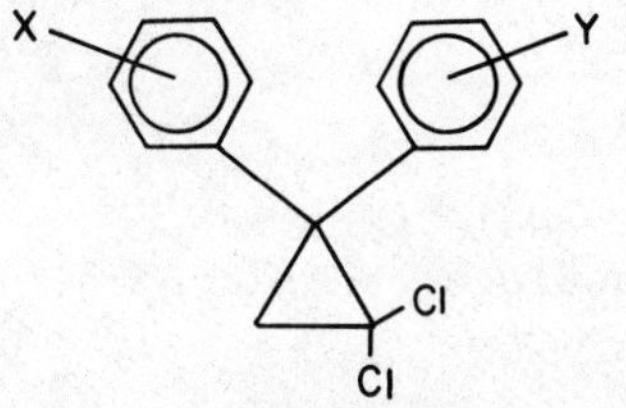

Number	X	Y	House fly LD_{50} (μg/g) Susceptible	House fly LD_{50} (μg/g) Resistant	*Aedes aegypti* LC_{100} (ppm)
8	4-Cl	4-Cl	0.12	0.47	0.013
9	4-OC_2H_5	4-OC_2H_5	0.13	0.25	0.06
10	2,4-di-Cl	4-Cl	0.41	-	0.2
11	4-OCH_3	4-OCH_3	0.3	-	0.09
DDT	-	-	0.17	400	0.13
12 (Methoxy-chlor)			0.33	1,000	0.25

[a]Data from Holan (1969).

pouch with a limited degree of flexibility and was not a rigid model. Maximum activity was expected when there was maximum interaction between the receptor site and four key substituents, X, Y, L and Z, in the generalized DDT structure given below.

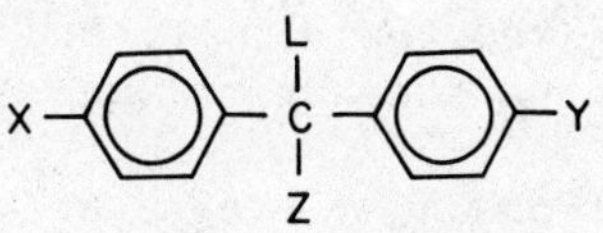

For maximum interaction the overall dimension or size of the DDT molecule, as expressed by the dimensions of X, Y, L and Z, was found to be critical, and any deviation from this size was expected to result in reduced interaction and reduced insecticidal activity.

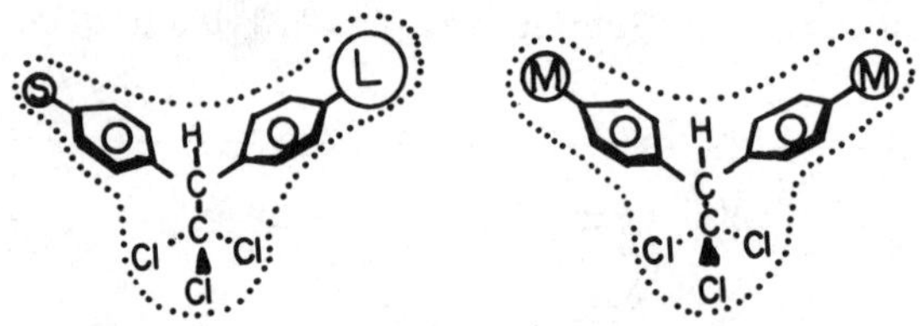

Figure 2. Hypothetical models showing the fit with the DDT receptor site of DDT analogs containing different size ring substituents.

The model is illustrated in Figure 2, in which M is a substituent approximately the size of chlorine, S is smaller than M, and L is larger than M. This model informs us that good fit with the receptor is not restricted to symmetrical molecules but that it also may be obtained with unsymmetrical analogs, as long as the overall dimension of the molecule is within the flexible framework of the receptor site. According to the model, as the summation of the size of X, Y, L and Z increases, interaction with the receptor increases, eventually reaching a maximum from which point increase in substituent size results in decreased interaction.

Using this model, the relationship between the structure and insecticidal activity of a large series of DDT analogs was examined in house flies and mosquito larvae by multiple regression analyses of activity and a variety of free energy parameters, including E_s, σ^*, σ and π. For a series of 25 compounds of the structure below,

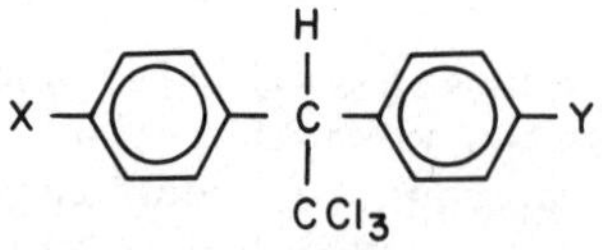

in which X and Y are different substitutents, the following best equation relating synergized toxicity to house fly toxicity with E_s (Taft's steric substituent constant) was obtained:

$$\log LD_{50} = 2.69 + 1.85\,\Sigma E_s + 0.85\,\Sigma E_{s^2} \qquad (1)$$

$$n = 25 \qquad r = 0.874 \qquad s = 0.31$$

A similar relationship (equation 2) was obtained for synergized house fly toxicity where the L and Z substituents were varied:

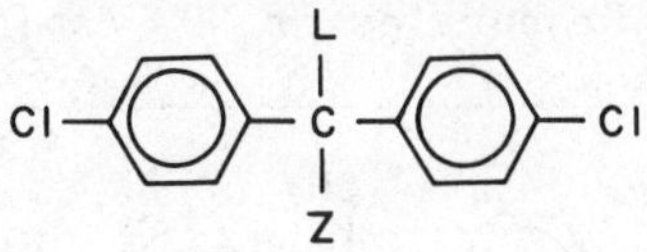

$$\log LD_{50} = 14.14 + 9.36\ \Sigma\ E_s + 1.12\ \Sigma\ E_s2 + 4.31\ \Sigma\sigma^* \qquad (2)$$

$$n = 9 \qquad r = 0.856 \qquad s = 0.31$$

Analogous equations were obtained for larvicidal activity in mosquitoes.

The reasonably satisfactory correlation observed between toxicity and the steric substituent constant E_s, a constant that gives an estimate of the substituent's size, provided support for a flexible receptor site. This model was subsequently used for the design of new, insecticidally active DDT analogs having structures which suggested that they might also be effective against DDT-R insects (Abu-El-Haj et al., 1979). According to the model, any change in L must be compensated for by a change in Z. Therefore, if the α-hydrogen atom in DDT is replaced by a methyl moiety, resulting in a quaternary α-carbon atom, the side-chain trichloromethyl moiety must be reduced in size to maintain insecticidal activity. Several compounds of this type were prepared and data for their effectiveness against S house flies and mosquito larvae are presented in Table 3. The data reveal that DDT analogs containing a quaternary carbon atom are insecticidally active, and one of the compounds (17) was equal to DDT against house flies and mosquito larvae. Against the R_{SP} strain of house flies, the LD_{50} of 17 was 50 µg/g, a value that is at least 100-fold smaller than the >5000 µg/g for this strain. Since the high tolerance of the R_{SP} strain to DDT is attributed to rapid dehydrochlorination of DDT to DDE, compound 17, which cannot be dehydrochlorinated because of the methyl group, would be expected to be effective. However, the LD_{50} value of 50 µg/g suggests that 17 may be detoxified to a significant extent by at least one other mechanism, probably an oxidative process, since the LD_{50} of 17 synergized with piperonyl butoxide, an inhibitor of the mixed-function oxidases, was 7.5 µg/g for the S-NAIDM strain.

Overall, the results illustrate the usefulness of models in designing DDT-type compounds having activity against R insects.

MODIFICATION OF METHYLCARBAMATE INSECTICIDES

In recent years a large amount of work has been carried out on the "derivatization" of methylcarbamate insecticides in an attempt to convert them into materials with favorable properties of selectivity. Carbamate insecticides are notorious for their generally

Table 3. Insecticidal Properties of DDT Analogs of the Structure[a]

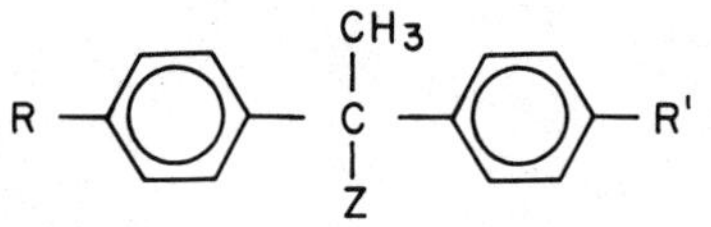

No.	R	R'	Z	House fly LD_{50} (μg/g)	*Culex quinquefasciatus* LC_{50} (ppm)
13	Cl	OCH_3	$CHCl_2$	72	0.115
14	Cl	OCH_3	CHBrCl	68	0.145
15	Br	OCH_3	$CHCl_2$	87	0.23
16	Br	OCH_3	CHBrCl	98	0.18
17	OC_2H_5	OC_2H_5	$CHCl_2$	10	0.06
DDT		-	-	14	0.07

[a]Data from Abu-El-Haj et al. (1979).

high mammalian toxicity and derivatization by substitution of the nitrogen proton of the carbamyl moiety by different types of functional groups has been highly successful in generating compounds with significantly reduced mammalian toxicity but high insecticidal activity (Fukuto, 1976; Fahmy et al., 1978). While the principal objective of this approach has been the design of compounds with an improved order of selectivity, in some cases the products have demonstrated effectiveness against R insects.

The high mammalian toxicity of methylcarbamate insecticides may be attributed to the fact that they are direct inhibitors of acetylcholinesterase, and therefore, they do not possess a "delay factor" in the intoxication process such as is provided by P=S to P=O conversion with many organophosphorus insecticides. A classical example of the delay factor is found in the safe organophosphorus insecticide malathion, for which it was suggested that slow metabolic activation to the anticholinesterase malaoxon provided the opportunity for carboxylesterases present in mammals to degrade malathion to nontoxic products (Krueger and O'Brien, 1959). In malathion-R house flies, however, increased phosphotriesterase activity appeared to be one of the mechanisms responsible for the degradation of the organophosphorus insecticide (Matsumura and Dauterman, 1964). Therefore, it

Table 4. Toxicological Properties of Some *N*-Dialkoxyphosphinothioyl-*N*-methylcarbamates[a]

No.	R	I_{50} (M) Fly AChE	LD_{50} mg/kg House fly S_{NAIDM}	R_{MIP}	Mouse (oral)
	m-Isopropylphenyl-OC(=O)N(CH_3)R				
18	H (MIP)	3.7×10^{-7}	41	125	16
19	$P(S)(OCH_3)_2$	6.6×10^{-5}	32.5	33.5	760
20	$P(S)(OC_2H_5)_2$	2.9×10^{-6}	67.5	165	400-550
	2-Isopropoxyphenyl-OC(=O)N(CH_3)R				
21	H (Propoxur)	6.9×10^{-7}	22	45	24
22	$P(S)(OCH_3)_2$	8.4×10^{-6}	32	37	1400
23	$P(S)(OC_2H_5)_2$	1.4×10^{-5}	37	38	1700

[a]Data from Fahmy et al. (1970).

was anticipated that substitution of the proton on the *N*-methyl moiety of carbamate esters that are toxic to both mammals and insects by a dialkoxyphosphinothioyl moiety might give derivatives of comparable insecticidal activity and low mammalian toxicity (Fahmy et al., 1970). In essence, the dialkoxyphosphinothioyl group attached to the carbamyl nitrogen was expected to provide the "delay factor" that allowed the carbamate to be degraded to nontoxic metabolites in mammals. In insects, however, owing to phosphotriesterase action in vivo, regeneration of the parent methylcarbamate was expected, resulting in intoxication of the insect. Several dialkoxyphosphinothioyl derivatives of toxic methylcarbamate insecticides were prepared and toxicological data for some of these are presented in Table 4. The data clearly reveal a decrease in mammalian toxicity, along with retention of insecticidal activity. Furthermore, the derivatized methylcarbamates were, in some cases, more toxic to a

Table 5. Toxicity of Carbaryl and Its *N*-Acylated Derivatives to the Egyptian Cotton Leafworm[a]

No.	Compound	LD_{50} mg/g to the Egyptian cotton leafworm			Resistance ratio (F_4)
		Susceptible	Carbaryl-resistant		
			F_1	F_4	
24	Carbaryl	0.6	3.8	6.2	10.3
25	*N*-Formyl carbaryl	0.2	2.7	2.5	12.5
26	*N*-Acetyl carbaryl	9.0	3.4	5.3	0.6
27	*N*-Butyryl carbaryl	1.7	3.2	3.5	2.1
28	*N*-Benzoyl carbaryl	1.4	2.6	4.5	3.2

[a]Data from Fahmy and Fukuto (1970).

carbamate-R strain of house flies (R_{MIP}) than the parent methylcarbamate: compounds 19 and 23 were virtually as toxic to the R_{MIP} strain as they were to the S-NAIDM strain. Although none of the above dialkoxyphosphinothioyl derivatives has been examined for comparative metabolism in S and R house flies, the toxicological data indicate that equal amounts of the parent methylcarbamate are present at the target site of both S and R strains after treatment with the derivative. Evidently, the presence of a dialkoxyphosphinothioyl moiety in the molecule reduces the rate of metabolic detoxification of the methylcarbamate ester (perhaps by providing an alternate pathway for metabolism, e.g., P=S to P=O conversion), and sufficient amounts of the parent methylcarbamate are generated at or near the target site. A study of the metabolism of the dimethoxyphosphinothioyl derivative of carbofuran in S house flies has shown that 4 hr after topical application, 60% of the applied dose of 8 μg/g was either carbofuran, the derivatized carbofuran, or an intact metabolite of carbofuran (Krieger et al., 1976).

Although a large number and variety of derivatives of methylcarbamate insecticides have been synthesized and examined for toxicity against insects and mammals, relatively few have been

Table 6. Synergism of Carbaryl and Two of Its *N*-Acylated Derivatives by Piperonyl Butoxide in a Constant Dose of 20 μg Piperonyl Butoxide/Larvae on a Carbaryl-resistant Strain of the Egyptian Cotton Leafworm[a]

No.	Compound	LD_{50} mg/g		Degree of synergism
		Alone (A)	+PB (B)	(A)/(B)
24	Carbaryl	22.5	2.6	8.7
25	Carbaryl *N*-formyl	5.3	0.8	6.6
26	Carbaryl *N*-acetyl	6.0	3.0	2.0

[a]Data from Fahmy and Fukuto (1970).

evaluated against resistant insects. The toxicity of different acylated derivatives of carbaryl to S and R Egyptian cotton leafworms, *Spodoptera littoralis*, has been determined (Fahmy and Fukuto, 1970) and the results are presented in Table 5. The data show that the *N*-acylcarbaryl derivatives were generally less toxic to the S but more toxic to the R strains than carbaryl. Variations in toxicity to the R strains, however, were small compared to the S strain. The results are consistent with the belief that an initial enzymatic deacylation step is required for intoxication since the deacylation enzyme, perhaps an esterase or amidase, could be more abundant in the R than the S strain.

Of interest is the higher toxicity of *N*-acetylcarbaryl to the R strain compared to the S strain (resistance ratio of 0.59), a case of negative correlation. This example serves as additional evidence for the increased presence of a deacylation enzyme in the R strain. Evidently, the balance between deacetylation to form carbaryl near the target enzyme and detoxication of either carbaryl or the *N*-acetyl derivative results in greater toxicity against the R than against the S strain.

Table 6 presents data concerning the synergism of carbaryl, *N*-formyl-(25), and *N*-acetylcarbaryl (26) by piperonyl butoxide on a highly resistant strain obtained from the field (42-fold more resistant to carbaryl). The significant level of synergism obtained by the addition of piperonyl butoxide to the acylcarbamates with the R strain supports the hypothesis that the deacylation process is different from the processes responsible for the direct detoxification of carbaryl. As in the case of the less resistant strain, the two *N*-acyl derivatives (25 and 26) were also more toxic than carbaryl on

Table 7. Toxicity of Carbofuran and *N,N'*-Thiodicarbamate Analogs to *Culex quinquefasciatus* Say Larvae[a]

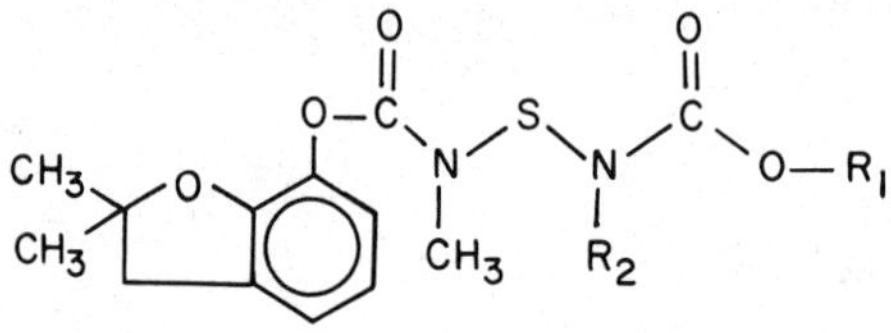

Compound	R_1	R_2	LC_{50} (ppm) Susceptible strain	Resistant strain	RR[b]
29 Carbofuran	-	-	0.0520	1.25	24
30	CH_3	CH_3	0.0230	0.41	17.8
31	C_2H_5	CH_3	0.0160	0.25	15.6
32	C_3H_7	CH_3	0.0090	0.12	13.3
33	$CH(CH_3)_2$	CH_3	0.0088	0.15	17.1
34	C_4H_9	CH_3	0.0043	0.068	15.8
35	C_5H_{11}	CH_3	0.0022	0.02	9.1
36	C_7H_{15}	CH_3	0.0010	0.011	11.0
37	C_8H_{17}	CH_3	0.0008	0.01	12.5
38	$C_{10}H_{21}$	CH_3	0.0012	0.011	9.2

[a]Data from Lawson (1979).
[b]RR = Resistance ratio = LC_{50} resistant strain/LC_{50} susceptible strain.

this highly resistant strain without the use of a synergist. Use of a synergist, however, minimized the difference in the activity of the three compounds, suggesting that synergism in these cases could be attributed to protection of the carbamate after in vivo liberation.

More recently, a large number of *N,N'*-thiodicarbamate derivatives of toxic methylcarbamate insecticides have been examined for toxicity to insects and mammals (Fahmy et al., 1978). The majority of these compounds were outstanding insecticides with the added benefit of being relatively safe to mammals. Subsequent examination of some of these derivatives to a carbamate-R strain of mosquito larvae, *Culex quinquefasciatus*, under constant selection pressure with propoxur revealed that many of the derivatives were substantially more toxic to the R strain than the parent methylcarbamate (Table 7) (Lawson, 1979). Similar results were obtained with the *N,N'*-thiodicarbamate analogs of propoxur and *m*-isopropylphenyl methylcarbamate. An increase in toxicity was generally related to the increase in the chain length of the alkoxy moiety (R_1), suggesting that toxicity was dependent on the lipophilic properties of the molecule. In the case of these derivatives, the greater toxicity of the longer chain derivatives may be attributed to faster penetration of the material from the aqueous test solution into the relatively lipophilic mosquito larvae. Subsequent cleavage of the N-S bond would lead to intoxicating amounts of the toxic methylcarbamate inside the mosquito larvae.

MODIFICATION OF ORGANOPHOSPHORUS INSECTICIDES

Glutathione *S*-Transferase

Glutathione *S*-transferase-catalyzed dealkylation has been implicated as a possible mechanism for the resistance of house flies (Motoyama and Dauterman, 1972) and phytoseiid mites, *Amblyseius fallacis* (Motoyama et al., 1971, 1977), to the organophosphorus insecticide azinphosmethyl or *O,O*-dimethyl *S*-[4-oxo-1,2,3-benzotriazin-3(4H)-ylmethyl] phosphorodithioate. Data showing the toxicity of azinphosmethyl and several of its analogs to S and R strains of house flies and *A. fallacis* are given in Table 8. According to the data, the R house flies and phytoseiid mites were able to tolerate very large dosages of azinphosmethyl (39), but the level of resistance was markedly reduced with the oxon (40) and the corresponding ethyl analog (41). The R strains, however, were substantially more tolerant (11- to 42-fold) to compounds 40 or 41 than the S strains. Owing to the much greater susceptibility of *O,O*-dimethyl esters to dealkylation compared to *O,O*-diethyl esters, the toxicological results suggest the involvement of glutathione *S*-transferase in the resistance mechanism.

Support for this mechanism was provided by studies on the metabolism of azinphosmethyl in R and S strains of house flies and phytoseiid mites. Incubation of azinphosmethyl with subcellular fractions of abdomens obtained for S and R house flies resulted in a faster rate of azinphosmethyl demethylation by the R than the S strain (Motoyama and Dauterman, 1972). Furthermore, assay for

Table 8. Toxicity of Azinphosmethyl and Analogs to Susceptible (S) and Resistant (R) Strains of House Flies and *Amblyseius fallacis*

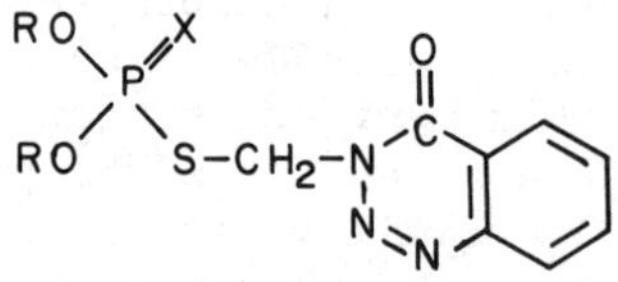

Substitution No.	R	X	House fly[a] LD$_{50}$ (μg/♀ fly) S	R	Resistance factor R/S	*Amblyseius fallacis*[b] LD$_{50}$ (μg/cm^2) S	R	Resistance factor R/S
39	CH_3	S	0.12	200	1,667	0.127	100688	79287
40	CH_3	O	0.07	0.89	13	0.851	35.960	42
41	C_2H_5	S	0.12	3.04	25	0.750	7.960	11
42	*n*-C_3H_7	S	-	-	-	3.959	42.890	11
43	*iso*-C_3H_7	S	-	-	-	>4210	>4210	-
44	*n*-C_4H_9	S	-	-	-	>4210	>4210	-

[a]Data from Motoyama and Dauterman (1972).
[b]Data from Motoyama et al. (1977).

glutathione transferase showed a higher titer of this enzyme in R than in S house flies. Use of [methoxy-^{14}C]azinphosmethyl revealed the formation of *S*-methyl-glutathione as well as desmethyl azinphosmethyl, thus confirming the involvement of the transferase in azinphosmethyl degradation. Related metabolism studies with the phytoseiid mites, both in vitro and in vivo, gave similar results (Motoyama et al., 1971), i.e., higher titers of transferase and faster rates of azinphosmethyl demethylation in R than in S strains.

Overall, the findings suggest the possibility of substitution by other alkoxy groups for the methoxy moieties, e.g., ethoxy, to counteract resistance brought about by the action of the transferase enzymes. The data in Table 8 provide support for this counter-

Table 9. Toxicity of Malathion and Analogs of Susceptible (S) and Malathion-resistant (R) Strains of Blowflies (*Chrysomya putoria*) and House Flies (*Musca domestica*)[a]

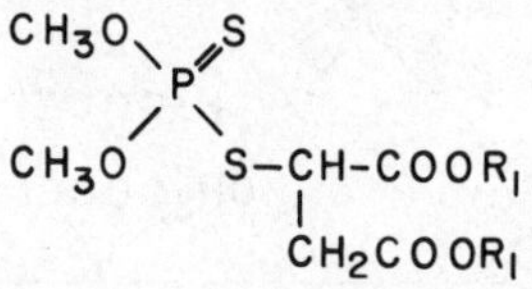

No.	R_1	Blowflies LD_{50} (μg/g) S	Blowflies LD_{50} (μg/g) R	Blowflies Ratio R/S	House flies LD_{50} (μg/g) S	House flies LD_{50} (μg/g) R	House flies Ratio R/S
45	Me	5.9	850	144	21	3066	146
46[b]	Et	9.6	2496	260	27	4239	157
47	*n*-Pr	13	>2366	>182	19	4750	250
48	*i*-Pr	5.9	>4720	>800	19	2090	110
49	*n*-Bu	24	>4800	>200	21	3822	182

[a]Data from Townsend and Busvine (1969).
[b]Malathion.

measure. Unfortunately, azinphosmethyl analogs with alkyl substituents larger than ethyl resulted in compounds of intrinsically poor insecticidal activity.

Carboxylesterase

One of the most vulnerable places in the malathion molecule for metabolic degradation is the ethoxycarbonyl moiety, and therefore, development of resistance to malathion by virtue of increased carboxylesterase activity in insects is not surprising. As indicated earlier, the low toxicity of malathion to mammals is attributed to carboxylesterase-mediated detoxification of malathion into malathion monoacid. Resistance development attributable to increased carboxylesterase activity has been suggested for house flies (Welling and Blaakmeer, 1971), *Culex tarsalis* mosquitoes (Matsumura and Brown, 1961), *Chrysomya putoria* blowflies (Townsend and Busvine, 1969) and *Tetranychus telarius* spider mites (Matsumura and Voss, 1964, 1965).

Table 10. Toxicity of Malathion and Analogs to Susceptible (S) and Resistant (R) Strains of Blowflies and House Flies[a]

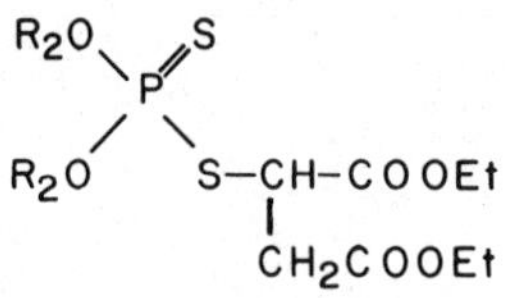

No.	R_2	Blowflies LD_{50} (μg/g) S	Blowflies LD_{50} (μg/g) R	Blowflies Ratio R/S	House flies LD_{50} (μg/g) S	House flies LD_{50} (μg/g) R	House flies Ratio R/S
46	Me	9.6	2496	260	27	4239	157
50	Et	3.8	38	10	6.5	26	4
51	*n*-Pr	12	36	3	65	390	6
52	*i*-Pr	14	70	5	27	108	4
53	*n*-Bu	81	243	3	400	3200	8

[a]Data from Townsend and Busvine (1969).

Since malathion resistance in insects and mites can be attributed to their ability to degrade malathion by hydrolysis of the ethoxycarbonyl moiety, and assuming that the ethyl ester is a specific substrate for the degrading enzyme, a possible countermeasure for resistance lies in the substitution of the ethyl group with another alkyl group. A variety of malathion analogs, with variations in R_1 of the structure shown in Table 9, were examined for toxicity to S and R blowflies, house flies, mosquito larvae and spider mites. In general, this kind of substitution had very little effect in counteracting resistance, as exemplified by the data in Table 9 for blowflies (Townsend and Busvine, 1969) and house flies (Matsumura and Dauterman, 1964). Similar results were obtained with malaoxon and analogs, although in this case, the degree of resistance observed was significantly less; e.g., R/S ratio values of 10 to 30.

In a few cases, the malathion derivatives were more effective against mosquito larvae (Plapp et al., 1965; Dauterman and Matsumura, 1962) and spider mites (Voss et al., 1964). For example, the toxicity (LC_{50}) of the methoxycarbonyl analog of malathion (45) to S and R *Tetranychus telarius* was 0.15% and 0.3%, respectively. Also, compound 45 was equally toxic to S and R *Culex tarsalis* larvae; however, the other analogs (47-49) were substantially less effective against R than against S strains.

In contrast to the ineffectiveness of carboxylester modification in counteracting resistance, modification of the dialkoxyphosphinyl moiety of malathion was more successful, as is evident from the toxicity data presented in Table 10 for compounds of the structure indicated. Although these analogs were much more toxic to the R strain than malathion (46), they were, however, 3- to 10-fold less effective against the R than the S strain. While the explanation for the relative effectiveness of the dialkoxyphosphinyl malathion analogs was based on the poor fit of these compounds on the carboxylesterase enzyme (Matsumura and Dauterman, 1964), it is possible that other resistance mechanisms were contributing to the high tolerance of these insects to malathion. The importance of glutathione *S*-transferase in insecticide resistance during this period was not recognized, and in light of the relative effectiveness of the dialkoxyphosphinyl analogs it seems more likely that the transferases were contributing strongly to resistance in these cases.

SUMMARY

In the preceding discussion, examples were provided that described attempts to suppress metabolic resistance by structure modification, including modification of DDT, carbamate and organophosphorus insecticides. Although a number of examples were given in which structure modification successfully diminished resistance caused by increased metabolic detoxification, cases where this was accomplished intentionally have been rare. The early work on deutero-DDT analogs represents one of the few bona fide cases of intentional structure modification based on anticipated stability of the modified compound to metabolic degradation. In general, the primary intent in the other examples presented was not in counteracting resistance but in the discovery of new and effective insecticides. High activity against resistant insects was incidental but in some cases, e.g., the dichlorocyclopropane and quaternary DDT analogs, effectiveness was readily explainable on chemical grounds.

In recent years, substantial progress has been made in the design of selectively toxic insecticides by taking advantage of known differences in the metabolism of insecticides in insects and mammals. The same kind of approach should be used for the design of compounds that might be effective against insecticide-resistant insects. However, insects are highly adaptable animals and there is no guarantee that they will remain susceptible to the new material.

REFERENCES

Abu-El-Haj, S., Fahmy, M. A. H., and Fukuto, T. R., 1979, Insecticidal activity of 1,1,1-trichloro-2,2-bis(*p*-chlorophenyl)-ethane (DDT) analogues, *J. Agric. Food Chem.*, 27:258.

Barker, R. J., 1960, Syntheses of the aliphatic deuterium analogs of DDT and TDE and their toxicity and degradation when applied to adult houseflies, *J. Econ. Entomol.*, 53:35.

Dauterman, W. C., and Matsumura, F., 1962, Effect of malathion analogs upon resistant and susceptible *Culex tarsalis* mosquitoes, *Science*, 138:694.

Fahmy, M. A. H., and Fukuto, T. R., 1970, Insecticidal activity of *N*-acylcarbamates on susceptible and carbaryl-resistant strains of the Egyptian cotton leafworm, *J. Econ. Entomol.*, 63:1783.

Fahmy, M. A. H., Fukuto, T. R., Myers, R. O., and March, R. B., 1970, The selective toxicity of new *N*-phosphorothioylcarbamate esters, *J. Agric. Food Chem.*, 18:793.

Fahmy, M. A. H., Fukuto, T. R., Metcalf, R. L., and Holmstead, R. L., 1973, Structure-activity correlations in DDT analogs, *J. Agric. Food Chem.*, 21:585.

Fahmy, M. A. H., Mallipudi, N. M., and Fukuto, T. R., 1978, Selective toxicity of *N,N'*-thiodicarbamates, *J. Agric. Food Chem.*, 26:550.

Fukuto, T. R., 1976, Carbamate insecticides, *in:* "The Future for Insecticides: Needs and Prospects," R. L. Metcalf and J. J. McKelvey, Jr., eds., *Adv. Environ. Sci. Technol.*, 6:313-346, Wiley-Interscience, New York.

Hennessy, D. J., Fratantoni, J., Hartigan, J., Moorefield, H. H., and Weiden, M. H. J., 1961, Toxicity of 2-(2-halogen-4-chlorophenyl)-2-(4-chlorophenyl)-1,1,1-trichloroethanes to normal and to DDT-resistant house-flies, *Nature*, 190:341.

Holan, G., 1969, New halocyclopropane insecticides and the mode of action of DDT, *Nature*, 221:1025.

Holan, G., 1971a, Rational design of insecticides, *Bull. Wld. Hlth. Org.*, 44:355.

Holan, G., 1971b, Rational design of degradable insecticides, *Nature*, 232:644.

Krieger, R. I., Lee, P. W., Fahmy, M. A. H., Chen, M., and Fukuto, T. R., 1976, Metabolism of 2,2-dimethyl-2,3-dihydrobenzofuranyl-7-*N*-dimethoxyphosphinothioyl-*N*-methylcarbamate in the house fly, rat and mouse, *Pestic. Biochem. Physiol.*, 6:1.

Krueger, H. R., and O'Brien, R. D., 1959, Relationship between metabolism and differential toxicity of malathion in insects and mice, *J. Econ. Entomol.*, 52:1063.

Lawson, M. A., 1971, "Toxicology and Mode of Action of *N,N'*-Thiodicarbamates in Susceptible and Resistant *Culex quinquefasciatus* Say Larvae," Ph. D. dissertation, University of California, Riverside.

Lipke, H., and Kearns, C. W., 1960, DDT-dehydrochlorinase, *Advan. Pest Control Res.*, 3:253.

Matsumura, F., and Brown, A. W. A., 1961, Biochemistry of malathion resistance in *Culex tarsalis*, *J. Econ. Entomol.*, 54:1176

Matsumura, F., and Dauterman, W. C., 1964, Effect of malathion analogues on a malathion-resistant housefly strain which possesses a detoxication enzyme, carboxylesterase, *Nature*, 202:1356.

Matsumura, F., and Voss, G., 1964, Mechanism of malathion and parathion resistance in the two-spotted spider mite *Tetranychus urticae*, *J. Econ. Entomol.*, 57:911.

Matsumura, F., and Voss, G., 1965, Properties of partially purified malathion carboxylesterase of the two-spotted spider mite, *J. Insect Physiol.*, 11:147.

Metcalf, R. L., and Fukuto, T. R., 1968, The comparative toxicity of DDT and analogues to susceptible and resistant houseflies and mosquitoes, *Bull. Wld. Hlth. Org.*, 38:633.

Metcalf, R. L., Kapoor, I. P., and Hirwe, A. S., 1971, Biodegradable analogues of DDT, *Bull. Wld. Hlth. Org.*, 44:363.

Moorefield, H. H., Weiden, M. H. J., and Hennessy, D. J., 1962, Relationship of the insecticidal and the free radical activities of DDT, *Contrib. Boyce Thompson Inst.*, 21:481.

Motoyama, N., and Dauterman, W. C., 1972, In vitro metabolism of azinphosmethyl in susceptible and resistant houseflies, *Pestic. Biochem. Physiol.*, 2:113.

Motoyama, N., Rock, G. C., and Dauterman, W. C., 1971, Studies on the mechanism of azinphosmethyl resistance in the predaceous mite, *Neoseiulus* (T.) *fallacis* (Family: Phytoseiidae), *Pestic. Biochem. Physiol.*, 1:205.

Motoyama, N., Dauterman, W. C., and Rock, G. C., 1977, Toxicity of *O*-alkyl analogues of azinphosmethyl and other insecticides to resistant and susceptible predaceous mites, *Amblyseius fallacis*, *J. Econ. Entomol.*, 70:475.

Mullins, L. J., 1956, Structure of nerve cell membranes, *Amer. Inst. Biol. Sci. Publ.*, 1:123.

Pillai, M. K. K., Hennessy, D. J., and Brown, A. W. A., 1963, Deuterated analogues as remedial insecticides against DDT-resistant *Aedes aegypti*, *Mosq. News*, 23:118.

Plapp, P. W., Jr., Orchard, R. D., and Morgan, J. W., 1965, Analogs of parathion and malathion as substitute insecticides for the control of resistant house flies and the mosquito *Culex tarsalis*, *J. Econ. Entomol.*, 58:953.

Townsend, M. G., and Busvine, J. R., 1969, The mechanism of malathion resistance in the blowfly *Chrysomya putoria*, *Entomol. Exp. Appl.*, 12:243.

Tsukamoto, M., 1961, Metabolic fate of DDT in *Drosophila melanogaster*, III. Comparative studies, *Botyu-Kagaku*, 26:74.

Voss, G., Dauterman, W. C., and Matsumura, F., 1964, Relation between toxicity of malathion analog and organophosphate resistance in the two-spotted spider mite, *J. Econ. Entomol.*, 57:808.

Welling, W., and Blaakmeer, P. T., 1971, Metabolism of malathion in a resistant and a susceptible strain of houseflies, *in:*

"Proceedings of the 2nd International IUPAC Congress of Pesticide Chemistry, Vol. II," A. S. Tahori, ed., pp. 61-75, Gordon and Breach, New York.

SUPPRESSION OF ALTERED ACETYLCHOLINESTERASE OF THE GREEN RICE LEAFHOPPER BY *N*-PROPYL AND *N*-METHYL CARBAMATE COMBINATIONS

Izuru Yamamoto, Yoji Takahashi* and Nobuo Kyomura*

Department of Agricultural Chemistry
Tokyo University of Agriculture
Setagaya-ku, Tokyo 156, Japan

INTRODUCTION

The green rice leafhopper, *Nephotettix cincticeps* Uhler, has developed cross resistance to many carbamate insecticides in Japan and has caused serious problems in the protection of rice plants from the diseases that the hopper transmits. The worst problem is that this resistance has developed in areas where conventional organophosphorus insecticides already have become useless. The resistance is mainly due to the reduced sensitivity of the hopper's acetylcholinesterase (AChE) to carbamates and this finding has initiated research into the development of a new group of carbamates that would control the resistant green rice leafhopper.

DISCOVERY OF *N*-PROPYL CARBAMATES AS POTENT INHIBITORS OF ALTERED ACETYLCHOLINESTERASE

Discovery of new insecticide chemicals usually occurs during trial and error approaches by the screening of many randomly synthesized compounds. However, whenever available, any biological, physiological, biochemical or physicochemical information on targets and chemicals should be incorporated into the design of pesticidal compounds and such information is more abundant for organophosphorus and carbamate insecticides than for any other type of pesticide. Both of these insecticides inhibit AChE, and metabolism is often an important factor in the efficacy. In relation to resistance, the

*Central Research Laboratories, Mitsubishi Chemical Industries Ltd., Midori-Ku, Yokohama, Kanagawa 227, Japan.

increased degradation of malathion by increased activity of carboxylesterase made the compound ineffective. Other organophosphorus insecticides also became ineffective, probably by increased degradation, although the exact mechanisms have not been studied.

The mechanism of carbamate resistance in the green rice leafhopper is quite different from those mentioned above. It is principally due to the reduced sensitivity of AChE to carbamates in the resistant hopper (Hama and Iwata, 1971; Yamamoto et al., 1977): AChE activity remains at the same level or becomes slightly higher in the resistant (R) strain. Aliesterase activity also becomes higher in the R strain, but aliesterase does not seem to be involved in the resistance mechanism. The degree of metabolism of carbamates is different in the R and susceptible (S) strains (Asakawa and Kazano, 1978), but it is only a minor part of the resistance mechanism. The degree of resistance correlates well with the degree of difference in I_{50}s for AChE from S and R strains.

Thus, carbamate resistance is due to insensitive AChE, which may arise from modification of the AChE structure. With this biochemical information in hand, we tried to find carbamate structures that would increase the interaction with such insensitive, modified or altered AChE. Therefore, emphasis was given to the anti-AChE activity of the synthesized compounds.

CHEMICAL MODIFICATIONS OF CARBAMATE STRUCTURE

Many modifications were attempted (Kyomura, 1979). None of the compounds modified at Site 1 (Figure 1) to thio, thiol or dithio approached the corresponding carbamates in anti-AChE and insecticidal activity (Metcalf, 1971), although modifications at Site 2 had produced a variety of effective carbamates in the past. We repeated the studies and added more modifications, but none gave good inhibition of resistant AChE. Substitution of hydrogen with dimethoxyphosphinothioyl and other groups at Site 3 sometimes produced a good insecticidal effect with certain R insects that lethally synthesized the original carbamates by increased metabolic activity (Fukuto, 1976). This approach, however, was not successful with the present

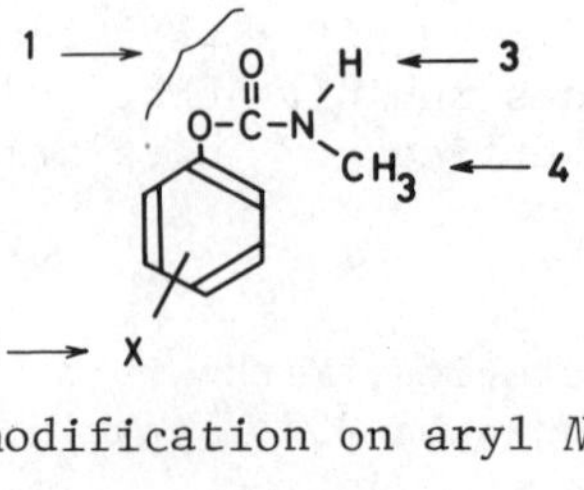

Figure 1. Sites of modification on aryl *N*-methyl carbamates.

problem, because the metabolic activity in the R strain was not particularly high, and even if the parent carbamate was produced by metabolism, the resistant AChE was not affected.

With Site 4, it is well known that *N*-methyl and *N,N*-dimethyl carbamates are potent anti-AChE compounds. The inactivity of other *N*-substituted carbamates was thought to be established. For example, the change from *N*-methyl to *N*-higher alkyl, phenyl, or benzyl results in a dramatic loss of anti-AChE activity. Even a small change to *N*-ethyl severely reduced AChE inhibition (Yu et al., 1972).

INHIBITION OF ALTERED ACETYLCHOLINESTERASE BY *N*-PROPYL CARBAMATES

In the past, the interaction of *N*-methyl carbamates with normal AChE (positive), *N*-methyl carbamates with altered AChE (negative), and *N*-higher alkyl carbamates with normal AChE (negative) were examined. Still to be studied was the interaction of *N*-higher alkyl carbamates with altered AChE.

Three strains were used as the enzyme source: S, a susceptible strain collected at Warabi, Saitama, and maintained at the National Institute of Agricultural Sciences, and RN-N and RN-4, the Nakagawara strains collected in 1969 and 1974, respectively. The relationship

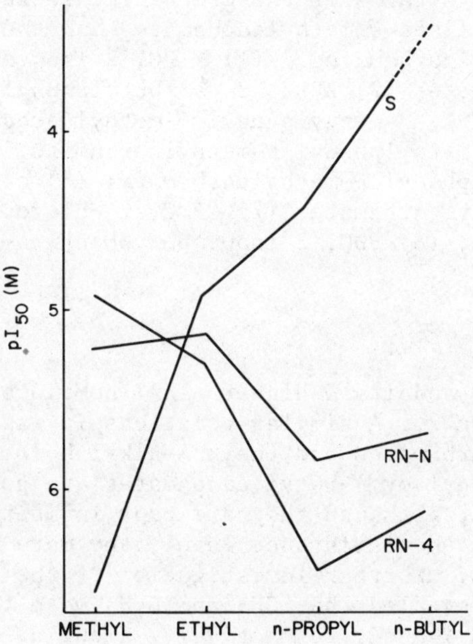

Figure 2. Inhibition of AChE from susceptible (S), resistant (RN-N) and more resistant (RN-4) strains of green rice leafhopper.

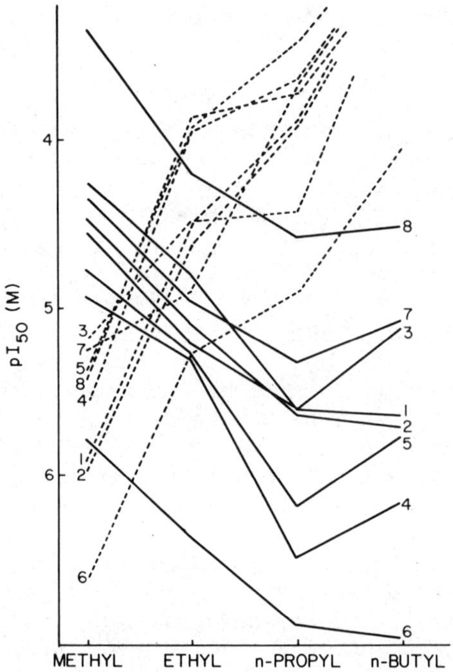

Figure 3. Inhibition of AChE from susceptible (S) and resistant (RN-4) strains of the green rice leafhopper by eight commercialized *N*-methylcarbamates and their corresponding *N*-alkylcarbamates. (1) MIPC, 2-isopropylphenyl *N*-methylcarbamate; (2) BPMC, 2-*sec*-butylphenyl *N*-methylcarbamate; (3) MTMC, 3-methylphenyl *N*-methylcarbamate; (4) MPMC, 3,4-dimethylphenyl *N*-methylcarbamate; (5) XMC, 3,5-dimethylphenyl *N*-methylcarbamate; (6) carbaryl, 1-naphthyl *N*-methylcarbamate; (7) CPMC, 2-chlorophenyl *N*-methylcarbamate; (8) PHC, 2-isopropoxyphenyl *N*-methylcarbamate.

as found for MPMC and its *N*-higher alkyl homologs (*N*-alkyl MTMC) is shown in Figure 2. A similar relationship was found for seven other *N*-methyl carbamates and their *N*-alkyl homologs (Figure 3). In general, *N*-propyl or *N*-butyl carbamates are potent inhibitors of resistant AChE, although they are poor inhibitors of susceptible AChE. The higher the resistance level, the more effective the *N*-propyl carbamates. Further investigation of the *N*-substituted naphthyl carbamates revealed that CH_3 and C_3H_7 were the best substituents for S-AChE and RN-N-AChE, respectively, among H, CH_3, C_2H_5, ClC_2H_4, C_3H_7, i-C_3H_7, $CH_2{=}CHCH_2$, $CH_3CH{=}CH$, $CH_3C{\equiv}C$, C_4H_9, i-C_4H_9, C_5H_{11}, C_6H_{12}, CH_3O, CH_3SCH_2, and CH_3OCH_2.

Table 1. Synergism of Insecticidal Activity to the Resistant Green Rice Leafhopper (RN-4) Between *N*-Methyl and *N*-Propyl Carbamates[a]

Treatment	Mortality after 24 hrs		
	200	100	50ppm
BPMC	56.7	3.3	0%
MTMC	100	35	13.3
BPMC + *N*-Propyl MTMC (1:1)	100	100	78.3
MTMC + *N*-Propyl MTMC (1:1)	100	100	90.0

[a]Applied as wettable powder.
BPMC = 2-*sec*-butylphenyl *N*-methylcarbamate, MTMC = 3-methylphenyl *N*-methylcarbamate.

INSECTICIDAL ACTIVITY - SYNERGISM BETWEEN *N*-PROPYL AND *N*-METHYL CARBAMATES

N-Propyl carbamates are potent inhibitors of resistant AChE, but still have only a limited effect against the R green rice leafhopper. Conventional synergists are ineffective in increasing insecticidal activity; however the combination of *N*-propyl carbamates with *N*-methyl carbamates gives a synergism that results in good control of R as well as S strains in both the laboratory (Tables 1 and 2) and the field. A 1:1 ratio is the most effective and synergism is also found for *N*-methyl carbamates with *N*-ethyl and *N*-butyl carbamates, although *N*-propyl is the most effective. Such synergism is found not only between *N*-methyl and *N*-propyl carbamates having the same leaving group, but also with those having different leaving groups (Table 3).

ALTERED ACETYLCHOLINESTERASE

Mechanism of Insensitivity of Altered Acetylcholinesterase to *N*-Methyl Carbamates

Reduced sensitivity of insect AChE was first recognized by Schuntner and Roulston (1968): the time required for 50% inhibition of AChE by diazinon became longer in the R *Lucilia cuprina* fly. Hama and Iwata (1971) provided definite evidence for reduced sensitivity of AChE to carbamate insecticides as a resistance mechanism in the green rice leafhopper. For indicating AChE sensitivity to

Table 2. Synergism of Insecticidal Activity to the Resistant Green Rice Leafhopper (RN-4) Between *N*-Methyl and *N*-Propyl Carbamates[a]

Treatment	Mortality after 24 hrs
	250 mg/pot
BPMC	0%
MTMC	7.5
MTMC + *N*-Propyl MTMC (1% + 1%)	97.5
BPMC + *N*-Propyl MTMC (1% + 1%)	97.5

[a]Applied as dusts.

the inhibitors, I_{50} values have been used as the parameter. Tripathi and O'Brien (1973) first introduced a more detailed analysis, finding that the reduced sensitivity of AChE to tetrachlorvinphos (stirofos), an organophosphorus insecticide, as a resistance mechanism for the house fly was due to the reduction of affinity of AChE to tetrachlorvinphos.

For AChE inhibition by carbamates, it is generally accepted that the inhibition occurs according to the following reaction sequence:

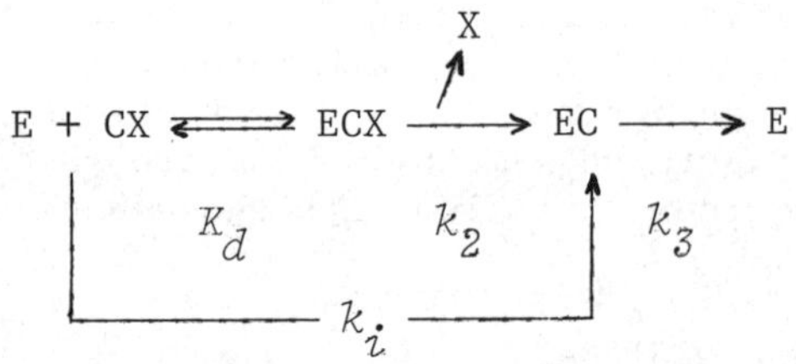

in which E is AChE; CX is a carbamate; C is the carbamyl group; X is the leaving group; ECX is a complex controlled by a dissociation constant, K_d; EC is the carbamylated AChE; k_2 and k_3 are the carbamylation and decarbamylation rate constants, respectively; and k_i = (k_2/k_3), the bimolecular reaction constant.

Yamamoto et al. (1977) analyzed the inhibition of the hopper's AChE by carbamates. Table 4 shows their results in comparison with results from the house fly. It is apparent that in both cases, an

Table 3. Insecticidal Synergism Between *N*-Propyl MTMC and *N*-Methyl Carbamates to Resistant Green Rice Leafhopper (RN-4)

Treatment	Topical LD_{50} (μg/g)		
	Single compound	1:1 (W/W) mixture with *N*-Propyl MTMC	Co-toxicity coefficient
N-Propyl MTMC	50		
MTMC	61	19	2.9
MIPC	155	25	3.0
BPMC	148	18	4.2
Propoxur	200	28	2.8
CPMC	54	23	2.2
MPMC	41	19	2.4
XMC	66	22	2.6
Carbaryl	44	19	2.4

increase in the K_d value or a reduced affinity of AChE to the inhibitors was the major cause of reduced inhibition. Thus, the AChE from such R strains is called "altered AChE." Such alteration is significant in terms of the enzyme's interaction with the inhibitors, but there is only a small difference in the AChE affinity to the substrate, acetylthiocholine, as shown in Table 5.

Multiplicity of Altered Acetylcholinesterase and Mechanism of Synergism

Insecticides, including carbamates, are synergized by such compounds as methylenedioxyphenyls, organothiocyanates, arylpropynyl ethers, propynyl phosphonates, aryloxyalkylamines, benzothiadiazoles, and imidazoles. Synergists often broaden the insecticidal activity in R strains. The mechanisms of methylenedioxyphenyl synergists are known: they inhibit the mixed-function oxidase system and save the insecticide from detoxication. Some insecticidally inactive carbamates synergize carbamate insecticides. In some cases this analog synergism may be due to the preferential detoxication of the synergist, thus sparing the insecticide. The synergistic action of naphthyl *N*-propyl carbamate (*N*-propyl carbaryl) for synthetic

Table 4. Inhibition Constants from Susceptible and Resistant Insects

	House fly, Tetrachlorvinphos[a]		Green rice leafhopper, BPMC[b]	
	Susceptible	Resistant	Susceptible	Resistant
K_d (M)	4.8×10^{-8}	2.8×10^{-5}	1.5×10^{-6}	5.0×10^{-4}
k_2 (min^{-1})	0.6	1.6	1.3	0.9
k_i ($M^{-1}min^{-1}$)	1.2×10^{7}	6.0×10^{4}	9.0×10^{5}	1.7×10^{3}

[a]Tripathi and O'Brien (1973).
[b]Yamamoto et al. (1977).

pyrethroids is due to its inhibition of hydrolytic detoxication (Jao and Casida, 1974).

However, the synergism between *N*-propyl and *N*-methyl carbamates does not involve the inhibition of detoxication. Rather it seems to involve the AChE (Table 6). A higher synergism occurs at I_{90} than I_{50}. Also, the increase in AChE inhibition by the increased concentration of the inhibitor is not efficient with individual *N*-methyl or *N*-propyl carbamates, but is quite efficient with a combination of the two (Takahashi et al., 1978).

Such synergism is based on the multiplicity of the resistant AChE as evidenced by electrophoresis and kinetic studies (Takahashi et al., 1978). The resistant AChE consists of a normal inhibition site that is sensitive to *N*-methyl carbamates and an additional altered inhibition site that is sensitive to *N*-propyl but not *N*-methyl carbamates.

A kinetic analysis giving a log v - t relation (v is the velocity of substrate hydrolysis catalyzed by AChE after inhibition for t min) indicates the uniformity or multiplicity of AChE sensitivity to the inhibitors. Figure 4 shows the results obtained with the strains of the green rice leafhopper: S, RN-4 and R-M. The latter is a highly R strain selected from RN-4 by selection pressure in the laboratory with MTMC for 16 generations (Table 7). With MTMC, a rapid straight-line inhibition occurs for the S strain; for RN-4 the inhibition almost stops after 40% inhibition, resulting in a curved line. According to Main (1969), such a curve in the rate plot is a reflection of the inhibition of a multiple AChE system. For R-M there is almost no inhibition. In contrast, inhibition by

Table 5. Affinity of Susceptible and Resistant AChE to Acetylthiocholine and the Inhibitors

	House fly[a]			Green rice leafhopper[b]		
	Susceptible(S)	Resistant(R)	R/S	Susceptible(S)	Resistant(R)	R/S
Acetylthiocholine (*Km*)	9.5×10^{-6}	3.3×10^{-5}	3.4	5.6×10^{-5}	1.0×10^{-4}	1.9
Insecticides (*Kd*)						
Tetrachlorvinphos	4.8×10^{-8}	2.8×10^{-5}	573			
BPMC				1.5×10^{-6}	5.0×10^{-4}	346

[a]Tripathi and O'Brien (1973).
[b]Yamamoto et al. (1977).

Table 6. Synergism at Different Inhibition Levels of AChE from Resistant Green Rice Leafhopper Between *N*-Methyl and *N*-Propyl Carbamates

Treatment	I_{50}	I_{70}	I_{90}
MTMC	1.2×10^{-4}	5.0×10^{-4}	4.0×10^{-3}
N-Propyl MTMC	4.0×10^{-5}	2.8×10^{-4}	4.8×10^{-3}
MTMC + *N*-Propyl MTMC (1:1)	1.4×10^{-5}	3.1×10^{-5}	1.2×10^{-4}
Co-toxicity coefficient	4.3	11.6	37.0

N-propyl MTMC is almost absent with the S strain, it almost stops after 60% inhibition for RN-4, resulting in a curved line, and it progresses rapidly for R-M. For the RN-4 strain, an *N*-methyl or *N*-propyl carbamate individually produces a curved line, but a 1:1 mixture produces a strong inhibition, with a straight line relationship to dosage.

Furthermore, by combining S- and R-M-AChE in various ratios, we can reproduce the curved line in accordance with the ratio (Figure 5). For a mixture consisting of S- and R-M-AChE in a 3:1 ratio, the inhibition stops at 75% with MTMC and 25% with *N*-propyl MTMC, and vice versa for a 1:3 enzyme mixture. With a 1:1 AChE mixture, either MTMC or *N*-propyl MTMC gives ca. 50% inhibition, and their combination affords a good inhibition with a straight line. Conversely, we can estimate the ratio of normal to altered sites in an AChE preparation.

These are indications that S-AChE consists of an *N*-methyl carbamate-S site and R-M-AChE consists of an *N*-propyl carbamate-S site, while RN-4-AChE consists of two kinds of site, one being sensitive to *N*-methyl and the other being sensitive to *N*-propyl carbamates. Figure 6 illustrates the nature of AChE and the mechanism of joint inhibition of resistant AChE by *N*-methyl and *N*-propyl carbamates.

INSECT POPULATION AND ALTERED ACETYLCHOLINESTERASE

Selection of a green rice leafhopper population with *N*-methyl carbamates results in the development of reduced susceptibility of the population to these *N*-methyl compounds. The population thus

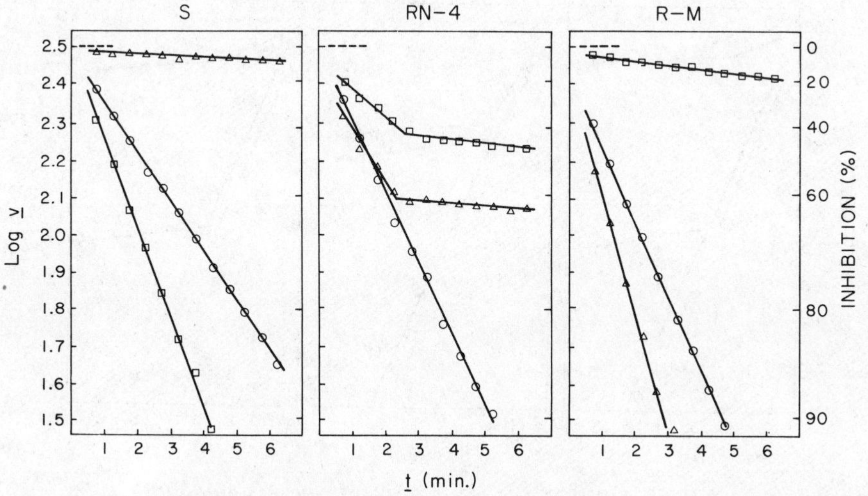

Figure 4. Rate of inhibition of AChE from susceptible (S) and resistant (RN-4, R-M) strains of the green rice leafhopper by *N*-methyl, *N*-propyl MTMC and their 1:1 mixture. □, MTMC, 1.2 x 10^{-4}M; Δ, *N*-propyl MTMC, 1.2 x 10^{-4}M; o, MTMC + *N*-propyl MTMC, 1.2 x 10^{-4} (1:1).

Table 7. Susceptibility to Carbamates of Three Strains of the Green Rice Leafhopper

	LD_{50} (μg/g) after 24 hrs		
	Susceptible	Resistant	
	S	RN-4	R-M
MIPC	2.0	143	565
BPMC	2.6	162	358
MTMC	5.1	82	128
Carbaryl	1.1	46	151

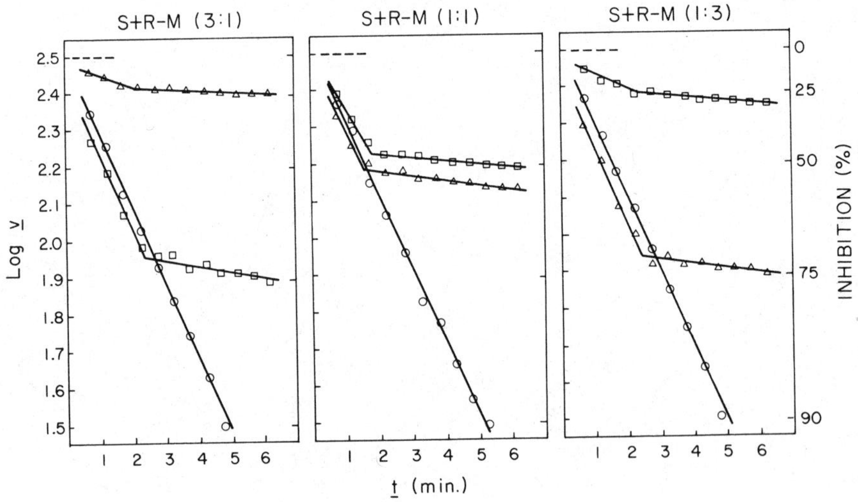

Figure 5. Rate of inhibition obtained by mixing AChE with different sensitivities to the inhibitors. □, MTMC, 1.2 x 10^{-4}M; Δ, *N*-propyl MTMC, 1.2 x 10^{-4}M; o, MTMC + *N*-propyl MTMC, 1.2 x 10^{-4}M (1:1).

obtained is susceptible to *N*-propyl and *N*-methyl carbamate combinations. By further selection with *N*-methyl carbamates, a population like R-M, which is susceptible to *N*-propyl but resistant to *N*-methyl carbamates, is obtained. The population becomes susceptible to *N*-methyl carbamates by selection with *N*-propyl carbamates. Thus, we can shift the insect population back and forth by alternate use of the two carbamate groups (Koyama et al., 1979).

A question arises as to whether an R strain like RN-4 consists of S- and R-M strains or of individuals having AChE that contains both normal and altered inhibition sites. The log v - t examination of AChE from individual insects gives the results shown in Figure 7. Individual carbamate sensitivity is plotted in terms of frequency of inhibition after 3 min inhibition time. In the S strain, the whole population consists of individuals having MTMC-sensitive but *N*-propyl MTMC-insensitive AChE. In the R-M strain, the situation is almost the opposite. Between the two extremes, the RN-4 strain consists of individuals having AChE of S-type sensitivity and incomplete sensitivity to both *N*-methyl and *N*-propyl carbamates. Figure 8 illustrates the development of resistance from the view point of the log v - t relation.

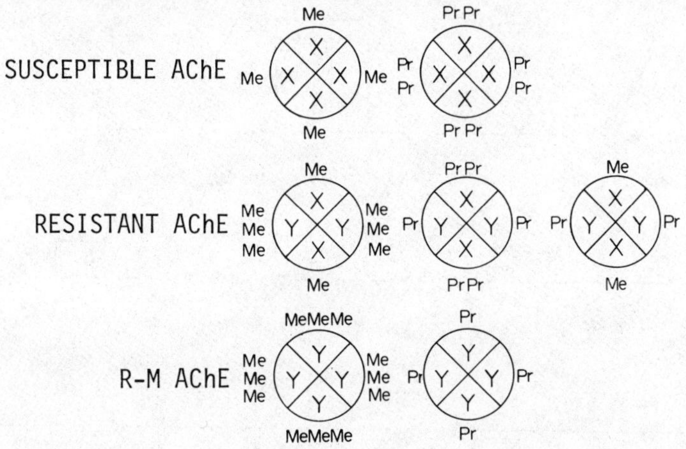

Figure 6. A circle represents AChE divided into inhibition sites, and numbers of Me and Pr outside the circle indicate the relative concentrations of *N*-methyl and *N*-propyl carbamates required for the enzyme inhibition. The susceptible AChE is depicted with a single type of inhibition site (X), which is sensitive to Me, but less sensitive to Pr. The resistant AChE may be a single enzyme with multiple inhibition sites (X and Y) or a mixture of enzymes with X or Y sites. The Y site is sensitive to Pr, but is less sensitive to Me. The R-M AChE has only Y sites. Pr gives better inhibition of resistant AChE than Me, but it is still insufficient. When Me and Pr are combined, each inhibitor interacts with the preferred inhibition site, thus achieving synergism.

PROSPECT

Resistance due to reduced sensitivity of AChE to the inhibitors has been found in mites (Smissaert, 1964; Voss and Matsumura, 1964), ticks (Lee and Batham, 1966; Roulston et al., 1968), house flies (Devonshire, 1975; Triphati and O'Brien, 1973), mosquitoes (Ayad and Georghiou, 1975) and the green rice leafhopper (Hama and Iwata, 1971). AChE insensitivity may evolve in response to an extremely severe selection pressure that exceeds a population's capacity for metabolic defense (Georghiou and Taylor, 1977). Regardless of the mechanism, resistance due to AChE insensitivity may increase in the future if the current use of AChE inhibitors is continued.

Carbamate insecticides are generally selective and have rather narrow spectra of insecticidal action. This may arise from metabolic

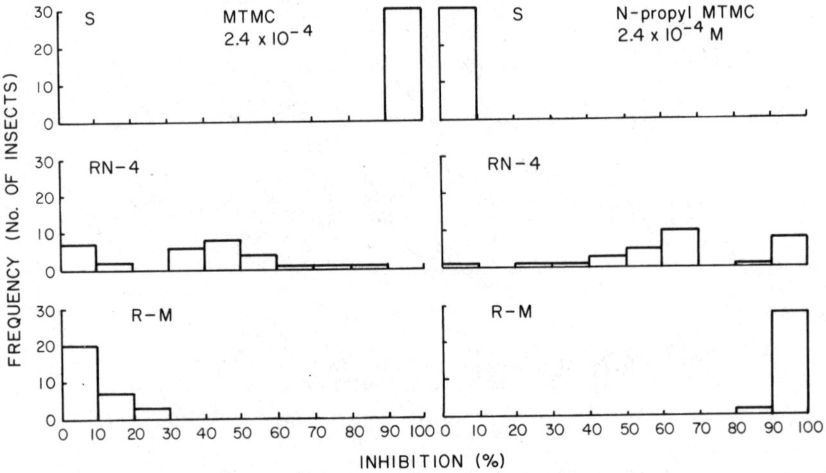

Figure 7. Individual carbamate-sensitivity of AChE from susceptible (S) and resistant (RN-4, R-M) strains of the green rice leafhopper, n = 30.

differences among species, but it also may be due to the differences in AChE in terms of normal and altered inhibition sites. A certain natural tolerance of insects to AChE inhibitors might be due to such a phenomenon.

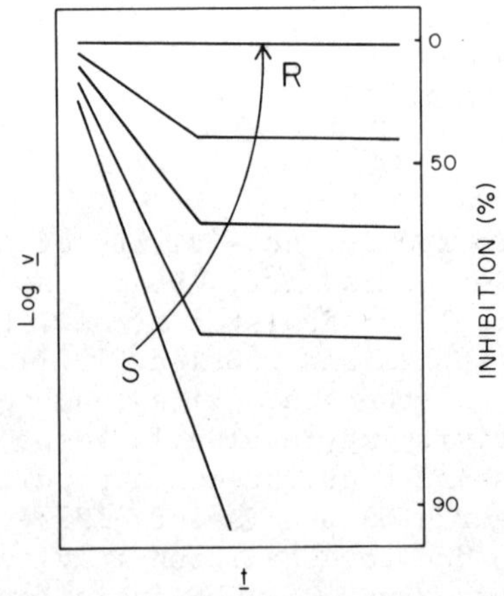

Figure 8. Resistance development and rate of AChE inhibition.

N-Propyl carbamates could provide a unique biochemical tool for investigating such problems. Recognition of the multiplicity of AChE sensitivity in the green rice leafhopper and the synergistic insecticidal action of combinations of *N*-propyl and *N*-methyl carbamates provides the rational basis for the use of combined AChE inhibitors and suggests a new screening system for finding effective insecticides.

REFERENCES

Asakawa, M., and Kazano, H., 1978, Reports on insecticide-resistance research, Japan Plant Protection Association (1978) p. 10.

Ayad, H., and Georghiou, G. P., 1975, Resistance to organophosphates and carbamates in *Anopheles albimanus* based on reduced sensitivity of acetylcholinesterase, *J. Econ. Entomol.*, 68:295.

Devonshire, A. L., 1975, Studies of acetylcholinesterase from houseflies (*Musca domestica* L.) resistant and susceptible to organophosphorus insecticides, *Biochem. J.*, 149:463.

Fukuto, T. R., 1976, Carbamate insecticide, *in:* "The Future for Insecticides, Needs and Prospects," R. L. Metcalf and J. J. Mckelvey, Jr., eds., pp. 313-342, John Wiley & Sons, New York.

Georghiou, G. P., and Taylor, C. E., 1977, Pesticide resistance as an evolutionary phenomenon, Proc. XV Intr. Congr. Ent., pp. 759-785, Entomol. Soc. Amer.

Hama, H., and Iwata, T., 1971, Insensitive cholinesterase in the Nakagawara strain of the green rice leafhopper, *Nephotettix cincticeps* Uhler (Hemiptera: Cicadellidae), as a cause of resistance to carbamate insecticides, *Appl. Ent. Zool.*, 6(4):183.

Jao, L. T., and Casida, J. E., 1974, Esterase inhibitors as synergists for (+)-*trans*-chrysanthemate insecticide chemicals, *Pestic. Biochem. Physiol.*, 4:456.

Koyama, Y., Takahashi, Y., and Yamamoto, I., 1979, unpublished data.

Kyomura, N., 1979, Aryl *N*-propyl carbamates, a potent inhibitor of acetylcholinesterase from the resistant green rice leafhopper, *Nephotettix cincteceps*, Ph. D. thesis, Tokyo University of Agriculture, Department of Agricultural Chemistry, Tokyo, Japan.

Lee, R. M., and Tatham, P., 1966, The activity and organophosphate inhibition of cholinesterase from susceptible and resistant ticks (Acari), *Ent. Exp. Appl.*, 9:13.

Main, A. R., 1969, Kinetic evidence of multiple reversible cholinesterase based on inhibition by organophosphates, *J. Biol. Chem.*, 244:829.

Metcalf, R. L., 1971, Structure-activity relationships for insecticidal carbamates, *Bull. Wld. Hlth. Org.*, 44:43.

Roulston, W. J., Schnitzerling, H. J., and Schuntner, C. A., 1968, Acetylcholinesterase insensitivity in the Biarra strain of the cattle tick *Boophilus microplus*, as a cause of resistance to organophosphorus and carbamate acaricides, *Aust. J. Biol. Sci.*, 21:759.

Schuntner, C. A., and Roulston, W. J., 1968, A resistance mechanism in organophosphorus-resistant strains of sheep blowfly (*Lucilia cuprina*), *Aust. J. Biol. Sci.*, 21:173.

Smissaert, H. R., 1964, Cholinesterase inhibition in spider mites susceptible and resistant to organophosphate, *Science*, 143:129.

Takahashi, Y., Kyomura, N., and Yamamoto, I., 1978, Mechanism of joint action of *N*-methyl and *N*-propyl carbamates for inhibition of acetylcholinesterase from resistant green rice leafhopper, *Nephotettix cincticeps*, *J. Pestic. Sci.*, 3:55.

Tripathi, R. K., and O'Brien, R. D., 1973, Insensitivity of acetylcholinesterase as a factor in resistance of houseflies to the organophosphate Rabon, *Pestic. Biochem. Physiol.*, 3:495.

Voss, G., Matsumura, G., 1964, Resistance to organophsophorus compounds in the two-spotted spider mite: Two different mechanisms of resistance, *Nature*, 202:319.

Yamamoto, I., Kyomura, N., and Takahashi, Y., 1977, Aryl *N*-propylcarbamates, a potent inhibitor of acetylcholinesterase from the resistant green rice leafhopper, *Nephotettix cincticeps*, *J. Pestic. Sci.*, 2:463.

Yu, C. C., Kearns, C. W., and Metcalf, R. L., 1972, Acetylcholinesterase inhibition by substituted phenyl *N*-alkyl carbamates, *J. Agri. Food Chem.*, 20:537.

SUPPRESSION OF RESISTANCE THROUGH SYNERGISTIC COMBINATIONS WITH EMPHASIS ON PLANTHOPPERS AND LEAFHOPPERS INFESTING RICE IN JAPAN

Kozaburo Ozaki

Fuchu Branch, Kagawa Agricultural Experiment Station

9177, Fuchu-cho, Sakaide, Kagawa-ken 762, Japan

INTRODUCTION

Resistance of insect pests to insecticides has been increasingly a worldwide problem, especially with regard to the major insect pests of a country's main crops. Such major insect pests have developed resistance due to exposure to the specific insecticides used for controlling them. Examples include parathion resistance in the rice stemborer, *Chilo suppressalis* Walker, and malathion resistance in both the green rice leafhopper, *Nephotettix cincticeps* Uhler, and the smaller brown planthopper, *Laodelphax striatellus* Fallen. These phenomena are generally considered to be related to the amount of parathion or malathion previously applied (Kimura, 1965; Ozaki, 1962, 1966; Yokoyama and Ozaki, 1968). On the other hand, the possibility exists that these insect pests might have multifarious genes.

Development of insecticide resistance by insect pests may be avoided by reducing the continued usage of individual insecticides in the field. The ideal would be to establish techniques for "Integrated Pest Management" and in fact many specialists are working toward this goal, but to achieve it, various ecological, physiological and toxicological studies must be carried out. It will take some time before such integrated pest management programs can be established. In the meantime, it is important to investigate plans to suppress the development of insecticide resistance under the conditions in which chemical control predominates.

DEVELOPMENT OF RESISTANCE IN THE SMALLER BROWN PLANTHOPPER

Development of resistance to malathion, fenitrothion, carbaryl, MTMC (*m*-toyl methylcarbamate) and methomyl was studied using a susceptible (LE) strain of the smaller brown planthopper. This strain exhibited low esterase activity toward β-naphthyl acetate and was not resistant to organophosphate (OP), chlorinated hydrocarbon or carbamate insecticides (Okuma and Ozaki, 1969; Ozaki et al., 1973a). Selection was conducted over successive generations by the contact method at 70-75% mortality.

A rapid increase in resistance of the malathion-selected strain (Rm) was observed during the first five generations, followed by a slow increase for the next few generations, and a rapid increase appeared again after more than ten generations. Resistance levels of this strain at the 5th, 9th and 12th generations were 7-, 8- and 80-fold, respectively. It attained 250-fold resistance at the 24th generation. A fenitrothion-selected strain (Rf) showed some increase in the level of resistance under the same conditions as above, but despite rigorous selection, no change in the degree of resistance was observed in the later generations. This is explained by the fact that the resistant factor(s) had attained homozygosity. The resistance level was only about 60-fold, a small increase compared to the 250-fold resistance of the Rm strain at the 24th generation of selection.

When the LE strain of the smaller brown planthopper was selected successively with carbaryl for 25 generations, the LD_{50} values were somewhat smaller than those of the original LE strain. The slope of the ld-p line at the final generation of selection was not as steep, which could be explained by assuming that intensive selection with carbaryl in the laboratory caused a certain percentage of the smaller brown planthoppers to become more susceptible to the insecticide.

In the MTMC-selected strain, a small increase in resistance was found after the first five generations of selection. The resistance level at the 12th generation was 6-fold, but despite rigorous selection, no change was observed in later generations. The degree of resistance development was much smaller than that of the Rm and Rf strains, and it seems that the resistance that can be induced in the strain by successive selection with MTMC in the laboratory has only a limited potential to reach the MTMC resistance level actually encountered in the field.

No changes in LD_{50} values of methomyl were found during ten generations of selection with methomyl (methomyl-selected strain), but a sharp increase occurred at the 11th generation, which was followed by a slow increase in later generations. The resistance level at the 11th generation was 4-fold, and it was 7.5-fold at the 20th generation. Thus, the development of resistance in the

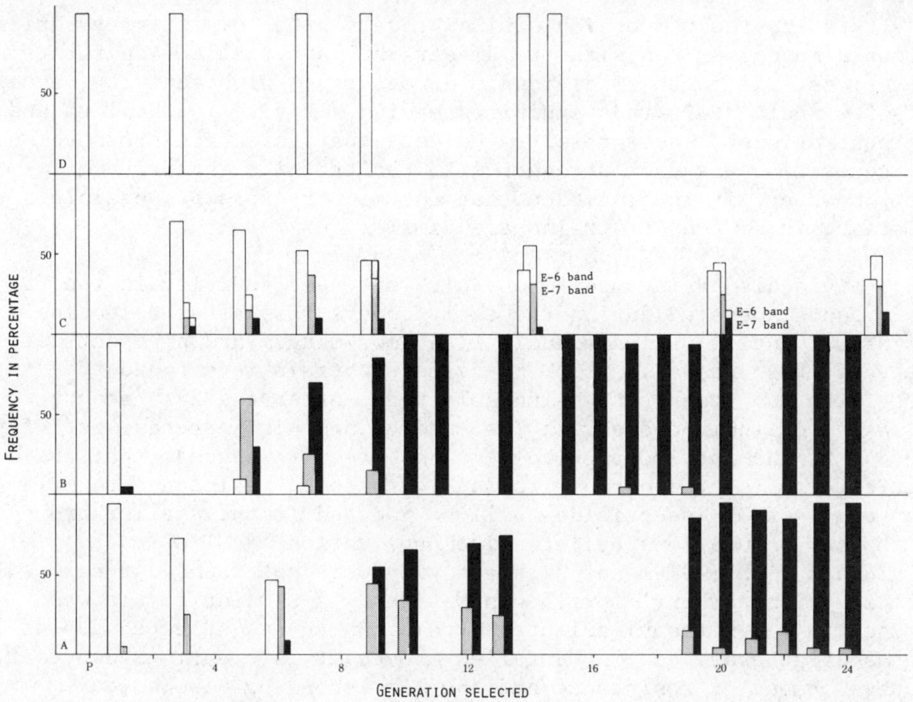

Figure 1. The percentage frequency of individuals with low, medium and high bands of esterase activity in the strains of the smaller brown planthopper selected successively with malathion (A), fenitrothion (B), MTMC (C) and carbaryl (D). White bars indicate low activity, grey bars indicate medium activity, and black bars indicate high activity. (From Okuma and Ozaki, 1969; Ozaki et al., 1973a.)

methomyl-selected strain was small, being about the same level as that in the MTMC-selected strain.

Ozaki and Kassai (1970) reported that the Busshozan population of the smaller brown planthopper collected from a field in Busshozan, Kagawa Prefecture, showed a significantly higher esterase activity to β-naphthyl acetate in the E_7 band following thin layer electrophoresis. The esterase activity in the E_7 band from every generation of the strains selected by malathion, fenitrothion, MTMC, carbaryl and methomyl was examined by the same technique. Figure 1 shows the frequency of individuals with low, medium and high esterase activity in the E_7 band. Intergeneration differences were shown clearly in both the Rm and Rf strains. In the Rm strain, the ratio of individuals with low to medium to high activity in the E_7 band at the 6th generation was 48:42:10, and 0:5:95 at the 24th generation. In the Rf strain, the proportion of individuals with low activity rapidly decreased in comparison with the Rm strain. After selection for ten generations, all of the individuals demonstrated high activity, except for a few individuals that exhibited medium activity in the 17th and 19th generations. These results indicate that selection pressure on each generation was much stronger with fenitrothion than with malathion; however, it is concluded that the main factors responsible for resistance in Rm and Rf strains are similar.

Individuals having medium and high esterase activity in the E_6 and E_7 bands appeared during the early generations after selection with MTMC. Ratios of individuals with low, medium and high activity in the esterase bands in the MTMC-selected strain were found to change with the generation being selected. After the 24th generation, the ratios of individuals with low, medium and high esterase activity were 35, 50 and 15%, respectively. These results suggest that the genetic factor(s) responsible for an increase in resistance to malathion or other OP insecticides will be induced in the smaller brown planthopper populations by intensive application of MTMC in the field. Individuals with medium and high activity were not found during selection with carbaryl for 15 generations. A similar result was obtained in the methomyl-selected strain, which was selected for as long as 25 generations. As detailed above, it is clearly shown that the development of resistance and the corresponding biochemical characteristics vary remarkably, depending on the insecticide.

Ozaki and Kassai (1970) studied the relationship between malathion resistance and esterase activity in the E_7 band of the LE and HE (malathion-resistant) strains and their F_1, F_2 and backcross offspring and suggested that esterase activity in the E_7 band and malathion resistance depend on the same co-dominant autosomal factor. From these results, one may assume that individuals having low, medium and high activity in the E_7 band are susceptible, semi-resistant (hybrid) and resistant to malathion, respectively.

SUPPRESSION OF RESISTANCE DEVELOPMENT IN THE SMALLER BROWN PLANTHOPPER

Efficiency of Alternate Application of Two Insecticides

When insect pests have developed resistance to insecticides, they can be controlled by using another insecticide to which the resistant (R) insects have not shown cross resistance. Since 1964, when the green rice leafhopper developed resistance to OP insecticides, green rice leafhoppers, smaller brown planthoppers and other planthoppers have been controlled with various carbamate insecticides, but the green rice leafhopper became resistant to carbamate insecticides in Ehime Prefecture in 1969. This specific case indicated that the alternative insecticides were unexpectedly short-

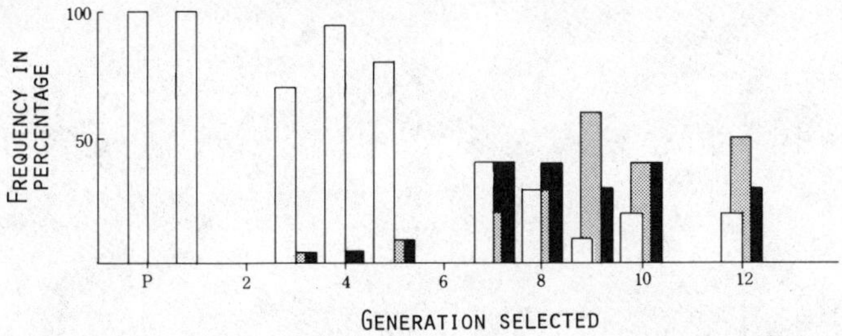

Figure 2. Changes in the percentage frequency of susceptible, semi-resistant (hybrid) and resistant individuals to malathion during selection with malathion and carbaryl on alternate generations. White bars indicate susceptible individuals; grey, semi-resistant, and black, resistant individuals. (From Ozaki et al., 1973b.)

lived in effectiveness against R insect pests. The population of insect pests in the field seemed to have a potential to achieve a large number of genetic variations. The author believes that a solution to this problem cannot be obtained by simply substituting the insecticide being used with other compounds.

Generally, it has been said that alternate applications with more than two insecticides differing in their mode of action would be effective in repressing the development of insecticide resistance. It is clear, as mentioned in the above section, that the mechanism of resistance to OP insecticides is not the same as that of carbamate resistance in the planthopper. The LE strain of the smaller brown planthopper was selected alternately with malathion and carbaryl in every other generation (Ozaki et al., 1973b). This was conducted in the laboratory by the contact method at 70-75% mortality, and the development and pattern of resistance were then investigated. LD_{50} values of carbaryl did not change significantly during the 12 generations of selection with alternate treatments of malathion and carbaryl, and no increase in malathion resistance was found during the first four generations, although it did develop after further selection, attaining 11-fold at the 12th generation. In spite of the absence of malathion selection after the 12th generation, the resistance level increased slightly, but the cause of this phenomenon was not determined. In the strain selected alternately with malathion and carbaryl, some malathion-R individuals (high activity in the esterase bands) appeared in early generations, with continued selection resulting in increasing numbers of such individuals. The percentages of individuals susceptible (S) (low activity in the esterase bands), semi-R (hybrid) (intermediate activity) and R to malathion at the 12th generation were 20, 50 and 30, respectively (Figure 2).

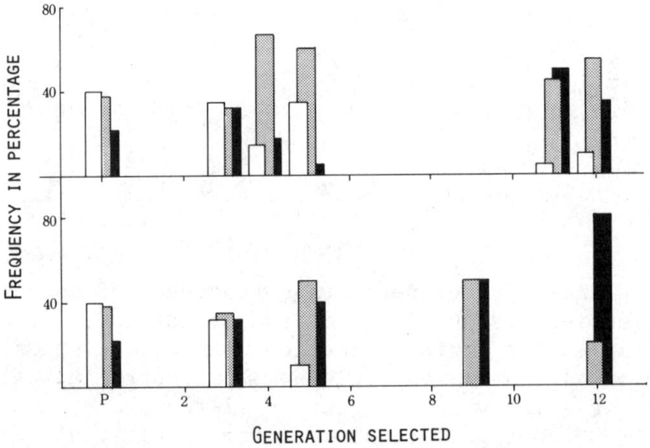

Figure 3. Changes in the percentage frequency of susceptible, semi-resistant (hybrid) and resistant individuals to malathion during successive selection with malathion (lower) and alternate selection with malathion and carbaryl at every third generation (upper). White bars indicate susceptible individuals; grey, semi-resistant, and black, resistant individuals. (From Ozaki, 1969.)

The Busshozan population, which showed only slight resistance to malathion, was also selected successively with malathion or alternately with malathion and carbaryl in every three generations (Ozaki, 1969). In the former selection, individuals resistant to malathion increased rapidly. No S individuals were found in the 9th generation, and the R individuals were 80% at the 12th generation. An interesting result was obtained from the alternate selection with malathion and carbaryl in every three generations: Individuals resistant to malathion that appeared after selection with malathion noticeably decreased in percentage during carbaryl selection for three generations. Thus, the rate of increase in percentage of R individuals was much slower in the strain selected alternately with malathion and carbaryl as compared with the strain selected successively with malathion only. The percentages of S, semi-R (hybrid) and R individuals to malathion at the 12th generation were 10, 55 and 35, respectively (Figure 3).

The strain selected successively with malathion was 202-fold resistant to malathion and 69-fold resistant to phenthoate at the 15th generation. In contrast, the strain selected alternately with malathion and carbaryl over three-generation periods was only about 20-fold resistant to malathion and 10-fold resistant to phenthoate by the 12th generation. LD_{50} values of carbaryl did not change significantly during selection for 12 generations. These results suggest that alternate applications of two insecticides different in their modes of resistance remarkably suppress the development of resistance in the insect pest and that efficient pest control may be achieved in alternate applications over long periods of time.

Efficiency of Mixtures of More Than Two Chemicals

The LE strain of the smaller brown planthopper was selected successively with a mixture of malathion and carbaryl (MN) or MTMC (MT) in the ratio of 1:1 by the contact method at 70-75% mortality (Ozaki et al., 1973b). A small increase in LD_{50} values of MT was found during selection with MT for eight generations, but in the following generations of selection the LD_{50} value decreased and became the same as that of the original LE strain by the 23rd generation. In contrast, the LD_{50} value of MTMC somewhat increased in the 8th generation and became about 4.5-fold in the 23rd generation. The semi-R (hybrid) and R individuals to malathion appeared in the early generations selected with MT. As shown in Figure 4, the ratio of S, semi-R and R individuals to malathion changed with each generation being selected. The percentages of semi-R and R individuals in the 4th generation were 20 and 7, and in the 20th generation, 60 and 40, respec-

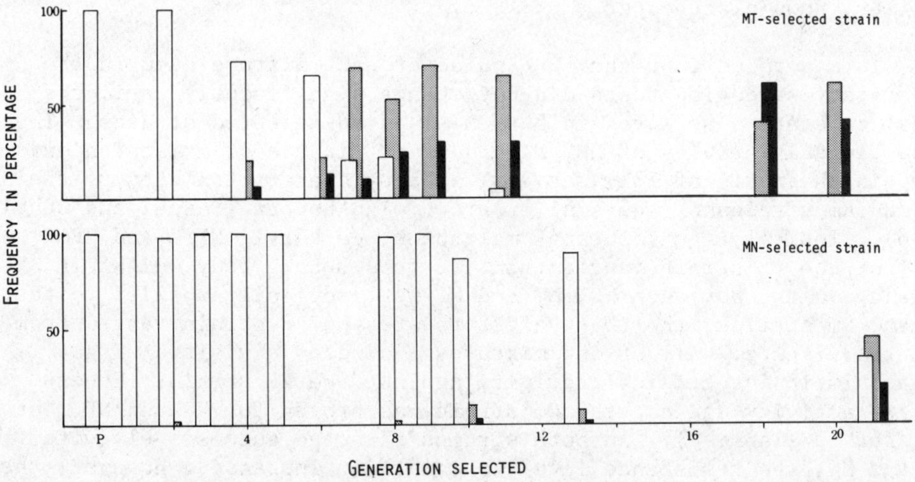

Figure 4. Changes in the percentage frequency of susceptible, semi-resistant (hybrid) and resistant individuals to malathion during selection with MN (malathion-carbaryl mixture) and MT (malathion-MTMC mixture). White bars indicate susceptible individuals; grey, semi-resistant, and black, resistant individuals. (From Ozaki et al., 1973b.)

tively. In the MT-selected strain the LD_{50} values of malathion were assayed after the final (23rd) generation of selection. The results showed that 50-fold resistance to malathion was attained in the 23rd generation.

A similar result was obtained in the strain selected successively with MN (MN-selected strain). The susceptibility to MN at the final (21st) generation was the same as that of the original LE strain. The LD_{50} values of carbaryl did not change significantly during the selection for 21 generations. However, a small increase in the LD_{50} value of malathion was found at the 15th generation, and thereafter, the level of resistance to malathion increased with each generation being selected, becoming 13-fold in the 21st generation.

Although individuals semi-R to malathion first appeared during the 2nd generation of selection with MN, no further increase in the percentage of semi-R or R individuals was found until the 9th generation. They then began to increase in the 10th generation, and at the 21st generation, the percentages of the S, semi-R and R individuals were 35, 45 and 20, respectively. It is suggested that malathion resistance develops gradually with each generation; however, the rate of development was much slower in the strain selected succesively with MN than in the strain selected alternately with malathion and carbaryl. The results indicate that MN is much more effective than MT in suppressing the occurrence of insecticide resistance in the smaller brown planthopper.

It was found that the LD_{50} value of each mixture used in the laboratory selection tests did not change significantly until the final generation selected in both MT- and MN-selected strains. The insecticidal activity of MN, MT and MP (a mixture of malathion and MPMC in the ratio of 1:1) was evaluated by contact toxicity to the LE and Rm strains of the smaller brown planthopper (Sasaki and Ozaki, 1976). Topical LD_{50} values of malathion, carbaryl, MTMC and MPMC against the LE strain ranged from 1.2 to 3.3 μg/g body weight in female adults, but when MN, MT and MP were topically applied to the LE and Rm strains, the LD_{50} values toward the LE strain ranged from 3.2 to 3.7 μg/g. All of the mixtures evaluated exhibited slight antagonistic insecticidal activity against the LE strain. Synergistic activity against the Rm strain was proven for MN and MP, but not for MT (Table 1). In both strains selected successively with MT and MN, the resistance level to malathion increased; however, the degree of malathion resistance development was lower in the strain selected with MN than in that selected with MT. From these results it is suggested that highly efficient suppression of insecticide resistance may be obtained by application of a mixture of insecticides showing synergistic activity against the R strain of the insect pest and that the result of applying a mixture without synergism would not be as good.

Table 1. Insecticidal Activity of Mixtures of Malathion and Carbamate Insecticides Against Susceptible (LE) and Malathion-resistant (Rm) Strains of the Smaller Brown Planthopper

Mixture[a]	LE strain		Rm strain	
	LD_{50}[b]	Co-toxicity coefficient	LD_{50}[b]	Co-toxicity coefficient
Malathion-carbaryl	3.4	48	5.6	112
Malathion-MTMC	3.7	43	9.3	80
Malathion-MPMC	3.2	54	9.2	127
Malathion	1.2		70.9	
Carbaryl	2.8		3.3	
MTMC	2.7		3.9	
MPMC	3.3		6.4	

[a]Mixed ratio of 1:1.
[b]μg/g body weight of the female adult.
Data in Sasaki and Ozaki (1976).

The LE strain was selected successively with a mixture of fenitrothion, carbaryl and methomyl in the ratio of 1:1:1 (FNMe) (Ozaki et al., 1973b). No elevated LD_{50} values of FNMe were found until the 8th generation, but at the 11th generation, the LD_{50} value of FNMe was twice as high as that of the original LE strain. Despite rigorous selection, no change in the LD_{50} values occurred in later generations. When the LD50 values for fenitrothion, carbaryl and methomyl were assayed in later generations of selection, slightly elevated LD_{50} values for fenitrothion and methomyl were observed, and at the 23rd generation they were 3 to 4 times higher than those of the original LE strain. The LD_{50} values of carbaryl did not change significantly during 23 generations of selection. The FNMe showed synergistic activity against the LE strain, and at the final generation, the LD_{50} value for FNMe was lower than those for fenitrothion, carbaryl and methomyl combined as a mixture. This indicates that the joint toxic action of the combined insecticides against the smaller brown planthopper became stronger with successive applications in the field. In the FNMe-selected strain, individuals having medium and highly active esterase bands were still not found in the 9th generation, but they did appear in the 10th generation, and then increased with each successive generation selected. The percentages of individuals with low, medium and high activity esterase bands at the 9th, 10th, 21st and 23rd generations were 100:0:0, 75:5:20, 60:20:20 and 50:21:29, respectively. Following further selection with FNMe, slightly elevated LD_{50} values were found with fenitrothion and methomyl, but their rate of increase was much lower than that of malathion in both strains selected successively with MT or MN. It is suggested that the efficiency in suppression of insecticide resistance development in the smaller brown planthopper is much higher with FNMe than with MT and MN because the suppression is achieved with a lower pressure exerted by the individual insecticides when they are combined in a mixture.

To further understand the mechanism by which mixtures of more than two insecticides suppress the development of insecticide resistance, the LE strain of the smaller brown planthopper was selected with malathion and carbaryl in parallel experiments (Ozaki et al., 1973b). In this test, the fourth and fifth instar nymphs of the smaller brown planthopper were selected respectively with malathion and carbaryl at 70-75% mortality. The surviving nymphs were placed at a ratio of 1:1 in a glass vessel containing rice seedlings and were then allowed to copulate freely. This process was continued in every generation. Their susceptibility to malathion and carbaryl and the esterase bands of individuals were investigated in each generation.

The LD_{50} values of carbaryl did not change significantly during parallel selection with malathion and carbaryl over 12 generations. Malathion resistance developed after the 3rd generation, and the resistance level was shown to be 15-fold in the 12th generation. The resistance level was the same as that at the 21st generation in the strain selected successively with MN. Figure 5 shows the frequency of occurrence of S, semi-R (hybrid) and R individuals to malathion. Individuals semi-R and R to malathion appeared as early as the 2nd generation at 10% and 5%, respectively. They did not increase until the 5th generation, but a rapid increase was found thereafter. The percentages of semi-R and R individuals in the 9th generation were 40 and 20, and in the 12th generation, 50 and 37, respectively. It was ascertained that development of semi-R and R individuals was much faster in the parallel selection with malathion and carbaryl than in the successive selection with MN.

It is generally accepted that the suppression of the development of malathion resistance in the smaller brown planthopper is caused by the lower pressure of the respective insecticides of malathion and carbaryl. However, from analysis of the results of successive selection with MN and the parallel selection with malathion and carbaryl, and from the testing of the insecticidal activity of

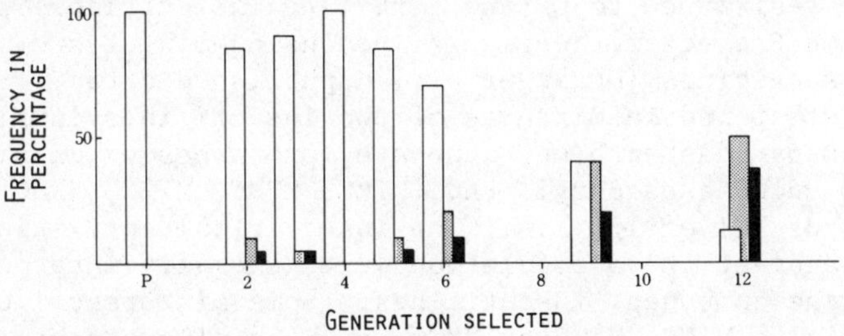

Figure 5. Changes in the percentage frequency of susceptible, semi-resistant (hybrid) and resistant individuals during parallel selections with malathion and carbaryl. White bars indicate susceptible individuals; grey bars, semi-resistant, and black bars, resistant individuals. (From Ozaki et al., 1973b.)

MN on the Rm strain, suppression of the development of malathion resistance by MN seems to be mainly caused by intercepting at an early stage of selection the appearance of smaller brown planthoppers having malathion-R factor(s). Furthermore, the LD_{50} values of MN did not increase during selection over 21 generations. This phenomenon is due to the percent frequency of the malathion-R individuals being kept low. Also, a new factor(s) that would cause resistance to MN in addition to malathion and carbaryl did not appear in the intensive selection with MN.

Moorefield (1968) reported that selection with carbaryl rapidly produced house flies virtually immune to the compound, and within a few generations the chemical was completely ineffectual. However, when the house fly was selected with mixed doses of carbaryl and piperonyl butoxide (1:5 ratio), the summation of altered susceptibility was only approximately 5-fold after 50 generations of pressure. He suggested that, in addition to supplementing the activity of the carbamate, the addition of a synergist apparently suppressed the development of carbamate resistance. Metcalf (1967) suggested that the use of a synergist with an insecticide may well provide the most feasible way to prevent the development of R strains of pests. Consequently, it may be concluded that the application of mixtures of two insecticides, each inhibiting the development of resistance to the other and each having a significantly different mode of action, can suppress the development of insecticide resistance.

SYNERGISTIC ACTIVITY AGAINST THE GREEN RICE LEAFHOPPER WITH MIXTURES OF TWO CHEMICALS

Toxicities to the S population of the green rice leafhopper collected from Sendai, Miyagi Prefecture, and the Rop-c population (multiple resistance to OP and carbamate insecticides) from Nakagawara, Ehime Prefecture, were examined using mixtures of OP and carbamate insecticides (Ozaki et al., unpublished data). Synergistic activity was found in mixtures of two insecticides consisting of an OP such as pyridaphenthion, acephate, propaphos or diazinon, and a carbamate such as carbaryl, PHC, MTMC, MPMC, MIPC, BPMC and XMC, in the ratio of 1:1 or 1.0:1.5. The insecticidal activities of these mixtures against the S population were generally intermediate between those of the component insecticides. Some mixtures, i.e., pyridaphenthion plus MTMC, MPMC or XMC, acephate plus carbaryl, XMC or BPMC, propaphos plus carbaryl, and diazinon plus carbaryl, showed synergistic activity against the S population. Although the insecticidal activities of mixtures of pyridaphenthion, acephate, propaphos, or diazinon plus a carbamate insecticide were less active against the Rop-c population than against the S population, synergistic action against the Rop-c population was demonstrated except for mixtures of acephate plus MTMC or XMC. Hama and Iwata (1973) reported that mixtures consisting of PHC plus methyl parathion, diazinon, phenthoate, malathion or dimethoate and those of carbaryl plus methyl parathion, diazinon or malathion in the ratio of 1:1 had shown synergistic activity against the Rop-c population of the green rice leafhopper collected from Nakagawara, Ehime Prefecture. These mixtures were much less toxic to the Rop-c population, i.e., LD_{50} values against the S population collected in Sendai, Miyagi Prefecture, ranged from 0.9 to 3.0 μg/g body weight of female adults, while the LD_{50} values against the Rop-c population ranged from 25 to 350 μg/g.

In the green rice leafhopper, Hama and Iwata (1972) also reported that aliesterase (esterase hydrolyzing methylbutyrate) was higher in activity in the Rop-c population than in the S population, and that the aliesterase in the Rop-c population was about 100 times more sensitive to PHC than that in the S population. They assumed that synergistic activity of the mixtures of carbamate and OP insecticides was related to interference with the OP resistance factor in the green rice leafhopper by carbamates.

As shown in Table 2, the mixture of propaphos and carbaryl showed synergistic activity against the Rop-c population of the green rice leafhopper. Yoshioka (1976) reported that when mixtures of propaphos and carbamate insecticide in the ratio of 1:1 were applied topically to the Rop-c population synergism was detected. An <u>in vitro</u> experiment was conducted on propaphos inhibition of a modified ChE (ChE insensitive to carbamate insecticides) from the carbamate-R (Rmc) strain established by laboratory selection and a normal ChE from the S strain (Hama, 1975). The modified ChE was inhibited by propaphos more than the normal ChE, i.e., the I_{50} values of propaphos against modified ChE and normal ChE were 7.5 x 10^{-5}M and 7.5 x 10^{-3}M, respectively. It is suggested that the synergistic activity in the mixtures of propaphos and carbamate insecti-

Table 2. Insecticidal Activity of Mixtures of Organophosphorus and Carbamate Insecticides Against Susceptible and Rop-c Populations of the Green Rice Leafhopper

Mixture	Susceptible (Sendai)		Rop-c (Nakagawara)	
	LD_{50}[a]	Co-toxicity coefficient	LD_{50}[a]	Co-toxicity coefficient
Diazinon-carbaryl[c]	0.47	171	28.1	214
Propaphos-carbaryl[c]	0.33	200	10.1	169
Ofunack-carbaryl[b]	1.41	80	32.2	217
Ofunack-BPMC[b]	2.52	75	53.7	300
Acephate-carbaryl[c]	1.30	109	17.8	130
Acephate-MIPC[b]	5.64	87	16.9	148
Acephate-BPMC[b]	2.50	157	20.5	120
Diazinon	0.67		127	
Propaphos	0.46		8.87	
Ofunack	1.42		164	
Acephate	6.70		13.4	
Carbaryl	0.93		44.4	
MIPC	3.84		180	
BPMC	2.77		158	

[a]µg/g body weight of the female adult.
[b]Mixed ratio of 1:1.
[c]Mixed ratio of 1.0:1.5.
Data in Ozaki et al. (in preparation).

cides might have been due to the fact that the modified ChE was inhibited by propaphos.

Yoshioka et al. (1975) demonstrated that Kitazin P (fungicidal S-benzyl diisopropyl phosphorothiolate) greatly elevated the tox-

Table 3. Insecticidal Activity of Mixtures of Kitazin P and Organophosphorus or Carbamate Insecticides Against Rop-c Population of the Green Rice Leafhopper

Mixture[a]	LD_{50}[b]	Co-toxicity coefficient
Kitazin P - phenthoate	12	991
Kitazin P - malathion	10	3656
Kitazin P - dimethylvinphos	13	252
Kitazin P - diazinon	23	357
Kitazin P - fenthion	195	249
Kitazin P - carbaryl	78	83
Kitazin P - BPMC	78	188
Kitazin P	465	
Phenthoate	68	
Malathion	300	
Dimethylvinphos	17	
Diazinon	45	
Fenthion	510	
Carbaryl	35	
BPMC	87	

[a]Mixed ratio of 1:1.
[b]μg/g body weight of the female adult.
Data in Yoshioka et al. (1975).

icity of OP insecticides to the S and Rop-c populations of the green rice leafhopper. Kitazin P was less toxic to the green rice leafhopper, as shown in Table 3, but simultaneous application with phenthoate, malathion or dimethylvinphos showed higher toxicity against both the S and Rop-c populations than did each insecticide alone. Kitazin P was also a good synergist with diazinon and fenthion, but the LD_{50} values of these mixtures were higher than those of Kitazin P plus either malathion or phenthoate. Miyata et al. (1979) investigated both the metabolism of C^{14}-methyl-labeled malathion by several populations of the green rice leafhopper and the inhibition of carboxyesterase by Kitazin P. Degradation of malathion was inhibited 80-96% by 10^{-5}M Kitazin P, and higher inhibition was observed in the R populations than in the S population. According to these results, the authors suggested that the inhibition of carboxyesterase by Kitazin P was the main factor responsible for synergism between malathion and Kitazin P against the green rice leafhopper.

SUPPRESSION OF RESISTANCE DEVELOPMENT IN THE GREEN RICE LEAFHOPPER

The Rop-c population collected from Nakagawara, Ehime Prefecture, was selected successively with a mixture of malathion and Kitazin P in the ratio of 1:1 (MK) by the spraying method at 50-90% mortality, and the change in susceptibility to MK and malathion was investigated (K. Yoshioka, personal communication). In the MK-selected strain, the LD_{50} values of MK increased with each generation selected; however, the increase in the LD_{50} value was very low, i.e., at the 19th generation the LD_{50} value was only 1.5 times higher than the relative value of the original population. The LD_{50} value of malathion did not change significantly during selection for 19 generations, as shown in Figure 6. The results are similar to those with the smaller brown planthopper strain that was selected with MN. In this case, changes in the biochemical characteristics during successive selection with MK were not checked. The mechanism for suppression of insecticide resistance development by use of mixtures of malathion and Kitazin P seemed to be the same here as in the case of the smaller brown planthopper that had been selected successively with MN. This follows the suggestion of Miyata et al. (1979) that synergistic activity of a malathion-Kitazin P mixture toward the green rice leafhopper was caused by an increase in carboxyesterase inhibition by Kitazin P.

The results mentioned above suggest that a mixture of propaphos and a carbamate insecticide would be able to suppress the development of insecticide resistance in the green rice leafhopper. The Rop-c population collected from Yoshida, Hiroshima Prefecture, was selected successively with a mixture of propaphos and carbaryl (PN) in the ratio of 1:1 by the spraying method at 7-60% (average 34%) mortality (Hosoda and Fujiwara, 1977). The LD_{50} values of propaphos, carbaryl and PN against the original population were 6.1, 17.1 and 7.7 μg/g body weight of female adults, respectively, and those of propaphos and PN at the 24th generation of selection were 12.0 and 11.5 μg/g, respectively. Susceptibility to these insecticides did not change significantly during selection for 24 generations, and the selected strain showed a gradual decrease in carbamate resis-

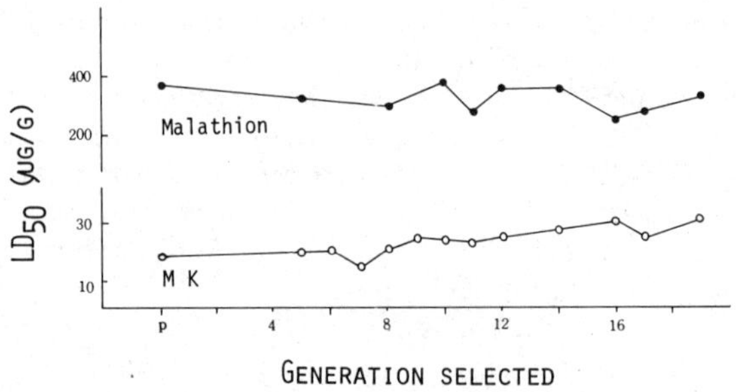

Figure 6. Change in susceptibility to malathion and MK (malathion-Kitazin P mixture) during selection with MK of the green rice leafhopper. (From K. Yoshioka, personal communication.)

tance to the same degree as the standard strain reared without insecticide pressure. When the Yoshida population was selected successively with propaphos (mortality ranging from 5-77%, averaging 26.0%) for 24 generations, the green rice leafhopper became 4.1-fold resistant to propaphos. Thus, the authors suggested that PN might suppress the development of insecticide resistance in the green rice leafhopper.

Since 1975 the Japan Plant Protection Association (JPPA) has been taking countermeasures against insecticide resistance in rice insect pests, and many cooperative studies have been carried out in several Prefectural Agricultural Experiment Stations to determine the efficiency in suppression of the development of resistance in the green rice leafhopper by applications of mixtures of propaphos plus BPMC (PB) in the ratio of 1:1 (JPPA, 1975 to 1978). Full particulars of the obtained results will be given by the committee in JPPA, but an example follows.

The Rop-c population collected from Nakagawara, Ehime Prefecture, was selected successively with propaphos and PB by the spraying method at 50-90% mortality. In contrast to the lower degree of development of propaphos resistance in the strain from Hiroshima Prefecture that was selected successively with propaphos, the LD_{50} value of propaphos increased steadily during successive selection with propaphos, and at the 13th generation the LD_{50} value was 15 times higher than the relative value of the original population. Following further selection, however, no change in the LD_{50} value was found. In the PB-selected strain a slight decrease in resistance to BPMC was observed during the first six generations, but it again increased after the 8th generation. After prolonged selection, the 21st generation was as resistant to BPMC as the original population. The LD_{50} value of PB was twice as high as that of the original population at the 5th generation. Following further selection the rate of increase in the LD_{50} value was much lower, and the value was only 4.5 times higher at the 21st generation. This strain also showed an increase in resistance to propaphos, but the rate of increase in this resistance was much lower than that of the propaphos-selected strain.

As mentioned above, it is expected that a mixture of propaphos and a carbamate insecticide will suppress the development of insecticide resistance in the green rice leafhopper, and at the present time, these are practical insecticides for the control of the green rice leafhopper, the smaller brown planthopper and other planthoppers in many parts of Japan. Topical toxicities of propaphos, carbaryl and PN against the green rice leafhopper populations from Hiroshima and Kagawa Prefectures were investigated. LD_{50} values of these three insecticides against the green rice leafhopper in 1974 were 7.7, 17.1, and 6.1 μg/g body weight of female adults, respectively, and in 1978, they were 4.2-10.1, 21.7-66.2 and 5.7-8.9 μg/g, respectively. In both Hiroshima and Kagawa Prefectures, the level of green rice leafhopper resistance to carbaryl increased significantly over five years; however, no significant change in susceptibility to propaphos and PN was found. This means that the observations made in laboratory selection tests are similar to those made in fields, except that the resistance level to carbaryl increased. It is thought that increases in the level of resistance of the green rice leafhopper to carbaryl were correlated with the selection pressure of carbamate insecticides used for the control of the brown planthopper, *Nilaparvata lugens* Stal, in autumn.

SUPPLEMENTAL REMARKS

As a countermeasure toward insecticide resistance in insect pests, it has been considered important that a priority be given to the selection of chemicals effective against resistant insects and to the examination of their methods of application. This is necessary as a tentative measure, but as was pointed out previously, a substitute insecticide was unexpectedly short-lived in effectiveness against R insects, although this alternative insecticide proved very effective initially. At the present time, the commercial development of new insecticides takes a lot of time and money; therefore, new ways have to be devised so that an insecticide can remain in practical use for longer periods of time.

It has been mentioned before that alternate applications of two or more insecticides with different modes of action would be very effective in the prevention of insecticide resistance; however, results from laboratory selection tests suggest that it is not such an efficient method. In contrast, it was found that the suppression of insecticide resistance development in the smaller brown planthopper could be achieved by simultaneous application of two or more insecticides, such as a mixture of malathion and carbaryl in the ratio of 1:1. Highly efficient suppression of the development of resistance can be achieved by mixtures that have synergistic insecticidal activity against strains resistant to the individual insecticides in the mixture. The author and his group call these mixtures "Fukugōzai" (multi-insecticide). Although the definition of multi-insecticide has not been officially discussed, we have defined it as a mixture of two or more chemicals that can suppress the development of insecticide resistance in insect pests by synergistic action. One could develop multi-insecticides as follows:

First, establish various R strains by successive selection with a particular chemical, using various chemicals that differ in their insecticidal mode of action. Next, make a thorough examination for synergistic action against the established R strains using a mixture of two or more chemicals. Although the establishment of R strains is time-consuming, this approach is worth pursuing in order to help solve the problem of insecticide resistance.

Because of the diversity of modes of resistance to insecticides, one would naturally expect that resistance to multi-insecticides would develop after successive applications. To cope fundamentally with the problem of insecticide resistance, every material and technique must be integrated, taking full advantage of the ecological, physiological, genetic and biochemical nature of the insect.

REFERENCES

Hama, H., 1975, Toxicity and antiacetylcholinesterase activity of propaphos against the resistant green rice leafhopper, *Nephotettix cincticeps* Uhler, *Botyu-Kagaku,* 40:14.

Hama, H., and Iwata, T., 1972, Sensitive ali-esterase to a carbamate insecticide, propoxur, in the resistant strains of the green rice leafhopper, *App. Ent. Zool.,* 7:177.

Hama, H., and Iwata, T., 1973, Synergism of carbamate and organophosphorus insecticides against insecticide-resistant green rice leafhopper, *Nephotettix cincticeps* Uhler, *Jap. J. Appl. Ent. Zool.,* 17:181.

Hosoda, A., and Fujiwara, A., 1977, Changes of susceptibility in the carbamate insecticide-resistant green rice leafhopper, *Nephotettix cincticeps* Uhler, during the continuous selection with diazinon, propaphos and propaphos:NAC mixture, *Bull. Hiroshima Pre. Agr. Exp. Sta.,* 39:21.

Japan Plant Protection Association, 1975-1978, Committee for the countermeasure to insecticide resistance, Studies on the green rice leafhopper, pp. 25, 27-29.

Kimura, Y., 1965, Resistance to malathion in the smaller brown planthopper, *Laodelphax striatellus* Fallén, *Jap. J. Appl. Ent.*

Zool., 9:251.

Metcalf, R. L., 1967, Mode of action of insecticide synergists, *Ann. Rev. Ent.*, 12:229.

Miyata, T., Sakai, H., Saito, T., Yoshioka, K., Tsuboi, A., Ozaki, K., and Sasaki, Y., 1979, Synergistic action of malathion and Kitazin P in green rice leafhoppers and mice, *Ann. Meet. Jap. Soc. Appl. Ent. Zool.*, (abst.).

Moorefield, H. H., 1960, Insect resistance to the carbamate insecticides, *Misc. Publ. Ent. Soc. Amer.*, p. 145-152.

Okuma, M., and Ozaki, K., 1969, Development of resistance to sumithion and malathion in the smaller brown planthopper, *Laodelphax striatellus* Fallèn, *Proc. Assoc. Pl. Prot. Shikoku*, 4:45.

Ozaki, K., 1962, Resistance to parathion in the rice stem borer, *Chilo suppressalis* Walker, *Botyu-Kagaku*, 27:81.

Ozaki, K., 1966, Some notes on the resistance to malathion and methyl parathion of the green rice leafhopper, *Nephotettix cincticeps* Uhler, *Appl. Ent. Zool.*, 1:189.

Ozaki, K., 1969, Resistant insect pests of rice plant and countermeasures for their control, *Agr. and Hort.*, 44:213.

Ozaki, K., and Kassai, T., 1970, Biochemical genetics of malathion resistance in the smaller brown planthopper, *Laodelphax striatellus*, *Ent. Exp. & Appl.*, 13:162.

Ozaki, K., Sasaki, Y., and Ueda, M., 1973a, The development of resistance to carbamate insecticides in the smaller brown planthopper, *Laodelphax striatellus* Fallèn, *Botyu-Kagaku*, 38:216.

Ozaki, K., Sasaki, Y., Ueda, M., and Kassai, T., 1973b, Results of the alternate selection with two insecticides and the continuous selection with mixtures of two or three ones of *Laodelphax striatellus* Fallèn, *Botyu-Kagaku*, 38:222.

Sasaki, Y., and Ozaki, K., 1976, Evaluation of mixtures of two insecticides for control of the susceptible, malathion- and fenitrothion-resistant strains of smaller brown planthopper, *Laodelphax striatellus* Fallèn, *Botyu-Kagaku*, 41:177.

Tanaka, H., Kitakata, S., Kojima, K., Umeda, K., and Kitagaki, Y., 1967, Cross resistance in malathion-resistant green rice leafhopper, *Ann. Meet. Jap. Soc. Appl. Ent. Zool.*, (abst.).

Yokoyama, Y., and Ozaki, K., 1968, Resistance of the smaller brown planthopper to organophosphorus insecticides in Kagawa Prefecture, *Proc. Assoc. Pl. Prot. Shikoku*, 3:35.

Yoshioka, K., 1976, Insecticidal effectiveness of various pesticidal compounds to diazinon-resistant green rice leafhopper, *Ann. Meet. Jap. Soc. Appl. Ent. Zool.*, (abst.).

Yoshioka, K., Matsumoto, M., Bekku, I., and Kanamori, S., 1975, Synergism of IBP and insecticides against insecticide resistant green rice leafhoppers, *Proc. Assoc. Pl. Prot. Shikoku*, 10:49.

INSECT GROWTH REGULATORS:

RESISTANCE AND THE FUTURE

Thomas C. Sparks and Bruce D. Hammock*

Department of Entomology
Louisiana Agricultural Experiment Station
Louisiana State University
Baton Rouge, Louisiana 70803

INTRODUCTION

In 1967 Carroll Williams summarized several years of campaigning for a revolution in pesticide chemistry with the publication of an article entitled "Third Generation Pesticides." His lectures on the golden oil of *Cecropia* helped to inspire the identification of the first juvenile hormone (JH) structure (Röller et al., 1976) as well as the synthesis of potent and relatively inexpensive JH mimics (Law et al., 1966; Bowers, 1968, 1969). Among the many potential advantages of the third generation pesticides was one predicted in the statement, "Now insect hormones promise to provide insecticides that are not only more specific but also proof against the evolution of resistance" (Williams, 1967). This statement was seriously questioned (Ellis, 1968; Schneiderman, 1971, 1972), and shortly thereafter it was shown to be invalid by the publication of two papers demonstrating that strains resistant (R) to pesticides could be cross-resistant to JHs (Dyte, 1972) and their mimics (Cerf and Georghiou, 1972). That resistance could develop should come as no surprise if one considers that: (a) an insect has the ability to regulate its own hormones by metabolism, and (b) exogenous molecules are usually treated as foreign compounds (xenobiotic) and therefore are subject to the normal inactivation processes often associated with resistance to conventional pesticides. This chapter, in part, catalogs a series of failures in which insects of many orders were shown either to be

*Division of Toxicology and Physiology, Department of Entomology, University of California, Riverside, California 92521.

Table 1. Cross Resistance to Insect Growth Regulators (IGR)

Species	IGR	Insecticide-resistant strain	Level of cross resistance[a]	Reference
		JUVENOIDS		
DIPTERA				
Ae. aegypti	methoprene	DDT	+	Rongsriyam and Busvine, 1975
	R-20458	DDT	+	
	Mon-0585	DDT	+	
Ae. taeniorhynchus	methoprene	malathion	0	Dame et al., 1976
An. gambiae	methoprene	DDT/dieldrin	+++	Kadri, 1975
		dieldrin	++	
		DDT	±	
	methoprene	DDT	+	Rongsriyam and Busvine, 1975
	R-20458	DDT	+	
	Mon-0585	DDT	+	
An. quadrimaculatus	methoprene	DDT	+	Rongsriyam and Busvine, 1975
	methoprene	DDT	0	Dame et al., 1976
	R-20458	DDT	+	Ronsriyam and Busvine, 1975
	Mon-9585	DDT	+	

Cx. quinquefasciatus	methoprene	OP-R[b]	+	Georghiou et al., 1975
Cx. fatigans	methoprene	DDT	+	Rongsriyam and Busvine, 1975
	R-20458	DDT	+	
	Mon-0585	DDT	+	
M. domestica	methoprene	dimethoate	+++	Cerf and Georghiou, 1972, 1974a
		OMS-15[c]	+++	
		OMS-12[d]	+++	
		fenthion	++	
		Chlorthion	+	
		SK-R[e]	+	
		parathion	±	
		DDT/lindane	±	
	methoprene	dimethoate	++	Jakob, 1973
		diazinon	+	
		malathion	+	
		DDT	0	
	methoprene	Baygon	+++	Plapp and Vinson, 1973
		R-Fc[f]	++	
	methoprene	O[g]	++	Rupes et al., 1977
		O_{II}[h]	++	
		OS[g]	+	
		H-DDT[i]	0	
		B[g]	−	
		O_v[j]	−	
	methoprene	OP-R[g]	+++	Arevad, 1974
	hydroprene	Baygon	++	Plapp and Vinson, 1973
		R-Fc	±	

(continued)

Table 1. (Cont'd)

Species	IGR	Insecticide-resistant strain	Level of cross resistance[a]	Reference
M. domestica	R-20458	dimethoate	+++	Cerf and Georghiou, 1974a
		OMS-15	+++	
		fenthion	++	
		OMS-12	++	
		DDT/lindane	+	
		Chlorthion	0	
		paration	0	
		SK-R	0	
	R-20458	Baygon	++	Plapp and Vinson, 1973
		R-Fc	+	
	Ro-7-9767	dimethoate	+++	Cerf and Georghiou, 1974a
		OMS-15	+++	
		OMS-12	+++	
		fenthion	+++	
		DDT/lindane	++	
		Chlorthion	+	
		parathion	+	
		SK-R	+	
	ENT-70033	Baygon	+++	Plapp and Vinson, 1973
		R-Fc	++	
	NIA 23509	dimethoate	+++	Cerf and Georghiou, 1974a
		fenthion	+++	
		OMS-15	++	
		parathion	++	

M. domestica	NIA 23509	DDT/lindane	++	Cerf and Georghiou, 1974a
		Chlorthion	+	
		OMS-12	+	
		SK-R	+	
	sesamex	dimethoate	+++	Cerf and Georghiou, 1974a
		fenthion	+++	
		OMS-15	+++	
		OMS-12	+++	
		Chlorthion	++	
		DDT/lindane	±	
		parathion	±	
		SK-R	±	
COLEOPTERA				
T. castaneum	JH I	R^k	+	Dyte, 1972
	9 juvenoids	malathion	0	Amos et al., 1974
	epofenonane	malation	0	Hoppe, 1976
	Ro-20-3600	malathion	±	
	methoprene	malathion	0	Amos et al., 1977
	hydroprene	malation	0	
T. confusum	methoprene	malathion	0	Amos et al., 1977
HOMOPTERA				
T. maculata	hydroprene	malathion	±	Hrdy, 1974
	kinoprene	malathion	–	

(continued)

Table 1 (Cont'd.)

Species	IGR	Insecticide-resistant strain	Level of cross resistance[a]	Reference
T. vaporariorum	kinoprene	DDT/malathion	0	Wardlow et al., 1976
	triprene	DDT/malathion	±	
LEPIDOPTERA				
H. virescens	JH II	CS[l]	±	Benskin and Vinson, 1973
	ENT-70033	CS	-	
S. littoralis	hydroprene	fenitrothion	+	El-Guindy et al., 1975
		aminocarb	0	
		endrin	-	
	R-20458	fenitrothion	+	El-Guindy et al., 1975
		aminocarb	+	
		endrin	+	
	ENT-34070	fenitrothion	++	El-Guindy et al., 1975
		aminocarb	+	
		endrin	+	
P. interpunctella	JH II	malathion (St, Co)[m]	++	Silhacek et al., 1976
		malathion (Se)[n]	+	
		BENZOYLPHENYL UREAS		
DIPTERA				
Ae. aegypti	diflubenzuron	DDT	+	Rongsriyam and Busvine, 1976
	PH 60-38	DDT	+	

Ae. taeniorhynchus	diflubenzuron	malathion	0	Dame et al., 1976
An. quadrimaculatus	diflubenzuron	DDT	0	Dame et al., 1976
	diflubenzuron	DDT	+	Rongsriyam and Busvine, 1975
	PH 60-38	DDT	±	
An. gambiae	diflubenzuron	DDT	+	Rongsriyam and Busvine, 1975
	PH 60-38	DDT	+	
Cx. fatigans	diflubenzuron	DDT	+	Rongsriyam and Busvine, 1975
	PH 60-38	DDT	+	
Cx. quinquefasciatus	diflubenzuron	OP-R[b]	±	Georghiou et al., 1975
M. domestica	diflubenzuron	dimethoate	+++	Cerf and Georghiou, 1974b
		OMS-15	+++	
		OMS-12	+++	
		fenthion	+++	
		DDT/lindane	+++	
		SK-R	+++	
		Chlorthion	++	
		parathion	++	
	diflubenzuron	diazinon (Nic)[o]	++	Oppenoorth and van der Pas, 1977
		malathion	+	
		diazinon (Fc)[o]	+	
		trichlorphon	+	
	diflubenzuron	MRF[p]	+	Keiding, 1977
	diflubenzuron	OMS-12	+++	Pimprikar and Georghiou, 1979
		methoprene	++	

(continued)

Table 1. (Cont.d)

Species	IGR	Insecticide-resistant strain	Level of cross resistance[a]	Reference
COLEOPTERA				
T. castaneum	diflubenzuron	malathion/lindane	-	Carter, 1975
		OP[q]	+	
S. oryzae	diflubenzuron	OP[q]	-	Carter, 1975
HOMOPTERA				
T. vaporariorum	diflubenzuron	DDT/malathion	±	Wardlow et al., 1976
	PH 60-38	DDT/malathion	-	

[a]Cross-resistance levels at the LD_{50}, ED_{50} or EC_{50}; high (>15X) +++; medium (5-15X) ++; low (>2-5X) +; tolerance (>1-2X) ±; none (1X) 0; negative (<1X) -.
[b]Organophosphate multiresistant field strain.
[c]Resistant to *m*-isopropylphenyl methylcarbamate.
[d]Resistant to *O*-ethyl-*O*-(2,4-dichlorophenyl) phosphoramidothioate.
[e]A resistant strain from the field.
[f]Resistant to DDT, carbamates and phosphates.
[g]A field strain resistant to DDT and to organophosphates.
[h]A substrain of O carrying chromosome II.
[i]Resistant to DDT.
[j]A substrain of O carrying chromosome V.
[k]Resistant to DDT, lindane, bromodan, 4 carbamates and 22 organophosphates.
[l]An insecticide-R field strain.
[m]Malathion-R strains; St = Statesboro; Co = Cottonwood.
[n]Malathion-R strain; Se = Sechreist.

[o]Diazinon and other organophsophate-R strains; high oxidase activity.
[p]Multiresistant field population.
[q]A non-specific organophosphorus-R strain.

Table 2. Induced Resistance to Insect Growth Regulators (IGR)

Species	IGR	Number of generations selected	Level of resistance	Reference
		JUVENOIDS		
DIPTERA				
Cx. pipiens	methoprene	8	8	Brown and Brown, 1974
	methoprene	40	218	Brown et al., 1978
	triprene	27	10	
	hydroprene	8	0	
	Ro-20-3600	14	0	
	R-20458	19	0	
Cx. quinquefasciatus	methoprene	20	0	Schaefer and Wilder, 1973
	Mon-0585	20	0	Hsieh et al., 1974
Cx. tarsalis	methoprene	62	86	Georghiou et al., 1974
D. melanogaster	Ro-20-3600	--	1.4-21.5	Arking and Vlach, 1976
M. domestica	methoprene	62	2000	Georghiou et al., 1978
COLEOPTERA				
T. confusum	methoprene	11	4	Brown et al., 1978
	hydroprene	11	7	
	R-20458	11	1.6	
	Ro-20-3600	5	0	

HEMIPTERA				
O. fasciatus	methoprene	11	6	Brown et al., 1978
	kinoprene	9	3	
	R-20458	18	3	
	BENZOYLPHENYL UREAS			
DIPTERA				
Cx. pipiens	diflubenzuron	5	7	Brown et al., 1978
M. domestica	diflubenzuron	10	50	Oppenoorth and van der Pas, 1977
	diflubenzuron	--	1000	Pimprikar and Georghiou, 1979
COLEOPTERA				
T. confusum	diflubenzuron	8	2	Brown et al., 1978

[a]Taken at the LD_{50}, ED_{50} or EC_{50}.

cross resistant or to have developed resistance to hormone mimics (Tables 1 and 2). There is, however, cause for optimism: thus far, resistance to hormone mimics and other insect growth regulators (IGRs) seems to be largely the result of reduced penetration and increased metabolism, while resistance at the site of action has not been demonstrated. Thus, Williams' predictions were correct, in part, because insects have not become truly resistant to their own hormones; they have simply evolved ways to avoid disruption of their development by exogenous hormones and hormone mimics.

Insect Growth Regulators

Some treatment of terminology is necessary before entering into a discussion of resistance to IGRs. The age of third-generation pesticides was initiated by mimics or analogs of insect JH (JHM or JHA). With mimics of the natural products pyrethrum and rotenone having been named "pyrethroids" and "rotenoids," respectively, Hideo Kamimura (1972) proposed the term "juvenoids" for JHMs. Since the implication of endocrine involvement might have led to regulatory problems, the term "insect growth regulator" was subsequently proposed (Stall, 1975). The strength and weakness of the term "IGR" lies in its analogy with the very successful plant growth regulators (PGR) (Lieberman, 1977). On strict analogy with PGR, IGR implies a group of compounds that stimulate the favorable regulation of insect growth, leading to large yields of high-quality insects. Due to this favorable connotation of IGRs, many compounds that act on systems other than the nervous system are now included, and the definition may become still broader in the future. The compounds in field use that currently fall under the IGR umbrella include the juvenoids and chitin synthesis inhibitors, both of which will be discussed in this chapter.

Extrapolation of the Third-generation Concept

An extension of Williams' definition of third-generation pesticides facilitates the use of this concept to explain current trends in pesticide chemistry. The first generation of pesticides might be characterized as those chemical means of pest control arising from many decades or even centuries of a trial-and-error approach to determining remedies. This group includes such chemicals as the arsenicals, oils and botanicals. The second generation of pesticides includes the organochlorines, carbamates and organophosphates, and arose from a vigorous synthesis program initiated largely by the discovery of DDT (Ordish, 1967). Williams (1967) seems to have defined third-generation pesticides as those based on insect hormones. If this procedure is extended every time a new mode of action is considered, one could label the resulting compounds as fourth, fifth or sixth-generation pesticides (Bowers, 1977a). A more useful concept of third-generation pesticides can be obtained by a careful examination of Williams' (1967) article and a closer analogy with Ordish's (1967) definition of second-generation pesticides. Rather than define the third generation of pesticides as an end product, the term could be used to define a process leading to bioactive compounds. Thus, a second-generation approach to pesticide development involves the process of screening and bioactivity-directed synthesis. A third-generation approach involves an appreciation of the physiology and ecology of target and nontarget species which directs a synthetic program to produce environmentally acceptable chemicals.

Most pesticides will obviously be hybrids resulting from the second and the third-generation approaches. The juvenoids are clearly our closest approximation to the result of the third-generation approach, but much of the structure optimization of juvenoids was generated by a second-generation approach (Siddall, 1976). Other IGRs, such as the anti-juvenile hormones (anti-JHs) and the benzoylphenyl ureas, were discovered mostly by a second-generation approach that used innovative screening methods. Among the best examples resulting from the third-generation approach are the biodegradable DDT mimics and the carbamates and organophosphates derivatized for increased safety and selectivity (Fukuto and Mallipudi, this volume). Although the parent compounds were obtained by a second-generation approach, their subsequent modification clearly involved a third-generation approach.

RESISTANCE TO INSECT GROWTH REGULATORS

Cross Resistance

Juvenoid Cross Resistance. The phenomenon of cross resistance to any IGR has been documented for 13 insect species from four orders (Table 1). Thus, IGR cross resistance is not an isolated event. The first report of such cross resistance was that of Dyte (1972) in which a highly insecticide-R strain of *Tribolium castaneum* (Herbst) displayed a 3X resistance toward JH I (Figure 1). Subsequent studies with malathion-R strains of *T. confusum* Jacquelin duVal and *T. castaneum*, however, have failed to detect cross resistance to several juvenoids, including methoprene, hydroprene and epofenonane (Figure 1) (Amos et al., 1974, 1977; Hoppe, 1976), although a very slight tolerance to Ro-20-3600 (Figure 1) was noted in *T. castaneum* (Hoppe, 1976).

An organophosphate-R strain of *Aedes nigromaculis* (Ludlow) was reported to have only a "vigor tolerance" to methoprene (Schaefer and Wilder, 1972), while dieldrin-R, dieldrin/DDT-R, and to a lesser extent, DDT-R strains of the *Anopheles gambiae* complex exhibited varying levels of cross resistance to methoprene (Kadri, 1975) (Table 1). DDT-R strains of the mosquitoes *Aedes aegypti* (L.), *An. quadrimaculatus* Say, *An. gambiae*, and *Culex fatigans* Weid. all exhibited either no resistance to methoprene (Dame et al., 1976) or only very low resistance (Rongsriyam and Busvine, 1975) to methoprene,

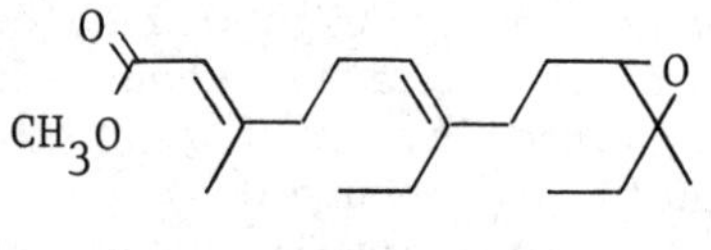

JH I

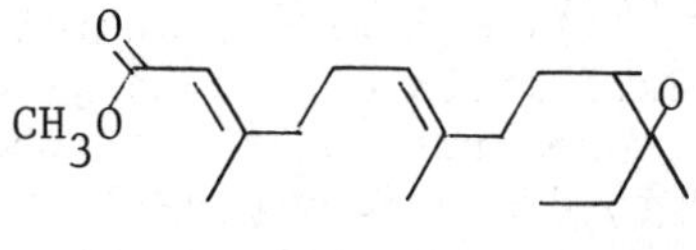

JH II

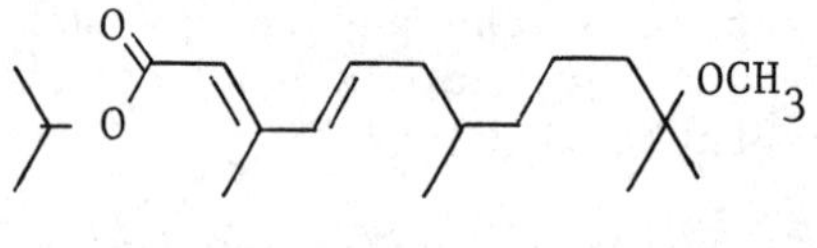

methoprene

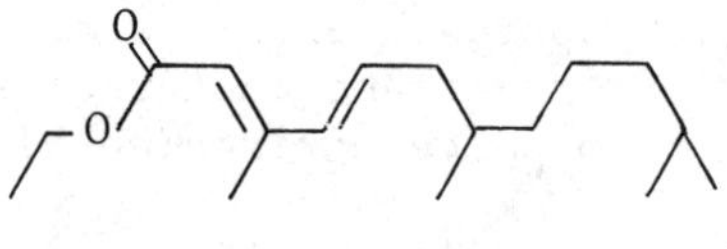

hydroprene

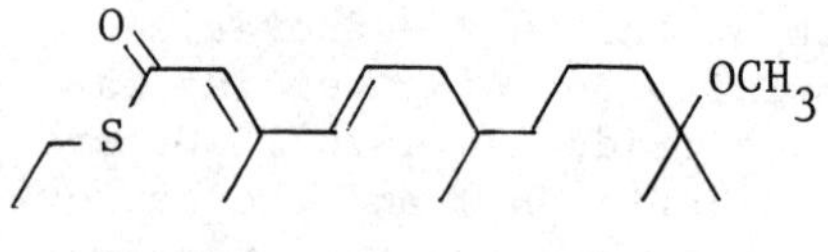

triprene

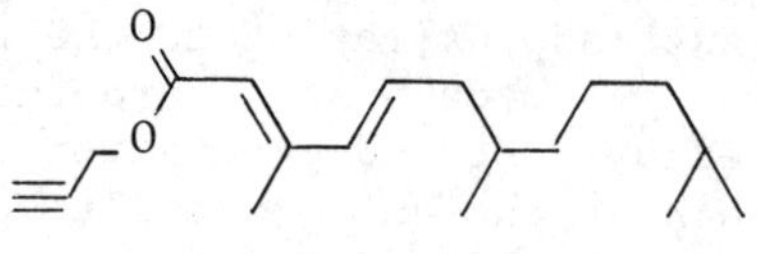

kinoprene

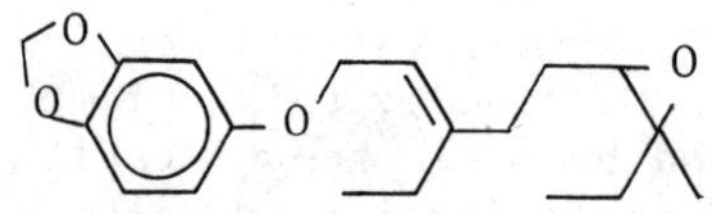

Ro-7-9767

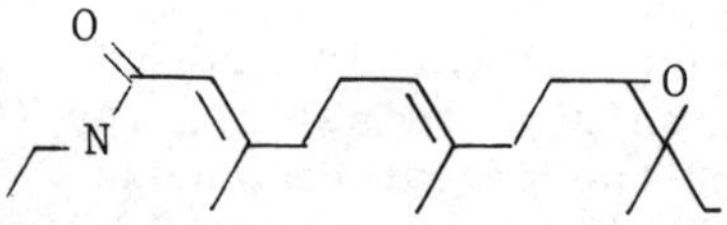

NIA 23509

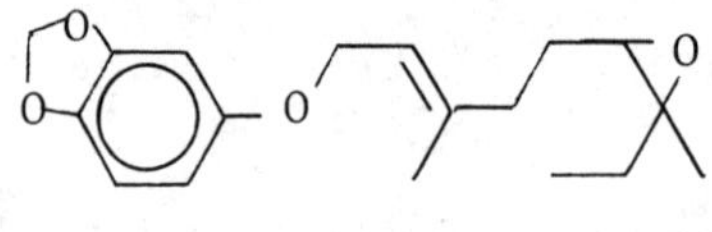

Ro-20-3600

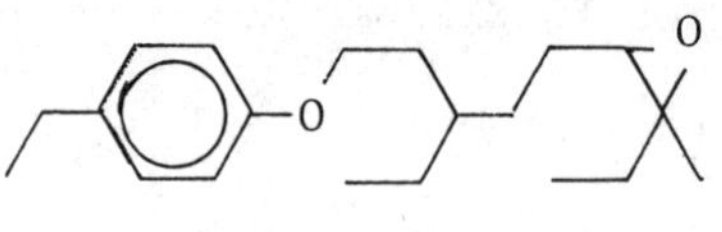

epofenonane

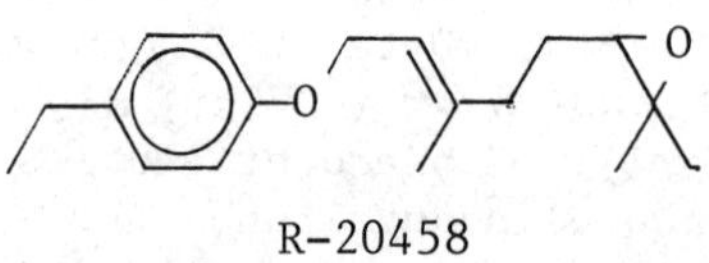

R-20458

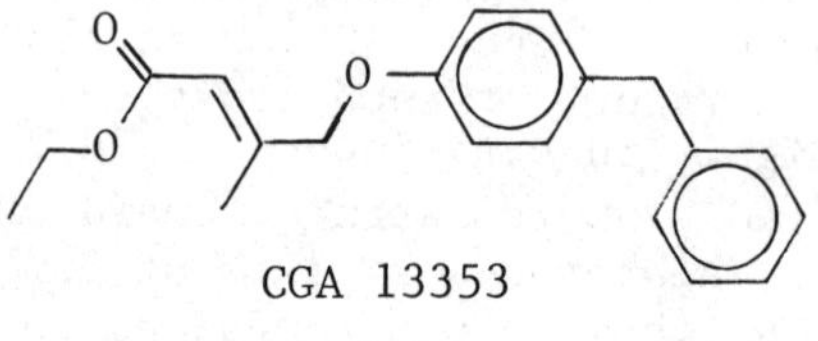

CGA 13353

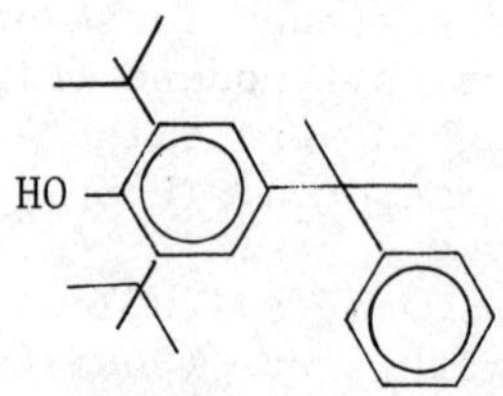

MON-0585

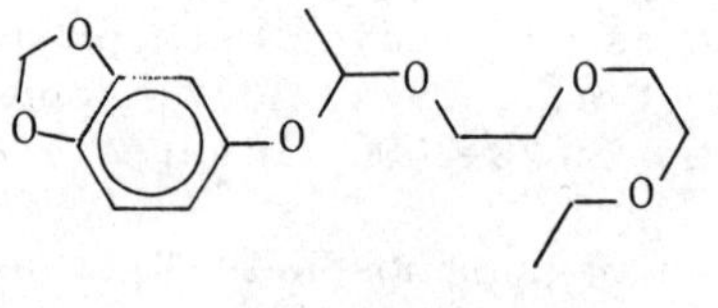

sesamex

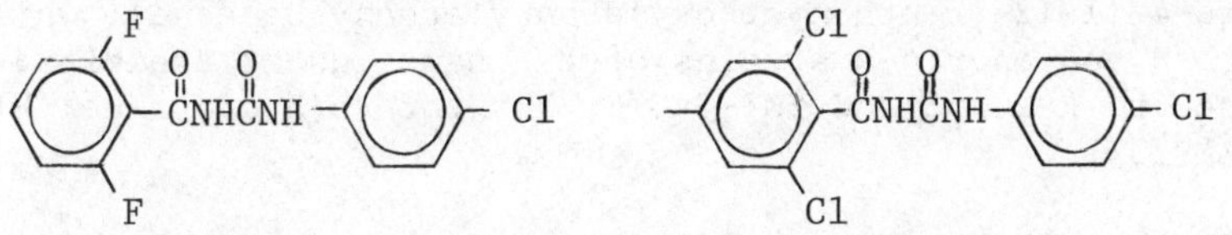

diflubenzuron PH 60-38

Figure 1. Structures of juvenile hormones and IGRs involved in resistance and cross resistance studies.

R-20458 and to the di-*t*-butylphenol, Mon-0585 (classed as a juvenoid for convenience) (Figure 1). Similarly, a malathion-R strain and a field strain of *Aedes taeniorhynchus* (Wied.) (Dame et al., 1975) displayed no resistance or only low resistance, respectively, to methoprene, while an OP-R strain of *Culex quinquefasciatus* exhibited a low level of resistance to methoprene (Georghiou et al., 1975).

Several insecticide-R strains of the house fly, *Musca domestica* L., most notably dimethoate-R and OMS-15-R (*m*-isopropylphenyl methylcarbamate), were found to possess cross resistance (39X) to methoprene (Cerf and Georghiou, 1972). Further studies by Cerf and Georghiou (1974a) demonstrated cross resistance to several other juvenoids (Table 1), and again, the dimethoate-R and OMS-12-R strains exhibited the highest cross resistance. Another dimethoate-R strain was also found to possess moderate cross resistance toward methoprene, while other R strains (malathion and diazinon) exhibited a very weak tolerance (Jakob, 1973). A low level (2.5X to 7.6X) of cross resistance to methoprene has also been observed in some strains of *M. domestica* resistant to bromophos-ethyl and fenitrothion (Rupes et al., 1976), while cross resistance to methoprene, hydroprene, R-20458 and ENT-70033 has been detected in two strains possessing resistance to DDT, carbamates and organophosphates (Plapp and Vinson, 1973). A high level of cross resistance to methoprene has also been noted for a field strain of *M. domestica* (Arevad, 1974).

A low level (1.4X) of cross resistance to hydroprene, but not kinoprene (Figure 1), was noted for a malathion-R strain of *Therioaphis maculata* (Buckton) (Hrdy, 1974). Likewise, a DDT/malathion-R strain of *Trialeurodes vaporariorum* (West.) was found to possess a low level of cross resistance to triprene and to kinoprene in the larval stage (Wardlow et al., 1976). An organophosphate-R strain of *Heliothis virescens* (Fab.) displayed a very low level of cross resistance to JH II (Figure 1) and an increased susceptibility to ENT-70033 (Benskin and Vinson, 1973; Vinson and Plapp, 1974), while three malathion-R strains of *Plodia interpunctella* (Hübner) were found to be cross resistant to JH I (Table 1) (Silhacek et al., 1976). Lastly, a fenitrothion-R strain of *Spodoptera littoralis* (Boisd.) displayed cross resistance to hydroprene, R-20458 and ENT-34070

(1-chloro-4-[1-[2-(2-ethoxyethoxy)ethoxy]ethoxy]benzene), while aminocarb-R and endrin-R strains of the same insect exhibited cross resistance to R-20458 and ENT-34070, but not to hydroprene (El-Guindy et al., 1975).

Benzoylphenyl Urea Cross Resistance. Although compounds of quite varied structures are known to inhibit chitin synthesis, only the benzoylphenyl ureas have been intensively studied. Medium to high levels of cross resistance to diflubenzuron (Figure 1) have been noted in several organophosphate-, carbamate-, organochlorine- and IGR-resistant strains of *M. domestica* (Table 1). Multi-R field populations of *M. domestica* have also displayed a low level of resistance (tolerance) to diflubenzuron (Keiding, 1977). Although no cross resistance to diflubenzuron was observed for DDT-R and malathion-R strains of *An. quadrimaculatus* and *Ae. taeniorhynchus*, respectively (Dame et al., 1976), a low level of cross resistance has been noted for an organophosphate-R strain of *Cx. quinquefasciatus* Say (Georghiou et al., 1975). Similarly, cross resistance (low) to diflubenzuron and PH 60-38 (Figure 1) has also been observed for DDT-R strains of *Cx. fatigans, Ae. aegypti, An. gambiae* and *An. quadrimaculatus*.

Organophosphate-R strains of *T. castaneum* and *Sitophilus oryzae* (L.) exhibited no cross resistance to diflubenzuron and were actually more susceptible (1.7X to 2.4X at the EC_{50}) than the non-resistant strain (Carter, 1975). However, a very low level of cross resistance to diflubenzuron was displayed by organophosphate-R and DDT/malathion-R strains of *T. castaneum* (Carter, 1975) and *Trialeurodes vaporariorum* (Wardlow et al., 1976), respectively, while the R strain of *T. vaporariorum* was found to be more susceptible to PH 60-38 during most of its larval and pupal development than the non-resistant strain (Wardlow et al., 1976).

Induced Resistance

Induced Resistance to Juvenoids. Initial attempts at laboratory selection of *Cx. quinquefasciatus* larvae with methoprene for 20 generations (Schaefer and Wilder, 1973) or Mon-0585 (Hsieh et al., 1974) failed to induce other than a transitory resistance (Table 2). However, a 13X resistance to methoprene was induced in *Cx. pipiens* L. after selection for only eight generations (Brown and Brown, 1974). This selected strain also demonstrated cross resistance to hydroprene (15X) and malathion (1.7X), but not to R-20458 or carbaryl (Table 3). Larvae of *Cx. tarsalis* Coquillet, selected with methoprene, displayed an 86X increase in the LC_{50} as compared to a susceptible laboratory strain (Georghiou et al., 1974). Further studies utilizing *Cx. pipiens* (Brown, 1977; Brown et al., 1978) resulted in the development of resistance when either methoprene or triprene (Figure 1) was used as the selecting agent. Furthermore, the methoprene-selected strain also demonstrated moderate to high levels of cross resistance to

Table 3. Cross Resistance to Other Insect Growth Regulators and Insecticides in IGR-selected Strains

Species	IGR used to select for resistance (level of resistance)	Compound	Level of cross resistance[a]	Reference
Cx. pipiens	methoprene (13X)	hydroprene	+++	Brown and Brown, 1974
		malathion	±	
		carbaryl	0	
		R-20458	–	
	methoprene (218X)	hydroprene	+++	Brown et al., 1978
		triprene	+++	
		CGA-13353	+++	
		R-20458	++	
		Ro-20-3600	+	
		carbaryl	±	
		dieldrin	±	
		temephos	±	
		malathion	0	
		parathion	–	
		DDT	–	
		diflubenzuron	–	
M. domestica	methoprene (1515X) (SK-R)	Ro-7-9767	+++	Georghiou et al., 1978
		R-20458	++	
		NIA 23509	+	
	methoprene (6300X) (dimethoate-R)	dimethoate	+++	Georghiou et al., 1978
		fenthion	+++	
		parathion	+++	

(continued)

Table 3. (Cont'd)

Species	IGR used to select for resistance (level of resistance)	Compound	Level of cross resistance[a]	Reference
M. domestica	methoprene (6300X) (dimethoate-R)	permethrin d-*trans*	+++	Georghiou et al., 1978
		permethrin d-*cis*	+++	
		isolan	++	
	methoprene (6000X)	diflubenzuron	++	Pimprikar and Georghiou, 1979

[a]Level of resistance: high (>15X), +++;
medium (5-15X), ++;
low (>2-5X), +;
tolerance (>1-2X), ±;
none (1X), 0;
negative (<1X), -.

hydroprene, triprene, CGA-13353 (Figure 1) and R-20458, but not to diflubenzuron or five conventional insecticides (Table 3). However, selection of *Cx. pipiens* larvae with hydroprene, Ro-20-3600 or R-20458 for 8, 14 and 19 generations, respectively, failed to result in the development of resistance (Brown et al., 1978).

After selection with methoprene for 75 generations, a resistant field strain (SK-R) of *M. domestica* possessed a 1515X level of resistance, while a dimethoate-R strain exhibited a 6300X resistance after 62 generations. The methoprene-selected SK-R strain also displayed cross resistance to several other juvenoids, including Ro-7-9767, R-20458 and NIA 23509 (Figure 1) (Georghiou et al., 1978). Concurrent with the observed increase in methoprene resistance, the methoprene-selected dimethoate-R strain exhibited a decrease in cross resistance to several carbamates and organophosphates, but a slightly increased tolerance was noted toward the pyrethroid permethrin. Selection of *Drosophila melanogaster* with Ro-20-3600 resulted in several strains that displayed resistance (1.4-21.5X) (Arking and Vlach, 1976).

Selection of *T. confusum* larvae for 11 generations with methoprene, hydroprene, R-20458 or Ro-20-3600 resulted in increased LC_{50} values by 4X, 7X, 1.6X and 0X, respectively (Brown, 1977; Brown et al., 1978). Similarly, *Oncopeltus fasciatus* Dallas nymphs selected with methoprene, kinoprene or R-20458 for 8, 9 and 18 generations resulted in 11X, 3X or 3X resistance, respectively (Brown, 1977; Brown et al., 1978).

Induced Resistance to Benzoylphenyl Ureas. Although prolonged selection of *Cx. tarsalis* with diflubenzuron failed to induce resistance (Georghiou, 1979), resistance (7X) to this compound was obtained in *Cx. pipiens* after only five generations (Brown et al., 1978) (Table 2). Very high levels of resistance to diflubenzuron have been induced in *M. domestica* (Oppenoorth and van der Pas, 1977; Pimprikar and Georghiou, 1979), and 2X resistance was observed in *T. confusum* after selection for eight generations (Brown et al., 1978).

JUVENILE HORMONE AND INSECT GROWTH REGULATOR METABOLISM

Metabolism of JH and Juvenoids

An overview of insect metabolism of JH and juvenoids must be presented before one can appreciate the implication of these metabolic pathways to resistance and the synthesis of improved compounds. It is not the purpose of this article to provide a comprehensive review of the subject; a review of JH and juvenoid metabolism by Hammock and Quistad (1976) covers the literature prior to 1975, and subsequent work on JH and juvenoids will be covered in reviews being prepared by deKort (1980) and Hammock and Quistad (1980).

Metabolism of Juvenile Hormone. Based on the work of Slade and Zibitt (1971, 1972) and two surveys that appeared shortly thereafter (White, 1972; Ajami and Riddiford, 1973), it has been generally accepted that epoxide hydration and ester cleavage are the major, primary routes of JH metabolism (Figure 2). The relative importance of these pathways in the insects examined is variable, with ester cleavage apparently of greatest importance in the Lepidoptera and epoxide hydration of major importance in many Diptera. For instance, the acid-diol is a major metabolite, but the diol is difficult to detect as a JH metabolite in *Manduca sexta* (L.) either in vivo (Slade and Zibitt, 1972) or in vitro (Hammock et al., 1976a). The diol is a major in vivo metabolite in *Sarcophaga bullata* Parker (Slade and Zibitt, 1972), and both the diol and the acid, but not the acid-diol, are important in vitro metabolites in *M. domestica* (Hammock et al., 1977a; Yu and Terriere, 1978a). In contrast, both epoxide hydration and ester cleavage are important in *Leptinotarsa decemlineata* (Say) (Kramer et al., 1977), *Tenebrio molitor* L. (Weirich and Wren, 1976), *Aphis mellifera* L. (Mane and Rembold, 1977) and *Locusta migratoria* L. (Erley et al., 1975).

Conjugation is undoubtedly an important route of JH metabolism, but it has not been well studied in insects and it may not be important in the regulation of JH titer. There is evidence that conjugation of JH diol with sulfate and glucose occurs in some insects (Slade and Zibitt, 1972; White, 1972; Slade and Wilkinson, 1974), and rat hepatocytes apparently form sulfate conjugates of the diol as well as both ether and ester glucoronides (Morello and Agosin, 1979). Based on our present knowledge of glutathione (GSH) transferase reactions, which comes mainly from mammalian studies (Chasseaud, 1974; Dauterman, this volume), JH is not a molecule that one would expect to be highly susceptible to GSH conjugation. This epoxide is not a highly reactive electrophile, and it is sterically hindered. Although significant GSH conjugation at the epoxide of JH I could not be demonstrated in vitro in S or R strains of *M. domestica* larvae or adults (Hammock, unpublished), the report by Morello and Agosin (1979) of high GSH transferase activity in liver homogenates, with JH I as the substrate and a mercapturic acid metabolite of JH I as a major metabolite of rat hepatocytes, suggests that insects should be closely examined for the presence of this metabolic pathway, especially since GSH has been implicated in some resistance mechanisms. The conjugated 2*E* olefin of JH (Figure 2) is susceptible to attack by biological nucleophiles, and there is a slow chemical reaction between JH and GSH at 37.5°C. GSH-S-alkene transferases are normally not active on sterically hindered substrates, so it was not surprising that neither *M. domestica* nor mouse liver enzyme preparations greatly accelerated the process (Hammock, unpublished). More recent work on JH metabolism has dealt largely with describing the primary routes of metabolism in insects not previously investigated, correlating JH metabolism with physiological changes, and investigating the

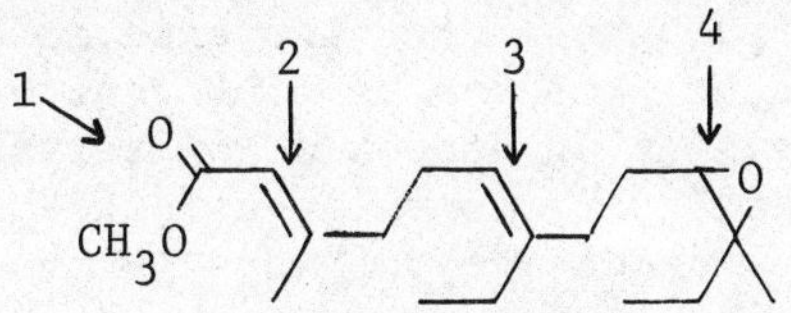

JH I

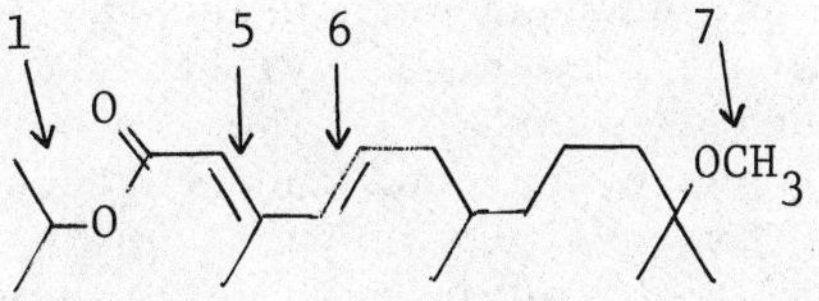

Methroprene

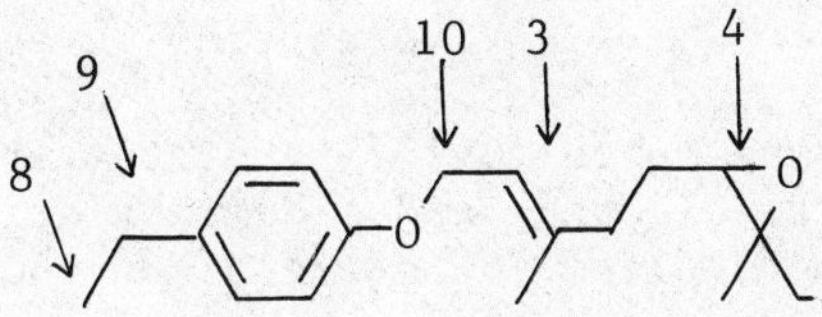

R-20458

Figure 2. Structures of JH I, methoprene and R-20458 showing sites of metabolism in insects. Ester cleavage at (1) for JH I and methoprene yields the acid. Epoxide hydration at (4) for JH I and R-20458 yields the diol. The conjugated 2*E*-olefin of JH I (2) may be subject to attack by nucleophiles. The 6,7 and 2,3-olefin of JH I and R-20458, respectively (3), can undergo epoxidation and oxidation. The 2,3-olefin of methoprene can be isomerized from 2*E* to 2*Z* (5) while the 4,5-olefin (6) can be oxidatively cleared to yield citronellic acid and citronellal derivatives. The methoxy group (7) can be oxidatively *O*-demethylated. R-20458 can undergo oxidation at the Omega and benzylic (9,10) positions.

regulation of JH metabolism (Kramer, 1978; Sparks and Hammock, 1979a; McCaleb and Kumaran, 1979).

The existence of oxidative pathways of JH metabolism was suggested by Ajami and Riddiford in 1973, but one would expect that the lipophilic JH molecule would be subject to allylic hydroxylation, 6,7-epoxidation and a host of other oxidase-mediated pathways. In general, oxidative routes of metabolism appear to be minor when compared to hydrolytic pathways. For instance, a close examination of JH metabolism by tissue homogenates of *M. sexta* failed to demonstrate significant levels of JH diepoxide or tetrahydrofuran diols, but as will be discussed later, some insecticide-R house fly strains rapidly metabolize JH by oxidative pathways (Hammock et al., 1977a; Yu and Terriere, 1978a).

Metabolism of Juvenoids. The metabolic pathways of only two groups of juvenoids have been examined in insects. One group includes the dienoate juvenoids, such as methoprene and hydroprene, while the other group includes the geranyl phenyl ether derivatives, such as R-20458 and epofenonane (Figure 2). The dienoate juvenoids are susceptible to ester cleavage, but the conjugated dienoate ester is much more stable than the ester of the natural JHs. Although ester cleavage has been reported to be an important metabolic pathway in several dipterous species for several dienoate ethyl, propynyl, and thioethyl esters (Yu and Terriere, 1975, 1977; Terriere and Yu, 1977), the isopropyl ester of methoprene has generally been found to be relatively resistant to esterases. Some in vivo ester cleavage has been reported in *T. molitor, O. fasciatus, M. domestica, Ae. aegypti* and *Cx. quinquefasciatus* (Solomon and Metcalf, 1974; Quistad et al., 1975). Surprisingly, both ester cleavage and *O*-demethylation were major routes of methoprene metabolism in *Solenopsis invicata* Buren (Bigley and Vinson, 1979a,b).

Oxidative pathways are also important in the metabolism of dienoate juvenoids. Solomon and Metcalf (1974) demonstrated that $^{14}C5$-methoprene was converted to $^{14}CO_2$ in *T. molitor* and that unknown metabolites were produced in vivo in both *T. molitor* and *O. fasciatus*, presumably as a result of oxidative pathways. Quistad et al. (1975) demonstrated the presence of methoxy and hydroxy citronellic acid and methoxy citronellal in *M. domestica, Ae. aegypti* and *Cx. quinquefasciatus*. These products presumably arose through oxidative cleavage of the terpene chain. A hydroxypyrone photoproduct of hydroprene has been identified (Hendrick et al., 1975) and has chromatographic properties similar to those of hydroprene acid. Such a metabolite might result from metabolism of methoprene via a peroxidase pathway, which could also result in backbone cleavage to yield methoxycitronellal. Oxidative *O*-demethylation has been shown to be an important pathway of methoprene metabolism in all insects examined (Solomon and Metcalf, 1974; Quistad et al., 1975; Hammock et al., 1977a; Brown and Hooper, 1979). However, some insects have very low mixed-function oxidase (MFO) activity, causing methoprene to be almost inert to metabolism, a fact that is probably more significant to insect biochemists than to field entomologists. The presence of other oxidative pathways, such as epoxidation, has also been demonstrated with hydroprene in three species of Diptera (Yu and Terriere, 1977) and with methoprene in *M. domestica* (Hammock et al., 1977a).

Table 4. Effect of Terminal Functionality on Juvenoid Resistance in *Musca domestica*

Compound	LD_{50} ng/larvae		Resistance factor (D/N)
	Dimethoate	NAIDM	
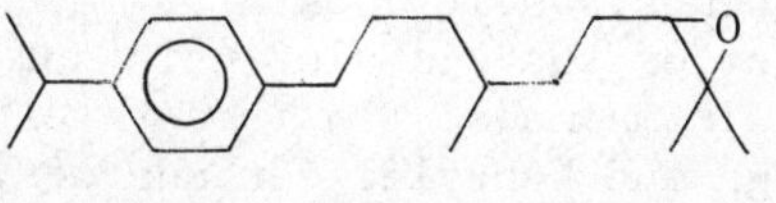	0.9	0.11	8.2
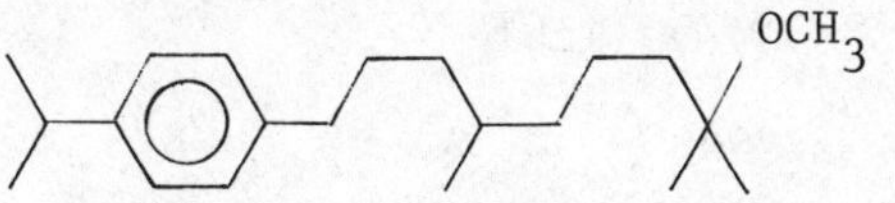	50	1.5	33
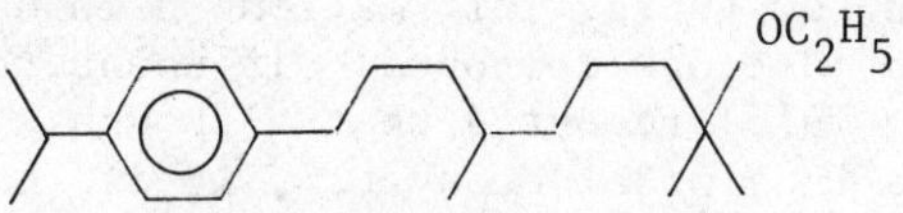	75	56	1.3
Methoprene	20	0.11	99

Arylterpenoid compounds are designated as AI3-36241, AI3-36206 (MV-678), and AI3-36093, respectively. LD_{50} values were determined by topically treating white larvae with the respective compounds and scoring as dead any insects that failed to emerge from their puparium.

Quistad et al. (1975) reported isomerization of 2*E* methoprene to 2*Z* methoprene in house flies. The existence of isomerized methoprene and hydroxymethoprene following application to *M. domestica* was confirmed by Hammock et al. (1977a); however, Yu and Terriere (1977) could not detect analogous isomerization of hydroprene. Since thiophenol-catalyzed isomerization of the 2,3-olefin is important in the synthesis of methoprene (Siddall, 1976), one might expect that such in vivo isomerization was due to the action of a biological nucleophile such as GSH. Incubation of 2*E* methoprene with GSH does result in a slow increase in the proportion of 2*Z* methoprene, but this reaction is not significantly enhanced by house fly enzymes. Although GSH reacts much more rapidly with methoprene than with JH, this Michael-like addition does not appear to be an important metabolic pathway, and, because GSH should isomerize both the 2*Z* and 2*E* olefins, this pathway does not totally explain Quistad's observation that 2*E* methoprene is converted to 2*Z* but that 2*Z* is not converted to 2*E*.

Epoxide hydration, as in the natural hormone, is an important route of metabolism for geranyl phenyl ether juvenoids such as R-20458 (Gill et al., 1972; Hammock et al., 1974a). When the epoxide is replaced by an alkoxide, *O*-dealkylation occurs, as it does with methoprene (Hammock et al., 1975b). Oxidation of the 2,3-olefin is not as important a pathway in insects as it is in mammals. In insects with high MFO activity, oxidation on the omega and benzylic carbons of the ethyl side chain are important. One would expect that steric hindrance would reduce benzylic oxidation on arylterpenoid compounds such as those developed by Schwarz et al. (1974) (Table 4).

Benzoylphenyl Urea Metabolism

Of the numerous benzoylphenyl urea insecticides that may be of commercial interest within the next few years, information on insect metabolism has emerged primarily on just one compound: diflubenzuron (Figures 1 and 3). Metcalf et al. (1975) reported that although diflubenzuron was moderately stable in a model ecosystem, it did not appear to be highly concentrated through the food chain. These workers attributed the degradation largely to the cleavage of the amidecarbonyl bonds. Verloop and Ferrell (1977) explained the relative stability of diflubenzuron and the closely related PH 60-38 (2,6-dichloro analog) on the basis of steric hindrance slowing the

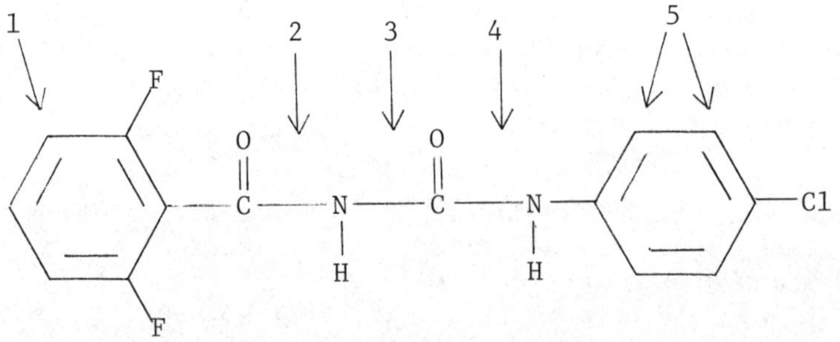

Figure 3. Structure of the benzoylphenyl urea diflubenzuron showing sites of metabolism in insects. Hydroxylation occurs at C3 of the difluorobenzamide ring (1) and at C2 of the *p*-chloroaniline ring (5) to yield phenols which may be conjugated. Cleavage at 2 yields 2,6-difluorobenzoic acid and *p*-chlorophenyl urea, cleavage at 3 yields 2,6-difluorobenzamide and cleavage at 4 yields *p*-chloroaniline. These primary metabolites may then be further metabolized.

formation of 2,6-dihalobenzoic acids from both PH 60-38 and diflubenzuron. Based on results with larvae of *Cx. quinquefasciatus* and *Estigmene acrea* Drury, Metcalf et al. (1975) concluded that diflubenzuron was refractory to metabolism in these insects because no metabolites of diflubenzuron could be found. Such metabolites were also absent from the larvae of *Pieris brassicae* L., *E. acrea* and *Anthonomus grandis grandis* Boheman (Verloop and Ferrell, 1977), further demonstrating the stability of diflubenzuron. However, in contrast to the other lepidopterous insects examined, Bull and Ivie (unpublished information) have found that late-instar *Heliothis* larvae metabolize diflubenzuron very rapidly, although interestingly, *Heliothis* is insensitive to the toxic effects of this compound. Still and Leopold (1975) found only unmetabolized diflubenzuron in the carcass and frass of *A. grandis grandis* either injected with or dipped into diflubenzuron solutions. Upon further examination of diflubenzuron metabolism in this species, Chang and Stokes (1979) confirmed that this chemical accounted for almost all of the radioactivity found in the intact insect four days following injection. In contrast to previous reports, these workers found that of the metabolites in the frass the great majority were apparently conjugates of diflubenzuron hydroxylated in either the 3 position of the difluorobenzamide ring or the 2 position of the chloroaniline ring. The fact that the conjugates were found without detecting the parent phenols in either the carcass or the frass indicates that conjugation must occur rapidly following hydroxylation. A further examination of the fate of diflubenzuron and three of its derivatives (*N*-methyl, 2-methoxy-*N*-methyl, 3-methoxy-*N*-methyl) in adult female *A. grandis grandis* (Bull and Ivie, 1980) indicated that diflubenzuron undergoes a limited amount of conjugation and metabolism. The diflubenzuron derivatives were much more rapidly and extensively metabolized than diflubenzuron and among the four compounds, the *N*-methyl derivative was most rapidly absorbed, metabolized and excreted, and diflubenzuron was acted upon least rapidly. Chang and Woods (1979a) also examined the metabolism of penfluron (2,6-difluoro-*N*-[[[4-(trifluoromethyl) phenyl] amino] carbonyl] benzamide) injected into adult *A. grandis grandis*. Although conjugates were present, unchanged penfluron accounted for virtually all of the injected radioactivity, and penfluron is about twice as stable as diflubenzuron in this species.

Three studies have dealt with the fate of diflubenzuron in flies. Ivie and Wright (1978) explored the phenomenon of reduced egg hatchability after treatment of adult *M. domestica* and *Stomoxys calcitrans* L. Following topical application of 2 μg of diflubenzuron to either males or females, unmetabolized diflubenzuron was found in the eggs of both *M. domestica* and *S. calcitrans*, supporting the hypothesis that diflubenzuron is an ovicidal agent in these insects rather than a chemosterilant. Diflubenzuron was found to be considerably more metabolically stable in *S. calcitrans* than in *M. domestica*: in *M. domestica*, the great majority of the material that was excreted, abraded, washed from the surface of the insect, or found in the carcass was unchanged parent material. The major route of metabolism to nonpolar metarials involved cleavage between the carbonyl and amide bonds, yielding trace levels of 2,6-difluorobenzamide, 4-chlorophenyl urea, and 2,6-difluorobenzoic acid. The expected 4-chloroaniline was probably rapidly acetylated since only 4-chloroacetanilide was found. Chang (1978) injected 10 μg of diflubenzuron into adult female *M. domestica* and found none of the metabolites listed above that would result from cleavage of the carbonylamide bond, and in contrast to the work of Ivie and Wright (1978), he reported that approximately 21% of the diflubenzuron present was in the form of polar metabolites in the excreta. Acid hydrolysis resulted in the identification of diflubenzuron hydroxylated in the 3 position of the difluorobenzamide ring, as was the case with *A. grandis grandis*. The discrepancies between these two studies in adult flies can probably be explained on the basis of strain, dose, route of administration, and analytical procedure.

Pimprikar and Georghiou (1979) examined the fate of low doses (10-50 ng) of diflubenzuron when topically applied to mature larvae of three strains of *M. domestica*. In contrast to the results of Ivie and Wright (1978) in which no hydroxylated metabolites were observed after large dosages of diflubenzuron were topically applied to adult flies, *M. domestica* larvae yielded hydroxylated metabolites identified as having phenols on either the difluorobenzamide or chloroanile rings. Larvae seemed to excrete polar "conjugates" less rapidly than adults. Enzymatic cleavage of these conjugates yielded largely 2,6-difluorobenzoic acid and 2,6-difluorobenzamide. The larvae contained much less parent diflubenzuron than the adult *M. domestica* (Ivie and Wright, 1978; Chang, 1978), which may be the result of enhanced abrasion due to the larval media, reduced penetration, dose effects, and/or more rapid metabolism. The work of these three laboratories supports the conclusion that diflubenzuron is only slowly metabolized in adult and larval *M. domestica*, and that several matabolic pathways are involved. Penfluron metabolism studies in adult female *M. domestica* (Chang and Woods, 1979b) demonstrated that this compound is much more stable (8X) than diflubenzuron. All of the recovered material in the excreta was apparently in the form of conjugates of phenolic penfluron metabolites.

It is apparent that a great deal more work is needed on the metabolism of benzoylphenyl urea insecticides in insects. Very little is known of the actual enzymes involved in their degradation and it is hoped that future studies will provide an indication of the role of metabolism in structure-activity relationships and species specificity of these interesting compounds.

MECHANISMS OF RESISTANCE TO INSECT GROWTH REGULATORS

Cross resistance to IGRs appears to be highest in those insecticide-R strains that possess high oxidase activity (Table 1). This trend seems to be especially applicable to *M. domestica* (Cerf and Georghiou, 1974; Plapp and Vinson, 1973; Vinson and Plapp, 1974; Rupes et al., 1976; Hammock et al., 1977a). IGR resistance often correlates with the presence of genes on chromosome II for high oxidase activity and does not appear to be associated with DDT-dehydrochlorinase activity, cyclodiene resistance genes, or Kdr (Plapp and Vinson, 1973; Vinson and Plapp, 1974; Rupes et al., 1976).

Resistance Mechanisms to Juvenoids

Three strains of *M. domestica* were used in studies of possible resistance mechanisms (Hammock et al., 1977a; Georghiou et al., 1978). These strains, with their methoprene ED_{50} values at the time of examination, included a susceptible S-NAIDM strain (0.0033 µg/white pupa), a dimethoate-R strain that was cross resistant to methoprene (0.13 µg), and a methoprene-selected strain (17 µg). The primary methoprene metabolite both in vitro and in vivo was found to be the 11-hydroxy compound resulting from *O*-demethylation of methoprene. Although 11-hydroxymethoprene is very active in some insects and *O*-demethylation could even be construed as an activation pathway (Solomon and Metcalf, 1974), it is much less active than methoprene in *M. domestica* (Henrick et al., 1976), and *O*-demethylation clearly leads to reduced biological activity. Whether measured by using model substrates or by methoprene metabolism in vitro, the dimethoate-R and methoprene-R strains exhibited much higher oxidase levels than the S-NAIDM strain. However, there were minor differences between the cross-resistant and resistant strains, and metabolism actually appeared to be a little faster in the dimethoate-R strain. Methoprene and the geranyl phenyl ether R-20458 penetrated more slowly into the methoprene-R strain than into either the dimethoate-R or S-NIADM strain, but there was no significant difference in the methoprene levels in the dimethoate-R and methoprene-R strains. Thus, it is likely that although this study explained most of the cross resistance, the greater resistance of the methoprene-R strain remains a mystery that cannot be totally explained on the basis of metabolism or penetration. There are, however, some possible explanations for the resistance observed: several workers have reported that target tissues for juvenile hormone, such as the imaginal disks, have very high levels of JH esterase and epoxide hydrolase (Chihara et al., 1972; Hammock et al., 1975a; Reddy et al., 1979). Perhaps a strain was selected that could rapidly degrade methoprene in the critical tissues. As both the SK-R and the dimethoate-R strains were selected with methoprene, the subsequent generations demonstrated less steep slopes of the dose-response lines. A reduced slope is often taken as an indication of slower penetration, which is clearly involved in this resistance, but insects are also known to have so-called sensitive periods in their development during which they are very susceptible to juvenoid application. For instance, when 0- to 6-hr-old *T. molitor* pupae were treated with juvenoids, the resulting probit lines were very steep. As more asynchronous populations were treated, the resulting probit lines decreased in slope, and by analogy one may speculate that the sensitive window of the methoprene-selected *M. domestica* has become more narrow. A final possibility is that a JH receptor has been selected that more specifically recognizes JH and excludes methoprene. Hopefully, current research on JH receptors will help to elucidate the mechanism of resistance to methoprene.

The mechanisms of methoprene resistance in *Cx. pipiens* have also been examined. Larvae selected with methoprene for over 30 generations possessed a resistance level of about 200X (Brown and Hooper, 1979). R larvae, examined 24 hr after treatment, had not only absorbed less methoprene, but more of that abosrbed was in the form of polar conjugates and less was present as the parent compound (Brown and Hooper, 1979). Although ester hydrolysis was not eliminated as a possible resistance mechanism, the major methoprene metabolite was found to be the hydroxy-ester, and piperonyl butoxide was found to reduce the resistance ratio from 213 to 136. Thus, oxidative metabolism appeared to play an important role in the ability of resistant *Cx. pipiens* larvae to withstand methoprene. However, such resistance also appears to be associated with faster excretion, poorer distribution in the tissues and, possibly, reduced food intake (Brown and Brown, 1980).

Mechanisms of Resistance to Diflubenzuron

The tremendous differences in the susceptibility of some lepidopterous larvae to diflubenzuron and the variations in sensitivity among instars invite a biochemical approach. To date, the only approach towards elucidating resistance mechanisms has involved the comparison of susceptible (S-NAIDM), cross-resistant (OMS-12-R, methoprene-R) and resistant (diflubenzuron-R) strains of *M. domestica* (Pimprikar and Georghiou, 1979). The methoprene-R strain was discussed earlier, the OMS-12-R strain had been under pressure with *O*-ethyl-*O*-(2,4-dichlorophenyl) phosphoramidothioate since 1960, and the diflubenzuron-R strain was selected for several years with diflubenzuron. The resistance ratios at the ED_{50} values relative to the S-NAIDM strain were 7.5X, 384X and >1000X for the methoprene-R, OMS-12-R and diflubenzuron-R strains, respectively, and the resistance appeared to be due to many factors. The application of diflubenzuron to mature larvae reduced chitin formation much more in S than in OMS-12-R or diflubenzuron-R pupae. This in vivo effect is undoubtedly the result of many interacting systems, and it would be very interesting to examine chitin synthesis in both R and S strains by the use of an in vitro system. Penetration was much slower in OMS-12-R and diflubenzuron-R strains than in the S-NAIDM strain, and it was slightly lower in the diflubenzuron-R than in the OMS-12-R strain. The very shallow dose response curves of the action of diflubenzuron on OMS-12-R and diflubenzuron-R strains are taken as a partial indication of reduced penetration and has been cited as a partial resistance mechanism in other *M. domestica* strains as well (Oppenoorth and van der Pas, 1977). The consequences of even slightly reduced penetration may have profound toxicological effects. Diflubenzuron, BAY SIR 8514 and several other benzoylphenyl ureas are quite insoluble in most organic solvents, which makes application of large amounts of material difficult in the laboratory. Numerous workers have noted that these compounds tend to crystallize on various surfaces, including insects (Verloop and Ferrell, 1978; Booth, this volume). Such crystalline materials generally have reduced bioavailability, and the actual crystal structure formed may greatly influence the resulting toxicity. In their study, Pimprikar and Georghiou (1979) noticed the appearance of fine needles on the cuticle of larvae treated topically with high doses of diflubenzuron. Thus, any resistance mechanism that necessitates a larger dose of diflubenzuron is likley to cause a disproportionate increase in the ED_{50} because it might enhance the formation of external microcrystals. According to Plapp (discussion during symposium) steep probit lines are obtained when larvae are exposed to diflubenzuron either in the diet or as thin films that avoid the formation of stable crystalline structures.

The use of various insecticide synergists proved very informative. Diethylmaleate (DEM) is known to deplete GSH by the Michael-like addition mentioned earlier, and *S,S,S*-tributylphosphorotrithioate (DEF) (Figure 4) is known to inhibit many insect esterases. Neither of these compounds strongly synergized diflubenzuron, indicating that GSH and DEF-sensitive esterases are not important in its metabolism. Synergism ratios (SR) were determined for diflubenzuron in the S-NAIDM, OMS-12-R and diflubenzuron-R strains with piperonyl butoxide (Figure 4) (SR = 2.6, 1.5, 77.6, respectively) and sesamex (SR = 14.8, 29.4, 36.3). These results clearly indicate that oxidative pathways are important. An interesting observation is that sesamex is clearly the better synergist in OMS-12-R larvae and piperonyl butoxide is better in diflubenzuron-R larvae. The "solvent" effect of the synergist at high doses cannot be ruled out; however, the difference in sesamex and piperonyl butoxide synergism between the strains indicates that these compounds are doing much more than simply facilitating penetration.

When larvae were analyzed for either the 4-chlorophenyl or 2,6-difluorobenzoyl-labeled diflubenzuron 24 hr after topical application, the percentage of unmetabolized, internal diflubenzuron was 18 and 22, respectively, for the S-NAIDM strain, 0.6 and 0.4 for the OMS-12-R strain, and 1.4 and 0.9 for the diflubenzuron-R strain. The polar

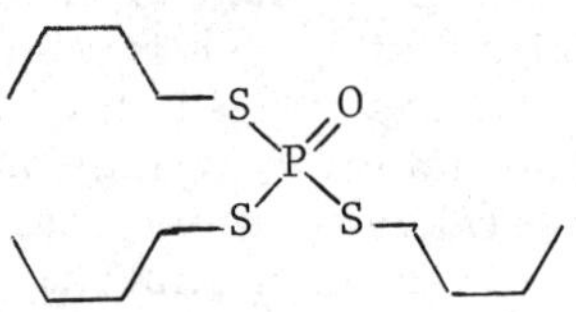

DEF

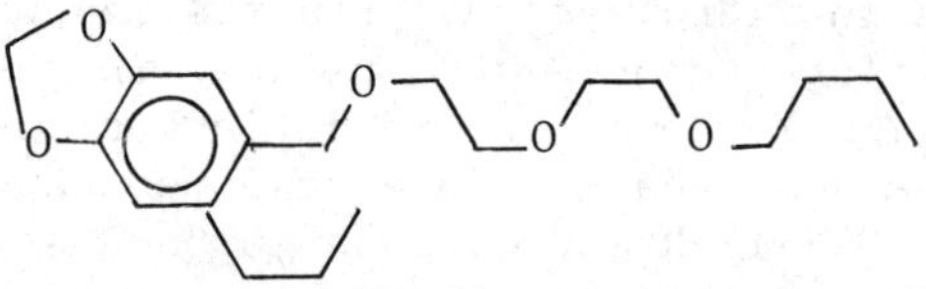

Piperonyl butoxide

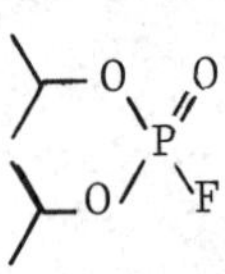

DFP

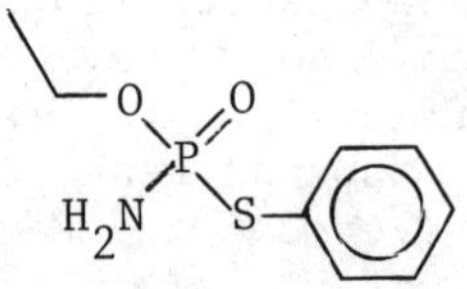

EPPAT

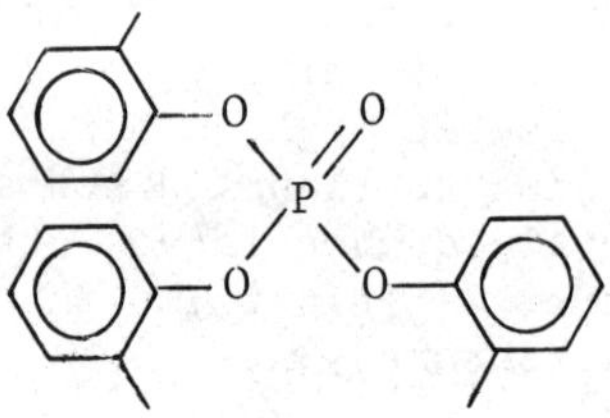

TOCP

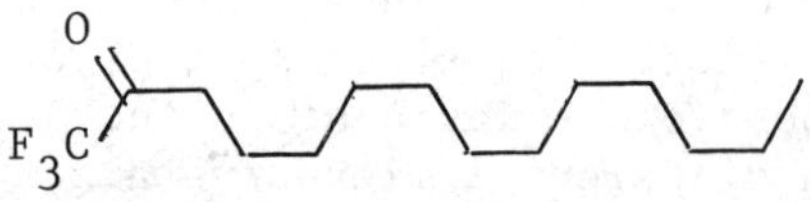

TFT

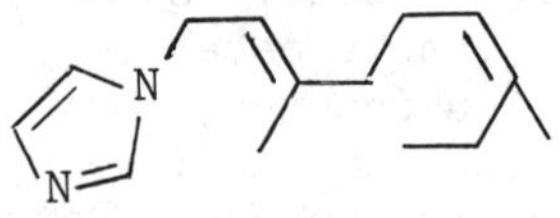

Methyl geranyl imidazole

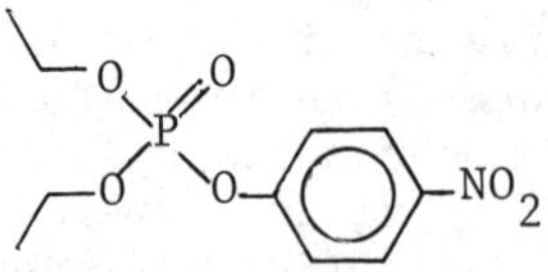

Paraoxon

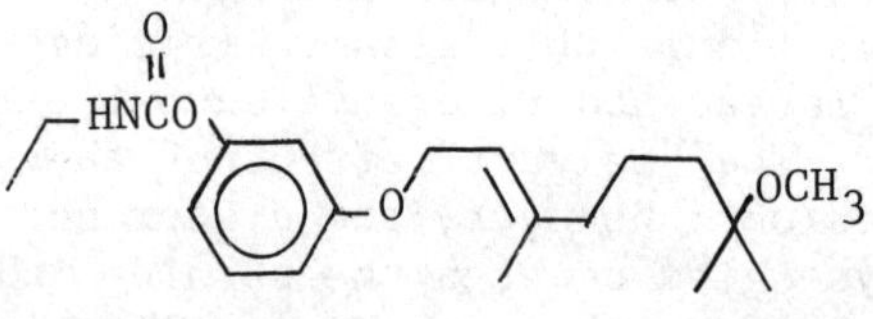

Geranyl phenyl ether carbamate

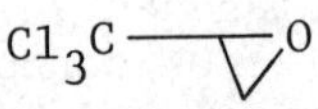

ETP

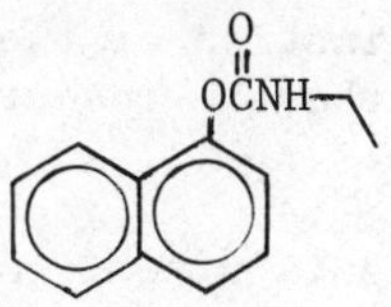

1-Naphthyl-*N*-ethyl carbamate

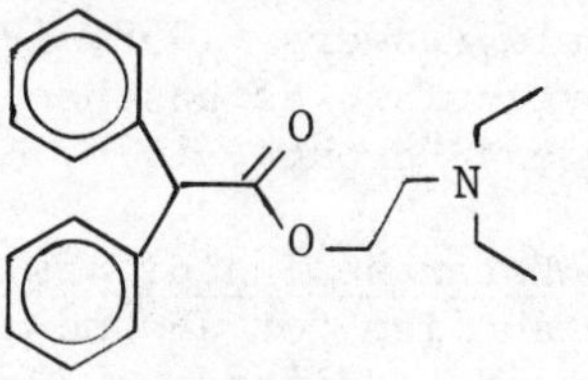

SKF-525A

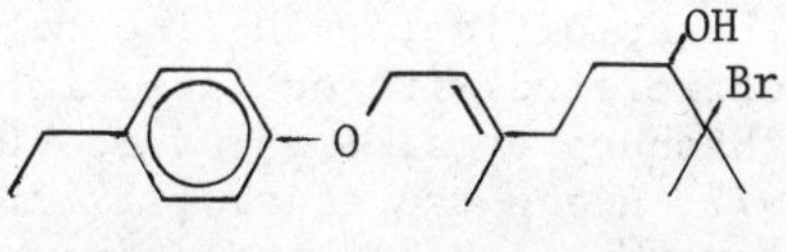

6-Hydroxy-7-bromo R-20458
geranyl phenoxy ether

Figure 4. Structures of compounds that can potentially synergize insect growth regulators.

metabolites detected in the external and cage washes indicated that all strains absorbed, metabolized and excreted diflubenzuron, and that this process was most extensive in the diflubenzuron-R larvae. Pimprikar and Georghiou (1979) have thus explained the cross resistance of the OMS-12-R strain and have partially explained the resistance of the diflubenzuron-R strain. However, the diflubenzuron levels inside the two R strains were so similar that other mechanisms must also be involved in the increased resistance following diflubenzuron selection.

COUNTERMEASURES TO IGR RESISTANCE

Countermeasures to resistance can involve a wide range of approaches; however, only those involving physiological/biochemical mechanisms will be considered in this chapter. There are several factors that could potentially play a role in resistance to IGRs. These factors include inactivation, penetration, transport, storage and length of sensitive period (window) (Georghiou, 1972; Hammock et al., 1977a). Obviously, countermeasures may not be possible for each one of these mechanisms; however, those for which some action can be taken will be examined.

Synergists

A potential method for countering the increased activity of detoxifying enzymes is the use of synergists; however, their use presents several problems, such as cost effectiveness, stability, chemical compatibility, registration, etc. Some of these problems may be simplified by incorporating the synergist into the IGR molecule (Bowers, 1968, 1969) or by derivatization of the IGR with the synergist, as has been done for some carbamate insecticides (Kuhr and Dorough, 1976).

Esterase Inhibitors. Ester hydrolysis is often a major route of metabolism for JH and some juvenoids (Slade and Zibitt, 1972; Ajami and Riddiford, 1973; Hammock and Quistad, 1976), especially in the hemolymph. During the last larval instar of many insects, esterases that seem to be specific for JH-like molecules appear in the hemolymph and are correlated with declines in the JH titer (Weirich et al., 1973; Sanburg et al., 1975a,b; Vince and Gilbert, 1977; Kramer et al., 1977; Hwang-Hsu et al., 1979; Sparks et al., 1979a). Thus, synergists that inhibit these enzymes may help stabilize JH and ester-containing IGRs.

DEF, an esterase inhibitor (Eto, 1974; Wilkinson, 1976), was found to stabilize methoprene in several stages of *S. invicta* (Bigley and Vinson, 1979b). Ester hydrolysis of several juvenoids, including hydroprene and kinoprene, has also been shown to be blocked by paraoxon (Figure 4) and, to a lesser extent, by piperonyl butoxide in *M. domestica* (Yu and Terriere, 1975). Similarly, ester hydrolysis of kinoprene was inhibited by DFP (diisopropyl phosphorofluoridate) and piperonyl butoxide in *Phorma regina* Meigen and *S. bullata* (Terriere and Yu, 1977). However, inhibition of ester hydrolysis by tri-o-cresylphosphate (TOCP) (Figure 4) in either *T. molitor* or *O. fasciatus* did not appear to be very effective in maintaining methoprene in vivo (Solomon and Metcalf, 1974), although SKF-525A (Figure 4), TOCP and piperonyl butoxide all reduced ester hydrolysis of JH I in both S and pyrethroid-R strains of *Sitophilus granarius* L. (Edwards and Rowlands, 1978). However, the observed inhibition of ester cleavage may also have been the result of a reduced rate of penetration as well as enzyme inhibition.

There may be at least two broad types of esterases acting on JH: general carboxylesterases of low specificity, which metabolize many esters, including α-naphthyl acetate (α-NA) and enzymes that appear to be very specific for JH (Whitmore et al., 1972; Sanburg et al., 1975a,b; Nowock and Gilbert, 1976; see also Sparks and Hammock, 1979b). Many of the aforementioned compounds are good in vitro inhibitors of general α-NA esterases (Sanburg et al., 1975b; Hammock et al., 1977b; Sparks and Hammock, 1980) but are relatively poor inhibitors of some specific JHEs.

Numerous compounds have been screened for their ability to block the ester hydrolysis of JH (Ajami, 1975; Pratt, 1975; Hooper, 1976; Hammock et al., 1977b; Kramer et al., 1977; Mumby et al., 1979; Sparks and Hammock, 1980). Several workers have shown that relatively nonspecific esterase inhibitors will stabilize JH; however, none of the potent JHE inhibitors have been thoroughly tested as potential synergists. Excluding their action against Diptera, which appears to be unique, several organophosphates, most especially *O*-ethyl-*S*-phenyl phosphoramidothiolate (EPPAT) (Figure 4), appear to be very potent inhibitors of JHEs (Hammock et al., 1977b; Sparks and Hammock, 1979b, 1980) and thus are potential JH synergists. Likewise, some trifluoromethyl ketones, such as 1,1,1-trifluorotetradecan-2-one (TFT) (Figure 4), also appear to be very effective inhibitors of the JHEs (Sparks and Hammock, 1980; Hammock et al., 1980). Unfortunately, compounds like EPPAT and TFT may be of limited use in vivo because EPPAT is relatively toxic to mammals (Sanborn and Fukuto, 1972), and TFT, possibly a transition state analog, is a reversible inhibitor (Hammock et al., 1980). However, the affinity of the trifluoromethyl ketone moiety for the JHEs may enhance the stability or facilitate the transport of juvenoids containing this functionality.

The JHEs of the Diptera appear to be unique: they are only poorly inhibited by the classical esterase inhibitors and the phosphoramidothiolates (Hooper, 1976; Sparks and Hammock, 1980). The best inhibition of the JHEs of *M. domestica* is provided by the *N*-alkyl carbamates of 1-naphthol and a geranyl phenyl ether (Figure 4) (Mumby et al., 1979; Sparks and Hammock, 1980). These compounds seem to be specific for the Diptera, and because some of them are relatively easy to synthesize, they hold promise as potential synergists. Thus, it is possible that a highly specific inhibitor of JH and juvenoid ester hydrolysis can be developed.

Although esterases with high substrate specificity are obviously important in the regulation of natural JH and some work has centered on inhibiting these enzymes, the practical utility of JHE inhibitors as synergists is questionable. Many active juvenoids contain no ester linkage (see Figure 1). The dienoate juvenoids are the only ester-containing compounds seriously being considered for development, and the conjugation of the ester with the Δ2, Δ4 diene makes these compounds even more stable to bases and presumably to esterases than JH itself, and most JHEs examined appeared highly specific for methyl esters. Although *M. domestica* can hydrolyze hydroprene, the isopropyl ester of methoprene is still the most stable ester (Yu and Terriere, 1975; Terriere and Yu, 1977), and ester hydrolysis has yet to be clearly demonstrated as a resistance mechanism toward juvenoids (Hammock et al., 1977a; Brown and Hooper, 1979). Thus, on the basis of present information, the usefulness of esterase inhibitors as practical juvenoid synergists is limited. JHE inhibitors, however, are useful research tools and they may be useful as stabilizers of JH or as direct inhibitors of insect development (Kramer et al., 1977; Sparks and Hammock, 1980). Here again there are potential difficulties because the JHEs of insects such as *Trichoplusia ni* (Hüber) are produced in such large amounts that a very high level of inhibition must be maintained in order to disturb development.

Epoxide Hydrolase Inhibitors. Epoxide hydration is another primary route of JH and juvenoid metabolism. The epoxide hydrolases (EHs) that act on JHs and juvenoids appear distinct from those acting on other substrates, such as HEOM (1,2,3,4,9,9-hexachloro-6,7-epoxy-1,4,4a,5,6,7,8,8a-octahydro-1,4-methano-naphthalene) or octene oxide (Slade et al., 1975; Mullin, 1979). Although the HEOM EHs of several insects are best inhibited by ETP (1,1,1-trichloropropane-2,3-epoxide) (Figure 4) (Brooks, 1973a,b, 1974; Slade et al., 1975; Craven et al., 1976; Yu and Terriere, 1978b), the EHs active on JH and juvenoids are best inhibited by more lipophilic molecules, such as 1-(6',7'-epoxy-8'-methyl geranyl) imidazole (Figure 4) (Slade, 1975), and are only poorly inhibited by ETP (Hammock et al., 1974a; Slade, 1975). Other compounds, especially those containing the methylene dioxyphenyl moiety, such as Ro-7-9767, Ro-20-3600 and piperonyl butoxide, also seem to be effective against the JH EHs (Slade and Wilkinson, 1973; Slade, 1975; Yu and Terriere, 1978a). The 7-bromo-6-hydroxy derivative of R-20458 (Figure 4) is also a weak inhibitor of *M. domestica* juvenoid EHs (Hammock et al., 1974a). SKF-525A reduced the formation of JH I 10,11-diol in both S and R strains of *S. granarius*, and TOCP slowed the EH activity in R but not S strains (Edwards and Rowlands, 1978). Kramer et al. (1977) reported that Triton X-100 seemed to inhibit EH as well as JHE activity in *L. decemlineata*. The greatest potential of the currently available juvenoids appears to be their use against the Diptera (Staal, 1975), but the JHEs do not seem to be as important as EHs in dipteran JH metabolism as they do in other insects (Slade and Zibitt, 1972; Hammock and Quistad, 1976; Wilson and Gilbert, 1978); thus, JH EH inhibitors could be potentially useful synergists in the Diptera, but unfortunately, the best inhibitors of JH or juvenoid EHs appear reversible and have I_{50} values of only about 10^{-5}M. There is a need to better understand the role of EHs in JH regulation and to develop more potent EH inhibitors. Such inhibitors will probably arise from a better understanding of the comparative mechanisms of insect and mammalian epoxide hydrolases.

Oxidase Inhibitors. Strains possessing elevated oxidase activity generally demonstrate the highest level of resistance. Thus, oxidase inhibitors appear to be good candidates for IGR synergists. Piperonyl butoxide (PB) has already been used as a general MFO inhibitor and as a synergist for pyrethroids and other insecticides (Casida, 1970; Wilkinson, 1976), and it has been found to increase methoprene's toxicity to several Diptera (Quistad et al., 1975) as well as to slow its oxidative metabolism in *S. invicta* (Bigley and Vinson, 1979b), *T. molitor* and *O. fasciatus* (Solomon and Metcalf, 1974). Unfortunately, few oxidase inhibitors other than PB have been examined for their ability to block oxidative metabolism or to synergize juvenoid action. Recently, PB and/or sesamex have also been shown to greatly increase the toxicity of diflubenzuron to house flies resistant to both diflubenzuron and OMS-12; DEF and DEM were found to be much less effective (Pimprikar and Georghiou, 1979). Likewise, chlordimeform appeared to be a good synergist for diflubenzuron in *H. virescens* larvae (Plapp, 1976b), but PB was much less effective as a synergist for diflubenzuron, PH 60-38 and other benzoylphenyl ureas in *Hylemya platura* (Meigen) (Vea et al., 1976). The very high activity of some methylene dioxyphenyl-containing juvenoids may imply a dual mode of action (Bowers, 1968, 1969). Unfortunately, these compounds are of poor field stability, are often very specific, and appear to be expensive to produce.

Designing IGRs to Circumvent Resistance

Site-of-action resistance may lie in the future of any insecticide, but resistance due to oxidative metabolism and/or reduced penetration is almost certain to occur in some species. It seems advisable to consider such resistance early in the commercialization of new compounds.

Juvenoids Resistant to Metabolism. It is reasonable to assume that the initial resistance problems encountered by IGRs will be due to previous selection with other pesticides, and, therefore, at least in part, to metabolism. Thus, the use of field strains or highly cross-resistant strains for screening work may prove informative. For instance, the bioassay data shown in Table 4 were obtained by applying juvenoids in 0.5 µl acetone topically to postfeeding larvae (white larvae) of *M. domestica*. When biological activities on the S-NAIDM strain were compared, methoprene demonstrated a clear advantage over the other compounds, notably the epoxide. Such epoxide-containing juvenoids were metabolized by insect epoxide hydrolases, and partly for this reason, alkoxide juvenoids have progressed further in development. However, the high oxidase levels in the dimethoate-R strain apparently resulted in resistance to all alkoxide-containing materials due to *O*-dealkylation, but epoxide hydrolase levels did not increase. In oxidatively resistant strains the epoxide-containing juvenoid demonstrated higher biological activity than methoprene (Table 4).

For use on field crops one needs a rather photostable skeleton. Thus, obvious sites of photoinstability and metabolic attack must be eliminated. One such site is the alkoxide moiety, because even if the resulting alcohol were very active biologically (Henrick et al., 1976), it would be likely to be susceptible to rapid conjugation. Hopefully, one could avoid such problems by using compounds with no terminal functionality, such as hydroprene, or possibly by returning to the epoxide, as with epofenonane (Figure 1). The epoxide is susceptible to both environmental degradation and to insect metabolism; however, no increase in epoxide hydrolase levels was found in several of the insecticide-R *M. domestica* strains examined, and it is possible that stability to normal levels of epoxide hydrolases could be increased by varying substituents about the epoxide moiety (Mori et al., 1975).

IGRs that are activated by metabolic enzymes could provide another method to overcome resistance, and such a concept might be illustrated by the juvenogens (biochemically activated juvenoids). These compounds combine an active juvenoid alcohol with a long chain alkyl acid to form a juvenogen ester. Once inside the insect the ester is cleaved, which releases the active juvenoid alcohol (Slama and Romanuk, 1976; Slama et al., 1978). However, to date none of these physiologically interesting compounds has been examined in insecticide-R or IGR-resistant insect strains. Such a strategy could be applied to biologically active metabolites or juvenoids with a functionality suitable for derivatization; other potential juvenoid metabolites are also relatively active for several insect species (Solomon and Metcalf, 1974; Hammock et al., 1974b; Henrick et al., 1976). To date, such compounds have not been tested on insecticide-R or IGR-resistant strains.

New Approaches for Benzoylphenyl Ureas. Neither resistance nor cross resistance appeared to be a serious problem that would restrict benzoylphenyl urea usage, but because several firms are considering the development of IGRs similar to difulbenzuron, an understanding of possible resistance mechanisms would be helpful in selecting specific compounds for development. If ring hydroxylation is generally found to be an important metabolic pathway in R insects, as suggested by Chang (1978), it may be useful to select aromatic systems more resistant to hydroxylation. Verloop and Ferrell (1977) indicated that there were three primary sites of hydrolysis in the benzoylphenyl ureas (Figure 3), and large 2,6-dichloro substituents greatly retarded the production of 2,6-dichlorobenzoic acid resulting from cleavage at one of these sites compared with 2,6-difluoro substituents on benzoylphenyl ureas. As the mechanism of insect resistance as well as environmental hydrolysis of the amide bonds in the benzoylphenyl ureas become better understood, compounds can probably be designed that will be of enhanced stability in insects without being overly persistent in the environment.

The study by Pimprikar and Georghiou (1979), as well as the sporadic activity of these compounds on some Lepidoptera, brings up the disturbing possibility of resistance by site insensitivity. At this point in our understanding of benzoylphenyl urea action, a screening of compounds on R strains is the only obvious recourse.

Whether resistance is due to reduced penetration, behavioral modification, altered distribution, enhanced metabolism or reduced sensitivity, one obvious short-term solution would be to get more active ingredient into the target insect. Such a course of action is not necessarily trivial with the benzoylphenyl ureas because many

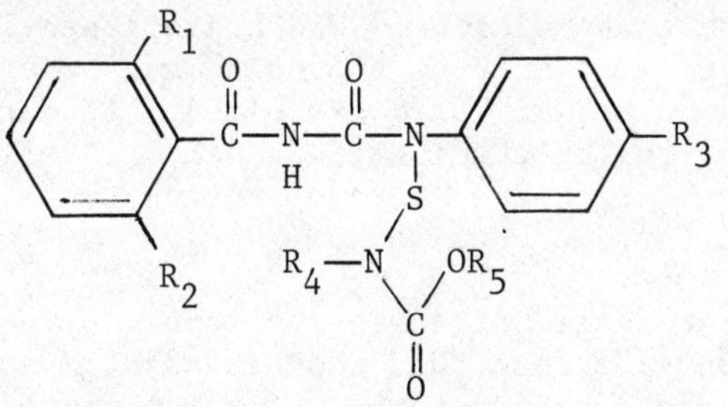

Figure 5. General structure of lipid-soluble sulfenyl-derivatives of benzoylphenyl ureas. R_1, R_2 = H, F, Cl; R_3 = C_2H_5, n-C_3H_7, i-C_3H_7, and R_5 = CH_3, n-C_3H_7, n-C_4H_9.

of these compounds are highly insoluble and thus difficult to formulate. They also tend to form very stable crystals that are apparently poorly absorbed. One possible solution to these problems is illustrated by the compounds shown in Figure 5, which were synthesized as potential haptens for immunochemical assays of the benzoylphenyl ureas (Sylwester et al., 1979). Wellinga et al. (1973) pointed out that when a third substituent was added to one or both nitrogens, the insecticidal activity of the benzoylphenyl ureas was greatly reduced, but three partial exceptions to this generalization involved compounds that were found to be unstable in aqueous ethanol yielding diflubenzuron upon hydrolysis. The studies of Fukuto and his co-workers on derivatizing insecticidal carbamates and phosphates (see Fukuto and Mallipudi, this volume) could logically be applied to this problem. Thus, the compounds shown in Figure 5 are soluble in hexane or carbon tetrachloride, although the parent benzoylphenyl urea is only marginally soluble in tetrahydrofuran. Some of these derivatives hydrolyze so readily on contact with water that the sulfenyl carbamate moiety could be considered as more of a formulating agent (Sylwester and Hammock, unpublished). Other derivatives are stable enough that the sulfenyl carbamate may actually aid in penetration. In all of these compounds, the toxicity of the derivatized material is very similar to that of the parent diflubenzuron or BAY SIR 8514 (*N*-[(4-trifluoro-methoxyphenyl)amino]carbonyl-2-chlorobenzamide) (Hammock, unpublished data). Such derivatives have been very successful in the past in increasing insect toxicity while decreasing mammalian toxicity (Fukuto, 1976). For the benzoylphenyl ureas, the derivatives seem potentially important as an aid in the formulation and in preventing the formation of a crystalline form of the compound, which may have reduced biological activity.

Development of Insect Growth Regulators with Novel Sites of Action

In theory, one advantage offered by insect growth regulators is that they act at new sites, thus "site-of action" cross resistance would be expected to be minimal. The benzoylphenyl ureas have illustrated the potential of such an approach, and hopefully, other new sites of action can be similarly exploited. As our knowledge of insect physiology and biochemistry advances, it will become increasingly possible to target new sites of action that are both critical to, and specific for, insects (Menn, 1980).

The endocrine system of insects offers many potential sites for such exploitation. Because the endocrine system is regulatory in function, small changes induced by a foreign compound will be amplified into major changes in the survival capacity of the insect, and the tremendous effect that foreign compounds can have on another regulatory system, the nervous system, has been exploited in the development of many classical pesticides, such as organochlorines, pyrethroids, organophosphates and carbamates.

The insect endocrine system appears to have evolved late enough that there are large differences between the endocrine biologies of insects and other organisms. For instance, terpenes have not yet been shown to have a regulatory function in any organism other than insects, and homoterpenes (such as JH I and JH II) have not been found in any other plant or animal. The variety of methods by which insects exercise endocrine control of their reproduction (Engelmann, 1970) indicates that the endocrine system is diverse even within the insects, and the tremendous selectivity that some juvenoids display for certain insect groups shows that variation in endocrine physiology extends to development as well as to reproduction.

Epithelial Endocrine System. Although ecdysone itself is not a promising control agent, the biosynthetic pathways leading to the critical molting hormones or the other insect steroids can possibly be exploited (Matolcsy et al., 1975; Slama, 1978; Lukovits et al., 1978). Another promising approach is the disruption of JH function or action. Compounds utilizing this mode of action have the potential advantage of being effective during the early larval stages of insect development and may therefore be better suited than the juvenoids to be agricultural insect control agents. Attempts to inhibit JH biosynthesis thus far have been only marginally successful, but such inhibitors could possess high specificity for insects (Matolcsy et al., 1975; Hammock, 1975; Pratt and Finney, 1977; Pratt and Bowers, 1977; Hammock et al., 1978; Hammock and Mumby, 1978). To date, only three anti-JHs have been discovered: the precocenes (Bowers, 1976a,b; 1977a,b; Bowers et al., 1976), A_{11} (methyl-6,7-dioxo-5α-10α-podocarpa-8,11,13-triene-15-oate) (Murakoshi et al., 1975; Slama, 1978), and ETB (ZR-2646, ethyl-4-[2-(*t*-butylcarbonyloxy)butoxy]benzoate) (Figure 6) (Staal, 1977; Riddiford and Truman, 1978; Sparks et al., 1979b). Although little is known about the anti-JH action of A_{11}, precocenes and ETB appear to have different types of action. The precocenes seem to be potent allatotoxins (Bowers, 1977a,b; Pener

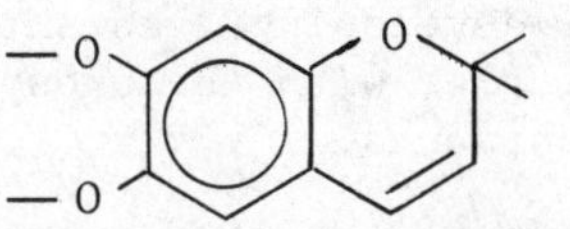

Precocene II

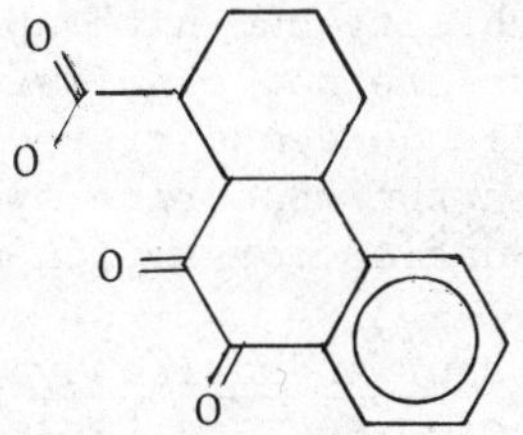

A_{11}

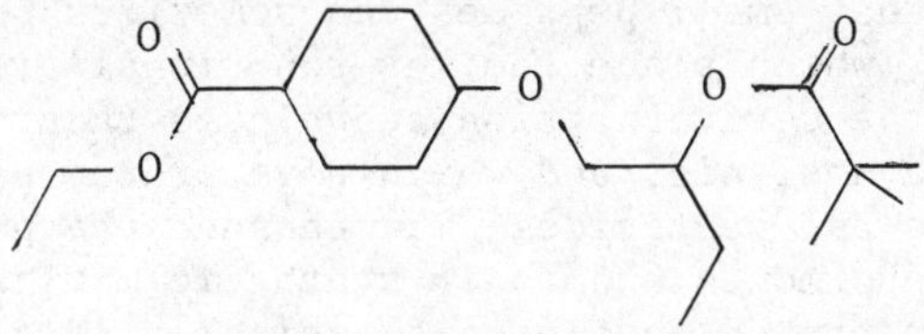

ETB

Figure 6. Structures of known anti-juvenile hormones.

et al., 1978), and several laboratories have hypothesized that the elusive precocene epoxide may destroy the corpora allata following its biosynthesis in situ. Preliminary evidence for this pathway has recently emerged (Jennings and Ottridge, 1979; Brooks et al., 1979). Other workers have investigated the basic endocrinology of juvenile hormone feedback regulation, and it has emerged from these studies that some juvenoids can inhibit JH production in vitro (Tobe and Stay, 1979) and depress JH titers in vivo (Schooley and Bergot, 1979). Such information supports the obvious hypothesis that the JHs have many sites of action and that the substrate specificity and sensitivity of the receptors involved may vary. One can thus envision two related modes of action for the anti-JHs represented by ETB (Stall, 1977), both of which are supported by experimental evidence. In one model for anti-JH action, ETB is predicted to bind nonproductively at a JH receptor and to prevent the productive binding of the natural hormones; it thus acts by the classical mammalian model of hormone agonist/antagonist (Sparks et al., 1979). Alternatively, ETB could productively bind with a receptor, which would result in feedback inhibition of JH biosynthesis and/or stimulation of JH metabolism and thus a lower JH titer (Schooley and Bergot, 1979). This model requires that ETB either fail to bind or that it bind nonproductively with other JH receptors controlling development. Both A_{11} and ETB contain ester moieties that should be easily metabolized, and the precocenes are subject to hydroxylation (Ohta et al., 1977; Burt et al., 1978) and potentially *O*-demethylation of MFOs.

Instability and high specificity insure that these particular compounds are not candidates for insect control agents, but precocene and ETB do demonstrate the concept of developing agents that would act on the endocrine system by interfering with the biosynthesis, degradation, or action of intrinsic hormones.

Neuroendocrine System. The neuroendocrine system probably offers more promise as a target than the epithelial endocrine system. It is a vital link between two regulatory systems, and many neurohormones in vertebrates are known to act at remarkably low levels. Most insect neurohormones are probably small peptides, which would be expected to penetrate poorly and be rapidly degraded; however, vertebrate work has demonstrated that small peptides can sometimes be mimicked by organic compounds, which place the key functionalities in the correct position. Also, the derivatization techniques that are used in peptide synthesis and analysis, and which have proven so useful for derivatizing classical pesticides, may enhance the penetration and stability of neurohormones and their synthetic mimics. Several studies have demonstrated that even classical pesticides could influence neurosecretion, and hopefully, careful industrial screening procedures may illustrate new compounds acting by this pathway (Maddrell and Reynolds, 1972; Granett and Leeling, 1972; Samaranayaka, 1974; Ruscoe, 1974; Gerolt, 1976). Although major advances in technology are needed for our understanding of insect neuroendocrinology and our ability to exploit this understanding, it is safe to predict that insect control agents based on the neuroendocrine system will emerge over the next few decades.

ADVANTAGES AND LIMITATIONS OF INSECT GROWTH REGULATORS IN CIRCUMVENTING RESISTANCE

Resistance and cross resistance to conventional insecticides is now widespread and continually advancing. Although IGRs have their own shortcomings, including resistance, they do possess many of the attributes necessary to function in a pest management program directed toward delaying or avoiding resistance.

A major problem often associated with the use of conventional second-generation insecticides is pest resurgence and secondary pest outbreaks (Smith, 1970; DeBach, 1974; Messenger et al., 1976). These pest resurgences or outbreaks must then be dealt with, most often by the further application of insecticides resulting in increased selection pressure, which hastens the development of resistance (DeBach, 1974). These pest problems result, in part, from the use of insecticides that not only kill the target pest insect, but also eliminate many of the nontarget beneficial predators and parasites that keep the pest population under control (Smith, 1970; DeBach, 1974).

In order to take the necessary steps toward maintaining the natural enemy population, and hence slow the development of insecticide resistance, compounds that are much more selective should be used. However, until recently most of the selectivity sought has been, quite naturally, between insects and mammals (Hollingworth, 1976). Among the conventional second-generation insecticides a high degree of specificity between insect species is rare. One reason for this is that most of the earlier toxicants cause lesions in biochemical or physiological systems that are similar throughout the animal kingdom. For instance, few neurotoxins have a high degree of specificity since the nervous system of all insects and even insects and mammals is quite similar. Metabolism, penetration, and other mechanisms rarely combine to give a selectivity ratio of over two orders of magnitude for most nontarget and target species (Winteringham, 1967), especially between a pest insect and its natural enemies (Croft and Brown, 1975).

Unlike the more conventional insecticides, the IGRs and in particular the juvenoids often display a high degree of selectivity. Toxicity ratios between insect species of over two orders of magnitude are quite common (Sorm, 1971; Jacobson et al., 1972; Slama et al., 1974; Sehnal, 1976) and as high as seven orders of magnitude have been observed (Slama et al., 1974). For example, some alicyclic juvenoids possess toxicity ratios between species of the Lepidoptera and species of the Hemiptera, Hymenoptera, Neuroptera and Diptera that span at least three orders of magnitude (Table 5). Many of the Lepidoptera are agricultural pest insects while the Hemiptera, Hymenoptera, Neuroptera and Diptera all contain natural enemies of lepidopterous pests. Thus, it is conceivable that selective IGRs can be developed for agricultural use that would have little effect on the natural enemies of the pest insects. By preserving the natural enemies, the possibility of pest resurgence or outbreak is reduced, which, by reducing the amount of insecticide needed for adequate control, would lower the selection pressure and delay the onset of insecticide resistance.

The use of nonpersistent insecticides or the application of the insecticide in a manner that leaves individuals untreated are but two of many possible approaches by which some members of a pest population may be purposely left relatively unselected (Georghiou and Taylor, 1977; Metcalf, 1980). By allowing unselected individuals to remain in a pest population the onset of resistance and the level of resistance attained could potentially be delayed and reduced, respectively (Georghiou and Taylor, 1977; Georghiou, this volume). Many IGRs possess characteristics that make them conducive to such a scheme, such as biodegradability and effectiveness only at specific intervals during insect development (Menn and Pallos, 1975; Staal, 1977; Riddiford and Truman, 1978). Due to the very short time they are present in the environment, and because not all individuals of a pest population are likely to be sensitive simultaneously, IGRs are more likely to allow individuals to remain relatively unselected, and hence susceptible, than most conventional insecticides.

Unfortunately, many of the qualities that make IGRs so desirable have prevented their use in agricultural pest control. Many IGRs are very unstable in the environment; however, better formulation technology and the advent of benzoylphenyl ureas and juvenoids of higher environmental stability (this chapter; Zurflueh, 1976) may yet lead to the increased use of IGRs in agriculture. Likewise, some of the currently available IGRs such as the benzoylphenyl ureas may be too broad in spectrum (Grosscurt, 1978), while the juvenoids act too slowly and only during specific intervals or windows (Slama et al., 1974; Staal, 1977). The selectivity of most highly active juvenoids would probably result in too small a market to support such efforts. Although one could argue that such an ideal compound would have an effective life of many years by avoiding resistance problems, the development of such compounds is not currently being considered. Perhaps placing a monetary value on deleterious environmental effects and upon worker safety, less expensive registration requirements for highly selective pesticides, and a re-evaluation of "profitability" by industrial marketing personnel could help change the situation. Until such a revolution of thought becomes practical and results in the calculated use of selective pesticides in well-planned pest management systems, the scientist may find his role largely restricted to describing and explaining rather than combating or circumventing resistance.

SUMMARY

The cross resistance and resistance of numerous insects to IGRs illustrates that such compounds do not offer the magic solution to our pest management problems. IGRs that are refractory to the development of resistance, however, can be selected for commercialization, and specific measures can be taken to delay the onset of resistance. The two factors that have most severely limited IGR deployment in pest control, selectivity and biodegradability, may be our best hope for delaying the development of resistance against compounds with a novel mode of action.

Acknowledgements: The original research presented in this chapter was supported by NIEHS grant ES01260-04 and the Louisiana and California Agriculture Experiment Stations. B. D. Hammock was supported by NIEHS Research Career Development Award 5 K04 ES00046-02. L. D. Newsom (Louisiana State University) and E. Hodgson (North Carolina State University) critically reviewed the manuscript.

REFERENCES

Ajami, A. M., 1975, Inhibitors of ester hydrolysis as synergists for biological activity of Cecropia juvenile hormone, *J. Insect Physiol.*, 21:1017.

Ajami, A. M., and Riddiford, L. M., 1973, Comparative metabolism of the Cecropia juvenile hormone, *J. Insect Physiol.*, 19:635.

Amos, T. G., Williams, P., Du Guesclin, P. B., and Schwarz, M., 1974, Compounds related to juvenile hormone: Activity of selected terpenoids on *Tribolium castaneum* and *T. confusum*, *J. Econ. Entomol.*, 67:474.

Amos, T. G., Williams, P., and Semple, R. L., 1977, Susceptibility of malathion-resistant strains of *Tribolium castaneum* and *T. confusum* to the insect growth regulators methoprene and hydroprene, *Ent. Exp. Appl.*, 22:289.

Arevad, K., 1974, Laboratory experiments with a juvenile hormone mimic against housefly larvae, *Danish Pest Inf. Lab. Ann. Rep.*, 1973:42.

Arking, R., and Vlach, B., 1976, Direct selection of mutants of *Drosophila* resistant to juvenile hormone analogues, *J. Insect Physiol.*, 22:1143.

Benskin, J., and Vinson, S. B., 1973, Factors affecting juvenile hormone analogue activity in the tobacco budworm, *J. Econ. Entomol.*, 66:15.

Bigley, W. S., and Vinson, S. B., 1979a, Degradation of [^{14}C]methoprene in the imported fire ant, *Solenopsis invicta*, *Pestic. Biochem. Physiol.*, 10:1.

Bigley, W. S., and Vinson, S. B., 1979b, Effects of piperonyl butoxide and DEF on the metabolism of methoprene by the imported fire ant, *Solenopsis invicta* Buren, *Pestic. Biochem. Physiol.*, 10:14.

Bowers, W. S., 1968, Juvenile hormone: Activity of natural and synthetic synergists, *Science*, 161:895.

Bowers, W. S., 1969, Juvenile hormone: Activity of terpenoid ethers, *Science*, 164:323.

Bowers, W. S., 1976a, Discovery of insect antiallatotropins, *in:* "The Juvenile Hormones," L. I. Gilbert, ed., pp. 394-408, Plenum Press, New York.

Bowers, W. S., 1976b, Hormone mimics, *in:* "The Future for Insecticides," R. L. Metcalf and J. J. McKelvey, eds., pp. 421-444, John Wiley and Sons, New York.

Bowers, W. S., Ohta, T., Cleere, J. S., and Marsella, P. A., 1976, Discovery of insect anti-juvenile hormones in plants, *Science*, 193:542.

Bowers, W. S., 1977a, Fourth generation insecticides, *in:* "Pesticide Chemistry in the 20th Century," ACS Symp. Ser. #37, J. R. Plimmer, ed., pp. 271-275, ACS, Washington, D.C.

Bowers, W. S., 1977b, Anti-juvenile hormones from plants: Chemistry and biological activity, *in:* "Natural Products and the Protection of Plants," G. B. Marini-Bettolo, ed., pp. 129-142, Elsevier Scientific Publishing Co., New York.

Brooks, G. T., 1973a, Insect epoxide hydrase inhibition by juvenile hormone analogues and metabolic inhibitors, *Nature,* 245:382.

Brooks, G. T., 1973b, The effects of metabolic inhibitors on insect epoxide hydrases, *Biochem. Soc. Trans.,* 1:1303.

Brooks, G. T., 1974, Inhibitors of cyclodiene epoxide ring hydrating enzymes of the blowfly, *Calliphora erythrocephala, Pestic. Sci.,* 5:177.

Brooks, G. T., Pratt, G. E., and Jennings, R. C., 1979, The action of precocenes in milkweed bugs, (*Oncopeltus fasciatus*) and locusts (*Locusta migratoria*), *Nature,* 281:570.

Brown, A. W. A., 1977, Epilogue: Resistance as a factor in pesticide management, *Proc. XV Inter. Cong. Entomol.,* 1976:816.

Brown, T. M., and Brown, A. W. A., 1974, Experimental induction of resistance to a juvenile hormone mimic, *J. Econ. Entomol.,* 67:799.

Brown, T. M., and Brown, A. W. A., 1980, Accumulation and distribution of methoprene in resistant *Culex pipiens pipiens* larvae, *Ent. Exp. Appl.,* in press.

Brown, T. M., and Hooper, G. H. S., 1979, Metabolic detoxication as a mechanism of methoprene resistance in *Culex pipiens pipiens, Pestic. Biochem. Physiol.,* 12:79.

Brown, T. M., DeVries, D. H., and Brown, A. W. A., 1978, Induction of resistance to insect growth regulators, *J. Econ. Entomol.,* 71:223.

Bull, D. L., and Ivie, G. W., 1980, Activity and fate of diflubenzuron and certain derivatives in the boll weevil, *Pestic. Biochem. Physiol.,* in press.

Burt, M. E., Kuhr, R. J., and Bowers, W. S., 1978, Metabolism of precocene II in the cabbage looper and European corn borer, *Pestic. Biochem. Physiol.,* 9:300.

Carter, S. W., 1975, Laboratory evaluation of three novel insecticides inhibiting cuticle formation against some susceptible and resistant stored products beetles, *J. Stored Prod. Res.,* 11:187.

Casida, J. E., 1970, Mixed-function oxidase involvement in the biochemistry of insecticide synergists, *J. Ag. Food Chem.,* 18:753.

Cerf, D. C., and Georghiou, G. P., 1972, Evidence of cross-resistance to a juvenile hormone analogue in some insecticide-resistant houseflies, *Nature,* 239:401.

Cerf, D. C., and Georghiou, G. P., 1974a, Cross resistance of juvenile hormone analogues in insecticide-resistant strains of *Musca domestica* L., *Pestic. Sci.,* 5:759.

Cerf, D. C., and Georghiou, G. P., 1974b, Cross-resistance to an inhibitor of chitin synthesis, TH 60-40, in insecticide resistant strains of the house fly, *J. Ag. Food Chem.,* 22:1145.

Chang, S. C., 1978, Conjugation, the major metabolic pathway of ^{14}C-diflubenzuron in the house fly, *J. Econ. Entomol.,* 71:31.

Chang, S. C., and Stokes, J. B., 1979, Conjugation: The major metabolic pathway of ^{14}C-diflubenzuron in the boll weevil, *J. Econ. Entomol.*, 72:15.
Chang, S. C., and Woods, C. W., 1979a, Metabolism of ^{14}C-penfluron in the boll weevil, *J. Econ. Entomol.*, 72:781.
Chang, S. C., and Woods, C. W., 1979b, Metabolism of ^{14}C-penfluron in the house fly, *J. Econ. Entomol.*, 72:482.
Chasseaud, L. F., 1974, The nature and distribution of enzymes catalyzing conjugation of glutathione with foreign compounds, *Drug Metabolism Review*, 2:185.
Chihara, C. J., Petri, W. H., Fristoom, J. W., and King, D. S., 1972, The assay of ecdysones and juvenile hormones on *Drosophila* imaginal disks in vitro, *J. Insect Physiol.*, 18:1115.
Craven, A. C. C., Brooks, G. T., and Walker, C. H., 1976, The inhibition of HEOM epoxide hydrase in mammalian liver microsomes and pupal homogenates, *Pestic. Biochem. Physiol.*, 6:132.
Croft, B. A., and Brown, A. W. A., 1975, Response of arthropod natural enemies to insecticides, *Ann. Rev. Entomol.*, 20:285.
Dame, D. A., Lowe, R. E., Wichterman, G. J., Cameron, A. L., Baldwin, K. F., and Miller, T. W., 1976, Laboratory and field assessment of insect growth regulators for mosquito control, *Mosq. News*, 36:462.
DeBach, P., 1974, "Biological Control of Natural Enemies," 323 p., Cambridge University Press, New York.
deKort, C. A. D., 1980, Regulation of juvenile hormone titer, *in:* "Ann. Rev. Entomol.," T. W. Mittler, ed., in preparation.
Dyte, C. E., 1972, Resistance to synthetic juvenile hormone in a strain of flour beetle, *Tribolium castaneum*, *Nature*, 238:48.
Edwards, J. P., and Rowlands, D. G., 1978, Metabolism of a synthetic juvenile hormone (JHI) in two strains of the grain weevil, *Sitophilus granarius*, *Insect Biochem.*, 8:23.
El-Guindy, M., Bishara, S. I., and Madi, S. M., 1975, Sensitivity to insect growth regulators (juvenile hormone analogues) in insecticide-resistant and -susceptible strains of *Spodoptera littoralis* (Boids.), *Z. PflKrankh PflSchutz.*, 82:469.
Ellis, P., 1968, Can insect hormones and their mimics be used to control pests, *Pest Articles and News Summaries*, 14:329.
Englemann, F., 1970, The Physiology of Insect Reproduction. International Series of Monographs in Pure and Applied Biology, Zoology Division, 44, G. A. Kerkut, ed., Pergamon Press, New York.
Erley, D., Southhard, S., and Emmerich, H., 1975, Excretion of juvenile hormone and its metabolites in the locust, *Locusta migratoria*, *J. Insect Physiol.*, 21:61.
Eto, M., 1974, "Organophosphorus Pesticides: Organic and Biological Chemistry," CRC Press, Cleveland, Ohio.
Fukuto, T. R., 1976, Carbamate insecticides, *in:* "The Future for Insecticides," R. L. Metcalf and J. J. McKelvey, eds., pp. 313-342, John Wiley and Sons, New York.

Georghiou, G. P., 1972, The evolution of resistance to pesticides, *Ann. Rev. Ecol. Syst.*, 3:133.
Georghiou, G. P., 1979, Status of development of alternative chemicals for control of resistant pests, *Proc. Papers 47 Annual Conf. of Calif. Mosq. Control Assoc. Jan. 28-31*, 1979:24.
Georghiou, G. P., and Taylor, C. E., 1977, Genetic and biological influences in the evolution of insecticide resistance, *J. Econ. Entomol.*, 70:319.
Georghiou, G. P., Lin, C. S., and Pasternak, M. E., 1974, Assessment of potentiality of *Culex tarsalis* for development of resistance to carbamate insecticides and insect growth regulators, *Proc. Papers 42 Annual Conf. of Calif. Mosq. Control Assoc., Feb. 24-27*, 1974:117.
Georghiou, G. P., Ariaratnam, V., Pasternak, M. E., and Lin, C. S., 1975, Organophosphorus multiresistance in *Culex pipiens quinquefasciatus* in California, *J. Econ. Entomol.*, 68:461.
Georghiou, G. P., Lee, S., and DeVries, D. H., 1978, Development of resistance to the juvenoid methoprene in the house fly, *J. Econ. Entomol.*, 71:544.
Gerolt, P., 1976, The mode of action of insecticides: Accelerated water loss and reduced respiration in insecticide-treated *Musca Domestica* L., *Pestic. Sci.*, 7(6):604.
Gill, S. S., Hammock, B. D., Yamamoto, I., and Casida, J. E., 1972, Preliminary chromatographic studies on the metabolites and photodecomposition products of the juvenoid 1-(4'-ethylphenoxy)-6,7-epoxy-3,7-dimethyl octene, *in:* "Insect Juvenile Hormones: Chemistry and Action," J. J. Menn and M. Beroza, eds., pp. 177-189, Academic Press, New York.
Granett, J., and Leeling, N. C., 1972, A hyperglycemic agent in the serum of DDT-prostrate American cockroaches, *Periplaneta americana*, *Ann. Entomol. Soc. Amer.*, 65(2):299.
Grosscurt, A. C., 1978, Diflubenzuron: Some aspects of its ovicidal and larval mode of action and an evaluation of its practical possibilities, *Pestic. Sci.*, 9:373.
Hammock, B. D., 1975, NADPH dependent epoxidation of methyl farnesoate to juvenile hormone in the cockroach, *Blaberus giganteus* L., *Life Sci.*, 17:323.
Hammock, B. D., and Mumby, S. M., 1978, Inhibition of the epoxidation of methyl farnesoate to juvenile hormone by corpora allata homogenates, *Pestic. Biochem. Physiol.*, 9:39.
Hammock, B. D., and Quistad, G. B., 1976, The degradative metabolism of juvenoids by insects, *in:* "The Juvenile Hormones," L. I. Gilbert, ed., pp. 374-393, Plenum Press, New York.
Hammock, B. D., and Quistad, G. B., 1980, Juvenile hormone analogs: Mode of action and metabolism, *in:* "Progress in Pesticide Biochemistry, Vol. 1," D. H. Hutson and T. R. Roberts, eds., John Wiley and Sons, Chichester, England, in preparation.
Hammock, B. D., Gill, S. S., and Casida, J. E., 1974a, Insect metabolism of a phenyl epoxygeranyl ether juvenoid and related

compounds, *Pestic. Biochem. Physiol.*, 4:393.

Hammock, B. D., Gill, S. S., and Casida, J. E., 1974b, Synthesis and morphogenetic activity of derivatives and analogs of aryl geranyl ether juvenoids, *J. Ag. Food Chem.*, 22:37.

Hammock, B., Nowock, J., Goodman, W., Stamoudis, V., and Gilbert, L. I., 1975a, The influence of hemolymph-binding protein on juvenile hormone stability and distribution in *Manduca sexta* fat body and imaginal disks in vitro, *Molec. Cell Endocrinol.*, 3:167.

Hammock, B., Gill, S. S., Hammock, L., and Casida, J. E., 1975b, Metabolic *O*-dealkylation of 1-(4'-ethylphenoxy)-3,7-dimethyl-7-methoxy or ethoxy-*trans*-2-octene, potent juvenoids, *Pestic. Biochem. Physiol.*, 5:12.

Hammock, B. D., Mumby, S. M., and Lee, P. W., 1977a, Mechanisms of resistance to the juvenoid methoprene in the house fly *Musca domestica* L., *Pestic. Biochem. Physiol.*, 7:261.

Hammock, B. D., Sparks, T. C., and Mumby, S. M., 1977b, Selective inhibition of JH esterases from cockroach hemolymph, *Pestic. Biochem. Physiol.*, 7:517.

Hammock, B. D., Kuwano, E., Ketterman, A., Scheffrahn, R. H., Thompson, S. N., and Sallume, D., 1978, Acute toxicity and developmental effects of analogs of ethyl α-(4-chlorophenoxy)-α-methylpropionate on two insects, *Oncopeltus fasciatus* and *Tenebrio molitor*, *J. Ag. Food Chem.*, 26:166.

Hammock, B. D., Lovell, V., Sparks, T. C., McLaughlin, J. K. D., 1980, Trifluoromethyl ketones: Potent transition state mimics of juvenile hormone esterases, in preparation.

Henrick, C. A., Willy, W. E., McKean, D. R., Baggiolini, E., and Siddall, J. D., 1975, Approaches to the synthesis of insect juvenile hormone analog ethyl 3,7,11-trimethyl-2,4-dodecadienoate and its photochemistry, *J. Org. Chem.*, 49:8.

Henrick, C. A., Willy, W. E., and Staal, G. B., 1976, Insect juvenile hormone activity of alkyl (2*E*,4*E*)-3,7,11-trimethyl-2,4-dodecadienoates. Variations in the ester function and in the carbon chain, *J. Ag. Food Chem.*, 24:207.

Hollingworth, R. M., 1976, The biochemical and physiological basis of selective toxicity, *in:* "Insecticide Biochemistry and Physiology," C. F. Wilkinson, ed., pp. 431-506, Plenum Press, New York.

Hooper, G. H. S., 1976, Esterase mediated hydrolysis of naphthyl esters, malathion, methoprene and Cecropia juvenile hormone in *Culex pipiens pipiens*, *Insect Biochem.*, 6:255.

Hoppe, T., 1976, Microplot trial with an epoxyphenylether (insect growth regulator) against several pests of stored wheat grain, *J. Stored Prod. Res.*, 12:205.

Hrdy, I., 1974, Effects of juvenoids on insecticide susceptible and resistant aphids (*Myzus persicae*, *Aphis fabae* and *Therioaphasis maculata*; Homoptera, Aphidae), *Acta ent. bohemoslovaca.*, 71:367.

Hsieh, M.-Y. G., Steelman, C. D., and Schilling, P. E., 1974, Selection of *Culex pipiens quinquefasciatus* Say for resistance to growth inhibitor, *Mosq. News,* 34:416.

Hwang-Hsu, K., Reddy, G., Kumaran, A. K., Bollenbacher, W. E., and Gilbert, L. I., 1979, Correlations between juvenile hormone esterase activity, ecdysone titer and cellular reprogramming in *Galleria mellonella, J. Insect Physiol.,* 25:105.

Ivie, G. W., and Wright, J. E., 1978, Fate of diflubenzuron in the stable fly and house fly, *J. Ag. Food Chem.,* 26:90.

Jacobson, M., Beroza, M., Bull, D. L., Bullock, H. R., Chamberlain, W. R., McGovern, T. P., Redfern, R. E., Sarmiento, R., Schwarz, M., Sonnet, R. E., Wakabayashi, N., Waters, R. M., and Wright, J. E., 1972, Juvenile hormone activity of a variety of structural types against several insect species, *in:* "Insect Juvenile Hormones," J. J. Menn and M. B. Beroza, eds., pp. 249-302, Academic Press, New York.

Jakob, W. L., 1973, Insect development inhibitors: Tests with house fly larvae, *J. Econ. Entomol.,* 66:819.

Jennings, R. C., and Ottridge, A. P., 1979, The synthesis of precocene I epoxide (2,2-dimethyl-3,4-epoxy-7-methoxy-2H-1-benzopyran), *J. Chem. Soc. Chem. Commun.,* 920.

Kadri, A. B. H., 1975, Cross-resistance to an insect juvenile hormone analogue in a species of the *Anopheles gambiae* complex resistant to insecticides, *J. Med. Entomol.,* 12:10.

Kamimura, H., Hammock, B. D., Yamamoto, I., and Casida, J. E., 1972, A potent juvenile hormone mimic, 1-(4'-ethylphenoxy)-6,7-epoxy-3,7-dimethyl-2-octene, labeled with tritium in either the ethylphenyl or the geranyl-derived moiety, *J. Ag. Food Chem.,* 20:439.

Keiding, J., 1977, Resistance in the housefly in Denmark and elsewhere, *in:* "Pesticide Management and Insecticide Resistance," D. L. Watson and A. W. A. Brown, eds., pp. 261-302, Academic Press, New York.

Kramer, S. J., 1978, Regulation of the activity of JH-specific esterases in the Colorado potato beetle, *Leptinotarsa decemlineata, J. Insect Physiol.,* 24:743.

Kramer, S. J., Wieten, M., and deKort, C. A. D., 1977, Metabolism of juvenile hormone in the Colorado potato beetle, *Leptinotarsa decemlineata, Insect Biochem.,* 7:231.

Kuhr, R. J., and Dorough, H. W., 1976, "Carbamate Insecticides: Chemistry, Biochemistry, and Toxicology," CRC Press, Cleveland, Ohio.

Law, J. H., Yuan, C., and Williams, C. H., 1966, Synthesis of a material with juvenile hormone activity, *Proc. Nat. Acad. Sci.,* 55:576.

Lieberman, M., 1977, Post harvest responses and plant growth regulators, *in:* "Pesticide Chemistry in the 20th Century," ACS Symposium Series #37, J. R. Plimmer, ed., pp. 280-292, ACS, Washington, D. C.

Lukovits, I., Tóth, B., Varjas, L., and Matolcsy, G., 1978, Quantitative relationship between structure and anti-ecdysone activity of triarimol analogues, *Acta Phytopath. Hung.*, 13:227.

McCaleb, D. C., and Kumaran, A. K., 1979, Control of juvenile hormone esterase activity in *Galleria mellonella* larvae, *J. Insect Physiol.*, in press.

Maddrell, S. H. P., and Reynolds, S. E., 1972, Release of hormone in insects after poisoning with insecticides, *Nature* (London), 236:404.

Mane, S. D., and Rembold, H., 1977, Developmental kinetics of juvenile hormone inactivation in queen and worker castes of the honey bee, *Apis mellifera, Insect Biochem.*, 7:463.

Matolcsy, G., Varjas, L., and Bordás, B., 1975, Inhibitors of steroid biosynthesis as potential antihormones, *Acta Phytopathologica Academiae Scientiarum Hungaricae*, 10:455.

Menn, J. J., 1980, Contemporary frontiers in chemical pesticide research, *J. Ag. Food Chem.*, 28:2.

Menn, J. J., and Pallos, F. M., 1975, Development of morphogenetic agents in insect control, *in:* "Insecticides of the Future," M. Jacobson, ed., pp. 71-88, Marcel Dekker Inc., New York.

Messenger, P. S., Biliotti, E., and van den Bosch, R., 1976, The importance of natural enemies in integrated control, *in:* "Theory and Practice of Biological Control," C. B. Huffaker and P. S. Messenger, eds., pp. 543-563, Academic Press, New York.

Metcalf, R. L., 1980, Changing role of insecticides in crop protection, *Ann. Rev. Entomol.*, 25:219.

Metcalf, R. L., Lu, P.-Y., and Bowlus, S., 1975, Degradation and environmental fate of 1-(2,6-difluorobenzoyl)-3-(4-chlorophenyl) urea, *J. Ag. Food Chem.*, 23:359.

Morello, A., and Agosin, M., 1979, Metabolism of juvenile hormone with isolated rat hepatocytes, *Biochem. Pharmacol.*, 28:1533.

Mori, K., Takigawa, T., Manabe, Y., Tominaga, M., Matsui, M., Kiguchi, K., Akai, H., and Ohtaki, T., 1975, Effect of the molecular chain length on biological activity of juvenile hormone analogs, *Agr. Biol. Chem.*, 39:259.

Mumby, S. M., Hammock, B. D., Sparks, T. C., and Ota, K., 1979, Synthesis and bioassay of carbamate inhibitors of the juvenile hormone hydrolyzing esterases from the house fly, *Musca domestica, J. Ag. Food Chem.*, 27:763.

Mullin, C. A., 1979, Purification and properties of an epoxide hydratase from the midgut of the southern armyworm (*Spodoptera eridania*), Ph.D. thesis, Cornell University, Ithaca, N. Y.

Murakoshi, S., Nakata, T., Ohtsuka, Y., Akita, H., Tahara, A., and Tamura, S., 1975, Appearance of three-moulters from larvae of the silkworm, *Bombyx mori* L., by oral administration of abietic acid derivatives, *Jap J. appl. Entomol. Zool.*, 19:267.

Nowock, J., and Gilbert, L. E., 1976, In vitro analysis of factors regulating the juvenile hormone titer of insects, *in:* "Invertebrate Tissue Culture," E. Kurstak and K. Maramorosch, eds., pp. 203-212, Academic Press, New York.

Ohta, T., Kuhr, R. J., and Bowers, W. S., 1977, Radiosynthesis and metabolism of the insect anti-juvenile hormone, precocene II, *J. Ag. Food Chem.*, 25:478.

Oppenoorth, F. J., and van der Pas, L. J. T., 1977, Cross-resistance of diflubenzuron in resistant strains of house fly, *Musca domestica*, *Ent. exp. Appl.*, 21:217.

Ordish, G., 1967, "Biological Methods in Crop Pest Control," Constable, London.

Pener, M. P., Orshan, L., and De Wilde, J., 1978, Precocene II causes atrophy of corpora allata in *Locusta migratoria*, *Nature*, 272:350.

Pimprikar, G., and Georghiou, G. P., 1979, Mechanisms of resistance to diflubenzuron in the house fly *Musca domestica* (L.), *Pestic. Biochem. Physiol.*, 12:10.

Plapp, F. W., 1976a, Biochemical genetics of insecticide resistance, *Ann. Rev. Entomol.*, 21:179.

Plapp, F. W., 1976b, Chlordimeform as a synergist for insecticides against the tobacco budworm, *J. Econ. Entomol.*, 69:91.

Plapp, F. W., and Vinson, S. B., 1973, Juvenile hormone analogues: Toxicity and cross-resistance in the house fly, *Pestic. Biochem. Physiol.*, 3:131.

Pratt, G. E., 1975, Inhibition of juvenile hormone carboxylesterase of locust haemolymph by organophosphates in vitro, *Insect Biochem.*, 5:595.

Pratt, G. E., and Bowers, W. S., 1977, Precocene II inhibits juvenile hormone biosynthesis by cockroach corpora allata in vitro, *Nature*, 265:548.

Pratt, G. E., and Finney, J. R., 1977, Chemical inhibitors of juvenile hormone biosynthesis in vitro, *in:* "Crop Protection Agents--Their Biological Evaluation," N. R. McFarlane, ed., pp. 113-132, Academic Press, New York.

Quistad, G. B., Staiger, L. E., and Schooley, D. A., 1975, Environmental degradation of the insect growth regulator methoprene V. Metabolism by houseflies and mosquitoes, *Pestic. Biochem. Physiol.*, 5:253.

Reddy, G., Hwang-Hsu, K., and Kumaran, A. K., 1979, Factors influencing juvenile hormone esterase activity in the wax moth, *Galleria mellonella*, *J. Insect Physiol.*, 25:65.

Riddiford, L. M., and Truman, J. W., 1978, Biochemistry of insect hormones and insect growth regulators, *in:* "Biochemistry of Insects," M. Rockstein, ed., pp. 307-357, Academic Press, New York.

Röller, H., Dahm, K. H., Sweely, C. C., and Trost, B. M., 1967, The structure of juvenile hormone, *Angew. Chem. internat. Edit.*, 6:179.

Rongsriyam, Y., and Busvine, J. R., 1975, Cross-resistance in DDT-resistant strains of various mosquitoes (Diptera, Culicadae), *Bull. ent. Res.*, 65:459.

Rupes, V., Zdarek, J., Svandova, E., and Pinterova, J., 1976, Cross-

resistance to a juvenile hormone analogue in wild strains of the house fly, *Ent. exp. Appl.*, 19:57.
Ruscoe, C. N. E., 1974, Exploitation of insect endocrine systems, *Chem. Ind.* (London), 16:648.
Samaranayaka, M., 1974, Insecticide-induced release of hyperglycaemic and adipolinetic hormones of *Schistocerca gregaria*, *Gen. Comp. Endocrinol.*, 24:424.
Sanborn, J. R., and Fukuto, T. R., 1972, Insecticidal, anticholinesterase, and hydrolytic properties of S-aryl phosphoamidothioates, *J. Ag. Food Chem.*, 20:926.
Sanburg, L. L., Kramer, K. J., Kezdy, F. J., and Law, J. H., 1975a, Juvenile hormone-specific esterases in the hemolymph of the tobacco hornworm, *Manduca sexta*, *J. Insect Physiol.*, 21:873.
Sanburg, L. L., Kramer, K. J., Kezdy, F. J., Law, J. J., and Oberlander, H., 1975b, Role of juvenile hormone esterases and carrier proteins in insect development, *Nature*, 253:266.
Schaefer, C. H., and Wilder, W. H., 1972, Insect development inhibitors: A practice evaluation as mosquito control agents, *J. Econ. Entomol.*, 65:1066.
Schaefer, C. H., and Wilder, W. H., 1973, Insect development inhibitors. 2. Effects on target mosquito species, *J. Econ. Entomol.*, 66:913.
Schneiderman, H. A., 1971, The strategy of controlling insect pests with growth regulators, *Mitt. Schweiz. Entomol. Ges.*, 44:141.
Schneiderman, H. A., 1972, Insect hormones and insect control, *in:* "Insect Juvenile Hormones: Chemistry and Action," J. J. Menn and M. Beroza, eds., pp. 3-27, Academic Press, New York.
Schooley, D. A., and Bergot, B. J., 1979, Biochemical studies on juvenile hormone antagonists, Paper 70, Pesticide Chemistry Div. 178th National American Chemical Society National Meeting, Sept. 9-14, Washington, D.C.
Schooneveld, H., Kramer, S. J., Privee, H., and Van Huis, A., 1979, Evidence of controlled corpus allatum activity in the adult Colorado potato beetle, *J. Insect Physiol.*, 25:449.
Schwarz, M., Miller, R. W., Wright, J. E., Chamberlain, W. F., and Hopkins, D. E., 1974, Compounds related to juvenile hormone. Exceptional activity of arylterpenoid compounds in four species of flies, *J. Econ. Entomol.*, 67:598.
Sehnal, F., 1976, Action of juvenoids on different groups of insects, *in:* "The Juvenile Hormones," L. I. Gilbert, ed., pp. 301-322, Plenum Press, New York.
Siddall, J. B., 1976, Insect growth regulators and insect control: A critical appraisal, *Environ. Hlth. Perspec.*, 14:119.
Silhacek, D. L., Oberlander, H., and Zettler, J. L., 1976, Susceptibility of malathion-resistant strains of *Plodia interpunctella* to juvenile hormone treatments, *J. Stored Prod. Res.*, 12:201.
Slade, M., 1975, unpublished information.
Slade, M., and Wilkinson, C. F., 1973, Juvenile hormone analogs: A possible case of mistaken identity, *Science*, 181:672.

Slade, M., and Wilkinson, C. F., 1974, Degradation and conjugation of Cecropia juvenile hormone by southern armyworm (*Prodenia eridania*), *Comp. Biochem. Physiol.*, 49B:99.

Slade, M., and Zibitt, C. H., 1971, Metabolism of cecropia juvenile hormone in lepidopterans, *in:* "Chemical Releasers in Insects. Proceedings of International IUPAC Congress on Pesticide Chemistry," A. S. Tahori, ed., 3:45, Gordon and Brench, New York.

Slade, M., and Zibitt, C. H., 1972, Metabolism of cecropia juvenile hormone in insects and mammals, *in:* "Insect Juvenile Hormones, Chemistry and Action," J. J. Menn and M. Beroza, eds., pp. 155-176, Academic Press, New York.

Slade, M., Brooks, G. T., Hetnarski, H. K., and Wilkinson, C. F., 1975, Inhibition of the enzymatic hydration of the epoxide HEOM in insects, *Pestic. Biochem. Physiol.*, 5:35.

Slama, K., 1978, The principles of antihormone action in insects, *Acta Entomologica Bohemoslovaca*, 75:65.

Slama, K., and Romanuk, M., 1976, Juvenogens, biochemically activated juvenoid complexes, *Insect Biochem.*, 6:579.

Slama, K., Romanuk, M., and Sorm, F., 1974, "Insect Hormones and Bioanalogues," 477p., Springer-Verlag, New York.

Slama, K., Kahovcova, J., and Romanuk, M., 1978, Action of some aromatic juvenogen esters on insects, *Pestic. Biochem. Physiol.*, 9:313.

Smith, R. F., 1970, Pesticides: Their use and limitations in pest management, *in:* "Concepts of Pest Management," R. L. Rabb and F. E. Guthrie, eds., pp. 103-118, North Carolina State University, Raleigh.

Solomon, K. R., and Metcalf, R. L., 1974, The effect of piperonyl butoxide and triorthocresyl phosphate on the activity and metabolism of Altosid (isopropyl 11-methoxy-3,7,11-trimethyldodeca-2,4-dienoate) in *Tenebrio molitor* L. and *Oncopeltus fasciatus* (Dallas), *Pestic. Biochem. Physiol.*, 4:127.

Sorm, F., 1971, Some juvenile hormone analogues, *Mitt. Schwiez ent. Gell.*, 44:7.

Sparks, T. C., and Hammock, B. D., 1979a, Induction and regulation of juvenile hormone esterases during the last larval instar of the cabbage looper, *Trichoplusia ni*, *J. Insect Physiol.*, 25:551.

Sparks, T. C., and Hammock, B. D., 1979b, A comparison of the induced and naturally occurring juvenile hormone esterases from last instar larvae of *Trichoplusia ni*, *Insect Biochem.*, 9:411.

Sparks, T. C., and Hammock, B. D., 1980, Comparative inhibition of the juvenile hormone esterases from *Trichoplusia ni*, *Musca domestica* and *Tenebrio molitor*, *Pestic. Biochem. Physiol.*, submitted.

Sparks, T. C., Willis, W. S., Shorey, H. H., and Hammock, B. D., 1979a, Haemolymph juvenile hormone esterase activity in synchronous last instar larvae of the cabbage looper, *Trichoplusia ni*, *J. Insect Physiol.*, 25:125.

Sparks, T. C., Wing, K. D., and Hammock, B. D., 1979b, Effects of the antihormone-hormone mimic ETB on the induction of insect juvenile hormone esterase in *Trichoplusia ni, Life Sci.*, 25:445.

Staal, G. B., 1975, Insect growth regulators with juvenile hormone activity, *Ann. Rev. Entomol.*, 20:417.

Staal, G. B., 1977, Insect control with insect growth regulators based on insect hormones, *in:* "Natural Products and the Protection of Plants," G. G. Marini-Bettolo, ed., pp. 353-383, Elsevier Scientific Publishing Co., New York.

Staal, G. B., 1979, Essential differences between natural juvenile hormones and juvenile hormone analogs elucidated by use of a substitution assay, *Pont. Acad. Scien. Scripta Varia,* The Vatican 41:353.

Still, G. G., and Lepold, R. A., 1978, The elimination of (*N*[[(4-chlorophenyl)amino]carbonyl]-2,6-difluorobenzamide) by the boll weevil, *Pestic. Biochem. Physiol.*, 9:304.

Sweeny, M., 1978, Fourth radioimmunoassay directory, Lab World, 29:48.

Sylwester, A., Wing, K. D., and Hammock, B. D., 1979, Immunochemical analysis of insecticides: Haptens and antigens for the insect growth regulator diflubenzuron, Paper 30, 63rd Pacific Branch meeting of the Entomological Society of America, June 26-28, Fresno, California.

Terriere, L. C., and Yu, S. J., 1973, Insect juvenile hormones: Induction of detoxifying enzymes in the house fly and detoxification by house fly enzymes, *Pestic. Biochem. Physiol.*, 3:96.

Terriere, L. C., and Yu, S. J., 1977, Juvenile hormone analogs: In vitro metabolism in relation to biological activity in blow flies and flesh flies, *Pestic. Biochem. Physiol.*, 7:161.

Tobe, S. S., and Stay, B., 1979, Modulation of juvenile hormone synthesis by an analogue in the cockroach, *Nature*, 281:481.

Vea, E. V., Yu, C.-C., Webb, D. R., Eckenrode, C. J., and Kuhr, R. J., 1976, Laboratory and field evaluation of insecticides and insect growth regulator for control of the seedcorn maggot, *J. Econ. Entomol.*, 69:178.

Verloop, A., and Ferrell, C. D., 1977, Benzoylphenyl ureas--a new group of larvicides interfering with chitin disposition, *in:* "Pesticide Chemistry in the 20th Century," ACS Symp. Ser. #37, J. R. Plimmer, ed., pp. 237-270, ACS, Washington, D.C.

Vince, R. K., and Gilbert, L. I., 1977, Juvenile hormone esterase activity in precisely timed last instar larvae and pharate pupae of *Manduca sexta, Insect Biochem.*, 7:115.

Vinson, S. G., and Plapp. F. W., 1974, Third generation pesticides: The potential for the development of resistance by insects, *J. Ag. Food Chem.*, 22:356.

Wardlow, L. R., Ludlam, A. G., and Bradley, L. F., 1976, Pesticide resistance in glasshouse whitefly (*Trialeurodes vaporariorum* West.), *Pestic. Sci.*, 7:320.

Weirich, G., and Wren, J., 1973, The substrate specificity of

juvenile hormone esterase from *Manduca sexta* haemolymph, *Life Sci.*, 13:213.
Weirich, G., and Wren, J., 1976, Juvenile hormone esterase in insect development: A comparative study, *Physiol. Zool.*, 49:341.
Weirich, G., Wren, J., and Siddall, J. B., 1973, Developmental changes of the juvenile hormone esterase activity in haemolymph of the tobacco hornworm, *Manduca sexta*, *Insect Biochem.*, 3:397.
Wellinga, K., Mulder, R., and van Daalen, J. J., 1973, Synthesis and laboratory evaluation of 1-(2,6-disubstituted benzoyl)-3-phenylureas, a new class of insecticides. I. 1-(2,6-dichlorobenzoyl)-3-phenylureas, *J. Ag. Food Chem.*, 21:348.
White, A. F., 1972, Metabolism of the juvenile hormone analogue methyl farnesoate 10,11-epoxide in two insect species, *Life Sci.*, 2:201.
Whitmore, D., Whitmore, E., and Gilbert, L. I., 1972, Juvenile hormone induction of esterases: A mechanism for the regulation of juvenile hormone titer, *Proc. Nat. Acad. Sci., USA*, 69:1592.
Wilkinson, C. F., 1976, Insecticide synergism, *in:* "The Future for Insecticides," R. L. Metcalf and J. J. McKelvey, eds., pp. 195-218, John Wiley and Sons, New York.
Williams, C. M., 1967, Third-generation pesticides, *Sci. Am.*, 217:13.
Williams, C. M., 1976, Juvenile hormone --- in retrospect and in prospect, *in:* "The Juvenile Hormones," L. I. Gilbert, ed., pp. 1-14, Plenum Press, New York.
Wilson, T. G., and Gilbert, L. I., 1978, Metabolism of juvenile hormone I in *Drosophila melanogaster*, *Comp. Biochem. Physiol.*, 60A:85.
Winteringham, F. P. W., 1969, Mechanisms of selective insecticidal action, *Ann. Rev. Entomol.*, 14:409.
Yu, S. J., and Terriere, L. C., 1975, Microsomal metabolism of juvenile hormone analogs in the house fly, *Musca domestica* L., *Pestic. Biochem. Physiol.*, 5:418.
Yu, S. J., and Terriere, L. C., 1977, Metabolism of [^{14}C]hydroprene (ethyl 3,7,11-trimethyl-2,4-dodecadienoate) by microsomal oxidases and esterases from three species of Diptera, *J. Ag. Food Chem.*, 25:1076.
Yu, S. J., and Terriere, L. C., 1978a, Metabolism of juvenile hormone I by microsomal oxidase, esterase, and epoxide hydrase of *Musca domestica* and some comparisons with *Phormia regina* and *Sarcophaga bullata*, *Pestic. Biochem. Physiol.*, 9:237.
Yu, S. J., and Terriere, L. C., 1978b, Juvenile hormone epoxide hydrase in house flies, flesh flies and blow flies, *Insect Biochem.*, 8:349.
Zurflueh, R. C., 1976, Phenylethers as insect growth regulators: Laboratory and field experiments, *in:* "The Juvenile Hormones," L. I. Gilbert, ed., pp. 61-74, Plenum Press, New York.

NATURAL ENEMY RESISTANCE TO PESTICIDES:

DOCUMENTATION, CHARACTERIZATION, THEORY AND APPLICATION*

B. A. Croft and K. Strickler

Pesticide Research Center
Department of Entomology
Michigan State University
East Lansing, Michigan 48824

INTRODUCTION

After arthropod pests first developed resistance to pesticides (e.g., DDT), entomologists questioned whether or not the natural enemies of arthropods would do the same (e.g., Pielou and Glasser, 1952; Wilkes et al., 1951; Spiller, 1958). In the early 1950s *Macrocentrus anyclivorus*, a braconid parasite of the Oriental fruit moth (*Grapholitha molesta*) was selected with DDT for 70 generations to see if a resistance potential was present and if it could be exploited by design. After six years and three million treated insects a disappointing maximum resistance of 12-fold resulted. When selection was discontinued, resistance regressed back to the original level within a few generations (Pielou and Glasser, 1952; Wilkes et al., 1951; Robertson, 1957). The failure to produce a resistant *M. ancylivorus* was typical of other attempts at selection of resistant beneficial arthropods in laboratory experiments during the period 1955-70 (e.g., Adams and Cross, 1967; Kot et al., 1971). There were no reports of significant resistance developing in field populations of natural enemies during this time period.

In the 1970s, interest in resistance of natural enemies to pesticides was rekindled by simultaneous reports of suspected organophosphate (OP) resistance in the predatory phytoseiid mites,

*Published as Michigan Agricultural Experiment Station Journal Article No. 9046. Supported in part by an EPA Grant No. CR-806277-02-2 to Texas A&M University and Michigan State University.

Table 1. Resistance to Pesticides Among Arthropod Parasites and Predators

Species (Family)	Pesticide	Fold resistance level	Crop	Conditions of selection[a]	Location	Selected references
Macrocentrus ancylivorus (Braconidae)	DDT	12	Peach	L	Canada	Pielou et al. 1952 Robertson 1958
Bracon mellitor (Braconidae)	DDT-Toxaphene	8	Cotton	L	U.S.A.	Adams and Cross 1967
Aphytis melinus (Aphelinidae)	Malathion Parathion	3–4 >10	Citrus	L F	Australia	Abdelrahman 1973 Strawn 1978
Trichogramma evanescens (Tricho-grammatidae)	Demeton-methyl	22	Several crops	L	Poland	Kot et al. 1977
Aphidoletes aphidomyza (Cecidomyidae)	Azinphosmethyl	[b]	Apple	F	U.S.A.	Adams and Prokopy 1977
Coleomagilla maculata (Coccinellidae)	Parathion	10–35	Cotton	F	U.S.A.	Chambers 1973
Amblyseius andersoni (Phytoseiidae)	Azinphosmethyl	[b]	Apple	F	Italy	Gambaro 1975

Species (Family)	Pesticide	Fold resistance level	Crop	Conditions of selection[a]	Location	Selected references
Amblyseius chilenensis (Phytoseiidae)	Phosmet	10	Apple	F	Uruguay	Croft et al. 1976a
	Azinphosmethyl	b/				
Amblyseius fallacis (Phytoseiidae)	Azinphosmethyl	100-1000	Apple	F	U.S.A.	Motoyama et al. 1970
	Carbaryl	25-77				Croft et al. 1976b
	Diazinon	119				Croft and Meyer 1973
	Parathion	102-152				
Amblyseius hibisci (Phytoseiidae)	Parathion	b/	Citrus	F	U.S.A.	Kennett 1970
Phytoseiulus persimilis (Phytoseiidae)	Parathion	>143	Greenhouse crops	F	Netherlands	Schulten et al. 1976
	Diazinon	292				
	Demeton-methyl	33				
Typhlodromus occidentalis (Phytoseiidae)	Azinphosmethyl	101-104	Apple	F	U.S.A.	Croft and Jeppson 1970
					Canada	Ahlstrom and Rock 1973
Typhlodromus pyri (Phytoseiidae)	Azinphosmethyl	10-42	Apple	F	U.S.A.	Watve and Lienk 1976
					New Zealand	Penman et al. 1976

a/ F = field, L = laboratory selection.

b/ Not calculated.

Typhlodromus occidentalis and *Amblyseius fallacis*. Whereas these predators previously were excluded from commercial apple orchards by OPs applied to control codling moth, *Laspeyresia pomonella*, and other key pests, they began to survive OP applications (Hoyt, 1969; Swift, 1970). Later, scientists documented these resistant (R) strains by comparison with susceptible (S) strains (Croft and Jeppson, 1970; Motoyama et al., 1970) and exploited them in integrated pest management (IPM) in several fruit-growing states (see reference in Table 5 and later discussion). Since these initial studies, additional documentations, characterizations, theoretical discussions and uses of insecticide-resistant natural enemies have been reported. Most of this work has been done with predatory phytoseiid mites and their tetranychid prey.

Documentation of Resistant Natural Enemy Species

Evidence of resistance in beneficial insects is less obvious than is resistance among pests, and entomologists rarely monitor for it (Croft and Brown, 1975). In spite of this, natural enemy resistance has been verified by study of a few populations under selection in the laboratory and field (see reviews by Croft and Brown, 1975; Croft, 1977; FAO, 1977). Table 1 contains an updated listing of these predators and parasites and the compounds to which resistance has developed. The frequencies of reported cases show some interesting trends. At the end of the DDT era (1952-1960) there were only a few documentations; after 10 to 20 years of OP use (1970-1978) there was an increase in verified cases. At present there are only 13 documented cases of resistant natural enemies, whereas 281 agricultural pest species are known to be resistant to at least one pesticide (FAO, 1977).

The first documented cases of R arthropod natural enemies were laboratory selections with *M. anyclivorus* (Pielou and Glasser, 1952; Wilkes et al., 1951; Robertson, 1957), *B. mellitor* (Adams and Cross, 1967), *T. evanescens* (Kot et al., 1971, 1977) and *Aphytis melinus* (Abdelrahman, 1973). As mentioned earlier, only relatively low levels of resistance then existed. Limited resistance in a field-selected parasite was demonstrated by Strawn (1978) for *Aphytis melinus* attacking California red scale (Table 1). Populations were taken from citrus orchards that were either unsprayed, moderately sprayed or heavily sprayed with methyl parathion. Parasites showed approximately a 10-fold increase in resistance at the moderately treated site, with lower resistance in both the untreated and heavily treated block. These observations may be explained by the food-limitation hypothesis discussed later. Developed resistance in field populations of parasites was suspected in *Aphelinus mali* (Spiller, 1958) and *Bracon mellitor* (Adams and Cross, 1967) but was never proven by S strain comparisons.

Among insect predators, resistance has been found in aphidophagous Coccinellidae and Cecidomyidae (Table 1). In Louisiana cotton fields, Atallah and Newson (1966) measured a 6-fold resistance to DDT in the ladybird beetle, *Coleomegilla maculata*. More recently, a 35-fold increase in resistance to methyl parathion, which has been used for two decades, was reported from Mississippe cotton fields (Chambers, 1973), and the cecidomyid, *Aphidoletes aphidimyza*, has developed some azinphosmethyl resistance in Massachusetts apple orchards (Adams and Prokopy, 1977). Similarly, OP-resistant strains of the coccinellid mite predator, *Stethorus punctum*, were believed to have developed in apple orchards of Pennsylvania in the mid 1960s (D. Asquith, personal communication), but this resistance has never been documented by comparison with S strains. Cases of incipient resistance in a dipterous predator of house fly larvae, a spider predaceous on cotton pests, and a corixid water bug were reviewed by Croft (1977).

Among the phytoseiid mites, resistance occurs in at least seven species, including four of national or international distribution. These mites are resitant to a variety of chemicals (Table 1) and have been widely managed and transferred for practical use in pest control programs (see later discussion). The most widespread and exploited species have been *T. occidentalis* and *A. fallacis* in North America. R strains of the species *Typhlodromus pyri* and *Phytoseiulus persimilis* are also increasing in both their geographic extent and levels of resistance, and should become important in IPM programs, as have the former two mites.

T. occidentalis was reported resistant or tolerant to OP compounds as early as 1952 by Huffaker and Kennett (1953). Thereafter, populations highly resistant or tolerant to several OP compounds, including azinphosmethyl, diazinon, phosmet, parathion, tepp and phosalone, were reported from the semiarid fruit-growing regions of western North America (see review and map in Croft and Brown, 1975). Resistance and cross resistance to other chemical groups (e.g., carbaryl) probably also occur, but as yet are undocumented.

In contrast to *T. occidentalis*, *T. pyri* has not developed high levels (>100 fold, see Table 1) of resistance to OPs even though such compounds are commonly applied in northwestern and northeastern North America, western and eastern Europe, and New Zealand where this species occurs. In New Zealand, Hoyt (1972) first reported a 10-fold resistance to azinphosmethyl, and Collyer and Geldermalsen (1975) later found populations in New Zealand with 14-fold resistance. In New York, Watve and Lienk (1976) reported an LC_{50} level that gave nearly a 20-fold resistance to azinphosmethyl when compared with the data of Hoyt (1972) for an S strain. At this level predators tolerated field applications and provided appreciable biological control of *Panonychus ulmi* in apple orchards. Most recently, Penman et al.

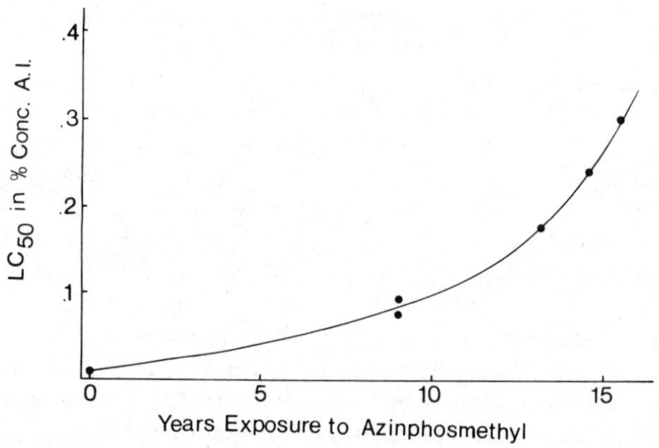

Figure 1. Relationship between LC_{50} values for field-collected strains of *T. pyri* and the number of years of exposure to azinphosmethyl. (Adapted from Penman et al., 1976.)

(1976) documented resistance to even higher levels in New Zealand but only after longterm exposure (Figure 1). Whereas Hoyt (1972) found resistance too low in 1970-71 to allow for integrated control, resistance in 1975-76 had increased 40-fold and was sufficient to provide for complete predator survival after 10 years of intensive azinphosmethyl application. *T. pyri* also is tolerant or has developed high levels of resistance to carbaryl (Van de Vrie, 1962; Watve and Lienk, 1976; Collyer and Geldermalsen, 1975). Contrast this with the situation for *A. fallacis:* Four years of exposure to Michigan and Indiana OP spray programs on apple led to 100-fold increases in azinphosmethyl resistance, but nine years of intensive selection were required for carbaryl resistance to reach levels of 22- or 70-fold (Croft and Meyer, 1973).

Resistance to azinphosmethyl and phosmet is present in *A. chilenensis* populations from Uruguayan apple orchards (Table 1). Selection continues since field rates applied for codling moth or leafroller control still cause some predator mortality, but biological control of the European red mite, *P. ulmi,* has been partially achieved in the area (Croft et al., 1976a). Recent reports (R. Weires and L. Campos, unpublished) indicate that OP resistance in this species is widespread in Chile. Italian populations of *Amblyseius andersoni* in peach orchards readily survive treatments with azinphosmethyl and other OPs, presumably due to developed resistance (Gambaro, 1975). *Amblyseius hibisci* occurring on citrus in California (parathion) and *T. caudiglans* in Canadian peach orchards (DDT) have only shown limited and isolated cases of resistance (Table 1).

A. fallacis strains resistant to DDT, carbaryl and several OP compounds (e.g., azinphosmethyl, parathion, phosmet, diazinon and stirofos) have been detected (Table 1), and OP cross resistance extends to at least 11 other insecticides (Croft and Nelson, 1972; Croft et al., 1976b). Whereas DDT and carbaryl resistance occur locally, OP resistance is present throughout midwestern and eastern North America (see map in Croft and Brown, 1975), to the extent that it is difficult to find S populations for experimental study (Croft, unpublished). Some success in developing a bi-resistant OP x carbaryl strain was reported by Croft and Meyer (1973), but attempts to establish, maintain and manage it in the field have been unsuccessful (Meyer, 1975; Croft and Hoying, 1975). OP resistance is easily developed and is stable in *A. fallacis* populations; however, carbaryl resistance develops only under intensive selection and is much less stable. Croft and Hoying (1975) concluded that if OP resistance develops among fruit pests other than mites, and if alternate pesticides are used, then resistance to carbaryl among *A. fallacis* populations could develop sufficiently to allow for integrated mite control only if carbaryl is used intensively.

Resistance to parathion and demeton-S-methyl in *P. persimilis* provides insights into the problems of developing laboratory-selected strains of resistant natural enemies from limited gene pools (Schulten et al., 1976). This tropical or subtropical mite was initially colonized from limited South American stocks. It has been released worldwide for controlling mite pests in greenhouses or in high-value vegetable and field crops (e.g., strawberries) (Oatman et al., 1968), especially in temperate areas where it does not overwinter. After colonization, selection for resistance in the laboratory and greenhouse met with little success (unpublished reports; Helle and Van de Vrie, 1974; Schulten et al., 1976). Schulten et al. (1976) recently found an R strain following recolonization from a greenhouse in the Netherlands. They reported a 100-fold resistance to parathion and a 30-fold resistance to demeton-S-methyl in one strain after eight years of selection, but resistance was entirely absent in another similarly treated strain from the same source population. These data suggest that resistance arose from a new mutation rather than from an R genotype originally present at a low frequency.

Characterization of Resistance

Genetics and toxicology of resistance and cross resistance in natural enemies has only been studied among phytoseiid mites. In Table 2, cross resistance in *A. chilenensis*, *A. fallacis*, *T. occidentalis* and *T. pyri* are sumarized for seven compounds used for deciduous fruit pest control. Azinphosmethyl resistance is sufficient to render field rates virtually innocuous to *A. fallacis* and *T. occidentalis*, and to *T. pyri* in some areas, but not to *A. chilenensis*. Phosmet (Imidan®) resistance is strongly linked to azinphosmethyl resistance, and lethal dosages of both compounds are comparable for each mite species. Parathion is very toxic to *A. chilenensis*

Table 2. Relative Toxicity of Field Application Rates and Cross Resistance Relationships of Seven Broadspectrum Insecticides Among Four Azinphosmethyl-resistant Phytoseiid Mites

Chemical compound	Species			
	Amblyseius chilenensis	*Amblyseius fallacis*	*Typhlodromus occidentalis*	*Typhlodromus pyri*
Azinphosmethyl	M[a]	L	L	L
Phosmet	M, (+)[b]	L, (+)	L, (+)	M, (+?)
Parathion	H, (-)	M, (+)	L, (-)	M, (+?)
Stirophos	L, (+)	M, (+)	M, (+)	NI[c]
Phosalane	NI	H, (-)	L, (+?)	L, NI
tepp	NI	H, (-)	L, (+?)	NI
Carbaryl	L, (-)	H, (-)	H, (-)	L, (-?)

[a]L = lowly, M = moderately, H = highly toxic.
[b](-) = no evidence of cross resistance to azinphosmethyl, (+) = positive evidence, ? = inferred from indirect data.
[c]NI = no information available.
(Adapted from Croft, 1977.)

and is not cross-related to azinphosmethyl resistance. For *A. fallacis,* however, parathion is moderately toxic and linked with azinphosmethyl resistance (Motoyama et al., 1970; Rock and Yeargan, 1971; Croft et al., 1976b). For *T. pyri,* parathion is moderately toxic and cross-resistant with azinphosmethyl (Watve and Lienk, 1975, 1976). *T. occidentalis* is inherently tolerant to parathion rather than resistant or cross resistant, as indicated by its immunity soon after parathion was registered for use (Huffaker and Kennett, 1953). Croft and Jeppson (1970) reported similar LC_{50} levels for parathion among two intensively selected strains (Washington and Utah) and two untreated strains of this species. More recently, tolerance to parathion was indicated in Australian populations of *T. occidentalis,* which are killed by field rates of azinphsomethyl but not parathion (Field, 1974, 1976).

Resistance to stirofos is cross-linked with azinphosmethyl resistance in all four species, but this compound is slightly toxic to *A. chilenensis* and highly toxic to *A. fallacis* and *T. occidentalis*

(Table 2). Azinphosmethyl-R strains of *A. fallacis* are not cross-resistant to phosalone and tepp, but resistant *T. occidentalis* are relatively immune to these two OPs, probably due to the mechanism conferring parathion tolerance. Carbaryl is almost nontoxic to *A. chilenensis* and *T. pyri*, but is extremely toxic to *A. fallacis* and *T. occidentalis*; there is no cross resistance between carbaryl and azinphosmethyl in any species.

In summary, *A. chilenensis, A. fallacis, T. occidentalis* and *T. pyri* vary greatly in their resistances, toxicities and cross resistances. Interspecific resistance differences within the family are as variable as those of their tetranychid prey (Helle and Van de Vrie, 1974).

Toxicology and Mechanisms of Resistance

Resistance mechanisms have been investigated in two populations of *A. fallacis* resistant to azinphosmethyl. Motoyama et al. (1971) reported that an OP-R strain from North Carolina degraded azinphosmethyl faster than an S strain, with less cholinesterase inhibition in vivo. No differences in biomolecular rate constants of the cholinesterase activities were detected between S and R strains. In the R strain, there was a 6-fold desalkylation of azinphosmethyl to a mono-desmethyl derivative, and a 3-fold increase in glutathione transferase activity. Since glutathione transferases are specific for methyl substrates, these data suggest that mites could be less resistant to OP compounds containing ethyl esters than to compounds containing methyl esters. Croft et al. (1976b) evaluated the differential cross resistance of a Michigan strain of *A. fallacis* to sixteen OP compounds. These mites were previously exposed mainly to azinphosmethyl and to limited applications of diazinon. Significantly high resistance ratios (R:S) to the methyl phosphorodithioates dimethoate, malathion and phosmet were observed, while significantly low resistance ratios to the ethyl phosphorodithioates carbophenothion and ethion were found. However, cross resistance to azinphosmethyl was relatively high. With respect to the methyl- versus ethyl-substituted phosphorothioates, resistance levels were appreciable for either group although the ethyl derivatives, parathion and diazinon, gave particularly high resistance ratios. Only two straight-chain phosphates, phosphamidon (methyl) and tepp (ethyl), were evaluated; cross resistance was moderate to the former and extremely low for the latter insecticide. Thus, levels of resistance were not consistently greater for methyl OP derivatives. However, because the mites had been exposed to some diazinon, this strain may have had other detoxification enzymes than glutathione transferase.

Motoyama et al. (1977) reported further studies of the role of glutathione transferase in OP resistance of *A. fallacis*. They compared the toxicity of *O*-alkyl analogs of azinphosmethyl to S and R strains and found that resistance decreased with ethyl and *n*-propyl analogs. This agrees with their previous findings that glutathione transferase confers azinphosmethyl resistance in the strain studied. They further predicted that R strains of *A. fallacis* would exhibit strong cross resistances, particularly to OP insecticides with *O*-methyl groups.

Genetics of Resistance

Croft et al. (1976b) evaluated the genetics of resistance to azinphosmethyl for a Michigan strain of *A. fallacis* using crossing and backcrossing methods with R and S mites. Offspring of reciprocal crosses showed almost identical dosage-mortality lines between parental lines, indicating partial dominance. When F_1 females were backcrossed with S males, the offspring showed a d-m line with long flexions at the 40-50% level, which is reasonably congruent with theoretical lines calculated for a 1:1 proportion of hybrid (R+) and S(++) genotypes. Data reflected clear segregation, indicating that the resistance factor is principally due to a single gene allele. Study of resistance in *P. persimilis* indicated that OP resistance probably is conferred by a dominant, single gene in this species as well (Schulten and van de Kloshorst, 1977).

Theory of Natural Enemy Resistance to Pesticides

Many factors influence pesticide resistance development in arthropod populations. Georghiou and Taylor (1977a,b) (Table 5) identified 21 factors influencing pests, categorized as genetic, biological and operational (i.e., those under the control of an operator or pest manager). Two additional factors (Croft and Brown, 1975) may help explain the greater frequency of resistance in pests as compared to natural enemies.

The Food Limitation Hypothesis

One factor influencing resistance in natural enemies is that these organisms depend on their hosts for survival and multiplication following an insecticide application. An insecticide may select R genotypes of both pests and natural enemies, but surviving pests have an unlimited food supply for reproduction whereas surviving natural enemies face a reduced prey or host population. With excessive pest mortality, the natural enemy starves, does not reproduce, or migrates out of the sprayed area to interbreed with a wild, non-resistant population (see more detailed discussion of this phenomenon in Huffaker, 1971; Croft, 1972; Georghiou, 1972; Newsom, 1974; Croft and Brown, 1975).

Morse and Croft (1980) tested the food limitation hypothesis with azinphosmethyl-S strains of the predaceous mite, *A. fallacis*,

Table 3. Biological Comparisons Between *Tetranychus urticae* and *Amblysieus fallacis*

Parameter	*T. urticae*	*A. fallacis*	Source[a]
number of developmental stages	4	4	van de Vrie et al. 1972 McMurtry et al. 1970
developmental period	6-10 days	5-11.6 days	van de Vrie et al. 1972 Ballard 1954 McClanahan 1968
preovipositional period	1 day (22-27 C)	1 day (26 C)	Bourdreaux 1954 Smith 1965
duration of oviposiiton	10.8-26.3 days	22 days	Caegle 1949 Ballard 1954
intrinsic rate of increase	.2585 (22C)	.279 (25 C)	Wrensch and Young 1975 Croft, unpubl.
mode of reproduction	arrhenotokous parthenogenesis	bisexual heteropycnosis	Helle and Bolland 1967 Helle et al. 1978
# chromosomes			
male (n)	3	4	Helle and Bolland 1967
female (2n)	6	8	Hansell et al. 1964
mating necessary for egg production	no	yes	Helle and Bolland 1967 Rock et al. 1976
sex ratio	varies 2.0-5.0	varies average 2.5	Overmeer and Harrison 1969 Burrell and McCormick 1964 Smith and Newsom 1970 See 1972

(cont'd)

Table 3 (cont.)

Parameter	*T. urticae*	*A. fallacis*	Source[a]
sex ratio (cont.)			Croft, unpubl. Rock et al. 1971 Ballard 1954
size (length by width)	450 by 300 (microns)	500 by 250 (microns)	
diapause	fertilized adult female	fertilized adult female	van de Vrie McMurtry et al. 1970

[a]Detailed literature source listing in Morse (1978).

and its prey, *T. urticae*. In Table 3, biological comparisons that have relevance to resistance development are made between these two species. This predator-prey system is ideal for studying the effects of prey density on development of predator resistance following pesticide treatment. Both species are similar in size, developmental stages and times, and intrinsic rates of increase. Both occupy plant substrates or treated surfaces the same way for all life stages even though the two species occupy different trophic positions (herbivore vs. predator) and have differing diets. The predator's movements are only slightly more extensive than the prey's. Increased predator movement may be important following a spray application because the predator may contact more pesticide as it searches for prey, whose density has been greatly reduced by the chemical.

Other comparisons between the two species that might affect resistance are in their genetic and toxicological properties. Their modes of reproduction and chromosome numbers differ somewhat: the spider mite can produce haploid males without mating whereas the predator mite cannot, and in both species, resistance to OPs, i.e., parathion in *T. urticae* (Herne and Brown, 1969) and azinphosmethyl in *A. fallacis* (Croft et al., 1976), appears to be conferred by a dominant or semi-dominant, single gene factor. In the study by Morse (1978), S strains of *T. urticae* and *A. fallacis* had similar slope values: 2.20 and 1.74 units probit mortality per 10-fold increase in concentration, respectively. One difference between the populations was their intrinsic susceptibility levels: an S strain of *T. urticae* had a baseline LC_{50} ca. 30 times greater than that for S populations of *A. fallacis* (see the differential susceptibility hypothesis discussed later in this chapter). Since the relative increase in resistance was of primary interest, this difference in the S strains was not considered critical in the following experiments.

Selections comparing the rates of development of resistance for the predator and prey were carried out in a greenhouse bean-plant environment in which both species occur naturally. Populations of pest mites and predators varied from 10^4 to 10^6 individuals at the time of the spray applications. Sprays to plant foliage were made with a compressed air sprayer using azinphosmethyl 50% w.p. Two treatments were compared: (1) S populations of predator or prey mites were selected separately with unlimited food following selection. Dosages were chosen to kill 75% of either predator or prey. Treatments were made every generation for 22 generations under similar environmental conditions and population densities. (2) To more closely simulate field conditions, S mites of both species were exposed on the same plants to concentrations giving 75% mortality of the prey, but only when the prey reached high population levels (comparable to densities requiring control under field conditions). Similar environmental conditions were maintained, as in the first two experiments. Predator-prey ratios fluctuated from 1:1 to 20:1 (prey:predator) during the test. In the initial phase of the experiment, S predator populations suffered mortality considerably in excess of 75% due to their greater susceptibility.

The LC_{50} level of the S populations of *T. urticae* and *A. fallacis* used in the two experiments were 0.192% and 0.0093% A.I. azinphosmethyl in water, respectively. These values are near the innate susceptibility level for each species, based on extensive testing with a group of possible source populations (Morse and Croft, 1981). In Figure 2, data on the two treatment comparisons are presented (note that the fold increase in resistance relative to an S strain is plotted in Figure 2a and the absolute change in LC_{50} values in Figure 2b): (1) With similar selection and population densities, and unlimited food, predatory mites developed resistance to azinphosmethyl as fast and possibly faster than did the prey mites when selected apart from the prey over 22 generations (Figure 2a). This indicated a comparable potential for resistance developemnt between the predator and pest populations tested. At the end of the test, the absolute LC_{50} level of each species was approximately equal because the rate of predator resistance development was slightly faster than that of the prey. (2) When selected together at 75% mortality of the prey (Figure 2b) prey mites developed resistance more slowly than in Experiment 1. This is because treatments were not made every generation, but only when the prey reached relatively high population levels (comparable to those conditions observed in the field). Predators that initially were selected at a level approaching 99.9% developed resistance rapidly when food was readily available, much faster than did the prey. When predators reached the resistance level of the prey (after the second selection),

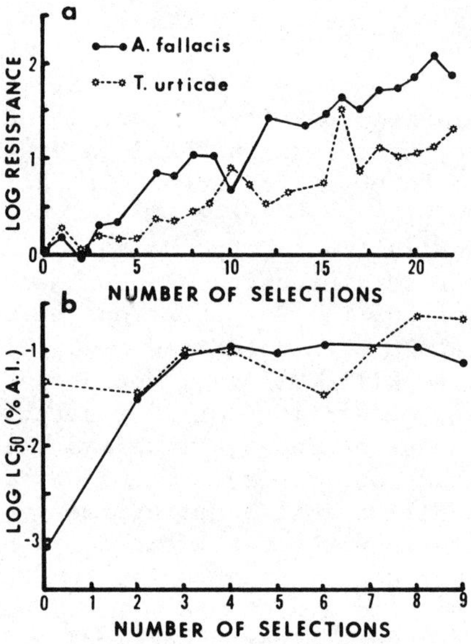

Figure 2. Developed resistance to azinphosmethyl in *Tetranychus urticae* (prey) and *Amblyseius fallacis* (predator) populations (adapted from Morse and Croft, 1980): (a) susceptible strains; separate selection; unlimited food; 75% selection, and (b) susceptible strains, coupled selection; unlimited food for prey; moderate food for predator; 75% selection of prey.

further resistance development slowed to about the same rate as that of the prey.

The pattern of azinphosmethyl resistance development in *A. fallacis* and spider mites in the field (Swift, 1970; Rock and Yeargan, 1971; Croft and McGroarty, 1973) may reflect a similar food limitation phenomenon. While it is difficult to reconstruct the timetable of resistance development, most fruit entomologists agree that resistance developed in the predator (*A. fallacis*) 3 to 5 years after it developed in the prey (*T. urticae* and *P. ulmi*) (B. A. Croft, unpublished survey data). These observations may be explained by the food limitation hypothesis as follows: When prey mites were selected with OP compounds (principally azinphosmethyl) from two to six times per season ($\bar{x} = 4$), resistance developed rapidly to the point that these chemicals were no longer recommended for mite control (1955-62). Predators during this period remained relatively susceptible, due in part to a scarcity of prey following OP treatments, but also because of excessive supression of pest mites with other highly effective acaricides. After 1962, OP compounds continued to be used for the control of other orchard pests (e.g., the codling moth, *Laspeyresia pomonella* L.) that had not developed resistance. Integrated mite control programs were more commonly used, and control with non-OP acaricides was not as intensive as during the 1955-62 period. As a result, predators developed R strains over the next 20+ selections as prey became more commonly available for reproductive exploitation following OP treatments.

These tentative hypotheses may have relevance to current tactics of chemical pest control and IPM. Pests that have effective natural enemies might be managed best by limiting insecticide use to only when necessary and even then to only give moderate kill of the prey/hosts or less than complete coverage of their habitat. Such action may result in greater resistance response in natural enemies and thus increase biological control as compared to current chemical control methods in which high kill of the pest is the norm. This assumes that growers will tolerate the short-term damage resulting from the higher pest population present initially (e.g., from foliage pests of apple, cotton, citrus).

There are ecological factors other than food limitation that influence the rate of development and stability of resistance in a natural enemy population, including development rates, exposure, vagility, refugia, etc. (Georghiou, 1972; Georghiou and Taylor, 1977a,b). It is this total complex plus certain genetic factors that ultimately account for the dynamics of resistance development in any natural enemy system. Some natural enemies (e.g., *Chrysopa* or a highly mobile, nonspecific inchneumonid parasite) may never develop R strains due to their genetic limitations, limited exposure, high vagility and mobility, etc.

The Differential Susceptibility Hypothesis

A second hypothesis to explain the discrepancy between the susceptibility and resistance observed in pests and natural enemies is that pests are more preadapted to handle conventional pesticides, as compared to predators and parasites. Gordon (1961) hypothesized that the larvae of certain polyphagous holometabolous insects were tolerant to pesticides because of biochemical stresses associated with the natural foods used by these insects over evolutionary time, e.g., plant secondary compounds. Krieger et al. (1971) found higher activity of aldrin epoxidase in the midgut tissues of oligophagous lepidopterous larvae than in monophagous species, and still higher activity in polyphagous species. They considered that enzyme activity by these species had been adjusted by natural selection to detoxify plant secondary compounds such as alkaloids, rotenoids and cyanides, etc. Brattsten et al. (1977) showed that mixed-function oxidases of the polyphagous insect larvae, *Spodoptera eridania*, were induced by a diversity of secondary plant substances in relatively small quantities. Following consumption of these compounds the larvae were less susceptible to dietary poisoning. Their data indicate that mixed-function oxidases (MFO) may play a major role in protecting herbivorous insects against many chemical stresses. These effects may act both within a generation (i.e., via dietary induction) and by genetic selection over a long-term, evolutionary period.

The relationship between the ability of phytophagous insects to detoxify both secondary plant compounds and pesticides raises questions about predatory and parasitic arthropods. Natural enemies encounter plant secondary compounds less frequently than do phytophages, and so would have less preadaptation to pesticides. Data from our laboratory comparing S strains of the phytophagous mite *T. urticae* and the predatory mite *A. fallacis* show the prey from two to 20 times less susceptible than the predator to a wide range of synthetic OP insecticides, and four to six times less susceptible to the plant toxin, nicotine (Strickler and Croft, unpublished). To carry the hypothesis one step further, predators, which in many cases have a mixed phytophagous and entomophagous habit, should be less susceptible to pesticides than are parasitic forms. Parasites, it is reasoned, are usually more specialized than predators, and therefore, may be even more limited in their exposure to plant secondary compounds.

Unfortunately, most comparisons of phytophage-natural enemy susceptibilities are confused by the history of insecticide exposure for each group of pests or natural enemies. The ideal comparisons should be made between S strains of each. However, there are some general trends and limited data that tend to confirm this hypothesis. For example, Croft and Brown (1975) reviewed LD, LC and LT_{50} data for a wide range of pests, predators and parasites associated with specific crops and found the following general trend of increasing susceptibility: pest < predators < parasites. Brattsten and Metcalf (1970) compared the synergistic ratio of carbaryl plus piperonyl butoxide (an MFO inhibitor) to indicate levels of mixed-function oxidases in a variety of insect species, and their data support the hypothesized relationship (Table 4). Although there was considerable variation within each group, pests on the average were approximately six times less susceptible to carbaryl than were predators, and 16 times less susceptible than parasites. A comparison of synergistic ratios of carbaryl and piperonyl butoxide (1:5) (SR, Table 4) from the study indicated an average 1.6 and 2.1 times greater MFO activity in pests than predatory and parasitic insects, respectively.

Differences in the susceptibilities of pests versus natural enemies and between predators and parasites may be further explained by differences in the relative importance of hydrolytic and oxidative detoxification enzymes (Plapp, 1980). Whereas both phytophagous and entomophagous arthropods may have hydrolytic enzymes of comparable

Table 4. Toxicity of Carbaryl (LD_{50}) and Synergistic Ratio (SR) of Carbaryl and Piperonyl Butoxide (1:5) to Three Classes of Insects (adapted from Croft and Morse, 1979).

	Insect Classification		
	Phytophage N = 25[a]	Predator n = 7	Parasite n = 8
LD_{50}	415.6 ± 245.7 μg/g[b]	70.3 ± 43.0 μg/g	25.9 ± 7.6 μg/g
SR	17.2 ± 9.0	10.6 ± 2.4	8.2 ± 1.7

[a]Represents mean value for 25 species.
[b]± standard error of the mean.
(Compiled from data presented by Brattsten and Metcalf, 1970).

efficiencies for metabolizing fats, proteins and other general foods (as well as insecticides that are detoxified hydrolytically), plant-feeding species may have more highly efficient oxidative detoxifying systems, maintained in response to certain secondary plant compounds (e.g., gossypol in cotton). If this is true, hydrolytically metabolized insecticides may be the most logical candidates for selective pesticides to be used in IPM.

The differential susceptibility hypothesis emphasizes the role of the MFOs in detoxifying both secondary compounds and insecticides; however, resistance in arthropods to many insecticides depends on other detoxification enzymes and mechanisms, e.g., hydrolases, glutathione transferases, and reduced sensitivity of target organs. Little is known about the mechanisms of detoxification in most plant secondary compounds. Future research should examine how natural enemies are adapted to detoxify secondary plant compounds and pesticides in a variety of ways (MFOs, hydrolases, etc.).

While the limited detoxification hypothesis suggests an explanation for differences in initial susceptibilities of phytophages, predators and parasites, it is less plausible in explaining differences in resistance development in these groups. To our knowledge, no studies have been made of the induction of MFO or other enzyme systems in predators and parasites, nor have studies been made on the rates at which enzyme systems can evolve in different groups of arthropods. Answers to these questions are needed before we can assess the role of diet in preadapting these arthropods to handle insecticides.

In summary, both the food limitation and the differential susceptibility hypotheses posed above are important areas for future research. No single hypothesis is likely to provide the complete explanation for differences observed between the susceptibility and resistance potential in plant-feeding arthropods and their entomophagous natural enemies; however, this discussion provides insight into the differential responses of these groups to pesticides.

Application of Resistant Phytoseiid Mites in IPM

Since insecticide-R phytoseiid mites were first discovered (Hoyt, 1969; Croft and Jeppson, 1970; Motoyama et al., 1970), they have been used in a variety of ways:

Endemic species. Exploiting endemic insecticide-R phytoseiid mites has been a widely employed tactic. It has been intensively used for deciduous tree fruit mites, especially on apple in North America. In the USA and Canada, R strains of *T. occidentalis* (western), *A. fallacis* (midwestern and eastern) and *T. pyri* (northeastern and eastern) are the principal species involved. The geographic range and extent of resistance in these species was referred to earlier (Table 1). The major integrated mite control programs based on these resistant mites are listed in Table 5 and include virtually every major apple-growing state or province in North America. To generalize, these systems are similar in that all employ (1) OP compounds (to which the predators are resistant) to control the major pests (e.g., codling moth, apple maggot, plum curculio, red-banded leaf roller), and (2) other selective insecticides, fungicides, miticides, herbicides, etc., to control other orchard pests. Insecticide-R phytoseiid mites provide biological control of spider mites under most conditions, although selective acaricides may be needed periodically to readjust predator-prey ratios in favor of the natural enemy. A detailed example of these mite management programs is given in Croft (1975) and Croft and McGroarty (1977).

Similar programs of integrated mite control are developing in South America and New Zealand (see below). In South America, *A. chilenensis* is widespread in apple-growing regions of Uruguay, Argentina and Chile. Resistance to OPs in this species also is relatively widespread (Croft et al., 1976a; R. Weires and L. Campos, personal correspondence). Although research on management of *A. chilenensis* has not yet been extensively developed, it is being used in localized areas with considerable success (e.g., in Uruguay, Chile and Argentina).

In New Zealand long-established strains of *T. pyri* (probably introduced initially from Europe with apple culture) show the highest resistance to OP compounds found to date (see earlier discussion, Table 1)· undoubtedly, this is related to the greater use of OPs for apple pest control on these islands (Penman et al., 1976) than in North America and Europe. Resistant *T. pyri* in New Zealand commonly persist in heavily sprayed orchards and provide biological control of *P. ulmi* if managed carefully with OPs and other selective pesticides (Penman et al., 1979). *T. pyri* populations from New Zealand are currently being exported for establishment in other fruit-producing areas where resistance in this species is not as high (see later discussion).

One additional native insecticide-R species that has only recently been discovered but that may have considerable potential for biological control is *A. andersoni,* which occure on deciduous tree fruits in the Italian and southern Swiss Alps of Europe (Gambaro, 1975). Apparently this species is the prominent natural enemy of orchard mites at the southerly distribution of *T. pyri.* Further study of the biology, resistance and management of this species may provide another species for introduction elsewhere.

The tactic of transferring and manipulating an insecticide-R mite within its native range while taking advantage of unique R features that may have developed locally is well established. It was first demonstrated by Croft and Barnes (1971, 1972), who established R strains of *T. occidentalis* from Washington and Utah in California orchards where this species occurred but was not resistant. Thereafter, an integrated mite control system was possible in this area (Croft and Barnes, 1971). There have been other unpublished reports of inter-orchard transfers of R phytoseiids within their distributional range; however, published reports are rare. One exception is the transfer of a multiple-R (OPs and carbamate) strain of *A. fallacis* in orchards of Michigan (Croft and Meyer, 1973; Croft and Hoying, 1975). Resistance to both chemicals was maintained in one strain by laboratory crossing experiments and selections; however, when released in the field and selected by both chemicals, resistance to OPs remained stable while carbamate resistance declined precipitously to a more susceptible level within a few generations.

Introduced species. The most successful case of establishment of an exotic, R phytoseiid mite to date has been with *T. occidentalis* in Australia (Readshaw, 1975). These mites from the USA initially were released in 1972 into a commercial orchard near Canberra and the results were spectacular. Despite six sprays of azinphosmethyl for codling moth control, predators successfully overwintered and maintained *T. urticae* well below damaging levels throughout the 1972-73 season and thereafter. Later, large numbers of *T. occidentalis* were released in small apple blocks throughout southeastern and southwestern Australia. Since their initial release, these predators have been recovered from all release sites. Only in the more humid areas of Australia, including Tasmania, has establishment of this mite not been successful. In these areas, work to establish

Table 5. Major Integrated Mite Control Programs on Apple in North America

State or Province	Principal Pest Controlled[a]	Principal Natural Enemies[b]	Research Reference[c]	Implementation Reference
Washington	Tm. Pu. As	To. Spp, Zm	Hoyt, 1969	Hoyt et al., 1970
New York	Pu, Tu	Af. Tp	Watve and Lienk, 1977	Tette, 1977
Michigan	Pu, Tu, As	Af, Sp. Ag, Zm	Croft & McGroarty, 1977	Croft, 1975
California	Tp, Tu, Pu	To, Ss	Croft & Barnes, 1971	- - -
Pennsylvania	Pu, Tu	Sp, Af	Asquith, 1971	Asquith, 1972
North Carolina	Pu, Tu, As	Af, Sp	Rock & Yeargan, 1971	Rock, 1972
Virginai, W Va	Pu, Tu	Af	Clancy & McAllister, 1968	- - -
Illinois	Pu, Tu	Af	Meyer, 1974	- - -
Oregon	Pu	To, Tp	Zwick, 1972	- - -
Missouri	Pu Tu	Af	Poe & Enn, 1969	- - -
Utah	Tu, Tm, Pu, As	To, Zm	Davis, 1970	- - -
Colorado	Tu Pu	To	Quist, 1974	- - -
New Jersey	Pu, Tu, As	Af	Swift, 1970	Christ, 1971
Ohio	Pu, Tu, As	Af, Zm, Ag	Holdsworth, 1968	Holdsworth, 1974

Massachusetts	Pu, Tu, As	Af	Prokopy et al., 1978	Hauschild et al., 1979
British Columbia	Tm, Pu, As	To, Tp, Zm	Downing & Molliet, 1971	Downing & Aarand, 1968
Quebec	Pu, Tu	Af	Parent, 1967	- - -
Nova Scotia	Pu, As	Several	Sanford & Herbert, 1970	Anon., 1970

[a]Pest importance in each area listed sequentially. Abbr.: Tm = *Tetranychus mcdanieli*; Pu = *Panonychus ulmi*; As = *Aculus schlechtendali*; Tu = *Tetranychus urticae*; Tp = *Tetranychus pacificus*.
[b]Natural enemy importance in each area listed sequentially. Abbr.: To = *Typhlodromus occidentalis*; Sp = *Stethorus punctum*; Zm = *Zetzellia mali*; Af = *Amblyseius fallacis*; Tp = *Typhlodroums pyri*; Ag = *Agistemus fleschneri*; Ss = *Scolothrips sexmaculata*; Spp = *Stethorus picipes*.
[c]References for this table are in Croft (1980).

R strains of *T. pyri* from New Zealand and *A. fallacis* from North America for control of the European red mite *P. ulmi* is currently underway.

In New Zealand, Penman et al. (1979) established strains of *A. fallacis* and *T. occidentalis* to augment resistant *T. pyri*, which probably were introduced along with *P. ulmi* from Europe. *A. fallacis* was first successfully introduced in 1973 from the USA. After evaluation for several seasons, researchers concluded that *A. fallacis* provided excellent control of *P. ulmi* under New Zealand weather conditions when placed directly into orchard trees, although its overwintering mortality in the ground and delayed movement into trees in early season limit its effectiveness when mites dispersed naturally. Researchers are considering introducing *A. fallacis* populations from a lower North American latitude, which may be better adapted to the mild winters of New Zealand than the original mites colonized from Michigan. Releases of insecticide-R *T. occidentalis* from North America have effectively controlled *T. urticae* populations in New Zealand; establishment of this predator appears likely as well. Unfortunately, *T. urticae* is only a pest in limited areas in New Zealand and therefore, work in the future will focus primarily on *T. pyri* or *A. fallacis*, which are better adapted as predators of *P. ulmi* than is *T. occidentalis*.

Development and utilization of R strains of *P. persimilis* in glasshouses in western Europe and the USSR is an example of management of an exotic phytoseiid that acquired resistance after leaving its native Chile (previous discussion). These mites are used in glasshouses for controlling *T. urticae* and other plant-feeding mites on a variety of crops in the USSR (Storozhkov et al., 1977) and in western Europe (Ann., 1977; van de Vrie, unpublished). Undoubtedly, these R mites will be transferred worldwide since this species is mass cultured widely for periodic and inundative release programs.

The last case history involved *T. pyri*, which is native to western Europe and probably was secondarily introduced into North America, Australia and New Zealand. In British orchards and most other areas of western Europe, *T. pyri* is the principal species associated with the European red mite, *P. ulmi*. However, in Europe there are fewer major insect pests of apple than in other areas of the world. The codling moth and various leafrollers are the only major pests, requiring minimal OP applications. Because of the low levels of OPs used and the relatively slow rate at which *T. pyri* acquires OP resistance (see earlier discussion; Penman et al., 1976), *T. pyri* populations have remained relatively susceptible in Europe.

Recently, UK scientists have reintroduced highly resistant strains of *T. pyri* from New Zealand back into Britain with success

(Cranham and Solomon, 1978, unpublished). These mites were readily established, overwintered, and tolerated normal field rates of insecticides that previously had eliminated the native strain each year. Will the resistance of these mites revert appreciably, given that selection pressure in British orchards is much less than the mites were previously exposed to in New Zealand? While reversion may occur, it is unlikely that resistance will decrease below that necessary for these mites to persist in the UK. Highly OP-resistant strains of *T. pyri* from New Zealand will probably be introduced throughout Europe and North America where less resistant strains of this mite now occur.

In summary, to our knowledge, the only major apple production regions (except for eastern Europe and China where published literature is difficult to obtain) where the potential for using insecticide-R phytoseiid mites has not been investigated are in South Africa and Japan. Undoubtedly, resistance will appear in endemic populations in these areas, or exotic species will be introduced in the near future. Phytoseiid mites should soon prove to be the major tool for apple mite control worldwide.

<u>Future uses</u>. Areas of new research on insecticide-R phytoseiid mites include: (1) the preselection of strains to new chemicals that might be used in crop protection (e.g., synthetic pyrethroids) (Hoy and Knop, 1979; Strickler and Croft, unpublished); (2) the utilization of multiple-R strains (Croft and Meyer, 1973; Croft and Hoying, 1975; Strickler and Croft, unpublished); (3) selection with certain compounds that confer multiple-factor resistances simultaneously (see example of dimethoate with the housefly, Keiding, 1977); and (4) the simultaneous favorable management of insecticide resistance in both pest mites and phytoseiids in the field (Croft, 1979; see later discussion).

Although use of insecticide-R phytoseiid mites is most well developed for deciduous fruits, it is expected that the approach will find application to other crops and other natural enemy groups. For example, *A. fallacis* is primarily adapted to low-growing vegetation including field crops, vegetables, cereals (corn and sorghum), cotton, soybeans and greenhouse crops; only secondarily does it invade fruit trees. It also occurs in North America from the temperate zones of Michigan and New York to as far south as the cotton and soybean fields of Mississippi and Louisiana. Among other natural enemies, certain insecticide-R aphelinid parasites that attack citrus scales may have a potential for manipulation and management, and a similar use of cecidomyid predators of aphids may be developed in the near future.

Management of Resistance in Pests and Natural Enemies

Considerable interest has developed in resistance management (Georghiou and Taylor, 1976, 1977a,b; Comins, 1977a,b; Croft, 1979) as a means to decrease the rate and level of the development of resistance in pests and/or to increase resistance of natural enemeis. A number of concepts underlie such an approach: First, it is known that a sequence of steps or phases are involved in resistance development (Brown and Pal, 1973; Croft, 1979) (see Figure 1). Initially, a gene or genes for resistance are usually associated with lowered fitness in an insect population. There may be a stage of stable or increasing susceptibility and a period of gene rearrangement, recombination and association of the resistance genes with fitness factors in a population, followed by a slow rate of resistance development. Later, a log increase in resistance to high LD_{50} levels may be observed among pests under intensive selection. Reversion to susceptible levels is rare once high resistance levels are reached, although reversion is more common if selection is discontinued during the initial phase of resistance development or before multiple resistance mechanisms are involved. To slow the rate of development one might monitor for resistance and change to a new class of compounds having no cross resistance with the earlier chemical used before the insect enters the log-increase resistance state (Keiding, 1967; Brown, 1977).

A second useful concept, which has emerged from resistance models incorporating the major factors affecting resistance development (Georghiou and Taylor, 1976; Comins, 1977a,b), is the resistance steady-state condition. A resistance steady-state occurs when the *operational* factors affecting resistance (e.g., application and selection thresholds, pest stage selected, mode of application, space-limited selection, and alternating chemicals) (see Georghiou and Taylor, 1976) are manipulated so that populations initially may show a limited increase in resistance to an insecticide but then revert back rapidly or frequently enough after selections that they never enter the log phase of resistance development.

At present, the operational factors that can be manipulated to keep an arthropod pest system in a steady-state or limited-resistance condition are unknown, but models show that such conditions can develop. Both sensitivity analyses with computer models and greenhouse experiments can be used to identify which parameters are most important for managing resistance. Studies suggest that leaving untreated refugia, high immigration rates and reduced selection pressure have a major impact in reducing resistance development of pests (Georghiou and Taylor, 1977a; Comins, 1977a) and natural enemies (Morse and Croft, 1981; reviewed earlier).

The question becomes, is resistance management practical, and what manipulations are feasible under current requirements for pest control (e.g., economic thresholds)? We believe resistance management is practical for some crops. In fact, it appears possible that a steady-state resistance condition may have been achieved unintentionally with apple pests. To understand the basis of this conclusion, a review of the resistance situation in the pest-natural enemy complex on apples is necessary.

Patterns of OP Resistance in North American Orchards

In apple orchards in North America, OP compounds, especially azinphosmethyl, have been widely used for 20 to 30 years. However, resistance to these insecticides has developed uniquely among certain secondary arthropod pests (i.e., mites, aphids and leafhoppers) and natural enemies, but not among the key pests (i.e., codling moth, plum curculio, apple maggot, red-banded leafroller). It seems highly improbable that all key pest species associated with apple lack the genetic potential to develop resistance. The codling moth developed resistance to DDT and lead arsenate in only 6 to 20 years (Hough, 1963; Morgan and Madsen, 1976). If selection pressures from OPs were similar to those of lead arsenate and DDT, then key pest species with more than one generation per season (i.e., codling moth, red-banded leaf roller) should currently exhibit resistance; yet in some parts of the United States, growers have used OPs extensively and continuously for the past 31 years without resistance problems in codling moth (e.g., Washington, S. C. Hoyt, personal communication). Species having only a single generation per season (apple maggot, plum curculio) would probably take longer to develop R strains.

Secondary pests of apple have developed different patterns of resistance than those described for key pests (Croft, 1979). Mites, aphids and leafhoppers developed R strains quickly, within the first 5 to 20 years of OP use. Fortunately, these species have predators and parasites that can provide significant biological control when not affected by chemicals used for key pest control. In the late 1950s and early 1960s when OPs were first used, there was a difficult transition period characterized by R secondary pests and greatly supressed beneficial natural enemies. However, with continued OP use in apple orchards (due to continued effectiveness on the key pests), there developed a number of pesticide-R natural enemies in the period from the mid-1960s to the present.

As noted earlier, the first evidence of widespread resistance involved predators of plant-feeding mites, including the phytoseiid mite species, *T. occidentalis* (Hoyt, 1969; Croft and Jeppson, 1970) and *A. fallacis* (Swift, 1970; Motoyama et al., 1970), and the cocinnellid beetle, *Stethorus punctum* LeConte (D. Asquith, personal communication). Thereafter, evidence that similar responses were occurring in natural enemy species associated with aphids (e.g., the cecidomyid, *Aphidoletes aphidimiza*) (Adams and Prokopy, 1977) and leafhoppers (Croft, unpublished data) was found. There may be other natural enemies associated with apple that have evolved or are in the process of evolving R strains, as yet unmonitored (e.g., *Aphelinus mali*). Evidence for this is the general stability observed in pest-natural enemy complexes associated with apple and OP usage.

The differences observed in resistance patterns of key pests, secondary pests and natural enemies may be related to differences between these groups in life history and other biological characteristics (Table 6). These characteristics contribute to greater OP selection and a faster buildup of populations after spraying for the secondary pests and certain R natural enemeis than for key pests and other natural enemies that have not developed resistance. Important characteristics contributing to these resistance development circumstances include differences in rate of reproduction, generations per year, life stages exposed, host specificity, and dispersal rates. For example, high dispersal rates and mobility of the key pests and some natural enemies increases the possibility that genes for resistance will be diluted by migration to and from wild alternate hosts or abandoned apple orchards.

In addition to biological characteristics of the pests and natural enemies, the general pattern of OP usage in apple orchards over the past 20 years may have favored a reduced probability that resistance would develop in key pests. During the first few years of OP use, compounds like azinphosmethyl were applied at relatively high rates (3-4.5 lbs. 50% w.p. per acre). In the mid-1960s the rate of application was lowered 20-25% because of high effectiveness and because phytoseiid mite predators were beginning to develop strains resistant to azinphosmethyl (Hoyt, 1969). More importantly, the emphasis in orchard pest control has been to reduce the number of broad-spectrum OP applications by some 25-50% on a seasonal basis in most major fruit-growing states (Croft, 1979). In general, we have continued to reduce selection pressure on key apple pests, and after many years without appreciable resistance developing in these key pests, one wonders if resistance will ever develop if present use patterns are maintained.

If we look closely at the features of OP use for apple pest control over the past 20+ years, we have serendipitously followed many tactics recommended for delaying or avoiding development of resistance (Keiding, 1967; Comins, 1977b): (1) many of these chemicals have a short residue life (e.g., parathion, diazinon); (2) selection is mainly directed at a single life stage, usually adults; (3) the application is localized rather than area-wide; (4) certain generations are left untreated (e.g., red-banded leafroller, codling moth), and (6) OP resistance may be unstable and reversion from resistance to susceptibility may be common among key apple pests (e.g., see review of OP reversion in Keiding, 1967).

Table 6. Biological Characteristics of Apple Pests and Natural Enemies That May Slow or Favor the Development of Resistance

NON-RESISTANT	RESISTANT
Key Pests	Secondary Pests
(codling moth, apple maggot, plum curculio, red-banded leafroller)	(mites, aphids, leafhoppers)
Natural Enemies	Natural Enemies
(syriphid flies, mirid bugs, chrysopids)	(phytoseiid mites, cecidomyid flies, *Stethorus* beetles)
Factors reducing selection pressure:	Factors increasing selection pressure:
1. Few life stages exposed to pesticides.	1. All active life stages exposed to pesticides.
2. Few generations per year (1-4).	2. Many generations per year (2-17).
Factors reducing population buildup, or diluting resistant genes:	Factors increasing population buildup, or preventing dilution of resistant genes:
3. Relatively slow rate of reproduction	3. Relatively rapid rate of reproduction.
4. Relatively high dispersal rates.	4. Relatively low dispersal rates.

Given that the key pests of apple have not developed resistance to OPs, whereas many secondary pests and their natural enemies have, a strategy of resistance management in each of these arthropod groups becomes feasible. Certain OP compounds (e.g., azinphosmethyl, phosmet) have become increasingly more selective to natural enemies. If their use is continued, more beneficial species may adapt to these compounds, while the probability of resistance among the key pests will hopefully lessen due to more limited use. Considerable research on the dynamics of resistance and on operational features available for manipulation must be completed if further management of resistance in apple systems is to be accomplished (Croft, 1979).

Based on these findings, researchers are opposed to extensive use of new chemicals (e.g., synthetic pyrethroids) (Croft and Hoyt, 1978; Croft, 1979) for apple pest control that do not have additional selectivity to natural enemies beyond that provided by the OPs, especially azinphosmethyl and phosmet. Though OPs may have limited effectiveness on a few secondary apple pests (e.g., leafrollers, leafminers, leafhoppers), the selective benefits of these compounds to natural enemies are sufficient to recommend their continued use for pest control until resistance developes in one or more of the key pests or until these compounds become unsuitable for use due to other factors (economic, human safety, environmental effects, etc.).

In summary, research scientists working with apple pest control have long sought to obtain selective chemicals that would provide control of the key pests of this crop, but would allow for survival of most secondary pests and their natural enemies. Such a selective chemical control system should optimally reduce pesticide inputs, yet provide effective chemical control of the target key pest species. The object of this selective control has not been to have a single selective chemical for each pest (e.g., see implementation problems associated with "species-specific" insecticide selectivity in Gruys, 1976), but rather to have a single chemical system that is selective to the entire group of key pests but nontoxic to secondary pests and natural enemies. Due to the unique development of OP resistance among apple pests in North America, we have begun to approximate this "ideal selectivity" and hopefully we will better approach this ideal in the future as a greater understanding of the factors influencing resistance in the entire arthropod complex is gained.

Acknowledgements; Appreciation is expressed to J. G. Morse and Dr. J. W. Vinson for their critical review of this manuscript.

LITERATURE CITED

Abdelrahman, I., 1973, Toxicity of malathion to the natural enemies of California red scale, *Aonidiella aurantii* (Mask.) (Hemiptera:

Diapidae), *Aust. J. Agric. Res.*, 24:119.

Adams, C. H., and Cross, W. H., 1967, Insecticide resistance in *Bracon mellitor*, a parasite of the boll weevil, *J. Econ. Entomol.*, 60:1016.

Adams, R. G., and Prokopy, R. J., 1977, Apple aphid control through natural enemies, *Mass. Fruit Note*, 42:6.

Ahlstrom, D. R., and Rock, G. C., 1973, Comparative studies on *Neosieulus fallacis* and *Metaseiulus occidentalis* for azinphos-methyl toxicity and effects of prey and pollen on growth, *Ann. Entomol. Soc. Am.*, 66:1109.

Anonymous, 1977, Ann. Rept. of Dept. Agric. Res. Rept. Royal Trop. Inst., The Netherlands, 72 pp.

Atallah, Y. H., and Newsom, L. D., 1966, the effect of DDT, toxaphene and endrin on the reproductive and survival potentials of *Coleomegilla maculata*, *J. Econ. Entomol.*, 59:1181.

Brattsten, L. B., and Metcalf, R. L., 1970, The synergistic ratio of carbaryl with piperonyl butoxide as an indicator of the distribution of multifunction oxidases in the Insecta, *J. Econ. Entomol.*, 36:101.

Brattsten L. B., Wilkinson, C. F., and Eisner, T., 1977, Herbivore-Plant interaction: Mixed-function oxidases and secondary plant substances, *Science*, 196:1349.

Brown, A. W., 1977, Epilogue: Resistance as a factor in pesticide management, p. 81624, Prox. XV Intern. Cong. Entomol. Public Entomol. Soc. Amer., College Pk., Md, 824 pp.

Brown, A. W. A., and Pal, R., 1973, Insecticide resistance in Arthropods, 2nd Ed. WHO Monograph Ser., 38, 483 pp.

Chambers, H. W., 1973, Comparative tolerance of selected beneficial insects to methyl parathion, Communication to Ann. Meet. Entomol. Soc. Amer., Nov. 28. p. 68.

Collyer, E., and Van Geldermalsen, M., 1975, Integrated control of apple pests in New Zealand, 1. Outline of experiments and general results, *N. A. J. Zool.*, 2:101.

Comins, H. N., 1977a, The development of insecticide resistance in the presence of migration, *J. Theoret. Biol.*, 64:177.

Comins, H. N., 1977b. The management of pesticide resistance,, *J. Theoret. Biol.*, 65:399.

Cranham, J. E., and Solomon, M. G., 1978, Establishment of predaceous mites in orchards, East Malling Res. Rept. for 1977, 110.

Croft, B. A., 1972, Resistant natural enemies in pest management systems, SPAN, 15:19.

Croft, B. A., 1975, Integrated control of apple mites, Ext. Bull. E-823 Mich. St. U., 12 pp.

Croft. B. A., 1977, Resistance in arthropod predators and parasites, p. 377-393, *in* "Pesticide Management and Insecticide Resistance," D. L. Watson and A. W. A. Brown, eds., Acad. Press, New York, 638 pp.

Croft, B. A., 1979, Management of apple arthropod pests and natural enemies relative to developed insecticide resistance, *Environ.*

Entomol., 8:583.

Croft, B. A., 1980, Introduction Cpt. 1, *in:* "Integrated Control of Deciduous Tree Fruit Pests, B. A. Croft and S. C. Hoyt, eds., Wiley Intersci., New York (in press).

Croft, B. A., and Barnes, M. M., 1971, Comparative studies on four strains of *Typhlodromus occidentalis*, III. Evaluations of releases of insecticide-resistant strains into an apple orchard ecosystem, *J. Econ. Entomol.*, 64:845.

Croft, B. A., and Barnes, M. M., 1972, Comparative studies on four strains of *Typhlodromus occidentalis*, VI. Persistance of insecticide resistant strains in an apple orchard ecosystem, *J. Econ. Entomol.*, 65:211.

Croft, B. A., and Brown, A. W. A., 1975, Responses of arthropod natural enemies to insecticides, *Ann. Rev. Entomol.*, 20:285.

Croft, B. A., and Hoying, S. A., 1975, Carbaryl resistance in native and released populations of *Amblyseius fallacis*, *J. Environ. Entomol.*, 4:895.

Croft, B. A., and Hoyt, S. C., 1978, Considerations in the use of pyrethroid insecticides for deciduous tree fruit pest control in the USA, *Environ. Entomol.*, 7:627.

Croft, B. A., and Jeppson, L. R., 1970, Comparative studies on four strains of *Typhlodromus occidentalis*, II. Laboratory toxicity of ten compounds common to apple pest control, *J. Econ. Entomol.*, 63:1528.

Croft, B. A., and McGroarty, D. L., 1977, The role of *Amblyseius fallacis* (Acarina:Phytoseiidae) in Michigan apple orchards. *Mich. Expt. Sta. Res. Rept.*, 333, 22 pp.

Croft, B. A. and Meyer, R. H., 1973, Carbamate and organophosphorus resistance patterns in populations of *Amblyseius fallacis*, *Environ. Entomol.*, 2:691.

Croft, B. A., and Morse, J. G., 1979, Recent advances on pesticide resistance in natural enemies, *Entomophaga*, 24:3.

Croft, B. A., and Nelson, E. E., 1972, Toxicity of apple orchard pesticides to Michigan populations of *Amblyseius fallacis*, *Environ. Entomol.*, 1:576.

Croft, B. A., Briozzo, J., and Carbonell, J. B., 1976a, Resistance to organophosphorus insecticides in the predaceous mite, *Amblyseius chilenensis*, *J. Econ. Entomol.*, 69:563.

Croft, B. A., Brown, A. W. A., and Hoying, S. A., 1976b, Organophosphorus resistance and its inheritance in the predaceous mite *Amblyseius fallacis*, *J. Econ. Entomol.*, 69:64.

F.A.O., 1977, Pest resistance to pesticides and crop loss assessment 1 Rept. of 1st F.A.O. Panel of Experts. F.A.O. Plant Prod. and Protect. Paper 6, 42 pp.

F.A.O., 1979, Pest resistance to pesticides, Rept. 2nd Panel of Experts Rome/Sept. 5-10, 1978, F.A.O. Plant Prot. Paper (in press).

Field, R. P., 1974 Occurrence of an Australian strain of *Typhlodromus* (Acarina:Phytoseiidae) tolerant to parathion, *J. Aust.*

Entomol. Soc., 13:255.
Field, R. P., 1976 Integrated pest control of Victorian peach orchards: The role of *Typhlodromus occidentalis* Nesbitt (Acarina:Phytoseiidae), *Aust. J. Zool.*, 24:565.
Gambaro, P. I., 1975, Organophosphorus resistance in a peach orchard population of predaceous mites, *Estrat. Inform. Fitopath.*, 25:21.
Georghiou, G. P., 1972, The evolution of resistance to pesticides, *Ann. Rev. Ecol. Syst.*, 3:133.
Georghiou, G. P., and Taylor, C. E., 1976, Pesticide resistance as an evolutionary phenomenon, Proc. 15th Int. Chg. Entomol. pp. 759-785.
Georghiou, G. P., and Taylor, C. E., 1977a, Genetic and biological influences in the evolution of insecticide resistance, *J. Econ. Entomol.*, 70:319.
Georghiou, G. P., and Taylor, C. E., 1977b, Operational influences in the evolution of insecticide resistance, *J. Econ. Entomol.*, 70:653.
Gordon, H. T., 1961, Nutritional factors in insect resistance to chemicals, *Ann. Rev. Entomol.*, 6:27.
Gruys, P., 1976, Development and implementation of an integrated control programme for apple orchards in the Netherlands, Proc. 8th Brit. Insecti. and Fungi. Conf., (1975) p. 823.
Helle, W., and van de Vrie, M., 1974, Problems with spider mites, *Outlook on Agric.*, 8:119.
Herne, D. H. C., and Brown, A. W. A., 1969, Inheritance and biochemistry of OP resistance in a New York strain of the two-spotted spider mite, *J. Econ. Entomol.*, 62:205.
Hough, W. S., 1963, Resistance to insecticides by codling moth and red-banded leaf roller, Virginia Ag. Expt. Sta. Tech. Bull., 166, 32 pp.
Hoy, M. A., and Knop, N. F., 1979, Studies on pesticide resistance in the phytoseiid *Metaseiulus occidentalis* in California, p. 8994, *in:* "Recent Advances in Acarology, Vol. 1," J. G. Rodriguez, ed., Acad. Press, New York, 631 pp.
Hoyt, S. C., 1969, Integrated chemical control of insects and biological control of mites on apples in Washington, *J. Econ. Entomol.*, 62:74.
Hoyt, S. C., 1972, Resistance to azinphosmethyl of *Typhlodromus pyri* from New Zealand, *N. Z. J. Sci.*, 15:16.
Huffaker, C. B., 1971, The ecology of pesticide interference with insect populations, p. 92107, Agricultural Chemicals Harmony or Discord for Food, People and the Environment, J. E. Swift, ed., Univ. Calif. Div. Agric. Sci. Publ., 151 pp.
Huffaker, C. B., and Kennett, C. E., 1953, Differential tolerance to parathion in two *Typhlodromus* predatory on cyclamen mite, *J. Econ. Entomol.*, 46:707.
Keiding, J., 1967, Persistence of resistant populations after the relaxation of the selection pressure, *World Rev. Pest Control*, 6:115.

Keiding, J., 1977, Resistance in the housefly in Denmark and elsewhere, p. 261302, *in:* "Pesticide Management and Insecticide Resistance," D. L. Watson and A. W. A. Brown, eds., Acad. Press, New York, 638 pp.

Kennett, C. E., 1970, Resistance to parathion in the phytoseiid mite *Amblyseius hibisci, J. Econ. Entomol.,* 63:1999.

Kot, J. T., Plewka, T., and Krudierek, T., 1971, Relationship in parallel development of host and parasite resistance to a common toxicant — Funal tecnical report PL-480, E21-Ent-19, F6-Po-203, *Inst. Ecol. Polish Acad. Sci.,* Wky Dz. Inf. PAN Zam., 25/71 66 pp.

Kot, J. T., Plewka, T., and Krukierek, T., 1977, Investigation on metasystox and DDT resistance of five populations of *Tricogramma evanescens* Westw. (Hymenoptera:Trichogrammatidae), *Polish Ecol. Stud.,* 1:175.

Krieger, R. I., Feeny, P. P., and Wilkinson, C. F., 1971, Detoxification enzymes in the guts of caterpillars: An evolutionary answer to plant defenses? *Science,* 172:579.

Meyer, R. H., 1975, Release of carbaryl-resistant predatory mites in apple orchards, *Environ. Entomol.,* 4:49.

Morgan, C. V. C., and Madsen, H. F., eds., 1976, Development of chemical, biological and physical methods for control of insects and mites, p. 256301, *in:* "History of Fruit Growing and Handling in United States of America and Canada," 18601972, Rogatta Press, Kelowna, B. C., Canada, 651 pp.

Morse, J. G., 1978, Development of resistance to azinphosmethyl, in the predatory mite, *Amblyseius fallacis,* and its prey, *Tetranychus urticae,* in greenhouse experiments, MS Thesis, Michigan State Univ., East Lansing, MI, 65 pp.

Morse, J. G., and Croft, B. A., 1981, A summary of resistance studies to azinphosmethyl in the predatory mite *Amblyseius fallacis* and its prey *Tetranychus urticae, Entomophaga,* (in press).

Motoyama, N., Rock, G. C., and Dauterman, W. C., 1970, Organophosphorus resistance in an apple orchard population of *Typhlodromus (Amblyseius) fallacis, J. Econ. Entomol.,* 63:1439.

Motoyama, N., Rock, G. C., and Dauterman, W. C., 1971, Studies on the mechanisms of azinphosmethyl resistance in the predaceous mite *Neoseiulus (T.) fallacis, Biochem. Physiol.,* 1:205.

Motoyama, N., Rock, G. C., and Dauterman, W. C., 1977, Toxicity to O-alkyl analogues of azinphosmethyl to resistant and susceptible predaceous mite, *Amblyseius fallacis, J. Econ. Entomol.,* 70:475.

Newsom, L. D., 1974, Predator insecticide relationships, *Entomophaga Mem. Ser.,* 7, 88 pp.

Oatman, E. R., McMurtry, J. A., and Voth, V., 1968, Suppression of the two-spotted spider mite on strawberry with mass releases of *Phytoseiulus persimilis, J. Econ. Entomol.,* 61:1517.

Penman, D. R., Ferro, D. N., and Waring, C. H., 1976, Integrated control of apple pests in New Zealand, VII. Azinphosmethyl resistance in strains of *Typhlodromus pyri* from Nelson, *N. Z. J. Expt. Agric.*, 4:377.

Penman, D. R., Wearing, C. H., Collyer, E., and Thomas, W. P., 1979, The role of insecticide-resistant phytoseiids in integrated mite control in New Zealand, Proc. V Intern. Cong. Acarol., 5 (in press).

Pielou, D. P., and Glasser, R. F., 1952, Selection for DDT resistance in a beneficial insecticide, *Science*, 115:117.

Plapp. F. W. Jr., 1980, Ways and means of avoiding or ameliorating resistance to insecticides, Proc. IV Intern. Cong. Plant Prot. Wash. D.C., 1979 (in press).

Readshaw, J. L., 1975, Biological control of orchard mites in Australia with an insecticide-resistant predator, *J. Aust. Inst. Ag. Sci.*, 41:213.

Robertson, J. G., 1957, Changes in resistance to DDT in *Macrocentrus ancylivorus* Rohw., *Can. J. Zool.*, 35:629.

Rock, G. C., and Yeargan, D. R., 1971, Relative toxicity of pesticides to organophosphorus-resistant orchard populations of *Neoseiulus fallacis* and its prey, *J. Econ. Entomol.*, 64:350.

Schulten, G. G. M., and van de Klashorst, G., 1977, Genetics of resistance to parathion and demeton-S-methyl in *Phyloseiulus persimilis* A. H. (Acari:Phytoseiidae), Proc. IV Cong. Acarol. Aust. (1974).

Schulten, G. G. M., van de Klashort, G., and Russel, V. M., 1976, Resistance of *Phytoseiulus persimilis* A. H. (Acari:Phytoseiidae) to some insecticides, *Zeit. Ang. Ent.*, 80:337.

Spiller, D., 1958, Resistance of insects to insecticides, *Entomologist*, 2:1.

Storozhkov, Yu. V., Chabanovaskii, A. G., Mosshukin, Yu. B., and Metreveli, N. P., 1977, The resistance of *Phytoseiulus* to pesticides, *Zashchita. Rastenii*, 10:26.

Strawn, A. J., 1978, Differences in response to four organophosphates in laboratory strains of *Aphytis melinus* and *Comperiella bifasciata* from citrus groves with different pesticide histories, M.S. Thesis, Univ. Calif. Riverside, 117 pp.

Swift, F. C., 1970, Predation of *Typhlodromus (A.) fallacis* on the European red mite as measured by the insecticidal check method, *J. Econ. Entomol.*, 63:1617.

van de Vrie, M., 1962, The influence of spray chemicals on predatory and phytophagous mites on apple trees in laboratory and field trials in the Netherlands, *Entomophaga*, 7:243.

Watve, C. M., and Lienk, S. E., 1975, Response of two phytoseiid mites to pesticides used in New York apple orchards, *Environ. Entomol.*, 4:797.

Watve, C. M., and Lienk, S. E., 1976, Resistance to carbaryl and six organophosphorus insecticides of *Amblyseius fallacis* and *Typhlodromus pyri* from New York apple orchards, *Environ. Entomol.*, 5:368.

Wilkes, A., Pielou, D. P., and Glasser, R. F., 1951, Selection for

DDT tolerance in a beneficial insect, *in:* "Conference on Insecticide Resistance and Insect Physiology," Nat. Acad. Sci. Publ. No. 219, pp. 78-81.

Zilbermints, I. V., 1975, Genetic change in the development and loss of resistance to pesticides, Proc. 8th Intern. Cong. Plant Protec., 2:85.

IMPLICATIONS AND PROGNOSIS OF RESISTANCE TO INSECTICIDES

Robert L. Metcalf

Department of Entomology
University of Illinois
Urbana-Champaign, Illinois

INTRODUCTION

Insect resistance to insecticides has been a scientifically described phenomenon for 65 years, since Melander (1914) first described the failure of lime sulfur sprays to control the San Jose scale, *Aspidiotus perniciosus*, on apple trees in the Clarkson Valley, Washington. Two generations of entomologists have watched the number of authentic cases of insect pest resistance grow at an exponential rate. Since 1948, the number of cases has doubled about every six years and has now passed 400 (see Georghiou, this volume). This strongly suggests that in another 10 years the number of examples may exceed 1,500 species. Since the number of important insect pests is generally estimated at about 600 in the United States and is perhaps about 1% of all described species of insects, it seems likely that by the turn of the century, man will have succeeded in producing some degree of insecticide resistance in nearly all pest insects. This is a monumental achievement and yet its accomplishment, toward which we are well on our way, is passing with relatively little notice by learned societies or governmental agencies, and with almost no notice by society in general.

It is sobering to contemplate that both the chemical and agricultural industries of the United States, faced with this impending debacle in insect control, have responded by increasing the rate of production of insecticides by 5-fold since 1948, an exponential growth rate with an average doubling time of about 12 years (Figure 1). From 1971 to 1976, for example, the farm use of insecticides in the United States increased from 153,800,000 lb. active ingredient (A.I.) to 162,000,000 lb., and the acreage treated rose by one-third from 49,000,000 acres to 66,000,000 acres (USDA, 1978).

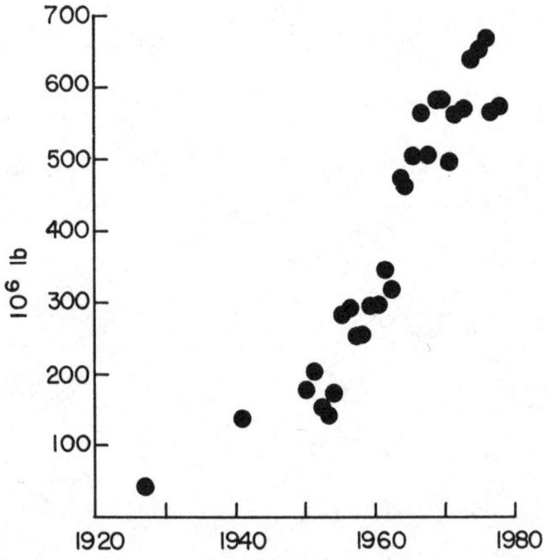

Figure 1. Production of insecticides in the United States (after Metcalf 1980, data from Pesticide Review).

From a biological viewpoint, insect resistance to insecticides is an evolutionary phenomenon brought about by intensive "natural selection" of the insect pest by massive applications of insecticides. The impact of these can perhaps be realized from a few statistics. Since 1950 in the United States more than 5,750,000,000 lb. (A.I.) of insecticides have been applied to the 400 million acres of cropland, or an average of more than 14 lb. per acre. In Egypt between 1961 and 1975, more than 811,000,000 lb. (A.I.) of insecticides were applied to about 4,000,000 acres of cotton to control the cotton leafworm, *Spodoptera littoralis*, the spiny bollworm, *Earias insulana*, and the pink bollworm, *Pectinophora gossypiella* (El-Sebae, 1977). This was an average of over 200 lb. per acre. In Illinois between 1953 and 1979, more than 100 million lb. (A.I.) of soil insecticides were applied to about 10 million acres of cropland to control the corn rootworms *Diabrotica longicornis* and *D. virgifera* (see later discussion), an average of about 10 lb. per acre. These insecticide residues have provided the massive selective processes responsible for the rapid increase in the number of resistant species.

Insecticide resistance seems to be the inevitable result of intensive selection of a large insect population having a substantial gene pool that incorporates, often at very low frequency, mutant alleles conferring fitness for survival under the modified environment contaminated by the insecticide. The classic example is, of course, the DDT-resistant house fly, *Musca domestica*, that appeared almost simultaneously in many parts of the world after the liberal use of DDT for residual fly control from 1946 to 1950. The resistant flies protected against the lethal action of DDT either by a detoxifying enzyme DDT'ase converting DDT to the inactive DDE, or by

Table 1. Number of Species of Insects and Mites Resistant to Insecticides.

Year	Agricultural Pests	Medical and Veterinary Pests	Total[a]	Reference
1908	1	-	1	(Brown 1971)
1918	3	-	3	ibid
1928	5	-	5	ibid
1938	5	2	7	ibid
1948	9	5	14	ibid
1955	13	12	25	(Metcalf 1955)
1957	37	33	70	(Brown 1958)
1960	65	72	137	(Brown 1961)
1967	127	97	185	(Brown 1968)
1969	119	105	224	(Brown 1971)
1976	225	139	364	(Georghiou & Taylor 1977)

[a]Metcalf (1955), Brown (1958, 1961, 1968, 1971), Georghiou and Taylor (1977).

a nerve axon insensitivity factor *kdr;* transferred these heritable characteristics to their offspring, giving rise to the astonishing phenomenon of house flies capable of living and reproducing normally in cages literally covered with white crystals of DDT. Insecticide resistance may result from a variety of mechanisms, including enhancement of physiological and biochemical barriers to intoxication, decrease in sensitivity of the target site of the insecticide, or modifications of behavior leading to reduced exposure (Georghiou, 1972a).

From a biological viewpoint, the prognosis for insecticide resistance is poor. The more than one million described species of Insecta comprise about two-thirds of all living animals and have aptly been described as "Man's Chief Competitors" (Flint and Metcalf, 1932). Their great genetic diversity, short life spans, and enormous reproductive capacities have enabled them to occupy nearly every available ecological niche in the terrestrial ecosystem. Man, by his rash single-factored selection by insecticides, is presently accelerating the evolutionary process and we already see evidence of new "monster" races of house flies, mosquitoes, cockroaches, bollworms, and rootworms that we are almost powerless to control.

Beyond the biological aspects, insect resistance to insecticides has given rise to a series of complex and interrelated economic and sociological problems that offer grave challenges to man. Aspects of these will be discussed in the balance of this paper, with the knowledge that "the past is prologue" and with the conviction that only by rigorous examination of the realities of past failure can we hope to present an optimistic prognosis for the future.

BRIEF HISTORY OF RESISTANCE

The history of insecticide resistance is discussed comprehensively elsewhere (Metcalf, 1955; Brown and Pal, 1971; Brown, 1971) and the emphasis here will be directed at exploring trends and making extrapolations for prognosis. Melander's (1914) report of San Jose scale resistance to lime sulfur received little attention and was followed by the reported resistance of California red scale *Aonidiella aurantii* and of black scale *Saissetia oleae* to hydrogen cyanide fumigation (Metcalf, 1955; Brown, 1971). This was greeted with some skepticism by many entomologists and various contradictory theories were proposed; however, the resistance was "real" and forced the abandonment of tree fumigation by the early 1940s because the higher dosages of HCN required were both phytotoxic and hazardous to the fumigator. By 1946, insecticide resistance was present in a total of 11 insect pests, including codling moth *Laspeyresia pomonella* and peach twig borer *Anarsia lineatella* to lead arsenate, cattle tick *Boophilus microplus* and blue tick *B. decoloratus* to sodium arsenite dip, and citrus thrips *Scirtothrips citri* and the gladiolus thrips *Taenothrips simplex* to tartar emetic. These examples resulted in the phasing out of two valuable control techniques, the sodium arsenite cattle dip and the tartar emetic sugar-bait sprays for thrips control. The prognosis for the future of insect control chemicals was poor. Insect resistance, however, received the scientific attention that it deserved only after the introduction of DDT in 1946 when this "miracle" insecticide failed to control the house fly *Musca domestica* in Sweden and Denmark in 1946, the mosquitoes *Culex pipiens* in Italy and *Aedes sollicitans* in Florida, the bedbug *Cimex lectularius* in Hawaii in 1946, and the human body louse *Pediculus corporis* in Korea and Japan in 1951 (Metcalf, 1955; Brown, 1971; Brown and Pal, 1971). This period, however, remained the golden age of synthetic insecticides, and it seemed that DDT could be readily replaced as required by methoxychlor, lindane, aldrin, dieldrin, heptachlor, chlordane, and toxaphene, and then by an unending stream of newer

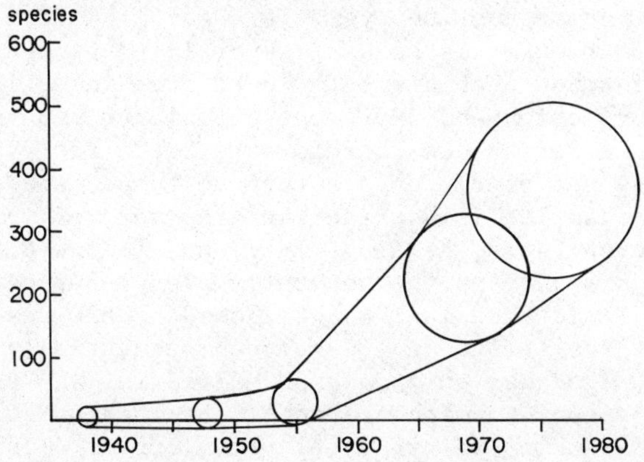

Figure 2. Rate of increase in insect pests with multiple resistance to insecticides. Area of circles is proportional to cases of multiple resistance (see text, data from Table 2).

organophosphorus insecticides. Instead, the use of these substitutes only served to increase the pace of development of insecticide resistance, and, as shown in Table 1, the numbers of examples of R pests in both agriculture and public health increased dramatically. The exponential nature of the rate of increase in numbers of insecticide-resistant pests is demonstrated in Figure 2.

Development of Multiple Resistance

The mere number of resistant (R) insect pest species does not adequately portray the impact of insecticide resistance upon applied entomology, crop production, or public health. *Cross resistance* enables these R species to survive exposure to closely related chemicals, e.g., DDT and methoxychlor, aldrin and heptachlor, or even lindane and dieldrin. Cross resistance arises because of a common detoxication system such as DDT'ase, or by target-site insensitivity (for example, mutant acetylcholinesterase) that is under the control of a single gene allele or of closely linked genes. Cross resistance limits the choice of available insecticides and can cause substantial economic problems, as in the case of the cyclodiene-R western corn rootworm *Diabrotica virgifera,* in which cross resistance to the soil insecticides aldrin, heptachlor and chlordane sharply decreased the benefit/cost ratio of soil insecticide treatments in corn (see later discussion). In early residual spray programs against *Anopheles gambiae* in West Africa, BHC was very effective, but the onset of resistance within two years brought with it cross resistance to dieldrin, thus phasing out two of the three available residual insecticides and placing sole reliance on DDT (see later discussion).

Multiple resistance, however, is a far more serious matter, as it results from the coexistence of several independent gene alleles producing resistance mechanisms for unrelated insecticides with different modes of action and different detoxication pathways (Sawicki, 1975). Multiple resistance therefore reflects the past history of insecticide use and precludes a return to those used previously. The following examples illustrate the disasters of multiple resistance. In Danish farms, Keiding (1977) has followed the development of multiple resistance in the house fly over a 30-year period beginning with DDT resistance in 1945, followed by BHC resistance in 1947. High levels of resistance have developed sequentially to the *organochlorines* DDT, lindane, chlordane and dieldrin, and to the *organophosphates* diazinon, coumaphos, ronnel, bromophos, fenitrothion, iodofenphos, malathion, dimethoate, trichlorfon, dichlorvos and tetrachlorvinphos. These strains were resistant to the *carbamates* propoxur, bendiocarb, dioxacarb and 4-benzothienyl N-methylcarbamate; to the *pyrethroids*, including pyrethrins and piperonyl butoxide, bioresmethrin, tetramethrin and decamethrin; and to the *juvenoid* methoprene. The cotton leafworm *Spodoptera littoralis* in Egypt has become resistant successively to the *organochlorines* toxaphene, DDT, lindane and endrin; the *organophosphates* trichlorfon, methyl parathion, fenitrothion, monocrotophos, dicrotophos, phospholan, mephosopholan, leptophos, chlorpyrifos, stirifos, methamidophos and azinphosmethyl; the *carbamates* carbaryl and methomyl; and the *chitin synthesis inhibitor* diflubenzuron. As a result, no new insecticide has remained effective for more than 2 to 4 years (El Sebae, 1977). In California's Central Valley, the mosquitoes *Aedes nigromaculis* and *Culex tarsalis* have developed multiple resistance successively to the *organochlorine* larvicides DDT, aldrin, dieldrin, and heptachlor, and to the *organophosphates* malathion, parathion, fenthion, EPN, chlorpyrifos and temephos.

In Australia the cattle tick *Boophilus microplus* has developed successive resistance to dips, first with sodium arsenite in 1937, and then to the *organochlorines* DDT, BHC, toxaphene, aldrin and dieldrin; the *organophosphates* diazinon, dioxathion, coumaphos, carbophenothion, ethion, bromophos, phosphamidon, chlorpyrifos and ronnel; and to *carbamates*, including carbaryl and 4-benzothienyl N-methyl carbamate (Shaw et al., 1968). Further selection has resulted in strains with altered acetylcholinesterase apparently insensitive to almost all organophosphorus insecticides. Multiple resistance in cattle tick control is especially serious because of the generally narrow margin between a lethal dosage to the tick and a toxic dose to the cattle, and because of problems with persistent residues in meat and milk (Wharton and Roulston, 1970). The problem is presently so serious that little hope remains for the continuation of successful chemical control.

Table 2. Development of Multiple Resistance to Insecticides[a]

Year	Resistant Species	2-Stage	3-Stage	4-Stage	5-Stage
1938	7	0	0	0	0
1948	14	1	0	0	0
1955	25	18	3	0	0
1969	224	42	23	4	0
1976	364	70	44	22	7[b]

[a]See Table 1 for sources.
[b]By 1979 at least 10 species showed 5-stage resistance: *Musca domestica, Culex pipiens, Myzus persicae, Heliothis virescens, Tribolium castaneum, Sitophilus granarius, S. oryzae, Boophilus microplus, Tetranychus urticae, T. telarius.*

The summation by Georghiou and Taylor (1977a) provides a record of the status of multiple resistance among the 364 species of R pests identified by 1976. The rate of development of multiple resistance (Table 2) can be surveyed from 1947 when *Musca domestica* was found to be resistant to both DDT and BHC and to the cyclodienes, to the present, using data provided by Metcalf (1955a), Brown (1969, 1971), Brown and Pal (1971), Georghiou and Taylor (1977a), and current literature. The various stages of multiple resistance can be classified generally as (1) arsenicals, (2) DDT and derivatives (3) BHC and cyclodienes, (4) organophosphates, (5) carbamates, (6) pyrethroids, (7) juvenoids, and (8) chitin synthesis inhibitors. A more vivid impression of the rate of development of multiple resistance is gained from Figure 2, which shows this as a function of the number of species of R pests that have been catalogued from 1938 to 1979 (Table 2). In Figure 2, the areas of the circles are proportional to the severity of multiple resistance, i.e., 1X for single-stage resistance and up to 5X for 5-stage resistance. It seems probable that the rate of development of insecticide-R species will slow within the next 10 to 20 years (Table 1) when most insects have become resistant, but the rate of development of multiple resistance will increase as our armamentarium of insecticides is expended in their control.

ECONOMIC ASPECTS OF INSECTICIDE RESISTANCE

Society is ruled by economic concerns and benefit/cost economics is increasingly dominated by consideration of externalities. The social economics of pest control inevitably must concern itself with quantification of such externalities as environmental pollution by pesticides, human health hazards of pesticide usage, destruction of natural enemies, costs of pesticide registration and regulation, and pest resistance and pest resurgences. Of the factors, the latter is perhaps the most intangible and has had little quantitative discussion.

Today, however, after 30 years of aggressive production of insecticide resistance, the entomologist, the sanitarian and the farmer must give sober consideration to the economic costs of resistance that now threaten and will soon jeopardize many of our present insect control and vector control practices.

Rising Costs of Insecticides

The early development of cross resistance in economic pests had little direct economic impact since the costs of the arsenical and organochlorine insecticides were extremely low. As recently as 1969, the wholesale prices per lb. in the United States were: lead arsenite, \$0.266; DDT, \$0.175; benzene hexachlorine, ca. \$0.13; toxaphene, \$0.22; and chlordane, \$0.59 (Pesticide Review, 1973). This price structure made it possible to use treatments with these insecticides as inexpensive forms of crop insurance and thus created both the technology of preplanting applications of soil insecticides and the pervasive philosophy of multiple applications that led to the "pesticide treadmill" of the cotton grower with as many as 50 to 60 applications in a single season (Vaughan and Gladys, 1977). This low cost and relative abundance, more than any other factor, was responsible for the pest eradication philosophy expressed in vast campaigns against the gypsy moth, spruce budworm, Japanese beetle, fire ant and the anopheline vectors of malaria. However, the switch from DDT to BHC, toxaphene or chlordane, or even to methyl parathion at \$0.55, parathion at \$0.59, or malathion at \$0.79 per lb. in 1969 as dictated by the development of pest resistance, caused little or no economic dislocation. Presently, however, because of global inflation, the prices of these "old reliable" insecticides are increasing rapidly, and from 1970 to 1977 the average wholesale price for these insecticides rose from \$0.69 per lb. to \$1.83 per lb., a rate of 21.7% per year (doubling time, 3.2 years). Increases reported over this period were: DDT, \$0.177 to \$0.34; toxaphene, \$0.22 to \$0.38; chlordane, \$0.59 to \$1.05; methyl parathion, \$0.55 to \$0.96; and malathion, \$0.79 to \$1.025 (Pesticide Review, 1977). This trend is inextricably linked to global inflation, and because of rapidly increasing petroleum prices, it seems likely to continue.

This rate of increase has profound implications to crop protection during the present period when the prices of such crops as cotton and corn are relatively static, compared to the increasing costs of insecticides. Even low level resistance necessitating a 2- to 4-fold increase in insecticide application rates in orchard or field can greatly affect benefit/cost ratios, as illustrated below with the corn rootworms (see later discussion).

The effects of multiple resistance are, however, much more severe. This selection process necessitates the steady development of new insecticides. Thus, on Danish farms from 1945 to 1977, more than 17 different insecticides were employed for fly control with an average effective lifetime of about three years. The R genes for DDT and cyclodienes have persisted for more than 20 years, and practical resistance to these compounds and to dimethoate reappeared within two months after reapplication (Keiding, 1977). In Egypt, between 1961 and 1975, 19 different insecticides were applied to cotton to control the cotton leafworm *Spodoptera littoralis*. No insecticide has remained effective for more than 2 to 4 years, and no signs of substantial reversion of resistance have occurred over 11 years (El Sebae, 1977).

It has become increasingly difficult to discover and develop new insecticides. In 1956, 1,800 compounds were screened per new product; in 1965, 3,600; in 1969, 5,040; and in 1972, 10,000 (Johnson and Blair, 1972). Most of the simple compounds have been investigated, but the costs of laboratory synthesis and screening are rising proportionately. Moreover, such newer products as the synthetic pyrethroids and insect growth regulators are chemically much more sophisticated: DDT can be produced in a one-step synthesis yet the synthetic pyrethroid allethrin requires 13 steps. These differences are reflected in greatly decreased yields of the final product and in much more elaborate synthetic plants, and registration requirements as mandated by EPA in 1979 are much more stringent than registration requirements with the USDA in 1956. The accumulated impact of all these factors including inflation are summed up in Figure 3, which shows the exponentially increasing costs of pesticide development from 1955 to 1977, a period in which costs rose about 20-fold. It has become fashionable to blame governmental overregulation for escalating R&D costs, yet present registration requirements represent less than one-third of the increase shown in Figure 3 (Campt, 1979). The accumulative effect of all these factors is reflected in the present prices of new types of insecticides, e.g., the synthetic pyrethroids, and the insect growth regulators are now selling for \$40 to \$50 per lb., or more than 100-fold higher than DDT and toxaphene. These new products are highly efficacious but the necessity for their use, brought about substantially because of multiple resistance, will drastically lower benefit/cost ratios from insecticidal employment. This is well demonstrated in the following examples.

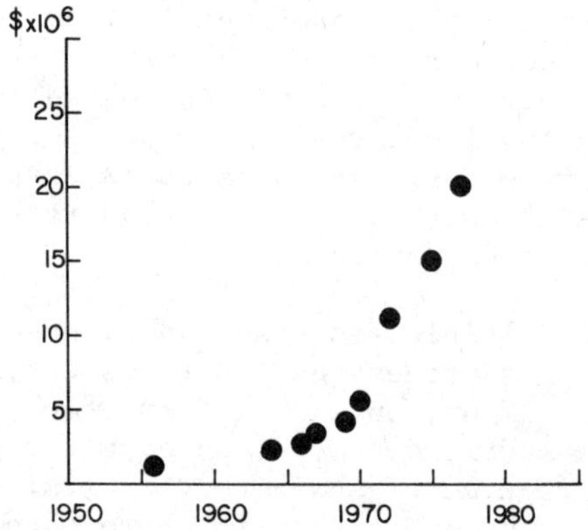

Figure 3. Rate of increase of developmental costs for new pesticides (after Metcalf, 1980; data from Goring, 1977; Hunter, 1974; Johnson and Blair, 1972; National Research Council, 1977; Secretary Health, Education, and Welfare, 1969; von Rumker et al., 1970; and Wellman, 1966).

IMPLICATIONS OF ANOPHELES RESISTANCE TO MALARIA ERADICATION AND CONTROL

The burgeoning resistance to insecticides by *Anopheles* vectors of malaria has the gravest implications to progress in malaria eradication and control, which have yet to be faced in a forthright manner (Sopper et al., 1961; Wright et al., 1972; WHO, 1976a). In simplest terms DDT, available until 1969 at $0.19 per lb., was the safest, most economical, and most effective residual insecticide for use on a variety of home building materials. Because of dwindling petroleum supplies, global inflation, escalating R&D costs, and insecticide resistance, it will never again be possible to achieve malaria control at the very modest cost of $0.19 to $0.43 per capita required for residual house spraying the the decade of the 1960s (Soper et al., 1961). Presently there are six alternative insecticides that may be used as substitutes for DDT in residual house spraying (Table 3). Dieldrin was originally about four times more effective than DDT and had comparable economics. However, dieldrin now costs almost seven times as much as DDT and is no longer recommended for residual spray programs because of poisoning among spray operators even when wearing protective clothing and under close supervision (WHO, 1972; Fontaine, 1976). Moreover, *Anopheles* resistance to dieldrin is even more widespread than DDT resistance, with 43 species of *Anopheles* now resistant to dieldrin vs. 24 for DDT. Lindane (99% gamma-isomer of BHC) was originally two to four times more effective than DDT on some building surfaces, but now costs 10 times as much as DDT (Table 3). Unfortunately, there is full cross resistance between lindane and dieldrin in *Anopheles*.

Propoxur (OMS-33) was developed under a WHO program to find substitutes for DDT in residual house spraying and has proved highly effective. However, propoxur is notably less persistent and must be rated as about 0.25 times as effective as DDT. Propoxur costs about 5 times as much as DDT and Wright et al. (1972) estimated annual insecticidal costs in residual house spraying at 20.4-fold that of DDT (Table 3). Unfortunately, *A. albimanus* has already demonstrated severe propoxur resistance, apparently due to widespread use of carbaryl and other agricultural insecticides in Central American cotton plantations.

Fenitrothion (OMS-43) is another residual insecticide developed by WHO as effective and apparently safe for residual house spraying. It is about 0.25 times as effective as DDT on a residual basis and presently residual spraying costs about 15.9 times as much as DDT treatment (Table 3, Fontaine et al., 1978).

The new synthetic pyrethroid decamethrin offers unprecedented activity as a residual insecticide and has been shown to be effective against *A. gambiae* at 0.05 g/m^2 for three months, or about 10-fold the effectiveness of DDT. However, decamethrin is presently priced at nearly 150 times as much as DDT, and it seems realistic to assume that the use of decamethrin would cost at least 10 to 20 times as much as DDT (Rishikesh et al., 1979).

Public Health Aspects of Anopheline Resistance

The consequences of multiple resistance by *Anopheles* vectors of malaria on human health are obviously of much greater import than the economic effects with which they are closely linked. DDT resistance was first observed in *A. sacharovi* in Greece in 1950 and was closely followed by the appearance of dieldrin resistance in about 1954 (Brown and Pal, 1971). The spread of resistance in *Anopheles* is shown in Table 4. The onset of resistance in *A. sacharovi* began a deterioration in malaria eradication that has continued for nearly 30 years, marked by sporadic epidemics such as the some 2000 cases of vivax malaria in 1956 (Brown and Pal, 1971). Resistance in this species is now found in Greece, Lebanon, Iran and Turkey, and by 1976 multiple resistance to malathion, fenitrothion, propoxur, and fenthion was observed in Turkey in an area where aerial spraying for agricultural pest control was practiced (WHO, 1976a,b; Pal, 1977).

Pronounced DDT resistance was found in *A. stephensi* in Iran and Iraq when full-scale residual spraying operations were begun in 1957-58, and dieldrin resistance appeared three years later. The malaria eradication program has struggled with the consequent operational difficulties for 20 years with malathion resistance appearing in Iran in 1975 after about six years of use (WHO, 1976a,b).

Table 3. Relative Costs of Insecticides for Residual House Spraying

Insecticide	Dosage g/m^2 (tech)	Approximate residual effect on mud - months	Cost per kg[a]	Cost per lb[b]	Relative cost per 6 months
DDT 75% wpd	2	6	\$0.33	\$0.34	1.0[a]
dieldrin[c] 50% wdp	0.5	6		\$2.34	1.7[c]
lindane[d] 50% wdp	0.5	3		\$3.45	5.1[c]
malathion[d] 50% wdp	2	3	\$0.89	\$1.02	5.3[a]
propoxur 50% wdp	2	3	\$3.40		20.4[a]
fenitrothion 40% wdp	2	3	\$2.63		15.9[a]
decamethrin 5% wdp	0.1	3		ca \$50	14.6[b]

[a]WHO data, Wright et al. (1972), Fontaine et al. (1978).
[b]Estimated from relative wholesale price of technical compound, Pesticide Review (1979).
[c]Use largely abandoned because of hazard to man and animals. Poisoning has occurred among spray operators even with full protective clothing.
[d]Disagreeable residual odor has interferred with use.

The best alternative presently is malathion, which now costs three times as much as DDT but is only about 0.25 to 0.33 times as effective on a variety of building surfaces (Table 3). Wright et al. (1972) calculated the cost ratio for insecticide in residual spraying at 5.3 times that of DDT. By 1976, malathion resistance had been demonstrated in *A. albimanus*, *A. culicifacies*, *A. hyrcanus*, *A. messeae*, *A. sacharovi*, and *A. sinensis* (WHO, 1966a). Recent experiences in India and Pakistan have shown that impure commercial grades of malathion can cause severe poisoning and even death in spray operators, when handled without protective clothing.

Table 4. Spread of Insecticide Resistance in Anopheles Mosquitoes[a]

Year	Number of species resistant to:			
	DDT	Dieldrin/lindane	Organophosphates	Carbamates
1951	1			
1957	4	6		
1959	7	23		
1962	12	34		
1969	15	37	1	1
1971	19	38	2	1
1975[b]	24	41	6	2
1976	24	43	6	2
1978[c]				

[a]Data from Brown and Pal (1971), Wright et al. (1972), WHO (1976a), Georghiou and Taylor (1977).
[b]Seventeen *Anopheles* vectors showed no resistance in 1975 (WHO 1976a).
[c]*A. albimanus* resistant to DDT, dieldrin, malathion, propoxur; *A. sacharovi* resistant to DDT, dieldrin, malathion, propoxur, fenthion, fenitrothion.

In Central America and the Caribbean, dieldrin spraying against *A. albimanus* was introduced in 1956 and widespread resistance was noticed in 1958. A return was made to DDT spraying and resistance to this compound, first noticed in 1958, was general by 1960. Propoxur spraying was begun in Guatemala, El Salvador, Honduras and Nicaragua in 1970-71, and pronounced resistance to this insecticide was present by 1974. Presently *A. albimanus* shows extensive multiple resistance to dieldrin, lindane, DDT, malathion and propoxur. The rapid onset of this multiple resistance has been markedly influenced by the vast quantities of insecticides used in cotton production. For example, Georghiou (1972b) reported the use in El Salvador from 1961-70 of 20,446,900 kg organophosphorus insecticides (93% parathion), 18,514,000 kg organochlorines, and 265,000 kg carbamates (8% carbaryl). This compares with 3,370,000 kg of DDT required for five years of residual house spraying to protect 3 million inhabitants in El Salvador.

The most dramatic public health consequences of vector resistance have been demonstrated by *A. culicifacies*. In Sri Lanka this species developed resistance to DDT in 1962 after spraying had continued for 15 years and eradication was in the consolidation stage with only 17 cases reported in 1963. An explosive epidemic occurred in 1967 with 3,466 cases of malaria and in 1968, there were more than 1,000,000 cases. *A. culicifacies* in India developed dieldrin resistance in 1958 and DDT resistance in 1959 after three years of spraying, but a widespread attack program was continued until 1965-66 when both DDT and BHC failed to control outbreaks of malaria in areas under consolidation and maintenance (Brown and Pal, 1971). Malathion was substituted in 1968 with some success, but widespread epidemics of malaria were reported with 4 million cases in 1975 compared to 125,000 in 1965. In Pakistan the experience was similar, with *A. culicifacies* demonstrating DDT resistance in 1963. The importance of this resistance was not recognized until malaria outbreaks began in the consolidated stage of operations in 1969 and neither DDT nor lindane was effective. Malaria cases in Pakistan were reported to number over 10 million in 1975, compared with 9,500 in 1961 (Time Magazine, 1975). *A. culicifacies* was reported to have developed malathion resistance by 1975.

Thus, in these areas where multiple resistance in *Anopheles* is widespread, malaria eradication and control efforts have almost completely broken down and there has been a massive resurgence of the disease, with some countries showing a 30- to 40-fold increase in malaria cases from 1968 to 1976 (Agarwal, 1979). The number of autochthonous cases of malaria in countries with control programs rose from 3,251,000 in 1972 to 7,517,000 in 1976, and it is believed that many more were unreported (WHO, 1978a).

Economic Effects of Multiple Resistance in Malaria Control

The global efforts of the World Health Organization to eradicate malaria by residual house spraying were predicated upon the use of DDT as a cheap, effective and safe household insecticide. DDT usage was standardized at 1 g per m^2 of wall surface, with applications at six-month intervals. WHO's experience indicated that the cost of residual house spraying ranged from $0.19 to $0.43 per capita, depending upon economic and geographic factors in the various countries involved (Soper et al., 1961). Even this meagre expenditure for a very important public health measure (malaria in 1955 was considered to be the "king of diseases" with about 200 million cases and 2 million deaths annually) was beyond the economic resources of most of the developing countries, and the program was supported for many years by hundreds of millions of dollars of foreign economic aid from the United States. It was originally anticipated that the total cost of the global eradication program would be approximately $1.3 billion (Soper et al., 1961).

The annual cost of spraying operations for the 70 million kg of DDT used in residual spraying operations in 1971 was estimated as \$60 million, but the annual cost for replacement by malathion was estimated at over \$180 million and by propoxur at over \$500 million (Wright et al., 1972). Because of the additional spraying rounds and safety equipment required with the substitute insecticides, the total cost for residual spraying per million population per six-month period was calculated at \$204,800 for DDT, \$637,000 with malathion, and \$1,762,000 with propoxur. Thus, to complete the malaria eradication program for the areas where *Anopheles* resistance is crucial with 256,111,000 inhabitants (WHO, 1976a) would require for the five years of spraying effort (10 applications) \$1.63 billion with malathion or \$7.67 billion with propoxur, the latter more than 5X the estimated \$1.3 billion cost for global malaria eradication. These estimates assume no further inflation or *Anopheles* resistance.

Malaria Control in Africa. This subject has remained enigmatic since the organization of global malaria eradication by WHO in 1955. Recent estimates place the population of tropical Africa at about 370 million and growing at the rate of 2.5 to 3% per annum (WHO, 1974). Malaria is endemic in much of Africa south of the Sahara and has been characterized as "stable malaria" of very high prevalence correlated with widely distributed and unusually efficient vectors, the *A. gambiae* complex and *A. funestus*. The malaria infection rate in holo- and hyperendemic regions may reach 90 to 95%, and it has been estimated that it is responsible for about one million deaths of infants and children annually.

A number of pilot malaria control projects using residual house spraying were conducted after 1950 and a considerable degree of control was obtained in Liberia, Cameroon and Uganda with the virtual disappearance of *A. funestus*. Results were not as promising from Senegal south to Rhodesia. It has never been possible to begin full-scale malaria eradication programs in continental Africa because of administrative and economic problems. Moreover dieldrin and lindane (BHC) resistance in the *A. gambiae* complex was detected in Nigeria in 1955 and is now widespread (Brown and Pal, 1971). DDT resistance in this complex was first observed in Upper Volta, and by 1975 was present in Benin, Cameroon, the Central African Republic, Congo, The Gambia, Guinea, Mauritius, Nigeria, Senegal, Sudan, Swaziland, Togo, and Upper Volta (WHO, 1976b). Each passing year exposes segments of the *A. gambiae* population to selection by a wide variety of agricultural insecticides used in Africa and hastens the onset of multiple resistance. DDT with its low cost, long residual effectiveness, and safety is particularly important to malaria control in rural Africa with its relatively low population density, shortages of trained labor, and poor communication. The countries of tropical Africa have the lowest per capita incomes in the world, e.g., Guinea, \$114; The Gambia, \$180; Mauritius, \$332; Nigeria, \$357; Senegal, \$284; Sudan,

$126; Uganda, $162; and Upper Volta, $66. The large-scale usage of the more expensive residual insecticides with shorter residual action, such as malathion, fenitrothion and propoxur, seems beyond the present capacity of these countries where the minimal cost of residual spray-int with DDT often represented 50-100% of the total medical and health budget per head per year (WHO, 1974). Therefore, it seems doubtful that malaria in Africa can be eradicated or even controlled by residual house spraying.

IMPLICATIONS OF WESTERN CORN ROOTWORM RESISTANCE

Diabrotica virgifera Le Conte, an important pest of corn *Zea mays*, was first described by Le Conte from specimens collected on "wild gourd" (probably *Cucurbita foetidissima*) near Fort Wallace, Kansas in 1867 (Smith and Lawrence, 1967). It was apparently sparsely distributed over the Colorado-New Mexico-Arizona area, and its original native host was thought to be *Tripsacum* (Smith, 1966). However, *D. virgifera* is strongly attracted to Cucurbitaceae and feeds avidly on wild bitter *Cucurbita* spp. (Howe et al., 1976). It is a compulsive feeder on the terpenoids cucurbitacins B and E (Metcalf, 1979). This is presumptive evidence of primary coevolution of the Diabroticites with the Cucurbitaceae and gradual host transference of several species, including *D. longicornia*, *D. undecimpunctata* and *D. virgifera* to the Graminaceae.

With the introduction of cultivated *Zea mays* by settlers migrating westward after the Civil War, a tempting new host plant became available in steadily increasing areas. *D. virgifera* (western corn rootworm) was first recorded as a pest of cultivated sweetcorn in 1909 near Fort Collins, Colorado and again in 1910 near Loveland, Colorado (Figure 4) (Gillette, 1912). It slowly spread across the western corn-growing area, producing injury in the five counties of southwest Nebraska in 1929 and heavy damage along the Platte River near Lexington in 1941, and extending eastward to Grand Island by 1945 (Tate and Bare, 1946). The western corn rootworm was first reported as injurious to corn in Kansas in Norton Co., Kansas in 1945, and by 1953 it was present as far east as Nemaha, Pottawattomie, Wabaunsee, Morris and Chase counties (Burckhardt and Bryson, 1955). The western corn rootworm crossed Nebraska to within 70 miles of the Missouri River by 1948, and by 1954 it was present all along the river from South Dakota to Missouri. Thus, from 1909 to 1948, the species traveled westward from Colorado to the Missouri River, about 470 miles, at an average of about 12 miles per year. In Kansas during the 1953 growing season it moved eastward 30 to 35 miles (Burckhardt and Bryson, 1955).

About this time, preplanting application of soil insecticides was introduced as a control measure for corn rootworms. BHC was recommended for this purpose in Nebraska in 1948 and large-scale

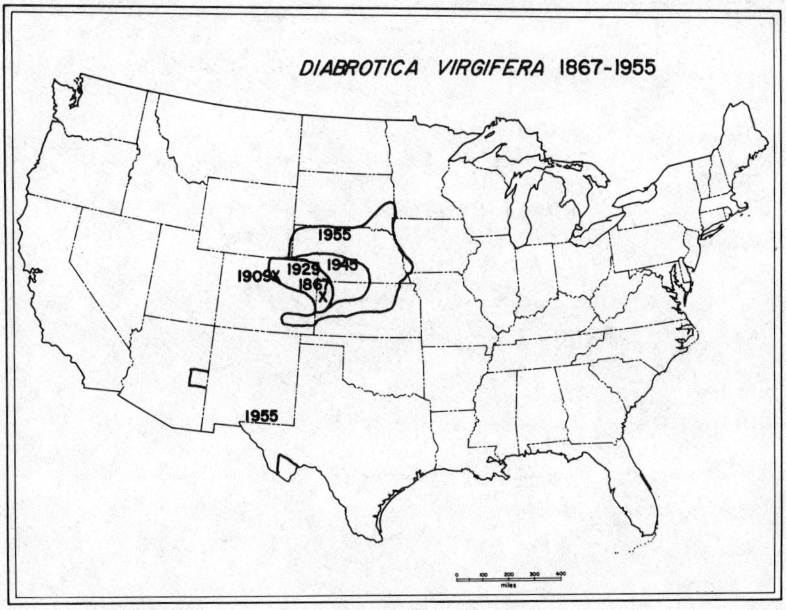

Figure 4. Distribution of western corn rootworm in the United States prior to onset of cyclodiene resistance.

applications were made in that state in 1949. Treatments with aldrin and chlordane were begun in 1952 and with heptachlor in 1954. The total area treated with these soil insecticides in Nebraska in 1954 was 1,740,000 acres (Ball and Weekman, 1962). Ineffective corn rootworm control was first noted in south central Nebraska in 1959 and became increasingly serious during 1960 and 1961. Determinations of topical LD_{50} values made in 1961 with western corn rootworm adults collected in the area of resistance in central Nebraska near Aurora were 38.02 μg for aldrin and 38.19 μg for heptachlor, compared with 0.38 μg and 0.43 μg respectively, in an area of susceptibility in eastern Nebraska near Ashland. Thus, the beetles in the R area had LD_{50} values about 100X those of the susceptible area (Ball and Weekman, 1962). An extension of these studies carried out in 1962 (Ball and Weekman, 1963) showed that the territory inhabited by the R beetles was expanding rapidly into an 80- to 100-mile-wide band along the Platte River in east central Nebraska and reaching into the western edges of South Dakota, Iowa and Missouri. The area in Nebraska near Kearney with LD_{50} values ranging up to 73.87 μg of aldrin per beetle coincided with that part of the state where the cyclodiene insecticides and BHC had been used the longest. For comparison of the 2-hr LD_{50} values, beetles from Scottsbluff in western Nebraska where few soil insecticides were used had an LD_{50} less than 0.05 μg for aldrin, and the differential of resistance was about 1500X.

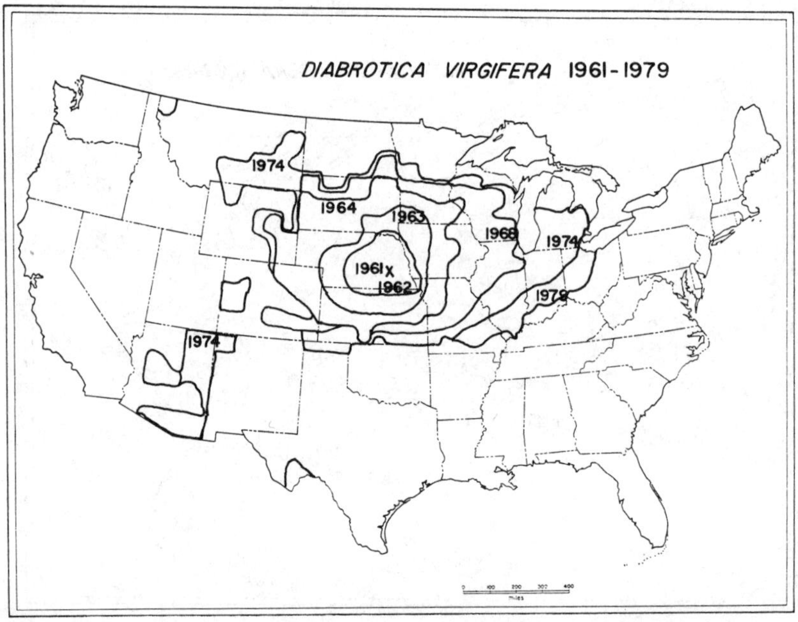

Figure 5. Distribtuion of western corn rootworm in the United States after onset of cyclodiene resistance.

A coordinated regional survey of the spread of the cyclodiene-R western corn rootworm was made in 1963 (Hamilton, 1965) and showed that the R beetles were present in western Iowa, northern Kansas, northwestern Missouri, southwestern Minnesota, and southeastern South Dakota. The relative susceptibility of the western corn rootworm ranged from LD_{50} 0.02 μg near Manhattan, Kansas to 15.5 μg (24-hr values) in northern Kansas near Belleville. The differential was about 775X.

Surveys made by the U.S. Department of Agriculture have delineated the continual spread of the cyclodiene-R western corn rootworm (Cooperative Economic Insect Report, 1956, 1965, 1969, 1975, 1977), as depicted in Figure 5. This provides perhaps the best record of the migration of a species in which resistance has been induced, apparently in a single locality in south central Nebraska. From 1961 to 1964 the R western corn rootworm traveled from near Grand Island, Nebraska to near Eau Clair, Wisconsin, about 360 miles or an average of 120 miles per year. Cyclodiene-R western corn rootworms reached northwest Indiana by 1968, a distance of about 500 miles, in seven years, or about 70 miles per year, and by 1979 they had almost completely covered all of the corn belt from Nebraska through Ohio, and Minnesota and Wisconsin through all of Illinois and most of Missouri (Figure 5). In Illinois in 1976 the topical LD_{50} values for the western corn rootworm were 51.37 μg for aldrin and 22.92 μg for heptachlor, or resistance of 1000X to 2500X primitive levels (Chio et al., 1978).

The astonishing change in the rate of migration of this species from 10 to 30 miles per year before onset of cyclodiene resistance to 70 to 120 miles per year after resistance seems likely to be the result of increased fitness of the R beetles and of a behavioral change associated with the R gene. Moreover, the R western corn rootworm is a superior competitor and through competitive displacement has become the dominant rootworm pest in an area where the northern corn rootworm, *Diabrotica longicornis* Say, that inhabits an almost identical ecological niche, was formerly the primary rootworm pest (Chiang, 1973). For example, in Illinois in 1967, the western corn rootworm comprised 9% of the corn rootworm population, 35% in 1975, and 65% in 1977, reaching 89-93% in some locations (Wedberg and Black, 1978).

Use Patterns of Soil Insecticides

The intensity of selection for resistance in the western corn rootworm has been enormous. The larvae of this insect feed on and tunnel into corn roots, directly reducing yields and causing corn plants to fall over and lodge after heavy rains and winds, which results in inefficient corn picking during mechanical harvesting. The feeding of a single corn rootworm larvae has been calculated to cause a loss of 0.72-0.87% of the yield (Apple, 1971; Kuhlman and Petty, 1971). The organochlorine soil insecticides BHC, chlordane, aldrin and heptachlor were relatively cheap, and when used in the soil at 1 lb. per acre (1.1 kg/hc) substantial decreases in loss were recorded. In Illinois between 1956 and 1962, aldrin applied to corn roots at 1 lb. per acre resulted in average savings of 8.5 bushels per acre (Bigger, 1963). From 1964 to 1973 in Illinois, soil treatments when needed for western corn rootworm control saved an average of 10%, or 10 bushels per acre (von Rumker and Horay, 1974).

Accurate statistics are available for Illinois about the extent of usage of soil insecticides for corn. They show the very rapid growth in the use of the organochlorines beginning in 1953 and reaching over 4 million acres by 1963. About 80% of this usage was aldrin and the average rate of use was about 1.3 lb. A.I. per acre. By 1967 the organochlorine insecticides were applied to 56% of the total corn acreage of Illinois. Resistant western corn rootworms were first found in northwest Illinois in 1964 and rapidly spread over the major corn-growing areas, which were almost completely infected by 1973. The State Extension Entomologist stopped recommending soil treatments with the organochlorines in 1970 because of widespread environmental pollution of milk, fish and humans with dieldrin and heptachlor expoxide and because they had become ineffective against

R corn rootworms. Aldrin and dieldrin were banned by the U.S. Environmental Protection Agency in 1974, and the use of heptachlor and chlordane was severely restricted in 1977. The use of organochlorines peaked in Illinois by 1967 at about 5,601,090 acres four years after R western corn rootworms had invaded and when almost half the state's corn acreage was infested (Figure 5), yet almost 1 million acres were still treated with organochlorines in 1976, when the R insects were present in 90% of the state's total corn acreage.

The emergence of cyclodiene resistance and emphasis on insecticide biodegradability resulted in the appearance of organophosphate insecticides, beginning with diazinon in 1964 and followed by phorate, fonofos, fensulfothion and terbufos. The carbamates metalkamate and carbofuran were introduced subsequently and the acreage treated with these materials in 1966 almost equaled the maximum treated with the organochlorines in 1967 and continues to rise. In 1972, 61% and in 1974, 59% of the total corn acreage in Illinois was treated with soil insecticides. It is worth noting that from 1953 to 1979 more than 100 million lb. (A.I.) of soil insecticides were applied to about 10 million acres of cropland, an average of about 10 lb. per acre. Yet this enormous rate of treatment has not appreciably ameliorated the corn rootworm problem nor halted the migration of R western corn rootworms that between 1964 and 1979 had infested all the corn-growing areas of the state.

Elsewhere in the corn belt, the use of soil insecticides for rootworm control was also monumental. USDA data indicate that 10,800,000 lb. (A.I.) of insecticides were used on corn in 1964; 17,500,000 lb. in 1966; 25,500,000 in 1971, and 31,978,000 lb. in 1976. The total corn acreage treated was 20,476,000 acres in 1971 (27.6% of the total acreage) and 31,966,000 acres in 1976 (38%) (USDA, 1976). In the latter year, the approximate quantities of soil insecticides used on corn were: carbofuran, 9,320,000 lb.; heptachlor, 1,741,000 lb.; diazinon, 1,100,000 lb.; chlordane, 1.026,000 lb.; and aldrin, 452,000 lb. (A.I.).

Selection Pressure on Adult Corn Rootworms

Several epidemic years of western corn rootworm populations have resulted in heavy damage by the adult beetles feeding on corn silks. This results in deficient pollination of the ears, or "scattercorn." Although this condition does not materially affect corn yields until beetle populations reach 10 or more per ear (Turpin et al., 1978), widespread aerial spray applications were made against adult corn rootwroms from 1976 to 1979. In Illinois approximately 1 million acres were treated in 1976; 768,140 acres in 1977, and 309,460 acres in 1978 (Black and Braness, 1979). In all the United States approximately 10,000,000 acres of corn were treated with carbaryl by air in 1976 (USDA, 1976). These massive applications of carbaryl to control adult corn rootworms have provided additional severe selection pressure for resistance.

Effects of Resistance on Economic Threshold

The economic injury level has been variously defined as "the lowest pest population density that will cause economic damage" (Stern et al., 1959), "a more critical density ..... where the loss caused by the pest equals in value the cost of the available control measures" (NAS, 1969), and the pest "population that produces incremental damage equal to the cost of preventing the damage" (Headley, 1972). Although these definitions were formulated during a period of relatively cheap and constant insecticide costs, it is apparent that the development of corn rootworm resistance necessitating the change to more expensive alternative insecticides can have a major effect on the use of soil insecticides for corn rootworm control.

The organochlorine insecticides provided cheap crop protection during the period of 1952 to 1962 when the average cost of application of 1.5 lb. per acre of aldrin or heptachlor granular was $2.20 per acre (von Rumker et al., 1975) and savings of corn as high as 8.5 bushels per acre (Bigger, 1963). At an average price for corn of $1.10 per bushel (USDA), the return to the grower, neglecting externalities, was about $4.25:$1.00. This cheap crop insurance, therefore, became an accepted grower practice, without regard to corn rootworm populations or economic thresholds (the action level immediately below the economic injury level) (Metcalf and Luckman, 1975).

Detailed records from Illinois indicate that from 1964 to 1973, the use of corn rootworm insecticides resulted in an average profit of $4.32 per acre (range: $3.50 to $5.00) and a total average profit for Illinois farmers of $24,978,810 (range: $16,364,500 to $32,540,535). Increasing corn rootworm resistance and the rapidly escalating costs of substitute insecticides have made this form of crop insurance less attractive. In 1968-70, the use of aldrin saved only 1.6 bushels per acre and the 1970 corn crop averaged $1.33 per bushel for a profitless $1.00:$1.00. During this period, however, the use of carbofuran at 1 lb. per acre costing $7.50 per acre saved an average of 13.7 bushels per acre (Kuhlman and Petty, 1971). Thus, in 1975 with corn averaging $2.54 per bushel, soil treatment with carbofuran costing $7.50 per acre saved as much as $34.80 worth of corn for a ratio of $4.09:$1.00 (von Rumker et al., 1975).

The prices of insecticides are presently rising at about 21.7% per year (Pesticide Review, 1977). In 1979 carbofuran treatment cost from $8 to $10 per acre, depending on spacing. In view of declining effectiveness, the $10 cost is more realistic. With corn at an average price of $2.20 per bushel the maximum benefit/cost was about $3.00:$1.00. This agrees well with Wedberg and Black's (1978)

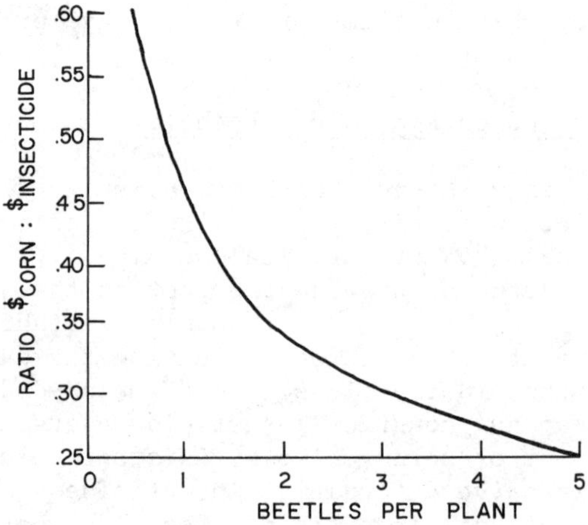

Figure 6. Economic threshold for corn rootworms based on ratio of price of corn per bushel to treatment cost per acre vs. number of adult rootworms per corn plant (after Taylor, 1975).

estimated return for Illinois in 1977 of \$3.42 per acre above treatment costs.

To evaluate the effects of escalating insecticide costs and variable corn prices upon the economic threshold for corn rootworm control, Taylor (1975) has devised a nomogram that determines the economic threshold (Figure 6) by relating the treatment threshold in numbers of adult beetles per corn plant in the fall to the ratio of the price of corn per bushel to the cost of treatment per acre. Applying this nomogram to the situations discussed above is informative: in 1956-62 with corn at \$1.10 per bushel and aldrin treatment costing \$2.20 per acre, the economic threshold was one adult beetle per plant; in 1975 with corn at \$2.54 per bushel and carbofuran treatment costing \$7.50 per acre, the economic threshold was three adult beetles per plant; in 1979 with corn at \$2.20 per bushel and carbofuran treatment at \$10.00 per acre, the economic threshold was greater than five adult beetles per plant. As Taylor indicates, this economic threshold is useful in determining alternative strategies for corn rootworm control, i.e., (1) rotation of corn and soybeans, (2) always apply soil insecticide, (3) never apply soil insecticide, and (4) integrated pest management (IPM) with observance of the economic threshold. As the costs of insecticides continue to rise, as resistance to alternative insecticides increases, and as crop prices lag further behind inflation, strategies (1) and (3), which cost the farmer little or nothing, will become increasingly attractive. Very few corn fields in Illinois average five adult beetles per plant (Taylor, 1975; Wedberg and Black, 1978), and careful evaluation of corn rootworm populations suggests that less than 20% of the corn acreage in Illinois needs soil treatment (Luckmann, 1978) and that in Indiana, less than 10% needs soil treatment (Turpin and Maxwell, 1976). A recent survey in Indiana (Turpin and Thieme, 1978) showed that corn rootworms caused economic loss in only 3.4% of untreated corn fields.

MIGRATIONS AND INTRODUCTIONS OF MULTIPLE-RESISTANT PESTS

The migration or inadvertent introduction of insect pests with dominant genes for multiple resistance represents a major threat to successful insect control and especially to established IPM programs. We are all familiar with the enormous damages that have resulted from the introduction of exotic pests into new geographic ranges; e.g., the gypsy moth *Lymantria dispar* into Massachusetts in 1892, the European corn borer *Ostrinia nubilalis* into Massachusetts about 1909, the Japanese beetle *Popillia japonica* into New Jersey about 1916, the Colorado potato beetle *Leptinotarsa decemlineata* into Germany about 1920, the red imported fire ant *Salenopsis invicta* into Alabama about 1933, and the spotted alfalfa aphid *Therioaphis maculata* into New Mexico in 1954. Today we have more or less reached some sort of accommodation with these invaders, largely by means of specific chemical control measures. Now the possibility of their reintroduction as multiple-resistance pests has the potentiality for even greater and more frequent economic disasters.

The migration of multiple-R forms of native or well-established pests from one area of the country to another or from one crop to another adds an additional dimension to this problem. We explored in detail (see previous section) the problems resulting from the immigration of the cyclodiene-R western corn rootworm *Diabrotica virgifera* through the U.S. corn belt. Each year its cousin, the southern corn rootworm, *D. undecimpunctata howardi*, migrates northward through the Mississippi Valley, bearing an assortment of genes for resistance to DDT, methoxychlor, cyclodienes, malathion and carbaryl selected by insecticides in the South (Chio et al., 1978). Newsom (Memorial Lecture, Entomological Society of America, Denver, Colorado, November 1979) described the challenge to soybean IPM in Louisiana resulting from the migration across the Gulf of Mexico of the soybean looper *Pseudoplusia includens* with multiple resistance to acephate, methomyl and pyrethroids, produced on cut flower crops in Florida. Malathion-R races of the flour beetles *Tribolium castaneum* and *T. confusum* have been shipped in grain supplies to nearly every port around the world. This phenomenon of the export and introduction of multiple-R insect pests is certain to become increasingly important. Its occurrence with important insect vectors of human diseases, e.g., the *Aedes aegypti* mosquito for yellow fever, *Anopheles gambiae* for falciparum malaria, and the human louse *Pediculus humanus*, could result in devastating epidemics against which our conventional insecticide remedies would be powerless.

PESTICIDE MANAGEMENT

Unmistakably, we have arrived at a time when the injudicious and unsophisticated employment of insecticides can no longer be justified or even tolerated. As we have seen, the cost of insecticides is increasing much more rapidly than the value of the crops they are used to protect, and the rise of multiple resistance necessitates the development of substitute insecticides that are increasingly more costly to produce, to market, and to apply. Moreover, the actual discovery of new insecticides that can circumvent, even temporarily, the widening pools of genes selected in multiple-R species is steadily becoming more difficult. Thus, insecticide resistance is not only a disaster to present methods of control but also prejudices the use of some of the most promising new insecticides even before they are fully developed. The examples of the cross-linkage between the *kdr* gene for DDT resistance and resistance to the synthetic pyrethroids (Keiding, 1977; Sawicki, 1978), and of the mutant cholinesterase of the Australian cattle tick (Wharton and Roulston, 1970) are particularly pertinent.

There is general agreement on the part of thoughtful scientists about practical measures to decrease the impact of insecticide resistance in control programs in both agriculture and public health (Brown, 1977; Elliott et al., 1978; Georghiou and Taylor, 1977; Keiding, 1977; Sawicki, 1975; Metcalf, 1980). The basic components are reduced selection pressure and insecticide management to choose the optimal sequence of insecticides used.

Reduced Selection Pressure

This is an obvious countermeasure that has had surprisingly little application in most control programs. The following components are generally agreed upon: (1) reduce the frequency of insecticide treatments by treating only when necessary, (2) reduce the extent of treatments, e.g., treat alternate rows or strips in field crops, alternate trees in orchards, (3) avoid insecticides with prolonged environmental persistence and slow release formulations, (4) reduce use of residual treatments, (5) avoid treatments that apply selection pressure on both larval and adult stages, e.g., in mosquitoes and corn rootworms, and (6) incorporate resistant crop varieties, source reduction, crop rotations, natural enemies, insect diseases and other non-chemical methods in control programs (Brown, 1977; Georghiou and Taylor, 1977; Keiding, 1977; Sawicki, 1975).

The combination of these principles is essentially a blueprint for integrated pest management. Conversely, these recommended practices have had almost no application today in crops such as cotton, corn or deciduous fruits; where economic thresholds are ignored, broad spectrum and relatively persistent insecticides are applied frequently and in excessive dosages to immense areas, slow release formulations are sought, both larvae and adults are treated either deliberately or accidentally, and source reduction by sanitation and crop rotation is seldom practiced (Georghiou and Taylor, 1977; Metcalf, 1980).

Management of Insecticides

The proper choice and methodology of use of insecticides in insect control and in IPM programs will be decisive in countermeasures for dealing with insect pest resistance; or more realistically, in preserving pest susceptibility. There is urgent need for much more basic knowledge about the genetic, biochemical and physiological factors involved in multiple resistance and in the cross-linkage of genetic factors not only for various types of resistance but also between resistance and behavioral traits. Today, it is reasonable to suggest the systematic application of the following practices: (1) monitor insect pest populations so that primitive susceptibility levels are understood (Chio et al., 1978) and so the early detection of resistance is possible (Brown, 1977). This should be carried out using the standardized and widely accepted methods for detecting resistance developed by FAO (Waterhouse, 1977) and WHO (1976a); (2) avoid the use of mixtures of insecticides, as this generally results in the simultaneous development of resistance to the components, since each component seems to develop the residual inheritance of the supporting genome for resistance to the other (Brown, 1977; Keiding, 1977); (3) extend the useful life of a satisfactory insecticide as long as possible, but by monitoring susceptibility, replace it before control fails; (4) choose a sequence of alternative insecticides based on genetic considerations of cross resistance and multiple resistance (Keiding, 1977; Sawicki, 1975). Carefully chosen remedial insecticides should always be available; e.g., methyl chlorpyrifos for temephos in the control of the *Simulium* vectors of onchocerciasis; chlorpyrifos for fenthion in larval control of *Culex fatigans*, the vector of filariasis; or carbaryl plus piperonyl butoxide synergist for malathion in the control of *Pediculus humanus*, the vector of endemic typhus (Brown, 1977). This procedure has worked reasonably well in the control of agricultural pests (perhaps more by serendipity than by careful planning); in the use of azinphosmethyl as a replacement for DDT in control of the codling moth; with carbofuran or phorate as replacements for aldrin and heptachlor in corn rootworm control, and with permethrin and fenvalerate as replacements for methyl parathion and mevinphos in the control of *Heliothis virescens* on cotton and lettuce.

Long-term studies of house fly resistance in Denmark (Keiding, 1977) have demonstrated that the incorrect choice of alternatives is likely to be damaging for future control. For example, DDT resistance as expressed in the "super" *kdr* mechanism has strong multiple resistance to the synthetic pyrethroids (Sawicki, 1978). For this reason, registration of residual formulations of pyrethroids in Denmark was denied (Keiding, 1978) with the hope of preserving the space spray effectiveness of the pyrethroids. Measures to consider include: (1) first, use insecticides with simple one-factor resistance and limited cross resistance, e.g., malathion; (2) avoid insecticides with complicated multiplicate resistance, e.g., diazinon; (3) avoid or delay use of insecticides that act as effective selectors of resistance for other insecticides, e.g., DDT or dimethoate for pyrethroids; and (4) exploit alternative treatments with insecticides without common major R factors and change insecticides before resistance develops (Brown, 1977; Keiding, 1977; Metcalf, 1980; Sawicki, 1975). Intensive genetic investigations of other insect and mite species with gross multiple resistance, e.g., the tobacco budworm *Heliothis virescens* (Wolfenbarger et al., 1977), the cotton leafworm *Spodoptera littoralis* (El Sebae, 1977), the green peach aphid *Myzus persicae* (Attia and Hamilton, 1978), the German cockroach *Blatella germanica* (Collins, 1976), and *Culex pipiens* (Georghiou et al., 1975), is urgently needed to determine if the genetic principles governing house fly multiple resistance have general applicability to a wide range of insect pests.

In summary, effective and rigorous strategies for pesticide management are now required that may extend the lifetime of insecticides presently available so that they may be useful in IPM programs. Such strategies will require biological wisdom, societal cooperation, and economic constraints. These can be achieved in large-scale public health programs (i.e., in the control of malaria, filariasis, onchocerciasis, etc.); but they will be much more difficult to implement in agriculture (Elliott et al., 1978). However, there is an equally urgent need for their development for agricultural pest control on heavily treated crops such as cotton, corn and deciduous fruits, where without proper pesticide management the resistance problem can only worsen (Metcalf, 1980).

Acknowledgements: Preparation of this paper was facilitated by research grants from the World Health Organization, Division of Vector Biology and Control, Geneva, Switzerland; and from the United States Department of Agriculture, Grant No. 5901-0410-8-0067, "Coevolutionary behavior of corn rootworms and cucumber beetles attacking corn and cucurbits," from SEA, Competitive Research Grants Office. Any opinions, findings, and conclusions or recommendations are those of the author and do not necessarily reflect the view of USDA.

REFERENCES

Agarwal, A., 1979, Pesticide resistance on the increase says UNEP, *Nature*, 279:280.

Apple, J. W., 1971, Gains from the use of carbofuran for northern corn rootworm control, *Proc. North Central Branch Entomol. Soc. Amer.*, 26:26.

Attia, F. L., and Hamilton, J. T., 1978, Insecticide resistance in *Myzus persicae* in Australia, *J. Econ. Entomol.*, 71:851.

Ball, H. J., and Weekman, G. T., 1962, Insecticide resistance in the adult western corn rootworm in Nebraska, *J. Econ. Entomol.*, 55:439.

Ball, H. J., and Weekman, G. T., 1963, Differential resistance of corn rootworms to insecticides in Nebraska and adjoining states, *J. Econ. Entomol.*, 56:553.

Bigger, J. H., 1963, Research on soil insecticides for field corn, *15th Illinois Custom Spray Operators Training School*, Urbana, Ill., pp. 21-25.

Black, K. D., and Braness, G. A., 1979, Insect situation and outlook and insecticide use, *31st Illinois Custom Spray Operators Training School.*, Urbana, Ill., pp. 110-130.

Brown, A. W. A., 1958, The spread of insecticide resistance in pest species, *Adv. Pest Control Res.*, 2:351.

Brown, A. W. A., 1961, The challenge of insecticide resistance, *Bull. Entomol. Soc. Amer.*, 7(1):6.

Brown, A. W. A., 1969, Insect resistance, *Farm Chemicals*, Sept., Oct., Nov.

Brown, A. W. A., 1968, Insect resistance comes of age, *Bull. Entomol. Soc. Amer.*, 14(1):3.

Brown, A. W. A., 1971, Pest resistance to pesticides, *in*: "Pesticides in the Environment," R. White-Stevens, ed., Dekker, N. Y., pp. 458-533.

Brown, A. W. A., 1977, Resistance as a factor in pesticide management, *Proc. 15th Interna. Cong. Entomol.*, Washington, D.C., pp. 816-824.

Brown, A. W. A., and Pal, R., 1971, Insecticide resistance in arthropods, *World Health Organization Monograph Series*, No. 38, Geneva, Switzerland, 491 pp.

Burckhardt, C. C., and Bryson, H. R., 1955, Notes on the distribution of the western corn rootworm, *Diabrotica virgifera* Lec. in Kansas, *J. Kansas Entomol. Soc.*, 28:1.

Campt, D. D., 1979, *Pestic. Toxic Chem. News*, Feb. 7, pp. 37-38.

Chiang, H. C., 1973, Bionomics of the northern and western corn rootworms, *Ann. Rev. Entomol.*, 18:47.

Chio, H., Chang, C.-S., Metcalf, R. L., and Shaw, J., 1978, Susceptibility of four species of *Diabrotica* to insecticides, *J. Econ. Entomol.*, 71:389.

Collins, W. J., 1976, German cockroach resistance: Propoxur selection induces the same resistance spectrum as diazinon, *Pestic. Sci.*, 7:1714.

Cooperative Economic Insect Report, U.S. Dept. Agr. 1956, 1965, 1969, 1975, 1977, Washington, D.C.

Elliott, M., Janes, N. F., and Potter, C., 1978, The future of pyrethroids in insect control, *Ann. Rev. Entomol.*, 23:443.

El Sebae, A. H., 1977, Incidents of local pesticide hazards and their toxicological interpretation, *Proc. Univ. California/ AID, Univ. Alexandria Seminar Pesticide Management,* Alexandria, Egypt, pp. 137-152.

Flint, W. P., and Metcalf, C. L., 1932, "Insects, Man's Chief Competitors," Williams & Wilkins, Co., Baltimore, Maryland.

Fontaine, R. E., 1978, House spraying with residual insecticides with special reference to malaria control, *World Health Organization,* VBC 78.704, Geneva, Switzerland, 28 pp.

Fontaine, R. E., Pull, J. H., Payne, D., Pradhan, G. D., Joshi, M. E., 1978, Evaluation of fenitrothion for control of malaria, *Bull. World Health Organ.,* 56:445.

Georghiou, G. P., 1972a, The evolution of resistance to pesticides, *Ann. Rev. Evol. Syst.,* 3:133.

Georghiou, G. P., 1972b, Studies on resistance to carbamate and organophosphorus insecticides in *Anopheles albimanus, Amer. J. Trop. Med. Hyg.,* 21:797.

Georghiou, G. P., Ariaratnam, V., Pasternak, M. E., and Lin, C. S., 1975, Organophosphorus multiresistance in *Culex pipiens quinquefasciatus* in California, *J. Econ. Entomol.,* 68:461.

Georghiou, G. P., and Taylor, C. E., 1977a, Pesticide resistance as an evolutionary phenomenon, *Proc. 15th Int. Cong. Entomol.,* Washington, D.C., pp. 759-785.

Georghiou, G. P., and Taylor, C. E., 1977b, Operational influences in the evaluation of insecticide resistance, *J. Econ. Entomol.,* 70:653.

Gillette, C. P., 1912, *Diabrotica virgifera* as a corn rootworm, *J. Econ. Entomol.,* 5:364.

Goring, C. A. I., 1977, The costs of commercializing pesticides, *in:* "Pesticide management and insecticide resistance," D. L. Watson and A. W. A. Brown, eds., Academic Press, New York, pp. 1-33.

Hamilton, E. W., 1965, Aldrin resistance in corn rootworm beetles, *J. Econ. Entomol.,* 58:296.

Headley, J. C., 1972, Economics of pest control, *in:* "Implementing practical pest management strategies," *Proc. National Extension Workshop,* Purdue Univ., LaFayette, Indiana, pp. 180-187.

Howe, W. L., Sanborn, J. R., and Rhodes, A. M., 1976, Western corn rootworm adults and spotted cucumber beetle associations with *Cucurbita* and cucurbitacins, *Environ. Entomol.,* 5:1042.

Hunter, R. C., 1974, Federal environmental pest control act of 1972 — what does it mean to the chemical industry? *Bull. Entomol. Soc. Amer.,* 20:103.

Johnson, J. E., and Blair, E. C., 1972, Cost, time, and pesticide safety, *Chem. Technol.,* (Nov.), pp. 666-669.

Keiding, J., 1977, Resistance in the housefly in Denmark and elsewhere, *in:* "Pesticide management and insecticide resistance," D. L. Watson and A. W. A. Brown, eds., Academic Press, New York, pp. 261-302.

Keiding, J., 1978, *Danish Pest Infestation Laboratory, Ann. Rept.*, p. 47.
Kuhlman, D. E., and Petty, H. B., 1971, Soil insect control demonstrations, *23rd Illinois Custom Spray Operators Training School*, Urbana, Illinois, pp. 32-45.
Kuhlman, D. E., Randell, R., and Cooley, T. A., 1973, Insect situation and outlook, 1973, *25th Illinois Custom Spray Operators Training School*, Urbana, Illinois, pp. 109-126.
Kuhlman, D. E., and Wedberg, J. L., 1978, Corn rootworms and their control in Illinois, *30th Illinois Custom Spray Operators Training School*, Urbana, Illinois, pp. 109-118.
Luckmann, W. H., 1978, Insect control in corn: Practices and prospects, *in:* "Pest Control Strategies," E. H. Smith and D. Pimentel, eds., Academic Press, New York, pp. 137-155.
Melander, A. L., 1914, Can insects become resistant to sprays? *J. Econ. Entomol.*, 7:167.
Metcalf, R. L., 1955, Physiological basis for insect resistance to insecticides, *Physiol. Rev.*, 35:197.
Metcalf, R. L., 1979, Plants, chemicals, and insects: Some aspects of coevolution, *Bull. Entomol. Soc. Amer.*, 25:30.
Metcalf, R. L., 1980, Changing role of insecticides in crop protection, *Ann. Rev. Entomol.*, 25:219.
Metcalf, R. L., and Luckmann, W. H., 1975, "Introduction to Insect Pest Management," John Wiley, New York, 587 pp.
National Academy Sciences, 1969, Principles of Plant and Animal Pest Control, Vol. 3, "Insect Pest Management and Control," Publ. 1965, Washington, D.C., 508 pp.
National Research Council, 1977, World Food and Nutrition Study, Supporting papers, Vol. 1, Washington, D.C., 318 pp.
Pal, R., 1977, Problems of insecticides resistance in insect vectors of disease, *Proc. 15th Inter. Cong. Entomol.*, Washington, D.C., pp. 801-811.
Pesticide Review, U.S. Dept. Agriculture, Agricultural Stabilization and Conservation Service, Washington, D. C.
Rishikesh, N., Clarke, J. L., Mathis, H. L., Pearson, J., and Obanewa, S. J., 1979, Stage V field evaluation of decamethrin against *Anopheles gambiae* and *Anopheles funestus* in a group of villages in Nigeria, *World Health Organ.*, VBC/79.712, Geneva, Switzerland, 25 pp.
Sawicki, R. M., 1975, Interactions between different factors or mechanisms of resistance to insecticides on insects, *in:* "Environmental Quality Safety, Suppl. Vol. 3, Pesticides," F. Coulston and F. Korte, eds., G. Thieme, Stuttgart, Germany, pp. 429-435.
Sawicki, R. M., 1978, Unusual response of DDT-resistant houseflies to carbinol analogues of DDT, *Nature*, 275:443.
Secreatry, Health, Education, and Welfare, 1969, Report Commission on Pesticides and their Relation to Environmental Health, Washington, D.C., 673 pp.

Shaw, R. D., Cook, M., and Carson, R. E., 1968, Developments in the resistant status of the southern cattle tick to organophosphorus and carbamate insecticides, *J. Econ. Entomol.*, 61:590.
Smith, R. F., 1966, The distribution of Diabroticites in western North America, *Bull. Entomol. Soc. Amer.*, 12:108.
Smith, R. F., and Lawrence, J. F., 1967, "Clarification of the status of the type specimens of Diabroticites," Univ. California Press, Berkeley and Los Angeles, 174 pp.
Soper, F. L., Andrews, J. A., Bode, K. F., Coatney, G. R., Earle, W. C., Keeney, S. M., Knipling, E. F., Logan, J. A., Metcalf, R. L., Quarterman, K. D., Russell, P. F., and Williams, L. L., 1961, Report and recommendations on malaria: A summary, *Amer. J. Trop. Med. Hyg.*, 10:451.
Stern, V. M., Smith, R. F., Van den Bosch, R., and Hagen, K. S., 1959, The integration of chemical and biological control of the spotted alfalfa aphid, *Hilgardia*, 29:81.
Tate, H. D., and Bare, O. S., 1946, Corn rootworms, *Nebr. Agr. Exp. Sta. Bull.*, 381:1.
Taylor, C. R., 1975, The economics of control of northern and western corn rootworms in Illinois, *Ill. Agr. Econ.*, July, p. 11.
Time Magazine, 1975, December 1, p. 63.
Turpin, F. T., and Thieme, J. M., 1978, Impact of soil insecticide usage on corn production in Indiana: 1972-74, *J. Econ. Entomol.*, 71:83.
Turpin, F. T., Leva, D., and Freeman, D., 1978, Impact of silk feeding by western corn rootworm beetles on corn grain yield, *30th Illinois Custom Spray Operators Training School*, Urbana, Illinois, pp. 101-102.
Turpin, F. T., and Maxwell, J. D., 1976, Decision making to use of soil insecticides by Indiana farmers, *J. Econ. Entomol.*, 69:359.
U.S. Department of Agriculture, 1976, Plant pest report No. 1 (1-4), pp. 22-36, Estimated losses and production costs attributed to insects and related arthropods, 1974, Washington, D.C.
U.S. Department of Agriculture, 1978, Farmers use of pesticides in 1976, Agr. Econ. Rept. No. 418, Washington, D.C.
Vaughan, M. A., and Gladys, L. Q., 1977, Pesticide management on a major crop with severe resistance problems, *Proc. 15th Int. Cong. Entomol.*, Washington, D.C., pp. 812-815.
Von Rumker, R., Guest, H. R., and Upholt, W. M., 1970, The search for safer, more selective and less persistant pesticides, *BioScience*, 20:1004.
Von Rumker, R., and Horay, F., 1974, Farmers pesticide use decisions and attitudes on alternate crop protection methods, U.S. Environmental Protection Agency, EPA 544/1-74-002, Office of Pesticide Programs.
Von Rumker, R., Kelso, G., Horay, F., and Lawrence, K. A., 1975, A study of the efficiency of the use of pesticides in agriculture, U.S. Environmental Protection Agency, EPA 540/9-75, Office of Pesticide Programs.

Waterhouse, D. F., 1977, FAO activities in the field of pesticide resistance, *Proc. 15th Inter. Cong. Entomol.*, Washington, D.C., pp. 786-793.

Wedberg, D. L., and Black, K. D., 1978, Insect situation and outlook and insecticide usage, *30th Illinois Custom Spray Operators Training School*, Urbana, Illinois, pp. 119-146.

Wellman, R. H., 1966, Industry's role in the development of pesticides, National Academy Sciences, Scientific Aspects of Pest Control, Publ. 1402, Washington, D.C., 470 pp.

Wharton, R. H., and Roulston, W. J., 1970, Resistance of ticks to chemicals, *Ann. Rev. Entomol.*, 15:381.

Wolfenbarger, D. A., Harding, J. A., and Davis, J. W., 1977, Isomers of (3-phenoxyphenyl)-methyl (±) *cis-trans*-3-(2,2-dichloroethenyl)-2,2-dimethylcyclopropanecarboxylate against boll weevils and tobacco budworms, *J. Econ. Entomol.*, 70:226.

World Health Organization, 1972, Vector Control in International Health, Geneva, Switzerland, 144 pp.

World Health Organization, Tech. Rept. Ser., 1974, Malaria control in countries where time limited eradication is impracticable at present, No. 537, Geneva, Switzerland, 66 pp.

World Health Organization, Tech. Rept. Ser., 1976a, Resistance of vectors and reservoirs of disease to pesticides, No. 585, Geneva, Switzerland, 88 pp.

World Health Organization, 1976b, Vector resistance to insecticides: A review of its operation significance in malaria eradication and control programs, WHO/VBC 76.634, Geneva, Switzerland, 19 pp.

World Health Organization, 1978a, The malaria situation in 1976, *WHO Chronicle*, 32:9.

World Health Organization, 1978b, Malaria control— a reoriented strategy, *WHO Chronicle*, 32:226.

Wright, J. W., Fritz, R. F., and Haworth, J., 1972, Changing concepts of vector control in malaria eradication, *Ann. Rev. Entomol.*, 17:75.

MANAGEMENT OF RESISTANCE IN PLANT PATHOGENS

J. D. Gilpatrick

New York State Agricultural Experiment Station
Cornell University
Geneva, New York 14456

INTRODUCTION

Crop losses from plant diseases caused by fungi, nematodes, bacteria and other agents have been estimated to be about 12% of the total world food production (Glass, 1976). The control of diseases with chemicals is an important component of modern crop production technology. These chemicals reduce crop losses from diseases and allow the production of crops in areas or times when they normally could not be grown economically. Of the pesticides used to control plant diseases, fungicides are largest in volume, followed be nematicides and bactericides.

There are many compelling reasons for the use of pesticides to control plant diseases, and no other disease control approach has been so thoroughly researched, tested in the field, and documented. Thus, chemical control is a safe and effective component of an overall management program and is often accepted as the best understood and most dependable method. Chemicals provide flexibility in disease control programs; they can be manufactured under precise quality control, packed in a variety of formulations, distributed quickly to the grower, and applied with reasonable assurance of success. They can be diluted and applied as prescribed dosages using well-tested methods, and can be restricted to geographic areas, fields or plant parts. They often provide the best control tactics for short-term decisions based upon disease forecasting systems.

In recent years, chemical control technology has been plagued with several problems, including: (1) the negative attitudes of the public and a segment of the scientific community toward pesticides;

(2) the impact of new government regulations and policies on the cost of development and registration of pesticides; (3) the fear that these new rules may lead to the deregistration of some of our more important pesticides; (4) a critical lack of education and basic research needed to support chemical control technology, and (5) the development of resistance in numerous plant pathogens to chemical control agents (Gilpatrick, 1979). Unquestionably, resistance of plant pathogens to fungicides and bactericides during the last decade has become a serious threat to the stability and success of chemical control of many important plant diseases. Several reviews of this subject have been written recently (Dekker, 1977a; Georgopoulos, 1976; Ogawa et al., 1977). Some have discussed methods of dealing with the resistance problem in plant pathogens (Anonymous, 1979; Dekker, 1976, 1977b; Gilpatrick, 1978; Sakurai, 1977).

Before considering how resistance to pesticides might be counteracted, one must first examine the history of pesticide usage for disease control and the various factors that are involved in the development of resistance.

HISTORY OF PESTICIDE USE

Concoctions and folk-art were used for plant disease control from earliest recorded history to the late 1800s. Between 1875 and 1950, compounds of sulfur and copper dominated the disease control scene, but preference has since shifted to the organic chemicals with a broad biochemical spectrum of activity, especially capton, dithiocarbamates, and several soil fumigants such as 1,3-D, EDB, DBCP, chloropicrin, and methyl bromide. Except in a few cases, resistance has not become a practical problem with sulfur, copper, or many organic multisite inhibitors after 20 to 100 years of use, but chemical molecules have been developed recently as therapeutants (systemics). Because these are much more biochemically selective than their predecessors, problems of resistance have occurred.

During the last decade, reports of resistance in plant pathogens have risen dramatically. A survey of the publication dates of papers referenced by one group of recent reviewers reveals a marked increase in reporting after the introduction of benzimidazoles in the late 1960s (Ogawa et al., 1977). Another survey conducted in 1976 by the FAO reported a total of 56 species of fungi and 10 of bacteria that had shown resistance in the laboratory or in practice (Anonymous, 1977). In 1978, the FAO noted that resistance had been found in one or more species to at least 52 different fungicides or bactericides (Anonymous, 1979). Most reports of resistance are for the benzimidazole fungicides for which 56 cases have been recorded. Only a few examples of resistance to nematicides are documented (Van Gundy et al., 1974).

FACTORS THAT AFFECT RESISTANCE IN PLANT PATHOGENS

The various factors involved in resistance include the genetics, fitness and biology of the pathogen; the chemistry of the pesticide and technology of its use; the nature of the disease, and the effect of the environment on these foregoing factors (Table 1).

Genetics

Resistance of plant pathogens to pesticides is a directional selection under repeated and often continuous pressure. It is an inherited and often stable trait with most examples of resistance considered to be the result of selecting individuals that have experienced mutations of chromosomal genes. Resistant (R) mutants may be either induced by exposing fungi to mutagens in the laboratory or isolated without benefit of mutagens from a population of apparently susceptible (S) isolates or from resistant field strains. Mutation frequency for R varies with the fungicide and fungus. R mutants for single-site fungicides, such as the benzimidazoles, occur relatively frequently, whereas those for multisite inhibitors, such as captan, are rare and probably will be lethal. The potential of a fungus to acquire fungicide resistance by artificial inducement using mutagens was measured as mutation frequency and degree of resistance in various fungi by Van Tuyl (1977). The limited-site inhibitors studied included a wide variety of chemical structures: benzimidazoles, carboxin, chloroneb, imazalil, cycloheximide and pimaricin. Induced resistance for each compound occurred in all cases in four or more different fungi. The mutation frequencies varied from 10^{-7} for pimaricin and benomyl to 2 x 10^{-4} for chloroneb. In Van Tuyl's study, mutation frequencies for R differed considerably among the fungal species: for example, against imazalil they were 2, 26, 82, 250 and 700 x 10^{-7} for *Aspergillus niger, Cladosporium cucumerinum, Aspergillus nidulans, Phialophora cinerescens* and *Penicillium expansum,* respectively.

The inheritance of fungicide resistance has been subjected to genetic analysis in several cases. Georgopoulos (1977) in discussing the chromosomal aspects of fungal resistance noted that in most cases studied, more than one locus for resistance to the fungicide was known: two loci were observed to control resistance of *Venturia inaequalis* to dodine, and as many as eight loci for cycloheximide in *Saccharomyces cerevisiae*. According to Van Tuyl (1977), the locus involved governs the degree of resistance in *Aspergillus nidulans*. He reported that resistance to benomyl and thiabendazole is determined by one main gene, *ben A*, that confers a high level of resistance, and by at least two other genes that give less pronounced resistance, *ben B* and *ben C*.

Table 1. Known or Suggested Factors Affecting the Selection of Resistance to Pesticides in Plant Pathogens

A.	Pathogen	- Genetics
		- Fitness of resistant mutants
		- Life cycle
		- Regeneration time
		- Inoculum potential
B.	Chemical	- Mode of action
		- Persistence
		- Application technology
		- Cross resistance
		- Control efficiency
C.	Disease	- Host and variety
		- Host range
		- Organs attacked
		- Tissues attacked
		- Organs treated
		- Economics
E.	Epidemiology	- Climate
		- Weather
		- Inoculum potential
		- Soil type

The fungi have well-defined nuclei that are typically very small. Little is known about the mitotic process, but meiosis in several fungi has been described as similar to that of higher organisms. In spite of a general lack of knowledge of nuclear division in fungi, the genetical consequences are similar to those in other organisms. Most fungi are in a haploid state most of the time, whereas the diploid phase is typically of short duration. Segregants are thus usually haploid and phenotypes are expressed directly.

Several mechanisms are available to fungi for genetic variability. If a sexual stage occurs, meiosis can provide genetic recombination. However, many fungi have no sexual phase and these must rely on mutations or other genetic mechanisms for variation. Often, fungi cells possess two or more genetically different nuclei, a condition known as heterokaryosis, which is unique to fungi and is responsible for increased genetic variability in this group (Day, 1974; Ross, 1979). Many plant pathogens, especially the Ascomycetes and fungi imperfecti (sexual cycle unknown), have multinucleate cells, and when asexual, uninucleate-celled spores are produced by heterokaryons, large numbers of variants can arise. Nuclei of heterokaryotic cells may combine into diploids, which with or without haploidization can also lead to variants. There can be migration of nuclei from one fungal segment to another through septal pores. Finally, the number of nuclei in heterokaryotic cells may vary from time to time depending on preconditions.

An example of a heterokaryotic fungus in which resistance to fungicides has occurred is the *Monilinia* spp., the cause of brown rot of stone fruits. Hall (1963) observed up to 40 nuclei in some hyphal cells of *M. fructicola* and Hoffman (1970) noted as many as 100 in tip cells of *M. fructigena* and *M. laxa*. The latter found that hyphal fusions commonly occurred in all three species. A similar situation exists in the fungus *Botrytis* for which more than 50 examples of benzimidazole resistance have been reported (Ogawa et al., 1977), and Meyer and Parmeter (1968) obtained resistance in *Thanatephorus cucumeris* to pentachloronitrobenzene after mixing susceptible asexual homokaryons.

Heritable resistance in the fungi should not be confused with physiological adaption to increasing dosages of fungicides in the laboratory on artificial media. This state is not stable, and the tolerance declines quickly when the fungus is grown on a fungicide-free medium. Such adaptions may occur in practice, but evidence for this as an important resistance mechanism is lacking.

The genetic mechanisms of resistance in plant pathogenic bacteria to bactericides is not well understood. In human and public health bactericides, genetic transfer of resistance by extrachromosomal plasmids (DNA) is considered to be an important mechanism for the transfer and storage of resistance factors, but the limited evidence available at this time suggests that cases of resistance to bactericides in plant pathogens is not under extrachromosomal control (Georgopoulos, 1977; Schroth et al., 1979).

Little is known about the inheritance mechanisms of plant parasitic nematodes as the subject was largely neglected by nematologists until recently (Day, 1974). It can be assumed that these pathogens possess genetic mechanisms similar to those of other animals.

Fitness

Whether or not an R mutant survives, increases, and eventually poses a problem in disease control will depend on its ability to function as a pathogen. Under the selective pressure of the pesticide, R mutants have an obvious ecological advantage over S forms and eventually will dominate a population if they have pathogenic capabilities. However, in the absence of pesticidal pressure, R strains may not succeed as well as S strains. The fitness of plant pathogens depends on several factors, including growth and sporulation rate, inherent pathogenicity, resistance to adverse conditions, and others. The fitness of a newly arisen R mutant may at first be inadequate, but under the continued selective pressure of the pesticide, the R strain could be selected for greater fitness. The common phenomenon of breaking host resistance by the evolution of new strains of plant pathogens indicates the enormous potential of these organisms to adapt to unfavorable conditions. The need of R mutants to adapt for fitness over a period of time may be a reason for the otherwise unexplainable variations in the time required for the appearance of resistance.

Laboratory-induced R mutants frequently are less virulent than S types from which they are derived; however, R strains obtained from the field often are as pathogenic as the S wild type, probably because selection has already occurred and only the most virulent mutants have survived. It has been suggested that the loss of pathogenicity in laboratory-induced R strains is a consequence of additional mutation from the mutagenic treatment.

Many reports of benzimidazole resistance in the field have indicated that R strains are as competitive as S strains when compared in laboratory and greenhouse tests. Selected examples include resistant isolates of *Venturia inaequalis* (Jones and Ehret, 1976; Tate and Samuels, 1976), *Botrytis cinerea* (Bertrand and Saulie-Carter, 1978; Bollen and Scholten, 1971; Polach and Molin, 1975), *Penicillium expansum* (Koffman et al., 1978), *Cladosporium cucumerium* (Van Tuyl, 1977), *Sphaerotheca fuliginea* (Schroeder and Provvidenti, 1969), *Cercospora* spp. (Georgopoulos and Dovas, 1973; Littrell, 1976), *Fusicladium effusum* (Littrell, 1976) and *Monilinia fructicola*

(Szkolnik and Gilpatrick, 1977). Other reports indicated that benzimidazole-resistant mutants of laboratory origin may be less fit than S types (Bertrand and Saulie-Carter, 1978; Van Tuyl, 1977).

Much information is now available on the fitness of R strains to fungicides and bactericides other than benzimidazoles. Recently, Dekker and Gielink (1979) studied the fitness of laboratory-selected strains of *Cladosporium cucumerinum* and *Fusarium oxysporum* f. sp. *narcissi* resistant to the antibiotic pimaricin. Increased resistance appeared to be associated with decreased fitness in vitro (growth and sporulation) and decreased pathogenicity. Pimaricin is a polyene macrolide antibiotic that complexes with sterols, and resistance to these antibiotics is probably due to a change in the quantity and type of sterol in the fungal membrane, which seems to induce a lower fitness in such R mutants. Several other promising new fungicides, e.g., triadimefon, phenapronil, biloxazol and CGA 64251 (1-[2-2,4-dichlorophenyl)-4-ethyl-1,3-dioxolan-2-yl-methyl]-1-H-1,2,4-triazole), are also believed to interfere with sterol biosynthesis. The association of resistance and decreased fitness in the case of sterol-inhibiting fungicides offers hope that resistance may be of much less importance for this group than for the benzimidazoles.

The best evidence for fitness of R strains is found when one examines their stability in the field following the removal of the selective pressure. Many workers observed that the level of resistance declined once the use of the pesticide in question was discontinued (Georgopoulos, 1977; Gilpatrick, unpublished; Kohmoto et al., 1974; Misato, 1975; Ruppel, 1975). In most cases, the resistance level again reached a high level if the pesticide was reintroduced, indicating an inferior fitness in the R population.

Warren et al. (1977) evaluated the potential for survival of tolerant strains of *Sclerotinia homeocarpa* on turf. In the absence of fungicides, benomyl-tolerant isolates declined to a small percentage of the population, but when benomyl was reapplied, R strains again dominated. However, cadmium-tolerant strains persisted at high levels even when cadmium use was curtailed. Three years after the use of benzimidazoles was discontinued, Dovas et al. (1976) found no indication that the level of resistance in the population of *Cercospora beticola* on sugar beets in Greece had declined. Nishimura et al. (1976) observed that strains of *Alternaria kikuchiana,* the causal agent for black spot of pear, resistant to polyoxin could be found in pear orchards 5 years after discontinuing use. Dodine-tolerant strains of *Venturia inaequalis* may still be recovered from the environs of apple orchards in which the fungicide has not been used for at least 10 years (Gilpatrick, unpublished). R strains of *Erwinia amylovora* appeared to be relatively stable and were detected in pear orchards 6 years after the termination of streptomycin application (Schroth et al., 1979).

As a general rule it may be stated that once resistance appears in practice, R strains will probably be stable and are likely to persist at least in low numbers for an indefinite period of time in the absence of the fungicide. With the reintroduction of the fungicide, the R population levels will quickly rise again.

Although the lack of fitness of R mutants may be a limiting factor in practice by delaying, reducing, or even preventing resistance, fungi have still been able to adapt to a wide variety of chemicals. Thus, inability to adapt for fitness does not seem to be a serious limiting factor so far in the development of resistance as a practical problem in agriculture.

Chemicals

Pesticides used in plant disease control may be classified as nonspecific or site-specific, according to their chemical mechanisms. The multisite inhibitors react with a variety of cellular sites, thus interfering with several metabolic functions, and the ability of the pathogen to develop viable R mutants to such chemicals is considered to be extremely low. The successful, wide use of captan, maneb and soil fumigants for 2 to 3 decades and copper and sulfur for almost 100 years demonstrates that these multisite inhibitors are not prone to develop resistance.

There are a few cases in which the use of a chemical that is classified as a multisite inhibitor has led to the development of resistance in fungi, e.g., dodine, pentachloronitrobenzene and organomercurials. This is probably due to the development of specific detoxification mechanisms (Georgopoulos, 1977). Resistance has also developed to members of the aromatic hydrocarbon group, especially in postharvest fungal pathogens, as discussed in detail by Eckert in another chapter of this book.

In spite of the above examples of resistance to chemicals that are indiscriminate in their mode of action, the most serious and widespread cases of resistance have occurred in those fungicides and antibiotics that are considered to be limited-site inhibitors. As stated previously, the most commonly occurring cases have been with the benzimidazoles. However, with less frequency, laboratory and field resistance has occurred with a wide variety of other chemicals as well. Notable examples include resistance to Kasugamicin in the rice blast fungus (*Piricularia oryzae*) (Ito et al., 1974), to streptomycin in the fire blight bacterium (*Erwinia amylovora*) on pear (Schroth et al., 1979), and to dimethirimol in the cucumber powdery mildew caused by *Sphaerotheca fuliginea* (Bent et al., 1971).

Cross Resistance and Multiple Resistance

As in other organisms, cross resistance between fungicides of similar biochemical modes of action occurs in the fungi. Again, most examples are found in the benzimidazoles, whose fungicidal activity is attributed primarily to a common breakdown product, carbendazim (MBC). The action of MBC is site-specific, interfering with mitosis, and cross resistance commonly occurs between benomyl, MBC, the thiophanates and thiabendazole (Dekker, 1977c). In practice, once resistance occurs to one benzimidazole, another benzimidazole cannot be substituted for it (Bollen and Scholten, 1971; Jones and Ekret, 1976).

Many examples of cross resistance in other fungicides are available, including *Ustilago maydis* resistance to the aromatic hydrocarbons dichloran, chloroneb, diphenyl, hexachlorobenzene, naphthalene, p-dichlorobenzene, pentachloronitrobenzene and sodium-o-phenylphenate (Tillman and Sisler, 1973; Van Tuyl, 1977). Van Tuyl (1977) concluded that resistance to PCNB, chloroneb and other aromatic hydrocarbons involves the same genetic factor. There are also examples of cross resistance between carboxin and oxycarboxin, i.e., oxathiin fungicides (Van Tuyl, 1977); between isoprothiolane and organophosphorus thiolates (Katagiri and Uesugi, 1977); among sterol inhibitors such as triforine, imazalil, fenarimol and triadimefon (Van Tuyl, 1977), and in derivatives of N-(3,5-dichlorphenyl) dicarboximides (Leroux et al., 1977; Sztejnberg and Jones, 1978). Although the mode of action of the dicarboximides, e.g., dichlozoline, vinclozolin, iprodione, dicyclidine, has not yet been fully explained, they do alter fungal lipid composition and one of them, iprodione, reduced sterol biosynthesis (Pappas and Fisher, 1979). Thus, the possibility of cross resistance between the sterol inhibitors and the dicarboximides must be considered, although Leroux et al. (1977) observed no such cross resistance between dicarboximides and sterol inhibitors in *Botrytis cinerea*. The sterol inhibitors and carboximides are the most likely candidates to replace the multisite inhibitors that are now being critically reconsidered by regulatory agencies as to their suitability for continued registration for use on crop plants. They are also the most likely replacements for the benzimidazoles whenever resistance develops.

Cross resistance is one of the most important aspects of the resistance problem in fungicides and may greatly inhibit the future use of limited-site inhibitors. This phenomenon is useful in understanding the mode of action of pesticides; conversely, mode of action studies can be useful in predicting the possibility of cross resistance occurring in practice.

In some cases, negatively correlated cross resistance has been reported for fungicides, for example, phosphoramidate and isoprothiolane in *Piricularia oryzae* (Uesugi et al., 1974). In such cases (Georgopoulos, 1977; Katagiri and Uesugi, 1977; Uesugi et al., 1974; Van Tuyl, 1977), the resistance to one fungicide is counteracted by an increased sensitivity to another, and if such fungicides could be developed for practical use, mixtures of compounds with negatively correlated cross resistance would be extremely valuable in limiting resistance in target pathogens.

The development of resistance to one fungicide does not appear to predispose a fungus to resistance to a chemical with a different chemical mode of action, but a few cases of multiple resistance are known. Strains of fungi resistant to benzimidazoles can also develop resistance to a second, nonrelated fungicide. Gilpatrick and Provvidenti (1973) reported that a strain of cucurbit powdery mildew, *Sphaerotheca fuliginea*, previously found resistant to benomyl, later showed resistance to the sterol inhibitors triforine and triarimol. This is an example of tolerance to two different modes of fungicidal action. Laboratory strains of *Botrytis* resistant to both dichloran and benomyl have also been reported (Chastagner and Ogawa, 1979; Geeson, 1978). Sakurai (1977) reported that some carbendazim-R strains of a *Botrytis* sp. were also resistant to chlorothalonil. Strains of *Venturia inaequalis* resistant to both benomyl and dodine have been isolated from New York apple orchards (Gilpatrick, unpublished) where the scab fungus developed resistance to dodine first and then to its replacement, benomyl.

Walmsley-Woodward et al. (1979) found isolates of *Erysiphe graminis* f. sp. *hordei* that were resistant to both ethirimol and tridemorph, Yaoita et al. (1979) reported triple-resistance to Blasticidin S, Kasugamycin and a phosphophorilate (IBP) in *Piricularia aryzae*, and Leroux et al. (1977) observed strains of *Botrytis cinerea* that had multiple resistance to dicarboximides, dichloran and pentachloronitrobenzene. These cases indicate that under continual selective pressure by more than one site-specific fungicide, the developemnt of tolerance to all fungicides used seems possible.

Pesticide Stability

The stability of a pesticide in a specific environment could greatly affect the duration of the selective process. Very little attention has been given to this aspect of resistance in plant pathogens, yet this could be an important factor in the development of resistance. Long persistence in soil has been related to the development of resistance in soil insects to soil insecticides (Harris, 1977), and this same phenomenon would apply to soil pathogens as well. Persistence in and on plant tissues would also be of importance in stem, foliar and fruit pathogens.

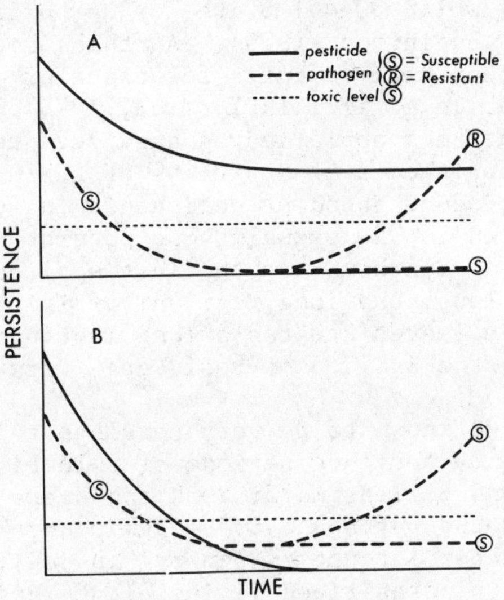

Figure 1. The relationship of chemical persistence in the environment to the dynamics of the pathogen. In A, the pesticide is present at a toxic level for a long duration and the selection pressure on the pathogen is high. In B, the chemical has a short residual life and the selection pressure is low.

Two theoretical situations relative to the persistence of pesticides and plant pathogens are given in Figure 1. In A, the relationship of the survival of the pathogen to the presence of a persistent pesticide is shown. In this case, the pesticide is always present at a toxic level; however, the pathogen is never completely eliminated and the selective pressure is maintained for a long period. This situation is favorable for the eventual appearance of R forms of the pathogen. In the second case (Figure 1-B), the pesticide has a short residual activity and quickly disappears from the environment. This removes the selective pressure, thus allowing S escapees to survive or proliferate. In this case, a few R forms of questionable fitness that must compete with a great number of S wild type escapees are not likely to survive and increase.

There is little evidence available to support the importance of chemical stability in plant pathogen resistance to pesticides, but a few examples may be mentioned. Oxycarboxin was found to be more stable than carboxin when applied to bean roots, as most of the carboxin was oxidized to a nonfungitoxic analog within 21 days after application, while 30-40% of the oxycarboxin could still be detected (Snel and Edgington, 1970). At this time, resistance to carboxin has not been noticed but a few cases of oxycarboxin resistance are known (Abiko et al., 1977; Iida, 1975). Persistence of benomyl in overwintering apple leaves has prevented the production of overwintering perithecia of S strains of *V. inaequalis*, but R isolates readily produce abundant perithecia in such leaves (Gilpatrick, unpublished). The sexual cycle occurs on fallen infected apple leaves about 5 months after the last application of benomyl to these leaves. Thus, the long residual activity of benomyl in overwintering apple leaves creates a long-lasting selective pressure that could lead to the buildup of resistance very quickly.

Benomyl is also known to be very persistent in soil, which offers an opportunity for long periods of selection against soil pathogens (Fuchs and Bollen, 1975; Vonk and Kaars Sijpesteijin, 1977) and above-ground pathogens by systemic uptake by the roots. Unquestionably the persistence of benomyl in soil, in decaying plant debris, and in parasitized living plants has been an important factor in the wide buildup of resistance to this fungicide in practice.

Translocation of systemic chemicals to above-ground parts and persistence there following seed and soil treatments afford unique protection of plants against infection by plant pathogens. The soil and seed act as a reservoir for the pesticide for an indefinite period and the pathogen is exposed to varying dosages of the toxicant because of variable uptake, dilution by growth of the host, and breakdown of the pesticide at both the application site and within the host. Such may have happened in the case of resistance in barley powdery mildew to ethirimol (Shepherd et al., 1975). Although there are many favorable reasons for using seed and soil treatments to control above-ground diseases, this approach probably provides a great risk of developing resistance to single-site inhibitors.

Disease, Epidemiology and Control Strategies

The nature of the disease, life cycle(s) of the pathogen(s), crop, weather and control tactics applied will have a marked influence on the developemnt of pesticide resistance. These are critical factors in determining the selective pressure applied to the pathogen by the pesticide and whether or not resistance will be expressed as a practical problem.

The Disease. Diseases that are highly destructive on economically important crops and where no suitable alternative control measure is available require intensive chemical control programs. Causal agents of such diseases are more likely to develop resistance than those of diseases where chemicals are required frequently. In the United States, foliage fungicide use is intensive on apples and other fruits, turf and ornamentals, vegetables, and peanuts (Gilpatrick, 1979). These are low acreage, high value crops in which multiple disease problems are intense, and R varieties or cultural means are often not satisfactory for control. It is on these crops that resistance problems have occurred most frequently and acutely. Use of pesticides for disease control on cereals, corn, soybeans and most other field crops of large acreage is at a low level in the United States and is greatly confined to seed treatments. With field crops, R varieties and other cultural means are the priamry tactic utilized in disease control and fungicide resistance has not been important. In other parts of the world, the situation is often similar, but in some areas there is an intensive use of chemicals to control cereal diseases on such crops as rice and barley, and resistance to fungicides has occurred.

The Pathogen. Resistance occurs mostly with those prolific pathogens that produce several cycles of reproductive propagules in large numbers each growing season and that require multiple application of the pesticide to contain the various cycles. In these cases the pathogens are repeatedly exposed to the selective pressure of the pesticide over a long period. Pathogens that possess infrequent cycles (one to a few cycles per season or one cycle in several seasons), that utilize chemically untreated alternate hosts for part of their life cycle, that have a host range that includes untreated hosts such as weeds, or that can survive as R structures for a long time even in the presence of the toxicant, are likely to be less prone to resistance than those pathogens that are not so characterized. In addition, the inherent genetic mechanisms of variability available to the pathogen will be important components of its propensity to mutate and adapt toward an R state.

The Control Tactics. The variety and sequence of chemical usage in disease control will depend on the blend and sequence of diseases present in the treated crop. If a resistance-prone chemical is used alone to control all diseases in the complex, the pressure to develop resistance is present in all pathogens. Pathogens that are exposed throughout the growing season will more likely develop resistance than those exposed for only a part of the season. Theoretically, if two fungicides of different modes of action that are toxic to the pathogen(s) present are used simultaneously, sequentially or alternately, resistance to one or both should be delayed if not prevented. In this case the chance for survival of a mutant resistant to both chemicals will be much smaller than for a mutant resistant to only one of them, because if an R mutant should arise to one of the chemicals, it is likely to be eliminated by the other chemical.

The use of multisite inhibitor simultaneously, sequentially or alternately with the resistance-prone, limited-site inhibitor is a promising possibility available to plant pathologists in combating resistance. It is commonly suggested that the two types of toxicants in combination be used at less than optimal dosage for each. With different practical modes of action, i.e., antisporulation, redistribution, eradication or protection, the two pesticides may actually complement each other. Even more compelling is that fact that at a reduced dosage the selective pressure with the resistance-prone chemical would be less intense, and R mutants would be inhibited by the broad-site inhibitor and disease control would be maintained at a good level. Furthermore, such R mutants might at first be less fit than the wild type and would not have a favorable environment for pathogenic adaption in the presence of the second fungicide. The foregoing applies only if the combinations are utilized as a disease control strategy prior to the development of detectable levels of resistance in the population and selection for fitness. Once resistance has occurred as a practical problem, the use of chemicals in mixtures, sequences or alternations seems to be of little value, especially if reduced rates of both chemicals are employed.

The greater the frequency and dosage of the pesticide required to control a disease, the greater the chances are that resistance will arise. Twenty or more sprays per season are used in some areas to control apple scab, but in other areas only one or two sprays may be required. In the latter case, resistance is much less likely to develop as a practical problem than in the former. Seed treatments with fungicides for the control of many soil pathogens are applied only once per growing season, and therefore only a small portion of the population of the pathogen is exposed to the fungicide. Thus, resistance of soilborne pathogens to seed-treatment fungicides should be rare.

<u>Epidemiological Factors</u>. The severity of disease, and hence selective pressure for resistance to pesticides, is greatly influenced by various epidemiological factors, such as climate, weather, soil type, inoculum potential and host susceptibility. The more acute the disease pressure, the greater is the call for chemical tactics, and the greater the pressure for selection and adaption for resistance in the pathogen. Many foliar diseases, such as apple scab, are much more important in moist, temperate climates, especially where the spring is cool and wet, than in dry temperate climates. Powdery mildew fungi are usually favored by warm, dry climates of low rainfall with high humidity at night. The success of soil pathogens is often determined by soil temperature, drainage and type. Root knot nematodes are generally more important pathogens in the lighter and warmer soils of the United States and seldom reach epidemic proportions in cool heavy soils. Diseases caused by

Pythium spp. are usually more destructive in wet soils than drier, whereas the reverse is true for *Rhizoctonia*.

The importance of host susceptibility in disease control with chemicals has been demonstrated by Fry (1977). The less intense the disease on a host cultivar or species, the less demanding will be the chemical control tactic needed, with a resulting reduction in selective pressure on the pathogen. Host susceptibility may be manipulated by genetic, nutritional and certain cultural means, including use of an R variety, proper fertilization, and escape techniques such as delayed planting or early harvesting.

Refugia

Entomologists consider refugia an important factor in the dynamics of insecticide resistance (Georghiou and Taylor, 1977). Plant pathologists have devoted little attention to it in considering the problems of pesticide resistance in plant pathogens, although it probably is of great significance.

Refugia has been defined as the purely accidental (fortuitous) survival of a certain proportion of individuals in a treated population (Georghiou, 1978). The most obvious examples are those individuals that somehow escape exposure to the chemical after treatment, but also involved are such factors as timing of the treatment, efficiency and extent of the coverage of the host part or soil treated, stability of the pesticide, availability of untreated host plants and presence of inoculum in a nonsusceptible state or non-treatable location. Thus, S forms of the pathogen may not be selected for resistance and will remain as important components of the entire population, offering competition to R forms by genetic and population dilutions. Because S forms are often well fitted for their roles as pathogens, they may continue to dominate the population if the selective pressure for resistance is not too intense.

On the other hand, refugia may offer a haven for R pathogens in the absence of the selective pesticide. Lines of fungi resistant to benzimidazoles have been found several years after their use was discontinued because of control failures: the pathogen appeared to somehow survive in the crop year after year. Wild or abandoned hosts adjacent to areas where resistance has occurred may harbor R forms of pathogens. In New York, lines of *Venturia inaequalis* resistant to dodine have been isolated from wild and abandoned orchards adjacent to orchards where dodine resistance caused control failures 10 years previously with no subsequent use of dodine in the area in the intervening period (Gilpatrick, unpublished). Lines resistant to benomyl were found on wild apple trees 0.5 Km distance from an isolated orchard where this fungicide failed to control scab two years earlier and where it had not been subsequently employed (Gilpatrick, unpublished).

CONTROL SCENARIOS AND RESISTANCE

The following examples illustrate different chemical control situations where several factors may interact to determine resistance in practice.

White Mold of Beans

The first example is a disease that is known to have only one reproductive cycle per season (Figure 2). White mold of beans is caused by the fungus *Sclerotinia sclerotiorum* (Abawi and Hunter, 1979), which overwinters as an R structure (sclerotium) in or on the soil. The sclerotium germinates during moist weather, producing a small mushroom-like structure (apothecium) from which spores produced sexually are released and are borne by wind to the bean flower. The flower is infected and rots, becoming covered with a fluffy white mycelial growth (white mold). The mycelium invades other tissues of the plant, including the pod, and may cause total destruction of the crop. No asexual secondary cycles are known for this organism. The sclerotia are produced on the dead tissue among the white mycelia, completing the life cycle as the sclerotia fall to the ground. Only

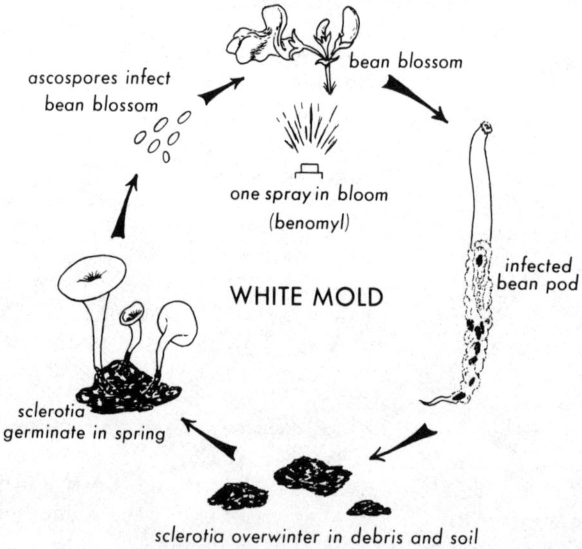

Figure 2. The white mold disease of bean caused by *Sclerotinia sclerotiorum* — life cycle and chemical control strategy.

a single, precisely timed spray of benomyl at flowering is required to give economic control of the disease by preventing infection of the flower. In this instance, the selection pressure on the pathogen by the fungicide occurs only once per season. In addition, beans may be rotated from one field to the other, and untreated hosts such as the dandelion may also harbor and provide benomyl-unselected inoculum for this disease, thus maintaining genetic diversity in populations of the pathogen. The probability of resistance developing in *S. sclerotiorum* from spraying fungicides on beans is not high, and in practice, no evidence of resistance has been observed in bean fields in New York even after 8 years of use of benomyl (Abawi, 1979).

Cedar-Apple Rust

Another example of a disease in which development of resistance in the causal agent is unlikely is cedar-apple rust (Agrios, 1978). The fungus *Gymnosporangium juniperi-virginianae* has a complicated biology requiring two different hosts, the red cedar and the apple, for the completion of its life cycle (Figure 3). The apple foliage and fruit are infected by wind-blown basidiospores produced in telial horns arising from galls on cedar foliage. The basidiospores are released from about the pink stage of apple phenology until the fruit is about 5 cm in diameter, a period lasting approximately 6 weeks. Water is required for the spores to germinate and infect the apple leaf or fruit, causing rust lesions. As fall approaches, another stage (aecial) appears on the apple lesions from which aeciospores

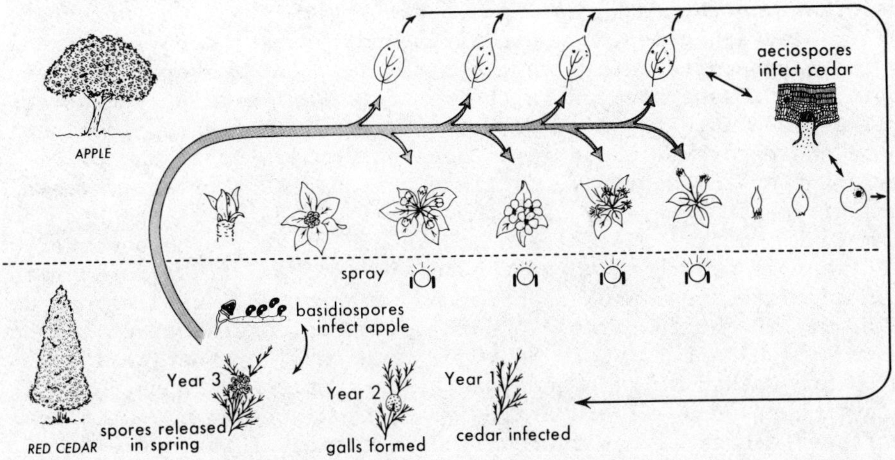

Figure 3. The cedar-apple rust disease caused by *Gymnosporangium juniperi-virginianae* — life cycle and chemical control strategy.

are produced and wind-blown to the cedar, causing infections in the presence of free moisture. The following year, small galls form on the cedar foliage, which increase in size, producing the telial horns in the third year. The fungus does not possess a recycling phase on either cedar or apple alone.

The disease is controlled on apple by avoiding the use of highly susceptible cultivars and attempting eradication of the cedar, but greatest reliance is placed on timely applications of fungicides. About four to six sprays of a fungicide are required on S cultivars of apple each year for economic control of this disease because the basidiospores are released over a prolonged period. Individauls in the fungus population are exposed to the fungicide only once in three years, even though a portion of the population is exposed each year. The pathogen receives no exposure to the fungicide on the alternate host, the cedar, and there is a significant influx of basidiospores from wild cedars, which have never been exposed to the fungicide. Thus, the influence of refugia as a genetic dilution factor may be great. To date, multisite inhibitors have been used to control rust on apple, and no resistance has been observed. The probability of developing resistance in this fungus to limited-site inhibitors appears to be extremely low, even though their use would be relatively intense each year.

Apple Scab

In the case of apple scab caused by *Venturia inaequalis*, fungicide use is very intense and resistance has been a problem with at least two fungicides — dodine and benomyl. This is a multicycle disease (Agrios, 1978) with abundant opportunity for selection for resistance in the pathogen. The fungus overwinters on infected fallen dead apple leaves where the sexual process occurs. Sexual spores (ascospores) are mature by the time that susceptible green apple leaf tissue appears in the spring. During rains the ascospores are released into the air and borne by wind to green apple tissue. These spores are released in increasing numbers until the bloom period of apple phenology, declining in numbers thereafter to an insignificant few by the time that the fruit has developed to about 5 cm in diameter (Gilpatrick and Szkolnik, 1978). The ascospores require free water on the surface of foliage and fruit for germination and infection, and depending on temperature, scab lesions develop within one to three weeks following infection. Each lesion bears a high number (up to 800,000) of asexual spores (conidia), which are washed by rain and dews to other tissues to recycle the infection process. In favorable years the inoculum potential can increase to high levels by repeated infection of the apple from primary spores and by repeated recycling of asexual spores. In the fall, the infected leaves fall to the ground to complete the annual cycle (Figure 4).

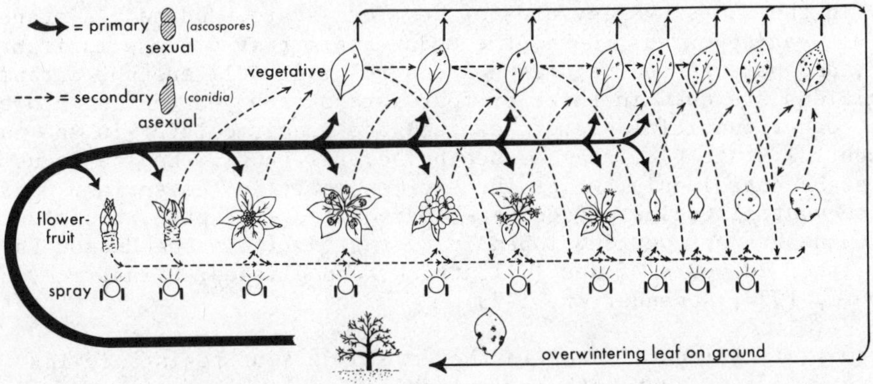

Figure 4. The apple scab disease caused by *Venturia inaequalis* — life cycle and chemical control strategy.

Control of the apple scab disease in a cool, wet climate is dependent primarily on the sequential spraying of fungicides on a weekly to biweekly basis, starting with the first appearance of green apple tissue (green tip) in the spring until the fruit has become resistant. This can occupy a period of three months or longer. In New York state, eight to twelve applications of appropriate fungicides are required each year. High selection pressure is placed on the sexual and asexual spores of the population by the fungicide. It is not surprising that resistance to dodine occurred in the apple scab fungus after about ten years of intensive and often exclusive use of this fungicide in western New York apple orchards (Szkolnik and Gilpatrick, 1969). The use of dodine was subsequently abandoned in western New York and replaced largely with benomyl. After similar use of benomyl, resistance developed to this fungicide in some orchards after only three years of use, which followed the pattern of benomyl resistance in *Venturia inaequalis* in other parts of the world (Jones and Walker, 1976; Wicks, 1974). Now growers in western New York have reverted to the use of multisite inhibitors such as captan, captafol, ethylene bis-dithiocarbamates and dichlone to combat scab. Mixtures of benomyl and multisite inhibitors such as captan and mancozeb have also been used. Whether or not this has effectively stemmed the increase of benomyl resistance is not yet known, but few cases of benomyl resistance have been reported, possibly because the weather has been unfavorable for the scab disease in recent years. In Michigan, where similar control practices have been used and where wetter growing seasons have been experienced than in western New York in recent years, the use of benomyl is no longer recommended because of increasing resistance in the scab fungus (Jones, 1979).

In the Hudson Valley area of New York state, dodine resistance has not yet appeared. Here, the cedar-apple rust disease is important, and because dodine does not control rust well enough, carbamate fungicides are used in three or four sprays each year during flowering. Other fungicides (e.g., captan) are required later in the summer to control fruit blemishes caused by several fungi. Dodine is used for scab control only two or three times early in the spring. Under the sequential regime of dodine -> carbamates -> captan, resistance to dodine has not yet developed in *V. inaequalis* in the Hudson Valley even after 20 years of use (Gilpatrick, unpublished; Gilpatrick and Blowers, 1974; Rosenberger, 1979).

One significant factor in the development of resistance in *V. inaequalis* is that both benomyl and dodine inhibit the sexual stage of S strains in the overwintering leaves in sprayed orchards by as much as 99% (Burchill, 1972; Gilpatrick, unpublished); however, R strains are not so affected (Gilpatrick, unpublished). Thus, significant shifts towards dominance of R strains can occur in the population from one season to the next.

Several strategies can be proposed to deal with the fungicide resistance problem in apples. These are presented in Table 2, and assume that such strategies are used each year from the time that the product is commercially introduced. In Strategy 1, a multisite inhibitor (M) is used in all sprays; the probability (P) of developing resistance is very low even after prolonged use. When a single-site inhibitor (S) prone to resistance is used in every spray (Strategy 2), P is very high, as proven in practice for benomyl. In Strategy 3, if M is used for one-half of the season and S for the other half, P for S is moderate. In Strategy 4, if two multisite inhibitors (M and N) are used in sequence with S so that use of S is infrequent, then P is low for S. In Strategy 5, when M and S are alternated throughout the season, P is moderate for S. In Strategy 6, M and S are combined in each spray at one-half the normally recommended rates of each, and P is low to moderate for S. Strategy 7 utilizes the single application technique (SAT) in which captafol is applied at a high rate and provides scab control until pink, which involves about one-half of the primary inoculum available each year. S (benomyl) is used for the remainder of the season, and P for benomyl has been found to be low in practice (Gilpatrick and Lienk, unpublished). In Strategy 8, a multisite inhibitor applied at a high dosage (D) is used early, followed by two sprays of S timed according to computer programming, followed by a postharvest spray of fungicide (A), which prevents the production of primary inoculum the following spring. In this case, P for S is low. P is also low in Strategy 9 if a single-site inhibitor (F) only slightly prone to resistance is used, but monitoring should be carried out to follow sensitivity levels in practice. In Strategy 10, if S and F are used in combination, resistance will not develop

Table 2. Spraying Strategies for the Prevention of Fungicide Resistance in the Apple Scab Fungus in a Cool, Moist Climate

No.	Fungicide at each apple tree growth stage											Probability of resistance occurring (P)
							Covers				Post-harvest	
	GrT	HIG	TC	P	B1	PF	1	2	3	4		
1	M	M	M	M	M	M	M	M	M	M	-	Very low (M)
2	S	S	S	S	S	S	S	S	S	S	-	Very high (S)
3	M	M	M	M	M	S	S	S	S	S	-	Moderate (S)
4	M	M	M	N	N	N	N	S	S	S	-	Low (S)
5	M	S	M	S	M	S	M	S	M	S	-	Moderate (S)
6	$\frac{M+S}{2}$	$\frac{M+S}{2}$	$\frac{M+S}{2}$	$\frac{M+S}{2}$	$\frac{M+S}{2}$	$\frac{M+S}{2}$	$\frac{M+S}{2}$	$\frac{M+S}{2}$	$\frac{M+S}{2}$	$\frac{M+S}{2}$	-	Low-to-moderate (S)
7	D	-	-	-	S	S	S	S	S	S	-	Low-to-moderate (S)
8	D	-	-	-	S	-	-	S	-	-	A	Low (S)
9	F	F	F	F	F	F	F	F	F	F	-	Low (F)
10	S+F	S+F	S+F	S+F	S+F	S+F	S+F	S+F	S+F	S+F	F	Low to-moderate (S) Low (F)

M,N = Multisite fungicides; e.g., captan, dithiocarbamates.
S = Single-site inhibitor; e.g., benomyl.
D = Multisite fungicide applied at high dosage; e.g., captafol.
A = Postharvest spray to trees to prevent overwintering of pathogen; e.g., triarimol, dodine.
F = Single-site inhibitor not prone to resistance; e.g., sterol inhibitors.
Monitor resistance level.

to F but eventually will to S; P for S resistance is considered lower than if S is used alone, as in Strategy 2.

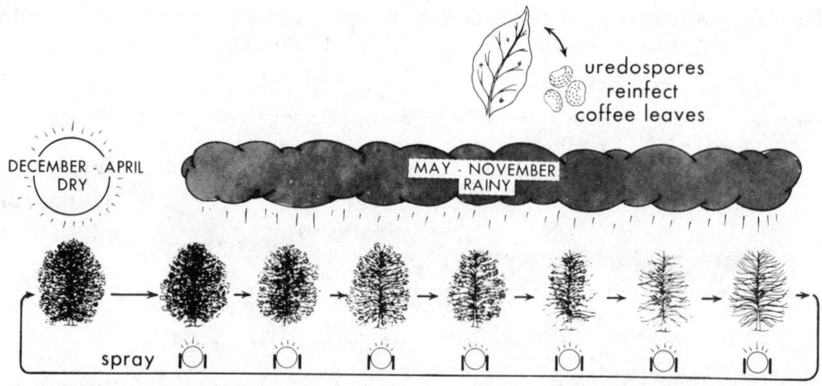

Figure 5. The coffee rust disease caused by *Hemileia vastatrix* — life cycle and chemical control strategy.

Coffee Rust

Many diseases are characterized by the apparent absence of a sexual cycle but with an asexual cycle that is repeated several times each year. Coffee rust caused by *Hemileia vastatrix* (Agrios, 1978) is an important disease in tropical climates. It is of only minor importance in dry months, but during the wet season the asexual stage recycles about every three weeks, leading to defoliation of the trees and berry destruction. The sexual stage does not play a role in the survival, proliferation or genetics of this fungus. The major source of inoculum is produced on infected berries throughout the fruiting season (Figure 5). Control has been obtained by spraying with a multisite fungicide (primarily copper) starting at the beginning of the rains and continuing at intervals during that season (Stover, 1977), so that from eight to eleven sprays applied at three- to four-week intervals may be needed each year. Resistance to copper fungicides has not yet been observed, but with the advent of fungicides with limited-site activity, one might expect resistance to develop rather quickly. This has happened with benzimidazoles for diseases caused by fungi of similar life cycles, e.g., *Cercospora, Botrytis, Penicillium* and powdery mildew fungi, where there has been extensive use of these fungicides.

Diseases Caused by *Cercospora* Species

Two dramatic examples of resistance are found in the case of leaf spots of sugar beets and peanuts caused by *Cercospora beticola* and *C. arachidicola*. Cercospora diseases are favored by warm, humid weather. The pathogens are seed-transmitted and spores can be blown long distances by the wind.

Table 3. Triphenyltin (TPT) Resistance Levels of Isolates of *Cercospora beticola* from Sugar Beets in Greece

Location	Years TPT used	% Isolates		
		Susceptible	Intermediate	Resistant
I	11	17	47	36
II	0	64	36	0
III	5	26	52	22

Sugar beets have been grown extensively in Greece since 1962 on a two-year rotation with cotton and corn. In one area (Location I) during 1962 and 1963, copper fungicides were applied for leaf spot control but were replaced with triphenyltin (TPT) in 1964, with five to eight sprays per season until 1970 (Giannopolitis, 1978). In 1971 benomyl replaced tin, but resistance developed to this fungicide in 1972. Since then, TPT has been used again, but resistance to tin fungicides occurred in sugar beet fields of Greece in 1976 and 1977. Giannopolitis compared the resistance to TPT isolates of *C. beticola* from the above location (I) to those from an area where no fungicides had been used (Location II) and to those from an area where fungicides were not used until 1970 (Location III). In the latter, TPT was used in 1970, followed by benomyl in 1971 and 1972, and then TPT from 1973 until the present. The resistance levels against TPT are shown in Table III.

Giannopolitis attributed the resistance in Locations I and III to the uniform and repeated application of TPT for many years. The discovery of intermediate resistance in an area where TPT has not been used (Location II) was unexpected, but this is an area where potatoes have been sprayed with TPT in the past, which may account for the resistance observed. In addition, sugar beet seeds are imported into Greece from areas that utilize TPT during seed production, and thus, long-distance spread of R strains by seeds is a possibility.

Cercospora leaf spot is an important disease of peanuts wherever the crop is grown. Rust (*Puccinia arachidis*) is increasing in importance worldwide and may reduce yields in some areas as much as Cercospora leaf spot. Benomyl controls leaf spot but not rust, while mancozeb controls rust but is relatively ineffective against leaf spot. In the state of Georgia in the United States, where leaf spot is severe and rust is of less importance, resistance to benomyl developed very quickly in *C. arachidicola* when this fungicide was used alone in six to nine sprays per year. In the state of Texas, however, rust is more severe than in Georgia, and fewer sprays are needed for *Cercospora*. Combinations of mancozeb and benomyl were used for both rust and leaf spot control, and under these conditions, resistance to benomyl has not developed in commercial peanut fields in Texas (Littrell, 1979; Smith et al., 1978).

Schiller and Indhaphun (1979) have demonstrated how the selection pressure by benomyl against *Cercospora* on peanuts can be further decreased by using pest management strategies. They observed that in Thailand, advancing the date of planting in relation to the July-to-October rainy season reduced the amount of disease from that occurring with planting closer to the rainy period. Using yield as a basis, it was concluded that with early plantings, the first sprays could be delayed until 50 days after sowing, followed by three additional sprays at 10-day intervals. There was no advantage in spraying beyond 80 days of sowing; thus, only four sprays of a combination of benomyl and mancozeb are needed in a typical rainfall year in the area of Thailand where this study was made. Without information on the effects of planting dates and spray timing on these diseases, unquestionably more sprays would be used to effect control. This reduction in sprays should still provide optimum disease control and minimal selection pressure from benomyl, thus lessening the probability of resistance occurring.

Brown Rot of Stone Fruits

The brown rot disease of stone fruits, caused by *Monilinia fructicola*, may be used to illustrate how use of a multiple number of fungicides could prevent resistance. The disease is characterized by two peaks of major activity, including a blossom disease phase and a pre- and postharvest period when the fruit is highly susceptible. Both periods are separated by a period of several weeks when the disease is relatively quiescent (Figure 6). Fungicides are used during bloom, before harvest, and as postharvest treatments for brown rot control. Sometimes fungicides are also applied during the lag phase between bloom and harvest, especially in the Eastern United States. One fungicide (e.g., triforine) could be used during bloom, a second in the lag phase (e.g., captan), a third at preharvest (e.g., vinclozolin), and a fourth at postharvest (e.g., benomyl). Triforine and benomyl have different modes of action, whereas that for vinclozolin is still not certain and is probably unique (Pappas and Fisher, 1979). The use of the multisite inhibitor captan, which is not highly effective against brown rot, should be adequate for light disease pressure during the lag phase. Using this regime, resistance to any of the fungicides used is unlikely to develop. This example illustrates the need to have a wide variety of pesticides available to combat resistance, but at this time triforine

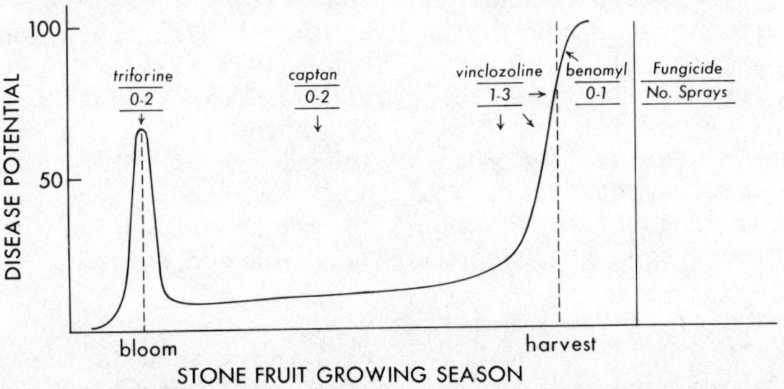

Figure 6. Proposed chemical control strategy for brown rot of stone fruit caused by *Monilinia fructicola,* utilizing four fungicides with different modes of action in order to avoid resistance in any one of the fungicides.

and vinclozolin are not approved for use on stone fruits in the United States.

COUNTERMEASURES

From the foregoing discussions, it can be concluded that the most probable way of preventing or alleviating the resistance problem is to reduce the selection pressure of the pesticide on the pest. This can probably best be achieved by a broad approach to disease control utilizing a chemical as one of the pest management tools. In this way, the probability of resistance can be minimized while still obtaining economic pest control. In such strategies, resistance-prone chemicals would be used as infrequently and as efficiently as possible.

The pest management approach requires considerable knowledge of (1) the chemical, e.g., mode of action, cross resistance, residue dynamics; (2) the biology of the pathogens; (3) epidemiological factors; (4) effects of crop culture on diseases; (5) threshold levels; and (6) economic factors. Often, much of this information is lacking, and consequently, a systems approach to disease control with chemicals is usually not possible at the time. A most common deficiency is information on the economic needs or benefits of the chemical control procedure. Thus, need cannot be accurately predicted at any point in time. The relationship of weather and other climatic factors on diseases and pesticide residue dynamics are little understood, and our ability to measure and predict weather in ways that are useful in spray forecasting is inadequate. We often lack knowledge of the biochemical mode of action of fungicides; thus, cross resistance between fungicides is often not predictable. Genetic mechanisms, propensity for resistance, fitness of R mutants, and the population dynamics of plant pathogens in the field are seldom clear. Research is needed in all of these areas if we are to effectively lower the selection pressure put on pathogen populations by pesticides. This information is the key to techniques of avoiding resistance, to more prolonged use of selective chemicals, and to the optimal utilization of these compounds in pest control and agricultural production. Finally, the value of any new technology must be demonstrated to be economically and ecologically sound.

In spite of the many deterrents to the application of pest management strategies to pest control, this method is the most promising way of solving the resistance problem. Unfortuantely, pest management cannot be easily legislated and regulated. Success will require the investment of a great amount of money, the ingenuity of the reseracher, the restraint of the pesticide manufacturer, and the understanding and cooperation of the farmer. Thus, knowledge, common sense and effort are the main vehicles to carry us to the solution.

There are several strategies that should prove useful in avoiding pesticide resistance in plant pathogens. The following relate to newly developing chemicals:

1) Determine the propensity of the pesticide for resistance and the potential fitness of R mutants in vivo and in vitro.

2) Determine the biochemical mode of action of the pesticide in order to predict cross resistance and to provide a guide for the synthesis of structures without a propensity for resistance.

3) Utilize pest management strategies to reduce the requirement for the chemical to a minimum and to use it most efficiently.

4) Avoid using resistance-prone chemicals in such a way as to put extensive selective pressure on the pathogen by repeated sprays or as seed or soil systemic treatments for foliar pathogens. Such chemicals should be used alone only when there are no alternatives or when there is a highly compelling need for the chemical. If intensive use is required, then the chemical should be used simultaneously, sequentially or alternately with chemicals of different biochemical modes of action, especially multisite inhibitors.

5) Monitor for resistance after the pesticide is put into practical use and alter strategies if resistance appears.

In addition to the above strategies, the following should be adopted once resistance has appeared as a practical problem:

1) The level and distribution of resistance should be confirmed by monitoring the pathogen population.

2) The use of the pesticide should be continued only with careful management. This might include its use only for emergencies or special uses. The use of the pesticide with reduced rates of other fungicides may be precarious if the extent of the resistance is unknown.

CONCLUSIONS

Plant pathologists have been fortunate in the past that many fungicides, nematicides and bactericides have not been as subject to resistance in pests as insecticides. This is because most plant disease control chemicals have been multisite inhibitors, whereas most of the modern insecticides and acaracides were relatively selective biochemically. After the introduction of site-selective compounds for plant disease control, especially fungicides, numerous cases of resistance occurred. Resistance in nematodes is still rare because unspecific soil fumigants are still predominantly used for control; but, indications are that as more specific compounds are utilized, resistance will develop in these pests as well (Anonymous, 1979).

The development of resistance by fungi to certain fungicides, especially the benzimidazoles, has been quick, widespread and disillusioning. Although all site-specific inhibitors of plant pathogens may be prone to resistance, indications are that with many of these, such factors as unfitness in R mutants and pesticide management may greatly temper the impact.

A variety of pesticides, ideally both site- and nonsite-specific, will be necessary to deal with resistance, and the development of new structures with a wide variety of biochemical actions should be encouraged. However, the regulatory and economic climate throughout the world is not conducive to such development of new pesticides. Virtually no new chemical structures useful in plant disease control are being approved at this time in the United States because of the extremely slow registration process (Gilpatrick, 1979). Furthermore, multisite inhibitors, including such widely used products as captan, ethylenebisdithiocarbamates, 1,2-dibromo-

3-chloropropane and pentachloronitrobenzene, are currently being considered for label restrictions or deregistration. Loss of any one of these would often remove the most important substitute or alternate for selective compounds and would thus lead to an increased use of site-selective pesticides. In addition, the availability of these multisite compounds for use together with selective chemicals in control programs would be prevented. The consequence would be an increase in the incidence of resistance.

A committee of the American Phytopathological Society (APS) recently advised that, at the present level of knowledge, it does not seem warranted for the United States Environmental Protection Agency to attempt to manipulate resistance by label directions or regulations (Gilpatrick, 1979). The APS report stated that any interference might only increase the resistance problem rather than improve it. The report concluded that such regulations, if imposed, should be based on well-proven scientific evidence and that label directions should be as flexible as possible so that scientists and extension workers in the field could make judgments on strategies as needed according to local situations.

A panel of experts of the FAO of the United Nations, meeting in Rome, Italy in 1978, were so concerned with the resistance problem in pesticides and the current regulatory climate throughout the world that they recommended that FAO draw the attention of member nations to the importance of this problem and the importance of retaining sufficient pesticides to provide adequate flexibility to deal with resistance when pesticide regulatory decisions are made (Anonymous, 1979).

To date, resistance in plant pathogens has not often caused serious crop losses except in local areas because the problem was discovered before it became widespread, and suitable alternate chemicals were available. Nevertheless, resistance has severely limited the use of some of our most effective fungicides and bactericides, and has reduced our flexibility in coping with numerous diseases. Yet, the potentials offered by systemics, which are usually single-site inhibitors, are too great to discourage their discovery, development and use because of the fear of resistance. These chemicals will be the major components of safe and effective disease control in the future. They promise a quality, variety and flexibility in control not previously possible and yet necessary for technological progress in agriculture.

What role resistance will play in plant disease control in the future is unknown, but it certainly will be a significant deterrent to the promiscuous use of many limited-site pesticides. Resistance has altered markedly the thinking of those designing chemical control programs for plant diseases, although there is considerable diversity of opinion as to how to prevent resistance. Much more research is needed on resistance before plant pathologists can develop and prove more effective strategies to deal with this disturbing aspect of crop production.

REFERENCES

Abawi, G. S., 1979, personal communication.

Abawi, G. S., and Hunter, J. E., 1979, White mold of beans in New York, New York's Food and Life Sciences Bulletin 77, p. 4.

Abiko, K., Kishi, K., and Yoshioka, A., 1977, Occurrence of oxycarboxin-tolerant isolates of *Puccinia horiana* P. Hennings in Japan, *Ann. Phytopathol. Soc. Japan,* 43:145.

Agrios, G. N., 1978, "Plant Pathology," Academic Press, New York, 2nd ed.

Anonymous, 1977, Pest resistance to pesticides and crop loss assessment, FAO Plant Production and Protection Paper, 6, Report to 1st Session of the FAO Panel of Experts, AGP:1976/M/10, p. 42.

Bent, K. J., Cole, A. M., Turner, J. A. W., and Woolner, M., 1971, Resistance of cucumber powdery mildew to dimethirimol, *Proc. 6th Br. Insectic. Fungic. Conf.*, 1:274.

Bertrand, P. F., and Saulie-Carter, J. L., 1978, The occurrence of benomyl-tolerant strains of *Penicillium expansum* and *Botrytis cinerea* in the mid-Columbia region of Oregon and Washington, *Plant Dis. Rep.,* 62:302.

Bollen, G. J., and Scholten, G., 1971, Acquired resistance to benomyl and some other systemic fungicides in a strain of *Botrytis cinerea* in cyclamen, *Neth. J. Plant Pathol.,* 77:83.

Burchill, R. T., 1972, Comparison of fungicides for suppressing ascospore production by *Venturis inaequalis* (Cke.) Wint., *Plant Pathol.,* 21:19.

Chastagner, G. A., and Ogawa, J. M., 1979, DCNA-benomyl multiple tolerance in strains of *Botrytis cinerea, Phytopathology,* 69:699.

Day, P. R., 1974, "Genetics of Host-Parasite Interaction," W. H. Freeman and Company, San Francisco.

Dekker, J., 1976, Prospects for the use of systemic fungicides in view of the resistance problem, *Proc. Am. Phytopathol. Soc.,* 3:60.

Dekker, J., 1977a, Resistance, *in:* "Systemic Fungicides," R. W. Marsh, ed., pp. 176-197, Longman, London.

Dekker, J., 1977b, The fungicide-resistance problem, *Neth. J. Plant Pathol.,* 83(Suppl. 1):159.

Dekker, J., 1977c, Effects of fungicides on nucleic acid synthesis and nuclear function, *in:* "Antifungal Compounds," Volume 2, M. R. Siegel and H. D. Sisler, eds., pp. 365-398, Marcel Dekker, Inc., New York.

Dekker, J., and Gielink, A. J., 1979, Acquired resistance to pimaricin in *Cladosporium cucumerinum* and *Fusarium oxysporum* f. sp.

narcissi associated with decreased virulence, *Neth. J. Plant Pathol.*, 85:67.

Dovas, C., Skylakakis, G., and Georgopoulos, S. G., 1976, The adaptability of the benomyl resistant population of *Cercospora beticola* in Northern Greece, *Phytopathology*, 66:1452.

Fry, W. E., 1977, Fungicides in perspective, *in:* "Antifungal Compounds," Volume 2, M. R. Siegel and H. D. Sisler, eds., pp. 19-50, Marcel Dekker, Inc., New York.

Fuchs, A., and Bollen, G. J., 1975, Benomyl, after seven years, *in:* "System-Fungicide," H. Lyr and C. Polter, eds., pp. 121-136, Academie-Verlag, Berlin.

Geeson, J. D., 1978, Mutational tolerance to carbendazim in *Botrytis cinerea*, *Ann. Appl. Biol.*, 90:59.

Georghiou, G. P., 1978, Strategies in the use of pesticides to delay or avoid development of resistance, Working Paper, FAO Panel of Experts on Pest Resistance to Pesticides, Rome, 1978, p. 16.

Georghiou, G. P., and Taylor, C. E., 1977, Genetics and biological influences in the evolution of insecticide resistance, *J. Econ. Entomol.*, 70:653.

Georgopoulos, S. G., 1976, The genetics and biochemistry of resistance to chemicals in plant pathogens, *Proc. Am. Phytopathol. Soc.*, 3:53.

Georgopoulos, S. G., 1977, Development of fungal resistance to fungicides, *in:* "Antifungal Compounds," Volume 2, M. R. Siegel and H. D. Sisler, eds., pp. 439-495, Marcel Dekker, Inc., New York.

Georgopoulos, S. G., and Dovas, C., 1973, A serious outbreak of strains of *Cercospora beticola* resistant to benzimidazole fungicides in Northern Greece, *Plant Dis. Rep.*, 57:321.

Giannopolitis, C. N., 1978, Occurrence of strains of *Cercospora beticola* resistant to triphenyltin fungicides in Greece, *Plant Dis. Rep.*, 62:205.

Gilpatrick, J. D., 1978, Strategies in the use of pesticides to delay or avoid development of resistance in fungi, A working paper for FAO Panel on Pest Resistance to Pesticides, Rome, Italy, 28 August - 1 September, 1978, p. 6.

Gilpatrick, J. D., ed., 1979, "Contemporary Control of Plant Disease with Chemicals: Present Status, Future Prospects, and Proposals for Action," EPA-68-01-3914, p. 169.

Gilpatrick, J. D., and Blowers, D. R., 1974, Ascospore tolerance to dodine in relation to orchard control of apple scab, *Phytopathology*, 64:649.

Gilpatrick, J. D., and Provvidenti, R., 1973, Resistance to fungicides by apple scab and cucurbit powdery mildew fungi in New York, Abstracts of Papers of 2nd International Congress of Plant Pathology, Minneapolis, Minnesota, #0780.

Gilpatrick, J. D., and Szkolnik, M., 1978, Maturation and discharge of ascospores of the apple scab fungus, *in:* "Proceedings

Apple and Pear Scab Workshop," A. L. Jones and J. D. Gilpatrick, eds., pp. 1-5, Special Report 28 of New York State Agricultural Experiment Station, Geneva.

Glass, E. H., ed., 1976, Research needs on pesticides and related problems for increased food supplies, Report to Science and Technology Office, National Science Foundation, Cornell University, Geneva, New York, p. 63.

Hall, R., 1963, Cytology of the asexual stage of the Australian brown rot fungus *Monilinia fructicola* (Wint.) Honey, *Cytologia*, 28:181.

Harris, C. R., 1977, Insecticide resistance in soil insects attacking crops, *in:* "Pesticide Management and Insecticide Resistance," D. L. Watson and A. W. A. Brown, eds., pp. 321-351, Academic Press, New York.

Hoffman, G. M., 1970, Kernverhaltnisse bei *Monilinia fructigena* und *M. laxa*, *Phytopathol. Z.*, 68:143.

Iida, W., 1975, On the tolerance of plant pathogenic fungi and bacteria to fungicides in Japan, *Japan Pestic. Information*, 23:13.

Ito, H., Miura, H., and Takahaski, A., 1974, Transition of the effectiveness of Kasugamycin at Shonai district in Yamagata, *Ann. Phytopathol. Soc. Japan*, 40:168.

Jones, A. L., 1979, personal communication.

Jones, A. L., and Ehret, G. R., 1976, Tolerance to fungicides in *Venturia* and *Monilinia* of tree fruits. *Proc. Am. Phytopathol. Soc.*, 3:84.

Jones, A. L., and Walker, R. J., 1976, Tolerance of *Venturia inaequalis* to dodine and benzimidazole fungicides in Michigan, *Plant Dis. Rep.*, 60:40.

Katagiri, M., and Uesugi, Y., 1977, Similarities between the fungicidal action of isoprothiolane and organophosphorus thiolate fungicides, *Phytopathology*, 67:1415.

Koffman, W., Penrose, L. J., Menzies, A. R., Davis, A. R., and Kaldor, J., 1978, Control of benzimidazole-tolerant *Penicillium expansum* in some pome fruit, *Scientia Horticulturae*, 9:31.

Kohmoto, K., Nishimura, S., and Udagawa, H., 1974, Distribution and chronological population shift of polyoxin-resistant strains of black spot fungi of Japanese pear, *Alternaria kikuchiana* in field, *Ann. Phytopathol. Soc. Japan*, 40:220.

Leroux, D., Fritz, R., and Gredt, M., 1977, Etudes en laboratoire de souches de *Botrytis cinerea* Pers., résistantes a la dichlozoline, au dicloran, au quintozene, à la vinchlozoline et au 26019 RP (on glycophene), *Phytopathol. Z.*, 89:347.

Littrell, R. H., 1976, Techniques for monitoring for resistance to plant pathogens, *Proc. Am Phytopathol. Soc.*, 3:90.

Littrell, R. H., 1979, personal communication.

Meyer, R. W., and Parmeter, J. R., Jr., 1968, Changes in chemical tolerance associated with heterokaryosis in *Thanatephorus cucumeris*, *Phytopathology*, 58:472.

Misato, T., 1975, The development of agricultural antibiotics in Japan, *Proc. 1st Intersect. Cong. of Intern. Assoc. Microbiol. Soc., 1974,* 3:589.

Nishimura, S., Kohmoto, K., and Udagawa, H., 1976, Tolerance to polyoxin in *Alternaria kikuchiana* Tanaka, causing black spot disease of Japanese pear, *Rev. Plant Protec. Res.,* 9:47.

Ogawa, J. M., Gilpatrick, J. D., and Chiarappa, L., 1977, Review of plant pathogens resistant to fungicides and bactericides, *FAO Plant Protection Bulletin,* 26:97.

Pappas, A. C., and Fisher, D. J., 1979, A comparison of the mechanisms of action of vinclozolin, procymidone, iprodione and prochloraz against *Botrytis cinerea, Pestic. Sci.,* 10:239.

Polach, F. J., and Molin, W. T., 1975, Benzimidazole-resistant mutant derived from a single spore culture of *Botryotinia fuckeliana, Phytopathology,* 65:902.

Rosenberger, D. A., 1979, personal communication.

Ross, I. K., 1979, "Biology of the Fungi," McGraw-Hill, New York.

Ruppel, E. G., 1975, Biology of benomyl-tolerant strains of *Cercospora beticola* from sugar beet, *Phytopathology,* 65:785.

Sakurai, H., 1977, Methods of determining the drug-resistant strains in phytopathogenic bacteria and fungi and its epidemiology in the field, *J. Pestic. Sci.,* 2:177.

Schiller, J. M., and Indhaphun, P., 1979, Economic control of Cercospora leaf spot and rust in rainfed peanut production, *Protection Ecology,* 1:109.

Schroeder, W. T., and Provvidenti, R., 1969, Resistance to benomyl in powdery mildew of cucrubits, *Plant Dis. Rep.,* 53:271.

Schroth, M. N., Thompson, S. V., and Moller, W. J., 1979, Streptomycin resistance in *Erwinia amylovora, Phytopathology,* 69:565.

Shepherd, M. C., Bent, K. J., Woolner, M., and Cole, A. M., 1975, Sensitivity to ethirimol of powdery mildew for U.K. barley crops, *Proc. 8th Br. Insectic. Fungic. Conf.,* 1:59.

Smith, D. H., McGee, R. E., and Vesley, L. K., 1978, Isolation of benomyl tolerant strains of *Cercospora arachidicola* and *Cercosporidium personatum* at one location in Texas, Tenth Annual Meeting of the American Peanut Research and Education Association, Inc., Gainesville, Florida, 1978.

Snel, M., and Edgington, L. V., 1970, Uptake, translocation and decomposition of systemic oxathin fungicides in beans, *Phytopathology,* 60:1708.

Stover, R. H., 1977, Fungicidal control of plant diseases in the tropics, *in:* "Antifungal Compounds," M. R. Siegel and H. D. Sisler, eds., pp. 353-370, Marcel Dekker, Inc., New York.

Szkolnik, M., and Gilpatrick, J. D., 1969, Apparent resistance of *Venturia inaequalis* to dodine in New York apple orchards, *Plant Dis. Rep.,* 53:86.

Szkolnik, M., and Gilpatrick, J. D., 1977, Tolerance of *Monilinia fructicola* to benomyl in western New York State orchards, *Plant Dis. Rep.,* 61:654.

Sztejnberg, A., and Jones, A. L., 1978, Tolerance of the brown rot fungus *Monilinia fructicola* to iprodione, vinclozolin and procymidone fungicides, *Phytopathol. News*, 12:187.

Tate, K. G., and Samuels, G. J., 1976, Benzimidazole tolerance in *Venturia inaequalis* in New Zealand, *Plant Dis. Rep.*, 60:706.

Tillman, R. W., and Sisler, H. D., 1973, Effect of chloroneb on the growth and metabolism of *Ustilago maydis*, *Phytopathology*, 63:219.

Uesugi, Y., Katagiri, M., and Noda, O., 1974, Negatively correlated cross resistance and synergism between phosphoramidates and phosphorothiolates in their fungicidal actions on rice blast, *Agric. Biol. Chem.*, 38:907.

Van Gundy, S. D., Thomason, I. V., and Castro, C. E., 1974, Resistance in nematodes to nematicides, FAO Working Paper AGP:WPR/74:11, p. 13.

Van Tuyl, J. M., 1977, Genetics of fungal resistance to systemic fungicides, *Mededelingen Landouwhogeshool Wageningen*, 77(2): 1-136.

Vonk, J. W., and Kaars Sijpesteijn, A., 1977, Metabolism, *in:* "Systemic Fungicides," R. W. Marsh, ed., p. 160-175, Longman, London.

Walmsley-Woodward, D. J., Laws, F. A., and Whittington, W. J., 1979, The characteristic of isolates of *Erysiphe graminis* f. sp. *hordei* varying in response to tridemorph and ethirimol, *Ann. Appl. Biol.*, 92:211.

Warren, C. G., Sanders, P. L., Cole, H., Jr., and Duich, J. M., 1977, Relative fitness of benzimidazole- and cadmium-tolerant populations of *Sclerotinia homeocarpa* in the absence and presence of fungicides, *Phytopathology*, 67:704.

Wicks, T., 1974, Tolerance of the apple scab fungus to benzimidazole fungicides, *Plant Dis. Rep.*, 58:886.

Yaoita, T., Goh, N., Aoyagi, K., Iwano, M., and Sakurai, H., 1979, Studies on drug-resistant strain of rice blast fungus, *Pyricularia oryzae*, Part 1. The occurrence of the multi-resistant strains of *Pyricularia oryzae* and its epidemiology in the field, *J. Niigata Agr. Exp. Sta.*, 28:61.

MANAGEMENT OF RESISTANCE IN ARTHROPODS

George P. Georghiou

Division of Toxicology and Physiology
Department of Entomology
University of California
Riverside, California 92521

INTRODUCTION

Since the earliest days of their awareness of resistance, entomologists have been concerned with understanding the factors responsible for its development and with divising measures for its control. It is remarkable that in reporting the first case of resistance -- in the San José scale toward lime sulfur -- Melander (1914) recognized the role of incomplete coverage and genetic recessiveness and speculated that should the scale become resistant also to oil sprays "we might have to introduce a weak strain to cross with the immune and thus return to the normal susceptible population." Melander and other early pioneers in studies of resistance (Quayle 1922, Woglum 1925) may have been ahead of their time, however, for in the subsequent 30 years or so, resistance evolved slowly, affecting only 12 species of arthropods (review by Babers 1949). Interest in resistance intensified with the introduction of DDT and with the rapid development in cases of resistance to organochlorine, organophosphate, carbamate and most recently to pyrethroid insecticides. The phenomenon now involves at least 428 species of arthropods and every class of commonly available compound (Georghiou and Mellon, this volume).

Since Melander's first observations, significant advances have been made in our knowledge of the genetics, physiology and biochemistry of resistance, but little progress, if any, has been achieved in formulating practical countermeasures for retarding or forestalling its evolution. However, the recent discoveries of new insecticides and the increasing emphasis on integrated pest management (IPM) have rekindled interest in research on countermeasures for resistance and have raised hopes that some practical breakthroughs might be forthcoming.

Among these newly discovered insecticides are the juvenile hormone mimics, the chitin synthesis inhibitors, a number of derivatized organophosphates and carbamates, several synthetic pyrethroids, and certain new strains of bacterial toxins. These discoveries are very significant in that they provide us with some diversification in available chemicals, and enable us to avoid an overreliance on one specific compound. However, these new chemicals in themselves do not offer solutions, since none has been shown to be immune to the development of resistance. Since additional chemicals may be slow in forthcoming, there is great interest both in agriculture and in public health in safeguarding these available chemicals against the development of resistance.

The growing acceptance of the concept of IPM is also significant since in advocating the integration of chemical with nonchemical control, IPM contributes to a reduction in chemical selection pressure and thus to a delay in the evolution of resistance. IPM may in fact be the most important enemy of resistance, and vice versa. It is becoming increasingly evident that for an IPM program to remain viable, it must ensure that resistance does not evolve: an imposed change to a different insecticide is usually so disruptive to the biological control component of IPM that an entirely new control program must then be formulated. As we know, such programs require prolonged and costly experimentation.

Thus, interest in research on resistance has increased substantially in recent years, but with a sensibly different orientation: whereas in the past the challenge of continued effective pest control rested almost solely with the synthetic chemist, now it is shared at least equally by the pest management specialist who must ensure that resistance itself does not evolve or that it is sensibly delayed. The question that is now being asked, therefore, is: What knowledge and what technological innovations are still needed to ensure the effective use of new chemicals without their exposure to the risk of resistance?

A large number of scientists today believe that more permanent solutions to the resistance problem should be sought through measures aimed at reducing the degree of chemical selection pressure. I will not review here the concept of "integrated control" or IPM, as source reduction, cultural control, pheromones, biological control, and other components of IPM are subjects for individual discussion in their own right. I will concern myself briefly with recent progress on the dynamics of resistance and will discuss operational measures that have been considered as holding promise for resistance management.

DYNAMICS OF RESISTANCE

It is an established fact that the rate of development of resistance is extremely variable. Resistance has arisen slowly in some species and fast in others. In the same species, resistance has developed rapidly under one set of circumstances and slowly or not at all under another. It is evident that an essential prerequisite of a control program involving chemicals is to determine how much insecticidal pressure can be applied against the target population without leading to resistance. Putting it simply, what is the resistance risk in the target population?

Credit for our early knowledge of the dynamics of resistance belongs to population geneticists who approached resistance as an evolutionary phenomenon (Crow, 1952, 1957, 1966; Milani, 1957, 1959, 1964) and to numerous other researchers who examined the development, stability and regression of resistance in various species, particularly mosquitoes and house flies (see critical reviews by Brown, 1957, 1961a, 1968, 1971, 1976; Busvine, 1957; Georghiou, 1965, 1972; Georghiou and Taylor, 1976; Keiding, 1963, 1967; Metcalf, 1955). With the proliferation of cases of resistance in species representing a wider range of biological, genetic and ethological characteristics, it has gradually become more feasible to examine the dynamics of resistance in greater perspective. The advent of computer technology has generated considerable interest in modeling of the known and presumptive factors that influence the evolution of resistance, and as a result a gratifyingly large number of studies have recently been published (Comins, 1977; Curtis et al., 1978; Georghiou and Taylor, 1976, 1977a,b; Plapp et al., 1979; Taylor and Georghiou, 1979; see also review by Taylor, this volume).

There is general concurrence that the evolution of resistance is determined by a variety of genetic, biological and operational factors, which in concert determine the degree of selection pressure that is exerted in a given ecological situation. Recently, these factors have been listed systematically (Georghiou and Taylor, 1976) (Table 1), their respective influence illustrated graphically by computer simulation (Georghiou and Taylor, 1977a,b), and examples provided from field case histories (Georghiou, 1980; Georghiou and Taylor, 1976).

Factors in the genetic and biological categories are inherent qualities of the population, and therefore are beyond man's control, but their assessment is essential in determining the "risk for resistance" of a target population. The operational factors, on the other hand, are man-made and are thus within our control. They can be altered to any extent necessary and feasible, depending on the risk for resistance that is revealed by the genetic and biological factors.

Table 1. Known or Suggested Factors Influencing the Selection of Resistance to Insecticides in Field Populations[a]

A. Genetic

1. Frequency of R alleles
2. Number of R alleles
3. Dominance of R alleles
4. Penetrance; expressivity; interactions of R alleles
5. Past selection by other chemicals
6. Extent of integration of R genome with fitness factors

B. Biological

a. Biotic

1. Generation turn-over
2. Offspring per generation
3. Monogamy/polygamy; parthenogenesis

b. Behavioral

1. Isolation; mobility; migration
2. Monophagy/polyphagy
3. Fortuitous survival; refugia

C. Operational

a. The chemical

1. Chemical nature of pesticide
2. Relationship to earlier used chemicals
3. Persistence of residues; formulation

b. The application

1. Application threshold
2. Selection threshold
3. Life stage(s) selected
4. Mode of application
5. Space-limited selection
6. Alternating selection

[a]Adapted from Georghiou and Taylor, 1976.

Certain of these factors may be especially influential under specific situations and thus have been the subject of special studies in the search of practical approaches to management of resistance. Recent studies have examined especially the beneficial role of susceptible individuals immigrating into the treated environment (Comins, 1977; Curtis and Rawlings, 1980; Curtis et al., 1978; Muir, 1977; Taylor and Georghiou, 1979; Wool and Manheim, 1980), the effect of dosage on the "functional" dominance of the resistance gene (Comins, 1977; Curtis and Davidson, 1980; Georghiou and Taylor, 1976), and the role of decay rates of pesticide residues (Sutherst and Comins, 1979, Taylor and Georghiou, 1980). These studies have contributed much information that is essential in modeling for comprehensive resistance management. Although the attainment of that goal may still appear elusive due to the lack of quantitative data on many critical parameters (Greever and Georghiou, 1979), further rapid progress can be anticipated in studies involving major pest species. An outstanding example of such an advance is the modeling of optimal strategies for long-term curtailment of resistance within a broad program of integrated control of the cattle tick, *Boophilus microplus*, in Australia (Sutherst and Comins, 1979; Sutherst et al., 1979). The suggested strategies involve the use of tick-resistant Zebu-type cattle as a basis for tick control, and include stringent quarantine measures to slow down the spread of resistant ticks, careful timing and moderation in the frequency of use of acaricides to minimize the number of selections, and high concentrations of acaricide to reduce the probability of survival of resistant individuals.

RESISTANCE MANAGEMENT

Reduction of selection pressure has been recognized since the dawn of resistance as the surest way to delay or avoid the evolution of resistant strains. IPM now offers the opportunity for such reductions in chemical selection pressure by introducing greater reliance on multiple interventions involving natural enemies, insect diseases, cultural control practices, host-plant resistance, and other non-chemical measures.

In discussing the status of research on resistance, Brown (1976) stated that one of the best ways of circumventing the problem lies in "practicing the whole range of measures included in the terms *integrated control* or *pest management*," but he also stressed that since the bulk of agricultural crops depend on chemical insecticides for their protection, "it is *pesticide management* that is needed, and our deficiencies in this regard are the measure of the deficiencies of entomologists in dealing with insecticide resistance."

It is evident that optimization of IPM must inevitably involve the introduction of strategies of insecticide usage that contribute only minimally to the development of resistance. Insecticide susceptibility is being recognized increasingly as a "depletable natural resource" (Hueth and Regev, 1975) that is "extremely expensive to replace once lost" (Sutherst and Comins, 1979).

My presentation below applies to the numerous cases of insect control, including many IPM programs, in which the use of chemical insecticides is required. It constitutes an attempt to recognize

Table 2. Chemical Strategies of Resistance Management

A. Management by moderation

- Low dosages, sparing a proportion of susceptible genotypes
- Less frequent applications
- Chemicals of short environmental persistence
- Avoidance of slow-release formulations
- Selection directed mainly against adults
- Localized rather than areawide applications
- Certain generations or population segments left untreated
- Preservation of "refugia"
- Higher pest population threshold for insecticide application

B. Management by saturation

- Rendering R gene "functionally" recessive by higher dosages on target
- Suppression of detoxication mechanisms by synergists

C. Management by multiple attack

- Mixtures of chemicals
- Alternation of chemicals

contrasting or alternative chemical interventions and their anticipated impact on resistance. The emphasis is on usage practices rather than on the choice of alternative insecticides for specific resistance mechanisms, as the latter aspect is discussed in papers by Fukuto and Mallipudi, and by Yamamoto et al. in this volume.

Measures for resistance management will be recognized under three principal categories: (1) management by moderation, (2) management by saturation, and (3) management by multiple attack (Table 2). The terms *moderation* and *saturation* in resistance management were introduced by Sutherst and Comins (1979) to express the use of contrastingly low or high dosages such that the target population is either spared severe depletion of susceptible genes or is entirely annihilated. The term *multiple attack* is introduced here to signify the application of multidirectional chemical selection pressure either on a short- or long-term basis. As explained later, these approaches are not totally exclusive of each other, and ingredients of each may be utilized in the same control program.

MANAGEMENT BY MODERATION

This approach recognizes that susceptible genes are a valuable resource that must be conserved, and it attempts to accomplish this through reduction of the selection pressure.

The process of selection may be illustrated by frequency distribution curves for susceptible (SS), hybrid (RS) and resistant (RR) individuals (Fig. 1). Initially, the susceptible genotypes are by far the most common, with only rare occurrence of an individual that is heterozygous for resistance. The frequencies of R genes in unselected wild populations have been assumed, based on mutation rates, to range from 0.001 to 0.0001 (Georghiou and Taylor, 1977a). Insecticides are usually applied at dosages that are lethal to the susceptible individuals but spare the heterozygotes and the homozygous resistant insects, thus progressively shifting the relative frequency of genotypes in favor of the resistant.

By applying a lower dose than that which kills all susceptible individuals, say LD_{90} or less, sufficient susceptible genes are preserved in the population to delay the onset of resistance. Likewise, incomplete coverage allows susceptible individuals to survive in untreated pockets (called "refugia") (Georghiou and Taylor, 1977a,b). Furthermore, setting a higher population density threshold for insecticide application results in fewer treatments, and hence in lower overall selection pressure.

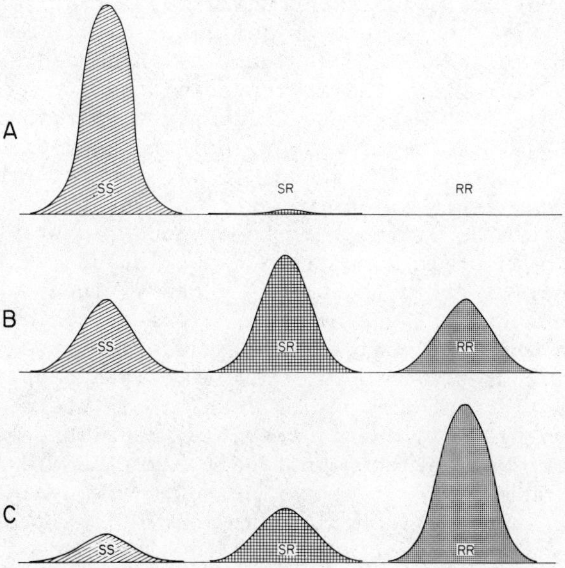

Figure 1. Diagrammatic representation of progressive changes in frequency of susceptible (SS), heterozygous resistant (SR) and homozygous resistant (RR) individuals with continued selection pressure (A to C).

These measures of resistance management by moderation (Table 2,A) may appear to be extreme and *in toto* they may be impracticable, but it should be remembered that the extent to which one or more may be necessary would depend on the "resistance risk" that has been ascribed to the target population through consideration of the genetic and biological factors listed in Table 1. Such measures would normally require the integration of a strong component of non-chemical pest control practices.

MANAGEMENT BY SATURATION

While management by moderation comes close to meeting environmental standards and is less destructive to biological controls, it may not be appealing where high-value crops are involved or where insect vectors of disease must be kept at very low densities. For these situations, a different approach to resistance management may be considered, namely "management by saturation." The term saturation here does not imply saturation of the environment by pesticides, but rather indicates the saturation of the defense mechanisms of the insect by dosages that can overcome resistance.

Rendering Resistant Genes Functionally Recessive

It has been shown by computer simulation (Georghiou and Taylor, 1977a) as well as by laboratory tests and selection experiments (Davidson and Pollard, 1958; MacDonald, 1959) that resistance develops rapidly if the R gene is dominant, but it is slowed considerably if the R gene is recessive (Georghiou and Taylor, 1977a). Management by saturation aims at rendering the R gene functionally recessive by applying dosages sufficiently high to be lethal to susceptible as well as heterozygous-resistant individuals (Curtis et al., 1978; Taylor and Georghiou, 1979). When the heterozygotes are killed, the R genes are eliminated and resistance does not evolve. It should be stated that homozygous-resistant genotypes are unlikely to exist in an untreated population due to the extremely low initial frequency of the R gene. Thus, while this approach appears promising against unselected populations, it would be inadvisable where selection has already given rise to homozygous-resistant individuals with a high degree of phenotypic resistance. Likewise, this approach may be applicable only where a high dose of a rapidly decaying pesticide is feasible, as with certain fumigants, or where a compound lacking significant mammalian toxicity, such as a juvenile hormone mimic or bacterial toxin, is available. There is certainly a need for technological innovations that would make possible the application of sufficiently high doses only to the target pest through microencapsulation, systemic uptake, use of attractants, or by some other means.

Suppression of Detoxication by Synergists

Synergists act by inhibiting specific detoxication enzymes and thus are capable of reducing or eliminating the selective advantage of individuals possessing such enzymes. This potential advantage of synergism was recognized early through studies on the synergism of DDT by chlorfenethol, a competitive inhibitor of dehydrochlorinase (March, 1952). It is perhaps unfortunate that this pioneering work was done on the house fly, a species possessing remarkable propensity for biochemical adaptation, since resistance to the combination of DDT and chlorfenethol was soon demonstrated (Moorefield and Kearns, 1955). Likewise, laboratory selection of house flies by carbaryl and the oxidase inhibitor piperonyl butoxide (p.b.) led gradually to the development of high resistance to the combination (Georghiou, 1962). Against the cattle tick, carbaryl synergized by p.b. was shown to be effective, even against the highly resistant *Biarra* strain (Schuster et al., 1974), but the relatively high cost of the synergist, problems with formulations, and the risk of reduction of the margin of mammalian safety have mitigated against its use (Nolan and Roulston, 1979).

The lasting utility of synergists as a means of inhibiting the evolution of resistance would obviously depend on the absence of an efficient, alternative mechanism of resistance in the target population. Recently, we demonstrated the application of this principle in the mosquito *Culex quinquefasciatus* using a strain that possesses high resistance to several organophosphates. The use of p.b. in combination with several of these insecticides, including oxon analogs, failed to produce synergism, thus indicating that this resistance was not due to an oxidase (Table 3). In contrast, treatment in combination with the esterase inhibitor DEF® (*S*,*S*,*S*-tributyl phosphorotrithioate) reduced resistance almost to the level found in the susceptible strain (Georghiou et al., 1975). Having thus determined that the strain possessed only esterases as a significant mechanism of resistance, we then showed by selection of substrains with temephos, temephos + p.b. or temephos + DEF® during 12 consecutive generations that resistance did not evolve when the insecticide was used jointly with DEF® but that it advanced to higher levels when the insecticide was used alone or in combination with p.b. (Table 4) (Ranasinghe and Georghiou, 1979).

Interest in identifying new inhibitors of resistance mechanisms has recently been revived. Perhaps the most significant new discovery of a synergist concerns IBP (Kitazin-P®, *O*,*O*-diisopropyl *S*-benzyl phosphorothiolate), a commercial fungicide employed in the control of the rice blast disease (Yoshioka et al., 1975). IBP has been shown to be a strong synergist of malathion in malathion-resistant strains through inhibition of carboxylesterase (Miyata et al., 1980). The lower synergistic effect of IBP on organophosphates that do not possess carboxylester groups would suggest that the

Table 3. Effect of Synergists p.b. and DEF® on Resistance to Organophosphates in *Culex quinquefasciatus* [a/]

	Susceptible Strain		Resistant Strain		Resistance Ratio at LC_{50} Insecticide + Synergist
	Insecticide alone LC_{50} (mg/L)	Insecticide + Synergist LC_{50} (mg/L)	Insecticide alone LC_{50} (mg/L)	Insecticide + Synergist LC_{50} (mg/L)	
		Synergist: piperonyl butoxide			
methyl parathion	0.005	0.0048	0.12	0.12	25.0
methyl paraoxon	0.025	0.024	0.4	0.31	12.9
chlorpyrifos	0.0023	0.0020	0.12	0.094	47.0
		Synergist: DEF			
methyl parathion	0.005	0.0012	0.12	0.0016	1.3
methyl paraoxon	0.025	0.0057	0.4	0.007	1.2
parathion	0.0041	0.0012	0.053	0.0028	2.3
chlorpyrifos	0.0023	0.0024	0.12	0.0013	0.54

[a/] Adapted from Georghiou et al., 1975

Table 4. Effect of Selection by Temephos Alone or in Combination with Synergists on Evolution of Organophosphate Resistance in *Culex quinquefasciatus*[a]

	Resistance Factor[b]				
Insecticide	Parental strain	Temephos-selected (F_{12})	Temephos + PB-selected (F_{12})	Temephos + DEF-selected (F_{12})	Not selected (F_{12})
temephos	117	322	122	3.1	3.1
chlorpyrifos	52	52	52	2.8	2.5
parathion	14	33	17	3.7	1.4
methyl parathion	24	36	28	1.3	2.8

[a] Adapted from Ransinghe and Georghiou, 1979.

[b] Resistance factor = LC_{50} of selected strain ÷ LC_{50} of susceptible strain.

compound may also inhibit such other detoxication pathways as those involving GSH S-transferase or phosphotriesterase.

MANAGEMENT BY MULTIPLE ATTACK

This group of chemical countermeasures aims at achieving control through the action of several independently acting forces such that the selection pressure by any one of them would be below that required for development of resistance. The concept is derived from the multi-site action of toxicants employed in earlier years for the control of insects and plant diseases, such as arsenicals, copper sulfate, etc. These chemicals, although not totally immune to development of resistance, nonetheless have demonstrated prolonged use-life, allegedly due to their effect on more than one biochemical site of action. Of course, we cannot return to the arsenicals, but an artificial means of achieving a multi-site attack is by the use of insecticides in mixtures, and on a longer-term basis, in rotations. Mixtures and rotations as a countermeasure for resistance are mentioned frequently in the literature (Brown, 1961a, 1971, 1973, 1976; Busvine, 1957; Crow, 1952, 1960) but for reasons explained below there is no concensus as to their practical usefulness.

Insecticide Mixtures

The concept of using mixtures as an anti-resistance measure assumes that the mechanisms for resistance to each member chemical are different and that they exist at such low frequencies that they do not occur together in any single individual within a given population. Thus, insects that survive one of the chemicals in the mixture are killed by another.

This statement, however, is an oversimplification of the concept, since there are a number of requirements that must be met if the mixture is to have even a remote probability of success. Synergistic action between the components of the mixture would reduce or eliminate the differential advantage of individuals possessing the respective resistance mechanisms and thus would enhance the probability of success. Such action would also have economic advantages: Nolan (*in* Nolan and Roulston, 1979) reported that in field trials against *Boophilus microplus*, a combination of ethion and pyrethroid required only 50% and 25% of the normal dose, respectively, for equivalent toxicity. Additionally, the components of the mixture must have approximately similar decay rates, or preferably possess short environmental stability. Use of the mixture should begin early, before resistance to one of the components has been selected. However, this requirement would not be applicable if the mixture consists of a pair of compounds that display negatively correlated toxicity (Ogita, 1958, 1961a,b; reviews by Brown, 1961b; Georghiou, 1966), i.e., resistance to one of the members is accompanied by enhanced susceptibility to the other and vice versa.

Perusal of pertinent literature reveals that there are more papers discussing the value of mixtures (as well as of rotations) than those that report actual research on the subject. This is probably because various mixtures of insecticides have been used for years to control multiple infestations, but have not caused any readily apparent delay in the appearance of resistance. It must be borne in mind, however, that the concept of mixtures as an anti-resistance measure requires considerably greater sophistication in the choice of compounds, formulations, and application methods than the mere combination of more than one chemical in the spray tank. The results of published studies and observations on mixtures show positive, negative, or no effect on resistance, but in a small number of cases in which chemicals with contrasting modes of action or detoxication pathways were employed, some delay in the onset of resistance or in the level attained was noted (Asquith, 1961; Burden et al., 1960; Graves et al., 1967; Ku, 1978; Ozaki et al., 1973; Takahashi, 1979). In the extreme case reported by Pimentel and Bellotti (1976), no resistance was obtained during selection of house flies in the laboratory with a mixture of six inorganic compounds, which contrasted sharply with the results obtained when each compound was used alone (Fig. 2).

Considerable interest in the concept of mixtures, as well as of rotations, is also apparent in other areas of plant and animal protection, as for example in the control of plant pathogens

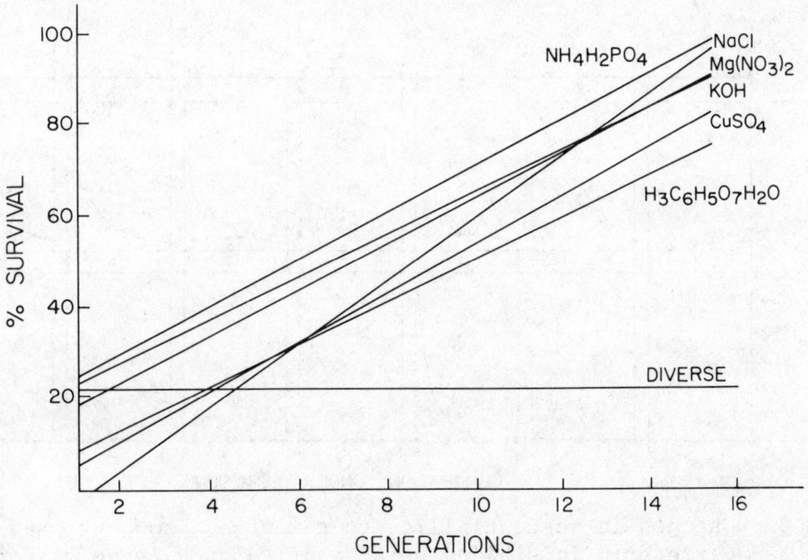

Figure 2. Results of laboratory selection of house fly colonies by inorganic chemicals alone and by their mixture. (After Pimentel and Bellotti, 1976.)

(Dekker, 1976; Delp, 1979; Kable and Jeffery, 1980; and papers by Gilpatrick and Ogawa et al. in this volume) and helminthic parasites of animals (Prichard et al., 1980).

The joint use of chemicals was recently investigated in our laboratory with three commercial insecticides that lack significant cross resistance: temephos, propoxur and permethrin. A strain of *Culex quinquefasciatus* was synthesized that contained genes for resistance to each of these compounds at the low frequency of 0.02. Following six generations of inbreeding, subcolonies of this strain were then placed under selection pressure with each compound separately and in various combinations. After nine generations each of the colonies selected by a single compound had developed high resistance to that compound, thus confirming that the respective R genes had been incorporated into the synthetic strain. However, the colonies selected by combinations developed some resistance only toward propoxur when this carbamate was part of the combination. Resistance to temephos and permethrin was suppressed by the combinations and dosages used (Fig. 3) (Lagunes and Georghiou, in prep.).

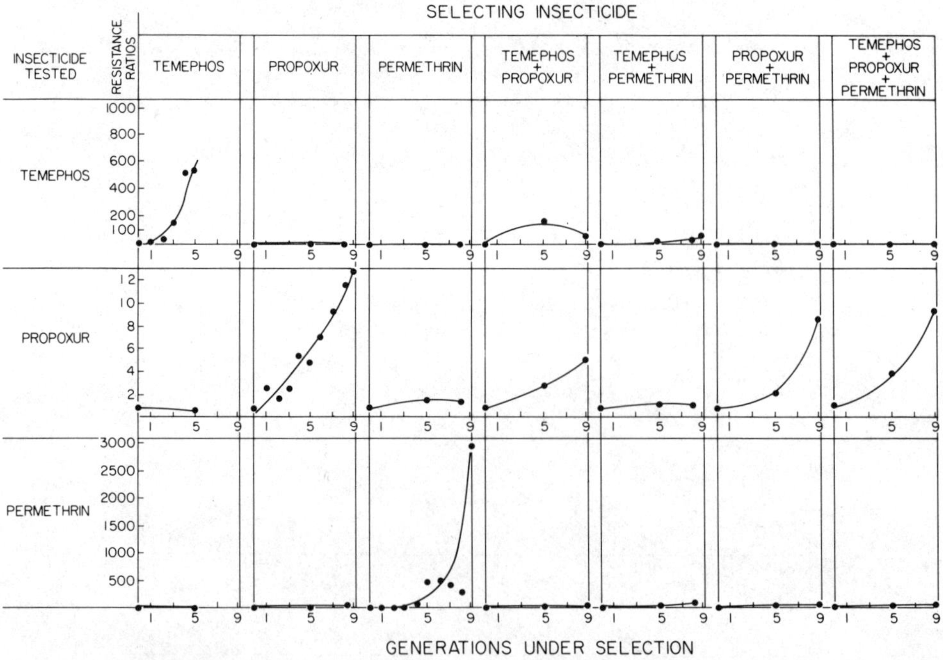

Figure 3. Changes in susceptibility to temephos, propoxur, and permethrin in sibling strains of *Culex quinquefasciatus* in the course of selection by these insecticides singly or in various combinations (data from Lagunes 1980, by permission).

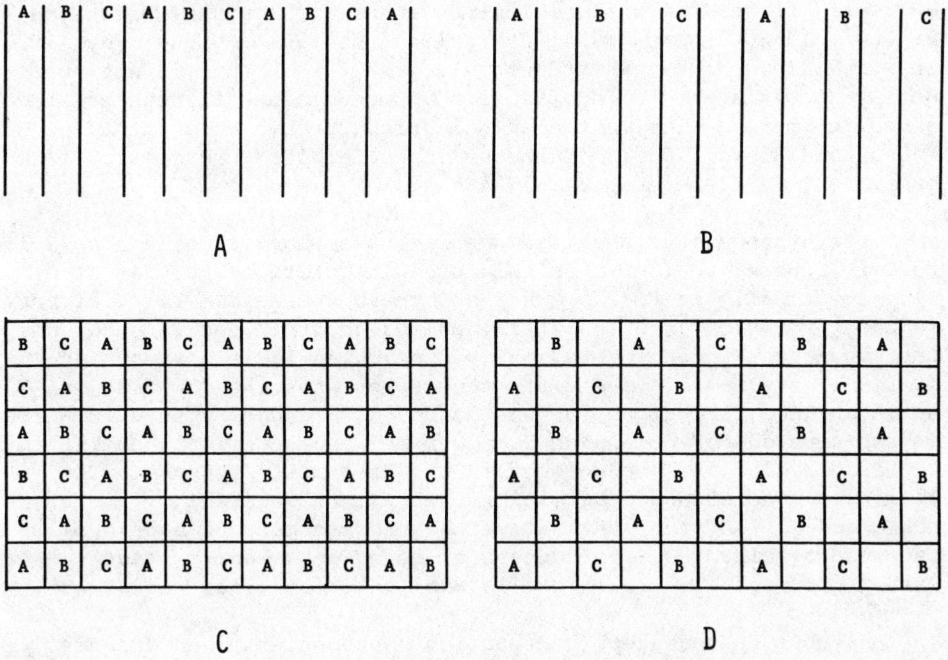

Figure 4. Application of insecticides in a fixed pattern of strips or blocks for malaria vector control. Unsprayed buffer strips or blocks may serve as refugia for "reversion" or may be subjected to non-chemical control measures (Muir, 1977, by permission).

An intriguing variation of the concept of mixtures that has potential use for malaria vector control is the application of unrelated insecticides in different sectors of a "mosaic" or grid pattern (Muir, 1977) (Fig. 4). The objective of this strategy is to avoid selection of the population for the same resistance mechanism in all regions of the treated area so that migrants that have not been killed in their sector of origin will be killed upon exposure to the insecticide used in the adjoining sector. The effectiveness of such strategy will depend on a reasonably high rate of exchange of migrants between sectors (Curtis and Rawlings, 1980). To ensure double exposure, the two insecticides may conceivably be applied on opposite walls of each house.

Insecticide Rotations

The concept of rotation of chemicals as an anti-resistance measure assumes that individuals that are resistant to one chemical have substantially lower biotic fitness than susceptible individuals, so that their frequency declines during the intervals between applications of that chemical. There are numerous studies indicating depressed fitness in insecticide-resistant arthropods (Bhatia and Pradhan, 1968; Brower, 1974; Ferrari and Georghiou, 1980; Rey, 1972; Shaw and Lloyd, 1969; Whitten et al., 1980), but this is not necessarily a consistent or invariable phenomenon since fitness has been shown to improve with continued selection through co-adaptation (Abedi and Brown, 1960; McEnroe and Naegele, 1968, reviews by Georghiou, 1972; Keiding, 1963, 1967).

The hypothetical oscillations in susceptibility of a population that is exposed to four chemicals used in rotation are presented diagrammatically in Fig. 5. Resistance to compound A rises slightly in the generation in which it is applied and then declines during the subsequent three generations during which it is not applied. It rises once more in the fifth generation when the compound is again applied, but drops in generations 6, 7 and 8. The same pattern is anticipated for compounds B, C, and D, each chemical oscillating one step behind its predecessor in the treatment sequence. The challenge consists of determining the optimal sequence of use of the chemicals and the stage at which a change must be made. As in the case of mixtures, the concept of rotations requires that the member chemicals are reciprocally unaffected by cross resistance.

Although the concept of rotation has been examined experimentally or empirically in only a small number of cases (Cutright, 1959; Graves et al., 1967; Ozaki, 1969; Ozaki et al., 1973), it recently has received increasing attention. Heather (1979) noted that the use of fumigants to eliminate infestations of grain pests that developed during storage has the effect of alternating chemical measures that are uncorrelated. He felt that the long-standing practice of fumigation, for example, had done much to slow the emergence of resistance to malathion, which had been applied to

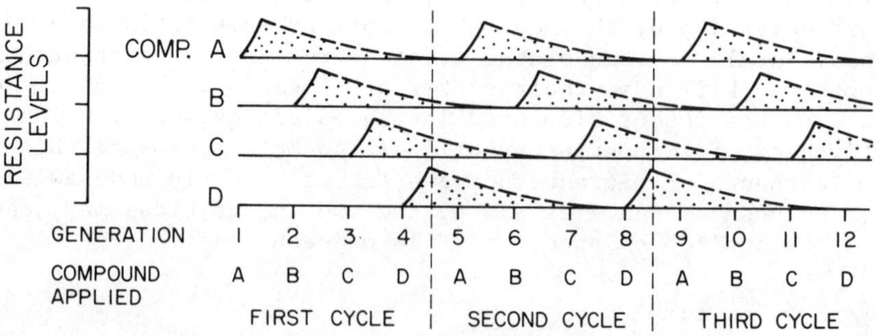

Figure 5. Hypothetical oscillations in susceptibility of a population that is exposed to four unrelated chemicals (A-D) used in rotation against succeeding generations (1-12) (Georghiou 1980, by permission).

grains on the farms. Against the western corn rootworm, *Diabrotica virgifera*, a rotation of carbofuran and organophosphate insecticides on an annual basis, was recommended as a means of arresting the development of carbamate resistance (Kantack et al., 1976).

In our laboratory work cited earlier (Fig. 3), the three strains that were selected by temephos, propoxur or permethrin showed distinctly different rates of loss of resistance when removed from selection pressure: temephos resistance declined rapidly, propoxur resistance slowly, and permethrin resistance declined at an intermediate rate (Lagunes and Georghiou, in prep.). Since the three strains had originated from the same "synthetic" population following a period of inbreeding and were thus presumed to have similar fitness qualities, the results suggest that the rate of reversion toward susceptibility may also be a function of the resistance gene itself.

In subsequent work, the rotation of selection by temephos, permethrin and propoxur was investigated on substrains of the synthetic strain in all six possible sequences. Each substrain was selected with the three compounds for one full cycle, in each case a change to the next compound being made after approximately five generations of selection or when substantial resistance to the selecting compound had developed.

The most interesting observation emerging from these selections was the rapid regression of resistance to temephos or permethrin when selection by one of these two chemicals was followed by selection by the other. Thus, temephos resistance regressed rapidly when temephos was replaced by permethrin as the selecting agent, and vice versa, and this regression appeared to be somewhat faster than was observed in strains that were removed completely from selection pressure. This reciprocal relationship did not appear to exist between propoxur and permethrin or between propoxur and temephos (Lagunes and Georghiou, in prep).

The rapid regression of temephos resistance during selection by permethrin is especially noteworthy since instances of negative cross resistance to pyrethroids have been observed in organophosphate-resistant *Culex quinquefasciatus* and *Cx. tarsalis* (Priester et al., 1980), *Cx. pipiens* (Weide Liu, personal communication), *Tetranychus urticae* (Chapman and Penman, 1979) and *Nephotettix cincticeps* (Ozaki, this volume).

CONCLUSION

It is obvious that generalized recommendations for solution of the resistance problem would be difficult if not impossible to expect in view of the considerable genetic, biological, and ecological diversity that exists in natural populations. For the near future, I visualize instead the elaboration of guidelines for the assessment of resistance risks based on the biological and ecological characteristics of the target population and the incorporation of resistance-delaying tactics as a vital component of integrated pest management programs. The essence of these tactics would be *moderation* in the use of pesticides, but other strategies as described under the principles of *saturation* and *multiple attack* should also prove useful under certain situations. The concepts of use of insecticides in mixtures, rotations or optimal sequences may be limited in many instances by economic or practical considerations. However, where control measures are applied on a large scale and are centrally coordinated, these concepts may present distinct advantages as means of delaying or averting the evolution of resistance, especially if supplemented by other integrated control measures.

My thoughts on the categories of inputs needed for optimization of resistance management are summarized diagrammatically in Fig. 6. There is need for new types of toxophores to provide greater diversity in available pesticides. Recent advances in natural product chemistry indicate that the discovery of new biochemical lesions need not depend on serendipity. There is also need for new synergists to serve as

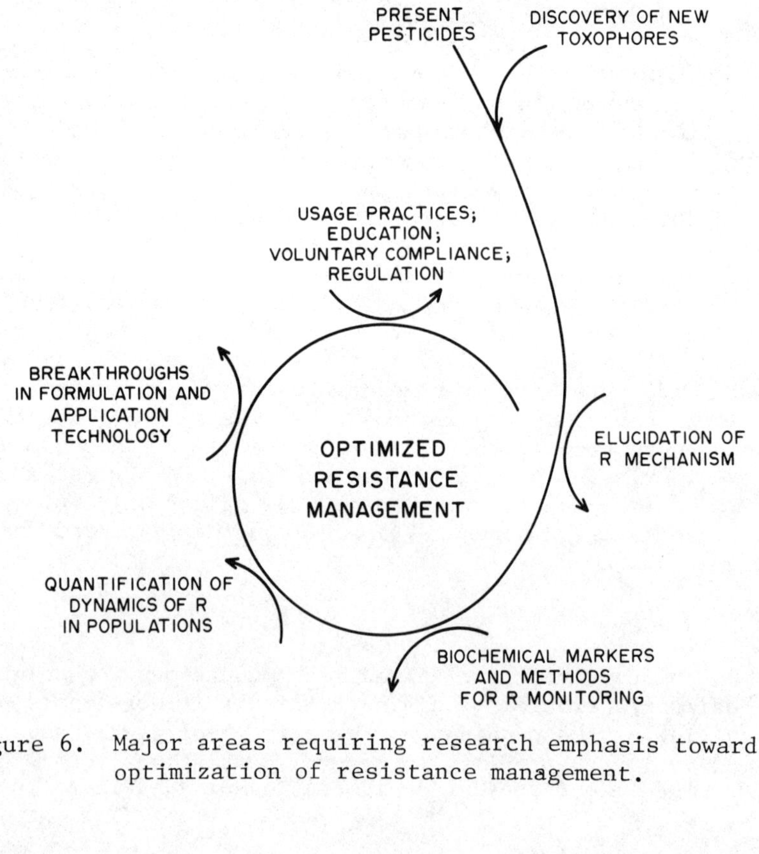

Figure 6. Major areas requiring research emphasis toward optimization of resistance management.

suppressors of resistance mechanisms. As new toxophores are discovered they must be assessed against reference strains that represent the presently known resistance mechanisms, and subsequently, through accelerated selection of representative populations, the type of potential resistance to these new chemicals can be determined. Simple, biochemical or toxicological tests for quick and reliable detection of resistance are needed for each type of pesticide based on the mechanisms of resistance involved. Recent discoveries of such simple tests for detection of organophosphate detoxifying esterases, and of "insensitive" acetylcholinesterase have improved the capability of detecting resistant genes at low frequencies.

There is need for breakthroughs in insecticide formulation and application technology so that the required "effective" dose can be placed on target in a manner consistent with the saturation concept described earlier. Such breakthroughs could also provide the means for joint application of insecticides or of insecticide/attractant or insecticide/synergist combinations at rates needed for optimal action and with stability characteristics that exclude unnecessarily prolonged selection of the target population.

There is no doubt that whether intentionally or not, pesticides are being abused and that many cases of resistance are the consequence of such abuse. Thus, any strategy for resistance management would require a strong advisory effort aimed at both the insecticide user and the insecticide marketer. Since some anti-resistance strategies may involve the acceptance of lower short-term benefits for the sake of long-term advantages, the question of voluntary compliance vs regulation will inevitably arise. Such potential complications and dilemmas emphasize even more the need for effective and rational resistance management strategies and ensure that the phenomenon of resistance will continue to challenge man's resourcefulness and ingenuity for many years to come.

REFERENCES

Abedi, Z. H., and Brown, A. W. A., 1960, Development and reversion of DDT resistance in *Aedes aegypti*, *Can. J. Genet. Cytol.*, 2:252.

Asquith, D., 1961, Methods of delaying selection of acaricide resistant strains of the European red mite, *J. Econ. Entomol.*, 54: 439.

Babers, F. H., 1949, Development of insect resistance to insecticides, *USDA, E-778*, 31 pp.

Bhatia, S. K., and Pradhan, S., 1968, Studies on resistance to insecticides in *Tribolium castaneum* (Herbst). III. Selection of a strain resistant to p,p' DDT and its biological characteristics, *Indian J. Entomol.*, 30:13.

Brower, J. H., 1974, Radiosensitivity of an insecticide-resistant strain of *Tribolium castaneum* (Herbst), *J. Stored Prod. Res.*,

10:129.
Brown, A. W. A., 1957, Insecticide-resistance and Darwinism, *Botyu Kagaku*, 22:277.
Brown, A. W. A., 1961a, The challenge of insecticide resistance, *Bull. Entomol. Soc. Amer.*, 7:6.
Brown, A. W. A., 1961b, Negatively correlated insecticides, *Pest Control*, 29:24.
Brown, A. W. A., 1968, Insecticide resistance comes of age, *Bull. Entomol. Soc. Amer.*, 14:3.
Brown, A. W. A., 1971, Pest resistance to pesticides, in: "Pesticides in the Environment," R. White-Stevens, ed., Vol. 1, Part 11, p. 457-552, Marcel Dekker, N.Y.
Brown, A. W. A., 1974, Insect resistance, IV. Countermeasures for resistance, *Farm Chemicals*, 127:1.
Brown, A. W. A., 1976, How have entomologists dealt with resistance? *Proc. Amer. Phytopath. Soc.*, 3:67.
Burden, G. S., Lofgren, C. S., and Smith, C. N., 1960, Development of chlordane and malathion resistance in the German cockroach. *J. Econ. Entomol.*, 53:1138.
Busvine, J. R., 1957, Insecticide-resistant strains of insects of public health importance, *Trans. Roy Soc. Trop. Med. Hyg.*, 51:11.
Chapman, R. B., and Penman, D. R., 1979, Negatively correlated cross-resistance to a synthetic pyrethroid in organophosphorus-resistant *Tetranychus urticae*, *Nature* (London), 281:298.
Comins, H. N., 1977, The development of insecticide resistance in the presence of migration, *J. Theoret. Biol.*, 64:177.
Crow, J. F., 1952, Some genetic aspects of selection for resistance, *Nat. Acad. Sci., Nat. Res. Council*, Publ. 219:72.
Crow, J. F., 1957, Genetics of insect resistance to chemicals, *Ann. Rev. Entomol.*, 2:227.
Crow, J. F., 1960, Genetics of insecticide resistance, general considerations, *Misc. Publ. Entomol. Soc. Amer.*, 2:69.
Crow, J. F., 1966, Evolution of resistance in hosts and pests, *Nat. Acad. Sci., Nat. Res. Council*, Publ. 1402:263.
Curtis, C. F., and Davidson, G., 1980, Possible means of inhibiting the evolution of insecticide resistance by mosquitoes, Unpublished document *Wld. Hlth. Org.*, WHO/EC/80.32, 12 p.
Curtis, C. F., and Rawlings, P., 1980, A preliminary study of dispersal and survival of *Anopheles culicifacies* in relation to the possibility of inhibiting the spread of insecticide resistance, *Ecol. Entomol.*, 5:11.
Curtis, C. F., Cook, L. M., and Wood, R. J., 1978, Selection for and against insecticide resistance and possible methods of inhibiting the evolution of resistance in mosquitoes, *Ecol. Entomol.*, 3:273.
Cutright, C. R., 1959, Rotational use of spray chemicals in insect and mite control, *J. Econ. Entomol.*, 52:432.
Davidson, G., and Pollard, D. G., 1958, Effect of simulated field deposits of gamma BHC and dieldrin on susceptible, hybrid and

resistant strains of *Anopheles gambiae*, *Nature*, 182:739.

Dekker, J., 1976, Prospects for the use of systemic fungicides in view of the resistance problem, *Proc. Amer. Phytopathol. Soc.*, 3:60.

Delp, C. J., 1979, Resistance to plant disease control agents: How to cope with it, *Proc. 9th Int. Congr. Plant Prot.*, (in press).

Ferrari, J. A., and Georghiou, G. P., 1980, Effect of insecticidal selection and treatment on reproductive potential of resistant, susceptible, and heterozygous strains of the southern house mosquito, *J. Econ. Entomol.*, (in press).

Georghiou, G. P., 1962, Carbamate insecticides: Toxic action of synergized carbamates against twelve resistant strains of the house fly, *J. Econ. Entomol.*, 55:768.

Georghiou, G. P., 1965, Genetic studies on insecticide resistance, *Adv. Pest Control Res.*, 6:171.

Georghiou, G. P., 1972, The evolution of resistance to pesticides, *Ann. Rev. Ecol. and Systematics*, 3:133.

Georghiou, G. P., 1980, Insecticide resistance and prospects for its management, *Residue Reviews*, 76:131.

Georghiou, G. P., and Taylor, C. E., 1976, Pesticide resistance as an evolutionary phenomenon, *Proc. XV Int. Congr. Entomol.*, p. 759.

Georghiou, G. P., and Taylor, C. E., 1977a, Genetic and biological influences in the evolution of insecticide resistance, *J. Econ. Entomol.*, 70:319.

Georghiou, G. P., and Taylor, C. E., 1977b, Operational influences in the evolution of insecticide resistance, *J. Econ. Entomol.*, 70:653.

Georghiou, G. P., and Taylor, C. E., 1979, Suppression of insecticide resistance by alteration of gene dominance and migration, *J. Econ. Entomol.*, 72:105.

Graves, J. B., Roussel, J. S., Gibbens, J., and Patton, D., 1967, Laboratory studies on the development of resistance and cross-resistance in the boll weevil, *J. Econ. Entomol.*, 60:47.

Greever, J., and Georghiou, G. P., 1979, Computer simulation of control strategies for *Culex tarsalis* (Diptera:Culicidae), *J. Med. Entomol.*, 16:180.

Heather, N. W., 1979, Aspects of the biology of malathion-resistant *Sitophilus oryzae* L., Ph.D. dissertation, Univ. of Queensland, 243 pp.

Hueth, D., and Regev, U., 1974, Optimal agricultural pest management with increasing pest resistance, *Am. J. Agric. Econ.*, 56:543.

Kable, P. F., and Jeffery, H., 1980, Selection for tolerance in organisms exposed to sprays of biocide mixtures: A theoretical model, *Phytopathology*, 70:8.

Kantack, B. H., Walgenbach, D. D., and Berndt, W. L., 1976, New emphasis on need to rotate chemicals for rootworm control, So. Dakota Insect Newsletter, So. Dakota Coop. Ext. Serv. No. 16, 2 p.

Keiding, J., 1963, Possible reversal of resistance, *Bull. Wld. Hlth. Org.*, Suppl. 29:51.

Keiding, J., 1967, Persistence of resistant populations after the relaxation of the selection pressure, *World Rev. Pest Contr.*, 6:115.

Ku, Te-yeh, 1978, Continued use of pesticides and its biological effects in rice, insect and disease control, (in press).

Lagunes, A. T., 1980, Impact of the use of mixtures and sequences of insecticides in the evolution of resistance in *Culex quinquefasciatus* Say (Diptera:Culicidae), Ph.D. dissertation, U. of Calif., Riverside, 209 p.

MacDonald, G., 1959, The dynamics of resistance to insecticides by Anophelines, *Riv. Parassitol.*, 20:305.

March, R. B., 1953, Synergists for DDT against insecticide-resistant house flies, *J. Econ. Entomol.*, 45:851.

McEnroe, W. D., and Naegele, J. A., 1968, The coadaptive process in an organophosphorus-resistant strain of the two-spotted spider mite, *Tetranychus urticae, Ann. Entomol. Soc. Amer.*, 61:1055.

Melander, A. L., 1914, Can insects become resistant to sprays? *J. Econ. Entomol.*, 7:167.

Metcalf, R. L., 1955, Physiological basis for insect resistance to insecticides, *Physiol. Rev.*, 35:197.

Milani, R., 1957, Ricerche genetiche sulla resistanza degli insetti alla azione delle sostanze tossiche, *Rend. Instituto Superiore Sanita*, Rome, 20:713.

Milani, R., 1959, Genetical considerations on insect resistance to insecticides, *Genetica Agratia*, 10:288.

Milani, R., 1964, Genetics and sanitary entomology, Symposia Genetica et Biologica Italica, Pavia, 13:178.

Miyata, T., Sakai, H., Saito, T., Yoshioka, K., Ozaki, K., Sasaki, Y., and Tsuboi, A., 1980, Mechanism of joint toxic action of Kitazin-P® with malathion in the malathion-resistant green rice leafhopper, *Nephotettix cincticeps* Uhler, *Appl. Ent. Zool.*, (in press).

Moorefield, H. H., and Kearns, C. W., 1955, Mechanism of action of certain synergists for DDT against resistant house flies, *J. Econ. Entomol.*, 48:403.

Muir, D. A., 1977, Genetic aspects of developing resistance of malaria vectors. 2. Gene flow and control pattern. Unpublished *Wld. Hlth. Org.* document WHO/VBC/77.659, 10 p.

Nolan, J., and Roulston, W. J., 1979, Acaricide resistance as a factor in the management of Acari of medical and veterinary importance, *Recent Adv. Acarol.*, 11:3.

Ogita, Z., 1958, A new type of insecticide, *Nature*, 182:1529.

Ogita, Z., 1961a, Relationship between the structure of compounds and negatively correlated activity: Genetical and biochemical studies on negatively correlated cross-resistance in *Drosphila melanogaster*, II, *Botyu Kagaku*, 36:20.

Ogita, Z., 1961b, Genetical studies on actions of mixed insecticides with negatively correlated substances: Genetical and biochemical studies on negatively correlated cross-resistance in *Drosophila melanogaster*. III, *Botyu Kagaku* 26:88.

Ozaki, K., 1969, Resistant insect pests of rice plant and countermeasures for their control, *Agr. and Hort.*, 44:213.

Ozaki, K., Sasaki, Y., Ueda, M., and Kassai, T., 1973, Results of the alternate selection with two insecticides and the continuous selection with mixtures of two or three ones of *Laodelphax striatellus* Fallen., *Botyu-Kagaku*, 38:222.

Pimentel, D., and Bellotti, A. C., 1976, Parasite-host population systems and genetic stability, *Amer. Natur.*, 110:877.

Plapp, F. Jr., Browning, C. R., and Sharpe, P. J. H., 1979, Analyses of rate of development of resistance based on a genetic model. *Environ. Entomol.*, 8:494.

Prichard, R. K., Hall, C. A., Kelly, J. D., Martin, I. C. A., and Donald, A. D., 1980, The problem of antihelmintic resistance in nematodes, *Austr. Vet. J.*, 56:239.

Priester, T. M., Georghiou, G. P., Hawley, M. K., and Pasternak, M. E., 1981, Toxicity of pyrethroids to organophosphate-, carbamate- and DDT-resistant mosquitoes, *Mosquito News*, 41:143.

Quayle, H. J., 1922, Resistance of certain scale insects in certain localities to HCN fumigation, *J. Econ. Entomol.*, 15:400.

Ranasinghe, L. E., and Georghiou, G. P., 1979, Comparative modification of insecticide resistance spectrum of *Culex pipiens fatigans* Wied. by selection with temephos and temephos/synergist combinations, *Pestic. Sci.*, 10:502.

Schuntner, C. A., Roulston, W. J., and Wharton, R. H., 1974, Toxicity of piperonyl butoxide to *Boophilus microplus*, *Nature* (London), 249:386.

Shaw, D. D., and Lloyd, C. J., 1969, Selection for lindane resistance in *Dermestes maculatus* DeGeer (Coleoptera:Dermestidae), *J. Stored Prod. Res.*, 5:69.

Srivastava, H. M. L., and Roy, R. G., 1972, Susceptibility studies of *Anopheles culicifacies* to DDT in Tamil Nadu, India, *J. Comm. Dis.*, 4:112.

Sutherst, R. W., and Comins, H. N., 1979, The management of acaricide resistance in the cattle tick, *Boophilus microplus* (Canestrini) (Acari:Ixodidae), in Australia, *Bull. Entomol. Res.*, 69:519.

Sutherst, R. W., Norton, G. A., Barlow, N. D., Conway, G. R., Birley, M., and Comins, H. N., 1979, An analyses of management strategies for cattle tick (*Boophilus microplus*) control in Australia, *J. Appl. Ecol.*, 16:359.

Takahashi, Y., 1979, Present status of insecticides for controlling the resistant green rice leafhopper, *Japan Pestic. Inf.*, 36:22.

Taylor, C. E., and Georghiou, G. P., 1979, Suppression of insecticide resistance by alteration of gene dominance and migration. *J. Econ. Entomol.*, 72:105.

Taylor, C. E., and Georghiou, G. P., 1980, The influence of pesticide persistence in the evolution of resistance, *Environ. Entomol.*, (in press).

Whitten, M. J., Dearn, J. M., and McKenzie, J. A., 1980, Field studies on insecticide resistance in the Australian sheep blowfly, *Lucilia caprina*, *Aust. J. Biol. Sci.*, (in press).

Woglum, R. S., 1925, Observations on insects developing immunity to insecticides, *J. Econ. Entomol.*, 18:593.

Wool, D., and Manheim, O., 1980, Genetically induced susceptibility to malathion in *Tribolium castaneum* despite selection for resistance, *Entomol. Exp. & Appl.*, 28:183.

Yoshioka, K., Matsumoto, M., Bekku, I., and Kanamori, S., 1975, Synergism of IBP and insecticides against insecticide-resistant green rice leafhopper, *Nephotettix cincticeps* Uhler, *Proc. Assoc. Plant Prot. Sikoku*, 10:49.

CONTRIBUTORS

Mitsuo Asada
Nippon Soda Co., Ltd.
Oiso-machi, Kanagawa, Japan

G. M. Booth
Brigham Young University
Provo, Utah

W. S. Bradshaw
Brigham Young University
Provo, Utah

S. D. Burton
Brigham Young University
Provo, Utah

B. A. Croft
Michigan State University
East Lansing, Michigan

Walter C. Dauterman
North Carolina State University
Raleigh, North Carolina

Joseph W. Eckert
University of California
Riverside, California

Minoru Fukada
Nihon Nohyaku Co., Ltd.
Osaka, Japan

T. Roy Fukuto
University of California
Riverside, California

George P. Georghiou
University of California
Riverside, California

J. D. Gilpatrick
Cornell University
Geneva, New York

Hiroshi Hama
National Institute of Agricultural Sciences
Tsukuba, Ibaraki, Japan

Bruce D. Hammock
University of California
Riverside, California

Akifumi Hayashi
Tokyo Medical and Dental University
Tokyo, Japan

C. R. Heaton
University of California
Davis, California

W. M. Hess
Brigham Young University
Provo, Utah

Yoshio Hisada
Sumitomo Chemical Co., Ltd.
Takarazuka, Hyogo, Japan

Ernest Hodgson
North Carolina State University
Raleigh, North Carolina

S. N. Irving
University of California
Riverside, California

Rokuro Kano
Tokyo Medical and Dental University
Tokyo, Japan

Toshiro Kato
Sumitomo Chemical Co., Ltd.
Takarazuka, Hyogo, Japan

Yasuo Kawase
Sumitomo Chemical Co., Ltd.
Takarazuka, Hyogo, Japan

Satoshi Kohno
Hyogo Prefectural Agricultural Center for Experiment
Kitaouji, Akashi, Japan

Akio Kudamatsu
Tokyo Medical and Dental University, and Nihon Tokushu Noyaku Seizo Co., Ltd.
Tokyo, Japan

Arun P. Kulkarni
North Carolina State University
Raleigh, North Carolina

Nobuo Kyomura
Mitsubishi Chemical Industries, Ltd.
Yokohama, Japan

J. R. Larsen
University of Illinois
Urbana, Illinois

Narayana M. Mallipudi
University of California
Riverside, California

B. T. Manji
University of California
Davis, California

Fumio Matsumura
Michigan State University
East Lansing, Michigan

Roni B. Mellon
University of California
Riverside, California

Robert L. Metcalf
University of Illinois
Urbana-Champaign, Illinois

T. A. Miller
University of California
Riverside, California

Tadashi Miyata
Nagoya University
Nagoya, Japan

Toshio Narahashi
Northwestern University Medical School
Chicago, Illinois

J. M. Ogawa
University of California
Davis, California

Kozaburo Ozaki
Kagawa Agricultural Experiment Station
Sakaide, Kagawa, Japan

J. Petri
Pennwalt Corp.
Monrovia, California

Frederick W. Plapp, Jr.
Texas A & M University
College Station, Texas

Steven R. Radosevich
University of California
Davis, California

L. M. Ross
Brigham Young University
Provo, Utah

Tetsuo Saito
Nagoya University
Nagoya, Japan

V. L. Salgado
University of California
Riverside, California

R. M. Sonoda
University of Florida
Ft. Pierce, Florida

Thomas C. Sparks
Louisiana State University
Baton Rouge, Louisiana

K. Strickler
Michigan State University
East Lansing, Michigan

Katsuhiro Tabata
Forestry and Forest Products Research Institute
Ushiku, Ibaraki, Japan

Yoji Takahashi
Mitsubishi Chemical Industries Ltd.
Yokohama, Japan

Charles E. Taylor
University of California
Riverside, California

Leon C. Terriere
Oregon State University
Corvallis, Oregon

Masuhisa Tsukamoto
University of Occupational and Environmental Health, Japan
Yahata-nishiku, Kitakyushu, Japan

Matazaemon Uchida
Nihon Nohyaku Co., Ltd
Osaka, Japan

Yasuhiko Uesugi
National Institute of Agricultural Sciences
Tsukuba, Ibaraki, Japan

T. C. Wang
Texas A & M University
College Station, Texas

D. J. Weber
Brigham Young University
Provo, Utah

Brian L. Wild
New South Wales Department of Agriculture
New South Wales, Australia

C. F. Wilkinson
Cornell University
Ithaca, New York

Tomio Yamada
Nippon Soda Co., Ltd.
Oiso-machi, Kanagawa, Japan

Izuru Yamamoto
Tokyo University of Agriculture
Tokyo, Japan

Kazuo Yasutomi
National Institute of Health
Tokyo, Japan

Hiromi Yoneda
Nippon Soda Co., Ltd.
Oiso-machi, Kanagawa, Japan

INDEX